Quantum Mathematical Physics

Springer
Berlin
Heidelberg
New York
Hong Kong
London
Milan
Paris
Tokyo

Physics and Astronomy ONLINE LIBRARY

http://www.springer.de/phys/

Walter Thirring

Quantum Mathematical Physics

Atoms, Molecules and Large Systems

Translated by Evans M. Harrell II

Second Edition
Corrected and Revised Second Printing
with Bibliographic Additions

With 63 Figures

Springer

Walter Thirring
Institute for Theoretical Physics
University of Vienna
Boltzmanngasse 5
1090 Vienna, Austria

Title of the original German edition:
Lehrbuch der Mathematischen Physik
Band 3: Quantenmechanik von Atomen und Molekülen
Band 4: Quantenmechanik großer Systeme

Library of Congress Cataloging-in-Publication Data applied for.

Die Deutsche Bibliothek – CIP-Einheitsaufnahme:
Thirring, Walter: Quantum mathematical physics : atoms, molecules and large systems /
Walter Thirring. Transl. by Evans M. Harrell. – 2.ed. –
Berlin ; Heidelberg ; New York ; Barcelona ; Hong Kong ; London ; Milan ; Paris ; Tokyo :
Springer, 2002
(Physics and astronomy online library)

ISBN 978-3-642-07711-1

Springer-Verlag Berlin Heidelberg New York
a member of BertelsmannSpringer Science+Business Media GmbH

http://www.springer.de

Cover design: Erich Kirchner, Heidelberg

Preface

This edition combines the earlier two volumes on Quantum Mechanics of Atoms and Molecules and on Quantum Mechanics of Large Systems, thus including in a single volume the material for a two-semester course on quantum physics. Since this volume is already quite heavy, I could not include many new results which show how lively the subject is. I just want to mention that inequality (IV:4.1.1.1) has been sharpened by T. Weidl by a factor 2π and the difficult problem 1 of (III:4.6) has been solved by A. Martin.

I have to thank N. Ilieva for the devotion in preparing this new edition.

Vienna, November 2001 *Walter Thirring*

Preface to the Second Edition: Quantum Mechanics of Atoms and Molecules

Ever since the first edition of this volume appeared in 1980 quantum statistical mechanics has florished. Innumerable results in many areas have been obtained and it would require a series of volumes to do justice to all of them. On the other hand the first edition was already rather crowded with many details so it would not be overburdened any more. Thus I added only one chapter on quantum ergodic theory where one can get the main notions across without too much pain. Nevertheless many subjects treated in the book had splendidely developed ever since and the only way out I could see is to add some recent references which the interested reader can consult.

For helpful advice I am indebted to many colleagues in particular to E.H. Lieb and H. Narnhofer. Furthermore I have to thank Z. Vakhnenko for resetting the complete book on the computer. Last but not least Nevena Ilieva took on the tedious task to fight with the program to get all the corrections across. We tried to do our best but if the reader still finds some faults, as I am sure he will, he might blame it onto the moods of the computer.

Vienna, December 1999 *Walter Thirring*

Preface to the First Edition: Quantum Mechanics of Atoms and Molecules

In this third volume of *A Course in Mathematical Physics* I have attempted not simply to introduce axioms and derive quantum mechanics from them, but also to progress to relevant applications. Reading the axiomatic literature often gives one the impression that it largely consists of making refined axioms, thereby freeing physics from any trace of down-to-earth residue and cutting it off from simpler ways of thinking. The goal pursued here, however, is to come up with concrete results that can be compared with experimental facts. Everything else should be regarded only as a side issue, and has been chosen for pragmatic reasons. It is precisely with this in mind that I feel it appropriate to draw upon the most modern mathematical methods. Only by their means can the logical fabric of quantum theory be woven with a smooth structure; in their absence, rough spots would inevitably appear, especially in the theory of unbounded operators, where the details are too intricate to be comprehended easily. Great care has been taken to build up this mathematical weaponry as completely as possible, as it is also the basic arsenal of the next volume. This means that many proofs have been tucked away in the exercises. My greatest concern was to replace the ordinary calculations of uncertain accuracy with better ones having error bounds, in order to raise the crude manners of theoretical physics to the more cultivated level of experimental physics.

The previous volumes are cited in the text as I and 11; most of the mathematical terminology was introduced in volume I. It has been possible to make only sporadic reference to the huge literature on the subject of this volume—the reader with more interest in its history is advised to consult the compendious work of Reed and Simon [3].

Of the many colleagues to whom I owe thanks for their help with the German edition, let me mention F. Gesztesy, H. Grosse, P. Hertel, M. and T. Hoffmann-Ostenhof, H. Narnhofer, L. Pittner, A. Wehrl, E. Weimar, and, last but not least, F. Wagner, who has transformed illegible scrawls into a calligraphic masterpiece. The Englist translation has greatly benefited from the careful reading and many suggestions of H. Grosse, H. Narnhofer, and particularly B. Simon.

Vienna
Spring, 1981

Walter Thirring

Preface to the First Edition: Quantum Mechanics of Large Systems

In this final volume I have tried to present the subject of statistical mechanics in accordance with the basic principles of the series. The effort again entailed following Gustav Mahler's maxim, "Tradition = Schlamperei" (i.e., filth) and clearing away a large portion of this tradition-laden area. The result is a book with little in common with most other books on the subject.

The ordinary perturbation-theoretic calculations are not very useful in this field. Those methods have never led to propositions of much substance. Even when perturbation series, which for the most part never converge, can be given some asymptotic meaning, it cannot be determined how close the nth order approximation comes to the exact result. Since analytic solutions of nontrivial problems are beyond human capabilities, for better or worse we must settle for sharp bounds on the quantities of interest, and can at most strive to make the degree of accuracy satisfactory.

The last two decades have seen successful and beautiful treatments of many fundamental issues – I have in mind the ordering of the states (2.1), properties of the entropy (2.2), noncommutative ergodic theory (3.1), the proof of the existence of the thermodynamic functions (4.3), and the mathematical analysis of Thomas-Fermi theory (4.1.2), which provides an understanding of the stability of matter. The day is surely not far off when most of the remaining holes in the conceptual structure of quantum statistical mechanics will have been filled in and the questions that are not satisfactorily answered today will be added to the list of achievements.

The successful completion of this course of mathematical physics in a reasonable time required the fortunate conjunction of several circumstances. As with volume III, I had active support from several collaborators, and in particular I am greatly obliged to B. Baumgartner, H. Narnhofer, A. Pflug, and A. Wehrl. Countless other colleagues have helped indirectly by coping with other time-consuming duties for me. The English edition has again greatly benefited from the critical reading of B. Simon. The working conditions at the University of Vienna were invaluable for the completion of this project. Last but not least, the frictionless collaboration of Springer-Verlag in Vienna and my secretary and calligrapher F. Wagner enabled the books to appear quickly and at a reasonable price.

I am aware that the uncompromising way of mathematical physics is not the easiest. Yet I feel that it has been one of the greatest intellectual accomplishments of our era to cast the laws of Nature in a clear mathematical form with rigorously deducible consequences. No amount of labor is too high a price to have paid for this. Let me conclude by also acknowledging and expressing my thanks to the reader who has borne with me to the end of the course.

Walter Thirring

Contents

Symbols Defined in the Text

$\sigma, \sigma^z, \sigma^\pm$	spin components	(1.1.1)
$\mathcal{H}_F$	Fock space	(1.3.1)
Ⓢ, $\wedge$	symmetric and antisymmetric tensor product	(1.3.1)
$a^*(f), a(f)$	creation and annihilation operators	(1.3.2)
$[\,]_+$	anticommutator	(1.3.3;2)
$\mathcal{A}_B$	C^*-algebra for bosons	(1.3.3;3)
$\mathcal{A}_F$	C^*-algbera for fermions	(1.3.3;4)
$a(\mathbf{x})$	annihilation operator	(1.3.3;7)
Ω	cyclic vector	(1.3.5)
$\mathcal{A}_E, \mathcal{A}_G$	even and gauge-invariant algebras	(1.3.8)
Tr	trace	(1.4.10)
I,II,III	factors of type I, II,III	(1.4.16)
$\rho(n)$	sum of the first n eigenvalues	(2.1.9)
?	ordering of density matrices	(2.1.10;1)
S_α	α-entropy	(2.2.2)
$S(\rho)$	von Neumann entropy	(2.2.4)
$d\Omega_z^N$	Liouville measure on phase space	(2.2.7)
ρ_{qu}	quantum-theoretic phase-space density	(2.2.7)
ρ_{cl}	classical phase-space density	(2.2.7)
$S(\sigma\|\rho)$	relative entropy	(2.2.22)
$\varepsilon, \sigma, \rho$	energy, entropy and particle densities	(2.3.8)
T	temperature	(2.3.16)
C_V, c_V	total and specific heat capacity at constant volume	(2.3.17;3)
P	pressure	(2.3.21)
κ	compressibility	(2.3.22;3)
μ	chemical potential	(2.3.27)
v_u	smeared potential	(2.4.9)

$\varphi(T, \rho)$	Legendre transform of ε	(2.4.14)
$\mathcal{L}$	Legendre transformation	(2.4.15;2)
z	fugacity	(2.5.9)
tr	trace on the one-particle space	(2.5.10)
v^u	unsmeared potential	(2.5.17)
$F_\sigma(z)$	generalized ζ function	(2.5.20)
$\mathbf{m}$	magnetization per volume	(22.5.37)
$\mathcal{R}$	covariance algebra	(3.1)
U	unitary operators	(3.1)
B_{eff}	effective magnetic field	(3.1.1;4)
τ_t	time-automorphism	(3.1.2)
τ_t^*	dual time-evolution	(3.1.2)
$\eta, \eta(a), \eta(\sigma)$	invariant mean	(3.1.14),(3.1.15)
E_0	projection onto eigenvectors of H with eigenvalue 0	(3.1.16;1)
a_t	transformed operator	(3.2.16;2)
J	conjugate-linear operator	(3.2.1)
π'	conjugate-linear representation	(3.2.1)
$\tilde{A}$	algebra of analytic elements	(3.2.6;(v))
$\sim$	Fourier transform	(3.2.6;(v))
τ_t^h	perturbed time-automorphism	(3.3.4)
R_h	corresponding operator	(3.3.2)
F_{ab}, G_{ab}	correlation functions	(3.3.14)
$\mathbf{X}_k, Z_k$	coordinates and charges of the nuclei	(4.1.3;1)
$W(\mathbf{x})$	wall potential	(4.1.3;4)
H_n	Hamiltonian with an effective field	(4.1.6)
C_n	correlation correction	(4.1.6)
$\Xi(H)$	grand canonical partition function	(4.1.8)
$\|n\|_c$	c-norm	(4.1.10)
h_n	one-particle Hamiltonian	(4.1.17)
v_s	singular part of the potential	(4.1.18)
$a_{p,q}, \rho_{p,q}$	annihilation and density operators	(4.1.25)
K, A, R	contributions to the energy	(4.1.33)
$\Phi(\mathbf{x})$	potential	(4.1.36)

Symbols Defined in Earlier Volumes

$\mathcal{B}(\mathcal{H})$	bounded operators	(III:2.1.24)
$\mathcal{C}_1(\mathcal{H})$	trace-class operators	(III:2.3.21)
$\mathcal{C}_2(\mathcal{H})$	Hilbert-Schmidt operators	(III:2.3.21)
$\otimes$	tensor product	(I:2.4.5)
$\oplus$	direct sum of Hilbert spaces	(III:2.1.9;2)
$\mathcal{A}'$	commutant of $\mathcal{A}$	(III:2.3.4)
$\|a\|_p$	trace norm	(III:2.3.21)

Part I

Quantum Mechanics of Atoms and Molecules

1
Introduction

1.1 The Structure of Quantum Theory

The structure of quantum mechanics differs startlingly from that of the classical theory. In volume I we learned that in classical mechanics the observables form an algebra of functions on phase space (p and q), and states are probability measures on phase space. The time-evolution is determined by a Hamiltonian vector field. It would be reasonable to expect that atomic physics would distort the vector field somewhat, or even destroy its Hamiltonian structure; but in fact the break it makes with classical concepts is much more drastic. The algebra of observables is no longer commutative. Instead, position and momentum satisfy the famous **commutation relations**,

$$qp - pq = i\hbar. \tag{1.1}$$

Since matrix algebras are not generally commutative, one of the early names for quantum theory was matrix mechanics. It became apparent in short order, however, that the commutator (1.1) of finite-dimensional matrices can never be proportional to the identity (take the trace of both sides), so attempts were then made to treat p and q as infinite-dimensional matrices. This proved to be a false scent, since infinite-dimensional matrices do not provide an ideal mathematical framework. The right way to proceed was pointed out by J. von Neumann, and the theory of C^* and W^* algebras today puts tools for quantum theory at our disposal, which are polished and comparatively easy to understand. There do remain a few technical complications connected with unbounded operators, for which reason the **Weyl**

relation

$$e^{i\alpha q} e^{i\beta p} e^{-i\alpha q} = e^{i\beta(p-\alpha)} \tag{1.2}$$

(setting $\hbar = 1$) is a better characterization of the noncommutativity.

Admittedly, Schrödinger historically first steered quantum mechanics in a different direction. The equation that bears his name treats p and q as differentiation and multiplication operators acting on the **Schrödinger wave-function** ψ, which has the interpretation of a **probability amplitude**: It is complex-valued, and $|\psi|^2$ is the probability distribution in the state specified by ψ. Superposition of the solutions of the equation causes **probability interference effects**, a phenomenon that cannot be understood classically at all. Later, ψ was characterized axiomatically as a vector in Hilbert space, but the peculiar fact remained that one worked with a complex Hilbert space and came up with real probabilities.

At long last the origin of the Hilbert space was uncovered. A state would normally be required to be represented as a **positive linear functional**, where positivity means that the expectation value $\langle a^2 \rangle$ of the square of any real observable a must also be nonnegative. It turns out that to each state there corresponds a representation of the observables as linear operators on some Hilbert space. (It is at first unsettling to learn that each state brings with it its own representation of the algebra characterized by (1.2), but it also turns out that they are all equivalent.) The schema of quantum theory thus adds no new postulates to the classical ones, but rather omits the postulate that the algebra is commutative. As a consequence, quantum mechanically there are no states for which the expectation values of all products are equal to the products of the expectation values. Such a state would provide an algebraic isomorphism to the ordinary numbers, which is possible only for very special noncommutative algebras. The occurrence of nonzero fluctuations $(\Delta a)^2 \equiv \langle a^2 \rangle - \langle a \rangle^2$ is in general unavoidable, and gives rise to the indeterministic features of the theory. The extremely good experimental confirmation of quantum mechanics shows that the numerous paradoxes it involves are owing more to the inadequacy of the understanding of minds raised in a classical environment than to the theory.

Quantum theory shows us where classical logic goes awry; the logical maxim *tertium non datur* is not valid. Consider the famous double-slit experiment. Classical logic would reason that if the only and mutually exclusive possibilities are "the particle passes through slit 1" and the "the particle passes through slit 2," then it follows that "the particle passes through slit 1 and then arrives at the detector" and "the particle passes through slit 2 and then arrives at the detector" are likewise the only and mutually exclusive possibilities. **Quantum logic** contests this conclusion by pointing to the irreparable change caused in the state by preparing the system to test the new propositions. The rules of quantum logic can be formulated just as consistently as those of classical logic. Nonetheless, the world of quantum physics strikes us as highly counterintuitive, more so even than the theory of relativity. It requires radically new ways of thinking.

The mathematical difficulties caused by the noncommutativity have all been overcome. Indeed, the fluctuations it cases often simplify problems. For example, the fluctuations of the kinetic energy, the **zero-point energy**, have the effect of weakening the singularity of the Coulomb potential and eliminating the problem of the collision trajectories, which are so troublesome in classical mechanics. Quantum theory guarantees that the time evolution can be continued uniquely from $t = -\infty$ to $t = +\infty$ for (non-relativistic) systems with $1/r$ potentials. In a certain sense this potential energy is only a small perturbation of the kinetic energy, and free particles can be used as a basis of comparison. Calculations are sometimes much easier to do in quantum theory than in classical physics; it is possible, for instance, to evaluate the energy levels of helium with fantastic precision, whereas only relatively crude estimates can be made for the corresponding classical problem.

1.2 The Orders of Magnitude of Atomic Systems

One can come to a rough understanding of the characteristics of quantum-mechanical systems by grafting discreteness and fluctuations of various observables onto classical mechanics. Their magnitudes depend on **Planck's constant** $\hbar$, which is best thought of as a quantum of angular momentum, since quantum-mechanically the orbital angular momentum L takes on only the values $l\hbar$, $l = 0, 1, 2, \ldots$. Suppose an electron moves in the Coulomb field of a nucleus of charge Z; then the energy is

$$E = \frac{p_r^2}{2m} + \frac{L^2}{2mr^2} - \frac{Ze^2}{r}. \tag{1.1}$$

For circular orbits ($p_r = 0$), quantization of the angular momentum means that

$$E(r) = \frac{l^2\hbar^2}{2mr^2} - \frac{Ze^2}{r}. \tag{1.2}$$

At the radius

$$r = \frac{l^2\hbar^2}{mZe^2} \equiv \frac{l^2 r_b}{Z}, \tag{1.3}$$

where r_b is known as the **Bohr radius**, the energy is minimized, with the value

$$E = -\frac{(Ze^2)^2}{2}\frac{m}{l^2\hbar^2} = \frac{-Z^2}{l^2}\frac{e^2}{2r_b} \equiv -\frac{Z^2}{l^2}\ \text{(Rydberg} \equiv \text{Ry)} \tag{1.4}$$

(**Balmer's formula**). If $l = 0$, then we would find $r = 0$ and $E = -\infty$, except that the stability of the system is saved by the inequality for the fluctuations $\Delta p \Delta q \geq \hbar/2$, the **indeterminacy relation**, which follows from (1.1). This makes $\langle p_r^2 \rangle \geq (\Delta p_r)^2 \approx \hbar^2/r^2$, the **zero-point energy**, and hence this part of the kinetic energy contributes as much as a centrifugal term with $l = 1$. This argument actually gives the correct ground-state energy. The reasoning is of course not a mathematically rigorous deduction from the indeterminacy relation, as the average of $1/r$ could conceivably be large without Δr being small. We shall later derive generalizations of the inequality $\Delta p \Delta q \geq \hbar/2$, which will justify the argument.

The virial theorem states that the velocity v of an electron is given classically by

$$\frac{mv^2}{2} = -E = \frac{Z^2e^4m}{2l^2\hbar^2} \Rightarrow v = \frac{Z}{l}\frac{e^2}{\hbar}.$$

The universal speed $e^2/\hbar$ is about $1/137$ times the speed of light. As Z increases, the nonrelativistic theory rapidly loses its accuracy. Relativistic corrections, entering through the increase of the mass and magnetic interactions, are $\sim v^2/c^2 \approx 10^{-5}Z^2$. For small Z they show up as fine structure of the spectral lines, but their effect becomes pronounced for heavy nuclei, and when Z is sufficiently greater than 137 the system is not even stable anymore. The relativistic kinetic energy is $\sqrt{m^2c^4 + p^2c^2} - mc^2$, which for large momenta grows only as $cp \approx c\hbar/r$. Equation (1.2) is accordingly changed to

$$E(r) \approx \frac{c\hbar}{r} - \frac{Ze^2}{r} = \frac{c\hbar}{r}\left(1 - \frac{Z}{137}\right), \tag{1.5}$$

which is no longer bounded below when $Z > 137$. The question of what happens for such large Z can only be answered in the relativistic quantum theory, and lies beyond the scope of this book.

If a second electron is introduced to form a helium-like atom, then the repulsion of the electrons makes it impossible to solve the problem analytically. To orient ourselves and to understand the effect of the repulsion, let us provisionally make some simplifying assumptions. Since an electron cannot be localized well, we can suppose that its charge fills a ball of radius R homogeneously. Such an electronic cloud would produce an electrostatic potential

$$V(r) = \begin{cases} -\dfrac{3e}{2R} + \dfrac{e}{2R}\left(\dfrac{r}{R}\right)^2, & r \leq R \\ -\dfrac{e}{r}, & r \geq R \end{cases} \tag{1.6}$$

(Figure 1.1). The potential energy of one electron and the nucleus is consequently $ZeV(0) = -3Ze^2/2R$. We can gauge the kinetic energy by reference to the hydrogen atom, for which the following rule of thumb leads to the correct ground-state energy: An electron cloud having potential energy $-Ze^2/r_b$ requires a kinetic energy $\hbar^2/2mr_b^2$. We set the kinetic energy equal to $9\hbar^2/8mR^2$, since $R = 3r_b/2$ provides the same amount of potential energy.

If the second electron is also a homogeneously charged sphere coinciding with the first one, then the electronic repulsion is

$$-\frac{3}{4\pi R^3}4\pi e^2 \int_0^R r^2 dr\, V(r) = \frac{6e^2}{5R}. \tag{1.7}$$

Therefore we obtain the ratio

$$\frac{|\text{Attraction of the electrons to the nucleus}|}{\text{Repulsion of the electrons}} = \frac{2\cdot(3Ze^2/2R)}{6e^2/5R} = \frac{5Z}{2}, \tag{1.8}$$

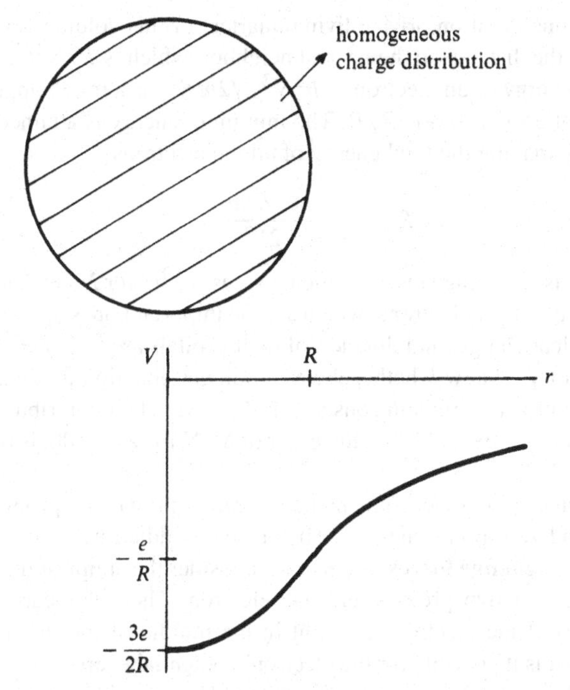

Figure 1.1. The potential of a homogeneous charge distribution.

and thus the total energy is

$$\begin{aligned} E(R) &= \text{kinetic energy} + \text{nuclear attraction} + \text{electronic repulsion} \\ &= 2\cdot\frac{9\hbar^2}{8mR^2} - 2\cdot\frac{3Ze^2}{2R}\left(1-\frac{2}{5Z}\right). \end{aligned} \tag{1.9}$$

This has its minimum at the value $R = R_{\min} = R_H/\left(Z-\frac{2}{5}\right)$, where

$$E(R_{\min}) = -Ry\cdot 2Z^2\left(1-\frac{2}{5Z}\right)^2. \tag{1.10}$$

If $Z = 2$, then $R_{\min} = 5R_H/8$, and the energy has the value $-2Ry\cdot\frac{64}{25} = -2Ry\cdot 2.56$. For such a primitive estimate, this comes impressively near to the experimentally measure $-2Ry\cdot 2.9$, and a helium atom is indeed only about half as large as a hydrogen atom. Actually, however, even if $Z = 1(H^-)$ the energy lies somewhat below $-Ry$ while (1.10) gives only $-\frac{18}{25}\,Ry$. In this case the picture of two equal spheres is not very apt, since the outer electron will travel out to large distances. Nevertheless, nonrelativistic quantum mechanics describes these systems very well.

If there are more than two electrons, then some of them must have spins in parallel, and **Pauli's exclusion principle** is of primary importance for the spatial configuration of atoms; it says that no two electrons may have the same position, spin, etc. An atom with N electrons and radius R has a volume of about R^3/N per particle. Electrons insist on private living quarters of this volume, so Δq will be on the order of the distance to the nearest neighbor, which is $R/N^{1/3}$. This makes the zero point energy of an electron $\approx \hbar^2 N^{2/3}/2mR^2$, as a rough approximation, and its potential energy $\approx -e^2 Z/R$. The minimum energy is attained at $R_{\text{min}} = \hbar^2 N^{2/3}/me^2 Z$, making the total energy of all the electrons

$$E(R_{\text{min}}) = -\frac{e^4 Z^2 m}{2\hbar^2} N^{1/3}. \tag{1.11}$$

The value R_{min} is an average radius, which goes as $N^{-1/3}$ for $N = Z$, making $E \sim N^{7/3}$. Yet the outermost electrons, which are the important ones for chemistry, see a screened nuclear charge, and the radii of their orbitals are $\approx \hbar^2/me^2$. Strangely enough, it is not yet know whether the Schrödinger equation predicts that these radii expand, contract, or remain constant as $Z \to \infty$. Their contribution of about 10 eV to the total energy (1.11), on the order of MeV for $Z \sim 100$, is rather slight, however.

Chemical forces also arise from an energetically optimal compromise between electrostatic and zero-point energies. History has saddled us with a misleading phrase for this, **exchange forces**. Let us now consider the simplest molecule, H_2^+, that is, a system of two protons and one electron. There is clearly a negative potential energy if the electron sits right in the middle of the line between the two protons. But is it possible for the electron's potential energy to be sufficiently negative to make the total energy less than that of H, or would its wave-function be too narrow, giving it an excessive zero-point energy? To be more quantitative about this question, let us again imagine that the electron is a homogeneously charged sphere with the potential (1.6). The radius R is chosen the same as for H, so there is no difference between this zero-point energy and that of hydrogen. If, as with H, we put one proton at the center of the cloud (Figure 1.2a), the potential energy is eV(0). Taking the Coulombic repulsion of the protons into account, we note that the second proton feels no potential as long as it is outside the cloud, but when it comes to within a distance $r < R$ its energy increases, because

$$V(0) + V(r) + \frac{e^2}{r} \geq V(0). \tag{1.12}$$

Hence there is no binding. However, if the two protons are placed diametrically across the center of the electron cloud, at radius r (Figure 1.2b), then the total potential energy

$$2V(r) + \frac{e^2}{2r} = -\frac{3e^2}{R} + \left(\frac{e^2}{R}\right)\left(\frac{r}{R}\right)^2 + \frac{e^2}{2r} \tag{1.13}$$

has the minimum

$$-\frac{3e^2}{2R}[2 - 2^{-1/3}] = -\frac{3e^2}{2R} \cdot 1.2 \tag{1.14}$$

at $r = 2^{-2/3} \cdot R$. This is more negative than $V(0)$, the energy with one proton outside the sphere, by a factor 1.2, and so we expect H_2^+ to be bound. If the total

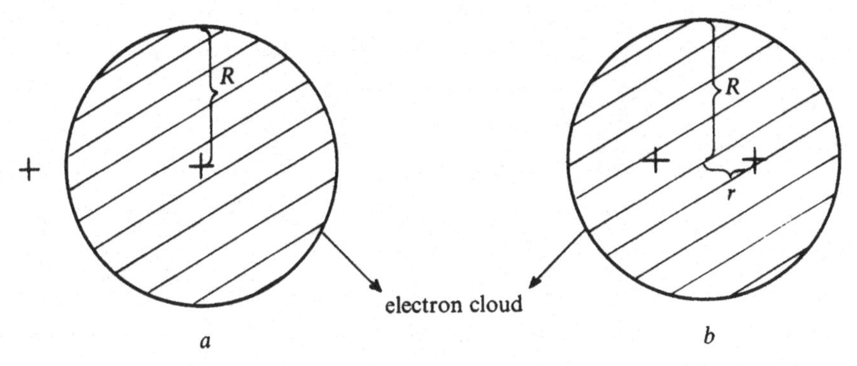

Figure 1.2a,b. Two electron distributions assumed for H_2^+.

energy is now minimized with respect to R, then $R_{\min} = R_{\mathrm{H}}/1.2$ and $E(R_{\min}) = -(1.2)^2 Ry$. The separation $2r$ of the protons at the minimum is $2^{1/3} R_{\min} = 1.57 r_b$, which is significantly smaller than the experimental value $2r_b$. The binding energy $((1.2)^2 - 1) Ry$ also amounts to more than twice the measured value, so the simple picture is not very accurate.

Finally, consider the molecule H_2, again assuming that the H atoms are spheres. If they do not overlap, then the electrostatic energy is twice that of a single H atom, and the two separate atoms exert no force on each other. As the spheres are pushed together, the energy first decreases, since the repulsion of the electrons is reduced (the energy of two uniformly charged spheres at a distance $r < 2R$ is less than e^2/r), while the other contributions to the energy remain unchanged. In order to find out how much energy can be gained by making the spheres overlap, let us superpose them and place the protons diametrically across their center at a distance r. As with the helium atom, the electronic repulsion is $6e^2/5r$, and hence the total potential energy is

$$V_{\mathrm{H}_2}(r) = -\frac{6e^2}{R} + 2\frac{e^2 r^2}{R^3} + \frac{e^2}{2r} + \frac{6e^2}{5R}. \tag{1.15}$$

The minimum at $r = R/2$ can now be compared with $2V(0)$:

$$V_{\mathrm{H}_2}\left(\frac{R}{2}\right) = -2\frac{3e^2}{2R} \cdot 1.1. \tag{1.16}$$

The minimum in R is now attained at $R_{\mathrm{H}}/1.1$, and the corresponding interprotonic distance $3r_b/2 \cdot 1.1 = 1.36 r_b$ is an excellent agreement with the actual distance. The resultant binding energy $2Ry((1.1)^2 - 1) \simeq 5.7$ eV is consequently also fairly close to the measured energy of dissociation 4.74 eV. Of course, it is necessary for the electrons in H_2 to have antiparallel spins, as otherwise the exclusion principle would restrict the room they have to move about in.

One lesson of these rough arguments is that delicate questions like that of stability depend on small energy differences. It will require highly polished calculational techniques to reach definitive conclusions.

2
The Mathematical Formulation of Quantum Mechanics

2.1 Linear Spaces

There are many surprising aspects to the infinitely many directions in an infinite-dimensional space. For this reason it is necessary to investigate carefully which of the familiar properties of finite-dimensional spaces carry over unchanged and which do not.

We begin by recollecting the basic definitions and theorems:

2.1.1 Definition

A **linear**, or **vector**, **space** $\mathbb{E} \ni v_i$ over the complex numbers $\mathbb{C} \ni \alpha_i$ is a set on which sums $\mathbb{E} \times \mathbb{E} \to \mathbb{E} : (v, u) \to v + u = u + v$ and products with scalars $\mathbb{E} \times \mathbb{C} \to \mathbb{E} : (v, \alpha) \to \alpha v$ are defined so that $\alpha_1(\alpha_2 v) = (\alpha_1\alpha_2)v$, $\alpha(v + u) = \alpha v + \alpha u$, $1 \cdot v = v$, and $(\alpha_1 + \alpha_2)v = \alpha_1 v + \alpha_2 v$.

2.1.2 Examples

1. Vectors in $\mathbb{C}^n$.
2. Complex $n \times n$ matrices.
3. Polynomials in n complex variables.
4. $C^r \equiv$ the r-times continuously differentiable functions.
5. Analytic functions.

Sums and products with α are defined in the usual way.

2.1.3 Remark

A subset $\mathbb{E}_1 \subset \mathbb{E}$ that is also a vector space is called a **subspace** of $\mathbb{E}$. For example, (2.1.2; 5) is a subspace of 4, and 3 is a subspace of 5. The **quotient space** $\mathbb{E}/\mathbb{E}_1$ consists of equivalence classes of vectors whose differences are elements of $\mathbb{E}_1$. In the absence of a scalar product there is no uniquely defined decomposition of vectors $v \in \mathbb{E}$ such that $v = v_1 + v_2$ with $v_1 \in \mathbb{E}_1$. However, if an $\mathbb{E}_2$ is also specified so that $\mathbb{E}_1 + \mathbb{E}_2 = \mathbb{E}$ and $\mathbb{E}_1 \cap \mathbb{E}_2 = \{0\}$, then there is such a decomposition with a unique $v_2 \in \mathbb{E}_2$; $\mathbb{E}$ is then the **sum** of $\mathbb{E}_1$ and $\mathbb{E}_2$, and $\mathbb{E}_2$ is a **complement** of $\mathbb{E}_1$. General sums of linear spaces can be defined in the same manner. According to the axiom of choice, it is always possible, by an inductive argument, to find a **Hamel basis** $\{e_\gamma\}$, $\gamma \in \mathbb{I}$, such that every vector can be written uniquely as

$$v = \sum_{\text{finite}} \alpha_i e_{\gamma_i}, \qquad \alpha_i \in \mathbb{C}.$$

Unfortunately, for infinite-dimensional spaces the set $\mathbb{I}$ is usually uncountable, and the Hamel basis is of little practical significance. The cardinality of $\mathbb{I}$ is known as the **algebraic dimension of the space**.

2.1.4 Definition

A **normed** linear space is a vector space on which there is defined a **norm** mapping $\mathbb{E} \to \mathbb{R}^+, v \to \|v\|$, such that $\|\alpha v\| = |\alpha|\|v\|$, $\|v+u\| \leq \|v\|+\|u\|$, and $\|v\| = 0$ iff $v = 0$.

2.1.5 Examples

1. $\mathbb{E} = \mathbb{C}^n \ni v = (v_1, v_2, \ldots, v_n)$, $\|v\|_p = \left[\sum_{i=1}^n |v_i|^p\right]^{1/p}$, $1 \leq p < \infty$, $\|v\|_\infty = \max_i |v_i|$.
2. $\mathbb{E} = n \times n$ matrices, $m = (m_{ij})$, $\|m\| = \left(\sum_{i,j} |m_{ij}|^2\right)^{1/2} = (\mathrm{Tr}\, mm^*)^{1/2}$.
3. $\mathbb{E} = n \times n$ matrices, $\|m\|^2 = \sup_{\sum_i |v_i|^2=1} \sum_i |\sum_k m_{ik} v_k|^2$.
4. Polynomials $P(z_i)$ for $z = (z_1, \ldots, z_n)$ in a compact set $K \subset \mathbb{C}^n$, $\|P\| = \sup_{z\in K} |P(z_i)|$.
5. The r-times continuously differentiable functions $f(z_i)$ on K,
$$\|f\| = \sup_{z\in K} |f(z_i)|.$$
6. Given a measure μ on K, it defines a norm $\|f\|_p = \left[\int d\mu |f|^p\right]^{1/2}$, $1 \leq p < \infty$. (We use the word **measure** to mean **positive measure**.) $L^p(K, \mu) \equiv \{f : \|f\|_p < \infty\}$.

2.1.6 Remarks

1. As $p \to \infty$, the norm $\|f\|_p$ approaches the norm of Example 5, which is denoted by $\|f\|_\infty$.
2. If μ is a sum of n point masses, then the space of Example 6 is the same as that of Example 1. If n is infinite, it is denoted by l^p.
3. As we see, different norms can be given to the same space, while, on the other hand, a space must sometimes be restricted for a norm to be finite on all of it.

2.1.7 Definition

If a norm on $\mathbb{E}$ satisfies the **parallelogram law** $\|u+v\|^2 + \|u-v\|^2 = 2\|u\|^2 + 2\|v\|^2$, then $\mathbb{E}$ is a **pre-Hilbert space**. In that case there exists a **scalar product**

$$\begin{aligned} \mathbb{E} \times \mathbb{E} \to \mathbb{C} &: (u, v) \to \langle u|v\rangle \\ &\equiv \frac{1}{4}(\|u+v\|^2 - \|u-v\|^2 - i\|u+iv\|^2 + i\|u-iv\|^2), \end{aligned}$$

which has the properties

$$\begin{aligned} \|v\|^2 &= \langle v|v\rangle, \langle v|u\rangle = \langle u|v\rangle^*, \\ \langle v|\alpha u\rangle &= \alpha\langle v|u\rangle, \langle u|v+w\rangle = \langle u|v\rangle + \langle u|w\rangle, \end{aligned}$$

and $\langle v|v\rangle = 0$ iff $v = 0$.

2.1.8 Examples

Of Examples (2.1.5), the only pre-Hilbert spaces (for $n > 1$) are Example 1 with $p = 2$, Example 2, and Example 6 with $p = 2$.

2.1.9 Remarks

1. Only the length of a vector is defined on a general normed linear space; on a pre-Hilbert space it is also known when two vectors are orthogonal. Pre-Hilbert spaces therefore conform better to our geometric intuition; by Problem 10,

   (i) $|\langle u|v\rangle| \le \|u\|\|v\|$ (the **Cauchy-Schwarz inequality**);

   (ii) $\langle u|v\rangle = 0 \Leftrightarrow \|u+v\|^2 = \|u\|^2 + \|v\|^2$ (**Pythagoras's law**).

2. If $\mathbb{E}_1$ and $\mathbb{E}_2$ are two pre-Hilbert spaces, then $\mathbb{E} \equiv \mathbb{E}_1 + \mathbb{E}_2$ can be made into a pre-Hilbert space, the **Hilbert sum** $\mathbb{E}_1 \oplus \mathbb{E}_2$, by setting $\langle (u_1, u_2)|(v_1, v_2)\rangle = \langle u_1|v_1\rangle + \langle u_2|v_2\rangle$. The vectors of $\mathbb{E}_1$ become orthogonal to those of $\mathbb{E}_2$ in the new space. Conversely, given a subspace $\mathbb{E}_1 \subset \mathbb{E}$ and defining $\mathbb{E}_1^\perp \equiv \{v \in \mathbb{E} : \langle v|u\rangle = 0 \text{ for all } u \in \mathbb{E}_1\}$, it follows that $\mathbb{E}_1 \cap \mathbb{E}_1^\perp = \{0\}$. It is tempting to single $\mathbb{E}_1^\perp$ out as the complement of $\mathbb{E}_1$. However, it can happen for infinite-dimensional spaces that $\mathbb{E}_1 \oplus \mathbb{E}_1^\perp \neq \mathbb{E}$: Let $\mathbb{E}_1 \subset l^2$ consist of the vectors having only finite many nonzero components; then $\mathbb{E}_1^\perp = \{0\}$ but $\mathbb{E}_1 \neq l^2$. This

is related to the fact, which we shall return to shortly, that in infinitely many dimensions not every linear subspace is topologically closed.

3. The tensor product $\mathbb{E}_1 \otimes \mathbb{E}_2$ and the antisymmetric tensor product $\mathbb{E}_1 \wedge \mathbb{E}_2$ can be defined as for finite-dimensional spaces (I: §2.4), and the scalar product in these constructions is multiplicative: $\langle v_1 \otimes v_2 | u_1 \otimes u_2 \rangle = \langle v_1 | u_1 \rangle \langle v_2 | u_2 \rangle$.
4. If two norms satisfy $\| \cdot \|_1 \leq a \| \cdot \|_2 \leq b \| \cdot \|_1$ for $a > 0, b > 1$, then they are said to be **equivalent**. They clearly produce the same topology (see below). Remarkably, all norms on finite-dimensional spaces are equivalent.
5. A mapping $a : \mathbb{E} \to \mathbb{F}$ satisfying $\|ax\| = \|x\|$ for all $x \in \mathbb{E}$ is called an **isometry**. We shall reserve the term **isomorphism** of normed spaces for a linear, isometric bijection.
6. Conversely, a scalar product $\langle u | v \rangle$ with the properties (2.1.7) defines a norm $\|x\|^2 = \langle x | x \rangle$ that obeys the parallelogram law.

Although the dimension of the space has only played a secondary role in the algebraic rules discussed above, infinite dimensionality disrupts the topological properties. These properties can be studied by using the norm (2.1.4), which induces a metric topology on a vector space with the distance function $d(u, v)$ defined as $\|u - v\|$. The neighborhood bases of vectors $v \in \mathbb{E}$ are $\{v' \in \mathbb{E} : \|v - v'\| \leq \varepsilon\}$. Definition (2.1.4) guarantees that addition and multiplication are continuous in this topology (Problem 3), i.e., the limit of sums or products equals the sum or product of the limits. There remains one obstacle to the use of the methods of classical analysis, in that not every **Cauchy sequence** v_n (i.e., for all $\varepsilon \to 0$ there exists an N such that $\|v_n - v_m\| \leq \varepsilon$ for all $n, m > N$) converges. In Example (2.1.5; 4), any continuous function is a limit of a Cauchy sequence of polynomials. Thus there are Cauchy sequences that do not converge in this space of polynomials. In order to exclude such difficulties with limits, we make

2.1.10 Definition

A normed space is **complete** iff every Cauchy sequence converges. A complete, normed, linear space (resp. pre-Hilbert space) is a **Banach** (resp. **Hilbert**) **space**.

2.1.11 Examples

Of Examples (2.1.5), only 1, 2, 3, 5 with $r = 0$, and 6 are complete.

2.1.12 Remarks

1. It is crucial that the limit exists as an element of the space in question. One can always complete spaces by appending all the limiting elements, but this can occasionally force one to deal with queer objects. For instance, if the polynomials (2.1.5; 4) are completed in the norm of (2.1.5; 6), then the resulting space $L^p(K, \mu)$ has elements that are not functions, but equivalence classes of functions differing on null sets.

2. One does not naturally have a good intuition about the concept of completeness, since finite-dimensional spaces are automatically complete. It should be distinguished from the notion of closure: Like every topological space, even an incomplete space is closed. It merely fails to be closed as a subspace of its completion, which is then its closure; in other words, it is dense in its completion.
3. Since convergent infinite sums are now defined and their limits exist, it is possible to introduce smaller bases than the Hamel basis. A set of vectors e_γ, $\gamma \in \mathbb{I}$ is said to be **total** provided that the set of its finite linear combinations is dense in $\mathbb{E}$. If $\mathbb{I}$ is countable, then $\mathbb{E}$ is **separable** (as a topological space).
4. By the axiom of choice, the e_γ can even be chosen orthonormal in a Hilbert space. If this has been done and $v = \sum_{\gamma \in \mathbb{I}} c_\gamma e_\gamma$, $c_\gamma = \langle e_\gamma | v \rangle$, then $\|v\|^2 = \sum_{\gamma \in \mathbb{I}} |c_\gamma|^2$, and the Hilbert space can be considered as $L^2(\mathbb{I}, \mu)$, where μ assigns every element of $\mathbb{I}$ the measure 1. If $\mathbb{I}$ is countable, then the Hilbert space is isomorphic to an l^2 space. If $\mathbb{I}$ is uncountable, then the countable sets and their complements constitute the measurable sets, and the resulting Hilbert space is not separable.
5. Every vector of a Hilbert space can be written in an orthogonal basis as a convergent infinite sum, $v = \sum_\gamma e_\gamma \langle e_\gamma | v \rangle$, and accordingly the sum (2.1.9; 2) of Hilbert spaces can easily be extended to infinite sums (though more care must be taken with the construction of infinite tensor products—see volume IV). However, if one approximates a vector v with an arbitrary total set $\{e_j\}$, say $v_n = \sum_{j=1}^n c_j e_j$, $\|v - v_n\| \le 1/n$, then it may be necessary to keep changing some of the c's substantially as $n \to \infty$, and the expansion $v = \sum_{j=1}^\infty c_j e_j$ may not exist. For instance, in l^2 the vectors

$$e_n = \left(1, \frac{1}{2^2}, \ldots, \underbrace{\frac{1}{n^2}}_{n\text{-th position}}, 0, 0, \ldots \right)$$

are total. If we expand $v = \lim_{n\to\infty} v_n \equiv \left(1, \frac{1}{2}, \ldots, 1/n, 0, 0, \ldots\right)$ then $v_n = -e_1 - e_2 - \cdots - e_{n-1} + n e_n$. Thus v can be approximated arbitrarily well by the e's, while the formal limit $v = -e_1 - e_2 - \cdots + \infty e_\infty$ does not make sense. In a general Banach space, where there is not an orthogonal basis at one's disposal, it is therefore unclear whether there exists a basis in which every vector can be written as a convergent sum. If there is a set of vectors in terms of which any vector can be written as a convergent sum, we shall call it **complete**. These distinctions may be somewhat unfamiliar, since for n vectors of $\mathbb{C}^n$, linearly independent $\Leftrightarrow$ total $\Leftrightarrow$ complete. In an infinite-dimensional space the implications go only one way; an infinite set of linearly independent vectors need not be total, and a total set need not be complete. For instance, $\{e^{inx}, n \in \mathbb{Z}\}$ is total and complete in $L^2((0, 2\pi), dx)$, but total and incomplete in the Banach space of continuous, periodic functions on $(0, 2\pi)$ with the sup-norm.

2.1.13 Definition

A **linear functional** w on a vector space $\mathbb{E}$ is a mapping $\mathbb{E} \to \mathbb{C} : v \to (w|v)$ such that $(w|v_1 + v_2) = (w|v_1) + (w|v_2)$ and $(w|\alpha v) = \alpha(w|v)$, for $a \in \mathbb{C}$.

2.1.14 Examples

In Examples (2.1.2) the linear functionals are

1. Scalar products with a vector.
2. Traces of the product of a matrix with some other matrix.

Linear functionals on the other examples include integrals of the functions by distributions and many other things. (See (2.2.19; 3).)

2.1.15 Remarks

1. The space of linear functionals on a vector space is called its **algebraic dual space**. It has a natural linear structure, $(w_1 + w_2|v) = (w_1|v) + (w_2|v)$ and $(\alpha w|v) = \alpha^*(w|v)$. The dual space of $\mathbb{R}^n$ can be identified with $\mathbb{R}^n$. However, infinite-dimensional spaces are not algebraically self-dual, and for that reason we introduce the abstract definition (2.1.13).
2. The concept defined in (2.1.13) is somewhat too general for our purposes, since the mapping $v \to (w|v)$ is automatically continuous only for finite-dimensional spaces (Examples 1 and 2). For example, consider $l^1 \equiv \{v = (v_1, v_2, v_3, \dots) : \|v\| \equiv \sum_i |v_i| < \infty\}$ with Hamel basis

$$\{e_i = (0, 0, \dots, \underbrace{1, 0,}_{i\text{-th position}} \dots, 0)\},$$

augmented with some other vectors $\bar{e}_\gamma$ to take care of vectors with infinitely many components. Every vector can be written as a finite sum, $v = \sum_{i,\text{ finite}} c_i e_i + \sum_{\gamma,\text{ finite}} \bar{c}_\gamma \bar{e}_\gamma$. If we define $(w|v) = \sum_{i=1}^{\infty} i c_i$, which converges because only finitely many c_i are nonzero, then w is obviously a linear functional, but it is not continuous. In fact, it is not even **closed**, i.e., there exists a sequence $v_n \to 0$ such that $(w|v_n) \to 1 \neq (w|0) = 0$; e.g., take

$$v_n = (0, 0, \dots, \underbrace{1/n,}_{n\text{-th position}} \quad 0, \dots).$$

This phenomenon can be understood as meaning that the steepness of w in the i-th direction is i; as i gets larger, it corresponds to a more nearly vertical plane. The formal reason for it is again that infinite-dimensional spaces can have nonclosed linear subspaces. The **kernel** of w, defined as $\{v : (w|v) = 0\}$, is a subspace, and if w were continuous, it would be closed, since it is the inverse image of the point zero. In this case, however, it contains all finite linear combinations of the vectors

$$v_{nm} = (0, \ldots, \underbrace{0, 1, 0, \ldots,}_{n\text{-th position}} \underbrace{-n/m,}_{m\text{-th position}} 0, \ldots),$$

and it is thus dense in l^1. It is desirable to exclude such pathologies, which is the motivation for

2.1.16 Definition

The linear space $\mathbb{E}'$ of the *continuous* linear functionals of a Banach space $\mathbb{E}$ is called its **dual space**.

2.1.17 Examples

As mentioned above, $\mathbb{C}^n$ and the space of $n \times n$ matrices are their own duals. More generally, all Hilbert spaces are self-dual; by a theorem of Riesz and Frêchet [3] any continuous linear functional on $\mathcal{H}$ can be written as a scalar product $v \to \langle w|v\rangle$ with a unique $w = \sum e_\gamma \langle w|e_\gamma\rangle \in \mathcal{H}$. Generalizing further, $(L^p(M, \mu))' = L^q(M, \mu)$ for $1/p + 1/q = 1$, $1 < p < \infty$; and $(L^1)' = L^\infty$, though, for infinite-dimensional spaces $(L^\infty)'$ is actually larger than L^1. The dual space of the continuous functions on a compact set, with the norm $\sup_{z\in K} |f(z)|$ consists of the (not necessarily positive) measures on K.

2.1.18 Remark

These statements depend critically on the completeness of the spaces. If we consider, for instance, the pre-Hilbert space $\mathbb{E}$ of the vectors of l^2 having finitely many nonzero components,then $(v_i) \to \sum_{i=1}^{\infty} v_i/i$ is a continuous linear functional that cannot be written as $\langle w|v\rangle$ for $w \in \mathbb{E}$, since

$$\left(1, \tfrac{1}{2}, \tfrac{1}{3}, \ldots\right) \notin \mathbb{E}.$$

The dual space $\mathbb{E}'$ is also a linear space, so the next task is to topologize it.

2.1.19 Definition

The neighborhood bases of vectors $w \in \mathbb{E}'$ will be defined alternatively by

$$U_{v,\varepsilon}(w) = \{w' \in \mathbb{E}' : |(w - w'|v)| < \varepsilon\}, \qquad v \in \mathbb{E}, \quad \varepsilon \in \mathbb{R}^+,$$

and by

$$U_\varepsilon(w) = \bigcap_{\|v\|=1} U_{v,\varepsilon}(w) \tag{2.1}$$

These produce respectively the weak * and the strong topology; the latter is equivalent to the topology given by the norm

$$\|w\| = \sup_{\|v\|=1} |(w|v)|, \tag{2.2}$$

which makes $\mathbb{E}'$ a Banach space (Problem 4). Its dual space is denoted $\mathbb{E}''$, and $\mathbb{E}'' \supset \mathbb{E}$. If $\mathbb{E}'' = \mathbb{E}$ (identifying elements of $\mathbb{E}''$ with those of $\mathbb{E}$ under the natural injection), then $\mathbb{E}$ is said to be **reflexive**.

2.1.20 Examples

Spaces with $\mathbb{E}' = \mathbb{E}$, such as Hilbert spaces, are clearly reflexive. As shown in Example (2.1.17), L^p is reflexive if $1 < p < \infty$, but not if $p = 1$ or ∞, since $\mathbb{E}$ cannot be reflexive unless $\mathbb{E}'$ is.

2.1.21 Remarks

1. It is also possible to topologize $\mathbb{E}$ weakly, by taking

$$U_{w,\varepsilon}(v) = \{v' \in \mathbb{E} : |(w|v - v')| < \varepsilon,\ w \in \mathbb{E}',\ \varepsilon \in \mathbb{R}^+\}.$$

   It is a corollary of the Hahn-Banach theorem that this is a Hausdorff topology. It is compatible with linearity in the sense that sums of vectors and multiplication by scalars are continuous mappings.
2. As its name suggests, the weak topology is weaker than the strong topology; in the weak topology the mapping $w \to \|w\|$ is not continuous, but only lower semicontinuous, as the supremum of continuous mappings, The weakening of the topology produces additional compact sets: in an infinite-dimensional Banach space the unit ball $\{v : \|v\| \leq 1\}$ fails to be norm-compact, but it is weak-* compact with respect to the space of which it is the dual (if this predual exists). Hence, if the Banach space is reflective, its unit ball is weak-* compact (cf. Problem 7).
3. The weak topologies do not have countable neighborhood bases, and they cannot be specified in terms of sequences; they require instead nets or filters. This means that the concepts of completeness and sequential completeness, and compactness and sequential compactness, are not identical. Hilbert spaces are weakly sequentially complete, but not weak complete. Another inconvenience is that not every point of accumulation is attainable as the limit of a convergent sequence (Problem 8). Fortunately, the **bounded sets**, i.e., $\{v : \|v\| \leq M\}$ in a Banach space with a separable dual space are a metrizable space when weakly topologized. For metric spaces the above notions coincide, and if only bounded sets are considered, these complications can be ignored.

Linear functionals are a special case of linear operators:

2.1.22 Definition

We let $\mathcal{L}(\mathbb{E}, \mathbb{F})$ denote the **space of continuous linear mappings** of the Banach space $\mathbb{E}$ into the Banach space $\mathbb{F}$. If $\mathbb{E} = \mathbb{F}$, define $\mathcal{B}(\mathbb{E}) \equiv \mathcal{L}(\mathbb{E}, \mathbb{E})$. The elements $a \in \mathcal{L}(E, F)$ are also called **operators**.

2.1.23 Examples

1. $\mathcal{L}(\mathbb{E}, \mathbb{C}) = \mathbb{E}'$.
2. $\mathcal{L}(\mathbb{C}^n, \mathbb{C}^m)$ consists of the $n \times m$ matrices.

2.1.24 Remarks

1. $\mathcal{L}(\mathbb{E}, \mathbb{F})$ is a vector space, as $\left(\sum \alpha_i a_i\right)x \equiv \sum \alpha_i a_i x$ for all $\alpha_i \in \mathbb{C}$ and $a_i \in \mathcal{L}(\mathbb{E}, \mathbb{F})$.
2. A linear mapping a is **bounded** iff it sends bounded sets to bounded sets, and thus $\|a\| \equiv \sup_{\|x\|=1} \|ax\|_{\mathbb{F}} < \infty$. For linear mappings the properties
   (i) continuous,
   (ii) continuous at the origin,
   (iii) bounded
   are all equivalent (Problem 11).
3. The transpose of a real, finite-dimensional matrix has an infinite-dimensional generalization: $a \in \mathcal{L}(\mathbb{E}, \mathbb{F})$ induces a mapping $a^* : \mathbb{F}' \to \mathbb{E}'$, known as the **adjoint operator**, since for $y' \in \mathbb{F}'$ the mapping $\mathbb{E} \to \mathbb{C}$ by $x \to (y'|ax)$ is continuous and linear, and consequently, it guarantees the existence of exactly one $x' \in \mathbb{E}'$ such that $(y'|ax) = (x'|x)$. Now define $x' \equiv a^* y'$. It is trivial to verify that the operator a^* is linear, and it is continuous in the norm topology (Problem 5).

   There are several ways to topologize $\mathcal{L}(\mathbb{E}, \mathbb{F})$.

2.1.25 Definition

The neighborhood bases of elements $a \in \mathcal{L}(\mathbb{E}, \mathbb{F})$ can be taken alternatively as

$$U_{y',x,\varepsilon}(a) = \{a' : |(y'|(a - a')x)| < \varepsilon\}$$

or

$$U_{x,\varepsilon}(a) = \{a' : \|(a - a')x\|_{\mathbb{F}} < \varepsilon\} = \bigcap_{\|y'\|=1} U_{y',x,\varepsilon}(a)$$

or

$$U_{\varepsilon}(a) = \bigcap_{\|x\|=1} U_{x,\varepsilon}(a).$$

The topologies are respectively called **weak**, **strong**, and **uniform**, and the associated kinds of convergence will be denoted by $\rightharpoonup$, $\rightarrow$, and $\Rightarrow$ (elsewhere often by w-lim, s-lim, and lim).

2.1.26 Remarks

1. The uniform topology corresponds to the norm $\|a\| = \sup_{\|x\|=1} \|ax\|_{\mathbb{F}}$, which makes $\mathcal{L}(\mathbb{E}, \mathbb{F})$ a Banach space (Problem 4).
2. Even though $\mathbb{E}$ and $\mathbb{F}$ are metrizable, the strong and weak topologies do not have countable neighborhood bases (cf. Problem 9), and only their restrictions to norm-bounded sets are metrizable. They are compatible with linearity, but not with the algebraic structure; multiplication is not necessarily a continuous mapping $\mathcal{L} \times \mathcal{L} \to \mathcal{L}$. However, it is sequentially continuous in the topologies $\mathcal{B}(\mathbb{E})$(weak) $\times$ $\mathcal{B}(\mathbb{E})$(strong) $\to$ $\mathcal{B}(\mathbb{E})$(weak) and $\mathcal{B}(\mathbb{E})$(strong) $\times$ $\mathcal{B}(\mathbb{E})$(strong) $\to$ $\mathcal{B}(\mathbb{E})$(strong). Of course, multiplication in one factor alone is continuous in all topologies.
3. For reflexive Banach spaces $\mathbb{E}$ and $\mathbb{F}$ the adjoint operation (2.1.24; 3) $\mathcal{L}(\mathbb{E}, \mathbb{F}) \stackrel{*}{\to} \mathcal{L}(\mathbb{F}', \mathbb{E}') : a \to a^*$ is a continuous mapping in the norm topology because $\|a\| = \|a^*\|$, and it is obviously continuous in the weak topology; yet it is not continuous in the strong topology. We shall later become acquainted with examples for which $\Omega_n \to \Omega$ but only $\Omega_n^* \rightharpoonup \Omega^*$.
4. The origin of many of the technical complications of quantum mechanics is that the norm topology of operators is too restrictive; one is often interested in a limiting operator of a sequence or family of operators that is not convergent in the norm topology. While weaker limits exist more frequently, the algebraic operations are not always continuous in the weak topologies, so great care must be taken in passing to a limit.
5. If $x_n \in$ a Hilbert space $\mathcal{H}$ converges weakly to $x \in \mathcal{H}$ and $\lim_{n\to\infty} \|x_n\| = \|x\|$, then the sequence is also strongly convergent: $\langle x_n - x | x_n - x\rangle = \|x_n\|^2 + \|x\|^2 - 2\,\mathrm{Re}\langle x_n | x\rangle \to 0$. Hence the strong and weak topologies are equivalent for unitary operators since for them $\|UX\| = \|X\|$. If unitary operators converge weakly but not strongly, the limit will not be unitary.

2.1.27 Problems

1. Show that the space l^∞ is not separable. (Hint: There exists an uncountable set of elements $v_i, i \in \mathbb{I}$, such that $\|v_i\| = 1$ and $\|v_i - v_j\| \geq 1$ whenever $i \neq j$.)

2. Show that the usual operator norm for operators on a Hilbert space satisfies the triangle inequality.

3. Prove the triangle inequality for the spaces L^p, $p \geq 1$. (Hint: the inequality $xy \leq x^p/p + y^q/q$ for $x, y \geq 0$ and $1/p + 1/q = 1$ implies **Hölders inequality**, $|\int fg\, d\mu| \leq \int |fg| d\mu \leq \|f\|_p \|g\|_q$, where $\|f\|_p \equiv (\int |f|^p d\mu)^{1/p}$. Next show that $\|f\|_p = \sup_{g:\|g\|_q=1} \int |fg| d\mu$ and conclude that $\|f + g\|_p \leq \|f\|_p + \|g\|_p$, which is known as **Minkowski's inequality**.

4. Let $\mathbb{E}$ and $\mathbb{F}$ be two Banach spaces. Show that the space of continuous linear mappings $\mathbb{E} \to \mathbb{F}$ (with the uniform topology) is also a Banach space. Moreover, show that if $\mathbb{F}$ is a normed space but not complete, then $\mathcal{L}(\mathbb{E}, \mathbb{F})$ is likewise not complete.

5. Let $a : \mathbb{E} \to \mathbb{F}$ be a continuous linear mapping of two Hilbert spaces. Show that $a^* : \mathbb{F} \to \mathbb{E}$ is also continuous.

6. Prove that on a Hilbert space $\mathbb{E} \|ax\| = \|a^*x\| = \|x\|$ for all $x \in \mathbb{E}$ iff $aa^* = a^*a = 1$.

7. Show that the unit ball in a separable Hilbert space $\mathcal{H}$ is weakly sequentially compact. Conclude that the Hilbert cube $\subset l^2 : \{v = (v_n) : |v_n| \leq 1/n\}$ is even strongly ($\equiv$ norm) compact.

8. Show that an infinite-dimensional Hilbert space is not metrizable in the weak topology. (Hint: Consider the vectors $x_n = (0, 0, \ldots, \sqrt{n}, 0, \ldots)$ in l^2. This set has a point of accumulation at 0, but it contains no convergent subsequences, which is impossible in a metric topological space).

9. Show that in the weak topology, compactness does not imply sequential compactness (except when the Hilbert space is separable).

10. Prove the Cauchy-Schwarz inequality $|\langle v_1|v_2\rangle| \leq \|v_1\|\|v_2\|$, and show that $|\langle v_1|v_2\rangle| = \|v_1\|\|v_2\|$ iff $v_1 = zv_2$ for some $z \in \mathbb{C}$ (and $v_i \neq 0$).

11. Show the equivalence of the properties of (2.1.24; 2).

2.1.28 Solutions

1. Let v_i be the vectors of the form $(c_1, c_2, \ldots, c_n, \ldots)$ with $c_n = 1$ or 0. This set has the power of the continuum, and $\|v_i\| = \sup |c_n| = 1$ (unless $v_i \equiv 0$) and $\|v_i - v_j\| \geq 1$, unless all the coefficients of v_i and v_j are equal. If there exists a countable dense set $A \subset l^\infty$, then for all v_i there would be an $a_i \in A$ with $\|v_i - a_i\| \leq \frac{1}{3}$. Since $\|v_i - v_j\| \geq 1$, the mapping $v_i \to a_i$ would be one-to-one, and the set v_i would have only the cardinality of a subset of A.

2. $\|a + b\| = \sup_{\|x\|=1} \|ax + bx\| \leq \sup_{\|x\|=1} \|ax\| + \sup_{\|x\|=1} \|bx\| = \|a\| + \|b\|$.

3. The inequality is trivial for $p = 1$, so assume $p > 1$. For $t \geq 0$, $t \leq t^p/p + 1/q$ (proof: find the minimum of the function $\varphi(t) = t^p/p - t$), and if $t = x/y^{q/p}$ this reads $xy \leq x^p/p + y^q/q$. Let f and g be two functions such that

$$\int |f|^p d\mu = \int |g|^q d\mu = 1.$$

Since

$$\left| \int fg\, d\mu \right| \leq \int |fg| d\mu \text{ and } \int |fg| d\mu \leq \frac{1}{p} \int |f|^p d\mu + \frac{1}{q} \int |g|^q d\mu = 1,$$

Hölder's inequality is proved in this special case. The general case follows by considering $f/\|f\|_p$ and $g/\|g\|_q$ in place of f and g. Furthermore,

$$\|f\|_p = \int |fg| d\mu \quad \text{with} \quad g = \frac{|f|^{p-1}}{\|f\|_p^{p/q}},$$

since

$$\int |fg| d\mu = \frac{\|f\|_p^p}{\|f\|_p^{p/q}} = \|f\|_p, \; \|g\|_q^q = \|f\|_p^{-p} \|f\|_p^p = 1.$$

Hence

$$\|f+g\|_p = \sup_{\|h\|_q=1} \int |(f+g)h| d\mu \leq \sup_h \int |fh| d\mu + \sup_h \int |gh| d\mu = \|f\|_p + \|g\|_p.$$

4. The vector-space properties are trivial. As for the norm, let $a : \mathbb{E} \to \mathbb{F}$,

$$\|a\| = \sup_{x \in \mathbb{E}} \frac{\|ax\|}{\|x\|}.$$

Then $\|\lambda a\| = |\lambda| \|a\|$ and $\|a + b\| \leq \|a\| + \|b\|$ as in Problem 2. Finally, $\|a\| = 0 \Rightarrow \|ax\| = 0$ for all $x \in \mathbb{E} \Rightarrow ax = 0 \Rightarrow a = 0$. As for completeness, let a_n be a Cauchy sequence; then $a_n x$ is also a Cauchy sequence in $\mathbb{F}$ for all $x \in \mathbb{E}$, and thus there exists a limit $\lim a_n x = ax \in \mathbb{E}$. This mapping is linear and bounded (since $\|a_n\| \leq C < \infty$ for all n, $\|a_n x\| \leq C\|x\|$, which implies $\|ax\| \leq C\|x\|$), and $\|a - a_n\| = \sup \|ax - a_n x\| / \|x\| \to 0$. The proof depends in an essential way on the completeness of $\mathbb{F}$. Remark: The Hahn-Banach theorem prevents $\mathcal{L}(\mathbb{E}, \mathbb{F})$ from being the trivial space $\{0\}$.

5. $\|a\| = \sup_{\substack{\|x\|=1 \\ x \in \mathbb{E}}} \|ax\| = \sup_{\substack{\|x\|=1, x \in \mathbb{E} \\ \|y\|=1, y \in \mathbb{F}}} |\langle y|ax\rangle| = \sup |\langle a^* y|x\rangle| = \|a^*\|.$

   (This is also true when $\mathbb{E}$ and $\mathbb{F}$ are only assumed to be Banach spaces.)

6. $aa^* = a^*a = 1 \Rightarrow \langle x|aa^*x\rangle = \|a^*x\|^2 = \langle x|a^*ax\rangle = \|ax\|^2 = \|x\|^2.$

$$\|ax\| = \|x\| \Rightarrow \langle x|a^*ax\rangle = \langle x|x\rangle \Rightarrow a^*a = 1,$$

and likewise

$$\begin{aligned} \|a^*x\| &= \|x\| \Rightarrow aa^* = 1. \\ (4\langle y|ax\rangle &= \langle x+y|a(x+y)\rangle - \langle x-y|a(x-y)\rangle \\ &\quad +i\langle x+iy|a(x+iy)\rangle - i\langle x-iy|a(x-iy)\rangle; \end{aligned}$$

it therefore follows from $\langle x|ax\rangle = 0$ for all x that $\langle y|ax\rangle = 0$ for all x and y, which implies that $a = 0$.)

7. Let $\{x_n\}$ be a total orthonormal set. Since the matrix elements $(x_n|a_k x_m)$ of a sequence $a_k \in \mathcal{B}(\mathcal{H})$ are bounded in absolute value by $\|x_n\|\|x_m\| \sup_k \|a_k\|$, for every n and m there is a point of accumulation a_{nm}. Let us define $a \in \mathcal{B}(\mathcal{H})$ by $ax_m = \sum_n a_{nm} x_n$ and note that $a_k \rightharpoonup a$, i.e., $(y|a_k x) \to (y|ax)$ for all $y, x \in \mathcal{H}$; this is because for all $\varepsilon > 0$, x and y can be written as $\sum_{\text{finite}} c_n x_n + \eta$ with $\|\eta\| < \varepsilon$, and the convergence of the finite sum follows by definition, so the convergence of general matrix elements is shown to within arbitrary accuracy, since $\sup_n \|a_n\| < \infty$. We next show that the strong and weak topologies are equivalent on the Hilbert cube: Let $v^{(n)} \rightharpoonup v$, and for any given ε choose r and N such that $\sum_{j=r+1}^{\infty} 1/j^2 < \varepsilon$ and if $n > N$ then

$$\sum_{j=1}^{r} |v_j^{(n)} - v_j|^2 < \varepsilon.$$

$\sum_{j=1}^{\infty} |v_j^{(n)} - v_j|^2 < 5\varepsilon$, and therefore $v^{(n)} \to v$. The Hilbert cube is thus strongly sequentially compact and, since the strong topology on a Hilbert space is a metric topology, also strongly compact.

8. A weak neighborhood of 0 has the form $U = \{x : |\langle v^{(1)}|x\rangle| + |\langle v^{(2)}|x\rangle| + \cdots + |\langle v^{(l)}|x\rangle| \le \varepsilon\}$. A neighborhood U necessarily contains some x_n, as otherwise $|\langle v^{(1)}|x_n\rangle| + \cdots + |\langle v^{(l)}|x_n\rangle| > \varepsilon$ for all n, which would mean that $|v_n^{(1)}| + \cdots + |v_n^{(l)}| > \varepsilon/\sqrt{n}$ and $\sum_{n=1}^{\infty} |v_n^{(1)}|^2 + \cdots + |v_n^{(l)}|^2 = \infty$, while, on the other hand, this sum $\le l\left(\sum_n |v_n^{(1)}|^2 + \cdots + \sum_n |v_n^{(l)}|^2\right) < \infty$. However, it is not true that there is an N such that $x_n \in U$ for all $n > N$. Despite the foregoing, there cannot be a weakly convergent subsequence of x_{n_k} of x_n; for consider the vector v whose n_{10^r}-th component is $1/r$ and whose other components are all 0. Then $\langle v|x_{n_k}\rangle = \sqrt{n_k}(1/r)$ if $k = 10^r$, and otherwise 0; but $n_k \ge k = 10^r$.
9. General theorems guarantee that the unit ball is always compact in the weak operator topology. Let us now investigate the nonseparable Hilbert space $\mathcal{H} = L^2([0,1], \mu)$, where μ is the measure assigning every point the measure 1. All of the multiplication operators multiplying any function in $\mathcal{H}$ by a function φ_n, a "saw-tooth" function going linearly from 0 to 1 in each interval $[k/10^n, (k+1)/10^n]$, $k \in \mathbb{Z} \cap [0, 10^n)$, have norm 1. But, even so, for each subsequence φ_{n_m} there exists a point x at which $\varphi_{n_m}(x)$ diverges, and consequently the sequence of operators is not weakly sequentially compact.
10. Let $v_2' = v_2 \exp(-i \arg\langle v_1|v_2\rangle)$. Then $|\langle v_1|v_2\rangle| = \langle v_1|v_2'\rangle$, and it follows from

$$\| \|v_1\| v_2' - \|v_2'\| v_1 \|^2 = 2\|v_1\|^2 \|v_2'\|^2 - 2\|v_1\| \|v_2'\| \langle v_1|v_2'\rangle \ge 0$$

that $\langle v_1|v_2'\rangle \le \|v_1\| \|v_2'\| = \|v_1\| \|v_2\|$. There can be equality only if $\|v_1\| v_2' - \|v_2'\| v_1 = 0$, i.e., $v_1 = z v_2$ with $z = (\|v_1\|/\|v_2\|) \exp(-i \arg\langle v_1|v_2\rangle)$.

11. (ii) $\Rightarrow$ (iii): Property (ii) implies that for all δ there exists ε such that $\|x\| < \varepsilon \Rightarrow \|ax\| < \delta$, which implies $\|a\| = \sup_{\|x\|=\varepsilon} \|ax\|/\|x\| < \delta/\varepsilon$, which $\Rightarrow$ (iii). (iii) $\Rightarrow$ (i): For all $\delta \exists \varepsilon = \delta/\|a\|$ such that for all $x' \in \mathbb{E}$,

$$\|x - x'\| < \varepsilon \Rightarrow \|ax - ax'\| \le \|a\| \cdot \|x - x'\| \le \delta.$$

(i) $\Rightarrow$ (ii) is trivial.

2.2 Algebras

C^* and W^* algebras are generalizations of algebras of matrices and functions. Their axioms are the basic algebraic and topological properties of these familiar algebras.

2.2.1 Definition

An **algebra** $\mathcal{A}$ is a vector space on which there is a mapping $\mathcal{A} \times \mathcal{A} \to \mathcal{A}$, called **multiplication**, having the properties

$$a(b_1 + b_2) = ab_1 + ab_2, \qquad a(bc) = (ab)c,$$

$$a(\alpha b) = \alpha ab, \qquad a, b, b_i \in \mathcal{A}, \qquad \alpha \in \mathbb{C}.$$

Additionally, we assume the existence of a **unit**, or **identity**, element $\mathbf{1}$ such that $a\mathbf{1} = \mathbf{1}a = a$ for all $a \in \mathcal{A}$; if this element should ever be lacking, we shall refer to $\mathcal{A}$ as an **algebra without a unit**. If $ab = ba$ for all a and $b \in \mathcal{A}$, then $\mathcal{A}$ is **Abelian**.

2.2.2 Examples

All of Examples (2.1.2) are algebras when multiplication is defined componentwise for vectors, pointwise for functions, and in the usual way for matrices. These multiplication rules make all of them Abelian except for the matrices. The spaces L^p, $p < \infty$, are not generally algebras; for examples, $x^{-1/2} \in L^1([0, 1], dx)$ but $x^{-1} \notin L^1([0, 1], dx)$. The spaces l^p are algebras, but they have no unit if $p < \infty$. The space $l^0 \equiv \{(v_1, v_2, \ldots) \in l^\infty : \lim_i |v_i| = 0\}$ is a subalgebra of l^∞ without a unit.

2.2.3 Remark

Every subspace of a vector space is the kernel of a homomorphism π, i.e., it is $\pi^{-1}(0)$. The kernels of homomorphisms of an algebra are only its **two-sided ideals**, i.e., subalgebras $\mathcal{B} \subset \mathcal{A}$ for which $a\mathcal{B} \subset \mathcal{B}$ and $\mathcal{B}a \subset \mathcal{B}$ for all, $a \in \mathcal{A}$. The quotient space with respect to a two-sided ideal is another algebra, known as the **quotient algebra**.

Since we work with the field of the complex numbers, there is another operation to axiomatize, complex conjugation:

2.2.4 Definition

A * **algebra** is an algebra on which there is a mapping $* : \mathcal{A} \to \mathcal{A}$, called **conjugation**, having the properties $(ab)^* = b^*a^*$; $(a + b)^* = a^* + b^*$; $(\alpha a)^* = \alpha^* a^*$ for $\alpha \in \mathbb{C}$; and $a^{**} = a$. The element a^* is known as the **adjoint** of a.

2.2.5 Examples

If * is complex conjugation or, in the case of matrices, Hermitian conjugation, then all of Examples (2.1.2) except for the analytic functions are * algebras.

2.2.6 Remark

It is at this point that complex numbers first become important. Anyone having philosophical objections to the occurrence of complex numbers in what ultimately pertains only to real physical measurements can just as well represent i, the square root of -1, as the real matrix

$$\begin{pmatrix} 0 & 1 \\ -1 & 0 \end{pmatrix}$$

or else postulate the existence of an abstract element $\mathbf{I} \in \mathcal{A}$ with the properties $\mathbf{I}^2 = -\mathbf{1}$, $\mathbf{I}^* = -\mathbf{I}$, and $\mathbf{I}a = a\mathbf{I}$ for all a.

Since matrices are the prototype of a * algebra, its elements are often referred to as operators, and the terminology follows that of matrices:

2.2.7 Definition

a is **normal** iff $aa^* = a^*a$
a is **Hermitian** iff $a = a^*$
a is **unitary** iff $aa^* = a^*a = \mathbf{1}$
a is a **projection**[1] iff $a = a^* = a^2$
a is **positive** iff $a = bb^*$ for some b
a is the **inverse** of b iff $ab = ba = \mathbf{1}$.

2.2.8 Remarks

1. The relationships among these sets of operators are depicted in the diagram below:

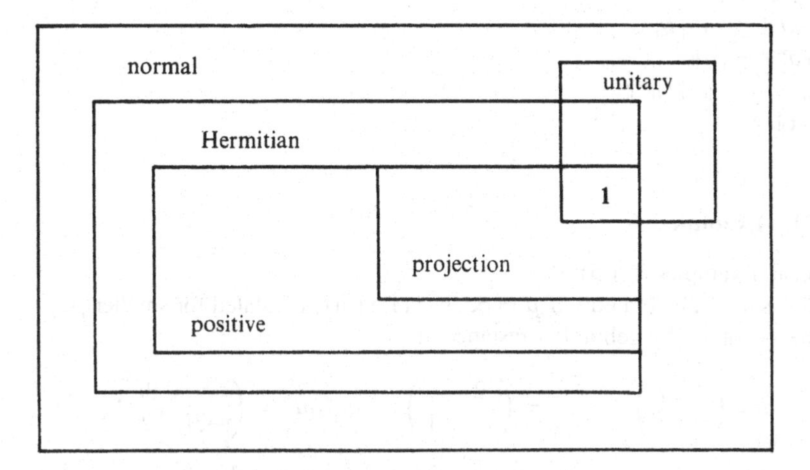

2. Although in a finite-dimensional space $ab = \mathbf{1}$ implies $ba = \mathbf{1}$, this is not true in general. A counterexample is given by the infinite matrix

$$a = \left| \begin{matrix} 0 & 1 & 0 & 0 & 0 & \dots \\ 0 & 0 & 1 & 0 & 0 & \dots \\ 0 & 0 & 0 & 1 & 0 & \dots \\ \dots & & & & & \end{matrix} \right. \quad \text{and} \quad b = a^*,\ aa^* = 1 \neq a^*a$$

Hence the property $ab = \mathbf{1}$ is not sufficient to make a the inverse of b.

The Definitions (2.2.7) easily imply the

[1] In this book the word "projection" will be understood as meaning "orthogonal projection."

2.2.9 Propostions

1. $(a^{-1})^{-1} = a$
2. $(ab)^{-1} = b^{-1}a^{-1}$
3. $(a^*)^{-1} = (a^{-1})^*$
4. The unitary elements form a subgroup of the group of invertible elements.

The next subject is the topology of the algebra, which must conform with the algebraic properties discussed above. This will allow us to generalize the analytic rules we are familiar with for matrices.

2.2.10 Definition

A C^* **algebra** is at the same time a * algebra and a Banach space, the norm of which satisfies
(i) $\|ab\| \leq \|a\|\|b\|$
(ii) $\|a^*\| = \|a\|$
(iii) $\|aa^*\| = \|a\|\|a^*\|$
(iv) $\|\mathbf{1}\| = 1$.

2.2.11 Examples

Recall Examples (2.1.5):
1. This is a C^* algebra only if $p = \infty$, because (iii) is violated for smaller p.
2. This is not a C^* algebra. For instance, if

$$a = \begin{pmatrix} 1 & i \\ 0 & 1 \end{pmatrix}, \quad aa^* = \begin{pmatrix} 2 & i \\ -i & 1 \end{pmatrix}, \quad aa^*aa^* = \begin{pmatrix} 5 & 3i \\ -3i & 2 \end{pmatrix},$$

then $\|aa^*\|^2 = \text{Tr}\, aa^*aa^* = 7 \neq \|a\|^2\|a^*\|^2 = (\text{Tr}\, aa^*)^2 = 9$.
3. This is a C^* algebra, as in fact is the more general $\mathcal{B}(\mathcal{H})$ (2.1.22), $\mathcal{H}$ a Hilbert space, and with the norm of (2.1.26; 1), because

$$\begin{aligned} \|a^*a\| &= \sup_{\|x\|=1=\|y\|} |\langle y|a^*ax\rangle| = \sup_{\|x\|=1} \langle x|a^*ax\rangle \\ &= \sup_{\|x\|=1} \|ax\|^2 = \|a\|^2, \end{aligned}$$

along with (i) and (ii), implies (iii).
4. The space of this example is not complete.
5. In this example the space is complete only if $r = 0$, in which case it is a C^* algebra.
6. L^p is not an algebra for $p < \infty$. If $p = \infty$, it is C^* algebra.

2.2.12 Remarks

1. Properties (i) and (ii) guarantee that multiplication and conjugation are continuous (Problem 3). Property (iii) roots the topology so deeply in the algebraic structure that (algebraic) homomorphisms of C^* algebras are norm preserving and thus automatically continuous (Problem 2). Property (iv) is just a convenient normalization.
2. It may happen that Property (iii) is satisfied by one norm and violated by another, although both norms produce the same topology. This occurs in Examples (2.2.11; 2) and (2.2.11; 3) as well as for the continuous functions on $[0, 1]$ with the norms $\|f\| = \sup_{x\in[0,1]} |f(x)|$ and

$$\|f\|_e = \sup_{x\in[0,1]} e^{-x}|f(x)|,$$

which are related by $\|\cdot\| \geq \|\cdot\|_e \geq e^{-1}\|\cdot\|$. The norm $\|\cdot\|$ yields a C^* algebra, but $\|\cdot\|_e$ does not, since $\|(e^x-1)\|_e^2 = (1-1/e)^2 \leq \|(e^x-1)^2\|_e = e+1/e-2$. In such situations we shall always choose the norm that satisfies (2.2.10).
3. $\mathbb{C}\backslash\{0\}$ is an open set, and $\{z \in \mathbb{C} : |z| = 1\}$ and $\{z \in \mathbb{C} : \operatorname{Im} z = 0\}$ are closed. Because both conjugation (*) and multiplication are continuous, these statements have the generalizations that the set of invertible elements of a C^* algebra is open, and that the unitary and the Hermitian elements form closed sets (Problem 4). Similarly, the sets of normal, positive, and projection operators are closed, and hence norm-limits of these types of operators are of the same types.

Given an operator a, it is always possible to get an invertible operator by adding some multiple of the identity $\mathbf{1}$ to it.

2.2.13 Definition

The **resolvent set** of $a \in \mathcal{A}$ is $\{z \in \mathbb{C} : (a - z)^{-1}$ exists$\}$, and its complement $\mathrm{Sp}(a)$ is known as the **spectrum**.

2.2.14 Examples

The spectrum of a matrix consists of its eigenvalues, and the spectrum of an ordinary function is its range.

2.2.15 Remarks

1. In (2.2.13) it is essential that $(a - z)^{-1}$ exist as an element of $\mathcal{A}$, and not in some other sense. Moreover, if $\mathcal{A}$ is a subalgebra of some other algebra $\mathcal{B}$, then one must specify whether $(a - z)^{-1}$ is to exist in $\mathcal{A}$ or $\mathcal{B}$. Fortunately, if $\mathcal{A}$ is a C^* algebra, then the inverse of $a - z$ belongs to the C^* algebra generated by a (that is, the norm-closure of the polynomials in a and a^*), and so one need not specify which algebra the inverse belongs to.

2. If $|z| > \|a\|$, then $(z - a)^{-1}$ can be expanded as a convergent series $z^{-1}\sum_{n=0}^{\infty}(a/z)^n$, and therefore $\mathrm{Sp}(a) \subset \{z \in \mathbb{C} : |z| \leq \|a\|\}$. In particular, all elements such that $\|a - \mathbf{1}\| < 1$ are invertible. The mapping $\mathbb{C} \to \mathcal{A} : z \to (a - z)^{-1}$ is actually analytic on the resolvent set, which is always open, by Problem 7.
3. It is easy to show that $\mathrm{Sp}(a^*) = \mathrm{Sp}(a)^*$ and $\mathrm{Sp}(P(a)) = P(\mathrm{Sp}(a))$ for any polynomial P and $a \in \mathcal{A}$. This implies (Problem 5) that the spectra of the unitary, Hermitian, positive, and projection operators lie respectively on the unit circle, the real axis, the positive real axis, and the set $\{0, 1\}$. If the operator is normal, the fact that the spectrum belongs to one of these sets implies that the operator belongs to the appropriate class (2.2.7) (Problem 5).
4. As the term "spectrum" suggests, the spectral values of an element represent the values it can attain in a certain sense; we shall see in (2.2.31; 2) that the convex combinations of the spectral values are all the possible expectation values of the element (2.2.18).

Positivity is a useful property in analysis, and it provides an algebra with an additional associative structure:

2.2.16 Definition

The algebra $\mathcal{A}$ has a **partial ordering** $a \geq b$ defined as meaning that $a - b$ is positive.

2.2.17 Remarks

1. As remarked in (2.2.15; 3), positivity is synonymous with having a positive spectrum (if $a^* = a$). According to (2.2.15; 4) the sum of two positive elements is positive, since expectation values are additive. Hence if $a \geq b$ and $b \geq c$, then $a \geq c$. If $a \geq 0$ and $-a \geq 0$, then $a = 0$, since 0 is the only Hermitian element a with $\mathrm{sp}(a) = \{0\}$. Thus $a \geq b$ and $b \geq a$ implies $a = b$. Since it is also true that $a \geq a$, the relationship $\geq$ is a partial ordering. Since positive operators are Hermitian, one might hope to extend the definition of $\geq$ to all Hermitian elements, but it fails to be a total ordering on this set: Consider:

$$\begin{pmatrix} 1 & 0 \\ 0 & 0 \end{pmatrix} \quad \text{and} \quad \begin{pmatrix} 0 & 0 \\ 0 & 1 \end{pmatrix}$$

which do not stand in this ordering relationship to each other.
2. Although it is true that $\geq$ is compatible with the linear structure of $\mathcal{A}$ in the sense that $a_i \geq b_i \Rightarrow \sum_i a_i \geq \sum_i b_i$, difficulties arise with products because the product of two Hermitian elements is not generally Hermitian. But even if it is, there remains the inconvenience that inequalities cannot be multiplied, $a \geq b \not\Rightarrow a^2 \geq b^2$ (Problem 10). Yet inverses of ordered positive operators have a definite ordering, $a \geq b > 0 \Rightarrow b^{-1} \geq a^{-1} > 0$, and as a consequence it is possible to show monotonicity of certain functions of operators with respect to

the ordering $\geq$ (Problem 11). Finally, note that $a \geq b$ clearly implies $c^*ac \geq c^*bc$ for any c.
3. The partial ordering is compatible with the topological structure; it commutes with the taking of limits.
4. Positivity is preserved by homomorphisms $\pi : \mathcal{A} \rightarrow \mathcal{B}$ of C^* algebras: $\pi(a^*a) = \pi(a)^*\pi(a) \geq 0$, and therefore $a \geq b \Rightarrow \pi(a) \geq \pi(b)$. Linear mappings of C^* algebras do not generally preserve positivity.

2.2.18 Definition

A linear functional f is **positive** iff $f(aa^*) \geq 0$ for all $a \in \mathcal{A}$. If, moreover, $f(\mathbf{1}) = 1$, then f is called a **state**, and $f(a)$ is the **expectation value** of a in the state f. If $f(aa^*) > 0 \, \forall \, a \neq 0$, f is called *faithful.*

2.2.19 Examples

1. Positive measures on function algebras are positive linear functionals. Probability measures are states.
2. The mapping $m \rightarrow \text{Tr } \rho m$ on $n \times n$ matrices m is positive iff ρ is positive in the sense that all of its eigenvalues are positive. If in addition $\text{Tr } \rho = 1$, then it is a state.
3. On the C^* subalgebra $\{v \in l^\infty : \lim_{i\rightarrow\infty} v_i \text{ exists}\}$ of l^∞, the functional $f(v) = \lim_{i\rightarrow\infty} v_i$ is a state.

2.2.20 Remarks

1. Definition (2.2.18) does not require continuity, i.e., the statement that there exists $M \in \mathbb{R}^+$ such that $|f(a)| < M\|a\|$ for all $a \in \mathcal{A}$, because it follows automatically. It is even true that $|f(b^*ab)| \leq \|a\| f(b^*b)$ and, as a generalization of the Cauchy-Schwarz inequality,
$$|f(b^*a)|^2 \leq f(b^*b)f(a^*a)$$
(Problem 8). It is consequently always possible to normalize a positive linear functional to be a state, which $\|f\| = \sup_{a\in\mathcal{A}} |f(a)|/\|a\| = 1$.
2. Convex combinations of states are states. States that cannot be written as convex combinations of other states are called **extremal**, or **pure**. In Examples (2.2.19), integrals with delta functions and traces with one-dimensional projections are pure states. A theorem of Krein and Milman [1, 12.15] says that our naive idea of convex, compact sets is valid for states; there must exist extremal points, and their convex combinations are dense in the space of states. Choquet's theorem allows any state to be written as an integral over pure states, though the measure used is unique only if the algebra is Abelian. For example, the state $m \rightarrow (1/n)\text{Tr } m$ of $n \times n$ matrices can be written as $(1/n)\sum_{k=1}^n \langle e_k|me_k\rangle$, where the e_k are any orthonormal system. The state $m \rightarrow \langle e_k|me_k\rangle$ (no sum) is pure, so there are many ways to write $(1/n)\text{Tr}$ as a convex combination of pure states. If the space

of states is pictured as a ball, then the pure states will constitute its surface. For Abelian algebras this ball becomes instead a simplex, only the corners of which are extremal. The extremal points of infinite-dimensional simplices may form a connected, closed set, like the surface of a ball. For instance, consider the Abelian C^* algebra of continuous functions on a compact set. The states are probability measures, and the extremal states are Dirac δ functions. They form a weak-* connected, closed set (see (2.2.28)), though their convex combinations are weak-* dense in the set of states.

3. There exist pure states for which the inequality $|f(a)| \le \|a\|$ of Remark 1 becomes an equality. This can be seen as follows: Given any $a \in \mathcal{A}$, one can construct a state for which $f(a^*a) = \|a\|^2$, by setting $f(\alpha+\beta a^*a) = \alpha+\beta\|a\|^2$ on the subspace spanned by $\mathbf{1}$ and a^*a. It is easy to convince oneself that this is a positive functional with $f(\mathbf{1}) = 1$ and $f(a^*a) = \|a\|^2$. According to theorems of Hahn and Banach and of Krein, the functional can be extended to all of $\mathcal{A}$ (but not necessarily uniquely, of course). Now let Z_a be the convex set of states such that $f(a^*a) = \|a\|^2$. The extremal points of Z_a are pure, since if $f_e = \lambda f_1 + (1-\lambda) f_2$, $0 < \lambda < 1$, then we would find $\|a\|^2 = \lambda f_1(a^*a) + (1-\lambda) f_2(a^*a)$, which implies that $f_i(a^*a) = \|a\|^2$, and f_e would not be extremal.
4. Another blemish afflicting positive linear functionals is that the supremum of linear functionals over many elements may not be the same as the linear functional of the maximal element using the partial ordering $\ge$. For example, if

$$v^{(n)} = (1, 1, \dots, \underbrace{1,}_{n\text{-th position}} 0, 0, \dots) \in l^\infty,$$

then with the state of Example (2.2.19; 3), $f(v^{(n)}) = 0$, but with $v \equiv \sup_n v^{(n)} = (1, 1, 1, \dots)$, $f(v) = 1$. (Of course, $v^{(n)} \not\rightarrow v$.)

The states suffering these afflictions can be set aside by

2.2.21 Definition

An **ascending filter** F is a norm-bounded subset of $\mathcal{A}$ in which any two elements are both exceeded (in the sense of $\ge$) by some element of F. The **supremum** $\sup F$ is the smallest element of $\mathcal{A}$ with $a \le \sup F$ for all $a \in F$. A state f is **normal** iff $\sup_{a \in F} f(a) = f(\sup F)$ for every ascending filter F.

If the supremum always exists in $\mathcal{A}$, and there are also sufficiently many normal states at hand, then the algebra has such nice properties that it merits a special name.

2.2.22 Definition

A W^* **algebra** is a C^* algebra in which
(i) every ascending filter achieves its supremum in $\mathcal{A}$; and
(ii) for all nonzero elements $a \in \mathcal{A}$, there exists a normal state f with $f(a) \neq 0$.

2.2.23 Examples

Matrices are W^* algebras, while the continuous functions on a compact set $\subset \mathbb{C}^n$ are not, because their supremum need not be continuous. The set of bounded, measurable functions $L^\infty(K, d^n x)$ is a W^* algebra. We saw earlier that it is a C^* algebra, and (i) is satisfied since monotonic, bounded sequences converge in L^∞. As for (ii), positive, normalized functions in $L^1 \subset (L^\infty)'$ provide the required normal states.

2.2.24 Remarks

1. Although the W^* property is defined with reference only to the ordering structure of the algebra, it will have both algebraic and topological consequences.
2. Integration theory relies on classes of functions that allow the taking of suprema. The permutability with integration is a fundamental characteristic of measures, distinguished them from such things as abstract averages. With W^* algebras much of measure theory can be generalized to the noncommutative case.
3. In atomic physics we shall primarily be concerned with the W^* algebra $\mathcal{B}(\mathcal{H})$, and the reader interested only in these problems need not worry much about the distinctions mentioned above. It is not until the fourth volume, *Quantum Mechanics of Large Systems*, that these notions will become important in the limit of infinite systems.

Because $\mathbb{C}$ is such a trivial space, the homomorphisms of C^* algebras into it are particularly simple. They are only of interest for Abelian C^* algebras, for which they completely determine the algebra's structure.

2.2.25 Definition

An algebraic *-homomorphism χ (i.e., $\chi(\alpha a + \beta b) = \alpha\chi(a) + \beta\chi(b)$, $\chi(ab) = \chi(a)\chi(b)$, and $\chi(a^*) = \chi(a)^*$ for all $a, b \in \mathcal{A}$ and $\alpha, \beta \in \mathbb{C}$) of an Abelian C^* algebra into $\mathbb{C}$ is called a **character**. The set of characters of $\mathcal{A}$ is denoted $X(\mathcal{A})$.

2.2.26 Examples

1. The characters of the algebra of $n \times n$ diagonal matrices are the maps $\chi_m : a \to a_{mm}$, $1 \leq m \leq n$; but the state $a \to \langle e|ae\rangle$ over this algebra is not necessarily a character for an arbitrary unit vector e.
2. The characters of the algebra $C(K)$ of continuous functions on a compact set $K \subset \mathbb{C}^n$ are of the form $\chi_z : f \to f(z)$, $z \in K$.

2.2.27 Remarks

1. Since the algebraic relationships are preserved by χ, the existence of $(a-z)^{-1}$ implies that of $(\chi(a)-z)^{-1}$. Hence, for all $\chi \in X(\mathcal{A})$, $\chi(a) \in \mathrm{Sp}(a)$, and thus $|\chi(a)| \leq \|a\|$.
2. Since $\chi(a^*a) = \chi(a)^*\chi(a) = |\chi(a)|^2 \geq 0$ and $\chi(\mathbf{1}) = 1$, every character is a state, which automatically makes the mapping $\chi : \mathcal{A} \to \mathbb{C}$ continuous. If χ were $\alpha\chi_1 + (1-\alpha)\chi_2$ with $\chi_1(a) > \chi_2(a)$ for some $a > 0$ such that $\chi(a) = c\chi_1(a)$ with $c < 1$ then $\alpha\chi_1(a^n) \leq \chi(a^n) = \chi(a)^n = c^n(\chi_1(a))^n \leq c^n\chi_1(a^n)$, $\forall_n$ and thus $\alpha = 0$. This shows that $\chi(a)$ consists of the extremal points of a convex set. Prop. (2.2.1.2) will imply that $X(\mathcal{A})$ contains all pure states; they provide irreducible representations of $\mathcal{A}$, which are one-dimensional if the algebra is Abelian, and therefore characters.
3. The kernel $\{a \in \mathcal{A} : \chi(a) = 0\}$ is a closed, two-sided ideal of $\mathcal{A}$. Since $\mathbb{C}$ has no proper ideals, the kernel is maximal in the sense that there are no larger proper ideals containing it. The converse of this statement is also true: to every maximal ideal there corresponds a character. Thus characters, pure states, and maximal ideals are bijectively related.
4. The set of characters $X(\mathcal{A})$ has the weak-* topology as a subset of $\mathcal{A}'$. Weak-* limits clearly preserve the algebraic characterization of $X(\mathcal{A})$ (for instance, $\chi_n(a) \to \chi(a)$ and $\chi_n(b) \to \chi(b) \Rightarrow \chi_n(ab) = \chi_n(a)\chi_n(b) \to \chi(a)\chi(b)$). Therefore $X(\mathcal{A})$ is a weak-* closed subset of the unit ball of $\mathcal{A}'$ and thus, according to Remark (2.1.21; 2), weak-* compact. By definition the mappings $X(\mathcal{A}) \to \mathbb{C} : \chi \to \chi(a)$ are weak-* continuous.

Since $\mathcal{A}$ is a subset of $\mathcal{A}''$, the elements $a \in \mathcal{A}$ can be considered as functions on $X(\mathcal{A})$, by setting $a(\chi) \equiv \chi(a)$. There is in fact a complete correspondence:

2.2.28 The Gel'Fand Isomorphism

Any Abelian C^* algebra $\mathcal{A}$ is isomorphic to the C^* algebra of the continuous functions $C(X(\mathcal{A})) : X(\mathcal{A})$ (with the weak- topology) $\to \mathbb{C}$.*

Proof: The mapping $\mathcal{A} \to C(X(\mathcal{A})) : a \to a(\chi)$ preserves all the algebraic properties such as $a_1a_2(\chi) = \chi(a_1a_2) = \chi(a_1)\chi(a_2) = a_1(\chi)a_2(\chi)$. Since $X(\mathcal{A})$ contains the pure states, Remark (2.2.20; 3) states that $\|a\| = \sup_{\chi \in X(\mathcal{A})} |a(\chi)|$, so the norms of $\mathcal{A}$ and $C(X(\mathcal{A}))$ are the same. It also follows from this that $a(\chi) = 0$ for all $\chi \Rightarrow a = 0$, and it only remains to show that $\mathcal{A}$ contains all the continuous functions on $X(\mathcal{A})$: A theorem of Weierstrass states that the polynomials in $z \in$ any compact set $\subset \mathbb{C}$ are dense in the continuous functions in the supremum topology. Stone [1, 7.3] generalized this to the statement that the norm-closure of any algebra of complex-valued functions with a unit and such that for all $\chi_1 \neq \chi_2$ there is an f with $f(\chi_1) \neq f(\chi_2)$ contains all continuous functions, and the $a(\chi)$ satisfy this requirement. Consequently $a \to a(\chi)$ is a bijection, preserving the algebraic and topological structure. □

2.2.29 Examples

1. In Example (2.2.26; 1), $X(\mathcal{A}) = \{\chi_1, \chi_2, \ldots, \chi_n\}$ (with the discrete topology), and $C(X(\mathcal{A})) = \{a_{mm} \in \mathbb{C}, m = 1, 2, \ldots, n : X_m \to a_{mm} X_m\}$ is the set of diagonal matrices.
2. In Example (2.2.26; 2), we already have a bijection between K and $X(C(K))$, $z \to \chi_z$. According to (2.2.28) the bijection is in fact a homeomorphism if $X(C(K))$ is equipped with the weak-* topology.

2.2.30 Remarks

1. The dual space of the continuous functions F is the space $M(F)$ of (not necessarily positive) measures. The following collection of continuous mappings into $\mathbb{C}$ summarizes the various identifications:

$$\mathcal{A} \xrightarrow{\mathcal{A}' \supset X(\mathcal{A})} \mathbb{C}$$

$$X(\mathcal{A}) \xrightarrow{C(X(\mathcal{A})) \equiv \mathcal{A}} \mathbb{C}$$

$$C(X(\mathcal{A})) \xrightarrow{M(C(X(\mathcal{A}))) \equiv \mathcal{A}'} \mathbb{C}.$$

2. These results for Abelian C^* algebras provide convenient representations of the normal elements a of any C^* algebra, when one simply considers the algebra generated by $\mathbf{1}$, a, and a^*.
3. Theorem (2.2.28) holds a fortiori for Abelian W^* algebras, which can also be represented as L^∞ functions on suitable measure spaces.

Since algebras of functions are easy to manipulate, (2.2.28) has a number of

2.2.31 Corollaries

1. The power series of $z \to (a(\chi) - z)^{-1}$ converges for all $\chi \in X(A)$ provided that $z > \sup_{\chi \in X(A)} |a(\chi)| = \|a\|$. As we see, if a is a normal element of a C^* algebra, then the radius of convergence of the series for $(a - z)^{-1}$ is exactly $\|a\|$. (This may be false if a is not normal: $\begin{pmatrix} z & 1 \\ 0 & z \end{pmatrix}$ is invertible for all $z \neq 0$, but $\|\begin{pmatrix} 0 & 1 \\ 0 & 0 \end{pmatrix}\| = 1$.) The spectrum of a is precisely the image of $X(A)$ under a.
2. Continuous functions $f(a)$ are defined on the range of the Gel'fand isomorphism as $f(a(\chi))$, and they exist for all normal a in any C^* algebra. More specifically, a Hermitian element can be decomposed into a positive and a negative part, and unique square roots can be taken of positive elements. In a W^* algebra, all the spectral projections $\theta(a - \alpha)$, $\alpha \in \mathbb{R}$, exist for every Hermitian element, since a step function is the supremum of continuous functions. It is always true that $\|f(a)\| = \sup_\chi |f(a(\chi))| = \sup_{\alpha \in \mathrm{Sp}(a)} |f(\alpha)|$.
3. Hermitian elements can be characterized by $-1 \leq a/\|a\| \leq 1$, and positive elements by $\|\mathbf{1} - a/\|a\|\| \leq 1$.

4. According to Remark (2.2.30; 1) the Gel'fand isomorphism maps a state w to a probability measure $d\mu_w$ on $C(X(\mathcal{A}))$: $w(a) = \int_{X(\mathcal{A})} d\mu_w(\chi)a(\chi)$ for a normal. The pure states are the point measures $d\mu_w(\chi) = \delta(\chi - \chi_0)$. $\chi_0 \in X(\mathcal{A})$, which $\Rightarrow w(a) = a(\chi_0)$. We again note that

$$\|a\| = \sup_{\chi \in X(\mathcal{A})} |a(\chi)| = \sup_{w \text{ pure}} |w(a)|.$$

5. Since a maps the compact set $X(\mathcal{A})$ continuously into the compact set $\mathrm{SP}(a)$, $a(\chi)$ can be introduced as a variable of integration as in Corollary 4, the integral being over the image measure $dw = a(d\mu_f)$:

$$\begin{array}{ccc} C(X(\mathcal{A})) & \xrightarrow{a} & C(\mathrm{Sp}(a)) \\ & \searrow \quad \swarrow & \\ & \mathbb{C} & \end{array}$$

$$\begin{array}{ccc} g(a(\chi)) & \longrightarrow & g(a) \\ \downarrow & & \downarrow \end{array}$$

$$\int_{X(\mathcal{A})} d\mu_w g(a(\chi)) = \int_{\alpha \in \mathrm{Sp}(a)} dw(\alpha) g(\alpha).$$

Thus every state w furnishes a probability measure on the spectrum of a normal element a, such that $f(a) = \int_{\alpha \in \mathrm{Sp}(a)} dw(\alpha)\alpha$. For Hermitian or unitary elements this becomes a measure on the real axis or, respectively, the unit circle in $\mathbb{C}$.

The mathematical framework that we have developed will now allow us to formulate the conceptual schema of quantum theory.

2.2.32 The Basic Assumption of Quantum Theory

The observables and states of a system are described by Hermitian elements a of a C^* algebra $\mathcal{A}$ and by states on $\mathcal{A}$. The possible outcomes of a measurement of a are $\in \mathrm{Sp}(a)$, and their probability distribution in a state w is dw, the probability measure induced on $\mathrm{Sp}(a)$ by w.

2.2.33 Remarks

1. Since physical measurements are always real numbers, observables are Hermitian operators, but they constitute a subalgebra (over the real numbers) only if $\mathcal{A}$ is Abelian.
2. In classical mechanics the observables were a real function algebra, and the spectrum of a function was its range. Assumption (2.2.32) generalizes the classical schema only by not requiring commutativity.
3. In this volume the C^*-algebra of observables will usually be $\mathcal{B}(\mathcal{H})$, and we will need to consider only the normal states over $\mathcal{B}(\mathcal{H})$.
4. For Abelian algebras we learned that maximal idea = character = pure state = point probability measure. These states are nondispersive for all observables,

i.e., the **mean-square deviation** $(\Delta_w(a))^2 \equiv w(a^2) - w(a)^2$ vanishes. If the algebra is noncommutative, nondispersive states do not normally exist, since the operator inequality

$$\left(\frac{a - w(a)}{\Delta_w(a)} + i\frac{b - w(b)}{\Delta_w(b)}\right)\left(\frac{a - w(a)}{\Delta_w(a)} - i\frac{b - w(b)}{\Delta_w(b)}\right) \geq 0$$

for an arbitrary state w implies the **indeterminacy**, or **uncertainty**, **relationship** (by taking w of the inequality above):

$$\left| w\left(\frac{ab - ba}{2i}\right)\right| \leq \Delta_w(a)\Delta_w(b).$$

A state that had no dispersion on any observable would yield zero for any commutator $[a, b] \equiv ab - ba$. The algebra of interest here will be $\mathcal{A} \equiv \mathcal{B}(\mathcal{H})$, on which that is not possible for normal states; there are increasing sequences of projection operators having **1** for supremum, but for which each one can be written as the commutator of two Hermitian operators (Problem 9).

5. Although it is obvious how a function of one observable is to be measured—take the function of the measured value of the observable—it is less clear how to measure the sum or product of noncommutating observables. The spectrum is certainly not just the sum or product of the original spectra; we shall see that the spectra of xp_y and yp_x are both $\mathbb{R}$, while their difference, the angular momentum, has spectrum $\mathbb{Z}$. That is, the only possible measured values of the angular momentum are integers, whereas measurements of x, y, p_x and p_y, or of the products xp_y and yp_x can yield any numbers whatsoever. This makes the algebraic structure of the observables rather problematic, for which reason there have been attempts to find alternative and more economically phrased axioms. Some of these will be discussed shortly and we shall see that they eventually lead back to the schema of (2.2.32), which is our justification for imbedding the observables in a C^* algebra.

It will not be possible here to unfurl the whole subject of the theory of measurement, so we shall merely describe the bare mathematical structures that have been proposed for the formulation of quantum theory.

2.2.34 Jordan Algebras

If one tries to invent an algebra containing nothing but observables, then one is confronted by the problem that, while sums of Hermitian elements are Hermitian, the same is not true of products. The symmetric product $a \circ b \equiv (a + b)^2 - a^2 - b^2$ results in a new Hermitian element, and can be used as an alternative binary relationship on an algebra of observables over $\mathbb{R}$. Abstracting from that the commutative and distributive laws for $\circ$, one can formulate the rules of a nonassociative algebra. It turns out that modulo a few topological assumptions, which are more or less convincing on physical grounds, these **Jordan algebras** can be imbedded in $\mathcal{B}(\mathcal{H})$, whereby $a \circ b = ab + ba$.

2.2.35 Propositional Calculi

In a propositional calculus the only observables are known as the **propositions** $p_i \in \mathcal{P}$, which corresponds to statements like "The particle is in region G," and can be tested by experiments having only *yes* and *no* as possible outcomes. The algebraic formulation represents the propositions as projection operators or, as above, the characteristic function of G for the statement just mentioned. Algebraic operations are avoided in favor of lattice-theoretical operations, which correspond to logical relationships and seem less burdened with the problematics of measurements. Next one postulates a partial ordering for $\mathcal{P}$ with a maximal element **1** and a minimal element **0**. In addition, there are assumed to exist

(i) $\inf\{p_1, p_2, \ldots, p_n\} \equiv p_1 \wedge p_2 \wedge \cdots \wedge p_n$,
(ii) $\sup\{p_1, p_2, \ldots, p_n\} \equiv p_1 \vee p_2 \vee \cdots \vee p_n$, and
(iii) a complementation$'$: $\mathcal{P} \to \mathcal{P}$ such that $p \wedge p' = \mathbf{0}$, $p'' = p$, and $p_1 \geq p_2 \Leftrightarrow p_1' \leq p_2'$.

From this it follows that $(p_1 \wedge p_2)' = p_1' \vee p_2'$, and thus $p \vee p' = \mathbf{1}$. The connection with logic is that a larger proposition makes a weaker statement, i.e., $p_1 \leq p_2$ means that $p_1 \Rightarrow p_2$; and the proposition **1** is always true and **0** always false. Thus $p_1 \wedge p_2$ (respectively $p_1 \vee p_2$) is the weakest (strongest) proposition that implies (is implied by) both p_1 and p_2. The proposition $p \wedge p' = \mathbf{0}$ means that there are no true propositions that imply both p and p'; a proposition cannot be true at the same time as its complement. In classical logic p' is the negation of p, $p_1 \wedge p_2$ means both p_1 and p_2, and $p_1 \vee p_2$ means either p_1 or p_2 (or both).

From the algebraic point of view, $\mathcal{P}$ is the set of projections with the ordering (2.2.16) and $p' = \mathbf{1} - p$. Classically, p_i are the characteristic functions of sets G_i in phase-space (in which case p_1' corresponds to the complementary set CG_1, $p_1 \wedge p_2$ to the intersection, and $p_1 \vee p_2$ to the union, of G_1 and G_2). These facts can, of course, also be expressed in terms of algebraic operations, and in the noncommutative case the product of characteristic functions generalizes to $p_1 \wedge p_2 = \lim_{n \to \infty} p_1(p_2 p_1)^n$.

On Hilbert space $\mathcal{H}$, the p_i are projections onto subspaces $\mathcal{H}_i$, and p_i' are projections onto the orthogonal subspaces $\mathcal{H}_i^{\perp}$, $p_1 \wedge p_2$ onto $\mathcal{H}_1 \cap \mathcal{H}_2$, and $p_1 \vee p_2$ onto the subspaces spanned by $\mathcal{H}_1$ and $\mathcal{H}_2$.

2.2.36 Remarks

1. The algebraic realization of the propositional calculus will require the W^* property to make the lattice-theoretical operators properly definable. All the projections then exist, as does the limit of the positive, decreasing sequence $p_1(p_2 p_1)^n$, $n \to \infty$.
2. Characteristic functions $\chi_I : \mathbb{R} \supset I \to 1$, $CI \to 0$, of observables are projections. $\chi_I(A)$ corresponds to the statement that some spectral value $a \in I$ has been measured, and $\chi_I(A)'$ means that the measured value is in CI.
3. In the commutative case the p_i are realizable as characteristic functions χ_i, and the distributive law $p_1 \wedge (p_2 \vee p_3) = (p_1 \wedge p_2) \vee (p_1 \wedge p_3)$ follows from

the correspondence with the set-theoretical operations. They are algebraically realized as follows: $p_1 \wedge p_2 \leftrightarrow \chi_1 \cdot \chi_2$, $p_1 \vee p_2 \leftrightarrow \chi_1 + \chi_2 - \chi_1 \cdot \chi_2$, and the distributive law states that $\chi_1 \cdot (\chi_2 + \chi_3 - \chi_2 \cdot \chi_3) = \chi_1 \chi_2 + \chi_1 \chi_3 - \chi_1 \cdot \chi_2 \cdot \chi_1 \cdot \chi_3$. If $p_1 > p_2$ for $p_i \in \mathcal{B}(\mathcal{H})$, then p_1 and p_2 commute, and the distributive law holds on the propositional subcalculus constructed from p_1, p_2, p_1', and p_2'. However, it does not hold in general, nor does the operator inequality $p_1 \vee p_2 \leq p_1 + p_2$.

2.2.37 Example

With the Pauli spin matrices

$$(\sigma_x, \sigma_y, \sigma_z) = \left(\begin{pmatrix} 0 & 1 \\ 1 & 0 \end{pmatrix}, \begin{pmatrix} 0 & -i \\ i & 0 \end{pmatrix}, \begin{pmatrix} 1 & 0 \\ 0 & -1 \end{pmatrix} \right)$$

on $\mathcal{H} = \mathbb{C}^2$, we form the one-dimensional projections

$$p_{\mathbf{n}} = \frac{1 + \boldsymbol{\sigma} \cdot \mathbf{n}}{2}, \qquad \mathbf{n} \in \mathbb{R}^3, \qquad |\mathbf{n}|^2 = 1.$$

Their physical interpretation is the statement, "A measurement of $\boldsymbol{\sigma}$ in the direction $\mathbf{n}$ definitely has the value 1." For all $\mathbf{n}_1 \neq \mathbf{n}_2$, $p_{\mathbf{n}_1} \wedge p_{\mathbf{n}_2} = 0$. Hence for different $\mathbf{n}_i$,

$$p_{\mathbf{n}_1} \wedge (p_{\mathbf{n}_2} \vee p_{\mathbf{n}_3}) = p_{\mathbf{n}_1} \wedge \mathbf{1} = p_{\mathbf{n}_1},$$

but

$$(p_{\mathbf{n}_1} \wedge p_{\mathbf{n}_2}) \vee (p_{\mathbf{n}_1} \wedge p_{\mathbf{n}_3}) = \mathbf{0} \vee \mathbf{0} = \mathbf{0}.$$

Hence, the classical conclusion: If the particle is in region G_1 and either in G_2 or G_3 then it is either in G_1 and G_2 or in G_1 and G_3, is invalid for noncommuting observables. The proposition "The spin points in the direction $\mathbf{n}_2$ as well as in $\mathbf{n}_3$" is certainly false ($p_{\mathbf{n}_2} \wedge p_{\mathbf{n}_3} = \mathbf{0}$). The complementary statement is the tautology "The spin points in some direction," and is the most restrictive statement implied by both $p_{\mathbf{n}_2}$ and $p_{\mathbf{n}_3}$ ($p_{\mathbf{n}_2} \vee p_{\mathbf{n}_3} = \mathbf{1}$). It does not imply that one of the measurements of $\boldsymbol{\sigma} \cdot \mathbf{n}_2$ or $\boldsymbol{\sigma} \cdot \mathbf{n}_3$ yields the value 1 with certainty, and thus $p_{\mathbf{n}_1} \wedge (p_{\mathbf{n}_2} \vee p_{\mathbf{n}_3})$ is to be read neither as "The spin has the direction $\mathbf{n}_1$ and $\mathbf{n}_2$ or $\mathbf{n}_3$" nor as "The spin has the direction $\mathbf{n}_1$ and $\mathbf{n}_2$ or $\mathbf{n}_1$ and $\mathbf{n}_3$." Therefore the classical distributive law fails in quantum mechanics.

It turns out that, up to technical assumptions, any propositional calculus in which the distributive law holds on (p_1, p_2, p_1', p_2') for $p_1 \geq p_2$ can be represented as a calculus of projections on Hilbert space, and for this reason the algebraic framework we have chosen seems to be the most appropriate one for quantum mechanics.

2.2.38 Problems

1. Show that the statement of (2.2.15; 2) is true.
2. Show that a *-homomorphism π of a C^* algebra is continuous.

3. Show that multiplication is continuous (in both factors simultaneously) in a C^* algebra.
4. Prove (2.2.12; 3).
5. Show that the statement of (2.2.15; 3) is true.
6. Consider a mixture of two states: Show that if $w = \alpha w_1 + (1-\alpha)w_2$, $0 < \alpha < 1$,then $(\Delta_w a)^2 \geq \alpha(\Delta_{w_1} a)^2 + (1-\alpha)(\Delta_{w_2} a)^2$, and that equality holds iff $w_1(a) = w_2(a)$.
7. Show that the resolvent set is open and that the spectrum is not empty.
8. Prove the inequality (2.2.20; 1).
9. Write the projection P_n onto the subspace of l^2 spanned by the first n basis vectors as i times the commutator of two Hermitian elements of $\mathcal{B}(l^2)$.
10. Find an example of 2×2 matrices for which $0 \leq a \leq b \not\Rightarrow a^2 \leq b^2$. (Hint: $a \geq 0$ iff $a = a^*$, Tr $a \geq 0$, and Det $a \geq 0$.)
11. Let $0 \leq a \leq b$. Show that (i) if a^{-1} exists, then $b^{-1} \leq a^{-1}$; (ii) if $\ln a$ exists, then $\ln a \leq \ln b$; and (iii) $a^\gamma \leq b^\gamma$ for $0 \leq \gamma \leq 1$. (Hint: Use

$$b \geq a \geq 0 \Rightarrow \int_0^\infty d\lambda\, \sigma(\lambda)(a+\lambda)^{-1} \geq \int_0^\infty d\lambda\, \sigma(\lambda)(b+\lambda)^{-1} \quad \text{for } \sigma \geq 0.$$

It can even be shown that all functions f for which $b \geq a \geq 0$ implies $f(b) \leq f(a)$ are of this form.)

2.2.39 Solutions

1. $(a-z)^{-1} = -(1/z)\sum_{n=0}^\infty (a/z)^n$, and the radius of convergence of this series is exactly $|z| = \lim \|a^n\|^{1/n} \equiv$ spr a (the **spectral radius**). It is always true that spr $a \leq \|a\|$, and for normal a, spr $a = \|a\|$ because $\|a^2\| = \|a\|^2$, etc. (cf. Corollaries (2.2.31; 1) and (2.2.31; 2)).
2. If $(a-z)^{-1}$ exists, then so does $(\pi(a)-z)^{-1}$, and consequently, by Problem 1, $\|\pi(a)\| \leq \|a\|$ for Hermitian a, and in general $\|\pi(a)\|^2 = \|\pi(a^*a)\| \leq \|a^*a\| = \|a\|^2$. With the aid of Remark (2.2.17; 4), one can also argue as follows:

$$a^*a \leq \|a\|^2 \cdot \mathbf{1} \Rightarrow \pi(a^*a) \leq \|a\|^2 \pi(\mathbf{1}) \Rightarrow \|\pi(a)\|^2 \leq \|a\|^2.$$

3. $\|(a+\delta a)(b+\delta b) - ab\| < \varepsilon$ for

$$\|\delta a\| \quad \text{and} \quad \|\delta b\| < (\varepsilon + ((\|a\| + \|b\|)/2)^2)^{1/2} - (\|a\| + \|b\|)/2.$$

4. $(a-\delta)^{-1} = a^{-1}(1-\delta \cdot a^{-1})^{-1} = a^{-1}\sum(\delta \cdot a^{-1})^n$; this series converges for all δ with $\|\delta\| < \|a^{-1}\|$. The mappings $a \to a^*a - aa^*$, $a \to a - a^*$, $a \to aa^* - \mathbf{1}$, $a \to a^*a - \mathbf{1}$, and $a \to a^2 - a$ are continuous, and hence the inverse image of $\mathbf{0}$ is closed in every case. The Gel'fand isomorphism allows positivity to be characterized by $\|a\| - \|\|a\| - a\| \geq 0$ for Hermitian a (2.2.31; 3). This function is continuous and thus the inverse image of $[0, \infty)$ is likewise closed.
5. It is trivial to see that $\mathrm{Sp}(a^*) = \mathrm{Sp}(a)^*$. Now suppose that $P(a) - \lambda = \alpha \prod_i (a - \lambda_i)$ for λ, λ_i and $\alpha \in \mathbb{C}$. Then $\lambda \in \mathrm{Sp}(a) \Leftrightarrow (P(a)-\lambda)^{-1}$ does not exist $\Leftrightarrow$ some $\lambda_i \in \mathrm{Sp}(a) \Leftrightarrow \lambda = P(\lambda_i) \in P(\mathrm{Sp}(a))$.

$$aa^* = a^*a = \mathbf{1} \Rightarrow \|a\| = \|a^{-1}\| = 1 \Rightarrow \mathrm{Sp}(a) \subset \{z \in \mathbb{C} : |z| = 1\}.$$

$$\begin{aligned} a &= a^* : \alpha + i\beta \in \mathrm{Sp}(a) \Leftrightarrow \alpha + i(\beta+\lambda) \in \mathrm{Sp}(a+i\lambda) \Rightarrow \alpha^2 + (\beta+\lambda)^2 \\ &\leq \|a+i\lambda\|^2 = \|a^*a + \lambda^2\| \leq \|a\|^2 + \lambda^2 \quad \text{for all } \lambda \in \mathbb{R} \Rightarrow \beta = 0. \end{aligned}$$

$$a^2 - a = 0, a = a^* : \mathrm{Sp}(a^2 - a) = (\mathrm{Sp}\, a)^2 - (\mathrm{Sp}\, a) = 0, \mathrm{Sp}(a) \subset \{0, 1\}.$$

If $a = b^*b$, then the proof of the positivity of the spectrum is a bit more involved, but it can be shown that it is possible to restrict to Hermitian b, for which positivity follows from $\mathrm{Sp}\, b^2 = (\mathrm{Sp}\, b)^2$. It follows from the Gel'fand isomorphism (2.2.28) that the spectral properties of normal operators satisfy the various operator relationships, since the corresponding facts for function algebras are obvious.

6. $(\Delta_w a)^2 - \alpha(\Delta_{w_1} a)^2 - (1-\alpha)(\Delta_{w_2} a)^2 = (\alpha - \alpha^2)[w_1(a) - w_2(a)]^2$.

7. The resolvent set is open because of the convergence of

$$(a-z)^{-1} = (a-z_0)^{-1} \sum_{n=0} (a-z_0)^{-n}(z-z_0)^n$$

$$\text{for } |z - z_0| < \|(a-z_0)^{-1}\|^{-1}$$

If the resolvent set were all of $\mathbb{C}$, then $(a-z)^{-1}$ would be an operator-valued, entire, bounded function, which would have to be constant by Liouville's theorem (cf. [7], IX.11).

8. With the method of (2.1.27; 10) it can be shown that $|f(a^*b)|^2 \leq f(a^*a)f(b^*b)$. If $a \geq \mathbf{0}$, then by (2.2.17; 2) and (2.2.31; 3) $b^*ab \leq \|a\| b^*b$ and $f(b^*ab) \leq \|a\| f(b^*b)$. For arbitrary a, $|f(bab^*)|^2 \leq f(baa^*b^*)f(bb^*) \leq \|aa^*\| f(bb^*)^2 = \|a\|^2 f(bb^*)^2$.

9. Let

$$S_n = \overbrace{\begin{pmatrix} 0 & 0 & \dots & 0 & 1 & 0 & 0 & \dots \\ 0 & 0 & \dots & 0 & 0 & 1 & 0 & \dots \\ 0 & 0 & \dots & 0 & 0 & 0 & 1 & \dots \end{pmatrix}}^{n\text{-th position}}.$$

Then $P_n = [S_n, S_n^\dagger] = (i/2)[S_n + S_n^\dagger, i(S_n - S_n^\dagger)]$.

10. $a \geq 0 \Leftrightarrow$ both eigenvalues $\geq 0 \Leftrightarrow$ the sum and product of the two eigenvalues ≥ 0, i.e., $\mathrm{Tr}\, a \geq 0$ and $\mathrm{Det}\, a \geq 0$. Now let

$$0 \leq a = \begin{pmatrix} 1 & 1 \\ 1 & 1 \end{pmatrix} \leq \begin{pmatrix} \lambda & 1 \\ 1 & 1 \end{pmatrix} = b, \qquad \lambda \geq 1.$$

$$\mathrm{Det}(b^2 - a^2) = \begin{vmatrix} \lambda^2 - 1 & \lambda - 1 \\ \lambda - 1 & 0 \end{vmatrix} = -(\lambda-1)^2 < 0 \Rightarrow b^2 - a^2 \not\geq 0.$$

11. (i) $0 < a \leq b \Rightarrow 0 < b^{-1/2}ab^{-1/2} \leq \mathbf{1} \Rightarrow \mathbf{1} \leq b^{1/2}a^{-1}b^{1/2} \Rightarrow b^{-1} \leq a^{-1}$.
(ii) $\ln b - \ln a = \int_0^\infty d\lambda [(\lambda + a)^{-1} - (\lambda + b)^{-1}]$.
(iii) $\int_0^\infty d\lambda\, \lambda^{-\gamma}(a+\gamma)^{-1} = \text{const.} \cdot a^{-\gamma}$ for $0 < \gamma < 1 \Rightarrow a^{-\gamma} \geq b^{-\gamma} \Rightarrow a^\gamma \leq b^\gamma$.

2.3 Representations on Hilbert Space

Algebras of matrices are typical C^* algebras, because any C^* algebra can be represented as an algebra of bounded operators on a Hilbert space.

The concepts of linear functional and character are generalized in

2.3.1 Definition

A **representation** π of a C^* algebra $\mathcal{A}$ is a *-homomorphism from $\mathcal{A}$ into $\mathcal{B}(\mathcal{H})$, that is $\pi(\lambda_1 a_1+\lambda_2 a_2) = \lambda_1\pi(a_1)+\lambda_2\pi(a_2), \pi(a_1 a_2) = \pi(a_1)\pi(a_2)$, and $\pi(a^*) = \pi(a)^*$ for all $a_i \in \mathcal{A}$ and $\lambda_i \in \mathbb{C}$. If $\pi(a) \neq 0$ whenever $a \neq 0$, then π is said to be **faithful**. Two representations π_1 and π_2 on $\mathcal{H}_1$ and $\mathcal{H}_2$ are **equivalent** iff there exists an isomorphism $U : \mathcal{H}_1 \to \mathcal{H}_2$ such that $\pi_2(a) = U\pi_1(a)U^{-1}$ for all $a \in \mathcal{A}$.

2.3.2 Examples

1. Matrix algebras represent themselves.
2. The continuous functions on a compact set K represent themselves as multiplication operators on $L^2(K, d\mu)$ if one defines $(\pi(a)\varphi)(x) = a(x)\varphi(x)$ for all $a \in \mathcal{A}$, $\varphi \in L^2$, and $x \in K$. ($\|\pi(a)\varphi\| \leq \|a\|\,\|\varphi\|$).

2.3.3 Remarks

1. It need not be required that π be continuous; it is automatically continuous because of positivity (2.2.17; 4): $0 \leq a^*a \leq \|a\|^2 \cdot \mathbf{1} \Rightarrow 0 \leq \pi(a^*)\pi(a) \leq \|a\|^2 \cdot \mathbf{1} \Rightarrow \|\pi(a)\| \leq \|a\|$. Note that $\|\pi(\mathbf{1})\| = 0$ or 1, since $\|\pi(\mathbf{1})\| = \|\pi(\mathbf{1})^*\pi(\mathbf{1})\| = \|\pi(\mathbf{1})\|^2$.
2. The kernel $\mathcal{K} = \pi^{-1}(0)$ is a closed, two-sided ideal of $\mathcal{A}$. Faithfulness of π means that $\mathcal{K} = \{0\}$, i.e., π is injective. The positivity argument of Remark 1 then also works for $\pi^{-1} : \pi(\mathcal{A}) \to \mathcal{A}$, and therefore π is faithful iff $\|\pi(a)\| = \|a\|$ for all $a \in \mathcal{A}$. If $\mathcal{A}$ has no proper two-sided ideals, it is said to be a **simple algebra**, and every nontrivial representation is faithful. More generally, π is always a faithful representation of the quotient algebra $\mathcal{A}/\mathcal{K}$. When topologized with the quotient norm, defined as inf $\|a + b\|$ for $b \in \mathcal{K}$, the representation of the quotient algebra is faithful and forms a C^* algebra [Bratelli and Robinson]. At any rate, $\pi(\mathcal{A})$ is itself a C^* algebra, and hence it is a norm-closed subalgebra of $\mathcal{B}(\mathcal{H})$.

Since π may fail to be either injective or surjective, the following terminology for subalgebras of $\mathcal{B}(\mathcal{H})$ is convenient:

2.3.4 Definition

Let $\mathcal{M}$ be a *-subalgebra of $\mathcal{B}(\mathcal{H})$. Then the *-subalgebra

$$\mathcal{M}' \equiv \{b \in \mathcal{B}(\mathcal{H}) : ba = ab \text{ for all } a \in \mathcal{M}\}$$

is its **commutant** (not to be confused with the dual of $\mathcal{M}$ as a linear space).
$\mathcal{M}' \cap \mathcal{M} \equiv \mathcal{Z}$ is its **center**.
If $\mathcal{M} \subset \mathcal{M}'$, then $\mathcal{M}$ is **Abelian**.
If $\mathcal{M} = \mathcal{M}'$, then $\mathcal{M}$ is **maximally Abelian**.
If $\mathcal{M} = \mathcal{M}''$, then $\mathcal{M}$ is a **von Neumann algebra**.
If $\mathcal{M}' = \{\lambda \cdot \mathbf{1}\}$, then $\mathcal{M}$ is **irreducible**.
If $\mathcal{Z} = \{\lambda \cdot \mathbf{1}\}$, then $\mathcal{M}''$ is a **factor**.

If $\mathcal{T}$ is a subspace of $\mathcal{H}$, then if $\mathcal{M} \cdot \mathcal{T} \subset \mathcal{T}$, $\mathcal{T}$ is an **invariant subspace**; if $\mathcal{M} \cdot \mathcal{T}$ is dense in $\mathcal{H}$, $\mathcal{T}$ is a **totalizer**. If the totalizer $\mathcal{T}$ is one-dimensional, then its vectors are said to be **cyclic** (with respect to $\mathcal{M}$).

2.3.5 Examples

1. Letting α and β take values in $\mathbb{C}$ or $\mathbb{C}^3$, some examples can be constructed with the Pauli spin matrices (2.2.37):

   (i) $\mathcal{M} = \{\alpha \cdot \mathbf{1} + \boldsymbol{\beta} \cdot \boldsymbol{\sigma}\} = \mathcal{M}''$; $\mathcal{M}' = \mathcal{Z} = \{\alpha \cdot \mathbf{1}\}$. This is irreducible, a factor, and non-Abelian. Every vector is cyclic, and there are no invariant proper subspaces.

   (ii) $\mathcal{M} = \{\alpha \cdot \mathbf{1} + \beta\sigma_z\} = \mathcal{M}''$, $\mathcal{M}' = \mathcal{M} = \mathcal{Z}$. This is reducible, not a factor, and maximally Abelian. $\begin{pmatrix} a \\ b \end{pmatrix}$ is cyclic only if a and b are both different from zero, while $\begin{pmatrix} a \\ 0 \end{pmatrix}$ and $\begin{pmatrix} 0 \\ b \end{pmatrix}$ are invariant subspaces.

   (iii) $\mathcal{M} = \{\alpha \cdot \mathbf{1}\} = \mathcal{Z} = \mathcal{M}''$, $\mathcal{M}' = \{\alpha \cdot \mathbf{1} + \boldsymbol{\beta} \cdot \boldsymbol{\sigma}\}$. This is reducible, a factor, and Abelian. There are no cyclic vectors, and every subspace is invariant.

2. $L^\infty(\mathbb{R}, dx)$, considered as multiplication operators on $L^2(\mathbb{R}, dx)$, is maximally Abelian. Every function in L^2 that is nonzero a.e. is a cyclic vector. Functions that vanish on some interval $I \subset \mathbb{R}$ are invariant subspaces. L^∞ is reducible, and not a factor.

2.3.6 Remarks

1. The following three conditions for irreducibility are equivalent (Problem 1):

   (i) $\mathcal{M}' = \lambda \cdot \mathbf{1}$.

   (ii) Every nonzero vector is cyclic.

   (iii) There are no invariant proper subspaces.

2. The direct sum $\pi_1 \oplus \pi_2$ and tensor product $\pi_1 \otimes \pi_2$ (representing $\mathcal{A} \times \mathcal{A}$) of two representations π_1 and π_2 are defined as for finite-dimensional spaces: If $x \equiv x_1 \oplus x_2 \in \mathcal{H}_1 \oplus \mathcal{H}_2 \equiv \mathcal{H}$ (respectively $x_1 \otimes x_2 \in \mathcal{H}_1 \otimes \mathcal{H}_2 \equiv \mathcal{H}$), then
$$\pi(a)x = \pi_1(a)x_1 \oplus \pi_2(a)x_2 \qquad \text{(respectively } \pi_1(a)x_1 \otimes \pi_2(a)x_2).$$
Sums of representations are reducible, and the $\mathcal{H}_i$ are invariant subspaces.
3. The commutant obviously has the properties:
$$\text{(i) } \mathcal{N} \supset \mathcal{M} \Rightarrow \mathcal{N}' \subset \mathcal{M}';$$
$$\text{(ii) } \mathcal{M}'' \supset \mathcal{M};$$
$$\text{(iii) } (\mathcal{M} \cap \mathcal{N})' \supset \mathcal{M}' \cup \mathcal{N}', (\mathcal{M} \cup \mathcal{N})' \supset \mathcal{M}' \cap \mathcal{N}'.$$
These imply that $\mathcal{M}''' = \mathcal{M}'$, since $(\mathcal{M}'')' \subset \mathcal{M}' \subset (\mathcal{M}')''$. It turns out that $\mathcal{M}''$ is the closure of $\mathcal{M}$ in both the strong and the weak topology (Problem 4). Strongly closed *-subalgebras of $\mathcal{B}(\mathcal{H})$ are the von Neumann algebras, and a theorem of Vigier (Problem 11) states that they have the properties of Definition (2.2.22), i.e., they are W^* algebras. Note that $\mathcal{M} \cap \mathcal{M}' = \lambda \cdot \mathbf{1} \Rightarrow \mathcal{M}' \cup \mathcal{M}'' = \{\lambda \cdot \mathbf{1}\}' \Rightarrow \mathcal{M}' \cap \mathcal{M}'' = \{\lambda \cdot \mathbf{1}\}'' = \lambda \cdot \mathbf{1}$, so that for a factor the center of $\mathcal{M}''$ is trivial.
4. For finite-dimensional spaces, $\mathcal{M} = \mathcal{M}''$, and $\mathcal{M}$ is
irreducible iff $= \mathcal{B}(\mathbb{C}'')$;
a factor iff $= \mathcal{B}(\mathbb{C}^n) \otimes \mathbf{1}$;
Abelian iff all $a \in \mathcal{A}$ are simultaneously diagonal; and
maximally Abelian iff to each pair of diagonal positions in the diagonal representation there exist elements with different eigenvalues.
5. If $\mathcal{M}$ contains a maximally Abelian subalgebra $\mathcal{N}$, then $\mathcal{M}' \subset \mathcal{N}' = \mathcal{N} \subset \mathcal{M}$, so $\mathcal{Z} = \mathcal{M}'$. In this case, being a factor is equivalent to being irreducible, though in general irreducibility implies being a factor but not vice versa.
6. $\mathcal{M}$ Abelian implies $\mathcal{M} = \mathcal{Z}$, so Abelian factors have the trivial form $\lambda \cdot \mathbf{1}$.
7. If $\pi(\mathcal{A})$ is reducible, then $s = s^* \in \pi(\mathcal{A})'$, $s \neq \lambda \cdot \mathbf{1}$ is said to induce a superselection rule. The Hilbert space decomposes into subspaces that are not connected by observables, and there exists a Hermitian operator s that assigns different quantum numbers to the various invariant subspaces. If $\pi(\mathcal{A})$ is a factor, then s does not belong to $\pi(\mathcal{A})$, and it is consequently not an observable, but rather a kind of hidden variable. There is no maximally Abelian subalgebra of $\pi(\mathcal{A})$, because s could always be added to any subalgebra.

In any representation π, every vector $x \in \mathcal{H}$, $\|x\| = 1$, produces a state $a \to \langle x|\pi(a)x\rangle$, $a \in \mathcal{A}$. We shall next show that, conversely, for every state there is a representation in which it is of this form. Since algebras have a linear structure, any $a \in \mathcal{A}$ can be represented as an operator on a linear space, namely the algebra itself, by $b \to ab$, $b \in \mathcal{A}$. For a C^* algebra, this linear space will only be a Banach space, but a state provides the scalar product needed to make the space a Hilbert space.

2.3.7 Lemma

If w is a state, then $\mathcal{N} \equiv \{a \in \mathcal{A} : w(a^*a) = 0\}$ is a closed, left ideal. The scalar product $\langle b|a\rangle = w(b^*a)$ makes the quotient space $\mathcal{A}|\mathcal{N}$ a pre-Hilbert space, and the canonical mapping $\mathcal{A} \to \mathcal{A}/\mathcal{N}$ is a continuous linear mapping of $\mathcal{A}$ (as a Banach space) onto $\mathcal{A}/\mathcal{N}$ (as a pre-Hilbert space).

Proof That $\mathcal{N}$ is a left ideal follows from (2.2.20; 1), as

$$w(a^*b^*ba) \leq \|b^*b\| w(a^*a),$$

and closure follows from continuity. The scalar product $\langle\,|\,\rangle$ on $\mathcal{A}/\mathcal{N}$ satisfies Postulates (2.1.7), and $|\langle b|a\rangle| = |w(b^*a)| \leq \|b\|\,\|a\|$ guarantees that the mapping is continuous. □

2.3.8 Remarks

1. Since $|w(a)|^2 \leq w(a^*a)$, the ideal $\mathcal{N} \subset \text{Ker}\, w = \{a \in \mathcal{A} : w(a) = 0\}$. Thus, in Example (2.3.5; 1) with $w(\cdot) = \binom{1}{0}(\cdot)\binom{1}{0}$, $\mathcal{N} = \left\{\begin{pmatrix} 0 & \alpha \\ 0 & \beta \end{pmatrix}\right\}$ and $\text{Ker}\, w = \left\{\begin{pmatrix} 0 & \alpha \\ \beta & \gamma \end{pmatrix}\right\}$ $\alpha, \beta, \gamma \in \mathbb{C}$.
2. Despite the norm-completeness of $\mathcal{A}$, the quotient $\mathcal{A}/\mathcal{N}$ may fail to be a Hilbert space. For example, let $\mathcal{A}$ be the continuous function in $x \in [0, 1]$ and $w(a) = \int_0^1 dx\, a(x)$; then $\mathcal{N} = \{0\}$, but $\mathcal{A}$ is strictly smaller than its completion $L^2([0, 1], dx)$.
3. Given the product of two algebras $\mathcal{A}$ and $\mathcal{B}$, i.e., each element of the product algebra is a linear combination of $a_i b_j = b_j a_i$, $a_i \in \mathcal{A}$, $b_j \in \mathcal{B}$, the Hilbert space constructed from a product state is the tensor product of the two Hilbert spaces gotten from $\mathcal{A}$ and $\mathcal{B}$.

2.3.9 Definition

The **Gel'fand–Naimark–Segal (GNS) representation** π_w of $\mathcal{A}$ on $\mathcal{B}(\mathcal{H})$, where $\mathcal{H}$ is the completion of $\mathcal{A}/\mathcal{N}$, corresponding to any state w is defined as the continuous extension of $\pi_w(a) : b \to ab$, $a \in \mathcal{A}$, $b \in \mathcal{A}/\mathcal{N}$ to all of $\mathcal{H}$.

2.3.10 Remarks

1. The elements of $\mathcal{A}/\mathcal{N}$ are equivalence classes of objects of the form $b + n$, $n \in \mathcal{N}$, though the mapping $\pi_w(a)$ is independent of the representative b, since $\mathcal{N}$ is a left ideal ($an \in \mathcal{N}$).
2. The general fact about continuity (2.2.20; 1) can be seen directly: $\|\pi_w(a)\| = \sup_{w(b^*b)=1}(w(b^*a^*ab))^{1/2} \leq \|a^*a\|^{1/2} = \|a\|$. Hence $\pi_w(a)$ is a continuous operator on $\mathcal{A}/\mathcal{N}$, and has a unique extension to $\mathcal{H}$.
3. $\text{Ker}\, \pi_w = \{a \in \mathcal{A} : w(b^*ac) = 0 \text{ for all } b, c \in \mathcal{A}\}$ is a closed, two-sided ideal contained in $\mathcal{N}$. It reduces to $\{0\}$ in the example of (2.3.8; 1), which shows

that the GNS representation may be faithful even if $\mathcal{N} \neq \{0\}$. The logical interrelationships are depicted below:

Ker w = linear space:	**$w(a) = 0$**
$\mathcal{N}$ = left ideal:	**$w(a^*a) = 0$**
Ker π_w = two-sided ideal:	**$w(bac) = 0$**

4. The vector corresponding to $\mathbf{1} \in \mathcal{A}/\mathcal{N}$ is cyclic.
5. π_w is irreducible iff w is pure (Problem 2).
6. If, conversely, we have a representation π with a cyclic vector Ω, then it defines a state $w(a) = (\Omega|\pi(a)\Omega)$, and π_w is then equivalent to π. By the axiom of choice, every representation is a sum of representations with a cyclic vector.
7. Since for all $a \in \mathcal{A}$ there is a state such that $w(a^*a) = \|a\|^2$, it is always possible to construct a faithful representation of any C^* algebra, by taking the sum of the representations for all possible w.
8. As we have seen, each vector Ω in the Hilbert space corresponds to a pure state, which corresponds to a ray in Hilbert space, i.e., $\{e^{i\alpha}\Omega, \alpha \in \mathbb{R}\}$. In wave-mechanics, this fact shows up as the **principle of superposition**, which states that the vector $\Omega = \alpha_1\Omega_1 + \alpha_2\Omega_2$, $|\alpha_1|^2 + |\alpha_2|^2 = 1$ describes the quantum-mechanical superposition of the states Ω_1 and Ω_2. Yet Ω contains information not contained in Ω_1 and Ω_2 taken separately, namely the relative phase of the vectors Ω_1 and Ω_2.

In order to study the form of the representation of Hermitian element a in more detail, consider the restriction to the C^* algebra generated by a. By the axiom of choice, we can choose $b_i \in \mathcal{H}$ such that $\mathcal{H}_i \equiv$ the completions of the sets of linear combinations of $a^n b_i$, $n = 0, 1, \ldots$ span all of $\mathcal{H}$. Each $\mathcal{H}_i$ provides a representation of the (Abelian) C^* algebra generated by a, and has b_i as a cyclic vector. By Corollary (2.2.31; 5), to the state

$$w_i : w_i(a^n) = \langle b_i | a^n b_i \rangle$$

there corresponds a measure μ_i on Sp(a) such that $w_i(\varphi(a)) = \int d\mu_i(\alpha)\varphi(\alpha)$. Taking the norm-closure of the polynomials next extends this to all continuous functions $\varphi \in C(\mathrm{Sp}(a))$; then the completion with the w_i norm extends this to $\mathcal{H}_i = L^2(\mathrm{Sp}(a), d\mu_i)$, on which $\pi(a)$ acts as the multiplication operator $\varphi(\alpha) \to \alpha\varphi(\alpha)$, $\alpha \in \mathrm{Sp}(a)$, $\varphi \in L^2(\mathrm{Sp}(a), d\mu_i)$. With (2.3.10;6) this yields

2.3.11 The Spectral Theorem

For any given Hermitian element $a \in \mathcal{A}$, every representation of $\mathcal{A}$ is equivalent to a representation $\mathcal{H} = \oplus\mathcal{H}_i$, for which $\mathcal{H}_i = L^2(\mathrm{Sp}(a), d\mu_i)$ and $\pi(a)_{|\mathcal{H}_i} : \varphi(\alpha) \to \alpha\varphi(\alpha)$. In this representation, a acts as a **multiplication operator**.

2.3.12 Remarks

1. Theorem (2.3.11) generalizes the statement that any finite-dimensional Hermitian matrix is diagonable with a unitary transformation. Of course, not all Hermitian elements of $\mathcal{A}$ are multiplication operators in this representation unless $\mathcal{A}$ is Abelian.
2. Although we made use of the GNS representation, by Remark (2.3.10; 6), the argument leading to (2.3.11) works just as well with any specified representation.
3. Theorem (2.3.11) shows that any Hermitian operator of $\mathcal{B}(\mathcal{H})$ can be transformed unitarily into a multiplication operator.
4. The scaling $\varphi \to 2^{n/2}\varphi$ is an isomorphism

$$L^2(\mathrm{Sp}(a), d\mu_n) \to L^2(\mathrm{Sp}(a), 2^{-n} d\mu_n);$$

   and hence $\mathcal{H}$ is also isomorphic to $\bigoplus_{n=1}^{\infty} L^2(\mathrm{Sp}(a), 2^{-n} d\mu_n)$ (assuming $\mathcal{H}$ is separable). Furthermore, since the μ_n are probability measures, $\mathcal{H}$ is also isomorphic to L^2 of a finite measure space. Incidentally, this shows that the μ_n are not in any way fixed uniquely.
5. The only fact that has been used so far is that the algebra generated by a and a^* is commutative, so all the same statements can be made for normal operators, except that $\mathrm{Sp}(a)$ would not be real, but just some subset of $\mathbb{C}$. If there are m mutually commuting operators $a_j = a_j^*$, then they can be represented simultaneously as multiplication operators on $L^2(\mathbb{R}^m, d\mu)$.

2.3.13 Examples

1. A Hermitian $n \times n$ matrix a with distinct eigenvalues α_i. The space $\mathbb{C}^n$ is isometric to $L^2(\mathbb{R}, d\mu)$ with $d\mu(\alpha) = \sum_{i=1}^{n} \delta(\alpha - \alpha_i) d\alpha$, $\langle w|v\rangle = \sum_{i=1}^{n} w_{\alpha_i}^* v_{\alpha_i}$,

$$\langle w|av\rangle = \sum_{i=1}^{n} w_{\alpha_i}^* \alpha_i v_{\alpha_i}.$$

2. $l^2(-\infty, \infty) = \{(v_n) : -\infty < n < \infty\}$, where $(av)_n = v_{n+1} + v_{n-1}$ is a Hermitian operator $\in \mathcal{B}(l^2)$. In order to rewrite it as a multiplication operator, map $l^2(-\infty, \infty)$ onto $L^2([-\pi, \pi], dx)$ by $(v_n) \to \sum_{n=-\infty}^{\infty} v_n e^{inx}$; then a becomes multiplication by $e^{ix} + e^{-ix} = 2\cos x$. Next write

$$L^2([-\pi, \pi], dx) = L^2([-\pi, 0], dx) \oplus L^2([0, \pi], dx)$$

and introduce the new variable $\eta = 2\cos x$, to make this isomorphic to $L^2([-2,2], d\eta/\sqrt{4-\eta^2}) \oplus L^2([-2,2], d\eta/\sqrt{4-\eta^2})$. On this space a has become the multiplication operator η.

We have found a representation on $L^2(\mathrm{Sp}(a), d\mu)$ of the C^* algebra generated by a, for which each element of the algebra corresponds to multiplication by a continuous function on $\mathrm{Sp}(a)$. The algebra does not, however, account for all multiplication operators on $L^2(\mathrm{Sp}(a), d\mu)$, as they constitute the much larger space $L^\infty(\mathrm{Sp}(a), d\mu)$ (Problem 5). Problem 4 shows that this space is obtained by strong closure and also has a purely algebraic characterization, as the bicommutant $\pi(a)''$. By taking strong limits one obtains a representation in which it is possible to describe the operator $f(a)$ for $f \in L^\infty$ more explicitly.

Once we know all the integrable functions of a Hermitian operator, and in particular the characteristic functions, the explicit form of the operator $f(a)$ can be written down in terms of the

2.3.14 Spectral Family

An element $a = a^* \in \mathcal{A}$ can be written as

$$a = \int_{-\infty}^{\infty} dP_a(\alpha)\alpha, \qquad P_a(\alpha) = \Theta(\alpha - a), \qquad \Theta(x) = \begin{cases} 1 & \text{if } x > 0 \\ 0, & \text{otherwise,} \end{cases}$$

and given any $f \in L^\infty$,

$$f(a) = \int_{-\infty}^{\infty} dP_a(\alpha) f(\alpha).$$

The set of projection operators $P_a(\alpha)$ is known as the **spectral family** *of a.*

2.3.15 Remark

The construction of (2.3.14) is a generalization of the Stieltjes integral to the case of operators. Just as for functions, it is defined as the limit of the sums

$$\begin{aligned} a &= \lim_{N\to\infty} \sum_{j=1}^{2^N} \|a\| \left\{ \frac{j-1}{2^N} \left[\Theta\left(a - \frac{\|a\|}{2^N} j\right) - \Theta\left(a - \frac{\|a\|}{2^N}(j-1)\right)\right] \right. \\ &\quad \left. - \frac{j}{2^N}\left[\Theta\left(a + \frac{\|a\|}{2^N} j\right) - \Theta\left(a + \frac{\|a\|}{2^n}(j-1)\right)\right]\right\}. \end{aligned}$$

Vigier's theorem (Problem 11) guarantees the existence of the strong limit, since the sums are a bounded, increasing sequence of operators.

There are many different ways to classify the spectra of Hermitian operators. The classification we shall make uses the Lebesgue decomposition of a measure on $\mathbb{R}$; any measure is the sum of a part $d\mu_{ac} = f(\alpha)d\alpha$, $f \geq 0$ and locally integrable, which is absolutely continuous with respect to Lebesgue measure $d\alpha$; a part $d\mu_p$ concentrated on some separate points, $d\mu_p = d\alpha \sum_n c_n \delta(\alpha - \alpha_n)$,

$\alpha_n \in \mathbb{R}$; and a remainder $d\mu_s$, the singular spectrum [2]. This last part is somewhat pathological (Problem 7) and will not occur in any of our applications (though there exist one-electron band models with $d\mu_s$). Each of the three pieces of the measure is concentrated on null sets with respect to the others, and there is an orthogonal decomposition of $L^2(\mathbb{R}, d\mu)$ as

$$L^2(\mathbb{R}, d\mu) = L^2(\mathbb{R}, d\mu_p) \oplus L^2(\mathbb{R}, d\mu_{ac}) \oplus L^2(\mathbb{R}, d\mu_s)$$

(Problem 6). By making the same decomposition of all the $d\mu_i$ of (2.3.11), one can decompose the space $\mathcal{H}$ according to the properties of any normal operator a into orthogonal subspaces invariant under a:

2.3.16 Definition

If the Hilbert space is decomposed as

$$\mathcal{H} = \mathcal{H}_p \oplus \mathcal{H}_{ac} \oplus \mathcal{H}_s,$$

as in Remark (2.3.15) for some normal a, then the **point spectrum, absolutely continuous spectrum**, and **singular spectrum** of a are defined by the restrictions of a to the subspaces,

$$\sigma_p(a) = \text{Sp}(a_{|\mathcal{H}_p}), \qquad \sigma_{ac}(a) = \text{Sp}(a_{|\mathcal{H}_{ac}})$$

and

$$\sigma_s(a) = \text{Sp}(a_{|\mathcal{H}_s}).$$

2.3.17 Examples

1. For the finite matrices (2.3.13; 1), $d\mu$ is a pure point measure, and $\mathcal{H}_{ac} = \mathcal{H}_s = \{0\}$, $\text{Sp}(a) = \sigma_p(a)$.
2. Let a be multiplication by α on $L^2([0, 1], d\alpha)$. Then $\mathcal{H}_p = \mathcal{H}_s = \{0\}$, and $\text{Sp}(a) = \sigma_{ac}(a)$.

2.3.18 Remarks

1. $\mathcal{H}_p$ is the space spanned by the eigenvectors. To see this, consider $\psi_n \in L^2(\mathbb{R}, d\mu_p)$ as above, so that $\psi_n(\alpha_n) = 1$, but $\psi_n(\alpha) = 0$ for other α. In $\mathcal{H}_p$, $\|\psi_n\| = 1$, but single-point sets have measure zero with respect to $d\mu_{ac}$ and $d\mu_s$, $\int d\mu_{ac}|\psi_n|^2 = \int d\mu_s|\psi_n|^2 = 0$. Therefore $\psi_n \in \mathcal{H}_p$, and $a\psi_n = \alpha_n\psi_n$. The vectors ψ_n form a basis for $\mathcal{H}_p$.
2. It is a natural question whether the decomposition of (2.3.16) depends on the choice of μ_i in (2.3.11). In fact the μ_i are unique up to the equivalence relation $\mu_i \to \mu_i f(\alpha)$, $f > 0$ and locally integrable, and equivalent measures effect the same decomposition of $\mathcal{H}$.
3. The sets σ_p, σ_{ac}, σ_s are closed, though they need not be disjoint, nor does the Lebesgue measure of σ_p or of σ_s have to be zero. Suppose, for example,

that α_n is a numbering of the rational numbers between 0 and 1, and $\mathcal{H} = L^2([0,1], d\alpha \sum_n \delta(\alpha - \alpha_n))$, and let a be the operator of multiplication by α. Then $\mathcal{H} = \mathcal{H}_p$ and $\sigma_p = [0, 1]$, because the spectrum is closed, but almost no point of σ_p is an eigenvalue. (I.e., the irrational points vastly outnumber the rationals.) Many authors define σ_p as just as the set of eigenvalues, which case $\sigma_p \cup a_{ac} \cup \sigma_s$ may be different from $\mathrm{Sp}(a)$.

4. The **essential spectrum** σ_{ess} comprises all points of $\mathrm{Sp}(a)$ other than isolated points of finite multiplicity, that is, having a finite-dimensional eigenspace. There is no essential spectrum on a finite-dimensional space, but in the infinite case every bounded Hermitian element has an essential spectrum.
5. Although there may not exist eigenvectors for every point α of σ_{ess}, the spectral representation contains sequences of functions that are more and more spectrally concentrated near α. This idea can be used to prove the following theorem: *For all $a = a^*$ and $\alpha \in \mathrm{Sp}(a)$, there exists a sequence $\{\psi_n\}_{n=1}^{\infty}$, $\|\psi_n\| = 1$, such that $\lim_{n\to\infty} \|(a-\alpha)\psi_n\| = 0$. $\alpha \in \sigma_{\mathrm{ess}}(a) \Leftrightarrow$ there exists such a set of orthogonal vectors, or, equivalently, a set of ψ_n, $\|\psi_n\| = 1$, that $\rightharpoonup 0$.*

The sum of the eigenvalues of a diagonable $n \times n$ matrix m is given by the trace

$$\mathrm{Tr}\, m = \sum_{i=1}^{b} \langle e_i | m e_i \rangle, \qquad \langle e_i | e_j \rangle = \delta_{ij}.$$

The trace is a unitary invariant, and hence independent of the basis $\{e_i\}$. If one attempts to define the trace of an element $a \in \mathcal{A}$ in some representation, the essential spectrum causes trouble. If the space is infinite-dimensional, then the question of whether $\sum_i$ converges must first be grappled with. One problem is that convergence in one basis does not necessarily imply convergence in another, even if the eigenvalues tend to zero. For example, if $a \in \mathcal{B}(l^2)$ is

$$a = \begin{bmatrix} 0 & 1 & & & & & \\ 1 & 0 & & & & & \\ & & 0 & \frac{1}{2} & & & \\ & & \frac{1}{2} & 0 & & & \\ & & & & 0 & \frac{1}{3} & \\ & & & & \frac{1}{3} & 0 & \\ & & & & & & \ddots \end{bmatrix},$$

then it has the absolutely convergent trace $\sum_{i=1}^{\infty} |a_{ii}| = 0$. Yet in a different basis a has the form

$$a = \begin{bmatrix} 1 & 0 & & & & & \\ 0 & -1 & & & & & \\ & & \frac{1}{2} & 0 & & & \\ & & 0 & -\frac{1}{2} & & & \\ & & & & \frac{1}{3} & 0 & \\ & & & & 0 & -\frac{1}{3} & \\ & & & & & & \ddots \end{bmatrix},$$

and $\sum_i a_{ii}$ is only conditionally convergent, which means that it can be rearranged (equivalent to a change of basis) so as to sum to any value whatsoever, or to diverge. This lack of definition is avoided if the operator is positive. In that case the worst possibility is divergence, but aside from that the sum has all the ordinary

2.3.19 Properties of the Trace

The mapping $m \rightarrow \operatorname{Tr} m = \sum_i \langle e_i | m e_i \rangle$, for $\langle e_i | e_j \rangle = \delta_{ij}$, sends the positive operators to $\overline{\mathbb{R}^+}$, and for $m_i \geq 0$,

(i) $\operatorname{Tr}(\alpha_1 m_1 + \alpha_2 m_2) = \alpha_1 \operatorname{Tr} m_1 + \alpha_2 \operatorname{Tr} m_2$, $\alpha_i \in \mathbb{R}^+$,

(ii) $\operatorname{Tr}\ U^{-1} m U = \operatorname{Tr} m$, *$U$ unitary,*

(iii) $m_1 \leq m_2 \Rightarrow \operatorname{Tr} m_1 \leq \operatorname{Tr} m_2$.

If m_i is not necessarily positive, but $\operatorname{Tr} |m_i| < \infty$, *where* $|m| \equiv (m^* m)^{1/2}$, *then* (i) *and* (ii) *are still true, and moreover*

(iv) $\operatorname{Tr}|m_1 + m_2| \leq \operatorname{Tr}|m_1| + \operatorname{Tr}|m_2|$,

(v) $(\operatorname{Tr}|m_1, m_2|)^2 \leq \operatorname{Tr}|m_1|^2 \operatorname{Tr}|m_2|^2$,

(vi) $\operatorname{Tr} ma = \operatorname{Tr} am$ *for all* $a \in \mathcal{B}(\mathcal{H})$.

Proof: Properties (i) and (iii) are trivial. For the other, see Problem 10. □

2.3.20 Remarks

1. The unitary invariance (ii) implies that the definition is independent of the choice of basis provided that $\operatorname{Tr} |m| < \infty$.
2. On an infinite-dimensional space, the trace is an unbounded, positive linear functional. This does not contradict Remark (2.2.20; 1), since the trace is not finite on a whole C^* algebra; for instance, $\operatorname{Tr} \mathbf{1} = \infty$.
3. For Property (iv) it was not necessary to assume that $\operatorname{Tr} |a| < \infty$, since $|\operatorname{Tr} am| \leq \|a\| \operatorname{Tr} |m|$. This can be shown most conveniently with a **polar decomposition**

$m = V|m|$ (see [3], VIII.9), where

$$V^*V = |m|^{-1}|m|^2|m|^{-1}$$

is the projection onto the space perpendicular to the null space of $|m|$, so $\|Vx\| \leq \|x\|$ for all $x \in \mathcal{H}$, and

$$|\text{Tr } am| = \left|\sum_i \langle |m|^{1/2} e_i | aV|m|^{1/2} e_i \rangle \right| \leq \sum_i \|a\| \| |m|^{1/2} e_i \|^2 = \|a\| \text{Tr } |m|.$$

4. Most trace inequalities valid for finite-dimensional matrices can be carried over to general Hilbert spaces, as will be discussed in the fourth volume.[2]
5. The trace has several technical advantages over the operator norm, which is only easy to work with in the spectral representation. Suppose $K \in L^2(\mathbb{R}^n \times \mathbb{R}^n, d^n x d^n x')$ is the kernel of a bounded integral operator on $L^2(\mathbb{R}^n, d^n x)$, $\psi(x) \to \int K(x, x')\psi(x') d^n x'$. Its norm is difficult to compute, while Tr $K^*K = \int d^n x\, d^n x'\, K^*(x, x')K(x', x)$, as can be seen by writing the operator in a basis,

$$K(x, x') = \textstyle\sum_{ij} K_{ij} e_i^*(x) e_j(x'),$$
$$\text{Tr } K^*K = \textstyle\sum_{ij} K_{ij}^* K_{ji} = \int d^n x\, d^n x'\, K^*(x, x')K(x', x).$$

Since Tr$|m|$ has the properties of a norm (2.1.4) by Property (2.3.19(iv)), it is interesting to set the operators of finite trace aside in a separate category. They bear a close resemblance to finite matrices.

2.3.21 Definition

Let $\mathcal{E} \subset \mathcal{B}(\mathcal{H})$ be the space of operators of finite rank, i.e., which map $\mathcal{H}$ to a finite-dimensional space. The completions of $\mathcal{E}$ in the norms $\|a\|_1 \equiv \text{Tr}|a|$, $\|a\|_2^2 \equiv \text{Tr } a^*a$, and $\|a\|_\infty \equiv \|a\|$ are denoted $\mathcal{C}_1$, the **trace class** operators; $\mathcal{C}_2$, the **Hilbert–Schmidt operators**; and $\mathcal{C}$, the **compact**, or **completely continuous, operators**.

2.3.22 Examples

In $\mathcal{B}(l^2)$, the matrices with only finitely many nonzero rows or columns belong to $\mathcal{E}$. Diagonal matrices with eigenvalues α_i belong to $\mathcal{C}_1$ provided that $\sum_i |\alpha_i| < \infty$; to $\mathcal{C}_2$ provided that $\sum_i |\alpha_i|^2 < \infty$; and to $\mathcal{C}$ provided that $\lim_{i\to\infty} \alpha_i = 0$.

2.3.23 Remarks

1. It follows from (2.3.19) that the $\| \ \|_p$ are norms. By Remark (2.3.18; 5), it is necessary to have $\sigma_{\text{ess}} = \{0\}$ for the trace to be finite, so the spectrum is purely

[2] Quantum Mechanics of Large Systems.

discrete. If $\alpha_i > 0$ are the eigenvalues of $(a^*a)^{1/2}$, then we conclude from $\sum_i \alpha_i^2 < \sum_i \alpha_i \sum_j \alpha_j$ that $\| \ \|_p \le \| \ \|_q$ for $p \ge q$, $p, q = 1, 2, \infty$. Hence a Cauchy sequence in $\| \ \|_q$ is also one in $\| \ \|_p$ for $p \ge q$, so we have the inclusions

$$\mathcal{E} \subset \mathcal{C}_1 \subset \mathcal{C}_2 \subset \mathcal{C} \subset \mathcal{B}(\mathcal{H}).$$

2. Let a be an operator such that $\|a\|_1 < \infty$ (resp. $\|a\|_2 < \infty$) and α_i are the eigenvalues of $(a^*a)^{1/2}$. The truncated operators $a_N \equiv P_N a P_N$, where P_N is the projection onto the first N basis vectors, obviously belong to $\mathcal{E}$ and converge to a in the $\| \ \|_1$ (resp. $\| \ \|_2$) norm. Hence the sets $\mathcal{C}_1$ and $\mathcal{C}_2$ contain all operators with finite $\| \ \|_1$ and respectively $\| \ \|_2$ norm. However, $\mathcal{C}$ is not all of $\mathcal{B}(\mathcal{H})$: $\|a\|$ is equal to $\sup_i |\alpha_i|$, and in this norm it does not generally suffice to have $\|a\| < \infty$ for a_N to converge in norm to a. (A simple counterexample is $a = \mathbf{1}$.) The correspondence with the l^p spaces is: $l^0 \leftrightarrow \mathcal{C}$; $l^1 \leftrightarrow \mathcal{C}_1$; $l^2 \leftrightarrow \mathcal{C}_2$; $l^\infty \leftrightarrow \mathcal{B}(\mathcal{H})$.
3. $\mathcal{E}$ is a two-sided ideal of $\mathcal{B}(\mathcal{H})$, and this is also a property of its completions $\mathcal{C}_p$, since $\|ab\|_p \le \min(\|a\|\|b\|_p, \|b\|\|a\|_p)$, $p = 1, 2, \infty$.
4. The essential spectrum of any operator of $\mathcal{E}$ is $\{0\}$. This property carries over to all of $\mathcal{C}$, and is a distinguishing characteristic of self-adjoint, compact operators (Problem 9).
5. An operator $a \in \mathcal{E}$ sends a bounded set $\mathcal{G} \in \mathcal{H}$ to a finite-dimensional, bounded set, which is necessarily relatively compact. The image of a bounded set remains relatively compact when one passes to the norm-completion: any $c \in \mathcal{C}$ can be written as $a + \delta$, where $a \in \mathcal{E}$ and for any $\varepsilon > 0$, $\|\delta\| \le \varepsilon / \sup_{v \in \mathcal{G}} \|v\|$, and $c\mathcal{G}$ is the relatively compact set $a\mathcal{G}$ added to a set of diameter less than ε. Relative compactness means that for any ε there exists a finite covering with balls of diameter less than ε, and this is also true of the image of $\mathcal{G}$ under $c \in \mathcal{C}$. This fact is the origin of the nomenclature for $\mathcal{C}$: compact operators carry bounded sets into relatively compact sets.
6. Completion of $\mathcal{E}$ in the strong topology yields all of $\mathcal{B}(\mathcal{H})$ (Problem 8), but the strong topology is not strong enough for Properties 4 and 5, to carry over to $\mathcal{B}(\mathcal{H})$.
7. $\mathcal{B}(\mathcal{H})$ is not a separable topological space (see (2.1.27; 1)) while $\mathcal{C}$ is separable when $\mathcal{H}$ is a separable Hilbert space.
8. The sets $\mathcal{C}_p$ can be defined for

$$1 \le p < \infty \quad \text{as} \quad \{a \in \mathcal{C} : \|a\|_p = (\mathrm{Tr}(a^*a)^{p/2})^{1/p} < \infty\}.$$

The $\mathcal{C}_p$ are complete, normed algebras with $\| \ \|_p$, but are not C^* algebras (see (2.2.11; 2)). $\mathcal{C}$ is one, and $\mathcal{B}(\mathcal{H})$ is even a W^*-algebra.

2.3.24 Problems

1. Show that the three conditions of Remark (2.3.6; 1) are equivalent under the restriction that in Condition (iii) the word "subspace" should be understood as "closed subspace."
2. Show that w is pure iff π_w is irreducible. (Hints: (i) w is pure iff for every positive linear functional w_1 such that $w_1 \le w$, $w_1 = \lambda w$ for some $\lambda \in (0, 1]$; and (ii) if

$w_1 \leq w$, then there exists a positive operator $t_0 \in \pi_w(\mathcal{A})'$, with $0 \leq t_0 \leq 1$, such that $w(b^* a) = \langle \pi_w(b)\Omega | t_0 \pi_w(a)\Omega \rangle$.)

3. Show that Ker $\pi_w = \{a : w(b^* a^* ab) = 0 \text{ for all } b \in \mathcal{A}\}$.

4. Let $\mathcal{A}(\ni \mathbf{1})$ be a C^* algebra of operators in $\mathcal{B}(\mathcal{H})$. Show that $\mathcal{A}''$ is both the weak and the strong closure of $\mathcal{A}$. This is known as von Neumann's density theorem. (The argument for why $\mathcal{A}''$ is contained in the strong closure of $\mathcal{A}$ proceeds by the following steps:
   (i) Let $x \in \mathcal{H}$. The projection P onto the closure of $\{ax : a \in \mathcal{A}\}$ belongs to $\mathcal{A}'$.
   (ii) Let $b \in \mathcal{A}''$. Then $P\mathcal{H}$ is stable under b, and thus for all $\varepsilon > 0$ there exists an $a \in \mathcal{A}$ such that $\|bx - ax\| < \varepsilon$.
   (iii) It remains to be shown that finite intersections of neighborhoods of b of the kind considered in step (ii) with various x_i also contain elements of the strong closure of $\mathcal{A}$. To do this, take n nonzero vectors $x_1, x_2, \ldots, x_n$, and consider the representation π of $\mathcal{A}$ on $\mathcal{H} \oplus \mathcal{H} \oplus \cdots \oplus \mathcal{H}$; For any a, $\pi(a) = a \oplus a \oplus \cdots \oplus a$, which is known as the amplification of a. $\pi(b) \in \{\pi(a)\}''$, and the same argument as before, with $x = x_1 \oplus x_2 \oplus \cdots \oplus x_n$, shows that there exists an $a \in \mathcal{A}$ such that $\sum_{i=1}^n \|(b - a)x_i\|^2 < \varepsilon^2$.)

5. Show that the strong closure of the operators acting on $L^2(\mathbb{R}^n, d^n x)$ by multiplication by continuous functions is L^∞. (Hint: If $f \in L^\infty$, consider the continuous functions $(\rho * f)(x) \equiv \int d^n x' \rho(x - x') f(x')$ for ρ continuous and $\int \rho d^n x = 1$, and then let ρ approach a delta function.)

6. Show that the sum in (2.3.16) is orthogonal.

7. Construct an operator with a purely singular continuous spectrum.

8. Show that $\mathcal{E}$ is strongly dense in $\mathcal{B}(\mathcal{H})$. (Use the fact that every vector is cyclic for $\mathcal{E}$.)

9. Show that compactness of a Hermitian operator on an infinite-dimensional Hilbert space is equivalent to: $\sigma_{ac} = \sigma_s = \emptyset$ and $\sigma_{ess} = \{0\}$.

10. Prove Properties (2.3.19).

11. Prove Vigier's Theorem: *Every bounded, increasing filter F of operators has a supremum s, i.e., there exists an operator s such that $a \leq s$ for all $a \in F$, and $a \leq s'$ for all $a \in F \Rightarrow s \leq s'$. The supremum s is unique and belongs to the strong closure of F.*

2.3.25 Solutions

1. (iii) $\Rightarrow$ (ii): Suppose $x \neq 0$. The closed subspace spanned by $\pi(a)x$, for $a \in \mathcal{A}$, is stable for all $\pi(a)$, and therefore identical to all of $\mathcal{H}$.
   (ii) $\Rightarrow$ (iii): Let $\mathcal{H}'$ be a nontrivial, closed, invariant subspace of $\mathcal{H}$ and let $x \in \mathcal{H}'$. Since $\pi(a)x \in \mathcal{H}'$ for all $a \in \mathcal{A}$, $\mathcal{H}'$ must be dense, so $\mathcal{H}' = \mathcal{H}$.
   (i) $\Rightarrow$ (iii): Let $\mathcal{H}' \subset \mathcal{H}$ be stable and P be the projection onto $\mathcal{H}'$. Then $P\pi(a)P = \pi(a)P$ for all $a \in \mathcal{A}$, so $P\pi(a^*)P = \pi(a^*)P$ for all $a \in \mathcal{A}$, which implies $P\pi(a) = \pi(a)P$, since $\pi(a^*) = \pi(a)^*$. But then $P = \mathbf{0}$ or $\mathbf{1}$.
   (iii) $\Rightarrow$ (i): If $a \in \pi(\mathcal{A})'$, then $a^* \in \pi(\mathcal{A})'$, and likewise for $b = a + a^*$ and $c = i(a - a^*)$. Hence all the spectral projections of b and c also belong to $\pi(\mathcal{A})'$. This means that a

is a multiple of the identity, since every projection in $\pi(\mathcal{A})'$ defines a stable, closed subspace.
Remark: It is also possible to show that for C^* algebras conditions (i)–(iii) are equivalent to algebraic irreducibility: The only invariant, closed *or unclosed* subspaces are $\{0\}$ and $\mathcal{H}$.

2. Proof of Lemma (i): *w is pure iff every $w_1 \leq w$ is of the form λw.*
$\Rightarrow$: $w = (w_1/w_1(1))w_1(1) + ((w - w_1)/(1 - w_1(1)))(1 - w_1(1))$ is a convex combination of two states unless $w_1 = \lambda w$.
$\Leftarrow$: w is not pure iff $w = \alpha w_1 + (1-\alpha)w_2$, $\lambda w \neq w_1 \leq w$.
Proof of Lemma (ii): *The mapping $(\pi_w(b), \pi_w(a)) \rightarrow w_1(b^*a)$ has a unique continuous extension to a positive bilinear form on $\mathcal{H} \times \mathcal{H}$, bounded by* 1. Hence there exists an operator t_0, $0 \leq t_0 \leq 1$, such that $w_1(b^*a) = \langle \pi_w(b)\Omega | t_0 \pi_w(a)\Omega\rangle$. (This is a direct corollary of the Riesz–Frêchet theorem (2.1.17), and is often referred to as the Lax–Milgram theorem). The substitution $a \rightarrow ca$ yields

$$\langle \pi_w(b)\Omega | t_0 \pi_w(c)\pi_w(a)\Omega\rangle = w_1(b^*ca) = w_1((c^*b)^*a)$$
$$= \langle \pi_w(c)^*\pi_w(b)\Omega | t_0\pi_w(a)\Omega\rangle = \langle \pi_w(b)\Omega | \pi_w(c)^* t_0 \pi_w(a)\Omega\rangle \quad \text{for all } a, b \in \mathcal{A}$$
$$\Rightarrow [t_0, \pi_w(c)] = 0 \quad \text{for all } c \in \mathcal{A} \Rightarrow t_0 \in \pi(\mathcal{A})'.$$

Proof of the theorem:
$\Rightarrow$: Let P be a projection operator $\in \pi_w(\mathcal{A})'$. The mapping $a \rightarrow \langle P\Omega | \pi_w(a) P\Omega\rangle$ is a positive linear functional $\leq w$, so $\langle P\Omega|\pi_w(a)P\Omega\rangle = \lambda\langle\Omega|\pi_w(a)\Omega\rangle$, and if a is replaced with b^*a, then $\langle P\pi_w(b)\Omega|\pi_w(a)\Omega\rangle = \langle\lambda\pi_w(b)\Omega|\pi_w(a)\Omega\rangle$, which implies that $P = \lambda \cdot \mathbf{1}$, since Ω is cyclic, and thus $P = \mathbf{0}$ or $\mathbf{1}$.
$\Leftarrow$: Suppose $0 \leq w_1 \leq w$. Then $w_1(a) = \langle\Omega|t_0\pi_w(a)\Omega\rangle = \lambda\langle\Omega|\pi_w(a)\Omega\rangle$, since $\pi_w(\mathcal{A})'$ consists only of scalars.

3. $w(b^*ac) = 0$ for all $b, c \in \mathcal{A} \Rightarrow w(b^*a^*ab) = 0$ for all b, using the substitution $b^* \rightarrow b^*a^*$, $c \rightarrow b$. Conversely, $w(b^*a^*ab) = 0$ for all $b \Rightarrow |w(b^*ac)|^2 \leq w(b^*b) \cdot w(c^*a^*ac)$, so $w(b^*ac) = 0$.

4. Since multiplication is continuous in one factor in any topology, it is easy to see that the strong closure of $\mathcal{A}$, which is contained in the weak closure of $\mathcal{A}$, is contained in $\mathcal{A}''$. Thus it suffices to show that $\mathcal{A}'' \subset$ the strong closure of $\mathcal{A}$.
(i) Let $\mathcal{K} = \{ax : a = a^* \in \mathcal{A}\}$. $a\mathcal{K} \subset \mathcal{K} \Rightarrow a\bar{\mathcal{K}} \subset \bar{\mathcal{K}}$, where $\bar{\mathcal{K}}$ is the closure of $\mathcal{K}$. For any $\bar{x} \in \mathcal{H}$, $PaP\bar{x} = aP\bar{x} \Rightarrow aP = PaP = (PaP)^* = Pa \Rightarrow [a, P] = 0$.
(ii) $bPb = bP = Pb \Rightarrow b\bar{\mathcal{K}} \subset \bar{\mathcal{K}} \Rightarrow bx \in \bar{\mathcal{K}}$.
(iii) Operators t on $\mathcal{H} \oplus \cdots \oplus \mathcal{H}$ can be considered as matrices (t_{ik}) the entries of which belong to $\mathcal{B}(\mathcal{H})$, in which case $(\pi(a))_{ik} = a\delta_{ik}$. $([t, \pi(a)])_{ik} = t_{ik}a - at_{ik}$, i.e., $\{\pi(a)\}'$ consists of all t such that $t_{ik} \in \mathcal{A}'$, and therefore $\pi(b) \in \{\pi(a)\}''$.

5. Since f and $\rho^* f$ are bounded, it suffices to show that $\|(f - \rho^* f)\varphi\|_2 \rightarrow 0$ for the dense set of $\varphi \in L^\infty$ and of compact support K. On K, f is also $\in L^2$, and

$$\|(f - \rho^* f)\varphi\|_2 \leq \|\varphi\|_\infty \|f - \rho^* f\|_2.$$

In Fourier-transformed space, $\|f - \rho^* f\|_2^2 = \int d^n k |\tilde{f}(k)|^2 |1 - \tilde{\rho}(k)|^2 \rightarrow 0$ if $\tilde{\rho}(k)$ tends monotonically to 1.

6. Let $a = \int dP_a(\alpha)\alpha$. To each vector x we can associate a measure $d\mu_x = d\langle x|P_a(\alpha)x\rangle$ and construct $\mathcal{H}_{ac} = \{x : d\mu_x$ is absolutely continuous with respect to $d\alpha\}$ and $\mathcal{H}_s' = \{x : d\mu_x$ is singular with respect to $d\alpha\}$. These two subspaces are orthogonal: Suppose

$x \in \mathcal{H}_{ac}$ and $y \in \mathcal{H}_s'$; then there is a set M of Lebesgue measure zero on which $d\mu_y$ is concentrated. With the notation $P(M) = \int_M dP_a(\alpha)$, $(\mathbf{1} - P(M))y = 0$, and so $\langle x|y\rangle = \langle x|P(M)y\rangle = \langle P(M)x|y\rangle = 0$. Now consider arbitrary $x \in \mathcal{H}$; $d\mu_x$ can be decomposed into a singular and an absolutely continuous part: $d\mu_x = d\mu_x^s + d\mu_x^{ac}$. (**Lebesgue decomposition**; see [1, 13.18.7].) Therefore there exists another set M of Lebesgue measure zero, on which $d\mu_x^s$ is concentrated, and thus $P(M)x \in \mathcal{H}_s'$ and $(\mathbf{1} - P(M))x \in \mathcal{H}_{ac}$. Since $\mathcal{H}_s' \perp \mathcal{H}_{ac}$ and $\mathcal{H}_s' + \mathcal{H}_{ac} = \mathcal{H}$, $\mathcal{H} = \mathcal{H}_s' \oplus \mathcal{H}_{ac}$. It is obvious that $\mathcal{H}_p \subset \mathcal{H}_s'$ and $\mathcal{H}_s \subset \mathcal{H}_s'$. On the other hand, if $x \in \mathcal{H}_s' \ominus \mathcal{H}_p$, then $x \in \mathcal{H}_s$, and the same argument as before shows that $\mathcal{H}_s' = \mathcal{H}_p \oplus \mathcal{H}_s$.

7. Let f be the **Cantor function**, defined as follows: The **Cantor set** $\mathcal{C}$ in $[0, 1]$ is the complement of $\left(\frac{1}{3}, \frac{2}{3}\right) \cup \left(\frac{1}{9}, \frac{2}{9}\right) \cup \left(\frac{7}{9}, \frac{8}{9}\right) \cup \left(\frac{1}{27}, \frac{2}{27}\right) \cup \cdots$. It is a closed set of Lebesgue measure zero. Now let $f = \frac{1}{2}$ on $\left(\frac{1}{3}, \frac{2}{3}\right)$, $\frac{1}{4}$ on $\left(\frac{1}{9}, \frac{2}{9}\right)$, $\frac{3}{4}$ on $\left(\frac{7}{9}, \frac{8}{9}\right)$, etc. (see figure below). The function f increases monotonically and has a unique extension to a continuous function. Let a be the multiplication operator on $\mathcal{H} = L^2([0, 1], df)$ defined by $\varphi(x) \to x\varphi(x)$. Then $P_a(\alpha)\mathcal{H} = \{\varphi : \varphi(x) = 0 \text{ for } x > \alpha\}$, and if $M \subset [0, 1]$, then

$$\|P(M)\varphi\|^2 = \int_M |\varphi(x)|^2 df.$$

In particular, $\|P(\mathcal{C})\varphi\|^2 = \|\varphi\|^2$, and hence $\mathcal{H}_{ac} = \{0\}$. The point spectrum is empty, since f has no discontinuities, so $\int_{\{\lambda\}} |\varphi(x)|^2 df = 0$ when the integral is over any one point set $\{\lambda\}$.

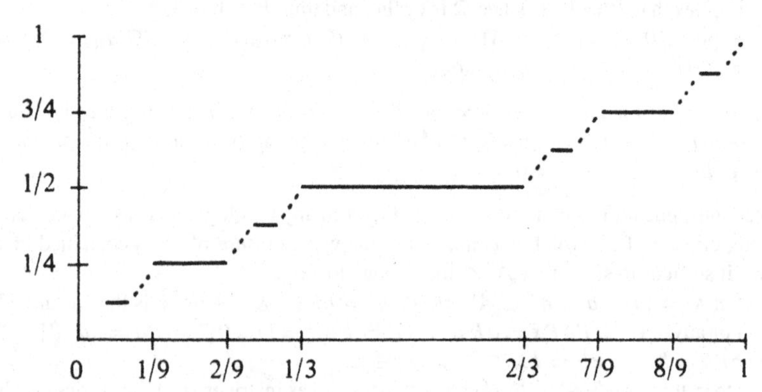

8. Let x and $y \in \mathcal{H}$. The operator $a : v \to \langle x|v\rangle y$, which maps x to y, is of finite rank and therefore compact. Thus every vector is cyclic for the compact operators. Consequently, the compact operators form an irreducible C^* algebra and are strongly dense in $\mathcal{B}(\mathcal{H})$ by von Neumann's density theorem.

9. $\sigma_{ac} \subset \sigma_{\text{ess}}$ and $\sigma_s \subset \sigma_{\text{ess}}$. It is not possible for each set to consist of one isolated point, so $\sigma_{\text{ess}} = \{0\}$ implies $\sigma_{ac} = \sigma_s = \emptyset$. Now let a be compact and $\lambda \in \sigma_{\text{ess}}(*a)$. There exists an orthonormal system $\{\psi_n\}$ such that $\|(a - \lambda)\psi_n\| \to 0$. The operator a sends the bounded set $\{\psi_n\}$ to a compact set, and hence $\{a\psi_n\}$ contains a strongly convergent subsequence $\{a\psi_{n_k}\}$. This implies that $\lambda = 0$, since no subsequence of $\{\psi_n\}$ is strongly convergent.

$\sigma_{\text{ess}}(a) = \{0\}$: Let $P_\varepsilon = P_a(\varepsilon) - P_a(-\varepsilon)$. Then $\dim(\mathbf{1} - P_\varepsilon)$ is finite for all $\varepsilon > 0$, so $a_n = \int_{-\infty}^{-1/n} dP_a(\alpha)\alpha + \int_{1/n}^{\infty} dP_a(\alpha)\alpha$ is of finite rank, and a fortiori compact. Since $\|a - a_n\| \leq 1/n$, a is compact.

10. (ii) $\operatorname{Tr} a = \sum_i \langle e_i | a e_i \rangle = \sum_i \|a^{1/2} e_i\|^2 = \sum_i \sum_k \langle U e_k | a^{1/2} e_i \rangle^2 = \sum_k \sum_i |\langle U e_k | a^{1/2} e_i \rangle|^2$ (since all summands are nonnegative), which $= \sum_k \|a^{1/2} U e_k\|^2 = \operatorname{Tr} U^* a U$.

 (iv) With the polar decompositions,

$$m_1 + m_2 = U|m_1 + m_2|, \qquad m_1 = V|m_1|, \qquad m_2 = W|m_2|,$$

$$\begin{aligned} \operatorname{Tr}|m_1 + m_2| &= \sum_k \langle e_k | U^*(m_1 + m_2) e_k \rangle \leq \sum_k (|\langle e_k | U^* V |m_1| e_k \rangle| \\ &\quad + |\langle e_k | U^* W |m_2| e_k \rangle|) \leq \operatorname{Tr}|m_1| + \operatorname{Tr}|m_2|. \end{aligned}$$

 For the final step, choose the e_k as the eigenvectors of $|m_1|$ and $|m_2|$, and observe that $\|V^* U e_k\| \leq 1$.

 (v) The trace is a positive linear functional on the $n \times n$ matrices, and (v) holds for matrices by (2.2.20; 1). Since each $m_i \in \mathcal{C}_2$ can be written as a finite matrix plus something of arbitrary small $\| \, \|_2$ norm, (v) holds for all $\mathcal{C}_2$ by Remark (2.3.20; 3): It can first be extended to $m_1 \in \mathcal{E}$ and $m_2 \in \mathcal{C}_2$, and then to both $m_{1,2} \in \mathcal{C}_2$.

 (vi) This follows from (ii), because any $a \in \mathcal{B}(\mathcal{H})$ is a linear combination of two Hermitian elements, each of which is in turn a linear combination of the positive elements $|a| \pm a$ or of the unitary elements $\|a\|^{-1}[a \pm i(\|a\|^2 - a^2)^{1/2}]$.

11. F is weakly relatively compact, and hence the set $\bigcap_{a \in F} \overline{\{b \in F : b \geq a\}}^{\text{weak}}$ is not empty, but must contain at least one element s. This $s \geq a$ for all $a \in \bar{F}^{\text{weak}}$, since the weak topology can be defined with the seminorm $\langle x | \cdot x \rangle$, and thus the weak closure preserves the ordering. If the set $\bigcap_a \cdots$ contained two elements $s_1 \neq s_2$, then there would exist some $x \in \mathcal{H}$ such that $|(x|s_1 x) - (x|s_2 x)| = \varepsilon > 0$, as well as $a_1, a_2 \in F$ such that $|(x|s_i x) - (x|a_i x)| < \varepsilon/2$, $i = 1, 2$. But then there would exist some $c \in F : c \geq a_1$ and $c \geq a_2$, so that $\varepsilon = |(x|s_1 x) - (x|s_2 x)| \leq |(x|s_1 x) - (x|cx)| + |(x|s_2 x) - (x|cx)| < \varepsilon/2 + \varepsilon/2 = \varepsilon$, which leads to a contradiction. The supremum $s \in \bar{F}^{\text{strong}}$ by the inequality

$$((b-a)x|(b-a)x) = ((b-a)^{1/2}x|(b-a)^{3/2}x) \leq (x|(b-a)x)^{1/2} \cdot \|x\| \|b\|^{3/2},$$

for all $b \geq a \geq 0$. $s' \geq a$ for all $a \in F \Rightarrow s' \geq a'$ for all $a' \in \bar{F}^{\text{weak}}$, which $\Rightarrow s' \geq s$.

2.4 One-Parameter Groups

Just as in classical mechanics, quantum-mechanical time-evolution is a one-parameter group. The group has a weaker sort of continuity than norm-continuity, which shows up in the unboundedness of the generators.

The dynamics of a closed system can described quantum-mechanically by an equation of the form

$$\frac{d}{dt} f = af, \tag{2.3}$$

where a is a time-independent operator. In this section we investigate the circumstances under which the formal solution,

$$f(t) = U_t f(0), \qquad U_t = \exp(at), \tag{2.4}$$

can be made sense of. In the applications f will be an element of a Banach space on which a acts linearly. From (2.4) we can abstract certain desiderata for an actual solution:

2.4.1 Definition

A mapping $\mathbb{R}^+ \to \mathcal{B}(\mathbb{E}) : t \to U_t$, is a **one-parameter semigroup** of operators on the Banach space $\mathbb{E}$ iff

(i) $U_{t_1+t_2} = U_{t_1} \cdot U_{t_2}$ for all $t_1, t_2 \geq 0$,

(ii) $U_0 = \mathbf{1}$.

If $\|U_t\| \leq 1$ (respectively $\|U_t \psi\| = \|\psi\|$), then we speak of **semigroups of contractions** (respectively **isometries**). If (i) and (ii) hold for all $t \in \mathbb{R}$ for a mapping $\mathbb{R} \to \mathcal{B}(\mathbb{E})$, then the semigroup is a group.

2.4.2 Remarks

1. Since $U_{t_1} U_{t_2} = U_{t_2+t_2} = U_{t_2} U_{t_1}$, all operators of a semigroup commute.
2. Groups of contractions are groups of isometries, since by definition $\|U_t\| \leq 1$ for all $t \in \mathbb{R}$, while $1 = \|U_t U_{-t}\| \leq \|U_t\| \cdot \|U_{-t}\| \Rightarrow \|U_t\| \geq 1$, so $\|U_t\| = 1$. On a Hilbert space, isometric groups are unitary groups, since for all $x \in \mathcal{H}$,

$$\|x\| = \|U^{-1} U x\| \leq \|U x\| \leq \|x\|,$$

so U and U^{-1} are both isometric $\Leftrightarrow$ U is unitary.
3. It is advisable to impose some continuity requirement on the mapping $\mathbb{R}^+ \to \mathcal{B}(\mathbb{E})$, as some crazy functions $\mathbb{R} \to \mathbb{R}$ are known which are linear but discontinuous. It only takes the weak topology on $\mathcal{B}(\mathbb{E})$ to guarantee that the norms are uniformly bounded on an interval (Problem 1): $\sup_{0 \leq t \leq \delta} \|U_t\| = M < \infty$. It then follows from the group property that

$$\|U_t\| \leq M^{t/\delta} \text{ for all } t \geq \delta,$$

so we may restrict ourselves to the study of the contractions $U_t M^{-t/\delta}$.

The strongest continuity property to require is that of the the norm topology on $\mathcal{B}(\mathbb{E})$. It in fact implies analyticity.

2.4.3 Theorem

For a semigroup, the following are equivalent:

(i) U_t *is norm-continuous;*

(ii) $\lim_{t\to 0} \|U_t - \mathbf{1}\| = 0$;

(iii) $\exists\, a \in \mathcal{B}(\mathbb{E})$, *such that* $\lim_{t\to 0} \|(1/t)U_t - \mathbf{1}) - a\| = 0$;

(iv) $U_t = \sum_{n=0}^{\infty} a^n (t^n/n!)$.

Because of (iv), *we write* $U_t = \exp(at)$. *The semigroup can be extended to a group, in which case* $\|U_t\| \leq \exp(\|a\|\,|t|)$ *for all* $t \in \mathbb{R}$.

Proof: It is obvious that (iv) $\Rightarrow$ (i) $\Rightarrow$ (ii) $\Leftarrow$ (iii) $\Leftarrow$ (iv), so it only remains to show that (ii) $\Rightarrow$ (iv). It follows from $U(0) = \mathbf{1}$ and norm-continuity at $t = 0$ that

$$\frac{1}{\tau}\int_0^\tau dt\, U_t$$

is close to the identity operator for small enough τ, and hence invertible. Therefore

$$a_\tau = \frac{U_\tau - 1}{\tau} \frac{1}{(1/\tau)\int_0^\tau dt\, U_t}$$

is well defined for small enough τ. This a_τ does not actually depend on τ, because

$$a_{n\tau} = \frac{U_\tau^n - 1}{\int_0^{n\tau} dt\, U_t} = \frac{(U_\tau - 1)(1 + U_\tau + \cdots + U_\tau^{n-1})}{\int_0^\tau dt\, U_t(1 + U_\tau + \cdots + U_\tau^{n-1})} = a_\tau.$$

Consequently, $a_{\tau'} = a_\tau$ whenever τ' is a rational multiple of τ and, by continuity, for all τ'. Since it is a constant, we may rename it a, and write

$$U_t = 1 + a\int_0^t ds\, U_s,$$

which leads to (iv) by iteration. Property (iv) implies the estimate $\|U_t\| \leq \exp(\|a\|\,|t|)$. □

2.4.4 Remarks

1. The exponential boundedness is the quantum version of a classical statement (I: 2.3.6; 1) for flows of bounded vector fields. Any faster growth, as for example for particles reaching infinity in a finite time, would violate the group structure.
2. It is of course possible for U_t to grow more slowly (e.g., for $a = \begin{pmatrix} 0 & 1 \\ 0 & 0 \end{pmatrix}$, $\exp(at) = 1 + at$) or even not at all ($a = i\begin{pmatrix} 0 & 1 \\ 1 & 0 \end{pmatrix}$, $\exp(at) = \cos t + a \sin t$).

We have seen that to each $a \in \mathcal{B}(\mathbb{E})$ there corresponds a U_t and vice versa. Now let us apply the methods of perturbation theory (I: §3.5) to the situation confronting us to evaluate the change in U_t if $a_0 \to a_0 + a_1$.

2.4.5 Theorem

Let $U_t = \exp(a_0, t)$, $V_t = \exp((a_0 + a_1)t)$, $a_i \in \mathcal{B}(\mathbb{E})$. *Then*

(i) $\|U_t - V_t\| \leq |t|\|a_1\| \exp(|t|(\|a_0 + \|a_1\|))$; *and, on the other hand, for all* $\lambda \geq \|a_0\| + \|a_1\|$,

(ii) $\|a_1\| \le (\|a_0\| + \lambda)(\|a_0 + a_1\| + \lambda) \int_0^\infty dt\, e^{-\lambda t} \|U_t - V_t\|$.

2.4.6 Remarks

1. Statement (i) is the precise analogue of the classical bound (I:3.5.4), and (ii) says that any perturbation has a noticeable effect after a rather short time.
2. Perturbation theory becomes quite inaccurate at large times, so it is not well suited as a tool for the study of the limit $t \to \infty$.

Proof:

(i) If we integrate

$$\frac{d}{d\lambda} \exp(a_0 t\lambda) \exp((a_0 + a_1)t(1 - \lambda))$$
$$= -t \exp(a_0 t\lambda) a_1 \exp((a_0 + a_1)t(1 - \lambda))$$

between 0 and 1, we obtain

$$V_t - U_t = t \int_0^1 d\lambda U_{\lambda t} a_1 V_{t(1-\lambda)}.$$

With Theorem (2.4.3) this gives the bound (i).

(ii) This follows from the identity

$$a_1 = (a_0 - \lambda) \int_0^\infty dt\, e^{-\lambda t} (U_t - V_t)(\lambda - a_0 - a_1),$$

in which we have assumed that $\lambda \ge \|a_0\| + \|a_1\| \ge \max\{\|a_0\|, \|a_0 + a_1\|\}$ so as to be certain the the integral exists. □

There are many ways to construct V_t from U_t and a_1. These constructions are subject to the complications typical of noncommuting operators.

2.4.7 Theorem

Let $a_1(t) \equiv \exp(-a_0 t) a_1 \exp(a_0 t)$. *Then*

(i) $\exp((a_0 + a_1)t) = \exp(a_0 t)(1 + \sum_{n=1}^\infty \int_0^t dt_1 \cdots \int_0^{t_{n-1}} dt_n a_1(t_1) \cdots a_1(t_n))$

(*the Dyson expansion*).

(ii) $\exp((a_0 + a_1)t) = \lim_{n\to\infty} (\exp(a_0 t/n) \exp(a_1 t/n))^n$

(*the Trotter product formula*).

2.4.8 Remarks

1. The sum and limit as $n \to \infty$ converge in the norm topology.
2. The Dyson expansion is identical to the classical formula (I: 3.5.7), and (ii) also has a formulation for flows.

3. With the time-ordering symbol $\mathbf{T}$, defined by

$$\mathbf{T}(a(t_1)a(t_2)\cdots a(t_n)) \equiv a(t_{i_1})a(t_{i_2})\cdots a(t_{i_n}),$$

where $t_{i_1} \geq t_{i_2} \geq t_{i_3} \geq \cdots \geq t_{i_n}$, Formula (i) can be written as

$$\exp(-a_0 t)\exp((a_0+a_1)t) = \mathbf{T}\exp\left(\int_0^t dt' a_1(t')\right)$$

Proof:

(i) $\frac{d}{dt}\exp(-a_0 t)\exp((a_0+a_1)t) = a_1(t)\exp(-a_0 t)\exp((a_0+a_1)t)$

$$\Rightarrow \exp((a_0+a_1)t)$$
$$= \exp(a_0 t)\left[1 + \int_0^t dt_1 a_1(t_1)\exp(-a_0 t_1)\exp((a_0+a_1)t_1)\right],$$

from which (i) follows by iteration.

(ii) Let

$$S_n = \exp((a_0+a_1)/n), \qquad T_n = \exp(a_0/n)\exp(a_1/n)$$

$$S_n^n - T_n^n = \sum_{m=0}^{n-1} S_n^m (S_n - T_n) T_n^{n-m-1}.$$

Since for all $k \leq n$, both $\|S_n^k\|$ and $\|T_n^k\| \leq \exp(\|a_0\| + \|a_1\|)$,

$$\|S_n^n - T_n^n\| \leq n\|S_n - T_n\|\exp(\|a_0\| + \|a_1\|).$$

However

$$\|S_n - T_n\| = \left\| \sum_{m=0}^{\infty} \frac{1}{m!}\left(\frac{a_0+a_1}{n}\right)^m - \left(\sum_{m=0}^{\infty}\frac{1}{m!}\left(\frac{a_0}{n}\right)^m\right) \times \left(\sum_{m=0}^{\infty}\frac{1}{m!}\left(\frac{a_1}{n}\right)^m\right)\right\| \leq \frac{c}{n^2},$$

so

$$\|S_n^n - T_n^n\| \to 0.$$

□

2.4.9 Example

Perturbation of the Larmor precession of a spin. Let $a_0 = i\sigma_x$ and $a_1 = ig\sigma_y$, with the σ's from (2.2.37). Since $(\mathbf{b}\cdot\boldsymbol{\sigma})^2 = |\mathbf{b}|^2$, it follows that

$$\exp((a_0+a_1)t) = \cos t\sqrt{1+g^2} + i(\sigma_x + g\sigma_y)\frac{1}{\sqrt{1+g^2}}\sin t\sqrt{1+g^2}.$$

This is an entire function of g for all $t \in \mathbb{R}$, and (2.4.7(i)) is its Taylor series. The latter is rather cumbersome, because the t dependence is greatly affected by g as it varies in $\mathbb{C}$. It is oscillatory if g is real, linear if $g = \pm i$, and otherwise it grows exponentially.

We must next confront the physically important case where U_t is only strongly continuous. The integral $(1/\tau)\int_0^\tau dt\, U_t$ will not converge uniformly to the operator $\mathbf{1}$, and might not be invertible even for very small τ. If we simply formally adopt the expression derived above for the generator

$$a = \frac{U_\tau - \mathbf{1}}{\tau}\left[\frac{1}{\tau}\int_0^\tau dt\, U_t\right]^{-1},$$

we find that it is not an element of $\mathcal{B}(\mathbb{E})$, but at least it is still true that a is defined on a dense set of $\varphi \in \mathbb{E}$: The inverse must certainly exist on vectors $\in \mathbb{E}$ of the form

$$\psi = \frac{1}{\tau}\int_0^\tau dt\, U_t \cdot \varphi, \quad \varphi \in \mathbb{E}, \quad \tau > 0 \quad \text{and} \quad a\psi = \frac{U_\tau - \mathbf{1}}{\tau}\varphi.$$

Since strong continuity means that $(1/\tau)\int_0^\tau dt\, U_t\varphi$ converges to φ as $\tau \to 0$, every vector φ of $\mathbb{E}$ can be approximated arbitrarily well by such a ψ. On ψ it is also true that

$$a\psi = \lim_{h\to 0} \frac{U_h - \mathbf{1}}{h}\psi,$$

because

$$(U_h - \mathbf{1})\int_0^\tau dt\, U_t = \int_\tau^{\tau+h} dt\, U_t - \int_0^h dt\, U_t = (U_\tau - \mathbf{1})\int_0^h dt\, U_t,$$

so

$$\frac{U_h - \mathbf{1}}{h} = \frac{U_\tau - \mathbf{1}}{\tau}\left[\int_0^\tau \frac{dt}{\tau} U_t\right]^{-1} \frac{1}{h}\int_0^h ds\, U_s.$$

Since the last factor converges strongly to $\mathbf{1}$ as $h \to 0$, we make

2.4.10 Definition

The **generator** a of a strongly continuous semigroup U_t is a linear mapping $D(a) \to \mathbb{E}$. Its **domain of definition**

$$D(a) = \left\{\psi \in \mathbb{E} : \exists \lim_{h\to 0} \frac{U_h - \mathbf{1}}{h}\psi \equiv a\psi\right\}$$

is dense in $\mathbb{E}$. The image of $D(a)$ is the **range** of a, $\mathrm{Ran}(a) \equiv aD(a) \subset \mathbb{E}$.

2.4.11 Example

$\mathbb{E} = l^2$, $\psi = (v_1, v_2, \ldots, v_n, \ldots)$, $U_t\psi = (e^{it}v_1, e^{2it}v_2, \ldots, e^{nit}v_n, \ldots)$ is strongly but not uniformly (= norm) continuous, $a\psi = i(v_1, 2v_2, \ldots, nv_n, \ldots)$.

$$D(a) = \left\{\psi \in l^2 : \sum_{n=1}^{\infty} |nv_n|^2 < \infty\right\}$$

is dense in l^2 but not equal to all of l^2.

2.4.12 Remarks

1. The condition of strong continuity may be weakened to weak continuity; as for instance the strong and weak topologies are the same for unitary operators on Hilbert space.
2. Furthermore, even weak measurability (i.e., $t \to \langle\psi|U_t\varphi\rangle$ is measurable for all φ and $\psi \in \mathcal{H}$) is equivalent to strong continuity for unitary groups on separable Hilbert spaces $\mathcal{H}$. This is not true for nonseparable Hilbert spaces: Let $\mathcal{H} = \bigoplus_x \mathcal{H}_x$, $\mathcal{H}_x = C$ for all $x \in \mathbb{R}$, where the uncountable sum is to be understood in the sense that only countably many of the components ψ_x of a vector ψ are nonzero, and
$$\|\psi\|^2 = \sum_x |\psi_x|^2.$$
If $(U_t\psi)_x = \psi_{x+t}$, then $t \to U_t$ is a unitary group that is weakly measurable but not weakly continuous: $\langle\psi|U_t\psi\rangle = \|\psi\|^2$ for $t = 0$ and is otherwise nonzero for only countably many t's. In this example, there exists no generator of any kind.
3. In order for $t \to U_t$ to be strongly differentiable, it is necessary for a to be $\in \mathcal{B}(\mathbb{E})$ and $D(a) = \mathbb{E}$. But then $t \to U_t$ is in fact analytic and strong differentiability is equivalent to the conditions of Theorem (2.4.3).
4. If a unitary group acts on a Hilbert space, ia must be Hermitian:
$$\langle ia\psi|\varphi\rangle = \langle\psi|ia\varphi\rangle \qquad \text{for all } \psi, \varphi \in D(a).$$
This can be seen just as for finite matrices,
$$\langle a\psi|\varphi\rangle = \lim_{h\to 0}\left\langle \frac{U_h - \mathbf{1}}{h}\psi\middle|\varphi\right\rangle = \lim_{h\to 0}\left\langle \psi\middle|\frac{U_{-h} - \mathbf{1}}{h}\varphi\right\rangle = -\langle\psi|a\varphi\rangle,$$
the only additional complication being to worry about the domains. Unitary semigroups can obviously be extended to unitary groups, even if only strongly continuous.

If U_t is only strongly continuous, then although $D(a)$ is dense in $\mathcal{H}$, it is not all of $\mathcal{H}$, that is, there are sequences $\varphi_n \in D(a)$ which converge, $\varphi_n \to \varphi$, but the limit $\varphi \notin D(a)$. Yet if $a\varphi_n$ converges to some $\psi \in \mathcal{H}$, then $\varphi \in D(a)$ and $\psi = a\varphi$ (Problem 2). This property will be important later on, for which reason we make the

2.4.13 Definition

The **graph** Γ of an operator $a : D \to \mathbb{E}$ is $\Gamma(a) \equiv \{(\psi, \varphi) \in D \times \mathbb{E} : \varphi = a\psi\}$. If $\Gamma(a)$ is a closed subspace of $\mathbb{E} \times \mathbb{E}$, a is said to be **closed**.

2.4.14 Examples

1. $\mathbb{E} = L^2([0,1], d\alpha)$, $(a\psi)(\alpha) = (1/\alpha)\psi(\alpha)$ with the domain

$$D_1(a) = \{\psi \in \mathbb{E} : \psi = 0 \text{ on some neighborhood of } 0\}.$$

The operator a is not closed, for consider $\psi_n = \alpha$ when $\alpha > 1/n$ and otherwise $\psi_n = 0$; then $\psi_n \to \psi(\alpha) = \alpha$ and $a\psi_n \to 1$, but $\psi \notin D_1(a)$.

2. Let a and $\mathbb{E}$ be as in Example 1, but take

$$D_2(a) = \left\{\psi \in \mathbb{E} : \int_0^1 \left|\frac{1}{\alpha}\psi(\alpha)\right|^2 d\alpha < \infty\right\}.$$

Since D_2 contains all ψ for which $a\psi \in \mathbb{E}$, a is closed on D_2.

3. Let $a\varphi = \mathbf{1} \cdot \varphi\left(\frac{1}{2}\right)$, $D(a) = \{\varphi \in L^2([0,1], d\alpha) \equiv \mathcal{H},\ \varphi \text{ continuous}\}$. This operator is not closed, since $\varphi_n \equiv \exp(-(\alpha - \frac{1}{2})^2 n^2) \to 0$ because $\|\varphi_n\| = O(1/n)$, but $a\varphi_n = 1 \not\to 0$.

2.4.15 Remarks

1. Note that Γ is required to be closed in $\mathbb{E} \times \mathbb{E}$ and not in $D \times \mathbb{E}$. By this definition with $\mathbb{E} = [0, \infty)$, the mapping $x \to 1/x$, $D = (0, \infty)$, is closed, while $x \to x$, $D = (0, \infty)$ is not.
2. Since the graph of a continuous mapping $a : F \to E$ is closed in $F \times E$, Γ is closed in $D \times \mathbb{E}$. Then a is closed iff D is closed (and therefore equal to $\mathbb{E}$).
3. Definition (2.4.13) is equivalent to the statement that $D \ni \psi_n \to \psi$, $a\psi_n \to \varphi \Rightarrow \psi \in D$, $\varphi = a\psi$, and to the statement that D is complete in the norm $\|\psi\|_a = \|\psi\| + \|a\psi\|$ (Problem 3).
4. If a is injective and $aD(a)$ is dense in $\mathbb{E}$, then a^{-1} is densely defined operator, and it is closed whenever a is, since $\Gamma(a^{-1}) = J\Gamma(a)$, where $J(x, y) = (y, x)$.
5. It might be imagined that whenever an operator is not closed, the domain has merely been chosen too small, and that by taking the closure $\bar{\Gamma}$ of Γ in $\mathbb{E} \times \mathbb{E}$ one would get a closed operator. This does not always work, as $\bar{\Gamma}$ might not be the graph of an operator. The trouble can be understood with $\theta(x) = 1$ for $x > 0$ and 0 for $x < 0$; the closure of the graph assigns the two values 0 and 1 to the point $x = 0$. It is also clear in Example (2.4.14; 3) that making $D(a)$ larger will not produce a closed operator.

The operator $a : D(a) \to \mathbb{E}$ of (2.4.10) is a discontinuous mapping. Continuity of a linear operator is equivalent to continuity at any single point and to boundedness. All of these conditions imply that there exists an $M \in \mathbb{R}^+$ such that $\|a\psi\| \le M\|\psi\|$ for all $\psi \in D(a)$. The connection between the notion of continuity and the notions of closure and domain of definition is:

2.4.16 Theorem

Any two of the three properties

(i) $D(a) = \mathbb{E}$;

(ii) *a is continuous; and*

(iii) *a is closed*

imply the third.

Proof:

(i) $\wedge$ (ii) $\Rightarrow$ (iii): Every graph of a continuous mapping is closed.

(ii) $\wedge$ (iii) $\Rightarrow$ (i): This was explained in (2.4.15; 2).

(i) $\wedge$ (iii) $\Rightarrow$ (ii): This follows from the closed-graph theorem [1, (12.16.11)], which is rather profound and cannot be proved here. □

2.4.17 Corollaries

1. If an operator is closed but not continuous, it cannot be defined everywhere.
2. If an operator is defined on all vectors and is discontinuous, then it is not closed (cf. (2.1.15; 2)).
3. If an operator is continuous, then it can be extended to all of $\mathbb{E}$, and it is thus closeable.

Since the difficulties attendant on the definition of a have to do with the inversion of an operator, a reasonable expectation would be that the resolvent $R_z \equiv (a-z)^{-1}$ of the operator a ought to belong to $\mathcal{B}(\mathbb{E})$ for $z \notin \mathrm{Sp}(a)$. If we write formally that $U_t = \exp(at)$, then

$$R_z = -\int_0^\infty dt\, e^{-tz} U_t.$$

In fact, for this formula there is

2.4.18 Theorem

Let U_t be a strongly continuous contractive semigroup, with generator a. Then for all $z \in \mathbb{C}$ with $\mathrm{Re}\, z > 0$,

$$R_z = -\int_0^\infty dt\, e^{-tz} U_t$$

maps $\mathbb{E}$ into $D(a)$. The resolvent satisfies $(a-z)R_z = \mathbf{1}$ *and* $\|R_z\| \leq (\mathrm{Re}\, z)^{-1}$.

Proof: The statement about the norm follows from $\|U_t\| \leq 1$. If a operates after R_z, then

$$aR_z = \lim_{h\to 0} \frac{U_h - 1}{h} R_z = \lim_{h\to 0} \left[\frac{1 - e^{hz}}{h} \int_0^\infty dt\, e^{-zt} U_t + \frac{e^{zh}}{h} \int_0^h dt\, e^{-zt} U_t \right].$$

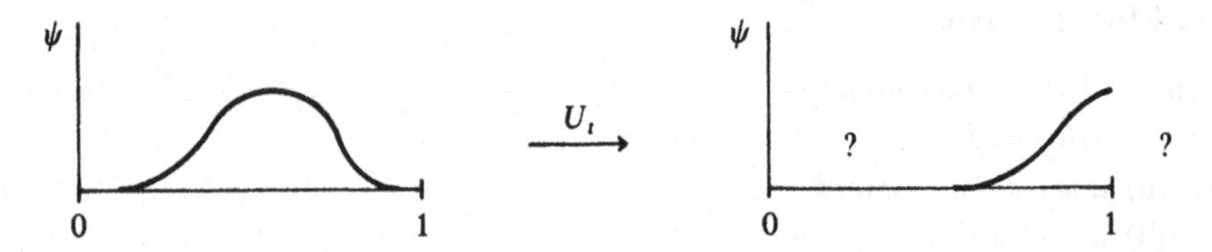

Figure 2.1. Unitary representation of the translation on [0, 1].

The first term converges uniformly to zR_z, and the second term converges strongly to $\mathbf{1}$. □

The problems that arise in physics usually go the other way: a is given and one tries to find U_t. It might be supposed that U_t could be defined as $\sum_{n=0}^{\infty}(t^n/n!)a^n$, but this often leads to disasters.

2.4.19 Example

Let $\mathbb{E} = L^2((0, 1), dx)$. Let us try to write the group of translations $\exp(itp)$ as a unitary family of operators by using the generator $p = -i(d/dx)$. So that all powers of id/dx will be well defined and Hermitian, we choose $D(p) = C_0^\infty(0, 1)$. These functions are supported within (0, 1), so

$$\left\langle \frac{d^n}{dx^n}\psi|\varphi\right\rangle = (-1)^n\left\langle \psi\left|\frac{d^n}{dx^n}\varphi\right.\right\rangle \quad \text{for all } \psi, \varphi \in D\left(\frac{d}{dx}\right),$$

and $\sum_{n=0}^{\infty}((it)^n/n!)(d^n/dx^n)$ is formally unitary. Unfortunately, the analytic functions on a complex neighborhood of (0, 1) for which this sum has a finite radius of convergence for all $x \in (0, 1)$ are not included in $D(-id/dx)$. Moreover, it is impossible to construct a unitary, finite translation this way, since it would have to translate part of the function out of (0, 1), and the missing part would affect the normalization integral:

$$\int_0^1 dx|\psi(x+t)|^2 = \int_t^1 dx|\psi(x)|^2 + \text{something unknown},$$

since $\psi(x)$ is not defined for $1 < x < 1 + t$ (see Figure 2.1).

Likewise, the attempt to write $U_t = \lim_{n\to\infty}(1 + at/n)^n$ opens the question of what $D(a^n)$ are for all $n \in \mathbb{Z}$. However, the use of the resolvent integral works without such difficulties, because it involves bounded operators only. It turns out that the properties we have found of a characterize the generators of semigroups; every such a determines a unique U_t.

2.4.20 The Hille–Yosida Theorem

Let a be a densely defined operator such that $(a - x)^{-1} : \mathbb{E} \to D(a)$ is bounded in norm as $\|(a - x)^{-1}\| \leq |x^{-1}|$ for all $x > 0$. Then there exists a unique contractive *semigroup U_t satisfying*

$$\lim_{h\to 0} \frac{U_h - \mathbf{1}}{h} \varphi = a\varphi \quad \textit{for all } \varphi \in D(a).$$

2.4.21 Remarks

1. It then follows from (2.4.18) that $(a - z)^{-1}$ exists for all z with $\text{Re}\, z > 0$, and is bounded in norm by $(\text{Re}\, z)^{-1}$.
2. Since $(a - x)^{-1}$ is defined on all of $\mathbb{E}$ and bounded, it is closed by Theorem (2.4.16). According to Remark (2.4.15; 4), $a - x$ and a are then also closed.
3. If $(a - x)^{-1}$ is defined only on a dense subspace, but is bounded there by $|x^{-1}|$, then it has a unique extension to all of $\mathbb{E}$, and Theorem (2.4.20) is still valid.
4. In the proof below we try to recover a from the resolvent by taking the limit $\lim_{x\to\infty}(-x - x^2(a - x)^{-1})$. It is also possible to work with $\exp(at) = \lim_{n\to\infty}(1 - at/n)^{-n}$.
5. Vectors φ on which $a^n\varphi$ is defined for all n and $\sum_n t^n \|a^n\varphi\|/n!$ converges for $|t| < t_0 > 0$ are called **analytic vectors**, or **entire vectors** if $t_0 = \infty$. The semigroup $\exp(at)$ is uniquely defined if a has a dense set of analytic vectors.

Proof: Let $a_x = -x - x^2(a - x)^{-1}$. For all $\varphi \in D(a)$, $a_x\varphi \to a\varphi$ as $x \to \infty$ (Problem 5). Consider the semigroup generated by $a_x \in \mathcal{B}(\mathbb{E})$. By Theorem (2.4.5), since the semigroups generated by a_{x_1} and a_{x_2} commute, for all x_1 and $x_2 > 0$ and $\varphi \in D(a)$,

$$\|(\exp(ta_{x_1}) - \exp(ta_{x_2}))\varphi\| \le c\|(a_{x_1} - a_{x_2})\varphi\|.$$

Because the vectors $a_x\varphi$ converge as $x \to \infty$, the vectors $\exp(ta_x)\varphi$ form a Cauchy sequence, which must always have a limit in a Banach space $\mathbb{E}$. Call the limit $U_t\varphi$. The operator U_t can be extended uniquely to $\mathbb{E}$, as the $\exp(ta_x)$ are uniformly bounded in x. To see that a is the generator of U_t, take the limit $x \to \infty$ of

$$\exp(a_x t)\varphi = \varphi + \int_0^t ds\, \exp(a_x s) a_x \varphi \quad \text{for } \varphi \in D(a).$$

Uniqueness follows from

$$\langle \psi | (a - x)^{-1} \varphi \rangle = -\int_0^\infty dt\, \exp(-xt) \langle \psi | U_t \varphi \rangle$$

and the fact that the Laplace transformation is injective on the continuous functions. □

2.4.22 Corollary (Stone's Theorem)

The operator ia is the generator of a unitary group on a Hilbert space $\mathcal{H}$ iff

(i) $\langle \psi | a\varphi \rangle = \langle a\psi | \varphi \rangle$ *for all* $\psi, \varphi \in D(a)$, *and*

(ii) $(a \pm i)D(a) = \mathcal{H}$.

Proof: It only needs to be shown that $(a - z)D(a) = \mathcal{H}$ for all z with $\operatorname{Im} z \neq 0$ when this holds for $z = \pm i$ and that $\|(a - z)^{-1}\| \leq |\operatorname{Im} z|^{-1}$. This is done in Problem 6. □

2.4.23 Examples

1. The a of example (2.4.11) obviously satisfies (i). It also satisfies (ii), since for any $\psi = (v_1, v_2, \ldots) \in l^2$ and $z \notin \mathbb{N}$,

$$\varphi = (a - z)^{-1}\psi = \left(\frac{v_1}{1-z}, \frac{v_2}{2-z}, \ldots, \frac{v_n}{n-z} \cdots\right) \in D(a)$$

and $(a - z)\varphi = \psi$.

2. In the troublesome Example (2.4.19), (i) is satisfied, as a never sends a vector out of $D(a)$. Since $C_0^\infty(0,1)$ is not all of $L^2((0,1), dx)$, (ii) is violated. $(-i(d/dx) - z)C_0^\infty(0,1)$ is not even dense in $L^2(0,1)$, since $\psi = e^{izx}$ is orthogonal to it: $\int_0^1 dx\, e^{-izx}(-i(d/dx) - z)\varphi(x) = 0$ for all $\varphi \in C_0^\infty(0,1)$.

In later sections we shall consider under what circumstances formal expressions for a can be interpreted so that the Hille–Yosida Theorem (2.4.20) applies, and whether the perturbation-theoretic formulas (2.4.5) and (2.4.7) also make some sense for only strongly continuous semigroups.

2.4.24 Problems

1. Show that the norms $\|U_t\|$ of a weakly continuous semigroup U_t are bounded for t in some interval $[0, \delta]$. Hint: argue as with the uniform boundedness principle.

2. Verity that if U_t is strongly continuous, then (cf. (2.4.10))

$$a = \frac{dU}{dt}, \quad D(a) = \left\{\varphi \in \mathbb{E} : \lim_{h \to 0} \frac{U_h - 1}{h}\varphi \text{ exists}\right\},$$

is a closed operator.

3. Show that an operator a on D is closed if D is complete in the norm $\|\psi\|_a = \|\psi\| + \|a\psi\|$.

4. Why is the graph of a continuous mapping $\mathbb{E} \to \mathbb{E}$ closed in $\mathbb{E} \times \mathbb{E}$?

5. Investigate the convergence of a_x from th e proof of (2.4.20).

6. Let a be a Hermitian operator. Show that the existence of $(a \pm i)^{-1}$ implies that of $(a - z)^{-1}$ for all z with $\operatorname{Im} z \neq 0$, and that $\|(a - z)^{-1}\| \leq |\operatorname{Im} z|^{-1}$.

2.4.25 Solutions

1. Let $p(\varphi) = \sup_{0 \leq t \leq \delta} |\langle\psi|U_t\varphi\rangle|$, fixing ψ for the moment. For any such ψ and $\varphi \in \mathbb{E}$, $p(\varphi)$ is the supremum of a continuous function on a compact set and consequently finite,

and the mapping $\varphi \to p(\varphi)$ is lower semicontinuous in the norm topology. Moreover, $p(\varphi_1+\varphi_2) \leq p(\varphi_1)+p(\varphi_2)$ and $p(\alpha\varphi) = |\alpha|p(\varphi)$. If p is bounded on any closed ball, then it is bounded everywhere because

$$p(\varphi_0 - \varphi) < M \text{ for all } \|\varphi\| \leq \varepsilon \Rightarrow p(\varphi) = \frac{\|\varphi\|}{\varepsilon}\, p\left(\frac{\varepsilon\varphi}{\|\varphi\|}\right) \leq \|\varphi\|\frac{p(\varphi_0)+M}{\varepsilon}.$$

Unboundedness of p on every ball would contradict lower semicontinuity: If p were unbounded on the ball K_{n-1}, then there would be $\varphi_n \in K_{n-1}$ with $p(\varphi_n) > n$. Since p is lower semicontinuous, there exist a ball $K_n \subset K_{n-1}$ containing φ_n and such that $p(\varphi) > n$ for all $\varphi \in K_n$. Since p must also be unbounded on the ball K_n, K_n must contain a φ_{n+1} such that $p(\varphi_{n+1}) > n+1$ and another ball $K_{n+1} \subset K_n$ such that $p(\varphi) > n+1$ for all $\varphi \in K_{n+1}$. The closed balls $K_1 \supset K_2 \supset \cdots$ are weakly compact, so there must exist some $\varphi \in \bigcap_n K_n$. Then $p(\varphi)$ would be greater than any n, which would contradict the finiteness of $p(\varphi)$ for all φ. Thus we conclude that $p(\varphi) < M_1\|\varphi\|$, and using the same argument for ψ, that

$$\sup_{0\leq t\leq\delta} \|U_t\| = \sup_{\substack{\|\psi\|=1\\ \|\varphi|=1}} \sup_{0\leq t\leq\delta} |\langle\psi|U_t\varphi\rangle| \leq M.$$

2. Let $\varphi_n \to \varphi$, $a\varphi_n \to \psi$;

$$\begin{aligned}\lim_{h\to 0}\frac{U_h-\mathbf{1}}{h}\varphi &= \lim_{h\to 0}\lim_{n\to\infty}\frac{U_h-\mathbf{1}}{h}\varphi_n = \lim_{h\to 0}\lim_{n\to\infty}\frac{1}{h}\int_0^h dt\, U_t a\varphi_n \\ &= \lim_{h\to 0}\frac{1}{h}\int_0^h dt\, U_t\psi = \psi \Rightarrow \varphi \in D(a), \quad a\varphi = \psi.\end{aligned}$$

3. A set (φ, ψ) in $\mathcal{H}\times\mathcal{H}$ is closed provided that it is complete in the norm $\|\varphi\| + \|\psi\|$. For a graph this norm is $\|\varphi\|_a = \|\varphi\| + \|a\varphi\|$.

4. Since $\|a\varphi\| \leq \|a\|\|\varphi\|$, the norm $\|\varphi\| + \|a\varphi\|$ is equivalent to $\|\varphi\|$, and the graph is closed if $D(a)$ is closed.

5. For all $\varphi \in D(a)$,

$$\lim_{x\to\infty}(1 + x(a-x)^{-1})\varphi = \lim_{x\to\infty}(a-x)^{-1}a\varphi = 0.$$

But if the bounded operators $-x/(a-x)$ converge on a dense set to $\mathbf{1}$, then they converge to $\mathbf{1}$ on the whole space, so

$$\lim_{x\to\infty}(-x(a-x)^{-1}a\varphi) = \lim_{x\to\infty} a_x\varphi = a\varphi$$

for all $\varphi \in D(a)$.

6. If a is Hermitian and $\mathrm{Ran}(a+i) = \mathcal{H}$, then $\|(a+i)^{-1}\| \leq 1$, since $(a+i)^{-1}x = y \Rightarrow \|x\|^2 = \|ay\|^2 + \|y\|^2$, and

$$\sup_x \frac{\|y\|^2}{\|x\|^2} = \sup_y \frac{\|y\|^2}{\|x\|^2} \leq 1.$$

Hence

$$(a+i+z)^{-1} = (a+i)^{-1}\sum_n(-z(a+i)^{-1})^n$$

has radius of convergence 1. The operators $a \pm 3i/2, a \pm 2i$, etc., have the same properties, and expansions around these points also converge up to the real axis, so the open half-planes can eventually be covered by such discs. For the second part, note that $(a+u+iv)^{-1}x = y, u, v \in \mathbb{R}, \Rightarrow \|x\|^2 = \|(a+u)y\|^2 + v^2\|y\|^2 \Rightarrow \|y\|^2/\|x\|^2 \leq 1/v^2$.

2.5 Unbounded Operators and Quadratic Forms

The generators of strongly continuous unitary groups are self-adjoint operators. Under the right circumstances, the domain of a formally Hermitian expression can be chosen so as to make it self-adjoint.

Typically a physicist is confronted with an unbounded Hermitian operator, and an important question is in what sense does such an operator generate a one-parameter group as the time-evolution. Since it is not always possible in classical mechanics to generate a satisfactory time-evolution—vector fields may fail to be complete, and a particle may reach infinity in a finite time—one must be prepared for trouble when doing quantum mechanics. Yet we shall discover later that the situation with $1/r$ potentials is much better quantum-theoretically than classically, and that the rather annoying question of the existence of collision trajectories in the classical three-body problem will cause no difficulty in quantum mechanics.

Our first task is to generalize the definition of the adjoint of an operator (2.1.24; 3) to cover unbounded operators, so as to ensure that all self-adjoint operators on Hilbert space generate unitary groups.

2.5.1 Definition

The **adjoint operator** a^* of an unbounded operator a having a dense domain $D(a)$ in $\mathcal{H}$ is defined on the domain

$$D(a^*) = \left\{ \varphi \in \mathcal{H} : \sup_{\psi \in D(a)} |\langle\varphi|a\psi\rangle| \|\psi\|^{-1} < \infty \right\}$$

by the formula $\langle\varphi|a\psi\rangle = \langle a^*\varphi|\psi\rangle$ for all $\psi \in D(a), \varphi \in D(a^*)$.

If $a = a^*$, then a is said to be **self adjoint**;

if $a^* = a^{**}$, then a is **essentially self adjoint**;

if $a^* \supset a$, then a is **Hermitian**.

The symbol $b \supset a$ means that b is an extension of a : $D(b) \supset D(a)$ and $b_{|D(a)} = a$.

2.5.2 Remarks

1. Since it is assumed that $D(a)$ is dense, (2.5.1) defines a^* uniquely.
2. The choice of the domain D for a fixes the domain of a^*; if $a^*\varphi \in \mathcal{H}$, then $\psi \to (a^*\varphi|\psi)$ is a continuous functional. $D(a^*)$ consists of all φ for which $\psi \to \langle\varphi|a\psi\rangle$ is a continuous functional $D \to \mathbb{C}$: it is thus the biggest possible domain.

3. If $D(a^*)$ is dense, then a^{**} is defined uniquely, and $a^{**} \supset a$. This is the case for Hermitian operators, where $D(a^*) \supset D(a)$, but in general $D(a^*)$ need be no larger than $\{0\}$.
4. If a is continuous, then $D(a^*) = \mathcal{H}$. This accords with Remark (2.1.24; 3), by which $a : D \to \mathcal{H}$ induces the mapping $a^* : \mathcal{H}^* = \mathcal{H} \to D^* = \mathcal{H}$. If, in addition, $D(a) = \mathcal{H}$, then the concepts of Hermitian and self-adjoint become synonymous.
5. $a \subset b \Rightarrow a^* \supset b^*$. In particular, a is Hermitian iff $a^* \supset a$, which implies that $a \subset a^{**} \subset a^*$; and a^* is Hermitian iff $a^{**} \supset a^*$. Therefore, if a and a^* are Hermitian, then a^* is self adjoint and a is essentially self-adjoint. Moreover, if b is a Hermitian extension of a, $a \subset b \subset b^* \subset a^*$; then b is determined by its domain, on which it must have the same action as a^*.
6. If a is essentially self-adjoint, then $a^* = a^{**} \supset a$ is the unique self-adjoint extension of a. $a \subset b = b^* \Rightarrow a^* \supset b^* = b^{**} \supset a^{**} = a^*$. The advantage of speaking of essential self-adjointness is the flexibility it allows in the choice of $D(a)$; a change in the domain $D(a)$ can leave a essentially self-adjoint, but it necessarily alters the statement that $a = a^*$.
7. The graph $\Gamma(a^*)$ (see (2.4.13)) can be described as follows: Let J be the unitary operator $(x, y) \to (y, -x)$ on $\mathcal{H} \oplus \mathcal{H}$. Then $\Gamma(a^*) = (J\Gamma(a))^\perp$ (Problem 3). Any subspace defined by orthogonality is closed, so a^* is always a closed operator. If a is Hermitian, then its closure is a^{**}, since $J^2 = -1$ and $(J(J\Gamma^\perp))^\perp = \bar{\Gamma}$. Hence Hermitian operators are always closeable, and we may assume them closed without loss of generality.

2.5.3 Examples

1. Let us recall Examples (2.4.14; 1) and (2.4.14; 2). Define $a_{1,2} \equiv a$ with the domains D_1 and D_2. The operator a_1 is not self-adjoint, since it is not closed. What is a_1^*? Its domain consists of all φ such that

$$\sup_{\psi \in D_1} \int_0^1 d\alpha \frac{1}{\alpha} \varphi^*(\alpha)\psi(\alpha) \left| \int_0^1 d\alpha |\psi(\alpha)|^2 \right|^{-1/2} < \infty,$$

so $(1/\alpha)\varphi^*(\alpha)$ must belong to $L^2([0, 1], d\alpha)$. Consequently $D((a_1)^*) = D_2$, $a_1^* = a_2 \supset a_1$, and we see that a_1 is Hermitian. It is also easy to see that a_2 is Hermitian and thus, according to (2.5.2; 5), self-adjoint: $a_2 \subset a_2^* \subset a_1^* = a_2$. This means that a_1 is essentially self-adjoint.
2. In Example (2.4.14; 3), $D(a^*)$ would be

$$\left\{ \varphi \in L^2([0, 1], dx) : \sup_{\psi \text{ continuous}} \int dx\, \varphi(x) \frac{\psi(\frac{1}{2})}{\|\psi\|} < \infty \right\}.$$

Since $\psi(\frac{1}{2})/\|\psi\|$ can be arbitrarily large, $D(a^*)$ is the subspace orthogonal to the constant function, and is not dense.
3. Let $a_1 : \psi(\alpha) \to i(d/d\alpha)\psi(\alpha) = i\psi'$, with $D(a) = \{\psi \in \mathcal{H} = L^2([0, 1], d\alpha) : \psi$ is absolutely continuous, $\psi' \in \mathcal{H}$, and $\psi(0) = \psi(1) = 0\}$.

Absolutely continuous functions are of the form

$$\psi(\alpha) = \int_0^\alpha d\alpha'\, g(\alpha') + \psi(0), \text{ with } g \text{ integrable.}$$

For such functions $\psi' = g$ a.e., so

$$\psi(\alpha) - \psi(0) = \int_0^\alpha d\alpha' \psi'(\alpha');$$

they are the functions for which integration by parts is justified. It is not restrictive enough to have merely functions that are continuous and have derivatives a.e. which are in L^2; there are continuous, strictly monotonic functions ψ for which ψ' vanishes almost everywhere (cf. (2.3.25; 7)). The boundary conditions ensure that a is Hermitian, for

$$\begin{aligned} \langle\varphi|a\psi\rangle &= i\int_0^1 d\alpha\, \varphi^*(\alpha)\frac{d}{d\alpha}\psi(\alpha) \\ &= i\,|\varphi^*(\alpha)\psi(\alpha)|_0^1 + \int_0^1 d\alpha \left(i\frac{d}{d\alpha}\varphi(\alpha)\right)^* \psi(\alpha) = \langle a\varphi|\psi\rangle \\ &\quad \text{for all } \varphi, \psi \in D. \end{aligned} \tag{2.5}$$

They are too strong for self-adjointness, however, since (2.5) is also valid without the requirement that $\psi(0) = \psi(1) = 0$. The other conditions are clearly necessary, so $D(a^*) = \{\varphi \in \mathcal{H} : \varphi$ is absolutely continuous and $\varphi' \in \mathcal{H}\}$. If we now calculate a^{**}, we are again led to (2.5), though this time it is necessary to reimpose the condition $\psi(0) = \psi(1) = 0$ to make $|\ldots|_0^1$ vanish. Therefore $a^{**} = a$, so a must be closed. The adjoint a^* is a proper extension of a and thus definitely not Hermitian, as $a^{**} \subset a^*$.

4. Consider again $a : \psi(\alpha) \to i(d/d\alpha)\psi(\alpha), D(a) = \{\psi \in \mathcal{H} = L^2([0,1], d\alpha) :$ ψ is absolutely continuous, $\psi' \in \mathcal{H}$, and $\psi(0) = \psi(1)e^{i\gamma}, \gamma \in \mathbb{R}\}$. It follows as in (2.5) that a is again Hermitian, but now $|\varphi^*(\alpha)\psi(\alpha)|_0^1 = 0$ requires that $\varphi(0) = \varphi(1)e^{i\gamma}$. Therefore $D(a) = D(a^*)$, and a is self-adjoint. This a is an extension of the a of Example 3 for any $\gamma \in \mathbb{R}$, so there is a one-parameter family of extensions of that a, and in fact it contains all possible self-adjoint extensions. (See (2.5.11).)
5. $a : \psi(\alpha) \to i(d/d\alpha)\psi(\alpha)$, $D(a) = \{\psi \in \mathcal{H} = L^2([0,\infty), d\alpha) : \psi$ is absolutely continuous $\psi' \in \mathcal{H}$, and $\psi(0) = 0\}$. This operator is also Hermitian, as it is easy to see that the upper limit contributes nothing upon integration by parts (Problem 4), and the equation

$$\langle\varphi|a\psi\rangle = \int_0^\infty d\alpha\, i\varphi^*\psi' = i\varphi^*(0)\psi(0) + \int_0^\infty d\alpha (i\varphi')^*\psi \stackrel{?}{=} \langle a\varphi|\psi\rangle$$

will hold provided that either $\varphi^*(0) = 0$ or $\psi(0) = 0$. This means that $D(a^*)$ lacks the condition $\psi(0) = 0$, and that $a^{**} = a$. It is not possible in this case to weaken the boundary condition to make $a^* = a$.
6. $a : \psi(\alpha) \to i(d/d\alpha)\psi(\alpha)$, $D(a) = \{\psi \in \mathcal{H} = L^2((-\infty,\infty) d\alpha) : \psi$ is absolutely continuous, and $\psi' \in \mathcal{H}\}$. The operator is Hermitian since, as in

Example 5, there is no contribution from $\pm\infty$ to the boundary term of the partial integration. In fact the operator is self-adjoint; there is no way to weaken the boundary conditions for $D(a^*)$. It is clear on reflection that the difficulties in Example (2.4.19) or with a unitary translation on $L^2(0,\infty)$ cannot arise on $L^2(-\infty,\infty)$, and nothing prevents $i(d/d\alpha)$ from being self-adjoint. Moreover, a^n is self-adjoint in this case on

$$D(a^n) = \left\{\psi \in \mathcal{H} : \psi, \ldots, \frac{d^{n-1}}{d^{n-1}\alpha}\psi \text{ absolutely continuous}, \right. \left. \frac{d^{n-1}}{d^{n-1}\alpha}\psi \in D(a)\right\}.$$

If a is Hermitian but not self-adjoint, then it may have any finite number of self-adjoint extensions, or an infinite number, or none at all. The adjoint a^* will not be Hermitian, and indeed it has complex eigenvalues. Returning to Example 3 (respectively 5), $\exp(-i\alpha z)$, $z \in \mathbb{C}$ (resp. $\operatorname{Im} z < 0$), is an eigenfunction of a^* with eigenvalue z. In these cases every point of $\mathbb{C}$ (resp. the lower half-plane) belongs to the point spectrum of a^*. This behavior is typical, as is shown by:

2.5.4 Theorem

Let a be a closed, Hermitian operator. Then

(i) $F_z = (a - z)D(a) = (\operatorname{Ker}(a^* - z^*))^\perp$, $(z = x + iy, y \neq 0)$, *is a closed subspace of* $\mathcal{H}$, *and* $(a - z)^{-1} : F_z \to D(a)$ *is a continuous bijection;*

(ii) $V : F_{-i} \to F_i : \psi \to (a-i)(a+i)^{-1}\psi$ *unitary, and* $\mathbf{1} - V = 2i(a+i)^{-1} : F_{-i} \to D(a)$ *is bijective, so that* $a\psi = i(\mathbf{1} + V)(\mathbf{1} - V)^{-1}\psi$ *for all* $\psi \in D(a)$;

(iii) $D(a^*) = D(a) + F_z^\perp + F_{z^*}^\perp$ *for all* $z \in \mathbb{C}$, $\operatorname{Im} z \neq 0$;

(iv) *if a is self-adjoint, then* $F_z = \mathcal{H}$ *for all* $z \in \mathbb{C}$, $\operatorname{Im} z \neq 0$; *and a is self-adjoint if* $F_z = \mathcal{H}$ *for some* $z \in \mathbb{C}$ *with* $\operatorname{Im} z > 0$ *and for some* z *with* $\operatorname{Im} z < 0$.

2.5.5 Remarks

1. Since all Hermitian operators are closeable, we have considered only closed operators a. If the assumption of closure were dropped, then Proposition (i)–(iv) would hold for the closure a^{**}. As a consequence, the resolvent $(a-z)^{-1}$ of an essentially self-adjoint operator a is densely defined, and since it is bounded, it has a unique extension $(a^{**} - z)^{-1}$ as a bounded operator defined on all of $\mathcal{H}$. (Recall Remarks (2.4.21).)
2. V is known as the **Cayley transform** of a.
3. These propositions are depicted schematically in Figure 2.2.

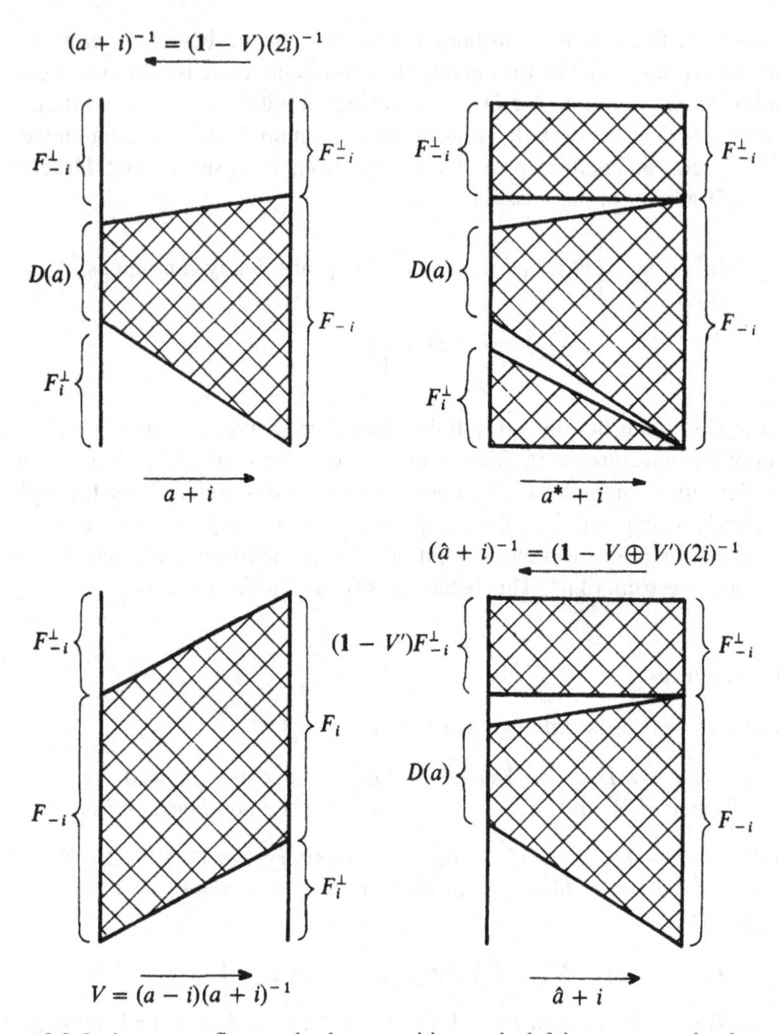

Figure 2.2. In the top two figures, the decomposition on the left is not assumed orthogonal. Although $D(a) \cap F^{\perp}_{\pm i} = \{0\}$, $D(a) \not\subset F_{\pm i}$. The extension $\hat{a}$, $a \subset \hat{a} \subset a^*$ and V' are defined in (2.5.10).

Proof:

(i) Because $\|(a - x - iy)\psi\|^2 = \|(a - x)\psi\|^2 + \|y\psi\|^2 \geq y^2\|\psi\|^2$, the resolvent $(a - z)^{-1}$ is bounded, and by assumption $\Gamma(a - z)$ and thus also $\Gamma((a - z)^{-1})$ are closed, so it follows from Theorem (2.4.16) that $D((a - z)^{-1}) = F_z$ is a Hilbert space, and thus a closed subspace of $\mathcal{H}$. Since $\langle\varphi|(a - z)\psi\rangle = \langle(a^* - z^*)\varphi|\psi\rangle$ for all $\varphi \in D(a^*)$, $\psi \in D(a)$, F_z is orthogonal to the eigenvectors of a^* of eigenvalue z^*. As a Hermitian operator, a cannot have any complex eigenvalues. Hence $a - z$ is injective and, by definition, surjective as a mapping $D(a) \to F_z$.

(ii) $\psi \in F_{-i} \Leftrightarrow \psi = (a+i)\varphi, \varphi \in D(a)$, and because $\|(a+i)\varphi\| = \|(a-i)\varphi\|$, the operators V and $V^{-1} = (a + i)(a - i)^{-1}$ are isometric. From $\varphi = (\psi - V\psi)/2i$ and a $\varphi = (\psi + V\psi)/2$, we conclude that $(\mathbf{1} - V)\psi = 0$ implies $(\mathbf{1} + V)\psi = 0$ and thus $\psi = 0$. Therefore $\mathbf{1} - V$ is invertible on F_{-i}, and $a = i(\mathbf{1} + V)(\mathbf{1} - V)^{-1}$.

(iii) Let $\psi \in D(a^*)$ and write $(a^* + i)\psi$ as a sum of vectors of F_{-i} and $F_{-i}^{\perp}$: $(a^* + i)\psi = (a + i)\eta + 2i\chi$, $\eta \in D(a)$, $\chi \in F_{-i}^{\perp} = \text{Ker}(a^* - i)$. Hence $2i\chi = (a^*+i)\chi \Rightarrow (a^*+i)(\psi-\eta-\chi) = 0$, because $a^* \supset a \Rightarrow a\eta = a^*\eta$. Therefore $\psi = \eta + \chi + \varphi$, $\varphi \in F_i^{\perp}$ is the required decomposition of any vector ψ of $D(a^*)$.

(iv) This follows from (iii), because if (2.4.25; 6) is generalized to unbounded operators, we see that $F_{z_0} = \mathcal{H} \Rightarrow F_z = \mathcal{H}$ for all z with $\text{Im}\, z/\text{Im}\, z_0 > 0$. □

2.5.6 Example

$a = i(d/d\alpha)$, $D(a)$ as in (2.5.3; 3)$\ni \varphi$; $F_{-i} \ni \psi(\alpha) = i(d/d\alpha + \mathbf{1})\varphi(\alpha)$ is orthogonal to $\text{Ker}(a^* - i) = \{ce^{\alpha}\}$, since

$$\int_0^1 d\alpha\, e^{\alpha}\left(\frac{d}{d\alpha} + 1\right)\varphi(\alpha) = e\varphi(1) - \varphi(0) = 0,$$

$$\varphi(\alpha) = -i\int_0^{\alpha} d\alpha' e^{\alpha' - \alpha}\psi(\alpha'),$$

so $\psi \to (a + i)^{-1}\psi$ is continuous, and

$$\begin{aligned}(V\psi)(\alpha) &= \left(i\left(\frac{d}{d\alpha} - 1\right)\varphi(\alpha)\right.\\ &= \int_0^1 d\alpha'[\delta(\alpha - \alpha') - 2\Theta(\alpha - \alpha')e^{\alpha'-\alpha}]\psi(\alpha')\\ &\equiv \left.\int_0^1 d\alpha' V(\alpha, \alpha')\psi(\alpha')\right).\end{aligned}$$

V is isometric on F_{-i}, since $\int_0^1 d\alpha' V^*(\alpha, \alpha')V(\alpha', \alpha'') = \delta(\alpha - \alpha'') - 2e^{\alpha+\alpha''}$.

2.5.7 Remarks

1. If a is not self-adjoint, then $a - z : D(a) \to \mathcal{H}$ with Im $z \neq 0$ is still injective, but it is not surjective, while $a^* - z : D(a^*) \to \mathcal{H}$ is surjective but not injective. Consequently, Sp(a) and Sp(a^*) each contain at least a half-plane, if Definition (2.2.13) is carried over for the spectrum of unbounded operators.
2. If a is self-adjoint, then $V : \mathcal{H} \to \mathcal{H}$ is unitary, so the spectral theorem extends to cover a. If $d\mu(\theta)$, $0 \leq \theta < 2\pi$, is one of the spectral measures (2.3.14) for V : $\text{Sp}(V) = \{e^{i\theta}\}$, $V(\psi)(\theta) = e^{i\theta}\psi(\theta)$, then the multiplication operator equivalent to a is $a : \psi(\theta) \to i(1 + e^{i\theta})/(1 - e^{i\theta})\psi(\theta)$. We saw earlier that V does not have the eigenvalue 1, so the measure of the point $\theta = 0$ is zero, and a is multiplication by an a.e. finite function. This form of the spectral theorem for unbounded self-adjoint operators is as general as possible. If $\lambda = \cot\theta/2$ is introduced as a new variable, then $L^2([0, 2\pi], d\mu(\theta))$ is mapped to $L^2((-\infty, \infty), d\mu(\lambda))$, and a becomes multiplication by λ. In analogy with (2.3.14), a may then be written as $\int_{-\infty}^{\infty} \lambda \, dP_\lambda$, which extends Theorem (2.3.14) to unbounded self-adjoint operators.
3. We shall understand convergence of a sequence of self-adjoint operators $a_n \to a$ to mean that all sequences of bounded functions f of the a_n converge: $f(a_n) \to f(a)$. It suffices to have convergence for either of the two classes of functions $f(a) = \exp(iat)$ for all $t \in \mathbb{R}$ and $f(a) = (a - z)^{-1}$ for all z with Im $z \neq 0$.
4. By Stone's Theorem (2.4.22), the self-adjoint operators are exactly the class of Hermitian operators that generate unitary groups. For instance, the a of Example (2.5.3; 4) generates the group of translations

$$(\exp(iat)\psi)(x) = \psi(x - t)$$

and answers the question that arose in (2.4.19) about the effect of the periodic boundary conditions: whatever is pushed past one end of the interval reappears at the other end with some constant change of phase (see Figure 2.3):
5. A converse of Theorem (2.5.4) can also be proved (Problem 7): If V is an isomorphism of a closed subspace F_- onto a subspace, such that $\mathbf{1} - V$ is a bijection of F_- onto a dense subspace of $\mathcal{H}$, then $i(\mathbf{1} + V)(\mathbf{1} - V)^{-1}$ is a Hermitian operator.

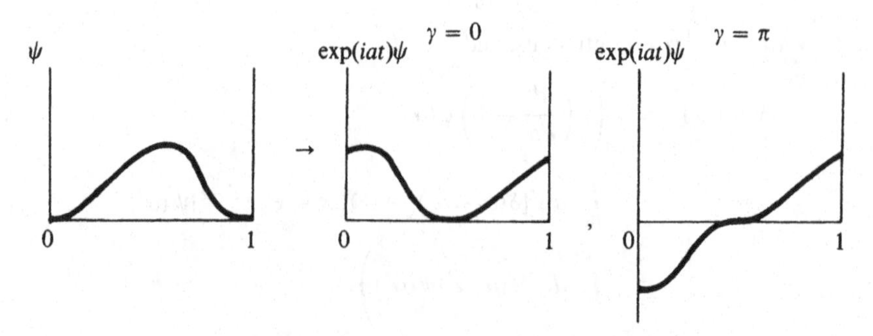

Figure 2.3. Two unitary representations of the translation on [0, 1].

6. The part of $D(a^*)$ not contained in $D(a)$ consists only of eigenvectors of a^* with complex eigenvalues. a^* acts on $D(a)$ like a, on $F_z^\perp$ (resp. $F_{z^*}^\perp$) it is z^* (resp. z). The sum of (iii) of (2.5.4) is of course not orthogonal.

Because of these facts, the most important criterion for self-adjointness is the absence of complex eigenvalues of a^*. To pursue this subject further, we make

2.5.8 Definition

$(m, n) \equiv \dim(\mathrm{Ker}(a^* \pm i)) = \dim(F_{\pm i}(a))^\perp$ are the **deficiency indices** of a.

2.5.9 Examples

1. a is essentially self-adjoint iff $(m, n) = (0, 0)$.
2. In Example (2.5.3; 3), $(m, n) = (1, 1)$.
3. In Example (2.5.3; 5), $(m, n) = (0, 1)$.

If one wishes to extend a Hermitian operator $a = i(\mathbf{1} + V)(\mathbf{1} - V)^{-1}$ to a self-adjoint operator $\hat{a} = i(\mathbf{1} + U)(\mathbf{1} - U)^{-1} \supset a$, then U must be a unitary extension of $V : F_{-i} \to F_{+i}$. It is clear that the extension must map the orthogonal complements $\mathrm{Ker}(a^* \mp i)$ unitarily onto each other, so U has to be an orthogonal sum $V \oplus V'$ for some V' acting on the orthogonal complements. This is possible only if $F_{\pm i}^\perp$ have the same dimension:

2.5.10 Theorem

A Hermitian operator a can be extended to a self-adjoint operator $\hat{a}$ iff the deficiency indices are equal. In that case, for every unitary mapping $V' : \mathrm{Ker}(a^* - i) \to \mathrm{Ker}(a^* + i)$ there exists a distinct extension

$$\hat{a} = i(\mathbf{1} + V \oplus V')(\mathbf{1} - V \oplus V')^{-1}, \qquad D(\hat{a}) = (\mathbf{1} - V \oplus V')\mathcal{H}.$$

2.5.11 Example

Let us return to (2.5.6), which has $(m, n) = (1, 1) : U(1, \mathbb{C})$ is multiplication by a phase factor, so there exists a one-parameter family $\{a_\beta\}$ of self-adjoint extensions of a. If $V'e^\alpha = e^{1-\alpha}e^{i\beta}$, then $U = V \oplus V'$ is defined on all of $\mathcal{H}$. With the procedure we have described above, $\varphi \in D(a)$ is written in the form

$$\begin{aligned}
\varphi(\alpha) &= ((U - \mathbf{1})\psi)(\alpha) \\
&= (U - \mathbf{1})\left(\psi(\alpha) - e^\alpha c \int_0^1 d\alpha' e^{\alpha'} \psi(\alpha') + e^\alpha c \int_0^1 d\alpha' e^{\alpha'} \psi(\alpha')\right) \\
&= -2 \int_0^\alpha d\alpha' e^{\alpha' - \alpha} \left(\psi(\alpha') - e^{\alpha'} c \int_0^1 d\alpha'' e^{\alpha''} \psi(\alpha'')\right) \\
&\quad + (e^{1-\alpha} e^{i\beta} - e^\alpha) c \int_0^1 d\alpha' e^{\alpha'} \psi(\alpha'),
\end{aligned}$$

where $c \int_0^1 d\alpha\, e^{2\alpha} = 1$, so φ satisfies the boundary conditions

$$\frac{\varphi(0)}{\varphi(1)} = \frac{e^{i\beta+1}-1}{e^{i\beta}-e} = \left(\frac{e^{1-i\beta}-1}{e^{-i\beta}-e}\right)^* = \left(\frac{e-e^{i\beta}}{1-e^{i\beta+1}}\right)^*$$
$$= \left(\frac{\varphi(1)}{\varphi(0)}\right)^* \Rightarrow |\varphi(0)| = |\varphi(1)|,$$

which makes a_β identical to the a of (2.5.3; 4).

2.5.12 Remarks

1. Theorem (2.5.10) reveals why it was not possible to weaken the boundary conditions in (2.5.3; 5) to make a self-adjoint. In that example, $m \neq n$.
2. Although the deficiency indices and extensions were defined with the special values $z = \pm i$, any other pair of complex conjugates z, z^*, Im $z \neq 0$, would have produced an equivalent definition.
3. If either m or n is zero, then the operator is maximal, i.e., it has no Hermitian extensions. If $m = n < \infty$, then every maximal Hermitian extension is self-adjoint, and if $m = n = \infty$, then there are maximal Hermitian extensions that are not self-adjoint.
4. If a Hermitian operator is real (see (3.3.18; 5)), that is, invariant under $i \to -i$, then $m = n$, and it always has a self-adjoint extension.

The delicate attribute of self-adjointness is not even preserved by the formation of linear combinations of operators. If a and b are self-adjoint, then $a + b$ (and $ab + ba$) may fail to be self-adjoint. The sum $a + b$ is a priori defined only on $D(a) \cap D(b)$, and $ab + ba$ only on $D(a) \cap D(b) \cap aD(a) \cap bD(b)$, and these sets might not be dense. Even if the intersection of the domains is dense in $\mathcal{H}$, it might be too small for the sum to be self-adjoint.

2.5.13 Example

A $1/r^2$ potential. Let $H = K + V$, $K : \psi(r) \to -(d^2/dr^2)\psi(r) : D(K) = \{\psi \in \mathcal{H} : \psi'$ is absolutely continuous, $\psi'' \in \mathcal{H}$, and $\psi(0) = 0\}$, where $\mathcal{H} = L^2([0,\infty), dr)$; $V : \psi(r) \to (\gamma/r^2)\psi(r)$, $\gamma \in \mathbb{R} : D(V) = \{\psi \in \mathcal{H} : (1/r^2)\psi \in \mathcal{H}\}$. With the a of Example (2.5.3; 5), $K = a^*a$, which is always self-adjoint. As in Example (2.5.3; 1), V is likewise self-adjoint. However, $D(H) = D(K) \cap D(V) = \{\psi \in \mathcal{H} : \psi'$ is absolutely continuous and $\in \mathcal{H}$, $\psi'' \in \mathcal{H}$, and $\psi(0) = \psi'(0) = 0\}$, and as in (2.5) the boundary condition does not show up in $D(H^*) : D(H^*) = \{\psi \in \mathcal{H} : \psi'$ is absolutely continuous and $\in \mathcal{H}$, and $(-\psi'' + \gamma\psi/r^2) \in \mathcal{H}\}$. It is possible for $D(H)$ to be a proper subset of $D(H^*)$, provided that neither ψ'' nor $\gamma\psi/r^2$ is in $\mathcal{H}$, but that their difference is, because of a cancellation of the singularities at 0. In order to understand when this happens, let us examine the solutions of the equation $\psi_\pm'' = (\gamma/r^2 \pm i)\psi_\pm$. The solutions that decrease when $r \to \infty$ as $\psi_\pm \sim \exp(-r(1 \pm i)/\sqrt{2})$ are linear combinations of

$r^{\pm\nu+1/2}$, $\nu = \sqrt{\gamma + 1/4}$, in the limit $r \to 0$. These functions are square-integrable only if $\nu < 1 \Leftrightarrow \gamma < \frac{3}{4}$, so if $\gamma \geq \frac{3}{4}$, then the deficiency indices are $(0, 0)$, and if $\gamma < \frac{3}{4}$, then they are $(1, 1)$. In the latter case there is a one-parameter family of self-adjoint extensions, which append $\psi_+ - \exp(2i\delta)\psi_i$, $\delta \in \mathbb{R}$, to $D(H)$ so that $H(\psi_+ - \exp(2i\delta)\psi_-) = i(\psi_+ + \exp(2i\delta)\psi_-)$. Even if $\gamma = 0$, $D(H)$ gets expanded by the inclusion of $\exp(-r(1+i)/\sqrt{2}) - \exp(2i\delta - r(1-i)/\sqrt{2})$, i.e., it can be characterized by the boundary condition $\psi(0)/\psi'(0) = \sqrt{2}/(\cot(\delta) - 1)$. Since $\varphi \equiv \psi/r$ in three-dimensional polar coordinates, $-\Delta\varphi$ becomes $\delta^3(x)\psi(0)/4\pi$, so this extension corresponds physically to the addition of a delta-function potential at the origin.

However, if b is small in comparison with a in a certain sense, then the addition of b to a does not affect its self-adjointness.

2.5.14 The Kato–Rellich Theorem

Let $a^* = a$, $b* \supset b$, and $D(b) \supset D(a)$, and suppose that there exist constants $0 \leq \alpha < 1$ and $\beta \geq 0$ such that $\|b\psi\| \leq \alpha\|a\psi\| + \beta\|\psi\|$ for all $\psi \in D(a)$. Then $a + b$ is self-adjoint on $D(a)$. If a is essentially self-adjoint on $D \subset D(a)$, then so is $a + b$.

Proof: In the spectral representation of a (see (2.5.7; 2)) we discover that

$$\|(a \pm i\eta)^{-1}\| \leq \eta^{-1} \quad \text{and} \quad \|a(a \pm i\eta)^{-1}\| \leq 1.$$

If η is large enough, then it follows that $\|b(a \pm i\eta)^{-1}\| \leq \alpha + \beta\eta^{-1} < 1$, so $\mathbf{1} + b(a + i\eta)^{-1}$ is bijection of $\mathcal{H}$. Consequently $(a + b \pm i\eta)D(a) = (\mathbf{1} + b(a \pm i\eta)^{-1})(a \pm i\eta)D(a)$ is either all of $\mathcal{H}$ or dense in $\mathcal{H}$, depending on whether $(a \pm i\eta)D(a)$ is all of $\mathcal{H}$ or only dense. □

2.5.15 Remarks

1. If b is bounded, then it is a fortiori relatively bounded, and $a + b$ is self-adjoint or essentially self-adjoint on $D(a)$ whenever a is.
2. Since $\sqrt{\alpha^2(1+\varepsilon) + \beta^2(1 + 1/\varepsilon)} \geq \alpha + \beta$ for all $\alpha, \beta, \varepsilon > 0$, Criterion (2.5.14) is equivalent to $\|b\psi\|^2 \leq \alpha^2\|a\psi\|^2 + \beta^2\|\psi\|^2$, or to $b^*b \leq \alpha^2 a^2 + \beta^2$.
3. For the statement about essential self-adjointness, α may be allowed to be 1.
4. For the physical systems that will concern us, b = a Coulomb potential is bounded relative to a = the kinetic energy. The Kato–Rellich Theorem is thus sufficient for our purposes to guarantee existence and uniqueness of the time-evolution.

It sometimes happens that formal Hamiltonians are not even strictly speaking operators, because they send every vector of $\mathcal{H}$ out of $\mathcal{H}$. However, knowledge of enough matrix elements is frequently sufficient to determine the time evolution, even in the absence of a well-defined operator.

2.5.16 Definition

A **quadratic form** q is a mapping $Q(q) \times Q(q) \to \mathbb{C} : (\varphi, \psi) \to \langle\varphi|q|\psi\rangle$, where $Q(q)$ is a dense subspace of $\mathcal{H}$ known as the **form domain**, such that $\langle\varphi|q|\psi\rangle$ is linear in ψ and conjugate-linear in φ. If $\langle\psi|q|\varphi\rangle^* = \langle\varphi|q|\psi\rangle$, then q is said to be **Hermitian**, and if $\langle\varphi|q|\varphi\rangle \geq 0$, it is **positive**. A positive quadratic form q is said to be **closed** iff $Q(q)$ is complete in the form norm $\|\varphi\|_q^2 = \langle\varphi|q|\varphi\rangle + \|\varphi\|^2$.

Gloss

Conjugate-linearity means that $\langle\lambda\varphi|q|\psi\rangle = \lambda^*\langle\varphi|q|\psi\rangle$. If $\langle\varphi|q|\varphi\rangle \geq -M\|\varphi\|^2$, then q is **semibounded**; equivalently, the form $\langle\varphi|q_1|\psi\rangle \equiv \langle\varphi|q|\psi\rangle + M\langle\varphi|\psi\rangle$ is positive.

2.5.17 Examples

1. Let a be a (densely defined, linear) operator. Then $(\varphi, \psi) \to \langle\varphi|a\psi\rangle$ is a quadratic form, and is Hermitian or positive iff a is.
2. Suppose a self-adjoint operator a has been written in a spectral representation on $\bigoplus_n L^2(\mathbb{R}, d\mu_m)$. Then $\langle\varphi|q|\psi\rangle = \sum_n \int_{-\infty}^{\infty} d\mu_n(\alpha)\alpha\varphi_n^*(\alpha)\psi_n(\alpha)$ for φ, $\psi \in Q(a) = \{\zeta \in \mathcal{H} : \sum_n \int_{-\infty}^{\infty} d\mu_n(\alpha)|\alpha||\zeta_n(\alpha)|^2 < \infty\}$ is Hermitian form, which is closed if $a \geq 0$. Observe that $Q(a)$ is different from $D(a) = \{\zeta \in \mathcal{H} : \sum_n \int_{-\infty}^{\infty} d\mu_n(\alpha)|\alpha|^2|\zeta_n(\alpha)|^2 < \infty\}$, but that rather $Q(a) = D(\sqrt{|a|})$.
3. For $\mathcal{H} = L^2(\mathbb{R}, d\alpha) \supset Q(q) = \{\psi \in \mathcal{H} : \psi$ is continuous at $0\}$, the "delta operator" $\langle\varphi|q|\psi\rangle = \varphi^*(0)\psi(0)$ is positive but not closed. The sequence $\psi_n = \exp(-n\alpha^2) \to 0$ in $\mathcal{H}$ is also a Cauchy sequence in the form norm, but without a limit in $Q(q)$: since the topology coming from $\| \ \|_q$ is finer than the one coming from $\| \ \|$, and the sequence tends to 0 in the latter topology, the only possible limit in $Q(q)$ would be 0. However, because $\langle\psi_n|q|\psi_n\rangle = 1$, the sequence does not tend to 0 in the $\| \ \|_q$ topology, and this fact is not changed by an enlargement of $Q(q)$. Hence q is not closed, and in fact not even closeable.

Thus Hermitian forms, in contradistinction to Hermitian operators, need not be closeable. However, if they are closeable, then they are always the quadratic form of some self-adjoint operator.

2.5.18 Theorem

If the form q is positive and closed, then it is the form of a unique self-adjoint, positive operator.

Proof: Consider $Q(q) \subset \mathcal{H}$ as a Hilbert space with the scalar product $\langle\varphi|\psi\rangle_q \equiv \langle\varphi|\psi\rangle + \langle\varphi|q|\psi\rangle$. The resulting topology is finer than that coming from $\langle \, | \, \rangle$, so the mapping $Q(q) \to \mathbb{C} : \psi \to \langle\varphi|\psi\rangle$ is continuous for all $\varphi \in \mathcal{H}$. As in (2.1.17), there exists a unique $\chi \in Q(q)$ such that $\langle\varphi|\psi\rangle = \langle\chi|\psi\rangle_q$. This defines an injection $c : \mathcal{H} \to Q(q) : \varphi \to \chi \equiv c\varphi$. If we consider c as a mapping $\mathcal{H} \to \mathcal{H}$, then it is Hermitian and bounded:

$$\langle\varphi|c\psi\rangle = \langle c\varphi|c\psi\rangle_q = \langle c\psi|c\varphi\rangle_q^* = \langle\psi|c\varphi\rangle^* = \langle c\varphi|\psi\rangle,$$

$$\|c\| = \sup_{\|\psi\|=1} \|c\psi\| \leq \sup_{\|\psi\|=1} \|c\psi\|_q = \sup_{\substack{\|\psi\|=1 \\ \|\varphi\|_q=1}} |\langle c\psi|\varphi\rangle_q| = \sup_{\substack{\|\psi\|=1 \\ \|\varphi\|_q=1}} |\langle\psi|\varphi\rangle| \leq 1.$$

Note that $c\mathcal{H}$ is dense in $Q(q)$ in the $\|\ \|_q$ norm (i.e., $\langle c\varphi|\psi\rangle_q = \langle\varphi|\psi\rangle = 0$ for all $\varphi \in \mathcal{H} \Rightarrow \psi = 0$), and hence it is also dense in $\mathcal{H}$ in the norm $\|\ \|$. Therefore c^{-1} is densely defined, and since c, being a bounded, Hermitian operator, is self-adjoint, c^{-1} is also self-adjoint on $c\mathcal{H}$ (Problem 10). The operator $c^{-1} - \mathbf{1}$ has the required property $\langle(c^{-1} - \mathbf{1})\varphi|\psi\rangle = \langle\varphi|q|\psi\rangle$ on the domain $c\mathcal{H} \subset Q(q)$. For uniqueness, see Problem 8. □

2.5.19 Example

We attempt to make sense of $d^2/dx^2 + V(x)$ with $V(x) = \lambda\delta(x)$ by extending the operator $H = -d^2/dx^2$ on $D(H) = \{\psi \in L^2((-\infty, \infty), dx) : \psi'$ is absolutely continuous, $\psi'' \in L^2$, and $\psi(0) = 0\}$. The Fourier transformation makes H a multiplication operator.

$$\begin{aligned}(H\psi)(k) &= k^2\psi(k), \\ D(H) &= \left\{\psi \in L^2\left((-\infty, \infty), \frac{dk}{2\pi}\right) : \int_{-\infty}^{\infty} dk|k^2\psi(k)|^2 < \infty, \right. \\ &\qquad \left. \int_{-\infty}^{\infty} dk\psi(k) = 0\right\}.\end{aligned}$$

It is closed since the graph norm (2.4.15; 3) is equivalent to $\|\psi\|_\Gamma^2 \equiv \langle\psi|\psi\rangle_\Gamma \equiv \int_{-\infty}^{\infty}(dk/2\pi)|\psi(k)|^2(1 + k^4)$, and

$$D(H) = \left\{\psi \in \mathcal{H} : \|\psi\|_\Gamma < \infty, \left\langle\psi\middle|\frac{1}{1+k^4}\right\rangle_\Gamma = 0\right\}$$

and $\|1/(1 + k^4)\|_\Gamma < \infty$. Clearly, since $\psi \in D(H) \Rightarrow \int dk\,\psi(k) = 0$,

$$(F_z)^\perp = \frac{1}{k^2 - z} \Rightarrow D(H^*) = D(H) + \left\{\frac{1}{k^2 - z}\right\}, \qquad z \in \mathbb{C} - \mathbb{R}^+,$$

$$H^* \frac{1}{k^2 - z} = z^* \frac{1}{k^2 - z}.$$

Therefore the deficiency indices are (1, 1), and there is a one-parameter family of self-adjoint extensions. Now consider H as quadratic form q. The domain $D(H)$ is not complete in the appropriate q-norm

$$\|\psi\|_q^2 = \int_{-\infty}^{\infty} \frac{dk}{2\pi}|\psi(k)|^2(1 + k^2) = \langle\psi|1 + q|\psi\rangle;$$

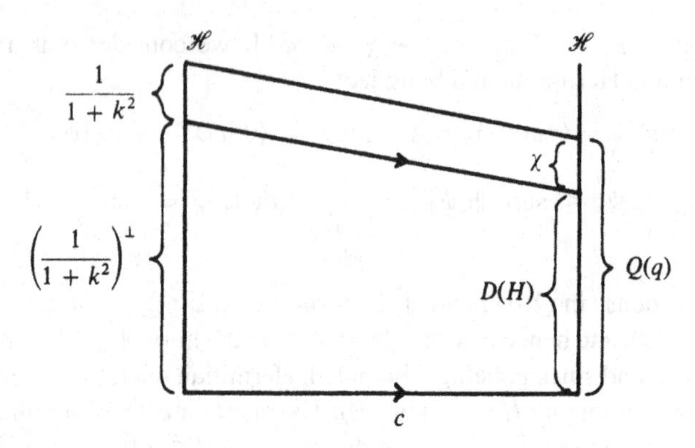

Figure 2.4. The domains involved in the extension of a quadratic form.

its completion contains ψ's which decrease only like $|k|^{-3/2-\varepsilon}$ as $k \to \infty$:

$$Q(q) = \left\{\psi \in L^2 : \|\psi\|_q < \infty, \int_{-\infty}^{\infty} dk\, \psi(k) = 0\right\}.$$

This is a subspace of $\{\psi : \|\psi\|_q < \infty\}$ closed in the norm $\|\ \|_q$, because

$$\int_{-\infty}^{\infty} dk\, \psi(k) = \left\langle \psi \middle| \frac{1}{1+k^2} \right\rangle_q \quad \text{and} \quad \left\| \frac{1}{1+k^2} \right\|_q = \frac{1}{\sqrt{2}};$$

therefore $\int_{-\infty}^{\infty} dk\, \psi(k)$ is a continuous linear functional on $Q(q)$ with the $\|\ \|_q$ norm (unlike the $\|\ \|$ norm). The injection c for the closed form $q = k^2$ with domain $Q(q)$ is determined by

$$\int_{-\infty}^{\infty} dk\, \varphi^*(k)\psi(k) = \int_{-\infty}^{\infty} dk (c\varphi(k))^* \psi(k)(1+k^2)$$

for all $\psi \in Q(q)$, $\varphi \in \mathcal{H}$, $c\varphi \in Q(q)$, which implies that

$$c : \varphi(k) \to \frac{\varphi(k)}{1+k^2} - \frac{1}{1+k^2} \int_{-\infty}^{\infty} \frac{dk'}{\pi} \frac{\varphi(k')}{1+k'^2}.$$

Hence c sends $(1/(1+k^2))^\perp = \{\varphi : \langle \varphi | 1/(1+k^2) \rangle = 0\}$ to $D(H)$ and $1/(1+k^2)$ to $1/(1+k^2)^2 - 1/2(1+k^2) \equiv \chi(k) \notin D(H)$, which means that $c^{-1}_{|D(H)} = 1+k^2$ and $c^{-1}\chi = 1/(1+k^2)$. The domain of H is enlarged by the inclusion of χ, on which the extension does not act as k^2 (see Figure 2.4). Since the deficiency indices are (1, 1), the domain of any self-adjoint extension is $D(H)+$ some one-dimensional space, so $D(H) + \{\chi\}$ is large enough to be $D(c^{-1})$. The operator $c^{-1} - \mathbf{1}$ is called the **Friedrichs extension** of H.

2.5.20 Remarks

1. If we choose $Q(\hat{q}) = \{\psi : \|\psi\|_q < \infty\} \supset Q(q)$, then

$$D(\hat{c}^{-1}) = \left\{\psi : \int dk |\psi(k)(1+k^2)|^2 < \infty\right\} \subset Q(\hat{q})$$

and $\hat{c}^{-1}$ is just multiplication by $1+k^2$, and thus different from c^{-1}, even though $\hat{q}_{|Q(q)} = q$ and both q and $\hat{q}$ are closed. In contrast, for self-adjoint operators, $a \subset \hat{a}$ necessarily implies $a = \hat{a}$.

2. If q arises from a positive operator a, $\langle\varphi|q|\psi\rangle = \langle\varphi|a\psi\rangle$, then q is always closeable to some $\hat{q}$ with $Q(\hat{q})$ the completion of $D(a)$ in the norm $\| \ \|_q$: To show this one has to verify that $Q(\hat{q})$ is a subspace of $\mathcal{H}$, i.e., to exclude that two Cauchy sequences $\{\varphi_n\}$ and $\{\varphi_n'\}$ converging to different elements in $Q(\hat{q})$ converge to the same element in $\mathcal{H}$. But if $\|\varphi - \varphi_n' - \psi\|_q \to 0$ then
$$\lim_{n,m\to\infty} \langle\varphi_n - \varphi_n'|\varphi_m - \varphi_m'\rangle_q = \lim_{m\to\infty}\left\{\lim_{n\to\infty}\langle\varphi_n - \varphi_n'|(a+1)(\varphi_m - \varphi_m')\rangle\right\} = 0.$$
Thus $\psi = 0$ and the trouble of (2.5.17) cannot happen.

3. In x-space the functions $(F_{\pm i})^{\perp}$ are of the form $\exp(-|x|(1\pm i)/\sqrt{2})$. The self-adjoint extensions append the function $\psi = \exp(-|x|(1+i)/\sqrt{2}) - \exp(2i\delta - |x|(1-i)/\sqrt{2})$ to $D(H)$ (Problem 5). The functions ψ satisfy $\lim_{\varepsilon\downarrow 0}\psi'(\varepsilon) - \psi'(-\varepsilon) = \lambda\psi(0)$, $\lambda = (\cot(\delta) - 1)/\sqrt{2}$, and so at $x = 0$,
$$\left(-\frac{d^2}{dx^2} + \lambda\delta(x)\right) \text{ is not singular.}$$
The form q defines the extension with $\lambda = \infty$, since χ has a discontinuous derivative at $x = 0$, but it vanishes at that point ($\int dk\, \chi(k) = 0$).

4. Since the norm on $Q(q)$ is weaker than the graph norm (2.4.15; 3) of H, the closure in $Q(q)$ produces an extension of the operator H, which is closed in its graph norm.

5. $Q(q)$ is closed with $\| \ \|_q$, but it is not all of $\{\psi \in \mathcal{H} : \|\psi\|_q < \infty\}$.

6. Whereas $\delta(x)$ of Example (2.5.17; 3) is not an operator, since its quadratic form is not closeable, $-d^2/dx^2 + \lambda\delta(x)$ is an operator.

2.5.21 Problems

1. Show that $a_{|D_2}$ of (2.4.14; 2) is closed.
2. Find an example of an operator l^2 with $D(a^*) - \ker a^* = \{\varphi : |\langle\varphi|a\psi\rangle| < c$ for all $\psi \in D(a)$, $\|\psi\| = 1$, $a^*|\varphi\rangle \neq 0\}$
3. Show that $\Gamma(a^*) = (J\Gamma(a))^{\perp}$.
4. Show that $a \subset a^*$ for the a of (2.5.3; 5).
5. What are the other self-adjoint extensions of H in (2.5.19)?
6. Determine the Friedrichs extension of H in (2.5.13).
7. Prove the claim made in (2.5.7; 5).
8. Show that the operator defined in (2.5.18) is unique.
9. Carry out the intermediate steps in the calculations of Examples (2.5.6) and (2.5.11).
10. Use the graph $\Gamma(a)$ to show that the inverse of a self-adjoint operator is self-adjoint whenever it exist.

2.5.22 Solutions

1.

$$\psi_n \to \psi, \quad \frac{1}{\alpha}\psi_n \to \varphi \Rightarrow \int_\varepsilon^1 \left|\frac{1}{\alpha}\psi - \varphi\right|^2 d\alpha = 0,$$

since

$$\le \int_\varepsilon^1 \left|\frac{1}{\alpha}\psi - \frac{1}{\alpha}\psi_n\right|^2 d\alpha + \int_\varepsilon^1 \left|\frac{1}{\alpha}\psi_n - \varphi\right|^2 d\alpha$$

for all $n \Rightarrow \int_\varepsilon^1 (1/\alpha^2)|\psi|^2 d\alpha \le \|\varphi\|^2$, i.e., $\psi \in D_2$ and $\|(1/\alpha)\psi - \varphi\| = 0$. Remark: The same argument works for any multiplication operator $\psi(\alpha) \to f(\alpha)\psi(\alpha)$, which is closed on the domain

$$\{\psi \in L^2 : f\psi \in L^2\}.$$

2. Let a be given in a matrix representation $a_{in} = (1/n)i$, and let $D(a)$ be

$$\{\psi = (\psi_1, \psi_2, \ldots) : \text{ only finitely many } \psi_i \text{ are nonzero}\}.$$

Then

$$\sup_{\|\psi\|\le 1} |\langle\varphi|a\psi\rangle| = \sup_{\|\psi\|\le 1} \sum_n \varphi_n \frac{1}{n} \sum_i i\psi_i = \infty \quad \text{for all} \quad \varphi \in l^2 \Rightarrow D(a^*) - \ker a^* = 0.$$

3. y is contained in the domain of a^* if there exists a $y^* \in \mathcal{H}$ such that $\langle y|ax\rangle = \langle y^*|x\rangle$ for all $x \in D(a)$, and if there is such a y^*, then $a^*y = y^*$. The equation $\langle y|ax\rangle = \langle y^*|x\rangle$ can be rewritten as $\langle (y, y^*)|(ax, -x)\rangle = \langle (y, y^*)|J(x, ax)\rangle = 0$, i.e., $\Gamma(a^*) = (J\Gamma(a))^\perp$.

4. It only needs to be shown that the upper limit contributes nothing to the integration by parts, that is $\varphi^*(\alpha)\psi(\alpha) \to 0$ as $\alpha \to \infty$. Since

$$\varphi^*(\beta)\psi(\beta) - \varphi^*(\alpha)\psi(\alpha) = \int_\alpha^\beta [\varphi^*(\alpha')\psi'(\alpha') + \varphi^{*\prime}(\alpha')\psi(\alpha')]d\alpha'$$

and because $[\ldots] \in L^1$, it follows first that $\varphi^*(\alpha)\psi(\alpha)$ is a Cauchy sequence and hence convergent, and secondly, because $\int_{-\infty}^\infty d\alpha|\varphi^*(\alpha)\psi(\alpha)| < \infty$, that the limit is 0.

5. Theorems (2.5.4) and (2.5.10) imply that every self-adjoint extension is of the form $\hat{H} : D(\hat{H}) = \{\psi = \eta + \varphi - V'\varphi, \eta \in D(H), \varphi \in F_i^\perp, V' \text{ an isometry } F_{-i}^\perp \to F_i^\perp\}$, $\hat{H}\psi = H\eta - i\varphi + iV'\varphi$. In our case, $\varphi = 1/(k^2 + i)$, and the most general V' acts by $V'^{-1}\varphi = e^{2i\delta}/(k^2 - i)$.

6. The form domain $Q(H)$ of $\langle\varphi|H|\varphi\rangle = \int_0^\infty dr(\varphi'^2 + \gamma\varphi^2/r^2)$ contains the operator domain $D_F(H)$ of the Friedrichs extension. The functions φ of $Q(H)$ must go to 0 faster than $r^{1/2}$ as $r \to 0$, so that $\int_0^\infty dr|\varphi|^2/r^2 < \infty$. Since functions of $D(H^*) \supset D_F(H)$ approach linear combinations of $r^{\pm\nu+1/2}$ in this limit, only $r^{+\nu+1/2}$ is possible, and that only if ν is real, i.e., $\gamma > -1/4$. Thus the Friedrichs extension amounts to appending the linear combination $\psi_+ - e^{2i\delta}\psi_-$, which behaves like $r^{\nu+1/2}$ as $r \to 0$. H is in fact a positive form only until the point $\gamma = -1/4$ since, by integration by parts,

$$\frac{1}{2}\int_0^\infty dr \frac{\varphi^2}{r^2} = \int_0^\infty dr \frac{\varphi'\varphi}{r} \le \left[\int_0^\infty dr\, \varphi'^2\right]^{1/2} \left[\int_0^\infty dr \frac{\varphi^2}{r^2}\right]^{1/2}$$

$$\Rightarrow \quad \int_0^\infty dr \frac{\varphi^2}{r^2} \leq 4 \int_0^\infty dr\, \varphi'^2.$$

Equality holds for $\varphi' =$ constant, which means $\varphi =$ (const.)r, though at large r, φ must somehow go to 0. The large-r dependence can be arranged so that when $\gamma < -\frac{1}{4}$, the form H is no longer positive. For the other extension H_δ with $D(H_\delta) \not\subset Q(H)$ we have $\langle\varphi|H_\delta\varphi\rangle \neq \langle\varphi|H|\varphi\rangle = \infty$ for all $\varphi \in D(H_\delta)$, $\varphi \notin Q(H)$.

7. V sends F_- to the dense subspace D, and the association given by

$$a : f = i(\mathbf{1} - V)g \to (\mathbf{1} + V)g$$

is linear. As to whether the operator is Hermitian: It is necessary to show that $\langle f'|af\rangle = \langle af'|f\rangle$ for f and $f' \in D$, i.e.,

$$\langle +i(\mathbf{1} - V)g'|(\mathbf{1} + V)g\rangle = \langle(\mathbf{1} + V)g'|i(\mathbf{1} - V)g\rangle.$$

This is true because V is an isometry.

8. $\langle(c^{-1} - 1)\varphi|\psi\rangle = \langle\varphi|q|\psi\rangle = \langle a\varphi|\psi\rangle$ for all $\varphi, \psi \in Q(q) \Rightarrow a \supset c^{-1} - \mathbf{1} \Rightarrow a = c^{-1} - \mathbf{1}$, since the latter operator is self-adjoint, and consequently cannot have any proper, Hermitian extensions.

9. Simple integration by parts.

10. Let $\bar{\mathcal{H}} = \mathcal{H} \oplus \mathcal{H} \supset \Gamma(a) = (J\Gamma(a))^\perp$, and let $U : \bar{\mathcal{H}} \to \bar{\mathcal{H}}$, $(x, y) \to (y, x)$.

$$\begin{aligned} \Gamma(a^{-1}) &= U\Gamma(a) \Rightarrow \Gamma((a^{-1})^*) = (J\Gamma(a^{-1}))^\perp \\ &= (JU\Gamma(a))^\perp = U(J\Gamma(a))^\perp = U\Gamma(a), \end{aligned}$$

i.e.,

$$\Gamma((a^{-1})^*) = \Gamma(a^{-1}), (a^{-1})^* = a^{-1}.$$

3
Quantum Dynamics

3.1 The Weyl System

Phase space is the arena of classical mechanics. The algebra of observables in quantum mechanics is likewise constructed with position and momentum, so this section covers the properties of those operators.

In classical mechanics, every function F on phase space generates a one-parameter group of diffeomorphisms $\exp(tL_{X_F})$ ($t \in \mathbb{R}$ and L_{X_F} is the Lie derivative with respect to the Hamiltonian vector field corresponding to F). Similarly, we learned in §2.4 that in quantum theory every observable a is associated with a one-parameter group of automorphisms $b \rightarrow \exp(iat)b\exp(-iat)$. One of the basic postulates quantum theory is that, in units with $\hbar = 1$, the groups generated by the Cartesian position and momentum coordinates $\mathbf{x}_j$ and $\mathbf{p}_j$ of n particles ($j = 1, \ldots, n$) are the same as classically, i.e., displacements respectively in the momenta and positions. Since $\mathbf{x}_j$ and $\mathbf{p}_j$ do not have bounded spectra, and hence cannot be represented by bounded operators, it is convenient to consider instead the bounded functions

$$\exp\left(i\sum_{i=1}^{n}\mathbf{x}_j \cdot \mathbf{s}_j\right) \quad \text{and} \quad \exp\left(i\sum_{j=1}^{n}\mathbf{p}_j \cdot \mathbf{r}_j\right), \qquad \mathbf{s}_j, \mathbf{r}_j \in \mathbb{R}^3,$$

so as not to have domain questions to worry about. The group of automorphisms can be written in terms of them as follows:

3.1.1 The Weyl Algebra

The operators

$$\exp\left(i\sum_{j=1}^{n}\mathbf{x}_j\cdot\mathbf{s}_j\right) \quad \text{and} \quad \exp\left(i\sum_{j=1}^{n}\mathbf{p}_j\cdot\mathbf{r}_j\right)$$

generate the **Weyl algebra** $\mathcal{W}$ with the multiplication law

$$\exp\left(i\sum_{j=1}^{n}\mathbf{p}_j\cdot\mathbf{r}_j\right)\exp\left(i\sum_{j=1}^{n}\mathbf{x}_j\cdot\mathbf{s}_j\right)\exp\left(-i\sum_{j=1}^{n}\mathbf{r}_j\cdot\mathbf{p}_j\right)$$
$$= \exp\left(i\sum_{j=1}^{n}(\mathbf{x}_j+\mathbf{r}_j)\cdot\mathbf{s}_j\right) \quad \text{for all } \mathbf{s}_j, \mathbf{r}_j \in \mathbb{R}^3.$$

3.1.2 Remarks

1. To simplify the notation, we consider $\mathbf{z}_j \equiv \mathbf{r_j} + i\mathbf{s}_j$ as a single vector in the Hilbert space $\mathbb{C}^{3n}$, with scalar product $(z|z') \equiv \sum_{j=1}^{n}\mathbf{z}_j^*\cdot\mathbf{z}_j$ and volume element $dz = d^3r_1\cdots d^3r_n d^3s_1\cdots d^3s_n$. Then we define the **Weyl operators** by

$$\begin{aligned} W(z) &\equiv \exp\left(-\frac{i}{2}\sum_{j=1}^{n}\mathbf{r}_j\cdot\mathbf{s}_j\right)\exp\left(i\sum_{j=1}^{n}\mathbf{r}_j\cdot\mathbf{p}_j\right)\exp\left(i\sum_{j}\mathbf{s}_j\cdot\mathbf{x}_j\right) \\ &= W^*(-z) = W^{-1}(-z). \end{aligned}$$

The multiplication law (3.1.1) can then be written compactly as

$$W(z)W(z') = \exp\left(\frac{i}{2}\operatorname{Im}(z|z')\right)W(z+z').$$

So the algebra $\mathcal{W}$ consists of linear combinations of $W(z)$ and a representation is characterized by $\langle u|W(z)u\rangle$ with u a cyclic vector.

2. We shall only be interested in representations for which $z \to W(z)$ is strongly continuous, so that we can recover $\mathbf{x}$ and $\mathbf{p}$ from knowledge of $W(z)$. In the representations we shall use, $\|W(z) - W(z')\| = 2$ whenever $z \neq z'$. Norm continuity is impossible, as $\mathbf{x}$ and $\mathbf{p}$ are always unbounded.
3. The C^* algebra gotten by taking the norm closure of $\mathcal{W}$ is too small for many purposes. In order to include all L^∞ functions of $\mathbf{x}$ and $\mathbf{p}$, it is necessary to take the strong closure $\overline{\mathcal{W}}$. The question then arises whether the isomorphism mentioned above, of the canonical and unitary transformations, can be extended to other coordinate systems involving L^∞ functions of $\mathbf{x}$ and $\mathbf{p}$. One cause for concern is that because of the noncommutativity of observables, a classical function $f(p,x)$ does not have a uniquely determined quantum mechanical version: Is the classical function p^2x^3 to be px^3p or $(p^2x^3 + x^3p^2)$, which by formal manipulation of (1.1.1) equals $-3x + px^3p$? It can even happen

that the product of operators simply fails to be defined because of the lack of a domain of definition (cf. (3.1.10; 5)). We shall not generally be able to settle the question of the proper quantum mechanical operators associated with all classical observables and find what groups they generate but shall instead consider successively more complicated special cases.

$L^1(\mathbb{C}^{3n})$ is the classical algebra of functions of phase space which we will now relate to the quantum operators.

3.1.3 Mapping of $L^1(\mathbb{C}^{3n})$ into $\overline{\mathcal{W}}$

Given a strongly continuous representation of $\mathcal{W}$,

$$W_f \equiv \int dz\, f(z) W(z) \in \overline{\mathcal{W}}$$

is well-defined for all $f \in L^1(\mathbb{C}^{3n})$, and:

(i) $W_{f+g} = W_f + W_g$;

(ii) $W_f^* = W_{\bar{f}},\ \bar{f}(z) = f^*(-z)$;

(iii) $W_{f*g} = W_f \cdot W_g,\ (f * g)(z) = \int dz' f(z - z') g(z') \exp((i/2)\mathrm{Im}(z|z'))$;

(iv) $W_{f_0} W(z) W_{f_0} = W_{f_0} \exp(-\frac{1}{4}(z|z))$ *for all* $z \in \mathbb{C}^{3n}$, $f_0(z) = (2\pi)^{-3n} \exp(-((z|z)/4))$;

(v) *The mapping* $L^1 \to \mathcal{B}(\mathcal{H}) : f \to W_f$ *is injective;*

(vi) $\|W_f\| \le \|f\|_1$.

Proof: Since $W(z)$ is strongly continuous, the integral is defined as a strong limit, but will not necessarily be in the norm closure of $\mathcal{W}$.

(i) and (ii) are obvious.

(iii) follows from (3.1.2; 1).

(iv) follows from (3.1.2; 1) and a Gaussian integral (Problem 5).

(v) If $W_f = 0$, then for all $z' \in \mathbb{C}^{3n}$ and $g, h \in \mathcal{H}$,

$$\begin{aligned} 0 &= \int dz\, f(z) \langle g | W(-z') W(z) W(z') h \rangle \\ &= \int dz\, f(z) \exp(i\, \mathrm{Im}(z|z')) \langle g | W(z) h \rangle. \end{aligned}$$

Now choose $h = W_{f_0} h_1,\ g = W_{f_0} g_1$ with $\langle g_1 | W_{f_0} h_1 \rangle \neq 0$. Then we learn that the Fourier transform of $f(z) \exp(-\frac{1}{4}(z|z))$ and thus f vanishes.

(vi) Since $W(z)$ is unitary, $\|W(z)\| = 1$ for all $z \in \mathbb{C}^{3n}$, and

$$\left\| \int dz\, f(z) W(z) \right\| \leq \int dz |f(z)| \|W(z)\|. \qquad \square$$

3.1.4 Consequences

Property (iv) implies that $W_{f_0} = W_{f_0}^*$ is a projection, and that $\mathcal{W} W_{f_0} \mathcal{H}$ is all of $\mathcal{H}$: The space orthogonal to it would be invariant under $\mathcal{W}$ and such that W_{f_0} would vanish on it, as $\langle y | W_{f_0} x \rangle = \langle W_{f_0} y | x \rangle = 0$ for all $x \in \mathcal{H} \Rightarrow W_{f_0} y = 0$. Thus in this subspace we would have a representation of $\overline{\mathcal{W}}$ with $W_{f_0} = 0$, which is impossible by Property (v). Therefore the subspace $W_{f_0} \mathcal{H}$ is a totalizer for $\mathcal{W}$, and in an orthogonal basis $\{u_j\}$ of $W_{f_0} \mathcal{H}$ with $W_{f_0} u_j = u_j$, the representation has the form:

$$\begin{aligned} \langle u_j | W(z) u_k \rangle &= \langle u_j | W_{f_0} W(z) W_{f_0} u_k \rangle \\ &= \langle u_j | W_{f_0} u_k \rangle \exp\left(-\frac{(z|z)}{4} \right) = \delta_{jk} \exp\left(-\frac{(z|z)}{4} \right). \end{aligned}$$

This argument proves

3.1.5 The Uniqueness of Theorem of Representations of W

Every strongly continuous representation of $\mathcal{W}$ is equivalent to a sum of cyclic representations by the state

$$\langle u | W(z) u \rangle = \exp\left(-\frac{(z|z)}{4} \right).$$

3.1.6 Remarks

1. If $\mathcal{H}$ is separable, then the sum is countable.
2. In the spectral representation of $\mathbf{x}$, the operator W_{f_0} projects onto

$$u(\mathbf{x}) = \pi^{-3n/4} \exp\left(-\sum_j \frac{x_j^2}{2} \right),$$

and

$$(W(z)u)(\mathbf{x}) = \exp\left(i \sum_{j=1}^{3n} s_j \left(x_j + \frac{r_j}{2} \right) \right) u(x_j + r_j)$$

(Problem 2).

3. If the assumption of strong continuity is left out, then there are more representations. Suppose $\mathcal{H}$ is as in (2.4.14; 2), and that, as before, $(W(z)\psi)_x = \exp(is(x + r/2))\psi_{x+r}$. This constitutes another representation of $\mathcal{W}$, on a non-separable $\mathcal{H}$. It is definitely not equivalent to (3.1.5), since $\exp(ixs)$ has a pure

point spectrum, every $z \in \mathbb{C} : |z| = 1$ is an eigenvalue. In this example the operator p does not even exist.

4. There are inequivalent representations of $\mathcal{W}$ for infinite systems ($n = \infty$), in fact uncountably many of them, even on separable Hilbert spaces. (See volume IV.)
5. If x-space is not infinite, but rather a torus (I: 2.1.7; 2), then there are infinitely many inequivalent representations of the Weyl relations. In that case $\exp(i \sum_{j=1}^{n} \mathbf{x}_j \cdot \mathbf{s}_j)$ is an observable only if $s_j \in (2\pi\mathbb{Z})^3$, and (3.1.1) is valid only for these s_j. The operator p is again formally the derivative $-i(d/dx)$, but according to (2.5.3; 4) this has a one-parameter family of self-adjoint extensions corresponding to the boundary conditions $\psi(1) = e^{-ij}\psi(0), 0 \leq j \leq 2\pi$, for $\psi \in L^2(0, 1) \cong L^2(T^1)$. Then $-(d/dx)$ has the eigenfunctions e^{ikx} with eigenvalues $k = 2\pi\mu - j, \mu \in \mathbb{Z}$. The representations are clearly inequivalent for different j and inequivalent to the representation (3.1.5), where the spectrum is absolutely continuous rather than pure point.
6. Theorem (3.1.5) gives what is known as a **ray representation** of $\mathbb{R}^{6n}$, which means that $W(z)W(z')$ equals $W(z+z')$ up to a phase factor. It may seem peculiar that the representation of $\mathcal{W}$ is essentially unique, even though every subgroup of $\mathbb{R}^{6n}$ is an invariant subgroup (a normal divisor), and a representation of any factor group is also a representation of $\mathbb{R}^{6n}$. The state of affairs can be understood as follows: for any $r \in \mathbb{R}$, the integral multiples $\{nr\}, n \in \mathbb{Z}$, constitute a normal divisor of $\mathbb{R}$, and $t \to \exp(2\pi i t/r)$ is the unique faithful representation of the factor group $\mathbb{R}/\{nr\}$. Hence there exists a one-parameter family of unfaithful representations of $\mathbb{R}$, and every (strongly continuous) representation is a sum or integral of them. However, the Weyl algebra is simple—it contains no nontrivial subideal—so that only the trivial representation fails to be faithful, and the irreducible faithful representations are all equivalent.

The self-adjoint generators $\mathbf{x}_j$ and $\mathbf{p}_j$ can be recovered from $W(z)$ by differentiation. Yet the problem remains of being precise about the commutation relations (1.1.1), since unbounded operators do not form an algebra. However, the fact that two operators commute can easily be interpreted in a mathematically reasonable way:

3.1.7 Definition

The statement that two unbounded, self-adjoint operators a and b **commute** will mean that $f(a)g(b) - g(b)f(a) \equiv [f(a), g(b)] = 0$ for all f and $g \in L^\infty$. We shall write this for simplicity as $[a, b] = 0$.

3.1.8 Remarks

1. For $[a, b]$ to equal 0, it suffices that $[\exp(iat), \exp(ibs)] = 0$ for all $t, s \in \mathbb{R}$, or that $[(a-z)^{-1}, (b-z')^{-1}] = 0$ for all $z, z' \in \mathbb{C}\backslash\mathbb{R}$ (Problem 6). A consequence of (3.1.1) is thus that $[\mathbf{x}_k, \mathbf{x}_j] = [\mathbf{p}_k, \mathbf{p}_j] = 0$ for all k, j, and $[\mathbf{x}_k, \mathbf{p}_j] = 0$ for $k \neq j$.

2. The Gel'fand isomorphism (2.2.28) is sill applicable to the C^* algebra generated by the bounded, continuous functions of such a and b, and it provides a spectral representation in which both a and b are multiplication operators. This is a generalization of the simultaneous diagonability of commuting Hermitian matrices.
3. It is tempting to conjecture that $[a, b] = 0$ whenever

   (i) there exists a dense domain D invariant under a and b;

   (ii) a and b are essentially self-adjoint on D; and

   (iii) $ab\psi = ba\psi$ for all $\psi \in D$.

   This is false; see the counterexample of Problem 4.

To make sense of $[x_k, p_j]$ for $k = j$, we need to find a $D \subset D(x_k) \cap D(p_i)$ such that $x_k D \subset D(p_j)$ and $p_j D \subset D(x_k)$. One such domain consists of the vectors that belong to $\mathcal{S}$ in the x-representation, where $\mathcal{S}$ is the space of C^∞ functions that decrease at infinity along with all their derivatives faster than any negative power of x. This space equals its Fourier transform $\hat{\mathcal{S}}$, and on $\mathcal{S}$, $x : f(x) \to xf(x)$, $p : f(x) \to i(\partial/\partial x)f(x)$, while on $\hat{\mathcal{S}}$, $x : f(p) \to i(\partial/\partial p)f(p)$ and $p : f(p) \to pf(p)$. On $\mathcal{S}$ we can write

3.1.9 The Heisenberg Commutation Relations

$$(x_k p_j - p_j x_k)\psi = i\delta_{jk}\psi \quad \textit{for all } \psi \in \mathcal{S}.$$

3.1.10 Remarks

1. The operators x and p are clearly Hermitian on $\mathcal{S}$ with deficiency indices $(0, 0)$, and thus essentially self-adjoint.
2. It is a natural question whether all representations of $[x_k, p_j] = i\delta_{jk}$ on dense domains D of essential self-adjointness lead to the Weyl relations (3.1.1). The answer is no. Additional assumptions are needed, such as that $\sum_{k=1}^{3n} x_k^2 + \sum_{k=1}^{3n} p_k^2$ be essentially self-adjoint on D. Otherwise, a variant of Problem 4 would provide a counterexample. Another possible condition is that $\prod_k (x_k + i)(p_k + i)D$ be dense in $\mathcal{H}$.
3. It has already been noted that finite matrices cannot satisfy (3.1.9). It is likewise impossible to represent p and x with bounded operators of any kind. Equation (3.1.9) also requires that $x^n p - px^n = inx^{n-1}$, so that $n\|x^{n-1}\| = \|x^n p - px^n\| \leq 2\|x^n\| \cdot \|p\|$, and thus $\|x\| \cdot \|p\| \geq n/2$ for all n.
4. There are many inequivalent representations of x and p if we allow matrices that satisfy $xp - px = i$ on some dense domain by formal manipulation; for

instance,

$$x = \begin{bmatrix} 0 & & & & \\ & 1 & & & \\ & & 2 & & \\ & & & 3 & \\ & & & & 4 \end{bmatrix}, \quad p = -i \begin{bmatrix} 0 & -1 & -\frac{1}{2} & -\frac{1}{3} & \dots \\ 1 & 0 & -1 & -\frac{1}{2} & \dots \\ \frac{1}{2} & 1 & 0 & -1 & \dots \\ \frac{1}{3} & \frac{1}{2} & 1 & 0 & \dots \\ & & & & \end{bmatrix}$$

$$[x, p] = -i \begin{bmatrix} 0 & 1 & 1 & 1 & 1 & 1 \\ 1 & 0 & 1 & 1 & 1 & 1 \\ 1 & 1 & 0 & 1 & 1 & 1 \\ 1 & 1 & 1 & 0 & 1 & 1 \\ 1 & 1 & 1 & 1 & 0 & 0 \end{bmatrix}.$$

On $D = \{(v) \in l^2 : \sum_i v_i = 0$, with only finitely many $v_i \neq 0\}$, with these matrices, $[x, p] = i$. This representation is not equivalent to the Weyl representation: $\mathrm{Sp}(x) = \mathbb{Z}^+$. The eigenvectors e_k of x do not belong to D, as otherwise there would be a contradiction, $(e_k|[x, p]e_k) = 0 = i(e_k|e_k)$.

5. One might hope that the Poisson bracket { } of classical mechanics goes over to the commutator in quantum mechanics not just for the Cartesian coordinates p and x, but also for generalized coordinates (cf. (3.1.2; 3)). Unfortunately, it does not. Consider p and x on the one-dimensional torus (circle) T^1; while x, $0 \leq x < 1$, is not a global coordinate, the equation $\{x, p\} = 1$ holds locally. Suppose that the quantum-mechanical Hilbert space is $\mathcal{H} = L^2(T^1, dx)$. Then the formal equation $[x, p] = i$ makes no sense as an operator equation, since p is defined only on the absolutely continuous functions on T^1 (which implies that $\psi(0) = \psi(1)$), while x maps functions out of this subspace. If the matrix elements are calculated with respect to the eigenfunctions $\psi_n = \exp(2\pi i n x)$, $n \in \mathbb{Z}$, then

$$\langle n|p|m\rangle = n\delta_{nm}, \qquad \langle n|x|m\rangle = \frac{i}{2\pi(n-m)}(1 - \delta_{nm}) + \frac{1}{2}\delta_{nm},$$

and if these matrices are multiplied, one finds that $[x, p] \neq i$. Hence (3.1.9) is not even valid in the sense of quadratic forms, so the representations (3.1.6; 5) cannot strict speaking be characterized by (3.1.9).

Following Remark (2.2.33: 3), the commutation relations (3.1.9) have as a consequence the

3.1.11 Indeterminacy Relations

$$\Delta x_i \Delta p_k \geq \frac{1}{2}\delta_{ik}.$$

3.1.12 Remarks

1. There are, of course, some domain questions to answer when Remark (2.2.33; 3) is extended to unbounded operators; but assuming that there is no real difficulty, (3.1.4) applies equally to all states, so there is no need for an index on Δ.
2. A natural question is whether there can be equality in (3.1.11), and, if so, for which states. The inequality of (2.2.33; 4) is of the form $\langle (a^* - \alpha^*)(a - \alpha) \rangle \geq 0$ for pure states, where we have let $a = x + 2ip(\Delta x)^2$ and $\alpha = \langle x \rangle + 2i\langle p \rangle (\Delta x)^2$. Equality would require that the state $| \rangle$ be an eigenvector of the (nonnormal) operator a with complex eigenvalue α. The operators a are the annihilation operators that are so important in the theory of many-particle physics.

In an x-representation ($p = (1/i)(\partial/\partial x$, $|\rangle = \psi(x) \in L^2(\mathbb{R}, dx)$), the a's can be used according to (3.1.12; 2) to construct

3.1.13 States of Minimal Uncertainty, or Coherent States

The equation

$$\Delta p = \frac{1}{2\Delta x}$$

holds only for the states

$$\psi(x) = \exp\left[-\frac{(x - \langle x \rangle - 2i\langle p \rangle (\Delta x)^2)^2}{4(\Delta x)^2} \right].$$

3.1.14 Remarks

1. If $(\Delta x)^2 = \frac{1}{2}$, then we get the states $W(-z^*)|u\rangle$ of (3.1.6; 2) with $z = \langle x \rangle + i\langle p \rangle$, which appeared in the GNS construction for the Weyl algebra. It follows that linear combinations of states of minimal uncertainty are dense in $\mathcal{H}$. The additional parameter Δx occurring here provides a standard of comparison between x and p, and was fixed earlier in the choice of f_0. States with different z are not orthogonal, even with the same Δx. There is, however, an analogue of the representation of the identity operator in an orthonormal system:

$$\mathbf{1} = \int \frac{dz}{2\pi} W(z)|\rangle\langle|W(-z)$$

This equation in fact holds for any normalized vector $|\rangle \in L^1 \cap L^\infty$ (Problem 7). The states $W(z)|\rangle$ are thus not only total, but moreover every vector can be written as an integral over them.
2. There is a strict inequality $(\Delta x)^2(\Delta p)^2 > \frac{1}{4}$ for impure states. Problem (2.2.38; 6) showed that $(\Delta x)^2$ and $(\Delta p)^2$ for a convex combination of two states are greater than or equal to the convex combinations of the two $(\Delta x)^2$ and the two $(\Delta p)^2$ of the constituent states, and equality holds only if the expectation values are the same in the two states. Since $\langle p \rangle$, $\langle x \rangle$, and

$\Delta x = 1/2\Delta p$ determine a unique coherent state, any genuine mixture of states will have $\Delta x \Delta p > \frac{1}{2}$: If states are averaged with a weight $p(\lambda) \geq 0$,

$$\int_0^1 d\lambda\, \rho(\lambda) = 1, \qquad W(a) = \int d\lambda\, \rho(\lambda) \langle a \rangle_\lambda,$$

then

$$\begin{aligned}(\Delta_W x)^2 (\Delta_W p)^2 &\geq \int_0^1 d\lambda (\Delta_\lambda x)^2 \rho(\lambda) \int_0^1 d\lambda' (\Delta_{\lambda'} p)^2 \rho(\lambda') \\ &\geq \left(\int d\lambda\, \rho(\lambda) \Delta_\lambda x \Delta_\lambda p \right)^2 \geq \frac{1}{4}.\end{aligned}$$

The last inequality is an equality only if $\Delta_\lambda x \Delta_\lambda p = \frac{1}{2}$, and the second one is an equality only if $\Delta_\lambda x = c \Delta_\lambda p$, so $\Delta_\lambda x = 1/2\Delta_\lambda p$ is independent of λ. But the first one is an equality only if all $\langle x \rangle_\lambda$ and $\langle p \rangle_\lambda$ are the same; so no genuine mixture makes all three equalities.

3.1.15 The Classical Limit

Until now, we have taken the microscopic standpoint and set $\hbar = 1$. In order to see how the operators turn into ordinary numbers in the classical limit $\hbar \to 0$, let

$$x_\hbar = x\sqrt{\hbar}, \qquad p_\hbar = p\sqrt{\hbar}, \qquad [x_\hbar, p_\hbar] = i\hbar.$$

If we used $W(z/\sqrt{\hbar})$ to cause a displacement by $r/\sqrt{\hbar}$ (respectively $s/\sqrt{\hbar}$) on $L^2(\mathbb{R}, dx)$ (resp. $L^2(\mathbb{R}, dp)$) at the same time as we let $\hbar \to 0$, then we would expect $x_\hbar$ (resp. $p_\hbar$) to converge to $r \cdot \mathbf{1}$ (resp. $s \cdot \mathbf{1}$). Indeed, the equation

$$W(z^* \hbar^{-1/2}) \exp(is(x_\hbar - r)) W(-z^* \hbar^{-1/2}) = \exp(isx_\hbar)$$

can be derived from (3.1.2; 1). As $\hbar \to 0$, $\exp(isx_\hbar) = (\exp(isx))^{\sqrt{\hbar}} \to 1$, so

$$W(z^* \hbar^{-1/2}) \exp(isx_\hbar) W(-z^* \hbar)^{-1/2}) \to \exp(isr),$$

and analogously for $\exp(itp_\hbar)$. In the sense of (2.5.7; 3), the operators $Wx_\hbar W^{-1}$ and $Wp_\hbar W^{-1}$ converge strongly to r and respectively s; the dilatation by $\hbar$ suppresses the fluctuations, and W translates the operators back to the proper positions.

If the particles are indistinguishable, then only the algebra $\mathcal{N}_s$ of symmetric functions of the $\mathbf{x}_i$ and $\mathbf{p}_i$ is observable. This algebra has a reducible representation on $\mathcal{W} \cdot u$: the unitary operator Π, permuting the indices $(1, 2, \ldots, n)$ to $(\pi_1, \pi_2, \ldots, \pi_n)$, $\Pi \mathbf{x}_i \Pi^{-1} = \mathbf{x}_{\pi_i}$, $\Pi \mathbf{p}_i \Pi^{-1} = \mathbf{p}_{\pi_i}$, then commutes with all observable. The set of permutations forms a representation of the symmetric group S_n. If it is decomposed into its irreducible parts, then $\mathcal{N}_s$ maps each part into itself, and there is a superselection rule (cf. (2.3.6; 7)). Unless we restrict ourselves to the identical or alternating representation of S_n, the algebra $\mathcal{N}_s$ contains no subalgebra maximal Abelian in $\mathcal{B}_\mathcal{H}$, and the Π become the hidden parameters mentioned in (2.3.6; 7). By definition they can never be observed, but they decompose the

Hilbert space since they do not simply multiply by ± 1. These superselection rules apparently do not exist in Nature.

The identical (respectively alternating) representation of S_n is obtained by restricting the tensor product of the Hilbert spaces belonging to individual particles to the symmetric (antisymmetric) subspace (cf. (I: 2.4.28)). As is well known, the symmetrization or antisymmetrization of ψ in the particle coordinates leads to Bose-Einstein or respectively Fermi-Dirac statistics. Relativistic quantum theory correlates these statistics with the spin of the particle, but in the framework of nonrelativistic quantum mechanics it appears as a special postulate:

3.1.16 The Connection between Spin and Statistics

For a system of indistinguishable particles with integral (respectively half-odd integral) spin, the representation must be restricted to the subspace of the identical (resp. alternating) representation of S_n.

3.1.17 Problems

1. Let $\mathcal{A} = \mathcal{B}(\mathbb{C}^2) \otimes \mathcal{B}(\mathbb{C}^2) = \{\sigma_1 \otimes \sigma_2\}''$. Construct a P such that $P\sigma_1 P = \sigma_2$ and $P = P^* = P^{-1}$
2. Verify the x-representation of u (3.1.6; 2).
3. According to (2.4.21; 5), a dense set of analytic vectors determines $\exp(iat)$ uniquely, and thus a Hermitian operator a defined on such a set is essentially self-adjoint. Show that the analytic vectors for $a = -i\, d/dx$ on the space $L^2((-\infty, \infty), dx)$ are (complex-valued) real-analytic functions.
4. Let M be the Riemann surface of $\sqrt{z}$ and $\mathcal{H} = L^2(M, dz = dx\, dy)$. Moreover, let the operators $a = -i\, \partial/\partial x$ and $b = -i\, \partial/\partial y$ be defined on $D = \{C^\infty$ functions with compact support not including $0\}$. Show that

   (i) a and b are essentially self-adjoint;

   (ii) D is mapped into itself by a and b;

   (iii) $ab\psi = ba\psi$ for all $\psi \in D$;

   but for the closures $\bar{a} = a^{**}$ and $\bar{b} = b^{**}$,

   (iv) $\exp(i\bar{a}t)\exp(i\bar{b}t) \neq \exp(i\bar{b}t)\exp(i\bar{a}t)$.
5. Verify (3.1.3(iv)).
6. Show that a sufficient condition for two operators to commute is that their exponentials or resolvents commute.
7. Verify (3.1.14; 1).

3.1.18 Solutions

1. $P = (1 + \sigma_1 \cdot \sigma_2)/2$

2.

$$W_{f_0} u = \int \frac{d^{3n} r d^{3n} s}{(2\pi)^{3n}} \exp\left(-\sum_j (|\mathbf{r}_j|^2 + |\mathbf{s}_j|^2)/4\right) \exp\left(i \sum_j \mathbf{s}_j \cdot (\mathbf{x}_j + \mathbf{r}_j/2)\right)$$

$$\times \exp\left(-\sum_j |\mathbf{x}_j + \mathbf{r}_j|^2/2\right) \pi^{-3n/4} = u.$$

3. If $\psi \in L^2$ is an analytic vector, then $\psi \in \bigcap_n D(a^n) \subset C^\infty$.

$$e^{iat} \psi(x) = \psi(x+t) = \sum \frac{(it)^n}{n!} a^n \psi(x).$$

The sum converges iff ψ is analytic.

4. Propositions (ii) and (iii) are obvious, as are $a \subset a^*$ and $b \subset b^*$. As for essential self-adjointness: Let $D_x \subset D$ be the set of all functions the support of which never contains the x-axis on any sheet of the Riemann surface of $\sqrt{z}$. D_x is dense. The operators $U(t) : \psi(x, y) \to \psi(x+t, y)$ are isometric and have dense ranges, so they have unique unitary extensions, which are strongly continuous in t. $U(t)$ is differentiable on D_x, and $dU(t)/dt = ia_{|D_x}$, so $a_{|D_x}$ is essentially self-adjoint (cf. Problem 3), and therefore so is a. The argument for b is similar; $ib_{|D_y} = dV(t)/dt$, $V(t) : \psi(x, y) \to \psi(x, y+t)$, and $D_y \subset D$ is the set of functions the support of which never contains the y-axis on any sheet.
(iv): Let ψ be a function supported in the circle centered at $(-\frac{1}{2}, -\frac{1}{2})$ on the first sheet and having some radius less than $\frac{1}{2}$. Then $U(1)V(1)\psi$ has its support on the first sheet and $V(1)U(1)\psi$ has its support on the second sheet, so $U(1)V(1) \neq V(1)U(1)$.

5. This calculation of a Gaussian integral will be entrusted to the reader.

6. The von Neumann algebra $\mathcal{A} = \{f(a) : f \in L^\infty\}$ is generated by

   (i) the exponential functions $\exp(iat)$: If $f \in L^1$, then the Fourier transform

$$\int f(t) \exp(iat) dt \in \mathcal{A}.$$

   But the Fourier transformation is a bijection $L^1 \cap L^\infty \to L^1 \cap L^\infty$, and $L^1 \cap L^\infty$ is weakly dense in L^∞.

   (ii) the resolvents $(a + x + iy)^{-1}$:

$$(a + x + iy)^{-1} + (a + x - iy)^{-1} = 2(a + x)((a + x)^2 + y^2)^{-1}$$

   and

$$(a + x + iy)^{-1} - (a + x - iy)^{-1} = -2iy((a + x)^2 + y^2)^{-1},$$

   and by the Stone-Weierstrass theorem, these functions generate all continuous functions vanishing at infinity, which is also a weakly dense set in L^∞. If there are two algebras $\mathcal{A}_0$ and $\mathcal{B}_0$ such that $\mathcal{A}_0'' = \mathcal{A}$ and $\mathcal{B}_0'' = \mathcal{B}$, and for which $[\mathcal{A}_0, \mathcal{B}_0] = 0$, then $[\mathcal{A}, \mathcal{B}]$ also $= 0$.

7. Let $\psi(x)$ be $|\rangle$ in the x-representation. Then $W(z)|\rangle$ becomes $e^{ixs}\psi(x+r)$ and $\int (dz/2\pi)W(z)|\rangle\langle|W^*(z)\varphi$ is

$$\int \frac{ds\,dr}{2\pi} e^{ixs}\psi(x+r)\psi^*(x'+r)e^{-ix's}\varphi(x')dx' = \int dr|\psi(x+r)|^2\varphi(x) = \varphi(x).$$

This formal manipulation can easily be justified for instance if $\psi \in L^1 \cap L^\infty$.

3.2 Angular Momentum

In quantum physics, just as in classical physics, the angular momentum **L** *is the generator of the group of rotations. This group is compact, so all of its irreducible representations are finite-dimensional. It is possible, however, for* **L** *to be unbounded in a reducible representation.*

In the earlier sections it was postulated that the group of canonical transformations in classical mechanics generated by p and x was represented in quantum theory by $\exp(irp)$ and respectively $\exp(isx)$. The next most simple group of transformations to study is the one generated by the angular momentum $\mathbf{L} = [\mathbf{x} \times \mathbf{p}]$. Classically, $\mathbf{L}$ generates the point transformations

$$\mathbf{x} \to M\mathbf{x}, \qquad MM^t = \mathbf{1}. \tag{3.1}$$

We consider here a single particle and use matrix notation; everything factorizes for systems of many particles. In quantum theory it is more convenient to work with the bounded Weyl operators, so we wish to find a unitary transformation U for which

$$U^{-1}W(z)U = W(Mz) \tag{3.2}$$

(cf. (3.1.2; 1)). Such a transformation must exist, since the operators $W(Mz)$ also satisfy (3.1.2; 1), and all irreducible representations of those relationships are equivalent. Following (3.1.6; 2) we can write (3.2) in the

3.2.1 Schrödinger Representation for the Rotations

The unitary transformation

$$(U\psi)(x) = \psi(Mx)$$

produces the automorphism (3.2) of the Weyl algebra.

3.2.2 Remark

The operator U is not fixed uniquely by (3.2); $e^{i\alpha}U$ would do just as well for any $\alpha \in \mathbb{R}$. This, however, is the extent of the arbitrariness in an irreducible representation of $W(z)$, as U is determined up to a unitary element of the commutant. Hence at this stage the U's only constitute a ray representation of $O(3)$ [cf. (3.1.6; 5)].

Yet is is possible to show that every strongly continuous ray representation of a compact group is derived from a representation of the universal covering group (see Problem 5). The universal covering group of $O(3)$ is $SU(2)$, which will be discussed in more detail later. The compactness is essential: The Weyl system (3.1.1) provides a ray representation of $\mathbb{R}^{6n}$, which is its universal covering group, but it is not a representation in the ordinary sense.

The generators of the one-parameter subgroups of rotations about the coordinate axes will be denoted $\mathbf{L}$, as a vector operator. We begin the study of the generators by determining a

3.2.3 Domain of Essential Self-Adjointness for L

The operator $\mathbf{L}$ *is essentially self-adjoint on the linear hull* D *of the vectors*

$$\psi_k = \exp\left(-\frac{|\mathbf{x}|^2}{2}\right) x_1^{k_1} x_2^{k_2} x_3^{k_3}, \qquad k_i = 0, 1, 2, \ldots.$$

Proof: D is dense (Problem 2) and obviously invariant under rotations. It is convenient to change to polar coordinates about the z-axis = the axis of rotation to check the differentiability of U on D. It is then a question of showing that

$$\lim_{\delta\to 0} \delta^{-2} \int_0^{2\pi} d\varphi |P(\sin(\varphi+\delta), \cos(\varphi+\delta)) \\ - P(\sin\varphi, \cos\varphi) - \delta P'(\cos\varphi, -\sin\varphi)|^2 = 0,$$

where P is a polynomial and P' is its derivative. Since the integral is over a compact set, its existence poses no difficulties. Taylor's formula allows the difference to be estimated with P'', which remains bounded in $[0, 2\pi]$. Since the integrand converges pointwise, differentiability follows from Lebesgue's dominated convergence theorem. Thus D is contained in the domain of the generators. It remains to be shown that it is large enough for essential self-adjointness. To this end, consider the finite-dimensional subspace D_k generated by $\{\psi_k \in D : k_1+k_2+k_3 \leq k\}$, which is invariant under rotations and therefore represents the L's by finite matrices. All vectors are entire for finite matrices, so D is a dense set of entire vectors. According to (2.4.21; 5), it determines U uniquely, which means that $\mathbf{L}$ is essentially self-adjoint on D.

The connection between U and L can be written explicitly as

$$U_{\delta\mathbf{e}} = \exp(i\delta\mathbf{L}\cdot\mathbf{e}),$$

where $\mathbf{e}$ is a unit vector in the direction of the axis of rotation. In polar coordinates it is apparent that $-i(x\,\partial/\partial y - y\,\partial/\partial x) = -i\,\partial/\partial\varphi$ has the same action as L_z on ψ_k, so on D,

$$\mathbf{L} = [\mathbf{x}\times\mathbf{p}]. \qquad \square$$

3.2.4 Remarks

1. U is not strongly differentiable on all of $L^2(\mathbb{R}^3)$—e.g., not on $\exp(-|\mathbf{x}|^2/2)\Theta(|\varphi| - \alpha)$, $0 < \alpha < \pi$. Hence $\mathbf{L}$ is unbounded in the representation (3.2.1).
2. Furthermore, D is invariant under x_j and p_j, which are essentially self-adjoint on it (Problem 3). It is contained in the intersection of the three distinct domains on which $\mathbf{x}$, $\mathbf{p}$, and $\mathbf{L}$ are self-adjoint.
3. The sets D_k also consist of entire vectors for the operator $|\mathbf{L}|^2 = L_1^2 + L_2^2 + L_3^2$, since there is no question that

$$\sum_{n=0}^{\infty} (|\mathbf{L}_{|D_k}|^2)^n \frac{t^n}{n!}$$

converges for all $t \in \mathbb{C}$. According to (2.4.21; 5), this means that $|\mathbf{L}|^2$ is essentially self-adjoint on D.

3.2.5 The Commutation Relations of L

Since $\mathbf{L}$ is the generator of the rotations, its commutators with other operators tell how much they change by infinitesimal rotations. Thus, on D,

$$[L_m, V_r] = i\varepsilon_{mrs} V_S, \quad \text{for } \mathbf{V} = \mathbf{L}, \mathbf{x} \text{ or } \mathbf{p}. \tag{3.3}$$

These relations can be derived by differentiating (3.2) or directly from (3.1.9). The operator $|\mathbf{L}|^2$, as a scalar, commutes with $\mathbf{L}$:

$$[L_m, |\mathbf{L}|^2] = 0, \qquad m = 1, 2, 3. \tag{3.4}$$

These relationships are all initially valid on vectors of D, and can then be extended to exponential functions in the sense (3.1.7), since the vectors of D are entire for $\mathbf{L}$, $|\mathbf{L}|^2$, $\mathbf{x}$, and $\mathbf{p}$. In this extended sense, it is also true that $[\mathbf{L}, |\mathbf{x}|^2] = [\mathbf{L}, |\mathbf{p}|^2] = 0$.

3.2.6 Parity

The group $O(3)$ has two separate parts, depending on whether $\text{Det}\, M = \pm 1$. The matrix $M = -\mathbf{1}$ belongs to the component not connected with $\mathbf{1}$, and cannot be attained by letting one-parameter, continuous subgroups act on $\mathbf{1}$. The division of the group in two parts corresponds to the **parity operator** P such that

$$P^{-1} W(z) P = W(-z).$$

The phase factor is fixed by the condition that

$$(P\psi)(\mathbf{x}) = \psi(-\mathbf{x}), \qquad \psi \in L^2(\mathbb{R}^3),$$

so that we have $P^2 = \mathbf{1}$ and $P^{-1} = P^* = P$. The parity operator changes the sign of both $\mathbf{x}$ and $\mathbf{p}$, and thus commutes with $\mathbf{L}$:

$$P\mathbf{L}P^{-1} = \mathbf{L}.$$

3.2.7 Remark

Just as with **L**, P can be constructed out of **x** and **p** (Problem 1). This is different from classical mechanics, where although every one-parameter group of canonical transformations has a function of **x** and **p** as its generator, the finite canonical transformation $\mathbf{x} \to -\mathbf{x}$, $\mathbf{p} \to -\mathbf{p}$, has no infinitesimal generator; it cannot be reached by a continuous path from the identity.

3.2.8 The Spectrum of L

Let us consider one of the components of **L**, say L_3, and the Abelian C^* algebra $\mathcal{L}_3$ generated by $\exp(i\delta L_3)$. Because of the Gel'fand isomorphism (2.2.28), every point of the spectrum corresponds to a character on $\mathcal{L}_3$. Since, from (3.2.1), $\exp(2\pi i L_3) = \mathbf{1}$, every character is of the form $\exp(i\delta L_3) \to \exp(im\delta)$, $m \to \mathbb{Z}$, and so the only possible spectral values of L_3 (or the component of **L** in any other direction) are whole numbers. The construction given below will show that all these possible values actually occur.

3.2.9 The Eigenvectors of L

Different components of **L** do not commute with one another, so their only common eigenvectors ψ must be eigenvectors of their commutators with eigenvalue 0. Since the commutator of any two orthogonal components of **L** is always the third component, $\mathbf{L}\psi$ must equal 0, and ψ is invariant under rotations ($\psi(x) = \psi(r)$).

However, $[|\mathbf{L}|^2, \mathbf{L}] = 0$ when acting on any vector, so it is possible to have common vectors $|l, m\rangle$ of $|\mathbf{L}|^2$ and L_3:

$$L_3|l, m\rangle = m|l, m\rangle, \qquad |\mathbf{L}|^2|l, m\rangle = l(l+1)|l, m\rangle$$

(letting the eigenvalues of $|\mathbf{L}|^2$ be $l(l+1)$, with the benefit of hindsight). In order to discover the possible values of the new eigenvalue $l \geq 0$ and its relationship to m, note that $l(l+1)$ must always be $\geq m^2$, because $\langle l, m||\mathbf{L}|^2|l, m\rangle = m^2 + \langle l, m|L_1^2|l, m\rangle + \langle l, m|L_2^2|l, m\rangle = l(l+1) \geq m^2$. As already remarked, equality can only hold for $\mathbf{L}|l, m\rangle = 0$, i.e., $l = m = 0$. Now, (3.2.5) may be rewritten as

$$[L_3, L_\pm] = \pm L_\pm, \qquad L_\pm = L_1 \pm iL_2,$$

so

$$L_3 L_\pm|l, m\rangle = (m \pm 1)L_\pm|lm, \rangle,$$

$$|\mathbf{L}|^2 L_\pm|l, m\rangle = l(l+1)L_\pm|l, m\rangle.$$

Consequently $L_\pm|l, m\rangle$ is a simultaneous eigenvector of L_3 and $|\mathbf{L}|^2$ whenever $|l, m\rangle$ is, and we may write

$$L_\pm|l, m\rangle = c|l, m \pm 1\rangle, \qquad c \in \mathbb{C}.$$

Supposing that $|l, m\rangle$ have been normalized, the normalization constants c can be calculated from the equation

$$|\mathbf{L}|^2 = L_3^2 \mp L_3 + L_\pm L_\mp$$

(Problem 4). The result is that

$$L_\pm |lm, \rangle = \sqrt{l(l+1) - m(m \pm 1)} |l, m \pm 1\rangle.$$

In order not to violate the condition that $m^2 \leq l(l+1)$, repeated applications of $L_\pm$ must eventually yield 0, which can happen only if $l \in \mathbb{Z}^+$. It follows that $L_+|l, l\rangle = L_-|l, -l\rangle = 0$. A classical description of the action of L_- (respectively L_+) is that the direction of the angular momentum vector is changed while its length is held constant, and L_3 varies from a maximum value l to a minimum $-l$ (respectively from the minimum to the maximum). The eigenfunctions $|l, m\rangle$ constitute a $2l+1$-dimensional representation of the algebra generated by $\mathbf{L}$. The representation is irreducible, since every vector is cyclic (2.3.6; 1). The operator L_+ can be used to construct $|l, l\rangle$, and all the other eigenvectors of a given representation can be gotten by applying L_- to it.

3.2.10 The Eigenfunctions in the x-Representation

To construct $|l, l\rangle$ algebraically by applying operators to $|0, 0\rangle$, which corresponds to a radially symmetric $\psi(r)$, we rely on the equations

$$[L_3, x_1 \pm i x_2] = \pm(x_1 \pm i x_2),$$

$$[L_\pm, x_1 \pm i x_2] = 0,$$

$$[|\mathbf{L}|^2 - L_3^2 - L_3, x_1 + i x_2] = [L_-, x_1 + i x_2] L_+$$

(Problem 4), which imply that $(x_1 + i x_2)$ sends $|l, l\rangle$ to $|l+1, l+1\rangle$, up to normalization. Hence, again up to normalization

$$|l, m\rangle = L_-^{l-m} (x_1 + i x_2)^l |0, 0\rangle.$$

In the Schrödinger representation (3.2.1), which gives the probability measure $d^3x |\psi(\mathbf{x})|^2$ on Sp($\mathbf{x}$), the vector $|0, 0\rangle$ depends only on the radius r, and $|l, m\rangle$ will have the form $Y_l^m(\theta, \varphi) f(r)$.

3.2.11 Simple Special Cases

1. $l = m = 0$. This state is rotationally invariant, and the probability distribution $|\psi(r)|^2$ it corresponds to is spherically symmetric.
2. $l = \pm m = 1$. $|\psi(\mathbf{x})|^2 \sim |x \pm i y|^2 \sim \sin^2\theta$, and the 1-2 plane takes on the characteristics of an orbital plane.
3. $l = 1, m = 0$. $|\psi(\mathbf{x})|^2 \sim \cos^2\theta$, corresponds to a superposition of orbits in the 1-3 and 2-3-planes.

4. $l = \pm m$. $|\psi(\mathbf{x})|^2 \sim \sin^{2l}\theta$, and the particle is strongly concentrated in the 1-2-plane for large l.

3.2.12 Remarks

1. It seems paradoxical that L_3 has a discrete spectrum while its constituents $x_1 p_2$ and $x_2 p_1$ each have continuous spectrum. However, since they do not commute, they cannot possess precise values at the same time, and the sum of separate measurements of the summands is not acceptable as a measurement of an eigenvalue of the sum L_3. By the axiom of linearity, it is nonetheless possible to determine the average value of L_3 by making separate measurements of $x_1 p_2$ and $x_2 p_1$ on several identical copies of the system.
2. The commutation relations (3.2.5) require a state that is nondispersive for L_3 and $\mathbf{p}$ to satisfy $\langle|\mathbf{p}|\rangle = 0$.
3. Note that $\langle l, m|L_{1,2}|l, m\rangle = 0$, $(\Delta L_{1,2})^2 = (l(l+1) - m^2)/2$. It is because of the quantum fluctuations of $L_{1,2}$ that $l(l+1)$ always exceeds m^2 unless $l = m = 0$. There are nonzero quantum fluctuations even when $m = \pm l \neq 0$, though their value in that case is the least possible according to (2.2.33; 4) because $[L_1, L_2] = iL_3$.
4. As in (3.1.13), it is possible to characterize the states of minimal indeterminacy of L_1 and L_2 as the eigenvectors of $L_1 - i(\Delta L_1/\Delta L_2)L_2$, because of Remark (2.2.33; 3).

3.2.13 Spin

Many particles, including electrons and protons, have an intrinsic angular momentum **S**, known as the **spin**, in addition to their orbital angular momentum **L**. The spin operators satisfy the commutation rules

$$[S_j, S_k] = i\varepsilon_{jkl} S_l$$

and compute with **x** and **p**. The algebra of observables for particles with spin is the product of the Weyl algebra and the spin algebra: According to (2.3.8; 3), the Hilbert space of any representation is the tensor product of the Hilbert space for the dynamical variables **x** and **p** and the Hilbert space for the spin variables. More interestingly, the unitary operators $\exp(i\mathbf{S}\cdot\mathbf{e}\delta)$ for electrons and protons (or any particles of half-odd integer spin) are simply ray representations of SO(3), i.e., according to (3.2.2), representations of the universal covering group.

3.2.14 Gloss

SO(3), that is, the real 3×3 matrices M such that $MM^t = \mathbf{1}$, Det $M = 1$, is connected as a topological space, but it is not simply connected. In other words, there are paths in SO(3) that cannot be contracted to a point without breaking. To see why it is not simply connected, map the group space into a ball in $\mathbb{R}^3$ by

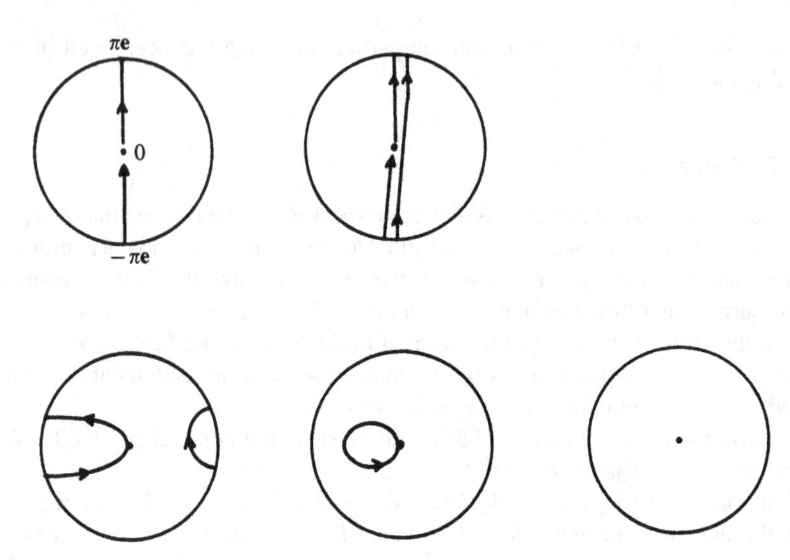

Figure 3.1. Homotopy of paths in SO(3).

associating a vector $\mathbf{e}\delta$ with any rotation, $\mathbf{e}$ specifying the axis of rotation and δ the angle. The angle may be restricted to the values $0 \leq \delta \leq \pi$, but then diametrically opposed points must be identified. For example, to rotate from 0 to 2π radians about the axis in the direction of $\mathbf{e}$, first go from 0 to $\pi\mathbf{e}$, which is equivalent to $-\pi\mathbf{e}$, and then return from there to the origin. There is no way to shrink this path down to the point 0, though a path that passed through the ball twice could be shrunk down. (See Figure 3.1.)

If the group space is doubled up like a two-sheeted Riemann surface, then it becomes simply connected and homeomorphic to the group SU(2). This new group comes into consideration as follows: For the spin matrices $\boldsymbol{\sigma}$ of (2.2.37), $-|\mathbf{v}|^2 = \operatorname{Det} \mathbf{v} \cdot \boldsymbol{\sigma}$ for any $\mathbf{v} \in \mathbb{R}^3$, and since any 2×2 matrix with trace 0 can be written as $\mathbf{v} \cdot \boldsymbol{\sigma}$

$$U^{-1} v_k \sigma_k U = M_{kl} v_l \sigma_k, \qquad U \in SU(2), \qquad M \in SO(3).$$

The homomorphism thereby set up from SU(2) to SO(3) is surjective but not injective, and indeed Schur's lemma shows that the identity in SO(3) corresponds to both $U = \pm\mathbf{1}$. If M is the rotation $\mathbf{e}\delta$, then it corresponds to $U = \exp(i\delta(\boldsymbol{\sigma} \cdot \mathbf{e})/2)$ (Problem 5), and letting δ increase from 0 to 2π brings one from $\mathbf{1}$ to $U = -\mathbf{1}$. Problem 5 also shows that SU(2) is simply connected, and is thus precisely the desired two-sheet universal covering group, so SO(3) is isomorphic to SU(2)/$\{1, -1\}$.

3.2.15 The Spectrum of S

The global properties are reflected in the spectrum. For SU(2) we only know that

$$\exp(4\pi i(\mathbf{S} \cdot \mathbf{e})) = \mathbf{1}$$

(4π rather than 2π), so the spectrum of any component may consist of both whole and half-odd integer values. This statement is consistent with our earlier construction of the representations, in which only $2l+1$ was required to be integral.

3.2.16 Representation of *S*

Since $|\mathbf{S}|^2$ commutes with $\mathbf{p}$ and $\mathbf{x}$, as do all the components of $\mathbf{S}$, it is a multiple of $\mathbf{1}$ in any irreducible representation. The experimental value found for electrons and protons is $\frac{1}{2}(\frac{1}{2}+1) = \frac{3}{4}$. The appropriate construction of a representation yields the matrices of (2.2.37):

$$\mathbf{S} = \frac{1}{2}\boldsymbol{\sigma}$$

(Problem 6). Thus, for n electrons, the overall Hilbert space is the antisymmetric tensor product of the Hilbert spaces for the individual electrons, each of which is a copy of $L^2(\mathbb{R}^3, d^3x) \otimes \mathbb{C}^2$.

3.2.17 Problems

1. Construct an explicit representation of the parity operator (3.2.6). (Hint: write p in the x-representation, decompose $L^2(\mathbb{R}^{3n}, d^{3n}x)$ as $\otimes_{k=1}^n L^2(\mathbb{R}^3, d^3x)$, introduce polar coordinates on $\mathbb{R}^3$, and see how P acts on the total set $\{f(r)Y_l^m(\theta, \varphi)\}$.)
2. Show that the ψ_k of (3.2.3) are total in $L^2(\mathbb{R}^3, d^3x)$.
3. Show that $\mathbf{x}$ and $\mathbf{p}$ are essentially self-adjoint on D (3.2.4; 3).
4. Verify the facts that (3.2.9) and (3.2.10).
5. Show that
   (i) SU(2) is simply connected; and
   (ii) every n-dimensional, continuous, unitary ray representation can be turned into an ordinary representation by an appropriate choice of phase. It is nontrivial, since there are irreducible ray representations of dimension greater than 1 for Abelian groups: see (3.1.6; 6).
   (iii) Find an irreducible ray representation of the (Abelian) Klein group of four, using spin matrices.
6. Show that the construction carried out in (3.2.9) produces $\mathbf{S} = \boldsymbol{\sigma}/2$ on a two-dimensional Hilbert space.

3.2.18 Solutions

1. $$\mathbf{x} \to -\mathbf{x} : r \to r, \theta \to \pi - \theta, \varphi \to \pi + \varphi,$$

   so

$$f(r)Y_l^m(\theta,\varphi) \to f(r)Y_l^m(\pi-\theta, \pi+\varphi) = (-1)^l f(r)Y_l^m(\theta,\varphi) = \exp(i\pi l) f(r)Y_l^m(\theta,\varphi);$$

which implies

$$P = \exp\left(i\pi\left(\sqrt{L^2 + \frac{1}{4}} - \frac{1}{2}\right)\right).$$

2. It suffices to consider the one-dimensional case. $\exp(-x^2/2 + itx)$ is the strong limit (s-lim) of $\exp(-x^2/2)\sum_{k=1}^{n}(itx)^k/k!$. Hence it follows from $\int \varphi(x)\exp(-x^2/2)P(x)dx = 0$ that $\int \varphi(x)\exp(-x^2/2 + itx)dx = 0$ for all t, and therefore $\varphi(x)\exp(-x^2/2) = 0$ a.e., so $\varphi = 0$.

3. All vectors in D are entire vectors for x_i and p_i.

4. This merely requires some differentiation.

5. (i) Any matrix $u \in \mathrm{SU}(2)$ is of the form

$$\begin{pmatrix} z_1 & z_2 \\ -z_2^* & z_1^* \end{pmatrix}, \qquad z_i \in \mathbb{C},\ |z_1|^2 + |z_2|^2 = 1.$$

The latter condition can be written as $\sum_{k=1}^{4}|x_k|^2 = 1$ with $z_1 = x_1 + ix_2$, $z_2 = x_3 + ix_4$, $x_k \in \mathbb{R}$, which shows that SU(2) is homeomorphic (and diffeomorphic) to the 3-sphere S^3. All n-spheres other than S^1, however, are simply connected, as can be seen with the following argument: Let

$$C : t \to x_k(t)(0 \le t \le 1; k = 1, 2, 3, 4; x_k(0) = x_k(1))$$

be a continuous, closed curve in S^3. By the Weierstrass approximation theorem there exist polynomials $P_k(t)$ such that $|x_k(t) - P_k(t)| < \varepsilon$ for all k, t, $P_k(0) = P_k(1) = x_k(0)$, and the curve $C_1 : t \to P_k(t)/\sqrt{\sum P_k^2(t)}$ is homotopic to the given curve C for ε small enough. By a theorem of Sard the set of all points of the curve C_1, as a differentiable mapping, has measure 0. Hence it is not possible for it to cover the whole 3-sphere. Therefore there exists a point $p \in S^3$ not on the curve. Since $S^3\backslash\{p\}$ is homeomorphic to $\mathbb{R}^3$, which is simply connected, C_1 can be continuously contracted to a point.

(ii) Let $u \to U(u)$ be a unitary, n-dimensional ray representation of SU(2), so $U(u)U(v) = \delta(u, v)U(uv)$ with $|\delta| = 1$. The associativity property implies that $\delta(u, v)\delta(uv, w) = \delta(u, vw)\delta(v, w)$, and it is obvious that $\delta(u, \mathbf{1}) = \delta(\mathbf{1}, u) = \mathbf{1}$. Since SU(2) is simply connected, $\sqrt[n]{\mathrm{Det}\, U(u)}$ is a well-defined number once $\sqrt[n]{\mathrm{Det}\, U(\mathbf{1})}$ has been fixed. by scaling $U(u) \to U'(u) = U(u)/\sqrt[n]{\mathrm{Det}\, U(u)}$ one obtains another ray representation with $U'(u)U'(v) = \delta'(u, v)U'(uv)$. However, since $\mathrm{Det}\, U'(u) = 1$, this means that $\delta'^n(u, v) = 1$ for all u and v, so $\delta'(u, v) = 1$ due to the simple connectedness of SU(2).

(iii) Klein's group of four contains four elements, e, a, b, and c, having the multiplication table

e	a	b	c
a	e	c	b
b	c	e	a
c	b	a	e

A ray representation can be obtained by setting $e \to \mathbf{1}$, $a \to \sigma_x$, $b \to \sigma_y$, and $c \to \sigma_z$ (cf. (2.2.37)).

6. The two vectors $|\uparrow\rangle$ and $|\downarrow\rangle$ such that $S_+|\uparrow\rangle = S_-|\downarrow\rangle = 0$ span the whole Hilbert space, and the matrix elements can be calculated as in (3.2.9).

3.3 Time-Evolution

As in classical mechanics the quantum-mechanical Hamiltonian generates the time-evolution, which is similar to its classical analogue, except that the influence of the noncommutativity must now be taken into account.

In the last two sections we have seen how to carry over the generation of the groups of translations and rotations from classical mechanics quantum mechanics. We now attempt the same feat for the time-evolution with a Hamiltonian H, and postulate a

3.3.1 Group of Automorphisms of the Time-Evolution

The algebra of observables evolves in time according to

$$a(t) = \exp(iHt)a\exp(-iHt) = \sum_{n=0}^{\infty} \frac{(it)^n}{n!} ad_H^n(a),$$

$$ad_H^0(a) \equiv a, \quad ad_H^n(a) \equiv [H, ad_H^{n-1}(a)], \quad a \in \mathcal{A}.$$

3.3.2 Remarks

1. Not every automorphism of a C^* algebra has this kind of representation. However, for our purposes, $\mathcal{A} = \mathcal{B}(\mathcal{H})$, for which every continuous, one-parameter group of automorphisms of the Jordan algebra (2.2.34) (i.e., it must be linear and preserve the symmetric product $\circ$) can be represented unitarily.
2. At this stage H is the classical Hamiltonian with p and q replaced by the operators of the Weyl system. However, because they do not commute, H is not uniquely defined, and in general even the question of a domain of essential self-adjointness is open. The systems we shall consider will not be so problematic, and self-adjointness will be taken care of by the Kato–Rellich theorem (2.5.14).
3. If a and H are bounded, then the series given in (3.3.1) converges because $\|ad_H^n(a)\| \leq 2^n \|H\|^n \|a\|$ for all t, and $t \to a(t)$ is continuous in norm. If H is unbounded, then the time-automorphism is still strongly continuous when $\exp(iHt)$ is, because

$$\begin{aligned}\|(\exp(iHt)a\exp(-iHt) - a)\psi\| &= \|(a\exp(-iHt) - \exp(-iHt)a)\psi\| \\ &\leq \|a(\exp(-iHt) - \mathbf{1})\psi\| + \|(\exp(-iHt) - \mathbf{1})a\psi\|.\end{aligned}$$

However, $da(t)/dt$ is not necessarily a bounded operator, and thus may not belong to $\mathcal{A}$. It is initially defined as the quadratic form $i[H, a]$ with $D(H)$ for its form domain. If a is itself unbounded, then the question of domain becomes more serious; under certain circumstances the Hermitian form $i[H, a]$ is not closeable, and can certainly not be the quadratic form of a self-adjoint operator.

Let us next investigate in some detail the time-evolution that will later serve as a standard of comparison.

3.3.3 Free Motion in Three Dimensions

The Hamiltonian for a free particle is

$$H = \frac{|\mathbf{p}|^2}{2m},$$

so in the spectral representation of the momentum H and its resolvent are

$$(H\psi)(\mathbf{p}) = \frac{|\mathbf{p}|^2}{2m}\,\psi(\mathbf{p}), \qquad D_p(H) = \{\psi \in L^2(\mathbb{R}^3, d^3p) : |\mathbf{p}|^2\psi \in L^2\},$$

$$(R(z)\psi)(\mathbf{p}) = \frac{1}{(|\mathbf{p}|^2/2m) - z}\,\psi(\mathbf{p}), \qquad (U(t)\psi)(\mathbf{p}) = \exp\left(-\frac{i|\mathbf{p}|^2}{2m}t\right)\psi(\mathbf{p}).$$

It is often desirable to have expressions for these quantities in the spectral representation of x, in which p is written as $-i\,\partial/\partial x$ and the two representations are related by the Fourier–Plancherel formula: The Fourier transformation maps $L^2(\mathbb{R}^3, d^3p)\cap L^1(\mathbb{R}^3, d^3p)$ isometrically onto $L^2(\mathbb{R}^3, d^3x)\cap L^\infty(\mathbb{R}^3, d^3x)$. Since both sets are dense in L^2, the Fourier transformation can be extended to a unitary transformation $L^2 \to L^2$. A calculation of the appropriate Fourier integral shows that

$$(H\psi)(\mathbf{x}) = -\frac{\Delta}{2m}\,\psi(\mathbf{x}),$$

$$(R(z)\psi)(\mathbf{x}) = \frac{m}{2\pi}\int d^3x'\,\frac{\exp(ik|\mathbf{x}-\mathbf{x}'|)}{|\mathbf{x}-\mathbf{x}'|}\,\psi(\mathbf{x}'), \qquad k = \sqrt{2mz},$$

$$(U(t)\psi)(\mathbf{x}) = \left(\frac{m}{2\pi i t}\right)^{3/2}\int d^3x'\,\exp\left(\frac{im|\mathbf{x}-\mathbf{x}'|^2}{2t}\right)\psi(\mathbf{x}')$$

(Problem 1).

3.3.4 Remarks

1. The Hamiltonian H is self-adjoint on the set of Fourier transforms $\tilde{\psi}$ of the vectors ψ of $D_p(H)$, and essentially self-adjoint on the Fourier transforms of the vectors of any set D that is dense in $D_p(H)$ in the graph norm. Examples of such states are the vectors of $\mathcal{S}$, the coherent states, and the domain of (3.2.3).
2. The vectors $\psi(\mathbf{x}) \in D_x(H)$ have some continuity properties because the integral kernel of the resolvent in $\mathbf{x}$-space is so nice. Furthermore, variants of Sobo-

lev's inequality show that functions whose derivatives have finite L^2-norms are bounded: Using the kernel for the resolvent we see that if $z = -\alpha^2$, $\alpha \in \mathbb{R}^+$, then

$$\begin{aligned} |\psi(\mathbf{x})| &= |(R(-\alpha^2)(|\mathbf{p}|^2 + \alpha^2)\psi(\mathbf{x})| \\ &= \int d^3x' \frac{\exp(-\alpha|\mathbf{x}-\mathbf{x}'|)}{4\pi|\mathbf{x}-\mathbf{x}'|}[2m(H\psi)(\mathbf{x}') + \alpha^2\psi(\mathbf{x}')] \\ &\leq (2m\|H\psi\| + \alpha^2\|\psi\|)(8\pi\alpha)^{-1/2}, \end{aligned}$$

by using the Cauchy–Schwarz inequality the fact that $\|\exp(-\alpha r)/4\pi r\|^2 = 1/8\pi\alpha$. One can also argue without using the kernel of the resolvent as follows: By the Cauchy–Schwarz inequality,

$$\begin{aligned} \left(\int |\tilde{\psi}(\mathbf{p})| d^3p\right)^2 &\leq \int \frac{d^3p}{(|\mathbf{p}|^2+\alpha^2)^2} \int (|\mathbf{p}|^2+\alpha^2)^2 |\tilde{\psi}(\mathbf{p})|^2 d^3p \\ &= \frac{\pi^2}{\alpha}\|(2mH+\alpha^2)\psi\|^2 \quad \text{for all } \alpha \in \mathbb{R}, \end{aligned}$$

which implies

$$\begin{aligned} |\psi(\mathbf{x})| &\leq (2\pi)^{-3/2}\int d^3p|\tilde{\psi}(\mathbf{p})| \leq \frac{\|(2mH+\alpha^2)\psi\|}{\sqrt{8\pi\alpha}} \\ &\leq \frac{(\alpha^2\|\psi\| + 2m\|H\psi\|)}{\sqrt{8\pi\alpha}}. \end{aligned}$$

Thus the functions $\psi(\mathbf{x})$ are bounded. Moreover, since $|\exp(i\mathbf{p}\cdot\mathbf{x}) - \exp(i\mathbf{p}\cdot\mathbf{x}')| \leq \min\{2, |\mathbf{p}||\mathbf{x}-\mathbf{x}'|\} \leq 2^{1-\gamma}|\mathbf{p}|^\gamma|\mathbf{x}-\mathbf{x}'|^\gamma$ for all $\gamma \in (0,1)$,

$$\begin{aligned} |\psi(\mathbf{x}) - \psi(\mathbf{x}')| &\leq \int \frac{d^3p|\mathbf{p}|^\gamma 2^{1-\gamma}}{(|\mathbf{p}|^2+\alpha^2)}(|\mathbf{p}|^2+\alpha^2)\tilde{\psi}(\mathbf{p})|x-x'|^\gamma \\ &\leq C(\gamma)|\mathbf{x}-\mathbf{x}'|^\gamma(\alpha^{(\gamma-1/2)}2m\|H\psi\| + \alpha^{(\gamma+3/2)}\|\psi\|) \\ &\qquad \text{for all } \gamma \in (0, \tfrac{1}{2}). \end{aligned}$$

So $\psi \in D_x(H)$ is in fact Hölder continuous with any exponent $< \frac{1}{2}$. Stronger properties such as C^1 cannot be hoped for (in three dimensions), since if $\psi = r^\gamma$, $1 > \gamma > \frac{1}{2}$, then $\psi'' \in L^2$ at small r but $\psi'(0) = \infty$. But at any rate, $D(|\mathbf{p}|^2) \subset L^\infty(\mathbb{R}^3)$.

3. The operator $U(t)$ also has a continuous integral kernel, and its effect is frequently to smooth functions. It describes how wave-packets damp out; the fact that they damp out is expressed by the weak convergence of $U(t)$ to 0 as $t \to \pm\infty$. For example, on the dense set of L^1 functions, $|(U(t)\psi)(\mathbf{x})| \leq (2\pi t/m)^{-3/2}\|\psi\|_1$. However, since $U(t)$ is invertible, the time-reversed motion is always possible.

4. The easiest way to see that H generates the classical time-automorphism $x \to x + pt$ for $m = 1$ is to use the Weyl operators:

$$\exp(ixs)\exp\left(-\frac{ip^2t}{2}\right)\exp(-ixs) = \exp\left(-\frac{i(p-s)^2t}{2}\right)$$
$$= \exp\left(-\frac{ip^2t}{2}\right)\exp\left(it\left(ps - \frac{s^2}{2}\right)\right)$$
$$\Rightarrow \exp\left(\frac{ip^2t}{2}\right)\exp(ixs)\exp\left(-\frac{ip^2t}{2}\right)$$
$$= \exp\left(it\left(ps - \frac{s^2}{2}\right)\right)\exp(ixs) = \exp(is(x+pt)).$$

This one-dimensional formula generalizes easily to vectors.

Most of the problems solved in introductory classical mechanics are also pretty easy in quantum theory:

3.3.5 Examples

1. Free fall. $H = p^2/2 + gx$, $L^2((-\infty, \infty), dx) \supset D \equiv$ the linear hull of $\{x^n \exp(-x^2)\}$. In the spectral representation of p, in which $x = i\, d/dp$, H can be defined as a self-adjoint operator on $D(H) \equiv \{\psi(p) \in L^2((-\infty, \infty), dp) :$ ψ is absolutely continuous and $(p^2/2 + ig\, d/dp)\psi \in L^2\}$. On D, $i[H, x] = p$ and $i[H, p] = -g$, so $\bar{x}(t) \equiv x + pt - gt^2/2$ and $\bar{p}(t) \equiv p - gt$ satisfy the same differential equations as $x(t)$ and $p(t)$. Since they agree at $t = 0$ and the vectors of D are entire for them, it follows that $\bar{x}(t) = x(t)$ and $\bar{p}(t) = p(t)$. The quantum fluctuations of these observables satisfy

$$\begin{aligned}(\Delta x(t))^2 &= \Delta x^2 + t^2(\Delta p)^2 + t(\langle xp + px\rangle - 2\langle x\rangle\langle p\rangle),\\ (\Delta p(t))^2 &= \Delta p^2.\end{aligned}$$

The damping out of the wave-packet does not depend on g; the uncertainty in p is constant while that of x grows linearly with Δpt. The spectrum of H is purely continuous, since on $D(H) H\psi = E\psi$ reads

$$H\psi = E\psi : ig\frac{d}{dp}\psi(p) = \left(E - \frac{p^2}{2}\right)\psi(p),$$

which

$$\Rightarrow \psi(p) = c\exp\left(-\frac{i(Ep - p^3/6)}{g}\right) \notin L^2((-\infty, \infty), dp).$$

2. The harmonic oscillator. $H = (p^2 + \omega^2 x^2)/2$; $D \subset L^2((-\infty, \infty), dx)$ as in Example 1 is invariant under H, and by the same argument the classical solution

$$x(t) = x\cos\omega t + \frac{p}{\omega}\sin\omega t,$$

$$p(t) = p\cos\omega t - \omega x \sin\omega t$$

again reproduces the correct quantum-mechanical time-evolution. The mean-square deviations are easily shown to satisfy

$$\begin{aligned}(\Delta x(t))^2 &= (\Delta x)^2 \cos^2 \omega t + (\Delta p)^2 \frac{1}{\omega^2} \sin^2 \omega t + \frac{1}{\omega} \cos\omega t \sin\omega t \\ &\times (\langle xp + px\rangle - 2\langle x\rangle\langle p\rangle).\end{aligned}$$

Wave-packets oscillate rather than decaying away. The last contribution cancels out for coherent states (3.1.13); moreover, if $\Delta x^2 = \Delta p^2/\omega^2$, then $(\Delta x(t))^2$ and $(\Delta p(t))^2$ are constant. $H = \omega(a^* a + 1/2)$, where

$$a = (\omega x + ip)(2\omega)^{-1/2}$$

(cf. (3.1.12; 2)), has a pure point spectrum, since $[a, a^*] = \mathbf{1}$ and $[H, a] = -a\omega$ (all acting on D), so $H\psi = E\psi \Rightarrow Ha\psi = (E - \omega)a\psi$. Since $H \geq \omega/2$, there is a vector ψ_0 such that $a\psi_0 = 0$, $H\psi_0 = (\omega/2)\psi_0$, and $H\psi_n = \omega(n + \frac{1}{2})\psi_n$, where $\psi_n = (a^*)^n (n!)^{-1/2}\psi_0$. In the spectral representation of x, $a(2\omega)^{1/2} = (d/dx) + \omega x$, $\psi_0 = \exp(-x^2\omega/2)$ (cf. (3.1.6; 2) and (3.1.12; 2)), and the ψ_n span D completely. Because D is dense in $\mathcal{H}$, H is self-adjoint on D and $\sigma_{ac}(H) = \sigma_s(H) = \emptyset$. Yet $\sigma(H)$ is not determined by the motion in configuration space alone; $\bar{H} = p_1 p_2 + x_1 x_2$ produces the same motion of the x_i as $H = (p_1^2 + p_2^2 + x_1^2 + x_2^2)/2$, (i.e., $\ddot{x}_i = -x_i$), but $\sigma(\bar{H}) = \mathbb{Z}$ and $\sigma(H) = \mathbb{Z}^+ + 1/2$.

3. A particle in a homogeneous magnetic field. The nonrelativistic version of the Hamiltonian (I: 5.1.9) with the magnetic field in the z-direction is

$$\begin{aligned} H &= \frac{|\mathbf{p} - e\mathbf{A}|^2}{2m} = \frac{m|\dot{\mathbf{x}}|^2}{2}, \qquad \dot{\mathbf{x}} = \frac{1}{m}\left(p_1 + \frac{eB}{2}x_2, p_2 - \frac{eB}{2}x_1, p_3\right) \\ &= i[H, \mathbf{x}], \qquad \left(\mathbf{A} = \frac{B}{2}(-x_2, x_1, 0)\right). \end{aligned}$$

Now the variables $(\dot{x}_1, \dot{x}_2)$ and $(\bar{x}_1, \bar{x}_2) = (x_1/2 + p_2/eB, x_2/2 - p_1/eB)$ are canonically conjugate pairs like (x_1, p_1) and (x_2, p_2), since on $D \times D$ as before $[\dot{x}_1, \dot{x}_2] = ieB/m^2$, $[\bar{x}_1, \bar{x}_2] = 1/ieB$, and $[\dot{x}_1, \bar{x}_k] = 0$ for all k and l. Writing $a = (\dot{x}_1 + i\dot{x}_2)\sqrt{m/2\omega}$, $\omega = eB/m$, we find that

$$H = \frac{p_3^2}{2m} + \left(a^* a + \frac{1}{2}\right)\omega, \qquad [a, a^*] = \mathbf{1}.$$

The Hilbert space is a tensor product $\mathcal{H}_{\bar{x}} \otimes \mathcal{H}_{\dot{x}} \otimes \mathcal{H}_{x_3}$ corresponding to the new pairs of conjugate observables, and H is the sum of the H of Example 1 with $g = 0$, acting on the last factor, and the H of Example 2 acting on the second factor. The time-evolution is accordingly

$$\begin{aligned} x_1(t) &= \bar{x}_1 - \frac{1}{\omega}(\dot{x}_2 \cos\omega t - \dot{x}_1 \sin\omega t), \\ x_2(t) &= \bar{x}_2 + \frac{1}{\omega}(\dot{x}_1 \cos\omega t + \dot{x}_2 \sin\omega t), \end{aligned}$$

$$x_3(t) = x_3 + \frac{p_3}{m} t.$$

Thus the constants $\bar{\mathbf{x}}$ function as the center of the orbit, and H, of course, is independent of them. The operator $\dot{x}_1^2 + \dot{x}_2^2$ therefore has an infinitely degenerate point spectrum, as it involves only one pair of conjugate variables. The operator H as a whole has continuous spectrum from $\omega/2$ to ∞, since it includes the kinetic energy in the 3-direction. As with the harmonic oscillator, the zero-point energy $\omega/2$ arises from the indeterminacy relation (3.1.11), according to which

$$H \sim \frac{1}{2}[(\Delta p^2) + \omega^2(\Delta x^2)] \sim \frac{1}{2}\left[\frac{1}{4(\Delta x^2)} + \omega^2(\Delta x^2)\right]$$

has its minimum $\omega/2$ when $\Delta x = (2\omega)^{-1/2}$. As in the classical case (see (I: §5.1)) it is important to distinguish between the canonical angular momentum $\hat{\mathbf{L}} = [\mathbf{x} \times \mathbf{p}]$ and the physical $\mathbf{L} = [\mathbf{x} \times m\dot{\mathbf{x}}]$. The former depends on the gauge, and is constant in the gauge chosen here. The physical angular momentum is independent of the gauge, but depends on time. In a magnetic field the ground state has $\langle 0|L_Z|0\rangle = -1$ (Problem 4), and for its orbit, $\Delta x \sim (\Delta p)^{-1} \sim \omega^{-1/2}$. As in the classical case H is invariant under translation combined with certain gauge transformations.

4. Radial motion in S states. Let $H = -\frac{1}{2}(d^2/dr^2) + V(r)$, where V and $V' \in L^\infty$, and $D(H) = \{\psi \in L^2([0,\infty), dr) : \psi \in C^1, \psi'$ is absolutely continuous $\psi'' \in L^2$, and $\psi(0) = 0\}$, making H self-adjoint according to (2.5.14) and (2.5.15). The operator $r : \psi(r) \to r\psi(r)$ is self-adjoint on $D(r) = \{\psi \in L^2 : r\psi \in L^2\}$. Its rate of change $\dot{r} = i[H, r]$ is at first defined as a quadratic form with the domain $D(H) \cap D(r)$, and is a restriction of the quadratic form associated with the Hermitian operator $p_r = -i\, d/dr$. $D(p_r) = \{\psi \in L^2 : \psi$ is absolutely continuous, $\psi' \in L^2$, and $\psi(0) = 0\}$. This is the operator studied in (2.5.3; 5). Thus Remark (2.5.13; 1) shows that the time-derivative of a self-adjoint operator need not have any self-adjoint extensions. An integration by parts reveals that the time-derivative of the form p_r is

$$\frac{d}{dt} p_r(\psi, \psi) = \int_0^\infty dr \left\{\left(-\frac{1}{2}\psi''^* + V\psi^*\right)\psi' - \psi^* \frac{\partial}{\partial r}\left(-\frac{1}{2}\psi'' + V\psi\right)\right\}$$

$$= -\int_0^\infty dr\, \psi^*(r)\psi(r)\frac{dV}{dr} + \frac{1}{2}|\psi'(0)|^2. \qquad \text{(a)}$$

This contains a noncloseable form in addition to the classical force, so $\dot{p}_r$ is not even an operator. Incidentally, (a) implies a relationship for the eigenvectors ψ of H that will be important later,

$$\left\langle \psi \left| \frac{dV}{dr} \psi \right.\right\rangle = \frac{1}{2}|\psi'(0)|^2. \qquad \text{(b)}$$

In applications we shall require

3.3.6 The Unitary Time-Evolution of a Time-Dependent Hamiltonian

The solution of

$$\frac{d}{dt}U(t,t_0) = -iH(t)U(t,t_0), \qquad U(t_0,t_0) = \mathbf{1}$$

is

$$\begin{aligned} U(t,t_0) &= 1+\sum_{n=1}^{\infty}(-i)^n \int_{t_0}^{t} dt_1 \int_{t_0}^{t_1} dt_2 \cdots \int_{t_0}^{t_{n-1}} dt_n H(t_1)H(t_2)\cdots H(t_n) \\ &\equiv T\left[\exp\left(-i\int_{t_0}^{t} dt' H(t')\right)\right]. \end{aligned}$$

3.3.7 Remarks

1. The operators U are not a one-parameter group, but it is still true that

$$U(t_2,t_1)U(t_1,t_0) = U(t_2,t_0)$$

(Problem 7).
2. If $H(t)$ is a step function, $H(t) = H_j$ for $t_{j-1} \le t \le t_j$, then

$$U(t_n,t_0) = \exp(-iH_n(t_n - t_{n-1}))\cdots\exp(-iH_1(t_1 - t_0)).$$

The sum (3.3.6) converges strongly in this case if there is a domain of entire vectors for all H_j, which is invariant under $\exp(-itH_j)$. Hence, passing to the continuous case, a sufficient condition for the series (3.3.6) to converge strongly would be the existence of a domain consisting of entire vectors for all $H(s)$, invariant under $\exp(-itH(s))$, and on which $H(s)$ is continuous enough so that the integrals applied in (3.3.6) to vectors of the domain can be strongly approximated by sums.

3.3.8 Examples

1. An oscillator with a spatially constant but time-varying force $f \in C^0(\mathbb{R})$. $H(t) = (p^2 + \omega^2x^2)/2 + xf(t)$. Since the equations of motion are linear,

$$U(t) = T\left[\exp\left(-\int_0^t dt'\, H(t')\right)\right]$$

produces the classical solution

$$U^{-1}(t)xU(t) = x\cos\omega t + \frac{p}{\omega}\sin\omega t + \xi(t),$$

$$U^{-1}(t)pU(t) = p\cos\omega t - \omega x\sin\omega t + \pi(t),$$

$$\xi(t) = -\frac{1}{\omega}\int_0^t dt'\sin\omega(t-t')f(t'), \qquad \pi(t) = \dot{\xi}(t).$$

Therefore the time-ordered product factorizes as

$$U(t) = \exp(-ip\xi(t)) \exp(ix\pi(t)) \exp\left(-\frac{it(p^2 + \omega^2 x^2)}{2}\right) \times \text{ a phase factor.}$$

Once again, the sum in (3.3.6) converges for all t on entire vectors for x, p and $p^2 + \omega^2 x^2$, such as the coherent states.

2. An oscillator with a changing frequency. $H(t) = (p^2 + \omega(t)^2 x^2)/2$, and the solution of the linear equation of motion $\dot{x} = p$, $\dot{p} = -\omega^2(t)x$ is the linear relationship

$$x(t) = \Omega_{11}(t)x + \Omega_{12}(t)p,$$

$$p(t) = \Omega_{21}(t)x + \Omega_{22}(t)p,$$

where in our notation the symplectic matrix $\Omega \in \mathrm{Sp}_2$ equals

$$T\left(\exp\left(\int_0^t dt' \begin{pmatrix} 0 & 1 \\ -\omega^2(t') & 0 \end{pmatrix}\right)\right).$$

The unitary transformation $U(t) = T[\exp(-i \int_0^t dt' H(t'))]$ that describes this time-evolution can be decomposed into factors as before: $\Omega \in \mathrm{Sp}_2 \Leftrightarrow \mathrm{Det}\,\Omega = 1$, so Ω has three free parameters and can be written as the product of the symplectic matrices

$$\begin{pmatrix} \cos \nu s & \frac{1}{\nu} \sin \nu s \\ -\nu \sin \nu s & \cos \nu s \end{pmatrix} \begin{pmatrix} e^{\beta} & 0 \\ 0 & e^{-\beta} \end{pmatrix}.$$

Since by Problem 8,

$$\exp\left(\frac{i\beta(px + xp)}{2}\right)(x, p) \exp\left(-\frac{i\beta(px + xp)}{2}\right) = (\exp(\beta)x, \exp(-\beta)p),$$

up to a phase factor $\exp(i\alpha)$,

$$U(t) = \exp(i\alpha) \exp\left(-\frac{is(p^2 + \nu^2 x^2)}{2}\right) \exp\left(\frac{i\beta(xp + px)}{2}\right).$$

It is not possible to write down the classical $\Omega(t)$ for an arbitrary $\omega(t)$, so the functions $\nu(t)$, $s(t)$, and $\beta(t)$ are also unknown. For some $\omega(t)$ a miracle happens and $\Omega(t)$ is an elementary function; for instance for $\bar{\omega}$, $\tau \in \mathbb{R}$,

$$\omega(t) = (t + \tau)\bar{\omega}\sqrt{1 - \frac{3}{4[\bar{\omega}(t + \tau)^2]^2}}, \qquad \begin{Bmatrix} c \\ s \end{Bmatrix} \equiv \begin{Bmatrix} \cos \\ \sin \end{Bmatrix} \bar{\omega}\frac{(t + \tau)^2}{2},$$

$$\Omega(t) = \sqrt{\frac{\tau}{t+\tau}} \begin{bmatrix} c + s/\bar{\omega}\tau^2 & s/\bar{\omega}\tau \\ \left(\frac{t+\tau}{\tau^2} - \frac{1}{t+\tau}\right)c - \left((t + \tau)\bar{\omega} + \frac{1}{\bar{\omega}\tau^2(t+\tau)}\right)s & \frac{t+\tau}{\tau}c - \frac{s}{\bar{\omega}\tau(t+\tau)} \end{bmatrix}.$$

3.3.9 Remark

The linear transformations of x and p leave the set K of states

$$\sim \exp\left(\frac{i(x-\gamma)^2}{2\alpha}\right), \quad \alpha, \gamma \in \mathbb{C}, \quad \operatorname{Im}\alpha < 0,$$

invariant: A transformation

$$x \to \Omega_{11}x + \Omega_{12}p + \xi,$$

$$p \to \Omega_{21}x + \Omega_{22}p + \pi,$$

changes $(x - \alpha p - \gamma)|\rangle = 0$ into $(x - \alpha' p - \gamma')|\rangle = 0$ with

$$\alpha' = \frac{\Omega_{12} + \alpha\Omega_{22}}{\Omega_{11} + \alpha\Omega_{21}} \quad \text{and} \quad \gamma' = \xi + \frac{\gamma + \pi(\Omega_{12} + \alpha\Omega_{22})}{\Omega_{11} + \alpha\Omega_{21}},$$

and hence $U|\rangle$ can be written as $\sim \exp(i(x-\gamma')^2/2\alpha')$. Although $\operatorname{Im}\alpha$ remains negative, $\operatorname{Re}\alpha$ does not remain equal to 0. Since the latter fact characterizes coherent states (3.1.13), linear transformations can affect the degree of indeterminacy (cf. (3.3.5; 1)), and Gaussian wave-packets may spread out.

If H is perturbed time-dependently to $H_1(t) = H + H'(t)$, then the eigenvalues of H_1 vary in time, therefore

$$H_1(t) \neq U^{-1}(t)H_1(0)U(t), \qquad U(t) = T\left[\exp\left(-\int_0^t dt' H_1(t')\right)\right].$$

As time passes, the family of projections $P_1(t)$ onto the eigenvectors of H_1 is more nearly transformed by $U(t)$ into itself the more slowly H' varies in comparison with the differences between energy levels. In other words, the transition probabilities approach zero in the limit of slow variation of H, even if the eigenvalues themselves change significantly.

3.3.10 Example

Recall Example (3.3.8; 1) and suppose that

$$H(s) = \frac{1}{2}\left(p^2 + \omega^2\left(x + \frac{f(s)}{\omega^2}\right)^2\right) - \frac{f^2(s)}{2\omega^2}, \quad 0 \le s \le 1, \quad f \in C^1.$$

The question is now whether the time-evolution according to $H(t/\tau), 0 \le t \le \tau$, transforms the projection onto the ground state of $H(0)$ into that of $H(1)$ as $\tau \to \infty$. The two ground-state eigenvalues are different, as the ground states satisfy

$$H(s)|E_0(s)\rangle = E_0(s)|E_0(s)\rangle, \qquad E_0(s) = \frac{\omega}{2} - \frac{f(s)^2}{2\omega^2},$$

$$a(t)U_\tau^*\left|E_0\left(\frac{t}{\tau}\right)\right\rangle \equiv (ip(t) + \omega x(t))U_\tau^*\left|E_0\left(\frac{t}{\tau}\right)\right\rangle = -\frac{f(t/\tau)}{\omega}U_\tau^*\left|E_0\left(\frac{t}{\tau}\right)\right\rangle,$$

$$a(t) = U^{-1}(t) a U(t).$$

As we saw earlier, the time-evolution of a is then

$$a(t) = a \exp(-i\omega t) - i \int_0^t dt' \exp(-i\omega(t - \tau')) f\left(\frac{t'}{\tau}\right),$$

or, after integration by parts,

$$\begin{aligned} a(\tau) &= a \exp(-i\omega\tau) - \frac{f(1)}{\omega} + \frac{\exp(-i\omega\tau)}{\omega} f(0) \\ &+ \frac{\exp(-i\omega\tau)}{\omega} \int_0^1 ds\, f'(s) \exp(i\omega\tau s). \end{aligned}$$

If the Fourier transform of f', which occurs in the last term, is denoted $\tilde{f}'(\omega\tau)$, then

$$a(\tau) + \frac{f(1)}{\omega} = \exp(-i\omega\tau)\left(a \frac{f(0)}{\omega} + \frac{\tilde{f}(\omega\tau)}{\omega}\right),$$

and, with (3.1.4),

$$|\langle E(0)|U_\tau E(1)\rangle| = \exp\left(-\left|\frac{\tilde{f}'(\omega\tau)}{2\omega}\right|^2\right).$$

If $f \in C^n$, then $\tilde{f}(\omega\tau) = O(\tau^{-n})$ as $\tau \to \infty$; the smoother the perturbation, the smaller the transition probability (cf. (3.3.7; 2)).

This behavior can be shown to occur more generally. Let $H(s)$, $0 \le s \le 1$, be a family of self-adjoint operators with a common domain D. An isolated eigenvalue $E(s)$ will be called **regular** iff $P(s)$, the projection onto its eigenvector, is finite-dimensional and continuously differentiable in s, as is $(H(s) - E(s))^{-1}(\mathbf{1} - P(s))$. Under these circumstances, there is a

3.3.11 Adiabatic Theorem

The probability of transition from a regular eigenvalue with the transformation

$$U_\tau \equiv T\left[\exp\left(i \int_0^\tau dt' H\left(\frac{t'}{\tau}\right)\right)\right]$$

goes as $O(\tau^{-1})$ in the limit $\tau \to \infty$.

3.3.12 Remarks

1. By assumption $E(s)$ is separated from all other eigenvalues by a nonzero distance for all s, so there is no question of crossing of eigenvalues. It is possible to show that the theorem remains valid when only a finite number of crossings can take place. However for degenerated eigenvalues the theorem fails.

2. The whole purpose of the domain assumptions is to ensure that U_τ is defined; they could be weakened in many ways.
3. Roughly speaking the theorem states that in the limit,

$$H\left(\frac{t}{\tau}\right) \to \sum_i E_i\left(\frac{t}{\tau}\right) U^{-1}(t) P_i U(t).$$

Proof: Let

$$\begin{aligned} P' &= \frac{d}{ds}P(s) : P^2 = P \Rightarrow P' = PP' + P'P \Rightarrow PP'P = 0, \Rightarrow P' \\ &= [[P', P], P]. \end{aligned}$$

Now write $P(s) = W(s)W^*(s)$, $P(0) = W^*(s)W(s) = W(0)$, where $W(s)$ is an isometry of the space of eigenvectors belonging to $E(0)$ onto that belonging to $E(s)$. Then $W(s) = P(s)W(s) = W(s)P(0)$, so $W' = P'W + PW'$ and $P' = W'W^* + WW^{*\prime} = [P'P]WW^* - WW^*[P', P]$ which is satisfied by $W' = [P', P]W$, which, because $PP'W = PP'PW = 0$ from the result above, implies that $P(s)W'(s) = 0$. The isometry $W(s)$ describes how the eigenvectors of $H(s)$ twist around as functions of s, and this must be compared with the time-evolution according to

$$V_\tau(s) = T\left[\exp\left(-i\tau\int_0^s ds'(H(s') - E(s'))\right)\right], \tag{a}$$

where a convenient phase factor has been included. From $V_\tau^{*\prime} = i\tau V_\tau^*(H - E)$, $(H - E)P = 0$, and the foregoing argument it follows that

$$(V_\tau^* W)' = i\tau V_\tau^*(H - E)PW + V_\tau^* W' = (i\tau)^{-1} V_\tau^{*\prime}(H - E)^{-1}(\mathbf{1} - P)W' \tag{b}$$

(writing H for $H(s)$, etc.). If Equation (b) is integrated by parts, then

$$\begin{aligned} V_\tau^*(1)W(1) - P(0) = (i\tau)^{-1} \Bigg\{ & V_\tau^*(H - E)^{-1}(\mathbf{1} - P)W'\Big|_0^1 \\ & - \int_0^1 ds\, V_\tau^* \frac{d}{ds}((H - E)^{-1}(\mathbf{1} - P)W') \Bigg\}. \end{aligned} \tag{c}$$

Since it has been assumed that the eigenvectors remain at least some positive distance apart, the operators $(H - E)^{-1}(\mathbf{1} - P)$ and $(d/ds)((H - E)^{-1}(\mathbf{1} - P))$ are uniformly bounded in s. The operator $\{\ldots\}$ is then also bounded, and (c) implies the adiabatic theorem

$$\|W(1) - V_\tau(1)P(0)\| = O(\tau^{-1}). \qquad \square$$

3.3.13 The Classical Limit

We saw in Examples (3.3.5; 1) through (3.3.5; 3) that the quantum-theoretical time-automorphism for linear equations of motion is the same as the classical one.

The connection between classical and quantum dynamics is not so easy in general, since it is possible that $\langle \dot{p} \rangle = -\langle V'(x) \rangle \neq V'(\langle x \rangle)$. Yet there is hope that as $\hbar \to 0$ the fluctuations can be neglected, leaving the classical time-evolution. In (3.1.15) we began with

$$W(\hbar^{-1/2}z^*)(x, p)W(\hbar^{-1/2}z^*) = (x, p) + \hbar^{-1/2}(\xi, \pi), \qquad z = \xi + i\pi,$$

in order to make the heuristic correspondence principle more precise. We ought to be able to show that quantum time-evolution converts this to

$$W(\hbar^{-1/2}z^*)(x(t), p(t))W(\hbar^{-1/2}z^*) \xrightarrow{\hbar \to 0} U_f^{-1}(t)(x, p)U_f(t)$$

$$+\hbar^{-1/2}(\xi(t), \pi(t)), \qquad z = \xi(0) + i\pi(0),$$

where $(\xi(t), \pi(t))$ are the solutions of the classical equations of motions with initial data $(\xi(0), \pi(0))$, and U_f gives the time-evolution of the equations of motion as linearized about the classical path (cf. (3.3.8; 2)):

$$U_f = T\left[\exp\left(-i\int_0^t dt' \left(\frac{p^2}{2m} + \frac{x^2}{2} V''(\xi(t'))\right)\right)\right]. \tag{3.5}$$

In the classical limit (3.1.15) this indeed reproduces the classical trajectory $(\xi(t), \pi(t))$:

$$\lim_{\hbar \to 0} W(\hbar^{-1/2}z^*)(x_\hbar(t), p_\hbar(t))W(\hbar^{-1/2}z^*) = (\xi(t), \pi(t)).$$

In other words, the diagram

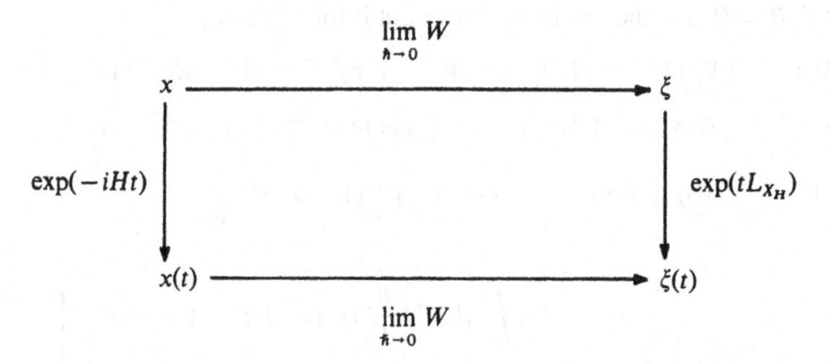

commutes. The mathematically precise statement of this fact uses the Weyl operators:

3.3.14 Theorem

Let $V \in C^3(\mathbb{R})$ and suppose that $D(V)$ and $D(|x^3 V'''|)$ contain K, the set of states $|\alpha|\sqrt{\pi/-\operatorname{Im}\alpha}\exp(i(x-\gamma)^2/2\alpha)$, $\alpha, \gamma \in \mathbb{C}$, and $\operatorname{Im}\alpha > 0$. Then for all t for which the classical trajectory $(\xi(t), \pi(t))$ continues to exist,

$$\lim_{\hbar \to 0} W(\hbar^{-1/2}z^*) \exp\left(i\frac{t}{\hbar}H_\hbar\right) \exp(i[r(x - \hbar^{-1/2}\xi(t)) + s(p - \hbar^{-1/2}\pi(t))])$$

$$\times \exp\left(-i\frac{t}{\hbar} H_\hbar\right) W(\hbar^{-1/2}z^*) = U_f^{-1}(t)\exp(i(rx+sp))U_f(t)$$

in the strong operator topology. The U_f in this formula is defined as in (3.5), $z = \xi(0) + i\pi(0)$, *and* $H_\hbar = H(x_\hbar, p_\hbar)$ *is any self-adjoint extension of the Hermitian operator* $(\hbar/2)p^2 + V(\hbar^{1/2}x)$ *on* K.

3.3.15 Remarks

1. Since $H_\hbar$ is a real, Hermitian operator (see (3.3.18; 5)), its deficiency indices are equal, and thus it has self-adjoint extensions. Any of these extensions serves to define $\exp(itH_\hbar/\hbar)$; in the limit they are all equivalent.
2. If $H_\hbar/\hbar$ is expanded around the classical trajectory in powers of $x_\hbar - \xi(t)$, $p_\hbar - \pi(t)$:

$$\begin{aligned}\frac{p^2}{2} + \hbar^{-1}V(\hbar^{1/2}x) &= \hbar^{-1}\left[\frac{\pi(t)^2}{2} + V(\xi(t))\right] \\ &\quad + \hbar^{-1/2}[\pi(t)(p - \hbar^{-1/2}\pi(t)) \\ &\quad + V'(\xi(t))(x - \hbar^{-1/2}\xi(t))] + \cdots \\ &\equiv \hbar^{-1}H_0 + \hbar^{-1/2}H_1 + H_2 + O(\hbar^{1/2}),\end{aligned}$$

then, in addition to $H_0(t)$, which is a multiple of **1**, there arises a linear term $H_1(t) = \hbar^{-1/2}(\pi(t)p + V'(\xi(t))x)$. This is precisely the generator of a displacement by $\hbar^{-1/2}(\xi(t) - \xi(0))$ and, in the momenta, $\hbar^{-1/2}(\pi(t) - \pi(0))$. The left side of (3.3.14) can therefore be written

$$U_\hbar(t)^* \exp(i(rx+sp))U_\hbar(t),$$

where

$$\begin{aligned}U_\hbar(t) &= W(\hbar^{-1/2}z^*)\left(T\left[\exp\left(-i\int_0^t dt'(H_1(t') + H_0(t'))\right)\right]\right)^* \\ &\quad \cdot \exp\left(-\frac{itH_\hbar}{\hbar}\right) W(\hbar^{-1/2}z^*).\end{aligned}$$

The theorem thus states that in the limit $\hbar \to 0$, the time-evolution according to $H_\hbar/\hbar$ differs from that according to H_1 by a factor

$$U_f(t) = T\exp\left(-i\int_0^t dt' H_2(t')\right),$$

where W has been used to translate the starting point back to the origin.

Proof: In order to show that $\lim_{\hbar\to 0} U_\hbar(t) = U_f(t)$ on K, consider the operator

$$\begin{aligned}U_\hbar(t_1, t_0) &\equiv W(\hbar^{-1/2}z^*)^*\left(T\left[\exp\left(-i\int_0^{t_1} dt' H_1(t')\right)\right]\right)^* \\ &\quad \cdot \exp\left(-\frac{i(t_1 - t_0)H_\hbar}{\hbar}\right) T\left[\exp\left(-i\int_0^{t_0} dt' H_1(t')\right)\right]\end{aligned}$$

$$\cdot W(\hbar^{-1/2}z^*)\exp\left(i\int_{t_0}^{t_1}dt' H_0(t')\right)$$

as in (3.3.6), and compare it with

$$U(t_1,t_0)\equiv T\left[\exp\left(-i\int_{t_0}^{t_1}dt'\left(\frac{p^2}{2}+\frac{V''(\xi(t'))x^2}{2}\right)\right)\right].$$

The set K is invariant under all the unitary factors that arise except possibly for $\exp(-itH_\hbar/\hbar)$; however, since $D(H_\hbar)\supset D(p^2)\cap D(V)\supset K$, the derivative by t_0 in the identity

$$U(t_1,0)-U_\hbar(t_1,0)=\int_0^{t_1}dt_0\frac{d}{dt_0}U_\hbar(t_1,t_0)U(t_0,0)$$

is justified on K. We find that

$$\begin{aligned}\frac{d}{dt_0}U_\hbar(t_1,t_0)U(t_0,0) &= iU_\hbar(t_1,t_0)\Bigg\{\hbar^{-1}V(\xi(t_0)+\hbar^{1/2}x)\\ &\quad -\hbar^{-1}V(\xi(t_0))-\hbar^{-1/2}V'(\xi(t_0))x\\ &\quad -\frac{V''(\xi(t_0))x^2}{2}\Bigg\}U(t_0,0).\end{aligned}$$

Now, with Taylor's formula, $\|\{\cdots\}\psi\|$, $\psi\in K$ is bounded by $\hbar^{1/2}\|x^3V'''\psi\|$, and since the $U_\hbar$ are unitary this goes to 0 as $\hbar^{1/2}\to 0$. □

3.3.16 Remarks

1. The examples looked at earlier show that when $V''>0$ the mean-square deviations oscillate about the classical trajectory, when $V''<0$ they are exponential functions of time, and when $V''=0$ they are linear in time. This corresponds exactly to the behavior of densities of finite spread according to classical stability theory.
2. Since $\hbar$ makes its original appearance only in $(\hbar^2/2m)\Delta$, the limit $\hbar\to 0$ is related for eigenvalues to the limit $m\to\infty$.
3. We have shown only that $U_\hbar$ converges. The conjugation $U\to U^*$ is not strongly continuous but only weakly so, which implies only the weak convergence of $U_\hbar^*$. However, since the limit is unitary and the weak and strong topologies are equivalent on the unitary operators, $U_\hbar^*$ also converges strongly. Finally, although the operator product is not strongly continuous, it is strongly sequentially continuous, so the proof of (3.3.14) goes through.

Classical trajectories generated by a Hamiltonian $H(x,p)=H(x,-p)$ satisfy $x(-t;x(0),p(0))=x(t;x(0),-p(0))$. Of course, $x\to x$, $p\to -p$ is not a canonical transformation, and it cannot be generated by a unitary transformation in quantum theory either; such a transformation would contradict $[x,p]=i$. The Weyl relations (3.1.2; 1) are nevertheless invariant under the

3.3.17 Antiautomorphism Θ of Reversal of the Motion

$$\Theta(\alpha A + \beta B) = \alpha\Theta(A) + \beta\Theta(B), \quad \alpha, \beta \in \mathbb{C} \quad \text{and} \quad A, B \in W,$$

$$\Theta(AB) = \Theta(B)\Theta(A), \qquad \Theta(W(z)) = W(-z^*).$$

3.3.18 Remarks

1. Θ preserves the structure of the Jordan algebra, i.e., $\Theta(A \circ B) = \Theta(A) \circ \Theta(B)$, and causes the transformation $\Theta(x) = x$, $\Theta(p) = -p$. If $\Theta(H) = H$ and $\Theta(A) = A$, then $\Theta(A(t)) = \Theta(\exp(iHt)A\exp(-iHt)) = \exp(-iHt)A \times \exp(iHt) = A(-t)$.
2. In the representation (3.1.4) for the W's, Θ is equivalent with complex conjugation to an operation Θ' such that

$$\Theta'(\alpha A + \beta B) = \alpha^*\Theta'(A) + \beta^*\Theta'(B),$$

$$\Theta'(AB) = \Theta'(A)\Theta'(B), \qquad \Theta'(W(z)) = W(z^*).$$

The operator Θ' also leaves the Weyl relations invariant, and it is easy to check that $\langle z_1|\Theta(W(z))|z_2\rangle = \langle z_2|\Theta'(W(z))|z_1\rangle^*$, where $|z_i\rangle \equiv W(z_i)|u\rangle$. Hence the matrix elements of Hermitian operators and consequently matrix elements of observables are the same with Θ' as with Θ. A bijection $K : \mathcal{H} \to \mathcal{H}$ is usually defined by

$$K\sum_i \alpha_i|z_i\rangle \equiv \Theta'\left(\sum_i \alpha_i W(z_i)\right)|u\rangle = \sum_i \alpha_i^*|z_i^*\rangle$$

and is known as **time-reversal**. Note that

$$\begin{aligned}\langle z_2|(K|z_1\rangle) &= \langle u|W(-z_2)W(z_1^*)|u\rangle = \langle u|W(-z_2^*)W(z_1)|u\rangle^* \\ &= \langle (K|z_2\rangle)|z_1\rangle^*.\end{aligned}$$

Since $\langle z_2|\Theta'(W(z))|z_1\rangle = \langle z_2|KW(z)K|z_1\rangle$, the operator Θ' is equivalent to this antilinear transformation o vectors.
3. Since $\Theta(\mathbf{L}) = -\mathbf{L}$, it would be reasonable to require that $\Theta(\boldsymbol{\sigma}) = -\boldsymbol{\sigma}$. We see incidentally that time-reversible operators $H = \Theta(H)$ must have at least doubly degenerate eigenvalues in the presence of spin. If w is the state associated with a certain eigenvalue of H, then the time-reversed state w_r defined by $w_r(a) = w(\Theta(a))$ is different from w, since for pure states $w(\boldsymbol{\sigma}) = -w_r(\boldsymbol{\sigma}) \neq 0$, although $w(H) = w_r(H)$.
4. It was possible to produce a spatial reflection with an element $\mathcal{P}$ of $\mathcal{W}$ that commutes with H if $H(x, p) = H(-x, -p)$, and thus $\mathcal{P}$ furnishes a constant of the motion (see (3.2.6) and (3.2.7)). However, reversal of the motion is not connected with a constant operator.
5. Hamiltonians that can be time-reversed in the sense of Remark 2 are real differential operators, which always have equal deficiency indices (recall (2.5.12; 4)).

Hence, by Theorem (2.5.10), there are self-adjoint extensions, which can be used to define a time-evolution. This is particularly interesting, since classically the existence of collision trajectories can destroy the one-parameter group of the time-evolution (cf. I: §4.5).

3.3.19 Problems

1. Verify (3.3.3).

2. How do the coherent states (3.1.13) evolve in time with the motions of (3.3.3) and (3.3.5; 2)?

3. (i) Calculate the resolvent $(H - z)^{-1}$ for the H of (3.3.5; 1) as a Fourier integral, and (ii) show that $\sigma_s(H)$ is empty and $\sigma_{ac} = \mathbb{R}$.
(Hints: For (i) make the ansatz that

$$(H - z)^{-1}\psi(p) = \int_{-\infty}^{\infty} \frac{dp'}{2\pi} \int_{-\infty}^{\infty} \frac{d\lambda}{\lambda - z} K(\lambda, p, p')\psi(p')$$

and determine K so that

$$\left(ig\frac{d}{dp} + \frac{p^2}{2} - z\right) \int_{-\infty}^{\infty} \frac{d\lambda}{\lambda - z} K(\lambda, p, p') = 2\pi\delta(p - p').$$

For (ii) use the formula

$$P(a, b) = s\text{-}\lim_{\varepsilon \to 0} \int_a^b \frac{dz}{2\pi i} [(H - z - i\varepsilon)^{-1} - (H - z + i\varepsilon)^{-1}],$$

where $P(a, b) = \int_a^b dP_H(\alpha)$ and $P_H(\alpha)$ is the spectral projection for H.)

4. Show that in (3.3.5; 3),

   (i) the canonical angular momentum $\hat{L}_3 = [\mathbf{x} \times \mathbf{p}]_3$ is constant, but that the physical $L_3 = [\mathbf{x} \times m\dot{\mathbf{x}}]_3$ is not; and

   (ii) $\langle 0|L_3|0\rangle = -1$ if $a|0\rangle = 0$.

5. Show that the addition of any vector potential, even one depending on the position, always causes an increase in the ground-state energy of a Hamiltonian with just an ordinary potential. This accounts for diamagnetism in hydrogen and helium atoms. (Actually, the statement can be generalized if the exclusion principle is taken into account.)

6. Show that even if ψ is an entire vector for a and b, it need not be an entire vector for $a + b$.

7. Prove the formula of (3.3.7; 1).

8. Prove that in (3.3.8; 2), $U(t) = \exp(i\alpha)\exp(-is(p^2 + v^2x^2)/2)\exp(i\beta(xp + px))$.

9. Interpret the adiabatic theorem for the soluble example (3.3.8; 2).

3.3.20 Solutions

1. $$(H\psi)(\mathbf{x}) = -\tfrac{\Delta}{2m}\,\psi(\mathbf{x}): \quad \text{this part is trivial.}$$

$$R(z)\psi : (R(z)\psi)(\mathbf{p}) = 2m\frac{\psi(\mathbf{p})}{|\mathbf{p}|^2 - k^2} \qquad k^2 = 2mz.$$

$$\begin{aligned}(R(z)\psi)(\mathbf{x}) &= \frac{2m}{8\pi^3}\int \frac{\exp(i\mathbf{p}\cdot\mathbf{x})}{|\mathbf{p}|^2 - k^2}\exp(-i\mathbf{p}\cdot\mathbf{x}')\psi(\mathbf{x}')dx'dp \\ &= \frac{m}{2\pi}\int \frac{\exp(ik|\mathbf{x}-\mathbf{x}'|)}{|\mathbf{x}-\mathbf{x}'|}\psi(\mathbf{x}')dx'.\end{aligned}$$

$$\begin{aligned}(U(t)\psi) : (U(t)\psi)(\mathbf{p}) &= \exp\left(-\frac{it|\mathbf{p}|^2}{2m}\right)\psi(\mathbf{p}) \Rightarrow (U(t)\psi)(\mathbf{x}) \\ &= \left(\frac{m}{2\pi i t}\right)^{3/2}\int d^2x' \exp\left(\frac{im|\mathbf{x}-\mathbf{x}'|^2}{2t}\right)\psi(\mathbf{x}').\end{aligned}$$

2. For the free time-evolution (in (3.1.13)), $\langle xp + px\rangle = 2\langle x\rangle\langle p\rangle)$, $\Delta(x(t))^2 = \Delta x^2 + t^2\Delta p^2 \neq 1/4\Delta(p(t))^2$, and

$$\psi_t(x) = (\Delta x^2 + t^2\Delta p^2)^{-1/4}\exp\left(-\frac{(x - \langle x\rangle - \langle p\rangle t)^2}{4(\Delta x^2 + t^2\Delta p^2)}\left(1 - \frac{it}{2\Delta x^2}\right) + i\langle p\rangle x\right).$$

For the oscillator with $\omega = 1$, Δx is constant, so the wave-packets do not spread themselves out to nothingness, and

$$\psi_t(x) = \exp\left[-\frac{(x - \langle x\rangle\cos\omega t - \langle p\rangle\sin\omega t + 2i\Delta x^2(\langle p\rangle\cos\omega t - \langle x\rangle\sin\omega t))^2}{4\Delta x^2}\right].$$

3. (i) $(H - z)^{-1}\psi(p)$
$= \int_{-\infty}^{\infty}(dp'/2\pi)\int_{-\infty}^{\infty} d\lambda\, \exp[-i(\lambda(p-p')/g - (p^3 - p'^3)/6g)]\psi(p')/(\lambda - z)$.
(ii) If $\psi \in L^1$, then

$$\begin{aligned}|\langle\psi|P(a,b)\psi\rangle| &= \Bigg|\lim_{\varepsilon\to 0}\int_a^b \frac{dz}{2\pi i}\int_{-\infty}^{\infty}\frac{d\lambda\,dp\,dp'}{2\pi} \\ &\quad\cdot\exp\left[-i\left(\frac{\lambda(p-p')}{g} - \frac{(p^3 - p'^3)}{6g}\right)\right]\psi^*(p)\psi(p') \\ &\quad\cdot\left(\frac{1}{\lambda - z - i\varepsilon} - \frac{1}{\lambda - z + i\varepsilon}\right)\Bigg| \\ &\le \lim_{\varepsilon\to 0}\int_a^b \frac{dz}{\pi}\int_{-\infty}^{\infty}\frac{d\lambda\,dp\,dp'}{2\pi}|\psi(p)\psi(p')|\frac{\varepsilon}{(\lambda - z)^2 + \varepsilon^2} \\ &\le \frac{(b-a)}{2\pi}\left(\int_{-\infty}^{\infty}|\psi(p)|dp\right)^2 \\ &\Rightarrow \psi \in \mathcal{H}_{ac} \Rightarrow \mathcal{H}_{ac} = \mathcal{H},\end{aligned}$$

since the vectors $\psi \in L^1$ are dense and $\mathcal{H}_{ac}$ is closed.

4. (i) $\hat{L}_3 = (m\omega/2)(\bar{x}_1^2 + \bar{x}_2^2) - (m/2\omega)(\dot{x}_1^2 + \dot{x}_2^2) \Rightarrow [L_3, H] = 0$, $L_3 = -(m(\dot{x}_1^2 + \dot{x}_2^2)/\omega) + m(\bar{x}_1, \dot{x}_2 - \bar{x}_2\dot{x}_1)$ is not constant.
   (ii) $a|0\rangle = 0 \Rightarrow \hat{L}_3|0\rangle = 0, 1, 2, \ldots$ For the orbit with the smallest radius, i.e., the least value of $\bar{x}_1^2 + \bar{x}_2^2$, it is also true that $(\bar{x}_1 - i\bar{x}_2)|0\rangle = 0$, which implies that $\hat{L}_3|0\rangle = 0$.

$$\langle 0|(\bar{x}_1\dot{x}_2 - \bar{x}_2\dot{x}_1)|0\rangle = 0 \Rightarrow \langle 0|L_3|0\rangle = -\frac{2}{\omega}\langle 0|H|0\rangle = -1.$$

5. Let $H_e = (p + eA(x))^2/2m + V(x)$ and $\psi(x) = R(x)\exp(iS(x))$, $R \geq 0$, S real. $\langle\psi|H_e|\psi\rangle = \int d^3x(|\nabla R + i(\nabla S + eA)R|^2/2m + R^2(x)V(x))$

$$\geq \int d^3x(\nabla R^2/2m + R^2 V) = \langle R|H_0|R\rangle \geq \text{ the ground-state energy with } e = 0.$$

6. Let $\mathcal{H} = L^2((-\infty, \infty), dx) \ni \psi(x) = 1$ when $0 \leq x \leq 1$, and otherwise 0, and let $a\psi(x) = \exp(x^2)\psi(x)$, $b\psi(x) = \psi(x+1) + \psi(x-1)$. b is bounded, so every vector is entire for it, and ψ is certainly entire for a since $\|a^n\psi\| \leq \exp(n)$. However,

$$\|(a+b)^n\psi\|^2 \geq \|ab^n\psi\|^2 \geq \int_n^{n+1} \exp(2x^2)dx \geq \exp(2n^2),$$

and $\sum_n \exp(2n^2)t^n/n!$ diverges for all $t > 0$.

7. Let $V(t) = U(t, t_1)U(t_1, t_0) - U(t, t_0)$. Then $dV/dt = -iHV$, which implies that $V(t) = 0$, since $V(t_1) = 0$.

8. $\frac{1}{2}(px + xp)$ is the generator of the group $U_\beta : \psi(x) \to \exp(\beta/2)\psi(\exp(\beta)x)$. This follows from the identity $i(xp + px)\psi(x) = 2\partial \exp(\beta/2)\psi(\exp(\beta)x)/\partial\beta_{|\beta=0}$, which holds for entire analytic functions ψ. It implies that

$$\begin{aligned} U_\beta x U_{-\beta}\psi(x) &= U_\beta(x\exp(-\beta/2)\psi(\exp(-\beta)x)) = \exp(\beta)x\psi(x), \\ U_\beta p U_{-\beta}\psi(x) &= U_\beta(-i\exp(-3\beta/2)\psi'(\exp(-\beta)x)) = \exp(-\beta)p\psi(x). \end{aligned}$$

9. If $t \gg \tau$, then $H(t)|0\rangle \sim (\omega(t)/2|0\rangle$, and the classical invariant E/ω becomes constant.

3.4 The Limit $t \to \pm\infty$

If particles escape to infinity, their time-evolution approaches that of free particles. In quantum theory this limit is achieved with great topological finesse.

The eigenvectors of H, which span the subspace $\mathcal{H}_p$ of (2.3.16), are related to classical trajectories that remain in compact regions indefinitely. The expectation value of an observable in this case is an almost periodic function $\sum_{j,k} \exp[it(E_j - E_k)]c_{jk}$, for which the time-average exists, but the time-limit does not. The operator $\exp(iHt)$ converges weakly on $\mathcal{H}_{ac}$, since in the spectral representation $\langle f|\exp(iHt)g\rangle = \int dh \exp(iht)f^*(h)g(h)$ approaches zero by Riemann–Lebesgue lemma. There is, of course, no chance for the unitary operators $\exp(iHt)$

to converge strongly to 0. In order to understand how some operator can converge strongly as $t \to \pm\infty$, it will be necessary to go more deeply into the ideas introduced with Theorem (2.5.14).

3.4.1 Definition

H' is said to be **bounded** (respectively **compact**) **relative** to H_0 iff $D(H') \supset D(H_0)$ and the mapping $H' : D_\Gamma(H_0) \to \mathcal{H}$ is continuous (compact). The space $D_\Gamma(H_0)$ is $D(H_0)$ topologized with the graph norm $\| \ \|_{H_0}$ (2.4.15; 3).

3.4.2 Remarks

1. Recall that continuous (respectively compact) linear mappings are those that send bounded sets to bounded (relatively compact) sets.
2. Relative boundedness is equivalent to the existence of a constant M such that $\|H'\psi\| \leq M(\|H_0\psi\| + \|\psi\|)$ for all $\psi \in D(H_0)$. Relative compactness in fact implies relative ε boundedness, which means that for all $\varepsilon > 0$ there exists an M such that $\|H'\psi\| \leq \varepsilon\|H_0\psi\| + M\|\psi\|$ for all $\psi \in D(H_0)$. As the Kato–Rellich theorem (2.5.14) showed, this implies that $(H\alpha) = H_0 + \alpha H'$ is self-adjoint on $D(H_0)$. Moreover, the H_0 and $H(\alpha)$ norms are then equivalent, so H' is also relatively compact with respect to all $H(\alpha)$.
3. If c is nonzero and outside the spectrum of H_0, then for all $\chi \in \mathcal{H}$,

$$\begin{aligned} \min\left\{1, \frac{1}{|c|}\right\} \|\chi\| &\leq \min\{1, |c|\} \left(\left\| \frac{H_0}{H_0 - c}\chi \right\| \frac{1}{|c|} + \left\| \frac{1}{H_0 - c}\chi \right\| \right) \\ &\leq \left\| \frac{1}{H_0 - c}\chi \right\|_{H_0} = \left\| \frac{H_0}{H_0 - c}\chi \right\| + \left\| \frac{1}{H_0 - c}\chi \right\| \\ &\leq [1 + (1 + |c|)\|(H_0 - c)^{-1}\|]\|\chi\|, \end{aligned}$$

which implies that the mapping $\mathcal{H} \to D_\Gamma(H_0) : \chi \to (H_0 - c)^{-1}\chi$ is continuous in both directions, and thus an isomorphism of these Hilbert spaces. Hence boundedness (respectively, compactness) of the mapping $H'(H_0 - c)^{-1}$:

$$\mathcal{H} \xrightarrow[\text{homeomorphic}]{(H_0-c)^{-1}} D_\Gamma(H_0) \xrightarrow[\text{continuous (resp., compact)}]{H'} \mathcal{H}$$

is equivalent to the relative boundedness (compactness) of H'. It must similarly be possible to extend the adjoint $(H_0 - c^*)^{-1}H'$ from $D(H')$ to a bounded operator on $\mathcal{H}$.

3.4.3 Examples

1. If two operators f and g commute, and thus have a common spectral representation on $\mathcal{H} = \bigoplus_i \mathcal{H}_i$, then g is bounded relative to f if there exists an M and $c > 0$ such that $|g_i(\alpha)| < M|f_i(\alpha)| + c$ for all i and α, where g_i and f_i are the multiplication operators on $\mathcal{H}_i$.

2. Let H_0 be the Hamiltonian of free motion (3.3.3) and H' be a multiplication operator $V(x) \in L^2(\mathbb{R}^3, d^3x)$. If the calculation is done in x space, then the Hilbert-Schmidt norm of $V(H_0 - c)^{-1}$ is

$$\begin{aligned} \|V(H_0 - c)^{-1}\|_2^2 &= \mathrm{Tr}(H_0 - c)^{-1} V^2 (H_0 - c)^{-1} \\ &= \int d^3\mathbf{x} V^2(\mathbf{x}) \int \frac{d^3p}{(|\mathbf{p}|^2 - c)^2} < \infty \quad \text{for all } -c \in \mathbb{R}^+. \end{aligned}$$

The operator $V(H - c)^{-1}$ is therefore in C_2 (see (2.3.21)) and consequently compact. In fact any potential V that falls off faster than $r^{-\varepsilon}$, $\varepsilon > 0$, at infinity and is not too singular at finite $\mathbf{x}$ is compact relative to H_0 (Problem 2). Roughly speaking, compact operators fall off in all directions in phase space, in both p and q.

3.4.4 Theorem

Let V be compact relative to H and bounded, and let P_{ac} be the projection onto the absolutely continuous spectrum of H. Then $V_t P_{ac}$ approaches zero strongly as $t \to \pm\infty$, where $V_t = \exp(iHt) V \exp(-iHt)$.

Proof: Let $c \notin \mathrm{Sp}(H)$, so that for any φ we can write $P_{ac}\varphi = (H - c)^{-1} P_{ac}\psi$. Then

$$\|V_t P_{ac}\varphi\| = \|V(H - c)^{-1} \exp(-iHt) P_{ac}\psi\|.$$

It was shown earlier that $\exp(iHt)P_{ac} \rightharpoonup 0$, and since $V(H - c)^{-1}$ is a compact operator it sends a weakly convergent sequence into a strongly convergent one. □

3.4.5 Corollaries

1. Functions that fall off as $r^{-\varepsilon}$ when $r \to \infty$ converge strongly to zero under free time-evolution.
2. Because of the resolvent equation

$$(H_0 + V - z)^{-1} = (H_0 - z)^{-1}(1 - V(H_0 + V - z)^{-1}), \qquad z \notin \mathbb{R},$$

any F that is compact relative to H_0 is also compact relative to $H_0 + V$ provided that V is relatively bounded by $H_0 + V$, which is always the case if V falls off as $r^{-\varepsilon}$ and is thus compact relative to H_0. The time-evolution with such potentials thus makes $F(t)P_{ac} \to 0$, where F is the characteristic function of any finite region in configuration space. This can be interpreted as meaning that the probability that a particle remains in the set given by F vanishes at large times: $\langle\psi_t|F\psi_t\rangle = \|F(t)\psi\|^2 \to 0$ for all $\psi \in P_{ac}\mathcal{H}$, $t \to \pm\infty$. In other words, the particle runs off to infinity. This distinguishes the absolutely continuous spectrum from the singular continuous spectrum; with the latter a particle may keep returning to near the origin again and again.

Now that the connection with classical physics has been looked into, let us proceed to find the quantum-mechanical analogies of the concepts of (I: §3.4).

3.4.6 Definition

The algebra $\mathcal{A}$ of the **asymptotic constants** is the set of operators a for which the strong limits

$$a_\pm = \lim_{t\to\pm\infty} \exp(iHt)a\exp(-iHt)$$

exists. The limits themselves form the algebras of $\mathcal{A}_\pm$, and we define $\tau_\pm$ as the (surjective) homomorphisms $\mathcal{A} \to \mathcal{A}_\pm : \tau_\pm(a) = a_\pm$.

3.4.7 Remarks

1. Since the product is not even sequentially continuous in the weak operator topology, the limit must be supposed to exist at least in the strong sense, in order that $\mathcal{A}$ and $\mathcal{A}_\pm$ be algebras and that $\tau_\pm$ be a homomorphism between them. Norm convergence is too much to ask for, as it would contradict the group structure of the time-evolution. If a_t were a Cauchy sequence in the norm $\| \ \|$, then for all ε there would exist a T such that

$$\|a_{t_1} - a_{t_2}\| = \|a - \exp(i(t_2 - t_1)H)a\exp(-i(t_2 - t_1)H)\| \le \varepsilon \quad \text{for all } t_1, t_2 > T,$$

and this is possible only if a_t is a constant.

2. It is immediately clear that $\mathcal{A} \supset \{H\}'$, and since

$$a_\pm = s\text{-}\lim_{t\to\pm\infty} \exp(i(t+\tau)H)a\exp(-i(t+\tau)H) = \exp(i\tau H)a_\pm\exp(-i\tau H) \quad \text{for all } \tau \in \mathbb{R},$$

$\mathcal{A}_\pm \subset \{H\}'$. Since furthermore, $\tau_{\pm|\{H\}'} = \mathbf{1}$, it follows that $\mathcal{A}_\pm = \{H\}' \subset \mathcal{A}$ and $\tau_\pm$ are endomorphisms.

3. As explained above, nothing converges on $\mathcal{H}_p$; if P_p is the projection onto this subspace, then $P_p a P_p$ belongs to $\mathcal{A}$ only if it is in $P_p\{H\}'P_p$.
4. If particles escape to infinity, then their momenta p ought to become nearly constant when they are far from any interaction. Consequently, a good candidate for an operator of $\mathcal{A}$ that is not in $\{H\}'$ would be $(\mathbf{1} - P_p)p(\mathbf{1} - P_p)$, or, better, some bounded function of p rather than p itself.

If the time-evolution becomes asymptotically equal to that of H_0, then it is a reasonable expectation that

$$\Omega_\pm = \lim_{t\to\pm\infty} \Omega(t) \equiv \lim_{t\to\pm\infty} \exp(iHt)\exp(-iH_0t)$$

exists. This raises the question of

3.4.8 Topologies in Which the Limit $\lim_{t\to\pm\infty}\Omega(t)$ Might Exist

1. **Norm convergence**. As remarked in (3.4.7; 1), there is no possibility of this kind of convergence, since

$$\|\Omega(t_1)-\Omega(t_2)\| = \|\exp(iH(t_1-t_2))-\exp(iH_0(t_1-t_2))\| < \varepsilon \qquad \text{for all } t_1, t_2 > T$$

implies that $H = H_0$. Physically, this means that without reference to a particular state, the times $\pm\infty$ are no better than any other times.

2. **Strong convergence**. This allows the possibility that the limit $\Omega_\pm$ of the unitary operators $\Omega(t)$ may not be unitary, since the equation $\Omega(t)\Omega^*(t) = \mathbf{1}$ is not necessarily preserved in the limit: As the mapping $a \to a^*$ is only weakly continuous, strong convergence of the Ω implies only weak convergence for the Ω^*. A product sequence $a_n b_n$ converges weakly to ab if $a_n \rightharpoonup a$ and $b_n \to b$. However, no statement can be made about the existence or value of the limit of $b_n a_n$. The following example on l^2 is illustrative of the different kinds of convergence:

$$\Omega_n = \begin{bmatrix} \overbrace{\ \ & & & & & 1}^{n} & & & \\ 1 & & & & & & & & \\ & 1 & & & & & & & \\ & & 1 & & & & & & \\ & & & 1 & & & & & \\ & & & & 1 & & & & \\ & & & & & & 1 & & \\ & & & & & & & 1 & \\ & & & & & & & & 1 \end{bmatrix} \to \Omega = \begin{bmatrix} & & & & & & & \\ 1 & & & & & & & \\ & 1 & & & & & & \\ & & 1 & & & & & \\ & & & 1 & & & & \\ & & & & 1 & & & \\ & & & & & 1 & & \\ & & & & & & 1 & \end{bmatrix}$$

$$\Omega_n^* = \begin{bmatrix} & 1 & & & & & & & \\ & & 1 & & & & & & \\ & & & 1 & & & & & \\ & & & & 1 & & & & \\ & & & & & 1 & & & \\ 1 & & & & & & & & \\ & & & & & & 1 & & \\ & & & & & & & 1 & \\ & & & & & & & & 1 \end{bmatrix} \rightharpoonup \Omega^* = \begin{bmatrix} & 1 & & & & \\ & & 1 & & & \\ & & & 1 & & \\ & & & & 1 & \\ & & & & & 1 \\ & & & & & \\ & & & & & \end{bmatrix}$$

The operator Ω_n^* converges only weakly, since $v_n \equiv \Omega_n^*(1,0,0,\ldots) = (\overbrace{0,0,\ldots,1}^{n},0,0,\ldots) \rightharpoonup 0$, while $\|v_n\| = 1$ for all n, so $v_n \not\to 0$. In this case we have $\mathbf{1} = \Omega_n^*\Omega_n \Rightarrow \Omega^*\Omega = \mathbf{1}$, but $\mathbf{1} = \Omega_n\Omega_n^* \not\rightharpoonup \Omega\Omega^* \neq \mathbf{1}$. The situation is the same for $\Omega_\pm$, since they are strong limits of unitary operators. Although $\Omega_\pm^*\Omega_\pm = \mathbf{1}$, since

$$\langle x|\Omega_\pm^*\Omega_\pm x\rangle = \|\Omega_\pm x\|^2 = \lim_{t\to\pm\infty}\|\Omega(t)x\|^2 = \|x\|^2 \quad \text{for all } x \in \mathcal{H},$$

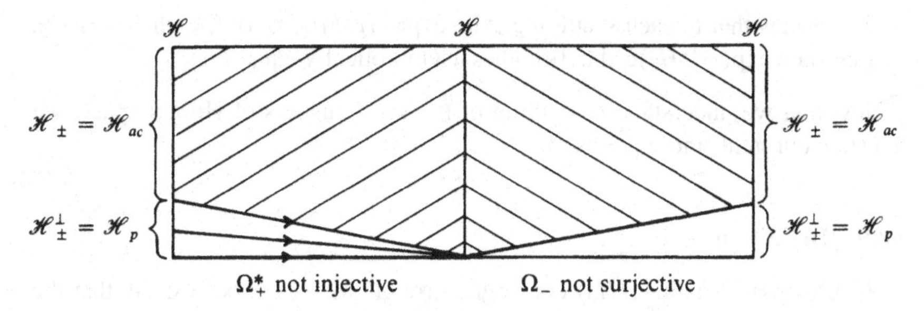

Figure 3.2. The domains and ranges of Ω^* and Ω.

this only tells us that $\Omega_\pm\Omega_\pm^* = \Omega_\pm\Omega_\pm^*\Omega_\pm\Omega_\pm^*$, i.e., that $\Omega_\pm\Omega_\pm^*$ is a projection. It projects onto a subspace $\mathcal{H}_\pm \subset \mathcal{H}$ on which $\mathcal{H}$ gets mapped unitarily by Ω, and which $\Omega_\pm^*$ maps back onto $\mathcal{H}$ (see Figure 3.2). These operators are related by $\Omega_{\pm|\mathcal{H}_\pm}^* = \Omega_{\pm|\mathcal{H}_\pm}^{-1}$. In analogy with (3.4.7; 2), $\Omega_\pm^* \exp(i\tau H)\Omega_\pm = \exp(i\tau H_0)$ for all $\tau \in \mathbb{R}$, so H acting on $\mathcal{H}_\pm$ is unitarily equivalent to H_0 on $\mathcal{H}$. Therefore the spectrum of H_0 is the same as a part of the spectrum of H. If H_0 has only an absolutely continuous spectrum ($H_0 : f(h) \to hf(h)$ on any summand of the spectral representation), and the eigenvectors of H are called $|E_i\rangle$, then

$$\langle f|\exp(itH_0)\exp(-itH)|E_i\rangle = \int d\mu(h)\exp(it(h - E_i))\langle f(h)|E_i\rangle \to 0$$

$$\text{as } t \to \pm\infty$$

by the Riemann–Lebesgue lemma. The bound states are thus in the kernel of $\Omega_\pm^*$. The strong convergence can be restated as: to every state $\varphi \in \mathcal{H}$ that evolves according to H_0, there exists a "scattering state" $\psi_\pm = \Omega_\pm\varphi$ such that the two states approach each other asymptotically:

$$\begin{aligned}\lim_{t\to\pm\infty} &\|\exp(iHt)\exp(-iH_0t)\varphi - \psi_\pm\| \\ &= \lim_{t\to\pm\infty} \|\exp(-iH_0t)\varphi - \exp(-iHt)\psi_\pm\| = 0.\end{aligned}$$

3. **Weak convergence**. The norm $\|a\| = \sup_{\|x\|=\|y\|=1} |\langle x|ay\rangle|$, as the supremum of weakly continuous functions, is weakly lower semicontinuous, so in any event we know that $\|\Omega_\pm\| \le 1$. Since the unitary operators are weakly dense in the unit ball, this is apparently the most that can be said. Weak convergence is not a very powerful property. The limit could simply be zero, as happens for $\exp(itH_0)$. If $\Omega(t)$ converges weakly but not strongly, then $\Omega_\pm$ cannot be unitary.

4. **Convergence of** Ω^*. If $\Omega(t) \to \Omega_\pm$, then we know that $\Omega^*(t)$ converges weakly to $\Omega_\pm^*$, and that $\Omega_\pm^*$ maps $\mathcal{H}_\pm$ unitarily onto $\mathcal{H}$ and sends everything else to 0. Since the weak topology and the strong topology are the same for the unitary operators (2.1.28; 5), $\Omega^*(t)$ converges strongly to $\Omega_\pm^*$ on $\mathcal{H}_\pm$, and converges weakly to 0 on its orthogonal complement. Strong convergence on

$\mathcal{H}_\pm$ means that to each scattering state $\exp(-itH)\psi$, $\psi \in \mathcal{H}_\pm$, there exists a free state $\exp(-itH_0)\varphi$ that becomes asymptotically equal to it.

Now that we understand something of the insidiousness of Hilbert space, we can state our goals more precisely.

3.4.9 Definition

(i) If $\exp(itH)\exp(-itH_0)$ converges strongly as $t \to \pm\infty$, we say that the **Møller wave operators** $\Omega_\pm = \lim_{t\to\pm\infty} \exp(itH)\exp(-itH_0)$ **exist**.

(ii) If $\mathcal{H}_\pm \equiv \Omega_\pm \mathcal{H} = \mathcal{H}_p^\perp$, then $\Omega_\pm$ are said to be **asymptotically complete**.

3.4.10 Remarks

1. The meaning of asymptotic completeness is that other than the bound states $\mathcal{H}_p$, every state approaches a free state as $t \to \pm\infty$. A simple classical example where this fails to be true has been provided by S. Sokolov: an otherwise free particle with an effective mass $\mathcal{M}(x) = \coth^2 x$, $H = p^2/\mathcal{M}(x)$, and $H_0 = p^2$. All incoming trajectories have a dead end at the origin, and the set of scattering trajectories is empty (cf. [4]). Pearson [4] has constructed a potential for which the analogous thing happens in quantum mechanics.
2. Invariance under reversal of the motion (3.3.17) does not suffice to guarantee that $\mathcal{H}_+ = \mathcal{H}_-$. We shall soon encounter many-channel systems for which $\mathcal{H}_+ \neq \mathcal{H}_-$.
3. Since

$$\frac{d}{dt}\Omega(t) = \exp(iHt)i(H - H_0)\exp(-iH_0t) = i\Omega(t)H_1(t),$$

$$H_1(t) \equiv \exp(iH_0t)(H - H_0)\exp(-iH_0t),$$

Ω_+ can also be written formally as

$$\Omega_+ = T\exp\int_0^\infty dt\, iH_1(t),$$

which, however, does not answer the question of the existence of the infinite integral.

3.4.11 Sufficient Conditions for the Existence and Completeness of $\Omega_\pm$

Let $H = H_0 + V$, $\sqrt{V} \equiv V/|V|^{1/2}$, $D(H) = D(H_0)$, and let χ_I be the characteristic function of an interval $I \subset \sigma(H_0)$. If

$$\sup_{\omega\in I}(\|\sqrt{V}\delta(H-\omega)\sqrt{V}\| + \|\sqrt{V}\delta(H_0-\omega)\sqrt{V}\|) < \infty,$$

then

$$\chi_I(H)\exp(iHt)\exp(-iH_0t)\chi_I(H_0) \quad \textit{and} \quad \chi_I(H_0)\exp(iH_0)\exp(-iHt)\chi_I(H)$$

converge strongly as $t \to \pm\infty$.

3.4.12 Remarks

1. In the case of a single channel, the spectrum normally has the properties

$$\sigma_{ac}(H_0) = \mathbb{R}^+ = \sigma_{ac}(H), \sigma_p(H_0) = \sigma_s(H_0) = \sigma_s(H) = \emptyset, \sigma_p(H) \subset \mathbb{R}^-.$$

It is convenient for technical reasons to use the projections χ_I to exclude the particles that move too slowly or too rapidly, by letting $I = (\varepsilon, 1/\varepsilon)$. If the supremum over $\omega \in I$ is finite for all $\varepsilon > 0$, then it follows that $\chi_I(H)\exp(itH)\exp(-itH_0)$ converges on a dense set in Hilbert space, and consequently on the whole space. Since $f(H)\Omega = \Omega f(H_0)$, we expect that

$$\chi_{I'}(H)\exp(iHt)\exp(-iH_0t)\chi_I(H_0) \to 0 \text{ for all } I' \cap I = \emptyset$$

and this is indeed verified in Problem 3. This equation shows that

$$\chi_I(H)\exp(itH)\exp(-itH_0)\chi_I(H_0)$$

has the same limit as

$$\exp(itH)\exp(-itH_0)\chi_I(H_0),$$

so (3.4.10) in fact implies what is required in Definition (3.4.9), viz., that $\exp(itH)\exp(-itH_0)$ converges on a dense set.

2. The operator $\sqrt{V}\delta(H-\omega)\sqrt{V}$ is to be interpreted as

$$\lim_{\varepsilon\to 0}\frac{1}{2\pi i}\sqrt{V}\left(\frac{1}{H-\omega-i\varepsilon}-\frac{1}{H-\omega+i\varepsilon}\right)\sqrt{V},$$

and this limit may exist even though $(H-z)^{-1}$ does not exist on the real axis. We shall soon discover that even compactness may survive the limit as $\varepsilon \downarrow 0$.

Proof: Let us make the abbreviations $\psi_I \equiv \chi_I(H_0)\chi$, $\varphi_I \equiv \chi_I(H)\varphi$. Then

$$\begin{aligned}
&\|\chi_I(H)(\exp(iHt_1)\exp(-itH_0t_1) - \exp(iHt_2)\exp(-iH_0t_2))\chi_I(H_0)\psi\| \\
&= \left\|\chi_I(H)\int_{t_1}^{t_2} dt\, \exp(iHt)V\exp(-iH_0t)\chi_I\right\| \\
&= \sup_{\|\varphi\|=1}\left|\int_{t_1}^{t_2} dt\langle\varphi|\chi_I(H)\exp(iHt)V\exp(-iH_0t)\psi_I\rangle\right| \\
&\le \sup_{\|\varphi\|=1}\int_{t_1}^{t_2} dt\|\sqrt{V}\exp(-iHt)\varphi_I\|\cdot\|\sqrt{V}\exp(-iH_0t)\psi_I\| \\
&\le \sup_{\|\varphi\|=1}\left[\int_{t_1}^{t_2} dt\|\sqrt{V}\exp(-iHt)\varphi_I\|^2\cdot\int_{t_1}^{t_2} dt\|\sqrt{V}\exp(-iH_0t)\psi_I\|^2\right]^{1/2}.
\end{aligned}$$

To show convergence, it thus suffices to show that as $t \to \pm\infty$, V_t converges in mean-square to zero (cf. (3.4.4)), whether it evolves in time according to H_0 or to H. To perform the time-integration, we use the generalization of Parseval's formula,

$$\int_{-\infty}^{\infty} dt \|f(t)\|^2 = \int_{-\infty}^{\infty} \frac{d\omega}{2\pi} \|\tilde{f}(\omega)\|^2,$$

for vectors in Hilbert space:

$$\int_{-\infty}^{\infty} dt \|\sqrt{V} \exp(-iHt)\varphi_I\|^2 = 2\pi \int_{-\infty}^{\infty} d\omega \|\sqrt{V}\delta(H-\omega)\varphi_I\|^2.$$

Now note that for any positive operators a and b, $\|\sqrt{b}a\varphi\|^2 \le \|\sqrt{b}\sqrt{a}\|^2 \times \|\sqrt{a}\varphi\|^2 = \|\sqrt{b}a\sqrt{b}\| \cdot \langle\varphi|a\varphi\rangle$, so:

$$\begin{aligned}
&2\pi \int_{-\infty}^{\infty} d\omega \|\sqrt{V}\delta(H-\omega)\varphi_I\|^2 \\
&\le 2\pi \int_{-\infty}^{\infty} d\omega \|\sqrt{V}\delta(H-\omega)\chi_I(H)\sqrt{V}\| \cdot \langle\varphi_I|\delta(H-\omega)\varphi_I\rangle \\
&\le 2\pi \sup_{\omega\in I} \|\sqrt{V}\delta(H-\omega)\sqrt{V}\| \int_{\infty}^{\infty} d\omega \langle\varphi_I|\delta(H-\omega)\varphi_I\rangle \\
&= 2\pi \sup_{\omega\in I} \|\sqrt{V}\delta(H-\omega)\sqrt{V}\| \|\varphi_I\|^2.
\end{aligned}$$

These relationships are still valid with H replaced by H_0, and show that the integral $\int_{t_1}^{\infty}$ gets arbitrarily small as $t_1 \to \infty$, since the integral $\int_{-\infty}^{\infty}$ exists. This implies the strong convergence of $\Omega(t)$ as $t \to \pm\infty$. □

3.4.13 Examples

1. A separable potential. Let $H_0 = |\mathbf{p}|^2$, $(V\varphi)(\mathbf{x}) = \lambda\rho(\mathbf{x}) \int d^3x' \rho^*(\mathbf{x}')\varphi(\mathbf{x}')$, in which $\int d^3x |\rho(\mathbf{x})|^2 = 1$, $\int d^3x d^3x' \rho(\mathbf{x})\rho^*(\mathbf{x}')/|\mathbf{x}-\mathbf{x}'| = M < \infty$, $\inf_{|\mathbf{p}|^2\in I} |\tilde{\rho}(\mathbf{p})| > 0$. Since $P = V/\lambda$ is a one-dimensional projection,

$$\begin{aligned}
(H-z)^{-1} &= (H_0-z)^{-1} - \lambda(H_0-z)^{-1}P(H_0-z)^{-1}D^{-1}(z), \\
D(z) &= 1 + \lambda \operatorname{Tr} P(H_0-z)^{-1}P
\end{aligned}$$

and

$$P(H-z)^{-1}P = P(H_0-z)^{-1}PD^{-1}(z).$$

By assumption,

$$\begin{aligned}
\operatorname{Tr} P(H_0-z)^{-1}P &= \int d^3p |\tilde{\rho}(\mathbf{p})|^2 (|\mathbf{p}|^2 - z)^{-1} \\
&= \int d^3x d^3x' \rho(\mathbf{x})\rho^*(\mathbf{x}') \frac{\exp(i\sqrt{z}|\mathbf{x}-\mathbf{x}'|)}{|\mathbf{x}-\mathbf{x}'|}
\end{aligned}$$

remains bounded by M for all $\sqrt{z}$. In addition, for all $y > 0$,

$$\begin{aligned}\lambda^{-1} \operatorname{Im} D(x+iy) &= \int d^3 p |\tilde{\rho}(\mathbf{p})|^2 \frac{y}{(|\mathbf{p}|^2 - x)^2 + y^2} \\ &\geq \inf_{|\mathbf{p}|^2 \in I} |\tilde{\rho}(\mathbf{p})|^2 \int_{|\mathbf{p}|^2 \in I} d^3 p \frac{y}{(|\mathbf{p}|^2 - x)^2 + y^2}\end{aligned}$$

is bounded below, uniformly in $x \in I$. Then

$$\sup_{x \in I} |D^{-1}(x+iy)| \leq \sup_{x \in I} \frac{|D(x)|}{|\operatorname{Im} D(z)|^2}$$

is also finite in the limit $y \to 0$, and

$$\lim_{y \to 0} \sup_{x \in I} \|\sqrt{V}(H-z)^{-1}\sqrt{V}\| < \infty.$$

2. Potentials $r^{-1-\varepsilon}$, $0 < \varepsilon < 1$. In momentum space,

$$\widetilde{r^{-\gamma}} = \int d^3 x \exp(i\mathbf{k} \cdot \mathbf{x}) \mathbf{r}^{-\gamma} = |\mathbf{k}|^{-3+\gamma} 4\pi \Gamma(2-\gamma) \sin(2-\gamma)\pi.$$

Consequently,

$$\begin{aligned}\operatorname{Tr}(\sqrt{V}\delta(H_0 - \omega)\sqrt{V})^n &= c(\varepsilon) \int \prod_{i=1}^{n} d^3 p_i \delta(|\mathbf{p}_i|^2 - \omega) |\mathbf{p}_i - \mathbf{p}_{i+1}|^{-2+\varepsilon} \\ &= \omega^{-n(1-\varepsilon)/2} \int \prod_{i=1}^{n} d\Omega_i |\mathbf{n}_i - \mathbf{n}_{i+1}|^{-2+\varepsilon},\end{aligned}$$

where by convention $\mathbf{p}_{n+1} = \mathbf{p}_1$, and $d\Omega_i$ stands for the solid angle element in the direction of the unit vector $\mathbf{n}_i \equiv \mathbf{p}_i/|\mathbf{p}_i|$. Now,

$$|\mathbf{n}_i - \mathbf{n}_{i+1}|^2 = 2(1 - \cos\theta_i), \qquad \theta_i = \measuredangle(\mathbf{n}_i, \mathbf{n}_{i+1}),$$

and

$$\int \prod_{i=1}^{n} d\Omega_i |\mathbf{n}_i - \mathbf{n}_{i+1}|^{-2+\varepsilon}$$

is smaller than $c(\varepsilon) < \infty$ for $n > 2/\varepsilon$ (Problem 5). Since $\| \; \| \leq \| \; \|_n$,

$$\sup_{\omega \in I} \|\sqrt{V}\delta(H_0 - \omega)\sqrt{V}\| \leq \sup_{\omega \in I} \omega^{-(1/2)+(\varepsilon/2)} c(\varepsilon) < \infty,$$

where we have taken I as a compact interval $\subset \mathbb{R}^+$. The Hölder continuity of $\sqrt{V}\delta(H_0 - \omega)\sqrt{V}$ in the norm $\| \; \|_n$ implies that the operator $\sqrt{V}(H_0 - x - iy)\sqrt{V}$ remains compact in the limit $y \to 0$ (see Problem 5). If from

$$\begin{aligned}&\sqrt{V}(H-z)^{-1}\sqrt{V} \\ &= \sqrt{V}(H_0 - z)^{-1}\sqrt{V} - \sqrt{V}(H-z)^{-1}V(H_0 - z)^{-1}\sqrt{V}\end{aligned}$$

we reason that

$$\sqrt{V}(H-z)^{-1}\sqrt{V} = \sqrt{V}(H_0 - z)^{-1}\sqrt{V}(\mathbf{1} + |V|^{1/2}(H_0 - z)^{-1}\sqrt{V})^{-1},$$

then we see that the operators $\sqrt{V}(H-z)^{-1}\sqrt{V}$ and $\sqrt{V}(H_0 - z)^{-1}\sqrt{V}$ differ only by the factor $(\mathbf{1}+|V|^{1/2}(H_0-z)^{-1}\sqrt{V})^{-1}$. Since

$$|V|^{1/2}(H_0-z)^{-1}\sqrt{V}$$

is compact and thus has a pure point spectrum with complex eigenvalues $\kappa_i(z)$ the only possible accumulation point of which is zero,

$$\|(\mathbf{1}+|V|^{1/2}(H_0-z)^{-1}\sqrt{V})^{-1}\| \leq \sup_i |(1-\kappa_i(z))^{-1}|.$$

The functions $z \to \kappa_i(z)$ are continuous, $z \to |V|^{1/2}(H_0-z)^{-1}\sqrt{V}$ is norm-analytic in $\mathbb{C}\backslash\mathbb{R}$, and it can be continued to $I \subset \mathbb{R}$. If the eigenfunctions decrease sufficiently fast at infinity, the values z_{ij} for which κ_i equals 1 are eigenvalues of H, because $|V|^{1/2}(H_0-z)^{-1}|V|^{1/2}\psi=0$ implies that $(H_0+V-z)|V|^{-1/2}\psi=0$. Hence, if $|V|^{-1/2}\psi \in L^2$, then z is an eigenvalue of H and for $\operatorname{Im} z \neq 0$ $\kappa_i(z)$ cannot be 1. A separate argument is necessary to exclude the values $\kappa_i=1$ for $z \in \mathbb{R}^+$. Thus if I is any compact set in $(0,\infty)\backslash\{z_{ij}\}$, (3.4.11) is satisfied.

3.4.14 Remarks

1. The analysis has been restricted to $\varepsilon<1$ so that the singularity at $r=0$ could not destroy the relative compactness of V. Since existence of Ω depends on the falling off of the potential as $r\to\infty$, it is clear that it exists for all V falling off faster that $1/r$, so long as the finite singularities are not strong enough to wreck the self-adjointness [4].
2. If $\varepsilon=0$, then $V_t^{1/2}\sim(pt)^{-1/2}$, which is not square-integrable in t, even if $p=0$ is excluded. This is to be expected, because $\Omega_\pm$ also fails to exist classically for the $1/r$ potential (I: 4.2.18; 2).
3. If $\varepsilon=1$, then the bound of $\|\sqrt{V}\delta(H_0-\omega)\sqrt{V}\|$ is independent of ω, so the supremum over ω in all $\mathbb{R}$ would be finite. This may seem surprising, since even in classical scattering theory the point $p=0$ has to be removed, since particles with $p=0$ never escape. In quantum mechanics, the diffusion of the wave-packets is enough to make $\langle r_t^{-2}\rangle$ square-integrable: With free time-evolution, $\Delta x_t^2\sim\Delta x_0^2+t^2/(\Delta x_0)^2$ (cf. (3.3.5; 1), and
$$\int_{-\infty}^{\infty} dt\langle r_t^{-\gamma}\rangle \sim \int_{-\infty}^{\infty} dt\left(\Delta x_0^2+\frac{t^2}{\Delta x_0^2}\right)^{-\gamma/2} \sim \Delta x_0^{2-\gamma}.$$
If $\gamma=2$, this is independent of Δx_0, so the bound ought to be independent of φ without the necessity of projecting out a neighborhood of $p=0$.
4. If there are bound states imbedded in the continuum, then $\exp(itH_0)\times\exp(-itH)$ cannot converge strongly on them, and it is necessary to project them out with a χ_I. With a potential $r^{-\gamma}, 0<\gamma<2$, then by the virial theorem to be proved in §4.1 they do not occur. This theorem states that an eigenvalue of the energy equals $(\gamma-2)/\gamma$ times the expectation value of the kinetic energy in the corresponding eigenstate. Since the latter quantity is positive, all eigenvalues are negative. If a potential oscillates, then Bragg reflection of waves can produce a bound state, even if in classical mechanics it would be energetically

possible for a particle to escape. For example, the function

$$\psi(r) = \frac{\sin r}{a + r - \frac{1}{2}\sin 2r} \in L^2((0,\infty), dr), \qquad a > 0,$$

satisfies the equation

$$\left(-\frac{d^2}{dr^2} + V(r) - 1\right)\psi(r) = 0,$$

$$V(r) = \frac{8\sin r}{(a+r-\frac{1}{2}\sin 2r)^2}(\sin r - (a+r)\cos r),$$

and it is thus an eigenfunction with eigenvalue $E = 1$ of a potential $V, |V(r)| < \varepsilon \min(1, 1/r)$, where ε can be taken arbitrarily small as a tends to $+\infty$. It can be shown that potentials that approach zero faster than $1/r$ as $r \to \infty$ have no positive eigenvalues ([3], §XIII).

3.4.15 Many-Particle Scattering

Different groupings are possible in a many-particle system as the particles go off to infinity, some remaining bound together while others get ever farther away from them. Formally, the Schrödinger equation for N particles acts on a $3N$-dimensional configuration space, and different ways of apportioning the particles into clusters correspond to different regions in $\mathbb{R}^{3N}$. A given distribution of $1, 2, \ldots, N$ into disjoint subsets, such as $(1,2), (3), (4,5,6), \ldots$, will be shorthand for the statement that particles 1 and 2 remain bound together, 3 approaches infinity by itself, 4 through 6 are bound, etc. Such a distribution is referred to as a **channel**. Pair potentials $V_{ij}(\mathbf{x}_i - \mathbf{x}_j)$ do not fall off in the directions where $\mathbf{x}_i$ and $\mathbf{x}_j$ are nearly equal, so the asymptotic time-evolution is not described by a unique Hamiltonian, but instead depends on the channel, i.e., on the direction in which the state goes to infinity. If we index the channels with a subscript α, then the interaction I_α between separate clusters goes to zero as $t \to \pm\infty$, and the time-evolution approaches that of $H_\alpha \equiv H - I_\alpha$.

3.4.16 Example

Consider three particles, and suppose either that the center of mass has been separated out or simply that one of the particles is infinitely heavy, as an approximation to a nucleus K and two electrons e_1 and e_2. Then the appropriate configuration space specifies the relative motion of the two coordinates $\mathbf{x}_1$ and $\mathbf{x}_2$ of the electrons, and there are four channels:

$(K)(e_1)(e_2)$. In this channel all the particles separate, and

$$H = \frac{|\mathbf{p}_1|^2}{2m_1} + \frac{|\mathbf{p}_2|^2}{2m_2} + V_1(\mathbf{x}_1) + V_2(\mathbf{x}_2) + V_{12}(\mathbf{x}_1 - \mathbf{x}_2)$$

breaks up into H_0 and $I_0 = V_1 + V_2 + V_{12}$.

$(K, e_1)(e_2)$. Particle 1 remains bound to the nucleus, while particle 2 escapes:

$$H_1 = \frac{|\mathbf{p}_1|^2}{2m_1} + \frac{|\mathbf{p}_2|^2}{2m_2} + V_1(\mathbf{x}_1), \qquad I_1 = V_2 + V_{12}.$$

$(K, e_2)(e_1)$. The same, with particles 1 and 2 switched.

$(K)(e_1, e_2)$. In this channel particles 1 and 2 remain bound together, which is of course impossible for electrons, but would be realistic in the scattering of a positron from a hydrogen atom. In this case,

$$H_{12} = \frac{|\mathbf{p}_1|^2}{2m_1} + \frac{|\mathbf{p}_2|^2}{2m_2} + V_{12}(\mathbf{x}_1 - \mathbf{x}_2), \qquad I_{12} = V_1 + V_2.$$

Once again, the existence of the Møller operators means that for each φ_α in which the clusters corresponding to a channel α are bound, and which evolves in time by $\exp(-itH_\alpha)$, there is a state ψ_α evolving by $\exp(-itH)$ and asymptotically approaching φ_α:

$$\| \exp(-itHt)\psi_\alpha - \exp(-itH_\alpha t)\varphi_\alpha \| \to 0.$$

Completeness of the Møller operators means that $\mathcal{H}_{ac}(H)$ is spanned by such ψ_α's.

3.4.17 The Møller Wave Operators for Many-Particle Scattering

If P_α is the projection onto the part of $\mathcal{H}_{ac}(H_\alpha)$ corresponding to the channel α, then the Møller operators

$$\Omega_{\alpha\pm} = \lim_{t\to\pm\infty} \exp(iHt)\exp(-iH_\alpha t)P_\alpha$$

are said to exist whenever the strong limit exists. In that event, the operators $Q_{\alpha\pm} \equiv \Omega_{\alpha\pm}\Omega^*_{\alpha\pm}$ are projections, and asymptotic completeness means that

$$\sum_\alpha Q_{\alpha\pm}\mathcal{H} = \mathcal{H}_{ac}(H).$$

3.4.18 Remarks

1. The operators P_α can be written as tensor products of the projections onto the bound states within the clusters of the channel and of identity operators in the relative coordinates. For instance, in (3.4.16) the projection P_α for $(K, e_1)(e_2)$ equals $P_p \otimes \mathbf{1}$, where P_p is the projection onto $\mathcal{H}_p(p_1^2/2m + V_1)$. The projections for different α will not generally be orthogonal, since they are related to different, and noncommuting, H_α. Although

$$\exp(itH)\exp(-itH_\alpha)$$

converges on all of $\mathcal{H}$, this limit is not terribly interesting.

2. The equation $\Omega^*\Omega = \mathbf{1}$ of (3.4.8; 2) has the generalization $\Omega^*_{\alpha\pm}\Omega_{\beta\pm} = \delta_{\alpha\beta}P_\alpha$ (Problem 4). As a result, the Q_α are orthogonal for different α:

$$Q_{\alpha\pm}Q_{\beta\pm} = \Omega_{\alpha\pm}\Omega^*_{\alpha\pm}\Omega_{\beta\pm}\Omega^*_{\beta\pm} = \delta_{\alpha\beta}Q_{\alpha\pm}.$$

This is to be expected, since all the Q_α involve the same H and commute with it:

$$\exp(iHt)Q_{\alpha\pm}\exp(-iHt) = \exp(iHt)\Omega_{\alpha\pm}\Omega^*_{\alpha\pm}\exp(-iHt)$$
$$= \Omega_{\alpha\pm}\exp(iH_\alpha t)\exp(-iH_\alpha t)\Omega^*_{\alpha\pm} = Q_{\alpha\pm}.$$

The physical significance of this is that the wave functions $\exp(itH)\psi_\alpha$, $\psi_\alpha = \Omega_{\alpha\pm}\varphi_\alpha$, turn into widely separated clusters after long times, so vectors corresponding to different channels are orthogonal. Since they all evolve according to $\exp(itH)$, this asymptotic orthogonality implies that they are orthogonal at all times.

3. The projections P_α and Q_α are rather unwieldy. $\sum_\alpha P_\alpha \neq P_{ac}$, and it is practically impossible to write Q_α explicitly. That is why it is more convenient to work with operators J_α, which approach P_α under the time-evolution of H_α, in place of the P_α themselves. Then $\Omega_{\alpha\pm}$ can be written as the limit of $\exp(iHt)J_\alpha\exp(-iH_\alpha t)$, since $\exp(iH_\alpha t)J_\alpha\exp(-iH_\alpha t) \to P_\alpha$ implies that $\exp(iHt)J_\alpha\exp(-iH_\alpha t) \to \Omega_{\alpha\pm}$. In Example (3.4.16), electron-hydrogen scattering, a good choice is

$$J_1 = \frac{|\mathbf{x}_1|^4 + |\mathbf{x}_2|^4}{1 + |\mathbf{x}_1|^4 + |\mathbf{x}_2|^4 + |\mathbf{x}_1|^8}, \qquad J_2 = \frac{|\mathbf{x}_1|^4 + |\mathbf{x}_2|^4}{1 + |\mathbf{x}_1|^4 + |\mathbf{x}_2|^4 + |\mathbf{x}_2|^8},$$

$J_{12} = 0$, $J_0 = 1 - J_1 - J_2$. If $\mathbf{x}_1$ remains in a finite region and $\mathbf{x}_2 \to \infty$, then J_1 goes to $\mathbf{1}$ and J_2 goes to 0, and vice versa. J_0 becomes $\mathbf{1}$ only if both particles go to infinity. It will be shown in §4.4 that this heuristic argument can actually be substanciated by a proof of strong convergence.

Criterion (3.4.11) for the existence and completeness of the Møller wave operators does not work for many-body systems, because the pair potentials V_{ij} are not compact relative to H. It is possible to write them as tensor products of a function of $\mathbf{x}_i - \mathbf{x}_j$ and the unit operator in the other coordinates, but a tensor product is compact only if both its factors are compact. This is where the functions J_α introduced above can be of use, for they decrease exactly in the directions in which I_α is constant, making $J_\alpha I_\alpha$ relatively compact and the methods of (3.4.11) applicable. This can be stated as a simple

3.4.19 Criterion for the Existence and Completeness of $\Omega_{\alpha\pm}$

Let J_α be positive operators for which s-lim $\exp(iH_\alpha t)J_\alpha\exp(-iH_\alpha t) = P_\alpha$ and $\sum_\alpha J_\alpha = \mathbf{1}$. If the strong limits of $\exp(iHt)J_\alpha\exp(-iH_\alpha t)$ and

$$\exp(iH_\alpha t)J_\alpha\exp(-iHt)P_{ac}(H)$$

exist at $t \to \pm\infty$, then $\Omega_{\alpha\pm}$ exist and are complete.

Proof: Since, by assumption, $\|(\exp(-iH_\alpha t)P_\alpha - J_\alpha\exp(-iH_\alpha t))\psi\| \to 0$ for all $\psi \in \mathcal{H}$, $\exp(iHt)\exp(-iH_\alpha t)P_\alpha$ converges strongly just as

$$\exp(iHt)J_\alpha\exp(-iH_\alpha t)$$

does and hence the latter operator converges to $\Omega_{\alpha\pm}$. Then, since

$$(\mathbf{1} - P_\alpha)\Omega^*_{\alpha\pm} = 0,$$

$$\begin{aligned} &s\text{-}\lim \exp(iHt)J_\alpha \exp(-iHt)P_{ac}(H) \\ &\quad = s\text{-}\lim \exp(iHt)\exp(-iH_\alpha t)(P_\alpha + (\mathbf{1} - P_\alpha)) \\ &\qquad \times \exp(iH_\alpha t)J_\alpha \exp(-iHt)P_{ac}(H) = Q_\alpha. \end{aligned}$$

Consequently,

$$\sum_\alpha Q_\alpha = s\text{-}\lim \exp(iHt)\sum_\alpha J_\alpha \exp(-iHt)P_{ac}(H) = P_{ac}(H). \qquad \square$$

3.4.20 Example

In the three-body system (3.4.16) let $V_{12} = 0$ and suppose that V_1 and V_2 are potentials such that the one-particle Møller operators ω_1 and ω_2 exist and are complete. The J_α have the following form in the different channels:
0-channel. Under the time-evolution according to

$$H_0 = \frac{|\mathbf{p}_1|^2}{2m_1} + \frac{|\mathbf{p}_2|^2}{2m_2},$$

J_1 and J_2 converge strongly to zero: If $\mathbf{x}_i \to \mathbf{x}_i + \mathbf{p}_i t$, then

$$J_i \to \frac{|\mathbf{p}_1|^4 + |\mathbf{p}_2|^4}{|\mathbf{p}_1|^4 + |\mathbf{p}_2|^4 + t^2|\mathbf{p}_i|^8}, \qquad i = 1, 2,$$

and this in turn approaches zero on the dense set of functions the support of which does not contain $\mathbf{p}_i = \mathbf{0}$. Therefore

$$\begin{aligned} s\text{-}\lim \exp(iHt)J_0 \exp(-iH_0 t) &= s\text{-}\lim \exp(iHt)\exp(-iH_0 t) \\ &\quad \times \exp(iH_0 t)J_0 \exp(-iH_0 t) = \omega_1 \otimes \omega_2 \end{aligned}$$

exists, and

$$\begin{aligned} s\text{-}\lim \exp(iH_0 t)J_0 \exp(-iHt)P_{ac}(H) &= s\text{-}\lim \exp(iH_0 t)J_0 \\ &\quad \times \exp(-iH_0 t)\exp(iH_0 t)\exp(-iHt)P_{ac}(H) = \omega_1^* \otimes \omega_2^*. \end{aligned}$$

1-channel. If particle 1 is bound, then

$$J_1 = \frac{|\mathbf{x}_1|^4 + |\mathbf{x}_2|^4}{1 + |\mathbf{x}_1|^4 + |\mathbf{x}_2|^4 + |\mathbf{x}_1|^8}$$

approaches **1** under the time-evolution by

$$H_1 = \frac{|\mathbf{p}_1|^2}{2m_1} + \frac{|\mathbf{p}_2|^2}{2m_2} + V_1(\mathbf{x}_1),$$

since $\mathbf{x}_2 \to \mathbf{x}_2 + t\mathbf{p}_2$ and $\mathbf{x}_1$ remains finite. If it is not bound, then by assumption the time-evolution becomes free, and $J_1 \exp(-iH_1 t)\psi$ approaches $J_1 \exp(-iH_0 t)\omega_1^* \otimes \mathbf{1}\psi$, which goes to 0. Thus

$$s\text{-}\lim \exp(iH_1 t)J_1 \exp(-iH_1 t) = P_1$$

and

$$s\text{-}\lim \exp(iHt)J_1 \exp(-iH_1 t) = \omega_2 \otimes \mathbf{1} \cdot P_1$$

The 2-channel works just like the 1-channel, and the 1-2 channel is empty. We see that in this trivial case, Criterion (3.4.19) reproduces the earlier results. We shall discuss more interesting examples later.

The operators $\Omega_{\alpha\pm}$ map the motion in channel α, i.e.,

$$a \to \exp(iH_\alpha t)a\exp(-iH_\alpha t),$$

to the actual motion as described by $\exp(iHt)$. Specifically, they produce the homomorphisms $\tau_\pm$ introduced in (3.4.6), and they send $\{H_\alpha\}'$ into $\mathcal{A}_\pm$. All $a \in \{H_\alpha\}'$ projected into channel α are in $\mathcal{A}$,

$$\begin{aligned}\tau_\pm(Q_\alpha a Q_\alpha) &\equiv \lim_{t\to\pm\infty} \exp(iHt)Q_\alpha a Q_\alpha \exp(-iHt)\\ &= \lim_{t\to\pm\infty} \exp(iHt)Q_\alpha \exp(-iH_\alpha t)(P_\alpha + \mathbf{1} - P_\alpha)a\exp(iH_\alpha t)Q_\alpha\\ &\qquad \times \exp(-iHt) = \Omega_{\alpha\pm}a\Omega^*_{\alpha\pm}.\end{aligned} \tag{3.6}$$

The set of constants of motion $\{H_\alpha\}'$ will contain the relative momenta of the individual clusters. They commute, and the vectors of their common spectral representation, denoted $|\alpha, k\rangle$, will be somewhat loosely referred to as eigenvectors of the momenta. The vectors $|\alpha, k\rangle$ are in the image of the projections P_α (3.4.17). The wave operators $\Omega_{\alpha\pm}$ transform $|\alpha, k\rangle$ into the eigenvectors of the asymptotic momenta $|\alpha, k, \pm\rangle \equiv \Omega_{\alpha\pm}|\alpha, k\rangle$ in such a way that, as in Remark (3.4.18; 2),

$$\langle \alpha, k, \pm|\tau_\pm(Q_\alpha a Q_\alpha)|\alpha, k, \pm\rangle = \langle \alpha, k|P_\alpha a P_\alpha|\alpha, k\rangle. \tag{3.7}$$

The states $|\alpha, k, \pm\rangle$ thus mean that the outgoing or, respectively, incoming particles have momenta k, and the transition probability from one such configuration to another can be measured macroscopically. As in classical mechanics (I: 3.4.9), this is the purpose of

3.4.21 Definition

$$S_{\alpha\beta} = \Omega^*_{\alpha+}\Omega_{\beta-}$$

is known as the S **matrix in the interaction picture**, and

$$S = \sum_\alpha \Omega_{\alpha-}\Omega^*_{\alpha+}$$

is the S **matrix in the Heisenberg picture**.

3.4.22 Remarks

1. The definition has been given in the form appropriate for a many-body system. One-particle scattering can be considered as a special case with only one channel.
2. The action of the $\Omega_{\alpha\pm}$ is depicted schematically in Figure 3.3: Since $\Omega_{\alpha+}\Omega^*_{\alpha-}|\beta, k, -\rangle = \delta_{\alpha\beta}|\alpha, k, +\rangle$, the transition probabilities can be expressed in terms of S as follows:

$$\langle \alpha, k', +|\beta, k, -\rangle = \langle \alpha, k'|S_{\alpha\beta}|\beta, k\rangle = \langle \alpha, k', +|S|\beta, k, +\rangle.$$

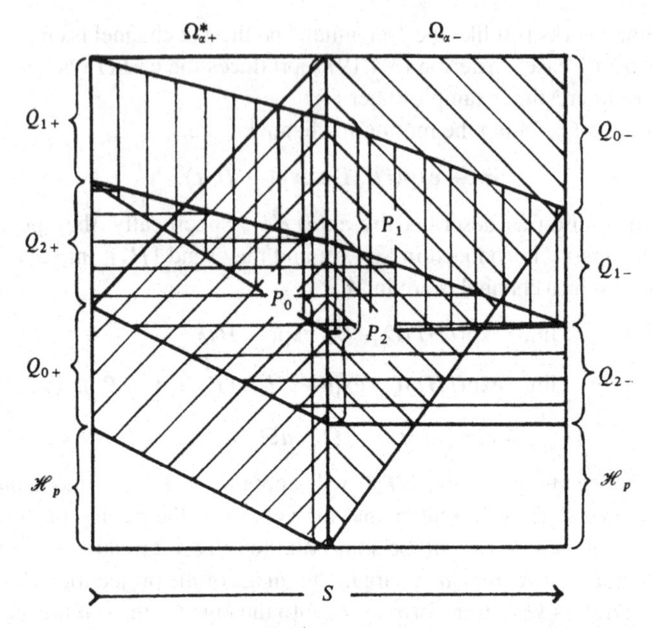

Figure 3.3. The domains and ranges of $\Omega^*_{\alpha+}$ and $\Omega_{\alpha-}$ in a system with several channels.

The operator S is thus a unitary transformation on $\mathcal{H}_{ac}(H)$, whereas $S_{\alpha\beta}$ maps nonorthogonal subspaces of $\mathcal{H}$ isometrically onto one another. Even so, $S_{\alpha\beta}$ is the more useful operator, since it is easier to make calculations with the states $|\alpha, k\rangle$ than with $|\alpha, k, \pm\rangle$.

3. Even if both $\sum_\alpha Q_{\alpha-}$ and $\sum_\alpha Q_{\alpha+}$ equal P_{ac}, it is still possible that $Q_{\alpha-}$ and $Q_{\alpha+}$ project onto different subspaces. For example, if a collision in (3.4.16) results in ionization, then one goes from Q_1 to Q_0. This illustrates the earlier remark that the mere existence of $s\text{-}\lim_{t\to\pm\infty} \exp(itH)\exp(-itH_0)$ does not guarantee that $\mathcal{H}_+$ and $\mathcal{H}_-$ are equal. It does not contradict invariance under reversal of the motion, as the operator K of (3.3.18; 2) just maps $\mathcal{H}_-$ onto $\mathcal{H}_+$.
4. From $\exp(iHt)\Omega_{\alpha\pm} = \Omega_{\alpha\pm}\exp(iH_\alpha t)$ it follows that $\exp(iHt)S\exp(-iHt) = S$ and $\exp(iH_\alpha t)S_{\alpha\beta}\exp(-iH_\beta t) = S_{\alpha\beta}$. The operator S does not in general commute with all constants of the motion, but only with

$$\bigcap_\alpha \{H_\alpha\}' \cap \{H\}'$$

(cf. (I: 3.4.11; 1)).
5. If $K \in \{H_\alpha\}'$ and $K = P_\alpha K$, then by (3.6) $Q_\alpha K Q_\alpha \in \mathcal{A}$, and

$$K_\pm \equiv \tau_\pm(Q_\alpha K_\pm Q_\alpha) = \Omega_{\alpha\pm} K \Omega^*_{\alpha\pm}.$$

Hence S transforms K_- into K_+:

$$K_+ = \Omega_{\alpha+}\Omega^*_{\alpha-}K_-\Omega_{\alpha-}\Omega^*_{\alpha+} = S^* K_- S.$$

For such observables, S gives the total change in time from $t = -\infty$ to $t = +\infty$.

6. If there is only one channel, then

$$S_{\alpha\beta} = s\text{-}\lim_{t\to\infty} \exp(iH_\alpha t)\exp(-2iHt)\exp(iH_\beta t)$$

can be written as

$$S_{00} = T\exp\left\{-i\int_{-\infty}^{\infty} dt\, H'(t)\right\}$$

(recall (3.4.10; 3)). The strong limit exists because $\exp(iH_0 t)\exp(-iHt)Q$ converges strongly and $(\mathbf{1} - Q)\exp(-iHt)\exp(iH_0 t)$ tends strongly to zero.

Scattering operators have been introduced by a comparison of the time-evolution with free motion of the clusters, as suggested by our experience with classical dynamics. It normally turns out, however, that S can be explicitly calculated only with methods that eliminate the time-variable. Sections 3.5 and 3.6 will be devoted to stationary methods.

3.4.23 Problems

1. Show that if H' is compact relative to H_0, then for every $\varepsilon > 0$ there exists a δ such that $\|H'\psi\| \le \varepsilon\|H_0\psi\| + \delta\|\psi\|$ for all $\psi \in D(H_0)$. (Hint: $a = H'(H_0 + i)^{-1}$ is compact. Let $P_n = \chi_{[-n,n]}(H_0)$, and show (i) that $\|a(\mathbf{1} - P_n)\| \to 0$ and (ii) that $H'P_n$ is bounded for all n.)

2. Show that $V = r^{-\varepsilon}$, $0 < \varepsilon < 2$, is compact relative to $H_0 = p^2$. (Show that $\mathrm{Tr}(V^{1/2}(H_0 + c^2)^{-1}V^{1/2})^n < \infty$ for $n \in \mathbb{Z}^+, n > 3/\varepsilon$.)

3. Let V be compact relative to H_0. Show that if $P(I)$ (respectively, $P_0(I)$) is the spectral projection for H (respectively, H_0) onto the interval I, then

$$(\mathbf{1} - P(I))\exp(itH)\exp(-itH_0)P_0(I) \to 0 \quad \text{as} \quad t \to \pm\infty.$$

4. Verify that $\Omega^*_{\alpha\pm}\Omega_{\beta\pm} = \delta_{\alpha\beta}P_\alpha$ for the system of three particles (3.4.16).

5. Let $A(\omega) \equiv \sqrt{V}\delta(H_0 - \omega)\sqrt{V}$, $V \sim r^{-1-\varepsilon}$ (cf. (3.4.13; 2)). Show that (a) $\|A(\omega)\|_n < c\omega^{-n(1+\varepsilon)/2}$, and (b) there exist $\rho > 0$, $\delta > 0$, and $c < \infty$ such that $\|A(\omega) - A(\omega')\|_n < c|\omega - \omega'|^\rho$ for all $|\omega - \omega'| < \delta$ and all $n > 2/\varepsilon$. From these two facts conclude that $\lim_{y\to 0}\sqrt{V}(H_0 - x - iy)^{-1}\sqrt{V} \in \mathcal{C}_n$.

3.4.24 Solutions

1. (i) Let $a_n = a^*(\mathbf{1} - P_n)a$. $\|a(\mathbf{1} - P_n)\|^2 = \sup_{\|\psi\|\le 1}\langle\psi|a_n\psi\rangle$. The mappings $\psi \to \langle\psi|a_n\psi\rangle$ are weakly continuous, because $\psi_i \rightharpoonup 0 \Rightarrow a\psi_i \to 0$, and

$$\psi \to \langle\psi|(\mathbf{1} - P_n)\psi\rangle$$

is strongly continuous. It follows that the sets $\{\psi : \langle\psi|a_n\psi\rangle \ge C\}$ are weakly closed. If $\|a(\mathbf{1} - P_n)\|^2$ were greater than $C > 0$ for all n, then the intersection of the decreasing sequence of weakly compact sets $\{\psi : \langle\psi|a_n\psi\rangle \ge C,\ \|\psi\| \le 1\}$ would not be empty, so there would exist a ψ such that $\|\psi\| \le 1$ and $\langle\psi|a_n\psi\rangle \ge C$ for all n. This is impossible, since $\mathbf{1} - P_n \to 0$.

(ii) $H'P_n = a\int_{-n}^{n}(\alpha+i)dP_\alpha$ is the product of two bounded operators. Therefore, if $\psi \in D(H_0)$ and n is sufficiently large, then

$$\begin{aligned} \|H'\psi\| &= \|a(H_0+i)\psi\| \\ &\leq \|a(1-P_n)(H_0+i)\psi\| + \|H'P_n\psi\| \leq \varepsilon\|(H_0+i)\psi\| + \|H'P_n\|\|\psi\|. \end{aligned}$$

2. With (2.3.20; 5) and (3.3.3),

$$\begin{aligned} &\mathrm{Tr}(V^{1/2}(H_0+c^2)^{-1}V^{1/2})^n \\ &= \int d^3x_1\cdots d^3x_n V(x_1)\frac{\exp(-c|x_1-x_2|)}{4\pi|x_1-x_2|} \\ &\quad\times V(x_2)\frac{\exp(-c|x_2-x_3|)}{4\pi|x_2-x_3|}\ldots V(x_n)\frac{\exp(-c|x_n-x_1|)}{4\pi|x_n-x_1|} \\ &= \int d^3y_1\ldots d^3y_n V(y_1)\frac{\exp(-c|y_2|)}{4\pi|y_2|}V(y_1+y_2)\frac{\exp(-c|y_3|)}{4\pi|y_3|}\cdots \\ &\quad\times V(y_1+y_2+\cdots+y_n)\frac{\exp(-c|y_2+y_3+\cdots+y_n|)}{4\pi|y_2+y_3+\cdots+y_n|}. \end{aligned}$$

The factor $\exp(-c|y_i|)$ takes care of the convergence of the integral by $dy_2\ldots dy_n$ at infinity, and the integral by dy_1 of

$$|y_1|^{-\varepsilon}|y_1+y_2|^{-\varepsilon}\cdots|y_1+y_2+\cdots+y_n|^{-\varepsilon},$$

converges provided that $n\varepsilon > 3$. The singularities at finite points are harmless, as long as $\varepsilon < 2$.

3. Since the operators are bounded in norm by 1 for all t, it suffices to show strong convergence on the dense set of $\varphi \in P_0(I')\mathcal{H}$, where I' is contained in the interior of I. On that set,

$$\begin{aligned} &(1-P(I))\exp(iHt)\exp(-itH_0t)P_0(I')\varphi \\ &= \frac{1}{2\pi i}\int_C dz(1-P(I))\exp(iHt)\left(\frac{1}{H_0-z}-\frac{1}{H-z}\right)\exp(-iH_0t)P_0(I')\varphi \\ &= \frac{1}{2\pi i}\int_C dz(1-P(I))\frac{1}{H-z}\exp(iHt)V\exp(-iH_0t)P_0(I')\frac{1}{H_0-z}P_0(I')\varphi, \end{aligned}$$

where C is a closed path of integration encircling I' but not cutting $\mathbb{R}\backslash I$ (see figure):

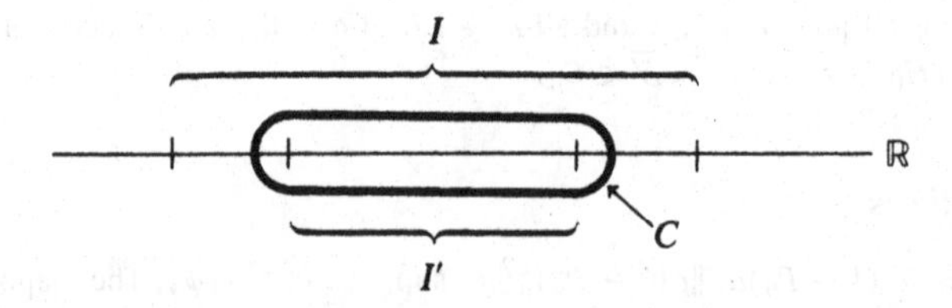

This makes the operators $(1-P(I))(H-z)^{-1}$ and $(H_0-z)^{-1}P_0(I')$ uniformly bounded on the path of integration. Theorem (3.4.4) then implies that the expression above converges strongly to zero. (Recall that $V_n \to 0$ and $\|a_n\| = 1$ for all $n \Rightarrow a_nV_n \to 0$.)

4. If $\alpha = \beta$, then this follows from the strong convergence of the operators (see (3.4.8; 2)), so it suffices to verify that

$$w\text{-}\lim P_\alpha \exp(iH_\alpha t)\exp(-iH_\beta t)P_\beta = 0$$

when $\alpha \neq \beta$. There are essentially just two cases:
$\alpha = 0, \beta = 1, 2, 3$:

$$\exp(iH_0t)\exp(-i(H_0 + V_\beta)t)P_\beta \rightharpoonup 0,$$

because

$$\exp(i|\mathbf{p}_1|^2t)\exp(-i(|\mathbf{p}_1|^2 + V_1(x_1))t)P_p(1) \rightharpoonup 0,$$

etc., since $P_p(1)$ contains only the eigenfunctions ψ_j, $H_1\psi_j = E_j\psi_j$, and

$$\exp(it(|\mathbf{p}_1|^2 - E_j))\psi_j \rightharpoonup 0.$$

$\alpha = 1, \beta = 2$:

$$P_p(1)\exp(i(|\mathbf{p}_1|^2 + V_1)t)\exp(-i|\mathbf{p}_1|^2t)\otimes\exp(i(|\mathbf{p}_2|^2t)\exp(-i(|\mathbf{p}_2|^2 + V_2)t)P_p(2) \rightharpoonup 0$$

for the same reason as above. Similarly for $\alpha = 1, \beta = 3$ and $\alpha = 2, \beta = 3$.

5. In (3.4.13; 2), the integral over $(S^2)^n$ is locally like an integral over $\mathbb{R}^{2n}$. Since the integrand depends only on the differences between the n_i, and a $2(n-1)$-fold integral over a homogeneous function of degree $n(-2+\varepsilon)$ is finite whenever $2(n-1) - n(2-\varepsilon) > 0$, it follows that $\|A(\omega)\|_n < \infty$ for $n > 2/\varepsilon$. As a consequence, $\|A(\omega) - A(\omega')\|_n$ is not only finite, but actually goes to zero Hölder-continuously as $\omega' \to \omega$. This guarantees the existence of the principal-value integral in

$$\lim_{y\to 0}\sqrt{V}(H_0 - x - iy)^{-1}\sqrt{V} = i\pi A(x) + P\int dz\,\frac{A(z) - A(x)}{z - x}$$

in the trace norm.

3.5 Perturbation Theory

Abrupt changes are the rule in infinite-dimensional spaces, but in physics a central question is under what circumstances eigenvalues are affected only slightly by perturbations.

Since most of the problem of physics cannot be solved analytically, it is the custom to approximate the solutions by carrying out Taylor expansions about suitably chosen, soluble limiting cases. The perturbed Hamiltonian is typically of the form $H(\alpha) = H_0 + \alpha H'$, which brings up the question of what quantities are analytic in α, and for what range of values. Of especial interest are the resolvent $R(\alpha, z) \equiv (H(\alpha) - z)^{-1}$, the isolated eigenvalues $E_k(\alpha)$ of $H(\alpha)$, and the projections onto them, which can be written

$$P_k(\alpha) \equiv \frac{1}{2\pi i}\int_{C_k(\alpha)} dz\, R(\alpha, z), \tag{3.8}$$

where $C_k(\alpha)$ is a closed path encircling $E_k(\alpha)$ and no other points of $\mathrm{Sp}(H(\alpha))$. Although $H(\alpha)$ is not diagonable for all complex values of α, for all α we know (Problem 1), if H_0 and H' are self-adjoint.

3.5.1 The Properties of the Projections

(i) $P_k(\alpha) = P_k^*(\alpha^*)$,

(ii) $P_i(\alpha)P_k(\alpha) = \delta_{ik} P_k(\alpha)$,

(iii) $[P_i(\alpha), R(\alpha, z)] = 0$.

(iv) *Except at the points α_s where the eigenvalues $E_k(\alpha)$ cross, the projections $P_k(\alpha)$ can be continued analytically in α such that* $\dim P_k(\alpha)\mathcal{H} = \mathrm{Tr}\, P_k(\alpha)$ *is constant throughout the region of analyticity.*

In most quantum mechanics books, operators are blithely manipulated as if they were finite-dimensional matrices. In the same spirit, let us warm up with 2×2 matrices (H_0 and H' are not necessarily Hermitian)

3.5.2 Examples

1. $H(\alpha) = \begin{pmatrix} \alpha & 0 \\ 0 & 0 \end{pmatrix}$, $E_{1,2}(\alpha) = \alpha, 0$. $R(\alpha, z) = \begin{pmatrix} 1/(\alpha - z) & 0 \\ 0 & -1/z \end{pmatrix}$,

   $P_1(\alpha) = \begin{pmatrix} 1 & 0 \\ 0 & 0 \end{pmatrix}$, $P_2(\alpha) = \begin{pmatrix} 0 & 0 \\ 0 & 1 \end{pmatrix}$.

2. $H(\alpha) = \begin{pmatrix} \alpha & 1 \\ 0 & 0 \end{pmatrix}$, $E_{1,2}(\alpha) = \alpha, 0$. $R(\alpha, z) = \begin{pmatrix} 1/(\alpha - z) & 1/z(\alpha - z) \\ 0 & -1/z \end{pmatrix}$,

   $P_1(\alpha) = \begin{pmatrix} 1 & 1/\alpha \\ 0 & 0 \end{pmatrix}$, $P_2(\alpha) = \begin{pmatrix} 0 & -1/\alpha \\ 0 & 1 \end{pmatrix}$.

3. $H(\alpha) = \begin{pmatrix} \alpha(\alpha+1) & \alpha \\ 0 & \alpha \end{pmatrix}$, $E_{1,2}(\alpha) = \alpha(\alpha+1), \alpha$.

   $R(\alpha, z) = \begin{pmatrix} 1/(\alpha(\alpha+1) - z) & -\alpha/(\alpha - z)(\alpha(\alpha+1) - z) \\ 0 & 1/(\alpha - z) \end{pmatrix}$,

   $P_1(\alpha) = \begin{pmatrix} 1 & 1/\alpha \\ 0 & 0 \end{pmatrix}$, $P_2(\alpha) = \begin{pmatrix} 0 & -1/\alpha \\ 0 & 1 \end{pmatrix}$.

4. $H(\alpha) = \begin{pmatrix} 1 & \alpha \\ \alpha & 0 \end{pmatrix}$, $E_{1,2}(\alpha) = \frac{1}{2}(1 \pm \sqrt{1 + 4\alpha^2})$.

   $R(\alpha, z) = \frac{1}{z(z-1)-\alpha^2} \begin{pmatrix} -z & -\alpha \\ -\alpha & 1 - z \end{pmatrix}$,

   $P_{1,2}(\alpha) = \pm\frac{1}{2}(1 + 4\alpha^2)^{-1/2} \begin{pmatrix} 1 \pm \sqrt{1 + 4\alpha^2} & 2\alpha \\ 2\alpha & 1 \pm \sqrt{1 + 4\alpha^2} \end{pmatrix}$.

These examples exhibit the

3.5.3 Singularity Structure of R, E_k, and P_k in the Finite-Dimensional Case

Let $H(\alpha)$ be a polynomial in α, and define $E_k(\alpha)$ as the poles (in z) of $R(\alpha, z)$.

(i) *The function $(\alpha, z) \to R(\alpha, z)$ is analytic except on $\bigcup_k \{z = E_k(\alpha)\}$.*

(ii) *The eigenvalues and projections $E_k(\alpha)$, $P_k(\alpha)$ are regular except at the crossing points α_s, at which the numbering of the eigenvalues changes. (In Examples* 1, 2, *and* 3, $\alpha_s = 0$, *and in Example* 4, $\alpha_s = \pm i/2$.) *At the points α_s the eigenvalues E_k and projections P_k may have algebraic singularities (Examples* 2, 3, 4), *but do not necessarily have them* (1, 2, *and* 3).

(iii) *In any event, $E_k(\alpha)$ is continuous in α. If $E_k(\alpha)$ has a branch point at α_s, then $\|P_k(\alpha)\| \to \infty$ as $\alpha \to \alpha_s$ (but not conversely; see Example* 2).

(iv) *If $\|P_k(\alpha)\|$ remains finite at α_s, then $H(\alpha_s)$ is diagonable (but not conversely; see Example* 3).

Proof:

(i) The singularities of $(H(\alpha) - z)^{-1}$ can only originate with zeroes in the denominator $[\mathrm{Det}(H(\alpha) - z)]^{-1} = \prod_k (E_k(\alpha) - z)^{-1}$.

(ii) $\mathrm{Det}(H(\alpha) - z) = (-z)^m + (-z)^{m-1}\mathcal{P}_i(\alpha) + \cdots$, where $\mathcal{P}_i$ are polynomials in α. Hence the $E_k(\alpha)$ are branches of the same algebraic functions, and as such have the desired properties. As a complex integral of the analytic function $R(\alpha, z)$, $P_k(\alpha)$ is analytic unless the contour C gets caught between two singularities, which can happen only at the points α_s. Since the integral (3.8) can be written in terms of the E_k and polynomials in α, the singularities at α_s are at worst algebraic.

(iii) The continuity of E_k follows from theorems on algebraic functions, and as a consequence, series expansions for the eigenvalues contain only positive powers of $(\alpha - \alpha_s)^{1/m}$. Suppose that this were also true for the P_k, so that $\|P_k\|$ would remain bounded. By continuing the $E_k(\alpha)$ along a circle above α_s, the E_k having a branch point there are permuted so that E_i becomes E_j for some $j \neq i$. By (3.8) the same thing happens with the P_k, so the first terms of $P_k(\alpha) = P_k(\alpha_s) + (\alpha - \alpha_s)^{1/m} P_k^{(1)} + \cdots$ would clearly have to satisfy $P_i(\alpha_s) = P_j(\alpha_s)$. Since $P_i^2(\alpha_s) = P_i(\alpha_s)$, $P_i(\alpha_s)P_j(\alpha_s) = P_j(\alpha_s)P_i(\alpha_s) = 0$, and $P_j^2(\alpha_s) = P_j(\alpha_s)$, this implies that $P_i(\alpha_s) = P_j(\alpha_s) = 0$.

(iv) $H(\alpha)$ is diagonable iff $H(\alpha) = \sum_k E_k(\alpha)P_k(\alpha)$. If E_k and P_k are continuous, then this equation can be continued analytically to α_s. □

3.5.4 Corollaries

1. As long as $H(\alpha)$ remains nondegenerate, everything is analytic, and H is diagonable.
2. If $H(\alpha)$ is Hermitian whenever α is real, and thus unitarily diagonable, then $\|P_k(\alpha)\| = 1$ on the real axis. Then it follows from (iii) that there can be no α_s on the real axis at which E_k has an algebraic singularity. This theorem, due to Rellich, is not trivial, as it may at first look, since it does not extend to the case

of two parameters: The eigenvalues $\alpha_1 + \alpha_2 \pm \sqrt{2}\sqrt{\alpha_1^2 + \alpha_2^2}$ of the matrix

$$\alpha_1 \begin{pmatrix} 2 & 1 \\ 1 & 0 \end{pmatrix} + \alpha_2 \begin{pmatrix} 0 & 1 \\ 1 & 2 \end{pmatrix}$$

have a branch point at $\alpha_1 = \alpha_2 = 0$.

3. All zeros of $\mathrm{Det}(H(\alpha) - z)$ are eigenvalues, for which reason analytic continuation of one of the $E_k(\alpha)$ always leads to another eigenvalue. This property is lacking on infinite-dimensional spaces. For instance, the eigenvalues of the hydrogen atom go as the square of the charge of the electron, and are thus entire functions in α. Yet they disappear when the charge becomes positive; their analytic continuation is not an eigenvalue.
4. Although $H(\alpha)$ is an entire function, it may happen that a power series for $E_k(\alpha)$ in α has only a finite radius of convergence. However, because of Corollary 2 the radius of convergence is necessarily greater than zero.

Let us now take up the question of how far these results carry over to the infinite-dimensional case. The set of eigenvectors will no longer span the whole Hilbert space, but instead there is the three-fold classification of spectra (2.3.16). It turns out, rather discouragingly, that the classification of spectra can be completely changed by arbitrarily small perturbations.

3.5.5 Theorem

The operators with pure point spectra are norm-dense in the set of Hermitian elements of $\mathcal{B}(\mathcal{H})$.

Proof: Given any $a = a^* \in \mathcal{B}(\mathcal{H})$ written in the spectral representation (2.3.11), $\mathcal{H} = \bigoplus_i \mathcal{H}_i$, $a_{|\mathcal{H}_i} : \psi(\alpha) \to \alpha\psi(\alpha)$, $\alpha \in \mathrm{Sp}(a)$. Define a_n such that $a_{n|\mathcal{H}_i} : \psi(\alpha) \to s_n(\alpha)\psi(\alpha)$, where $s_n(x) = m/n$ for $m/n \leq x < (m+1)/n$, $n \in \mathbb{Z}^+$, $m \in \mathbb{Z}$. Then $\|a - a_n\| \leq 1/n$, and $\mathrm{Sp}(a_n)$ is the set of values of s_n, i.e., $\{m/n : m \in \mathbb{Z}\}$, which is purely discrete. □

3.5.6 Remarks

1. More particularly, the theorem states that any operator with pure continuous spectrum can be converted into an operator with pure point spectrum by the addition of an arbitrarily small perturbation. Conversely, there are operators with continuous spectra and arbitrarily small norm, like $a_n\psi(\alpha) = (1/n)\sin\alpha\psi(\alpha)$, for which $\|a_n\| = 1/n$. These can convert the pure point spectrum of, for instance, the zero operator into a pure continuous spectrum.
2. Theorem (3.5.5) can be strengthened to state that the addition of an operator δ with trace norm (2.3.21) $\|\delta\|_p < \varepsilon$, $p > 1$, can render the spectrum discrete. The theorem does not hold for $p = 1$; if $H_0 = |\mathbf{p}|^2$ and $\|H_0 - H\|_1 < \infty$, then the Møller operators exist, and H_0 and HP_{ac} are unitarily equivalent.
3. The proof also works for unbounded self-adjoint operators.

4. Note that the eigenvalues of s_n have infinite multiplicity and hence belong to the essential spectrum. Nevertheless, the spectrum consists of isolated points, and the next theorem will show that a relatively compact perturbation cannot change a continuous spectrum into isolated points.

The essential spectrum σ_{ess} (2.3.18; 4) is less sensitive than the continuous spectrum. (From now on H_0 and H' will be supposed to be self-adjoint.)

3.5.7 Stability of the Essential Spectrum

If H' is compact relative to H_0, then $\sigma_{\text{ess}}(H_0 + H') = \sigma_{\text{ess}}(H_0)$.

Proof: The criterion of (2.3.18; 5) for the essential spectrum can be reformulated as follows: $\lambda \in \sigma_{\text{ess}}(H_0) \Leftrightarrow \exists\, \psi_n : \|\psi_n\| = 1,\ \psi_n \rightharpoonup 0,\ (H_0 - \lambda)\psi_n \to 0$. By Definition (3.4.1), $H'(H_0 - z)^{-1}$ is compact for all $z \notin \mathrm{Sp}(H_0)$, so

$$(H_0 + H' - \lambda)\psi_n = (H_0 - \lambda)\psi_n + H'(H_0 - z)^{-1}(H_0 - z)\psi_n \to 0,$$

since

$$(H_0 - z)\psi_n = (H_0 - \lambda)\psi_n + (\lambda - z)\psi_n \rightharpoonup 0,$$

and compact operators make weakly convergent sequences strongly convergent. We can then conclude that $\lambda \in \sigma_{\text{ess}}(H_0 + H')$, and switching H_0 and $H_0 + H'$ (cf. (3.4.5; 2)) yields the other direction of the theorem. □

3.5.8 Remarks

1. The addition of a relatively compact potential produces only finitely many bound states under $E_0 < 0$. A classical description would be that the volume of the phase space under E_0 is finite for such systems (cf. (3.5.36; 1)).
2. Compactness is essential. The addition of the bounded operator $\alpha \cdot \mathbf{1}$, $\alpha \in \mathbb{R}$, shifts the whole spectrum of any operator by α.
3. When applying this theorem, it should be remembered that if a is compact and b is bounded, then ab is compact, but $a \otimes b$ may not be.
4. If a Hilbert–Schmidt operator is added as in (3.5.6; 2) to an operator, changing a continuous spectrum to a purely discrete spectrum, then the new eigenvalues must be dense in the continuum of the original operator, since σ_{ess} is unchanged by the addition of a compact operator.

As is reasonable, the shift in the spectrum by α when one adds $\alpha \cdot \mathbf{1}$ is as great as possible with a perturbation by an operator of norm $\le \alpha$:

3.5.9 Theorem

If the distance from λ to the spectrum of H_0 satisfies $d(\mathrm{Sp}(H_0), \lambda) > \|H'\|$, then $\lambda \notin \mathrm{Sp}(H_0 + H')$.

Proof: The series

$$\frac{1}{H_0 + H' - \lambda} = \frac{1}{H_0 - \lambda} \sum_{n=0}^{\infty} [H'(\lambda - H_0)^{-1}]^n$$

is convergent in norm, because $\|(H_0 - \lambda)^{-1}\|^{-1} = d(\text{Sp}(H_0), \lambda)$. □

However, if H' is unbounded, then the addition of $\alpha H'$ can change any kind of spectrum in any way, no matter how small α is.

3.5.10 Examples

1. $H_0 = 0$, $H' \equiv \psi(x) \to x\psi(x)$ on $L^2((-\infty, \infty), dx)$. $\text{Sp}(H_0 + \alpha H') = \mathbb{R}$ for $\alpha \neq 0$, and $\{0\}$ for $\alpha = 0$.
2. $H_0 = -d^2/dx^2$, $H' = \alpha x^2$: $\sigma_{ac}(H_0) = \mathbb{R}^+$, $\sigma_p(H_0) = \sigma_s(H_0)$ is empty. $\text{Sp}(H_0 + \alpha H') = \sqrt{\alpha} \bigcup_{n=0}^{\infty} \{2n+1\}$, $\sigma_{ac} = \sigma_s$ is empty if $\alpha > 0$, and $\sigma_{ac}(H_0 + \alpha H') = \mathbb{R}$, $\sigma_p = \sigma_s$ is empty for $\alpha < 0$.

Most physically realistic perturbations are unbounded, so it may seem hopeless to conclude anything about $\text{Sp}(H_0 + \alpha H')$ from $\text{Sp}(H_0)$. Fortunately, the relevant condition is not that H' be small, but only that it be *small in comparison with* H_0.

3.5.11 Theorem

Let H' be bounded relative to H_0 (3.4.1) and $H(\alpha) = H_0 + \alpha H'$. Then the resolvent $R(\alpha, z) \equiv (H(\alpha) - z)^{-1}$ is analytic in the variables (α, z) throughout some region containing $\{0\} \times \{\mathbb{C} \backslash \text{Sp}(H_0)\}$.

Proof: If $z \notin \text{Sp}(H_0)$, then $H'(H_0 - z)^{-1}$ is bounded, so the series

$$\sum_{n=0}^{\infty} \alpha^n (H'(H_0 - z)^{-1})^n$$

for the resolvent converges for all α small enough. □

3.5.12 Remarks

1. The more precise form of the region of analyticity depends on the particulars of the operators. If say,

$$\text{Sp}(H_0) = \mathbb{R}^+, \quad \|H'\psi\| \leq a\|\psi\| + b\|H_0\psi\|,$$

then

$$\begin{aligned} \|H'(H_0 - z)^{-1}\| &\leq a\|(H_0 - z)^{-1}\| + b\|H_0(H_0 - z)^{-1}\| \\ &\leq \frac{a}{|\text{Im}\, z|} + \frac{b|z|}{|\text{Im}\, z|} \quad \text{for Re}\, z \geq 0, \end{aligned}$$

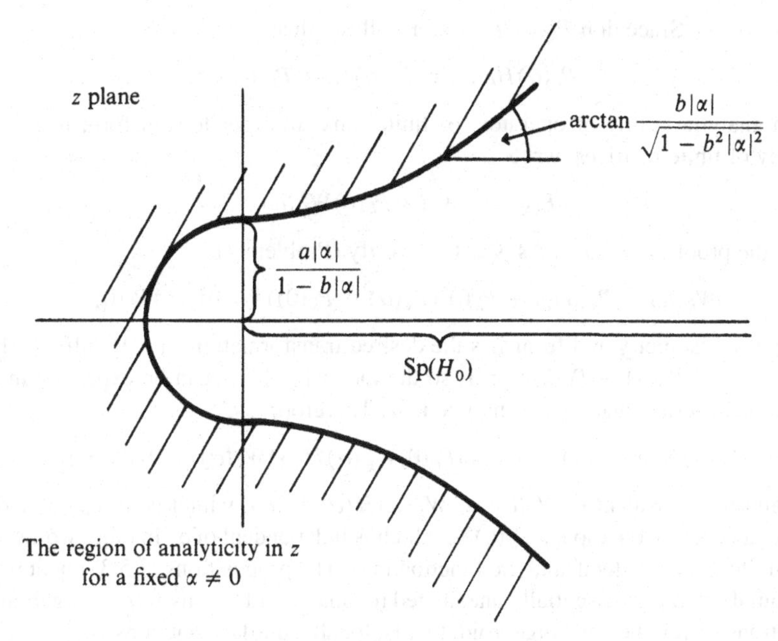

Figure 3.4. The region of analyticity of the resolvent.

and $\|H'(H_0 - z)^{-1}\| \le a/|z| + b$ for Re $z \le 0$, and so the series converges for

$$|\text{Im } z| \ \ge \ \frac{a|\alpha|}{1-b^2|\alpha|^2} + \frac{b|\alpha|}{\sqrt{1-b^2|\alpha|^2}}\sqrt{|\text{Re } z|^2 + \frac{a^2|\alpha|^2}{1-b^2|\alpha|^2}}, \qquad \text{Re } z \ge 0,$$

$$|z| \ \ge \ \frac{a|\alpha|}{1-b|\alpha|}, \qquad \text{Re } z \le 0.$$

2. The constants a and b of Remark 1 are not determined by H'. The constant b can be chosen smaller at the cost of increasing a. Hence it is difficult to formulate general statements, as in (3.5.9), about how much the spectrum is shifted.

The analyticity of the resolvent means that the results (3.5.3) about the eigenvalues remain valid away from the essential spectrum (see Figure 3.4).

3.5.13 Theorem

Let H' be bounded relative to H_0. Then the isolated eigenvalues of finite multiplicity of $H(\alpha) = H_0 + \alpha H'$, as well as the projections onto their eigenvectors, are analytical in α in a neighborhood of the origin.

Proof: For any isolated eigenvalue $E(0)$ of H_0, there exists a distance $d \in \mathbb{R}^+$ such that the circle $K = \{z \in \mathbb{C} : |z - E(0)| = d\}$ does not intersect the spectrum of H_0. If $\alpha_0^{-1} = \sup_{z\in K} \|H'(H_0 - z)^{-1}\|$, then $K \times (-\alpha_0, \alpha_0)$ is within the region of analyticity of the resolvent R, and $P_k(\alpha) = (1/2\pi i)\int_K dz\, R(z, \alpha)$ is analytic for $|\alpha| < \alpha_0$. Since dim $P_k(0)\mathcal{H} < \infty$, it follows that

$$P_k(\alpha)H_k(\alpha) \equiv P_k(\alpha)H(\alpha)P_k(\alpha)$$

is an analytic family of operators of finite rank. In order to transform it into a family of finite matrices, write

$$P_k(\alpha) = W_k(\alpha)P_k(0)W_k^*(\alpha),$$

as in the proof of (3.3.11). As is easy to verify (Problem 4),

$$W_k(\alpha) = P_k(\alpha)[1 + P_k(0)(P_k(\alpha) - P_k(0))P_k(0)]^{-1/2}P_k(0)$$

is a partial isometry and furnishes the desired transformation. If α is sufficiently small, then $\|P_k(\alpha) - P_k(0)\| < 1$, so the factor $[\ldots]^{-1/2}$ can be expanded in a convergent series, making W analytic in α. Therefore

$$P_k(\alpha)H(\alpha)P_k(\alpha) = W_k(\alpha)P_k(0)W_k^*(\alpha)H(\alpha)W_k(\alpha)P_k(0)W_k^*(\alpha)$$

is unitarily equivalent to $H_k(\alpha) \equiv W_k^*(\alpha)H(\alpha)W_k(\alpha)$, which is an operator of finite rank acting on a space $P_k(0)\mathcal{H}$, which is independent of α. In other words, it is a finite-matrix-valued analytic function in α. The propositions (3.5.3) about polynomials $H(\alpha)$ are essentially unchanged for analytic functions $H(\alpha)$—algebraic functions merely become algebroid, that is, locally similar, functions. □

The next subject is the derivation of explicit formulas for the change that a perturbation H' causes in an eigenvalue. Let us assume that an interval of $\mathbb{R}$ contains no essential spectrum, but only eigenvalues, and that the $E_k(\alpha)$ and their projections $P_k(\alpha)$ change continuously with α. This is always the case when H' is bounded relative to H_0, but may also happen otherwise. Next, rewrite $H(\alpha)$ as $H_0 + \alpha PH'P + \alpha(H' - PH'P)$, where $P = P_k(0)$ for the k of interest. If $E_k(0) + \alpha PH'P$ has been diagonalized, then the effect of $PH'P$ can be included in E_k, so without loss of generality we may assume that $PH'P = 0$. Since the eigenvector $|\alpha\rangle : (H(\alpha) - E(\alpha))|\alpha\rangle = 0$, $\langle\alpha|\alpha\rangle = 1$, varies continuously with α, let $|\alpha\rangle = c|0\rangle + |\perp\rangle$, $\langle 0|\perp\rangle = 0$, $|c|^2 = 1 - \langle\perp|\perp\rangle$, which $\neq 0$ for sufficiently small α. If this is substituted into the ei genvalue equation and the component parallel and perpendicular to $|0\rangle$ are separated, then

$$\begin{aligned} c(E(\alpha) - E(0))|0\rangle &= \alpha PH'|\perp\rangle, & P &= |0\rangle\langle 0|, \\ (H_0 - E(\alpha) + \alpha P_\perp H')|\perp\rangle &= -c\alpha H'|0\rangle, & P_\perp &= 1 - P. \end{aligned} \quad (3.9)$$

This produces

3.5.14 The Brillouin–Wigner Formulas

$$E(\alpha) = E(0) - \alpha^2\langle 0|H'(H_0 - E(\alpha) + \alpha P_\perp H'P_\perp)^{-1}H'|0\rangle,$$

$$|\perp\rangle = -c\alpha(H_0 - E(\alpha) + \alpha P_\perp H'P_\perp)^{-1}H'|0\rangle,$$

$$c^{-2} = 1 + \alpha^2\langle 0|H'(H_0 - E(\alpha) + \alpha P_\perp H'P_\perp)^{-2}H'|0\rangle.$$

3.5.15 Remarks

1. To ensure that the formal expressions make sense, we must assume that $|0\rangle \in D(H')$ and that $(H_0 - E(\alpha) - \alpha P_\perp H' P_\perp)^{-1}$ exists. If H' and consequently $P_\perp H' P_\perp$ is bounded relative to H_0 and and $E(\alpha)$ is isolated, then the series

$$(H_0 - E(\alpha))^{-1} \sum_{n=0}^{\infty} [\alpha P_\perp H' P_\perp (H_0 - E(\alpha))^{-1}]^n$$

converges on $P_\perp \mathcal{H}$ for α small enough, so (3.5.14) are well defined.

2. The eigenvalue $E(\alpha)$ is determined implicitly by (3.5.14). An explicit expression results from a comparison of the power series for both sides of the equation in α. The first few terms are fairly simple:

3.5.16 Lowest-Order Perturbation Theory

Up to $o(\alpha^2)$,

$$E(\alpha) = E(0) - \alpha^2 \langle 0|H' P_\perp (H_0 - E(0))^{-1} P_\perp H'|0\rangle,$$

$$|\perp\rangle = \alpha (H_0 - E(0))^{-1} H'|0\rangle,$$

$$c = 1 - \frac{\alpha^2}{2} \langle 0|H'(H_0 - E(0))^{-2} E'|0\rangle.$$

3.5.17 Remarks

1. An objective assessment of (3.5.16), which has been a daily tool for whole generations of physicists, is that it is unsatisfactory in several respects. Its shortcomings are that

   (i) if H' is unbounded, it is not obvious that $E(\alpha)$ should be analytic in α, and indeed it is not analytic in most of the standard examples of perturbation theory—the anharmonic oscillator, Stark effect, Zeeman effect, and hyperfine structure;

   (ii) even if the radius of convergence ρ is greater than zero, the n-th order terms get so complicated for large n that it is not easy to find out what ρ is;

   (iii) even the condition that $\alpha \ll \rho$ does not guarantee that (3.5.16) will be in close agreement to the true value. For example, the radius of convergence of $\sin(100\alpha)$ is infinite, but linear and parabolic approximations are not useful beyond a short range. If we wish to use (3.5.16), we ought to first show that the function $E(\alpha)$ does not have such wild oscillations.

2. The terms linear in α do not appear in (3.5.14), because $\alpha P H' P$ was defined away at the beginning. As a consequence, the **Feynman–Hellmann** formula

$$\left.\frac{\partial E}{\partial \alpha}\right|_{\alpha=0} = \langle 0|H'|0\rangle$$

holds for nondegenerate eigenvalues. If the eigenvalues are degenerate, one must first choose the right basis in the degeneracy space and be aware that the numbering of the analytic functions $E_k(\alpha)$ will not continue to order them by their magnitudes. For example, the eigenvalues $\pm\alpha$ of

$$H(\alpha) = \alpha H' = \begin{pmatrix} 0 & \alpha \\ \alpha & 0 \end{pmatrix}$$

are not the same as $\alpha\langle 0|H'|0\rangle$ with the eigenvector of $H(0)$

$$|0\rangle = \begin{pmatrix} 1 \\ 0 \end{pmatrix},$$

and the lowest eigenvalue $E_1(\alpha) = -|\alpha|$ is not differentiable at the point $\alpha = 0$.

3. Formulas (3.5.14) do not assume analyticity. Even without analyticity, (3.5.16) gives the correct asymptotic expansion under our assumptions:

$$\begin{aligned} E(\alpha) \quad &- \quad E(0) - \alpha^2\langle 0|H'P_\perp(H_0 - E(0))^{-1}P_\perp H'|0\rangle \\ &= \quad \alpha^2\langle 0|H'P_\perp(H_0 - E(0))^{-1}(E(\alpha) - E(0) + \alpha P_\perp H'P_\perp) \\ &\qquad \times(H_0 - E(\alpha) + \alpha P_\perp H'P_\perp)^{-1}H'|0\rangle = O(\alpha^3), \end{aligned}$$

and similarly at higher orders. However, in the absence of analyticity perturbation theory may lead to nonsense. It may happen that the series diverges for all $\alpha \in \mathbb{R}\backslash\{0\}$ although the discrete eigenvalues persist for all $\alpha \in \mathbb{R}$, or that the series converges, but to the wrong answer.

3.5.18 Examples

1. $H(\alpha) = p^2 + x^2 + \alpha^2 x^6$. Since the potential of this anharmonic oscillator goes rapidly to infinity for all $\alpha \in \mathbb{R}$ as $|x| \to \infty$, the spectrum remains discrete (see (3.5.36; 1)). Yet $R(\alpha, z)$ is not analytic:

$$-\left.\frac{\partial R}{\partial \alpha^2}\right|_{\alpha=0} = (p^2 + x^2 - z)^{-1} x^6 (p^2 + x^2 - z)^{-1}$$

is unbounded, since $\|x^3(p^2 + x^2 - z)^{-1}\varphi\|$ can get arbitrarily large.

2. $H(\alpha) = p^2 + x^2 - 1 - 3\alpha x^2 + 2\alpha x^4 + \alpha^2 x^6 = a^*a$, where $a = ip + x + \alpha x^3$. It is clear that $H(\alpha) \geq 0$ for all $\alpha \in \mathbb{R}$, the eigenvalues remain isolated for all $\alpha \in \mathbb{R}$, and $\sigma_{\text{ess}}(H(\alpha)) = \emptyset$. However, the eigenfunction $\exp(-x^2/2 - \alpha x^4/4)$ with the eigenvalue 0 belongs to

$$L^2((-\infty, \infty), dx)$$

only for $\alpha \geq 0$. Since perturbation theory produces an asymptotic series for the ground state $E(\alpha)$, which equals 0 for all $\alpha \geq 0$, all the perturbation coefficients must vanish. The series then also converges trivially for all $\alpha \leq 0$, although 0 is no longer an eigenvalue.

More precise information about the positions of the eigenvalues can be obtained with variational methods. They rely on the

3.5.19 Min-Max Principle

Let H be self-adjoint and bounded from below, and let the eigenvalues be $E_1 \leq E_2 \leq E_3 \leq \cdots \leq E_\infty$ (counting multiplicity), where by definition all E_k lying above the bottom of the essential spectrum E_∞ are set equal to E_∞, even if E_∞ is not an eigenvalue. Let D_n be an n-dimensional subspace of $D(H)$, let $D_n^\perp$ be its orthogonal complement in $D(H)$, and let Tr_{D_n} be the trace in D_n. Then

$$\sum_{k=i}^{i+j-1} E_k = \inf_{D_{i+j-1}} \sup_{D_j \subset D_{i+j-1}} \mathrm{Tr}_{D_j} H = \sup_{D_{i-1}} \inf_{D_j \subset D_{i-1}^\perp} \mathrm{Tr}_{D_j} H, \quad i \text{ and } j = 1, 2, \ldots$$

Proof: See Problem 5. □

3.5.20 Remarks

1. In particular, $\sum_{n=1}^{j} \langle \psi_n | H \psi_n \rangle$ for any orthonormal system $\{\psi_n\} \subset D(H)$ is always greater than the sum of the first j eigenvalues, counting the bottom of σ_{ess} as an infinitely degenerate eigenvalue. By the use of well chosen trial functions ψ_n provided with several parameters to adjust, excellent upper bounds on $\sum_{n=1}^{j} E_n$ can be obtained.
2. To get an upper bound for E_n itself, take an orthonormal $\{\psi_1, \ldots, \psi_n\} \subset D(H)$. The greatest eigenvalue of any $n \times n$ matrix $\langle \psi_i | H \psi_k \rangle$ is $\geq E_n$.

The astute reader will have realized from (3.5.16) that the second-order correction ($\sim \alpha^2$) for the ground state is always negative. More generally, (3.5.19) permits the proof of some

3.5.21 Concavity Properties of $E_n(\alpha)$

Let $H = H_0 + \alpha H'$, with $D(H') \supset D(H_0)$. Then $\sum_{n=1}^{j} E_n(\alpha)$, $j = 1, 2, \ldots$, are concave functions of α. For several α's it is jointly concave

3.5.22 Gloss

Concave functions $f : I \to \mathbb{R}$ are by definition those for which

$$f\left(\sum_{i=1}^{n} \alpha_i x_i\right) \geq \sum_{i=1}^{n} \alpha_i f(x_i), \qquad \alpha_i \geq 0, \quad \sum_{i=1}^{n} \alpha_i = 1, \quad x_i \in I.$$

The function $-f$ is said to be **convex**. Concave functions f have the following properties:

(i) f is continuous on the interior of I, has right and left derivatives at every point, and has first and second derivatives almost everywhere;

(ii) f'' is a negative distribution ($f''\, dx$ is a negative measure);

(iii) for $x > 0$, $f(x)$ is concave iff $xf(1/x)$ is concave;

(iv) if $f > 0$ and $1/f$ is concave, then f is convex;

(v) if the functions $f_i(x)$ are concave and $\alpha_i \geq 0$, then $\sum_i a_i f_i(x)$ is concave;

(vi) if $f_i(x)$ is concave, then $\inf_i f_i(x)$ is concave;

(vii) if the functions f_i are concave and $f_1' \geq 0$, then $f_1 \circ f_2$ is concave

Proof: The expression $\sum_{n=1}^{j} \langle \psi_n | H(\alpha) \psi_n \rangle$ is linear in α, so by Property (vi) its infimum over the ψ_n is concave. □

3.5.23 Remark

It is necessary that $D(H(\alpha)) = D(H_0)$, so that the infimum is taken over a set independent of α. For instance,

$$E_2(\alpha) = \inf_{\langle \psi | \psi_1(\alpha) \rangle = 0} \langle \psi | H(\alpha) \psi \rangle$$

is not necessarily concave.

Although the min-max principle guarantees that $E_1(\alpha)$ lies below any possible expectation value of $H(\alpha)$, it does not say how close to E_1 the expectation value comes. People who make variational calculations normally convince themselves that they come close by their faith in their pet trial functions. There are, however, a few criteria with which to gauge the accuracy.

3.5.24 Weinhold's Criterion of the Mean-Square Deviation

There is a spectral value of H in the interval $[\langle H \rangle - \Delta H, \langle H \rangle + \Delta H]$.

Proof: According to (3.5.19), there is a spectral value of $(H - \langle H \rangle)^2$ below $\langle (H - \langle H \rangle)^2 \rangle = (\Delta H)^2$, and hence H has a spectral value nearer to $\langle H \rangle$ than the distance ΔH. □

3.5.25 Remark

Criterion (3.5.24) can only be used after verifying that the eigenvalue in the interval is indeed the one wanted. For instance, it produces a lower bound for E_1 only if it is known that E_2 is greater than $\langle H \rangle + \Delta H$.

3.5.26 Duffin's Criterion of the Local Energy

Suppose that $H = |\mathbf{p}|^2 + V(\mathbf{x})$, $D(V) \supset D(|\mathbf{p}|^2)$ has isolated eigenvalues E_k. The eigenvector $\psi_1(\mathbf{x}) : H\psi_1 = E_1\psi_1$ can be assumed to be nonnegative. If $\psi(\mathbf{x}) > 0$ and $E(\mathbf{x}) \equiv H\psi(\mathbf{x})/\psi(\mathbf{x})$, then E_1 lies in the interval $[\inf_{\mathbf{x}} E(\mathbf{x}), \sup_{\mathbf{x}} E(\mathbf{x})]$.

Proof: Write $\psi_1(\mathbf{x}) = R(\mathbf{x})\exp(iS(\mathbf{x}))$, with R positive and S real (cf. (3.3.20; 5)); then

$$\begin{aligned}\langle\psi_1|H\psi_1\rangle &= \int dx(|\nabla R(\mathbf{x})|^2 + R^2(|\nabla S(\mathbf{x})|^2 + V(\mathbf{x}))) \\ &\geq \int dx(|\nabla R(\mathbf{x})|^2 + R^2 V(\mathbf{x})).\end{aligned}$$

For ψ_1 to give the lowest eigenvalue, S must be constant, and can be redefined as 0. This makes $\langle\psi_1|\psi\rangle > 0$, and

$$E_1\langle\psi_1|\psi\rangle = \langle H\psi_1|\psi\rangle = \int dx\, \psi_1(\mathbf{x})\psi(\mathbf{x})E(\mathbf{x}) \mathop{\geq \inf}^{\leq \sup} E(\mathbf{x})\cdot\langle\psi_1|\psi\rangle. \quad \square$$

3.5.27 Remark

This criterion does not involve integrals as in the calculation of expectation values, which is an advantage; but at least one of its error bounds is worse than the corresponding bound of (3.5.24) calculated with ψ:

$$\begin{aligned}(\Delta H)^2 &= \int dx(E(\mathbf{x}) - \langle H\rangle)^2\psi(\mathbf{x})^2 \\ &\leq \max\left\{(\langle H\rangle - \inf_{\mathbf{x}} E(\mathbf{x}))^2, (\langle H\rangle - \sup_{\mathbf{x}} E(\mathbf{x}))^2\right\}.\end{aligned}$$

If H is of the form $|\mathbf{p}|^2 + V(x)$, then, obviously, $E_1 \geq -\|V\|_\infty$. But even if $V \to -\infty$ somewhere, H may be bounded below. The uncertainty principle leads one to believe that this should always be the case when V approaches $-\infty$ more slowly than $-1/r^2$, which means that V is locally in L^p, $p > \frac{3}{2}$. This is in fact the case, as shown by the

3.5.28 General Lower Bound

In three dimensions,

$$|\mathbf{p}|^2 + V(\mathbf{x}) \geq -c_p\|V\|_p^{2p/(2p-3)}, \qquad p > \frac{3}{2},$$

$$c_p = \Gamma\left(\frac{2p-3}{p-1}\right)^{(2p-2)/(2p-3)}\left(\frac{p-1}{p}\right)^2(4\pi)^{-2/(2p-3)},$$

where the operator $|\mathbf{p}|^2 + V$ *is defined by the Friedrichs extension* (2.5.19).

Proof: The ground state ψ satisfies the equation

$$\psi(\mathbf{x}) = -\int d^3x' G(\mathbf{x}-\mathbf{x}')V(\mathbf{x}')\psi(\mathbf{x}'), \qquad G(\mathbf{x}) = \frac{\exp(-\sqrt{|E|}|\mathbf{x}|)}{4\pi|\mathbf{x}|}$$

because G is the Green function for $|\mathbf{p}|^2 - E$. Young's and Hölder's inequalities imply that if $p \geq 1$, then

$$\|\psi\|_2 \leq \|G\|_q \|V\|_p \|\psi\|_2, \qquad \frac{1}{p} + \frac{1}{q} = 1.$$

We thus calculate that

$$\begin{aligned} \|G\|_q &= \frac{1}{4\pi} \left[\int d^3x \frac{\exp(-q\sqrt{|E|}|\mathbf{x}|)}{|\mathbf{x}|^q} \right]^{1/q} \\ &= |E|^{(q-3)/2q} \cdot q^{(q-3)/q} \cdot \Gamma^{1/q}(3-q)(4\pi)^{-1/p}. \end{aligned}$$

Since this is finite up to the point $p = \frac{3}{2} \Leftrightarrow q = 3$, it can be substituted into the earlier inequality to get an upper bound on $|E_0|$ for the Friedrichs extension with $p > \frac{3}{2}, q < 3$. The argument does not work for arbitrary definitions of the sum of the operators $|\mathbf{p}|^2$ and V; the deficiency indices may be nonzero, and the lowest eigenvalue can be arbitrarily negative. □

If H' is positive, then lower bounds for E_k can be obtained from an eigenvalue problem restricted to some subspace (cf. (3.5.19) for contrast):

3.5.29 The Projection Method

Suppose that $H' \geq 0$ and $P = P^* = P^2$, so that $P(H')^{-1}P$ is bounded and invertible on $P\mathcal{H}$. Then the ordered sequence of eigenvalues of $H_0 + P(P(H')^{-1} \times P)^{-1}P$ are lower bounds of the ordered sequence of eigenvalues of $H_0 + H'$.

Application: For any projection Q, $Q \leq 1$, so $(H')^{1/2}Q(H')^{1/2} \leq H'$. If we take $Q = (H')^{-1/2}P(P(H')^{-1}P)^{-1}P(H')^{-1/2}$, then Q is a projection, since $Q = Q^* = Q^2$, and there results $H' \geq P(P(H')^{-1}P)^{-1}P$. Therefore, by the min-max principle, the ordered sequence of eigenvalues of

$$H_L \equiv H_0 + P(P(H')^{-1}P)^{-1}P$$

consists of lower bounds for the eigenvalues of $H = H_0 + H'$, since $H \geq H_L$, and so all expectation values satisfy $\langle\psi|H\psi\rangle \geq \langle\psi|H_L\psi\rangle$.. □

If $|L\rangle$ is an eigenvector of H_L and $P|L\rangle = 0$, then $|L\rangle$ must be one of the eigenvectors $|i\rangle$ of H_0, and thus E_i is eigenvalue of H_L. If $P|L\rangle \neq 0$, then we may write

$$P|L\rangle = -P(H')^{-1}P|\rangle,$$

which converts the eigenvalue equation into $(H_0 - E_L)|L\rangle = P|\rangle$, or the more convenient form

$$P|L\rangle = P(H_0 - E_L)^{-1}P|\rangle = -P(H')^{-1}P|\rangle.$$

3.5.30 Special Cases

1. Let $P = |\chi\rangle\langle\chi|$ be one-dimensional, so the comparison operator is $P \cdot \langle\chi|(H_0 - E_L)^{-1} + (H')^{-1}|\chi\rangle$. Since $E_i^{(0)}$ are trivial lower bounds, we need $\langle i|\chi\rangle \neq 0$, $i = 1, 2, \ldots, n$ in order to raise our estimates of the first n eigenvalues. If we let

$$|\chi\rangle = \sum_{i=1}^{n} c_i |i\rangle, \qquad \sum |c_i|^2 = 1,$$

then the problem becomes to solve the equation

$$\sum_i \frac{|c_i|^2}{E_i^{(0)} - E_L} + \sum_{i,k} c_i^* \langle i|(H')^{-1}|k\rangle c_k = 0.$$

Since $(H')^{-1} > 0$, there is always a solution for E_L between any $E_i^{(0)}$ and $E_{i+1}^{(0)}$. No one-dimensional projection can raise an eigenvalue above the next higher one. More specifically, $E_1^{(0)} + \langle 1|(H')^{-1}|1\rangle^{-1} \leq E_1 \leq E_1^{(0)} + \langle 1|H'|1\rangle$, whenever the left side is $\leq E_2^{(0)}$.

2. If we let $|\chi\rangle = c(H_0 - E_L)|\psi\rangle$, $\|\psi\| = 1$, then we need to find the least solution of

$$\langle\psi|H_0 - E_L + (H_0 - E_L)(H')^{-1}(H_0 - E_L)|\psi\rangle$$
$$= \langle\psi|H - E_L - (H - E_L)(H')^{-1}(H - E_L)|\psi\rangle = 0,$$

i.e.,

$$E_l = \langle\psi|H|\psi\rangle - \langle\psi|(H - E_L)(H')^{-1}(H - E_L)|\psi\rangle.$$

If $E_L \leq E_2^{(0)}$, then it will be a bound for E_1, and

$$\langle\psi|H|\psi\rangle - \langle\psi|(H - E_L)(H')^{-1}(H - E_L)|\psi\rangle \leq E_1 \leq \langle\psi|H|\psi\rangle,$$

provided that the left side is $\leq E_2^{(0)}$. In this case we are not confined to the use of eigenvectors of H_0, and may equip ψ with some parameters to vary and optimize the bounds. In addition, H can be written in various ways as a sum of H_0 and H'. If we let $H' = (E_2 - E_L) \cdot \mathbf{1}$ (why not, once we know that $E_L < E_2$), then

$$\langle\psi|(H - E_L)(E_2 - E_L) - (H - E_L)^2|\psi\rangle = 0,$$

which yields **Temple's inequality**,

$$E_1 \geq E_L = \langle H\rangle - \frac{(\Delta H)^2}{E_2 - \langle H\rangle}.$$

If $\Delta H < E_2 - \langle H\rangle$, this improves (3.5.24).

Once the E_k have been localized, the question can be raised of how well the trial function ψ approximates an actual eigenvector $|\rangle$. There is little chance for a general pointwise bound, but the accuracy in the L^2 norm i.e., in the mean-square sense, can be gauged in terms of inner products, for which there are useful estimates.

3.5.31 Bounds for the Overlap Integral

Let a constant be subtracted from H to adjust the lowest eigenvalue E_1 to 0, and let $|\rangle$ be its eigenvector. Then

$$1 - \frac{\langle H\rangle}{E_2} \leq |\langle\psi|\rangle|^2 \leq 1 - \frac{\langle H\rangle^2}{\langle H^2\rangle}$$

where $\langle\,\rangle$ is the expectation value in the state ψ. (These bounds are due respectively to Eckart and to Farnoux and Wang.)

Proof: Right side: $H\psi$ is orthogonal to $|\rangle$, so

$$|\langle|\psi\rangle|^2 \leq 1 - \frac{|\langle\psi|H\psi\rangle|^2}{\|H\psi\|^2}.$$

Left side: Let $P = 1 - |\rangle\langle|$. Then

$$\langle\psi|P(H - E_2)P|\psi\rangle = \langle\psi|H\psi\rangle - E_2(1 - |\langle|\psi\rangle|^2) \geq 0. \qquad \square$$

3.5.32 Remarks

1. It is not difficult to improve these bounds [5]. They show that the relevant facts for the accuracy of ψ in the L^2 sense are smallness of $\langle H\rangle - E_1$ and ΔH and a large isolation distance from E_2.
2. The upper bound holds only for the eigenvector of the ground state E_1, though similar lower bounds hold for excited states.

The motivation for the concepts that have been developed is the study of Hamiltonians of the form $H = |\mathbf{p}|^2 + V$, and we have assumed that there is only a discrete spectrum under the continuum on $\mathbb{R}^+$. The final topic of this section will be bounds on the number of bound states below a given energy; this in turn excludes σ_{ess} from below such an energy. The discussion begins with a lemma, which seems trivial for attractive potentials, but is surprising for potentials that are partially repulsive.

3.5.33 Monotony of $N(H)$ in the Coupling Constant

Let $H = |\mathbf{p}|^2 + \lambda V$ be such that $\sigma_{\text{ess}}(H)$ is contained in $\mathbb{R}^+$, and let $N(H) \equiv \operatorname{Tr}\Theta(-H)$ be the number of eigenvalues less than zero, counting multiplicity. Then $N(|\mathbf{p}|^2 + \lambda V)$ is a monotonically increasing function of λ for $\lambda > 0$.

Proof: $H_1 \leq H_2 \Rightarrow N(H_1) \geq N(H_2)$ and $N(\lambda H) = N(H)$ for all $\lambda > 0$, so

$$N(|\mathbf{p}|^2 + \lambda_1 V) \geq N\left(\frac{\lambda_1}{\lambda_2}(|\mathbf{p}|^2 + \lambda_2 V)\right) = N(\mathbf{p}|^2 + \lambda_2 V) \text{ for all } \lambda_1 \geq \lambda_2. \quad \square$$

The number of eigenvalues below $-c^2$, i.e., $N(H+c^2)$, can be estimated above by traces for potentials that are in some trace class relative to $|\mathbf{p}|^2$, that is, by certain integrals.

3.5.34 The Birman–Schwinger Bound

Let

$$|V|_- = \begin{cases} -V(\mathbf{x}), & \textit{where } V(\mathbf{x}) < 0 \\ 0, & \textit{otherwise.} \end{cases}$$

Then for all $p \geq 1$,

$$N(|\mathbf{p}|^2 + V + c^2) \leq \|(|\mathbf{p}|^2 + c^2)^{-1/2}|V|_-(|\mathbf{p}|^2 + c^2)^{-1/2}\|_p^p.$$

Proof: $N(|\mathbf{p}|^2 + V + c^2) \leq N(|\mathbf{p}|^2 - |V|_- + c^2)$, and all the eigenvalues of $|\mathbf{p}|^2 - \lambda|V|_- + c^2$ are continuous, decreasing functions λ. Hence $N(|\mathbf{p}|^2 - |V|_- + c^2)$ equals the number of values of $\lambda \leq 1$ for which $|\mathbf{p}|^2 - \lambda|V|_-$ has the eigenvalue $-c^2$ (see Figure 3.5). Since $|\mathbf{p}|^2 + c^2$ is invertible, it follows from $(|\mathbf{p}|^2 - \lambda|V|_-)\psi = -c^2\psi$ that

$$(|\mathbf{p}|^2 + c^2)^{-1/2}|V|_-(|\mathbf{p}|^2 + c^2)^{-1/2}\varphi = \frac{1}{\lambda}\varphi, \qquad \varphi = (|\mathbf{p}|^2 + c^2)^{1/2}\psi.$$

Thus $N(|\mathbf{p}|^2 - |V|_- + c^2) =$ [the number of eigenvalues $(1/\lambda_i) \geq 1$ of the operator $(|\mathbf{p}|^2 + c^2)^{-1/2}|V|_-(|\mathbf{p}|^2 + c^2)^{-1/2}$]

$$\leq \sum_i \left(\frac{1}{\lambda_i}\right)^p$$

$$= \mathrm{Tr}[(|\mathbf{p}|^2 + c^2)^{-1/2}|V|_-(|\mathbf{p}|^2 + c^2)^{-1/2}]^p. \qquad \square$$

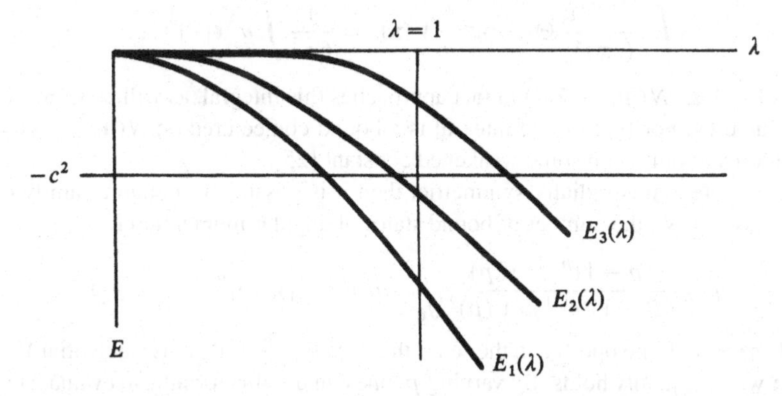

Figure 3.5. The eigenvalues as functions of the coupling constant.

3.5.35 Applications

1. Bound S-states. If V is radially symmetric, then one can ask about states of definite angular momentum l, where one thinks of the appropriate projection P_l onto an angular-momentum subspace as included in V. The operator $P_0(|\mathbf{p}|^2 + c^2)^{-1}P_0$ has an integral kernel

$$R(r, r') \equiv \frac{1}{crr'}[\sinh rc \exp(-r'c)\Theta(r' - r) + \sinh r'c \exp(-rc)\Theta(r - r')],$$

and, moreover, the bound involving $p = 1$,

$$\begin{aligned}\|(\mathbf{p}|^2 + c^2)^{-1/2}|V|_-(\mathbf{p}|^2 + c^2)^{-1/2}\|_1 &= \mathrm{Tr}|V|_-(|\mathbf{p}|^2 + c^2)^{-1} \\ = \int_0^\infty dr|V(r)|_- r^2 R(r, r) &= \int_0^\infty dr|V(r)|_- \frac{1 - \exp(-2rc)}{2c},\end{aligned}$$

may exist. For $c = 0$, this reduces to **Bargmann's bound**:

$$\text{(The number of bound } S\text{-states of } V(r)) \leq \int_0^\infty dr\, r|V(r)|_-.$$

2. In our discussion of free motion (3.3.3) in three dimensions, we saw that the integral kernel of $(|\mathbf{p}|^2+c^2)^{-1}$ was $\exp(-c|\mathbf{x} - \mathbf{x}'|)/4\pi|\mathbf{x} - \mathbf{x}'|$, which is infinite where $\mathbf{x} = \mathbf{x}'$. It is thus necessary to choose a larger exponent p; with $p = 2$, we get the bound of **Ghirardi and Rimini**:

$$\begin{aligned}N(|\mathbf{p}|^2 + V + c^2) \leq \left(\frac{1}{4\pi}\right)^2 \int d^3x\, d^3x' \frac{|V(\mathbf{x})|_-|V(\mathbf{x}')|_-}{|\mathbf{x} - \mathbf{x}'|^2} \\ \cdot \exp(-2c|\mathbf{x} - \mathbf{x}'|).\end{aligned}$$

3.5.36 Remarks

1. The classical analogue of $N(|\mathbf{p}|^2 + V)$ is the volume of phase space of negative energy,

$$\int \frac{d^3x d^3p}{(2\pi)^3}\Theta(-|\mathbf{p}|^2 - V(\mathbf{x})) = \frac{1}{6\pi^2}\int d^3x|V|_-^{3/2}.$$

As $\lambda \to \infty$, $N(|\mathbf{p}|^2 + \lambda V)$ in fact approaches this integral, as will be shown in volume IV. For finite λ the integral is a bound conjectured on $N(|\mathbf{p}|^2 + \lambda V)$, but proved only with some weakened constant [25].

2. If the potential is radially symmetric, then it is possible to obtain a family of bounds for N_l, the number of bound states of angular momentum l,

$$N_l \leq \frac{(p-1)^{p-1}\Gamma(2p)}{(2l+1)^{2p-1}p^p\Gamma(p)^2}\int_0^\infty dr\, r^{2p-1}|V(r)|_-^p, \qquad p \geq 1.$$

These bounds are optimal in the sense that for all $p \geq 1$, there is a potential $V_{l,p}$ for which equality holds. By varying p, one can use this formula to evaluate the number of bound states for most potentials to within a few percent.

3. The moments of the eigenvalues can be read off from N by [6]

$$\begin{aligned}\sum_i |E_i|^\gamma &= \mathrm{Tr}|H|^\gamma \Theta(-H) = \int_{-\infty}^0 dE|E|^\gamma \mathrm{Tr}\,\delta(E-H) \\ &= \int_{-\infty}^0 dE|E|^\gamma \frac{\partial}{\partial E} N(H-E) = \gamma \int_{-\infty}^0 dE|E|^{\gamma-1} N(H-E).\end{aligned}$$

3.5.37 Example

The Yukawa potential $V(r) = -\lambda \exp(-r)/r$, $\lambda > 0$. By (3.5.35; 1), the number of bound S-states is at most $\int_0^\infty dr|V(r)|r = \lambda$. If we use the trial function $\psi(\mathbf{x}) = u(r)/r$, $u(r) = (\lambda^3/8\pi)^{1/2} r \exp(-\lambda r/2)$ in (3.5.26), then $H\psi/\psi = -\lambda^2/4 + \lambda(1-\exp(-r))/r$, so we get the bounds $-\lambda^2/4 \le E_1 \le -\lambda^2/4 + \lambda$. The expectation values of $H = |\mathbf{p}|^2 + V$ and H^2 in the state ψ are

$$\langle \psi | H\psi \rangle = \frac{\lambda^2}{4} \frac{2\lambda + 1 - \lambda^2}{(1+\lambda)^2} = -\frac{\lambda^2}{4} + \lambda - \frac{3}{2} + O(1/\lambda)$$

and

$$\langle \psi | H^2 \psi \rangle = \frac{\lambda^4}{16}\left[5 - \frac{16\lambda + 12\lambda^2}{(1+\lambda)^2} + \frac{8\lambda}{2+\lambda}\right],$$

$$\Delta H = \frac{1}{2}\left(\frac{\lambda}{1+\lambda}\right)^2 \sqrt{\frac{2+3\lambda}{2+\lambda}}.$$

The min-max principle yields the upper bound $\langle \psi | H\psi \rangle$ for E_1, and (3.5.24) gives the lower bound $\langle \psi | H\psi \rangle - \Delta H$, once it is known that $E_2 > \langle \psi | H\psi \rangle + \Delta H$. Because $V > -\lambda/r$, $E_n > -\lambda^2/4n^2$ (see §4.1); consequently, for $\langle \psi | H\psi \rangle \le -\lambda^2/16$, we have definitely caught the eigenvalue E_1 between two bounds. For λ sufficiently large, the lower bound can be improved with Temple's inequality (3.5.30; 2), since $(\Delta H)^2 (E_2 - \langle H \rangle)^{-1} = O(\lambda^{-2})$. The projection method can make use of the exactly soluble case $H_0 = |\mathbf{p}|^2 - \lambda/r$ and $H' = \lambda(1-\exp(-r))/r$, yielding

$$\langle H'^{-1} \rangle = \frac{\lambda^2}{2}\int_0^\infty dr\, r^3 \frac{e^{-\lambda r}}{1-e^{-r}} = 3\sum_{n=0}^\infty \frac{\lambda^2}{(\lambda+n)^4} < \lambda^{-2}(3+\lambda),$$

so $-\lambda^2/4 + \lambda^2/(3+\lambda) \le E_1$. The general bound (3.5.28)

$$E \ge -\left(\int_0^\infty dr\, r^2 |V|^p\right)^{2/(2p-3)} \frac{\Gamma(3-q)^{2q/(3-q)}}{p^2}(p-1)^2$$

works only if $p < 3$, and thus never gives the actual asymptotic behavior $\sim \lambda^2$ as $\lambda \to \infty$; for instance, with $p = q = 2$, we get $-\lambda^4/16 \le E_1$. (See Figure 3.6.)

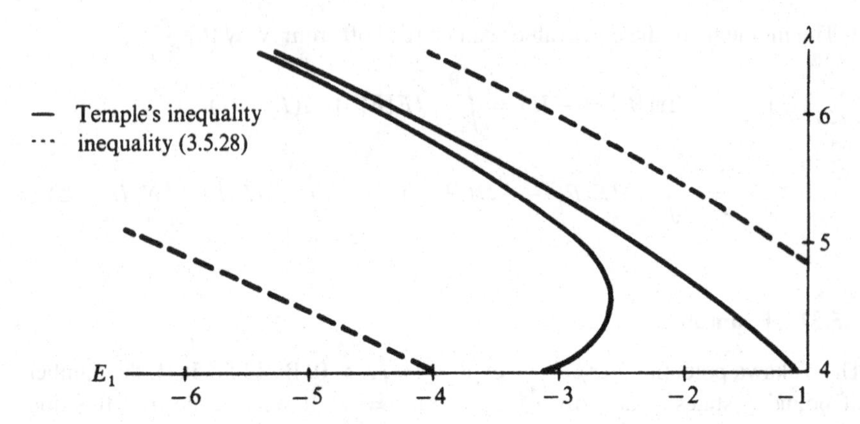

Figure 3.6. Bounds for the ground state with a Yukawa potential.

3.5.38 Problems

1. Show that $R(\alpha, z)$ has simple poles at E_k, and $P_k(\alpha) = P_k^*(\alpha^*)$, $P_i(\alpha)P_k(\alpha) = \delta_{ik}P_k(\alpha)$.

2. Show that a is compact iff $\psi_n \rightharpoonup 0 \Rightarrow a\psi_n \to 0$. (Hint: for the "only if" direction, recall that the strong and weak topologies are equivalent on strongly compact sets in Hilbert space).

3. Show that if H' is compact relative to H_0, then for all $\varepsilon > 0$ there exists a δ such that $\|H'\psi\| \leq \varepsilon\|H_0\psi\| + \delta\|\psi\|$ for all $\psi \in D(H_0)$. (The operator $a = H'(H_0 + i)^{-1}$ is compact. Let $P_n = \chi_{[-n,n]}(H_0)$, and show (i) that $\|a(\mathbf{1} - P_n)\| \to 0$, and (ii) that $H'P_n$ is bounded for all n.)

4. Show that for $|\alpha| < \delta$, $W(\alpha) = P(\alpha)[\mathbf{1} + P(0)(P(\alpha) - P(0))P(0)]^{-1/2}P(0)$, where W is as in the proof of (3.5.13).

5. Prove the min-max principle (3.5.19). (Use the unitary invariance of the trace and note that $\mathrm{Tr}_{D_1 \cup D_2} = \mathrm{Tr}_{D_1} + \mathrm{Tr}_{D_2}$ for D_1 orthogonal to D_2.)

6. Give an example of a 3×3 matrix for which the value $E_2 = \sup_{\|\psi\|=1} E_2(\psi)$, $E_2(\psi) = \inf_{\langle\varphi|\psi\rangle=0, \|\varphi\|=1}\langle\varphi|H\varphi\rangle$ is attained for some ψ other than ψ_1, the ground state.

3.5.39 Solutions

1. Let P_k be the projection onto the eigenvector for E_k. The Laurent series of $(H_0 - z)^{-1}$is $(E_k - z)^{-1}P_k + (\mathbf{1} - P_k)\times$ analytic factors.

$$2\pi i P_k(\alpha) = \oint_{c_k} dz\, R(\alpha, z) = -\oint_{-c_k} dz\, R(\alpha, z) = -\oint_{c_k} dz^*\, R(\alpha, z^*)$$
$$= -\left[\oint_{c_k} dz\, R(\alpha^*, z)\right]^*.$$

$$P_i(\alpha)P_k(\alpha) = (2\pi i)^{-2}\oint_{c_i}\oint_{c_k}\frac{dz\,dz'}{z - z'}[R(\alpha, z) - R(\alpha, z')].$$

$i = k$: $(2\pi i)^{-2} \int\int \cdots = (2\pi i)^{-1} \oint_{C_i} dz\, R(\alpha, z)$.

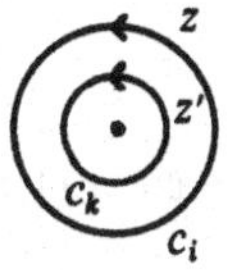

$i \neq k$: $(2\pi i)^{-2} \int\int \cdots = 0$.

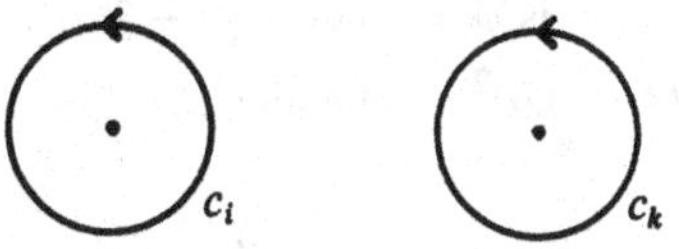

2. Lemma: Let $K \subset \mathcal{H}$ be strongly compact. If $M \subset K$ is weakly closed, then it is also strongly closed. Conversely, if M is strongly closed, then it is also strongly compact, hence weakly compact, hence weakly closed. (See also Problem (2.1.29; 7).)

   "only if": $\psi_n \rightharpoonup 0 \Rightarrow a\psi_n \rightharpoonup 0 \Rightarrow a\psi_n \to 0$ because of the lemma and because $\{\psi_n\}$ is bounded.

   "if": $(\psi_n \rightharpoonup 0 \Rightarrow a\psi_n \to 0) \Leftrightarrow (\psi_n \rightharpoonup \psi \Leftrightarrow a\psi_n \to a\psi)$.
   Now suppose X is a bounded set. To show that aX is strongly relatively compact, it suffices for every sequence $a\psi_n$, $\psi_n \in X$, to contain a strongly convergent subsequence. The sequence ψ_n contains a weakly convergent subsequence $\psi_{n_k} \rightharpoonup \psi \Rightarrow a\psi_{n_k} \to a\psi$.
3. (i) Let $a_n = a^*(\mathbf{1} - P_n)a$. $\|a(\mathbf{1} - P_n)\|^2 = \sup_{\|\psi\| \le 1} \langle \psi | a_n \psi \rangle$. If $\| \cdots \|^2$ were greater than some fixed positive C for all n, then the intersection of the decreasing sequence $\{\psi : \langle \psi | a_n \psi \rangle \ge C, \|\psi\| \le 1\}$ of weakly compact sets would be nonempty, so there would exist a ψ such that $\|\psi\| \le 1$ and $\langle \psi | a_n \psi \rangle \ge C$ for all n. This is impossible since $\mathbf{1} - P_n \to 0$.
   (ii) $H' P_n = a \int_{-n}^{n} (\alpha + i) dP_\alpha$ is the product of two bounded operators. Consequently, if $\psi \in D(H_0)$, then for n sufficiently large $\psi \in D(H_0)$,
   $$\begin{aligned} \|H'\psi\| &= \|a(H_0 + i)\psi\| \le \|a(\mathbf{1} - P_n)(H_0 + i)\psi\| + \|H' P_n \psi\| \\ &\le \varepsilon \|(H_0 + i)\psi\| + \|H' P_n\| \|\psi\|. \end{aligned}$$
4. $\|P(\alpha) - P(0)\| < 1$, $[\ \ ] > 0$, $[\ \ ]P_0 = P_0[\ \ ]$. $WW^* = P[\ \ ]^{-1/2} P_0 [\ \ ]^{-1/2} P = P P_0 [\ \ ]^{-1} P = P$. $W^* W = P_0 [\ \ ]^{-1/2} P [\ \ ]^{-1/2} P_0 = [\ \ ]^{-1/2} P_0 [\ \ ] P_0 [\ \ ]^{-1/2} P_0 = P_0$.
5. We consider only the infinite-dimensional case, so H has arbitrarily many eigenvalues greater than or equal to E_{i+j-1}. Let $H\psi_i = E_i \psi_i$, let $B_{i,j}$ be the subspace spanned by $\{\psi_i, \psi_{i+1}, \ldots, \psi_{i+j-1}\}$
   $$D_j^- = D_j \cap B_{1,i-1}, \quad D_j^0 = D_j \cap B_{i,j}, \quad D_j^+ = D_j \cap B_{i+j,\infty},$$
   and let d^α be the dimension of D_j^α. In the trace
   $$\mathrm{Tr}_{D_j} H = \mathrm{Tr}_{D_j^-} H + \mathrm{Tr}_{D_j^0} H + \mathrm{Tr}_{D_j^+} H$$
   the first contribution lies between $E_1 + E_2 + \cdots + E_{d^-}$ and $E_{i-d^- -1} + E_{i-d^-} + \cdots + E_{i-1}$, the second between $E_i + E_{i+1} + \cdots + E_{i+d^0-1}$ and $E_{i+j-d^0-1} + E_{i+j-d^0} + \cdots + E_{i+j-1}$, and the third is $\ge E_{i+j-1} d^+$. Hence
   $$\sup_{D_j \subset D_{i+j-1}} \mathrm{Tr}_{D_j} H \ge E_i + E_{i+1} + \cdots + E_{i+j-1}$$
   (equality holds for $D_{i+j-1} = B_{1,i+j-1}$); and
   $$\inf_{D_j \subset D_{i-1}^\perp} \mathrm{Tr}_{D_j} H \le E_i + E_{i+1} + \cdots + E_{i+j-1}$$

(equality for $D_{i-1} = B_{1,i-1}$).

6. $$H = \begin{bmatrix} 1 & 0 & 0 \\ 0 & 2 & 0 \\ 0 & 0 & 3 \end{bmatrix}, \quad \psi\left(\frac{1}{\sqrt{2}}, 0, \frac{1}{\sqrt{2}}\right) \neq \psi_1 = (1, 0, 0).$$

The general form of the φ's orthogonal to ψ is $\varphi = \left(\alpha, e^{i\varphi}\sqrt{1-2|\alpha|^2}, -\alpha\right)$, with $|\alpha| \leq \frac{1}{2}$, and $\langle\varphi|H\varphi\rangle = 2(1-2|\alpha|^2) + (3+1)|\alpha|^2 = 2$ for all α.

3.6 Stationary Scattering Theory

An explicit formula for the scattering operator S is obtainable with an Abelian limit, and analytic methods can be applied to it.

Historically, collision problems would be studied only with the methods of wave mechanics, and not with those of matrix mechanics. It has thus come to be believed that scattering theory should not be thought of as concerning the scattering of waves, and not of particles with observables **x** and **p**. More recently, the connections between the two points of view have become better understood, and as a result we shall be able to tie in directly with §3.4. Since we have by now learned which mathematical pitfalls are dangerous and which can be harmlessly circumvented, we shall indulge in formal manipulations without always pausing to investigate the finer points of rigor.

The Møller operators were introduced as time-limits of $\Omega(t) = \exp(iHt) \times \exp(-iH_0t)$. If they exist, then a fortiori (Problem 1) the limit as $\varepsilon \downarrow 0$ of $\varepsilon \int_0^\infty dt \exp(-\varepsilon t)\Omega(t)$ exists (cf. (I: 3.4.18)), which is an operator that no longer contains time explicitly. Since the integrand is an exponential function of t, the t-integration looks trivial at first sight, but because H and H_0 do not commute, it is not so simple. The difficulty can be eased with the partition of unity $\mathbf{1} = \int_0^\infty dE\,\delta(H_0 - E)$ given by the spectral representation of H_0, which we shall think of as $|\mathbf{p}|^2$. Then only the commuting variable E appears in the integral of the final exponential function, $\exp(-iH_0t) = \int_0^\infty dE \exp(-iEt)\delta(H_0 - E)$, and there is no further obstacle to the integration:

$$\begin{aligned} \Omega_\pm &= s\text{-}\lim_{\varepsilon\downarrow 0} \varepsilon \int_0^\infty dt \int_0^\infty dE \exp(\pm it(H - E \pm i\varepsilon))\delta(H_0 - E) \\ &= s\text{-}\lim_{\varepsilon\downarrow 0} \int_0^\infty \pm dE\, i\varepsilon(H - E \pm i\varepsilon)^{-1}\delta(H_0 - E) \\ &= \mathbf{1} - s\text{-}\lim_{\varepsilon\downarrow 0} \int_0^\infty dE(H - E \pm i\varepsilon)^{-1} V\delta(H_0 - E). \end{aligned} \tag{3.10}$$

3.6.1 Remarks

1. This means that $\Omega_\pm$ can be written in terms of the boundary values of the analytic function $z \to (H - z)^{-1}V$ on the branch cut $\mathbb{R}^+$. The V's that we shall deal

with are mapped by this function into compact operators. It will in addition be convenient to use the variable $k \equiv \sqrt{E}$ instead of E to simplify the integrals; the limits in $\Omega_\pm$ simply correspond to Im $k \downarrow 0$.

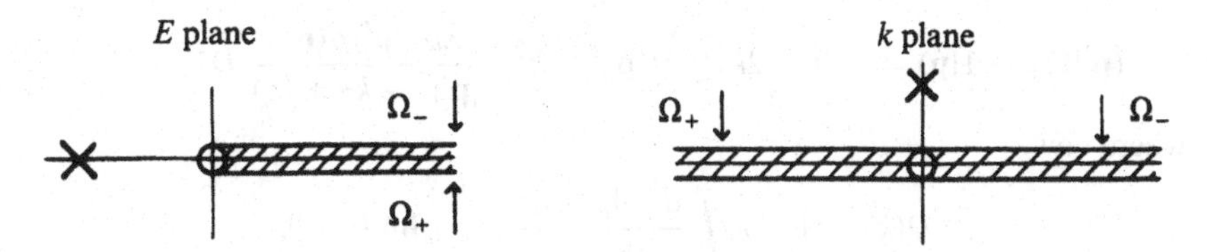

Under the right circumstances, the integrands can be continued analytically across the real axis:

$$\|V^{1/2}(H_0 - k^2)^{-1}V^{1/2}\|_2^2 = \int \frac{d^3x\, d^3x'}{(4\pi|\mathbf{x}-\mathbf{x}'|)^2} \times V(\mathbf{x})V(\mathbf{x}')\exp(i(k-k^*)|\mathbf{x}-\mathbf{x}'|)$$

is finite even if Im $k < 0$, provided that $V(\mathbf{x})$ falls off sufficiently rapidly. If we write

$$\Omega_\pm = s\text{-}\lim_{\varepsilon\downarrow 0} \int_0^\infty 2k\, dk\Omega(i\varepsilon \mp k)\delta(H_0 - k^2),$$

then ($V^{1/2} = V|V|^{-1/2}$)

$$\begin{aligned}\Omega(k) &\equiv 1 - (H - k^2)^{-1}V = (1 + (H_0 - k^2)^{-1}V)^{-1} \\ &= |V|^{-1/2}(1 + |V|^{1/2}(H_0 - k^2)^{-1}V^{1/2})^{-1}|V|^{1/2}\end{aligned}$$

has only poles, at the points where the compact operator in the denominator has the eigenvalue -1. The branch-point in the variable E at $E = 0$ disappears in the uniformizing variable k.

2. It is customary in wave mechanics to work outside the space L^2 and use plane waves $\varphi = \exp(i\mathbf{k}\cdot\mathbf{x})$ as eigenvectors of H_0. When multiplied by $\Omega_\pm$ they are turned into eigenvectors of H : $\psi_\pm = \Omega_\pm(\mathbf{k})\varphi$, which satisfy the **Lippmann–Schwinger equation**

$$\psi_\pm = \varphi - (H_0 - k^2 \pm i\varepsilon)^{-1}V\psi_\pm, \qquad k = |\mathbf{k}|,$$

because $\Omega(k) = (\mathbf{1} + (H_0 - k^2)^{-1}V)^{-1}$. In the x-representation, this reads

$$\psi_\pm(\mathbf{x}) = \exp(i\mathbf{k}\cdot\mathbf{x}) - \int \frac{d^2x' \exp(\mp ik|\mathbf{x}-\mathbf{x}'|)}{4\pi|\mathbf{x}-\mathbf{x}'|} V(\mathbf{x}')\psi_\pm(\mathbf{x}').$$

The new eigenvectors contain incoming, or respectively outgoing, spherical waves in addition to the plane waves.

3.6.2 Example

With the separable V of (3.4.13; 1), $\Omega(k) = \mathbf{1} - \lambda(H_0 - k^2)^{-1} P D^{-1}(k)$, or, written as an integral operator with a momentum-space kernel,

$$(\mathbf{p}'|\Omega_\pm - 1|\mathbf{p}) = -\int_0^\infty 2k\,dk\,\delta(|\mathbf{p}|^2 - k^2)\frac{\lambda\rho^*(\mathbf{p}')\rho(\mathbf{p})}{(|\mathbf{p}'|^2 - k^2 \pm i\varepsilon)} D^{-1}(\mp k),$$

where now

$$D(k) = 1 + \lambda \int \frac{d^3p|\rho(\mathbf{p})|^2}{|\mathbf{p}|^2 - k^2}, \qquad \operatorname{Im} k > 0.$$

If, say $\rho^2 = M^2/(|\mathbf{p}|^2 + M^2)|\mathbf{p}|^2$, $M > 0$, then $D(k) = 1 + (\lambda/4\pi)M^2/(M - ik)$. This function can be continued into the lower k-plane, equivalent to the second sheet in E, though it no longer equals the integral there, but instead develops a pole at $k = -iM$.

In the same way, the time-limit (3.4.22; 6) in S can be recast as an ε limit, the only difference being that the partition of unity is needed on both ends of the expression:

$$\begin{aligned} S &= s\text{-}\lim_{\varepsilon\downarrow 0} \varepsilon \int_0^\infty dE\,dE'\,\delta(H_0 - E) \\ &\quad \times \exp\left(-it\left(H - \frac{E+E'}{2} - i\varepsilon\right)\right)\delta(H_0 - E')dt \\ &= s\text{-}\lim_{\varepsilon\downarrow 0} \int_0^\infty dE\,dE'\,\delta(H_0 - E)\frac{-i\varepsilon}{H - (E+E')/2 - i\varepsilon}\delta(H_0 - E'). \end{aligned} \tag{3.11}$$

With the second iteration of the resolvent formula (suppose the domains of H_0 and H are equal),

$$(H - z)^{-1} = (H_0 - z)^{-1} - (H_0 - z)^{-1}[V - V(H - z)^{-1}V](H_0 - z)^{-1}$$

and the limit

$$\lim_{\varepsilon\downarrow 0} \frac{-i\varepsilon}{((E - E')/2 - i\varepsilon)((E' - E)/2 - i\varepsilon)} = 2\pi i\delta(E - E'),$$

these results

$$S = s\text{-}\lim_{\varepsilon\to 0} \int_0^\infty dE\{\mathbf{1} - 2\pi i\delta(H_0 - E)[V - V(H - E - i\varepsilon)^{-1}V]\}\delta(H_0 - E). \tag{3.12}$$

(As usual, S is in the interaction representation.) In order to discuss (3.12), we need another operator-valued analytic function of the uniformizing variable k (see (3.6.1; 1)):

$$\begin{aligned} t(k) &\equiv V\Omega_-(k) = V - V(H - k^2)^{-1}V \\ &= V^{1/2}[\mathbf{1} + |V|^{1/2}(H_0 - k^2)^{-1}V^{1/2}]^{-1}|V|^{1/2} \\ &= [V^{-1} + (H_0 - k^2)^{-1}]^{-1} = t^*(-k^*). \end{aligned} \tag{3.13}$$

The domains of definition, especially that of V^{-1}, will have to be checked later. As Im $k \downarrow 0$,

$$\begin{aligned} t^{-1}(k) - t^{-1}(-k) &= \lim_{\varepsilon\downarrow 0}[(H_0 - k^2 - i\varepsilon)^{-1} - (H_0 - k^2 + i\varepsilon)^{-1}] \\ &= 2\pi i \delta(H_0 - k^2) \end{aligned}$$

in the sense of convergence of quadratic forms, and we get

$$\mathbf{1} - 2\pi i \delta(H_0 - k^2) t(k) = (t^{-1}(k) - 2\pi i \delta(H_0 - k^2)) t(k) = t^{-1}(-k) t(k) :$$

3.6.3 The Spectral Representation of the S Matrix

$$S = \int_0^\infty 2k\, dk\, S(k) \delta(H_0 - k^2),$$

$$\begin{aligned} S(k) &= t^{-1}(-k) t(k) = s\text{-}\lim_{\varepsilon\downarrow 0}[\mathbf{1} + (H_0 - k^2 + i\varepsilon)^{-1} V] \\ &\quad \cdot [\mathbf{1} + (H_0 - k^2 - i\varepsilon)^{-1} V]^{-1}. \end{aligned}$$

From this there follow the

3.6.4 Unitary Properties of the S Matrix

(i) $S(k)S(-k) = \mathbf{1}$ on the domain of analyticity.

(ii) $\delta(H_0 - k^2) S(k)^* = S(-k) \delta(H_0 - k^2)$ for k real.

3.6.5 Example

Recall (3.6.2), with $V = \lambda P$, but here $\not\exists V^{-1}$. We find that $t(k) = \lambda P D^{-1}$ and $S(k) = \mathbf{1} - 2\pi i \delta(H_0 - k^2) \lambda P D^{-1}$. Since $\delta(H_0 - k^2) P \delta(H_0 - k^2) = \delta(H_0 - k^2) |\rho(k)|^2 \times (k/4\pi) P_0$, where P_l is the projection onto the states of angular momentum $|\mathbf{L}|^2 = l(l+1)$, and $D(k) - D(-k) = 2\pi i (k/4\pi) |\rho(k)|^2$, we find that

$$S = \int_0^\infty 2k\, dk\, \delta(H_0 - k^2) \left(P_0 \frac{D(-k)}{D(k)} + 1 - P_0 \right).$$

3.6.6 Remarks

1. Since $[S, H_0] = 0$ it follows that even though $S(k)$ maps functions off the energy shell $H_0 = k^2$, $S(k)\delta(H_0 - k^2)$ does not. The unitary relation $\delta(H_0 - k'^2) S(k')^* S(k) \delta(H_0 - k^2) = \delta(H_0 - k^2) \delta(k'^2 - k^2)$ then holds on the energy shell, and hence $S(k)\delta(H_0 - k^2)$ can be written[1] as $\exp(2i\delta(k))\delta(H_0 - k^2)$,

[1] It is unfortunately traditional to use the same letter for the phase-shift $\delta(k)$ and for Dirac's delta function. The reader should be alert for any possible confusion.

where $\delta(k) = \delta(k)^* = -\delta(-k)$. Making use of the spectral representation of $H_0 = |\mathbf{p}|^2$, we can write $\mathcal{H} = L^2(\mathbb{R}^+, 2k\,dk) \otimes L^2(S^2, d\Omega)$, and the operator $\delta(k)$ maps the angular part $L^2(S^2, d\Omega)$ onto itself. The operator $\delta \equiv \delta(\sqrt{H_0})$ then acts on all of $\mathcal{H}$, and $[\delta, H_0] = 0$. If V is spherically symmetric, $[\delta, \mathbf{L}] = 0$, so $\delta(k) = \sum_l \delta_l(k) P_l$, for $\delta_l(k) \in \mathbb{R}$. Then in the diagonal representation of $|\mathbf{L}|^2$, S becomes a multiplication operator in H_0:

$$S = \int_0^\infty dk^2 \delta(H_0 - k^2) \exp(2i\delta(k)) = \sum_l \int_0^\infty dk^2 \delta(H_0 - k^2) P_l \exp(2i\delta_l(k)).$$

2. The unitarity of S implies the **Low equation** for t (as before, defined with $\operatorname{Im} k \downarrow 0$):

$$\begin{aligned} t(-k) - t(k) &= 2\pi i V \Omega \delta(H_0 - k^2) \Omega^* V \\ &= 2\pi i t(k) \delta(H_0 - k^2) t(-k), \qquad k \in \mathbb{R}. \end{aligned}$$

3. If the Lippmann-Schwinger equation (3.6.1; 2) is written as $\psi_- = \varphi - (H_0 - k^2)^{-1} t(k)\varphi$, and if we use $\varphi = \exp(ik\mathbf{n} \cdot \mathbf{x})$, $\mathbf{k} = k\mathbf{n}$, noting that for $|\mathbf{x}| \gg |\mathbf{x}'|$,

$$\frac{\exp(ik|\mathbf{x} - \mathbf{x}'|)}{|\mathbf{x} - \mathbf{x}'|} \sim \frac{\exp(ikr)}{r} \exp(-ik\mathbf{n}' \cdot \mathbf{x}'), \qquad \mathbf{n}' = \frac{\mathbf{x}}{r},$$

then as $|\mathbf{x}| \to \infty$, in the x-representation,

$$\psi_-(\mathbf{x}) = \exp(i\mathbf{k} \cdot \mathbf{x}) + \frac{\exp(ikr)}{r} f(k; \mathbf{n}', \mathbf{n}),$$

$$f(k; \mathbf{n}', \mathbf{n}) \equiv \frac{-1}{4\pi} \int d^3x' d^3x'' \exp(-ik\mathbf{n}' \cdot \mathbf{x}') \langle x'|t(k)|x''\rangle \exp(ik\mathbf{x} \cdot \mathbf{x}'').$$

The angular dependence f of the outgoing spherical wave is thus determined by t in momentum space on the energy shell. Only this part of t shows up in $\exp(2\pi i\delta(k))\delta(H_0 - k^2) = (1 - 2\pi i\delta(H_0 - k^2)t(k))\delta(H_0 - k^2)$. In particular, if $[t(k), \mathbf{L}] = 0$, then by comparing coefficients (Problem 6),

$$\begin{aligned} f(k; \mathbf{n}', \mathbf{n}) &= \sum_l \langle \mathbf{n}'|P_l|\mathbf{n}\rangle \frac{\exp(2i\delta_l(k)) - 1}{2ik} \\ &= \sum_l \frac{2l+1}{k} P_l(\cos\theta) \exp(i\delta_l(k)) \sin\delta_l(k), \qquad \theta = \measuredangle(\mathbf{n}', \mathbf{n}). \end{aligned}$$

If the plane wave is expanded in spherical harmonics $\exp(i\mathbf{k} \cdot \mathbf{x}) = (\exp(ikr) - \exp(-ikr))/2ikr + \cdots$, then ψ_- becomes asymptotically $(\exp(i(kr + \delta(k))) - \exp(-i(kr + \delta(k)))/2ikr + \cdots$, which shows the significance of δ as the **phase-shift** of a spherical wave.

4. If there are several channels (see (3.4.22; 6)), then the generalization of (3.12) is

$$S_{\alpha\beta} = \int_0^\infty dE \left[\delta_{\alpha\beta} - 2\pi i \delta(H_\alpha - E) \left(V_\alpha - V_\alpha \frac{1}{H - E - i\varepsilon} V_\beta\right)\right] \delta(H_\beta - E).$$

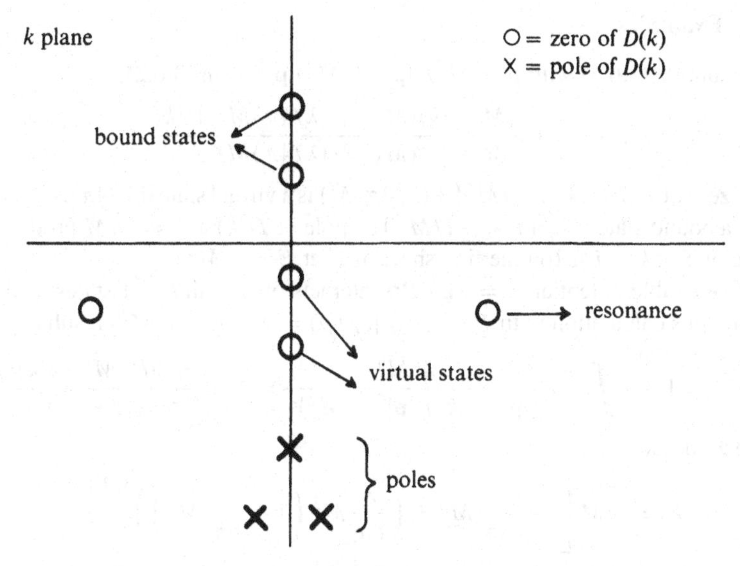

Figure 3.7. The configuration of the poles and zeroes of $D(k)$.

We shall assume in what follows that V decreases fast enough at infinity that the norm in (3.6.1; 1) remains finite for all k with $\mathrm{Im}\, k > \kappa_0 < 0$. Then for such k,

$$S(k) = V^{-1/2} D(-k) D^{-1}(k) V^{1/2}, \qquad D(k) = 1 + |V|^{1/2}(H_0 - k^2)^{-1}|V|^{1/2}$$

is a meromorphic function taking values in $\mathcal{B}(\mathcal{H})$. Our next topic is

3.6.7 The Configuration of the Poles of $S(k)$

The scattering operator S has a pole at any value of k at which either $D(k)$ has eigenvalue zero or $D(-k)$ has a pole. Both poles and zeroes occur for $-k^*$ whenever they occur for k. $D(k)$ has no poles in the upper half-plane, but only zeroes, and those are restricted to the imaginary axis.

As depicted in Figure 3.7, the terminology for these values of k is:

zeroes with $\mathrm{Im}\, k > 0$	**bound states**
zeroes with $\mathrm{Im}\, k < 0$ and $\mathrm{Re}\, k = 0$	**virtual states**
zeroes with $\mathrm{Re}\, k \neq 0$	**resonances**

Proof: $D(-k^*) = KD(k)K$ is the time-reversed version of $D(k)$ (cf. (3.3.18; 2)), so the two operators have the same poles and zeros. If $0 = D(k)\psi = \psi + |V|^{1/2}(H_0 - E)^{-1}|V|^{1/2}\psi$, then $(H_0 + V - E)\varphi = 0$, where $\varphi \equiv (H_0 - E)^{-1}|V|^{1/2}\psi$. If V decreases sufficiently fast, then φ is square-integrable whenever ψ is, and for such potentials the equation for φ can be solved in L^2 only if $E < 0$, i.e., for k purely imaginary. Complex zeroes and poles can appear after analytic continuation to the second sheet of E, which is the lower half-plane in k. □

3.6.8 Examples

1. In Example (3.6.5) with $\rho = M^2/(|\mathbf{p}|^2 + M^2)|\mathbf{p}|^2$ as in (3.6.2),
$$S(k) = P_0 \frac{(M - ik)(M(1 + (\lambda/4\pi)M) + ik)}{(M + ik)(M(1 + (\lambda/4\pi)M) - ik)} .$$
The zero of $D(k)$ at $k = -iM(1+(\lambda/4\pi)M)$ is a virtual state if $\lambda/4\pi > -1/M$, and a bound state if $\lambda/4\pi < -1/M$. The pole of $D(k)$ at $k = -iM$ produces a pole of S at $k = iM$ (on the first sheet of E at $E = -M^2$).
2. The separable potential $V = \lambda \mathbf{p} \cdot P\mathbf{p}$ interacts only with $l = 1$ states, and the analogous calculation with $P = |\rho\rangle\langle\rho|$, $\rho(p) = M^2/(p^2 + M^2)$ results in
$$D(k) = 1 + \lambda \int d^3p \frac{|\mathbf{p}|^2 M^4}{(|\mathbf{p}|^2 - k^2)(|\mathbf{p}|^2 + M^2)^4} = 1 + \frac{\lambda}{8\pi} \frac{M^2(M - 2ik)}{(M - ik)^2} .$$
The zeroes at
$$k = -iM \left[1 + \frac{\lambda}{8\pi} M^3 \pm \left(\frac{\lambda}{8\pi} M^3 \left(1 + \frac{\lambda}{8\pi} M^3 \right) \right)^{1/2} \right]$$
are virtual states if $\lambda > 0$, resonances if $-8\pi/M^3 < \lambda < 0$, and if $\lambda < -8\pi/M^3$ there is one bound and one virtual state.

3.6.9 Remarks

1. The poles of $D(k)$ were originally called spurious poles, since it was assumed that all poles of $S(k)$ on the first sheet of E should correspond to bound states. The poles of $D(k)$ have no physical significance, and serve only to show at what point analytic continuation makes the $\| \|_2$ norm in (3.6.1; 1) diverge.
2. $S(k)$ is determined by the phase of $D(k)$, and D approaches 1 at infinity in the upper half-plane. If we normalize the $\delta(k)$, defined in (3.6.6; 1) only modulo π, by setting $\delta(0) = 0$, then a well-known theorem of analytic function theory implies that $-\delta(\infty) = \pi$ times the number of bound states. The more general version of this fact is

3.6.10 Levinson's Theorem

Let V be compact relative to H_0, and suppose $\mathrm{Tr}|(H_0 - z)^{-1} - (H - z)^{-1}| \le M(z)$, *where* $M(z) \le O(|z|^{-1-\varepsilon})$ *as* $|z| \to \infty$ *and* $O(|\mathrm{Im}\, z|^{-1+\varepsilon})$ *as* $\mathrm{Im}\, z \to 0$, $\mathrm{Re}\, z > 0$, $\varepsilon > 0$. *Then* 2π *times the number of bound states equals* $i \lim_{k\to\infty} \ln \mathrm{Det}(S(k) - S(0)) = i \lim_{k\to\infty} \mathrm{Tr} \ln(S(k) - S(0))$ *if* $0 \notin \sigma_p(H)$, *so that* $S(0)$ *is well-defined.*

3.6.11 Gloss

$\ln(\mathbf{1} + A) \equiv \sum_{n=1}^{\infty}((-1)^n/n)A^n$ is defined for all A with $\|A\| < 1$, and $\mathrm{Det}(\mathbf{1} + A) \equiv \exp(\mathrm{Tr}\, \ln(\mathbf{1} + A))$ is defined for all A with $\|A\|_1 < \infty$. In general $\ln(AB) \neq \ln A + \ln B$, but $\mathrm{Det}(1 + A)(1 + B) = \mathrm{Det}(1 + B)(1 + A)$ for all $A + B + AB \in \mathcal{C}_1$, and $= \mathrm{Det}(1+A)\mathrm{Det}(1+B)$ for all $A, B \in \mathcal{C}_1$ [16]. If $A(z) : \mathbb{C} \to \mathcal{C}_1$ is analytic, then it follows that in the domain of analyticity

$$\mathrm{Tr}\,\frac{d}{dz}\ln(\mathbf{1} + A(z)) = \mathrm{Tr}(\mathbf{1} + A(z))^{-1}A'(z).$$

Proof of (3.6.10)**:** Let $Q_\pm(E) = \mathbf{1} + (H_0 - E \pm i\varepsilon)^{-1}V$, $S(E) = Q_+(E)Q_-^{-1}(E)$. Although $(H_0 - z)^{-1}V$ is compact, it is not trace-class. However, differences of two such terms with different z are trace-class, since $(H_0 - z_1)^{-1}V(H_0 - z_2)^{-1} = [(H_0 - z_1)^{-1} - (H - z_1)^{-1}][\mathbf{1} + (V - z_1 + z_2)(H_0 - z_2)^{-1}]$. Thus justifies the following formal manipulations:

$$\begin{aligned}
\mathrm{Tr}\,\frac{d}{dE}\ln S(E) &= \mathrm{Tr}\, Q_- Q_+^{-1}[Q'_+ Q_-^{-1} - Q_+ Q_-^{-1} Q'_- Q_-^{-1}] \\
&= \mathrm{Tr}(Q_+^{-1}Q'_+ - Q_-^{-1}Q'_-) \\
&= \mathrm{Tr}\{[\mathbf{1} + (H_0 - E + i\varepsilon)^{-1}V]^{-1}(H_0 - E + i\varepsilon)^{-2}V \\
&\qquad -(\varepsilon \leftrightarrow -\varepsilon)\} \\
&= \mathrm{Tr}\left[\frac{1}{H_0 - E + i\varepsilon} - \frac{1}{H - E + i\varepsilon} - (\varepsilon \leftrightarrow -\varepsilon)\right].
\end{aligned}$$

If we do the integration over E, then

$$\mathrm{Tr}\ln S(E) = \mathrm{Tr}\int_C dz((H - z)^{-1} - (H_0 - z)^{-1}),$$

where C, the contour of the complex integration, is as shown below:

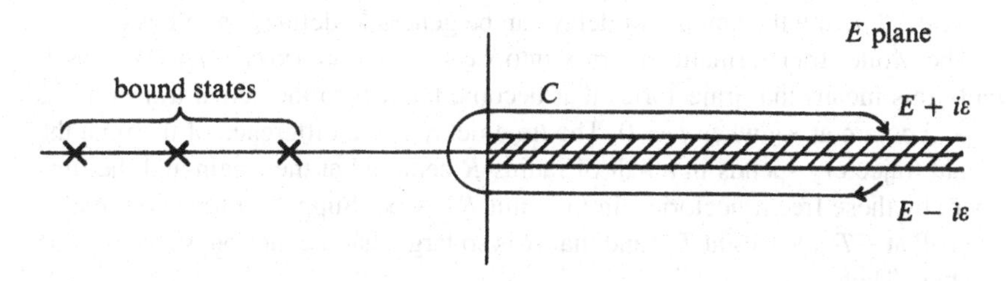

By assumption, the circle $K : |z| = E$ can be appended to C, since the extra contribution goes to 0 as $E \to \infty$. In this limit $C \cup K$ encircles all the poles of $(H - z)^{-1}$, but does not contain $\mathrm{Sp}(H_0)$. The proposition then follows from the residue theorem. □

3.6.12 Example

In the case of the separable potential (3.4.13; 1),

$$\begin{aligned}
\mathrm{Tr}((H_0 - z)^{-1} - (H - z)^{-1}) &= \lambda \mathrm{Tr}(H_0 - z)^{-1}P(H_0 - z)^{-1}D^{-1}(z) \\
&= D^{-1}(z)\frac{\partial}{\partial z}D(z).
\end{aligned}$$

In (3.6.2) we found that $D(z) = 1+(\lambda/4\pi)M^2/(M-i\sqrt{z})$ for $\rho^2(p) = M^2(|\mathbf{p}|^2+M^2)^{-1}$. Therefore

$$\frac{\partial}{\partial z} D(z) = \frac{i}{2\sqrt{z}} \frac{\lambda}{4\pi} \frac{M^2}{(M-i\sqrt{z})^2},$$

and the assumptions of (3.6.10) are satisfies. In fact

$$\delta(k) = \arctan \frac{kM^2\lambda/4\pi}{M^2(1+(\lambda/4\pi)M)+k^2}$$

has the limits $\delta(0) = 0$, $\delta(\infty) = 0$ if $1+(\lambda/4\pi)M > 0$, and $\delta(\infty) = -\pi$ if $1+(\lambda/4\pi)M < 0$. If, however, we pass to the limit $M \to \infty$, $\lambda \uparrow 0$, so that $\lambda_r \equiv M(1+(\lambda/4\pi)M)$ stays finite, then $S(k) = (\lambda_r - ik)/(\lambda_r + ik)$. In this case, $\delta(k) = -\arctan(k/\lambda_r)$ varies between 0 and $\pi/2$ times signum λ_r. For $\lambda_r < 0$ (a virtual state) as well as for $\lambda_r > 0$ (a bound state) (3.6.10) is violated, since $D(\infty) \neq 1$.

The classical scattering transformation (I: §3.4) for, say, a particle in a central potential in $\mathbb{R}^2$, is a canonical transformation that leaves p_r and L asymptotically invariant as $r \to \infty$, and as a consequence has an asymptotic generator $2\delta(p_r, L)$:

$$(r, \theta; p_r, L) \to \left(r - 2\frac{\partial\delta}{\partial p_r}, \theta - 2\frac{\partial\delta}{\partial L}; p_r, L\right).$$

Thus the generator contains information about the scattering angle $-2\partial\,\delta/\partial L$ and about $-2\partial\,\delta/\partial p_r$, the amount by which a particle evolving according to H outdistances one evolving according to H_0. This distance corresponds to a delay time $2(m/p)(\partial\delta/\partial p_r)$. Similarly, in quantum theory $\exp(-2i\delta(p))x\exp(2i\delta(p)) = x - 2\partial\,\delta/\partial p$, and the amount of delay can be generally defined as follows.

The Møller transformations turn $\mathbf{x}$ into $\mathbf{x}_\pm \equiv \lim_{t\to\pm\infty}(\mathbf{x}(t) - t\mathbf{p}(t))$. Classically this means that trajectories that become tangent to the actual trajectory as $t \to \pm\infty$ are at $\mathbf{x}_\pm$ when $t = 0$. The time-delay is the difference of the time the actual trajectory spends in a ball of radius R centered at the origin and the time spent by these free trajectories, in the limit $R \to \infty$. Suppose a trajectory enters the ball at $-T_-$, leaves at T_+, and that R is so large that the motion is free outside the ball. Then

$$\mathbf{x}(\pm T_\pm) = \mathbf{x}_\pm \pm T_\pm \mathbf{p}_\pm,$$

and if this equation is multiplied by $\mathbf{p}_\pm$, we find that

$$T_+ + T_- = \frac{|\mathbf{p}_+|\sqrt{R^2 - b_+^2} - \mathbf{x}_+ \cdot \mathbf{p}_+}{|\mathbf{p}_+|^2} + \frac{|\mathbf{p}_-|\sqrt{R^2 - b_-^2} + \mathbf{x}_- \cdot \mathbf{p}_-}{|\mathbf{p}_-|^2},$$

where $b_\pm$ are the smallest distances the free trajectories come to the origin. The times spent by the free trajectories in the ball are $2(\sqrt{R^2 - b_\pm^2}/|\mathbf{p}_\pm|)$, so with $R \to \infty$, the time-delay

$$D \equiv \text{time of actual trajectory} - \text{time of free trajectory}$$

$$= \frac{\mathbf{x}_- \cdot \mathbf{p}_- - \mathbf{x}_+ \cdot \mathbf{p}_+}{|\mathbf{p}_+|^2}.$$

It turns out that there is a direct relationship between D and the S matrix and the virial:

3.6.13 Definitions of the Time-Delay

(i) $D = \Omega_-(1/|\mathbf{p}|)(\mathbf{x}\cdot\mathbf{p}+\mathbf{p}\cdot\mathbf{x})(1/2|\mathbf{p}|)\Omega_-^* - \Omega_+(1/|\mathbf{p}|)(\mathbf{x}\cdot\mathbf{p}+\mathbf{p}\cdot\mathbf{x})(1/2|\mathbf{p}|)\Omega_+^*$;

(ii) $D = P_{ac}(1/\sqrt{H})\int_{-\infty}^{\infty} dt(2V_t + \mathbf{x}_t \cdot \nabla V_t)(1/\sqrt{H})P_{ac}$;

(iii) $D = w\text{-}\lim_{R\to\infty} \Omega_- \int_{-\infty}^{\infty} dt\{\exp(iHt)\Theta(R^2 - |\mathbf{x}|^2)\exp(-iHt)$
$- \exp(iH_0t)\Theta(R^2 - |\mathbf{x}|^2)\exp(-iH_0t)\}\Omega_-^*$;

(iv) $D = -i\Omega_- S^{-1}\int_0^{\infty} dE\, \delta(H_0 - E)(\partial S(E)/(\partial E)\Omega_-^*$;

where $H_0 = |\mathbf{p}|^2/2$ and $H = H_0 + V$.

3.6.14 Remarks

1. Definitions (i) and (ii) are possible whenever scattering theory works, i.e., for V and $x \cdot \nabla V$ falling off as $r^{-1-\varepsilon}$. In Definition (iv), however, it has so far been shown that $\partial S/\partial E$ is well-defined only for $r^{-4-\varepsilon}$ fall off.
2. It is clear because of its classical meaning that D should be independent of the choice of the point $\mathbf{x}$ on the trajectory. Therefore D should commute with H. This follows formally from (i), since

$$\begin{aligned} \exp(iHt)D\exp(-iHt) &= \Omega_- \frac{1}{|\mathbf{p}|}\{(\mathbf{x}+\mathbf{p}t)\cdot\mathbf{p} + \mathbf{p}\cdot(\mathbf{x}+\mathbf{p}t)\}\frac{1}{2|\mathbf{p}|}\Omega_-^* \\ &\quad -\Omega_+ \frac{1}{|\mathbf{p}|}\{(\mathbf{x}+\mathbf{p}t)\cdot\mathbf{p} + \mathbf{p}\cdot(\mathbf{x}+\mathbf{p}t)\}\frac{1}{2|\mathbf{p}|}\Omega_+^* \\ &= D. \end{aligned}$$

However, D is different for different trajectories; like S, it does not commute with spatial translations.
3. For repulsive potentials $\sim r^{-\nu}$,

$$D = (2-\nu)P_{ac}\frac{1}{\sqrt{H}}\int_{-\infty}^{\infty} dt\, V(t)\frac{1}{\sqrt{H}}P_{ac}.$$

If $\nu = 2$, then $D = 0$. If $\nu > 2$, then $D < 0$. This means that the actual trajectory spends less time in the ball than the straight trajectory does, since its path is shorter. If $\nu < 2$, then D is positive, and the dominant effect of V is to brake the particle. Note that from (iv), the phase-shift for such potentials is a monotonic function of E.
4. In the wave picture an incident wave $\exp(-ikr)$ is turned into $\exp(i(kr + 2\delta(k)))$. If we assume a wave-packet narrowly concentrated about k_0 and expand

$\delta(k) = \delta(k_0) + (k - k_0)(\partial\delta(k_0)/\partial k_0) + \cdots$, then the coefficient of k becomes $r + 2(\partial\delta(k)/\partial k)$ instead of r. Thus the center of the wave-packet is shifted from $r = k_0 t$ to $r = k_0(t - 2(\partial\delta/\partial E))$ after the scattering.

5. If there are resonances at $\pm k_r - ib$, then

$$S(k) = \frac{(-k - k_r + ib)(-k + k_r + ib)}{(k - k_r + ib)(k + k_r + ib)} \times \text{slowly varying factors.}$$

Ignoring the slowly varying parts,

$$-i\frac{\partial}{\partial k^2} \ln S(k) \cong \frac{b}{k}\left[\frac{1}{(k - k_r)^2 + b^2} + \frac{1}{(k + k_r)^2 + b^2}\right].$$

If $b \ll k_r$, then there is a sharp maximum $\sim 1/(bk_r)$ at the resonance energy, at which $\delta(k)$ passes rapidly through 90°. For this reason $1/(bk_r)$ can be thought as a lifetime, which can become so extremely long that there is hardly any difference between resonances and bound states. This happens for α particles radiating from nuclei.

6. If the potential is radial, then $2\delta_t(k)$ is the same as the classical generator of the scattering transformation, and (3.6.13) reduces to the classical formula.

The Equivalence of the Definitions of D. (i) $\Leftrightarrow$ (ii): Introduce the generator $G \equiv (\mathbf{x}\cdot\mathbf{p} + \mathbf{p}\cdot\mathbf{x})/2$ of dilatations. On the one hand

$$\begin{aligned} &i P_{ac}\frac{1}{\sqrt{H}}\int_{-T}^{T} dt\, \exp(iHt)[G, H]\exp(-iHt)\frac{1}{\sqrt{H}}P_{ac} \\ &= P_{ac}\frac{1}{\sqrt{H}}(\exp(-iHT)G\exp(iHT) \\ &\quad - \exp(iHT)G\exp(-iHT))\frac{1}{\sqrt{H}}P_{ac}, \end{aligned}$$

and on the other hand this equals (see (3.3.19; 8))

$$\begin{aligned} &P_{ac}\frac{1}{\sqrt{H}}\int_{-T}^{T} dt\, \exp(iHt)(\mathbf{x}\cdot\nabla V - |\mathbf{p}|^2)\exp(-iHt)\frac{1}{\sqrt{H}}P_{ac} \\ &= -4T P_{ac} + P_{ac}\frac{1}{\sqrt{H}}\int_{-T}^{T} dt\, \exp(iHt)(2V + \mathbf{x}\cdot\nabla V) \\ &\quad \cdot \exp(-iHt)\frac{1}{\sqrt{H}}P_{ac}. \end{aligned}$$

Now $G = \exp(-iH_0 t)G\exp(iH_0 t) + T|\mathbf{p}|^2$ and $P_{ac}(1/\sqrt{H})\exp(iHT)|\mathbf{p}|^2 \times \exp(-iHT)(1/\sqrt{H})P_{ac} \to 2$ as $T \to \pm\infty$. Thus the equality of the two expressions reduces to the equality of (i) and (ii) in the limit $T \to \infty$.

(i) $\Leftrightarrow$ (iv):

$$\text{(i)} = \frac{1}{2}\Omega_-\left(\frac{1}{\sqrt{H_0}}G\frac{1}{\sqrt{H_0}} - S^{-1}\frac{1}{\sqrt{H_0}}G\frac{1}{\sqrt{H_0}}S\right)\Omega_-^*$$

$$= \frac{1}{2}\Omega_- \frac{1}{\sqrt{H_0}} S^{-1}[S, G] \frac{1}{\sqrt{H_0}} \Omega_-^*, \text{ since } [S, H_0] = 0.$$

But

$$[G, S] = i\frac{\partial}{\partial\alpha}\int_0^\infty dE\delta(H_0\alpha^{-2} - E)S(E)|_{\alpha=1}$$
$$= 2i\int_0^\infty dE\delta(H_0 - E)E\frac{\partial S(E)}{\partial E},$$

since the angular part of $S(E)$ is unaffected by dilatations, so this means that (i) $\Leftrightarrow$ (iv).

(iii) $\Leftrightarrow$ (iv): This equivalence can be shown with the same methods but is slightly more involved. A proof will be sketched in Problem 2. □

The quantity related to S that is of interest in experimental physics is the cross-section σ. Following the classical theory (I: §3.4), we define the cross-section as the number of particles scattered into a given solid angle per unit area of incident particles. The momentum distribution of the incident particles is described by a wave-function $\varphi(\mathbf{k})$ in $L^2(\mathbb{R}^3, d^3k/(2\pi)^3)$. In reality a particle is never precisely aimed at the scattering target, but is rather a beam with momentum concentrated near $\mathbf{k}_0$, while its width in x-space will be macroscopic. The initial state is best described as a mixture

$$\frac{1}{F}\int_F d^2a|\exp(i\mathbf{a}\cdot\mathbf{k})\varphi(\mathbf{k})\rangle\langle\exp(i\mathbf{a}\cdot\mathbf{k})\varphi(\mathbf{k})|,$$

letting φ have compact support containing $\mathbf{k}_0 = (0, 0, \sqrt{E})$ and letting $\mathbf{a} = (a_1, a_2, 0)$ be a translation in the plane of the impact parameter, which is averaged over a surface F, the beam cross-sectional area. We next find the probability of measuring the momentum of the outgoing state in some cone C so far from $\mathbf{k}_0$ that $\varphi_{|C} = 0$, and there is no danger of measuring an unscattered particle. In this computation the $\mathbf{1}$ in S in (3.12) does not contribute, and with $\psi = -2\pi i\int dE\delta(H_0 - E)t\delta(H_0 - E)\varphi$, $t = -4\pi f$, we obtain

$$\frac{(2\pi)^{-3}}{F}\int_F d^2a\int_C d^3k|\psi(\mathbf{k}|^2 = \frac{(2\pi)^{-9}}{F}\int_F d^2a\int d^3kd^3k'd^3k''$$
$$\cdot\int_0^\infty dE\delta(|\mathbf{k}|^2 - E)\delta(|\mathbf{k}'|^2 - E)8\pi^2 f(\mathbf{k}, \mathbf{k}')8\pi^2 f^*(\mathbf{k}, \mathbf{k}'')\delta(|\mathbf{k}''|^2 - E)$$
$$\cdot\exp(i\mathbf{k}'\cdot\mathbf{a})\varphi(\mathbf{k}')\exp(-i\mathbf{k}''\cdot\mathbf{a})\varphi^*(\mathbf{k}'').$$

To get σ this has to be divided by the probability that the particle arrives through a unit area, i.e., $1/F$. Afterwards, we may let F become infinite, so $\int d^2a\exp(i(\mathbf{k}' - \mathbf{k}'')\cdot\mathbf{a}) = (2\pi)^2\delta^2(\mathbf{k}'_\perp - \mathbf{k}''_\perp)$, where $\perp$ denotes the projection into the 1-2 plane. Because $\delta^2(\mathbf{k}'_\perp - \mathbf{k}''_\perp)\delta(|\mathbf{k}'|^2 - |\mathbf{k}''|^2) = \delta^3(\mathbf{k}' - \mathbf{k}'')/2k'_3$ and $\int_0^\infty k^2dk\delta(k^2 - |\mathbf{k}'|^2) = |\mathbf{k}'|/2$, we get

$$\sigma\, d\Omega = d\Omega\int\frac{dk'^3}{(2\pi)^3}|\varphi(\mathbf{k}')|^2|f(\mathbf{k}, \mathbf{k}')|^2\frac{|\mathbf{k}'|}{|k'_3|}.$$

If φ is narrowly enough concentrated about $\mathbf{k}_0$ that we may set $|\mathbf{k}'|/|k_3'|$ to 1 and regard $f(\mathbf{k},\mathbf{k}')$ as a constant, then because of the normalization the detailed form of φ becomes irrelevant, and we obtain a formula for the

3.6.15 Scattering Cross-Section

$$\sigma(\mathbf{k},\mathbf{k}_0) = |f(\mathbf{k},\mathbf{k}_0)|^2, \qquad \sigma_t = \int d\Omega_k \sigma(\mathbf{k},\mathbf{k}_0).$$

3.6.16 Remarks

1. We have considered the probability of measuring a momentum $\mathbf{k}$ as $t \to \infty$. Since
$$s\text{-}\lim_{t\to\infty} \frac{\mathbf{x}(t)}{|\mathbf{x}(t)|} = s\text{-}\lim_{t\to\infty} \frac{\mathbf{p}(t)}{|\mathbf{p}(t)|}$$
(Problem 3), this equals the probability of measuring $\mathbf{x}$ in the same angular direction.
2. The scattering amplitude f is also the coefficient of the asymptotic spherical wave (3.6.6; 3). The complete wave-function $|\psi_-|^2$, however, is not asymptotically dominated by $|f|^2/r^2$, but instead by $|\varphi|^2$ and an interference factor $\sim 1/r$.
3. We learned in (II: §3.3 and 3.4) that the details of the exact wave-function ψ are quite complicated. For instance, $\sigma_t = \int d^3\Omega\sigma$ does not simply describe the shadow cast by an object, but rather refers to the asymptotic region in which the shadow dissolves (the Frauenhofer region of (II: 3.4.42)).

3.6.17 Properties of the Scattering Amplitude

If k is real, then

(i) $f(k;\mathbf{n}',\mathbf{n}) - f(k;\mathbf{n},\mathbf{n}')^* = (1/2\pi)\int d\Omega'' f(k;\mathbf{n}',\mathbf{n}'')f(k;\mathbf{n},\mathbf{n}'')^* ik;$

(ii) $f(k;\mathbf{n}',\mathbf{n}) = f(k;-\mathbf{n},-\mathbf{n}')$ *provided that* $KVK = V;$

(iii) $f(k;\mathbf{n}',\mathbf{n}) = f(k;-\mathbf{n}',-\mathbf{n})$ *provided that* $PVP = V.$

Time-reversal K and parity P were defined in (3.3.18; 2) *and* (3.2.6).

Proof:

(i) This follows from (3.6.6; 2), since for real k, $t(-k) = t(k)^*$.

(ii) If H and H_0 are invariant under K, then $KSK = S^*$, and so $Kt(k)K = t(k)^*$. From the rules $K^2 = \mathbf{1}$, $\langle a|Kb\rangle = \langle Ka|b\rangle^*$, $K\mathbf{p}K = -\mathbf{p}$ it follows that $\langle \mathbf{n}'|t(k)|\mathbf{n}\rangle = \langle \mathbf{n}'|Kt(k)^*K|n\rangle = \langle -\mathbf{n}'|t(k)^*|-\mathbf{n}\rangle^* = \langle -\mathbf{n}|(t(k)|-\mathbf{n}'\rangle$.

(iii) This proposition follows from $Pt(k)P = t(k)$ and $PpP = -p$. □

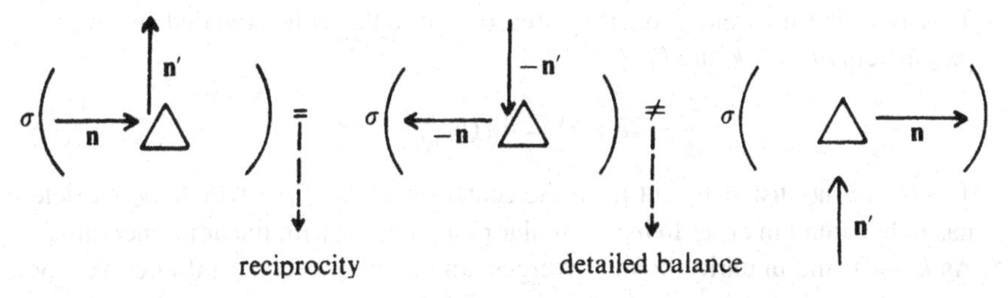

Figure 3.8. Scattering from a triangle.

3.6.18 Example

With the separable potential (3.6.5), $f(k; \mathbf{n}', \mathbf{n}) = 4\pi\lambda\rho^*(k\mathbf{n}')\rho(k\mathbf{n})D^{-1}(k)$. This satisfies (i), and invariance under K means that $\rho^*(\mathbf{k}) = \rho(-\mathbf{k})$, which implies (ii). Invariance under P means that $\rho(\mathbf{k}) = \rho(-\mathbf{k})$, which implies (iii).

3.6.19 Remarks

1. If $\mathbf{n} = \mathbf{n}'$, then (i) becomes the **optical theorem,** $\sigma_t(k) = 4\pi \operatorname{Im} f(k; \mathbf{n}, \mathbf{n})/k$. The information contained in the forward scattering amplitude includes the total scattering cross-section.
2. Proposition (ii) goes by the name of **reciprocity**. It states that if there is invariance under K, then the reversed motion is also possible: $\sigma(k; \mathbf{n}', \mathbf{n}) = \sigma(k; -\mathbf{n}, -\mathbf{n}')$.
3. Propositions (iii) and (ii) together imply that $\sigma(k; \mathbf{n}', \mathbf{n}) = \sigma(k; \mathbf{n}, \mathbf{n}')$, which is referred to as **detailed balance**. It is not valid for scattering from targets that are not invariant under reflections (see Figure 3.8).
4. For radial potentials the expression for f in (3.6.6; 3) makes

$$\sigma_t = \sum_l \sigma_l, \qquad \sigma_l(k) = \frac{4\pi}{k^2}(2l+1)\sin^2\delta_l(k).$$

The total cross-section is the sum of the contributions of all possible definite angular momenta, each of which is maximized by the **unitarity bound**

$$\sigma_l \le \frac{4\pi}{k^2}(2l+1).$$

This bound is attained at a resonance; for instance, with

$$\exp(2i\delta_l) = \frac{k - k_0 + ib}{k - k_0 - ib}$$

the l contribution would be

$$\sigma_l = \frac{4\pi}{k^2}(2l+1)\frac{b^2}{(k-k_0)^2 + b^2}.$$

This is four times the geometric area of a circular ring bounded by impact parameters $b_l = l/k$ and b_{l+1}:

$$\frac{4\pi}{k^2}(2l+1) = 4\pi(b_{l+1}^2 - b_l^2).$$

The b_l are the distances out from the center of the target at which the particle has to be aimed in order to have angular momentum l with linear momentum k.

5. As $k \to 0$ the unitarity bound diverges, and σ_l may become infinite. Yet for most potentials δ_l goes as k^{2l+1} as $k \to 0$, so only $l = 0$ contributes to σ_l. In terms of the **scattering length** $a \equiv -\lim_{k\to 0}\delta_0(k)/k = -f(0; \mathbf{n}, \mathbf{n})$,

$$\lim_{k\to 0}\sigma_l(k) = 4\pi a^2.$$

6. In classical physics potentials that extend to infinity, such as $r^{-\eta}$, always produce infinite total cross-section, since no matter how large the impact parameter b is, there is always a nonzero scattering angle $\sim b^{-\eta}$. This is no longer the case in quantum theory when the potential decreases faster than r^{-2} (Problem 4). The classical argument breaks down because the indeterminacy in the scattering angle should go as b^{-1}, which eventually dominates the classical scattering angle. This sort of reasoning cannot, however, explain why the critical value should be $\eta_c = 2$; especially as it depends greatly on the dimension ($\eta_c = (1+d)/2$).
7. Although S is a continuous function of V in the strong topology [8], f is not likewise continuous, since it involves matrix elements with plane waves, which are not square-integrable. It can thus happen that the forward scattering amplitude, and thus also σ_t, are finite for an r^{-2} potential cut off arbitrarily far away, but become infinite as the cut-off goes to infinity, even though the potentials are arbitrarily close in norm.

It remains to discuss how f can be calculated explicitly, or, if that is impossible, how to assess the accuracy of approximations to it. If two particles interact through a radial potential, the problem is to solve an ordinary differential equation, and $\delta(k)$ can be found by numerical integration. When there are more particles, however, we are confronted with a nontrivial partial differential equation. It is therefore advisable to survey the more general methods that are available. In the absence of better ideas, one frequently falls back on a series expansion in V, called the Born approximation. The hope is that at high energies, for which the kinetic energy overwhelms the potential energy, the result becomes accurate. Whether the hope is fulfilled depends on an

3.6.20 Error Estimate for the Born Approximation

Let $V \in L^1$, so $v_k \equiv |V|^{1/2} \exp(i\mathbf{k} \cdot \mathbf{x}) \in L^2$. The n-th Born Approximation $f^{(n)}$ to $f(k; \mathbf{n}', \mathbf{n})$,

$$-f^{(n)}(k; \mathbf{n}', \mathbf{n}) = 4\pi \langle v_{k'} | V^{1/2} |V|^{-1/2} \sum_{m=0}^{n-1} (|V^{1/2}|(k^2 - H_0)^{-1} V^{1/2})^m | v_k \rangle$$

satisfies

$$|f(k; \mathbf{n}', \mathbf{n}) - f^{(n)}(k; \mathbf{n}', \mathbf{n})| \leq \frac{\|K\|^n}{1 - \|K\|} \int d^3x |V(\mathbf{x})|,$$

where $K = |V|^{1/2}(H_0 - k^2)^{-1}|V|^{1/2}$. Since

$$\begin{aligned} \|K\|^4 &\leq (\mathrm{Tr}\, KK^*KK^*) \\ &= \int \prod_{i=1}^{4} \frac{d^3x_i}{(4\pi)^4} V(\mathbf{x}_j) \frac{\exp(ik(|\mathbf{x}_1 - \mathbf{x}_2| - |\mathbf{x}_2 - \mathbf{x}_3| + |\mathbf{x}_3 - \mathbf{x}_4| - |\mathbf{x}_4 - \mathbf{x}_1|))}{|\mathbf{x}_1 - \mathbf{x}_2||\mathbf{x}_2 - \mathbf{x}_3||\mathbf{x}_3 - \mathbf{x}_4||\mathbf{x}_4 - \mathbf{x}_1|} \\ &\equiv N(k) \end{aligned}$$

goes to zero as $k \to \infty$ by the Riemann–Lebesgue lemma, for all $n \geq 1$ and $\varepsilon > 0$ there exists an energy great enough that $|f - f^{(n)}| < \varepsilon$

3.6.21 Remark

If $N(0) < 1$ and the sign of V does not change, then the Born approximation converges for all k, since $N(k) \leq N(0)$. This can only occur, however, when there are no bound states (see §3.5). In essence, the power-series expansion has a chance only if V is small or E large.

3.6.22 Example

By setting $D(k)$ to 1 in (3.6.5), one gets the first Born approximation. If, say, $\rho(k) = M^2/(k^2+M^2)$ as before, the function D becomes $1+(\lambda/4\pi)M^2/(M-ik)$, so

$$|D|^2 - 1 = \frac{\lambda}{2\pi} \frac{M^3}{M^2 + k^2} \left(1 + \frac{\lambda M}{8\pi}\right).$$

When $\lambda M \gg 1$, the error is on the order of one percent for $k > M^2\lambda$.

If $H \geq 0$, it is straightforward to use the projection method (3.5.29) to obtain a fairly accurate upper bound for $t - V$ at $E = 0$ with the inequality $H^{-1} \geq P(PHP)^{-1}P$. If H has n bound states, then the correction to the first Born approximation may be positive, or even infinite, when there happens to be a bound state at $E = 0$. If the energies of bound states were known exactly the negative parts of H^{-1} could be projected out. If they are only approximately known, then the following lemma reduces the bound to the inversion of a finite-dimensional matrix:

3.6.23 Lemma

Suppose that an invertible, Hermitian operator a has n negative eigenvalues, and is positive on the subspace perpendicular to their eigenvectors. For any n-dimensional projection P such that $Pa^{-1}P < 0$, $a \geq P(Pa^{-1}P)^{-1}P$.

Proof: If $\chi \in a^{-1}P\mathcal{H}$, then $\langle\chi|a|\chi\rangle = \langle\chi|P(Pa^{-1}P)^{-1}P|\chi\rangle$ follows trivially. If $\chi \notin a^{-1}P\mathcal{H}$ and Q is the projection onto $P\mathcal{H} \cup \{\chi\}$, then we must have $\text{Det } Qa^{-1}Q/\text{Det } Pa^{-1}P \geq 0$, as otherwise $Qa^{-1}Q$ would have $n+1$ negative eigenvalues (since $Pa^{-1}P$ has n, $Qa^{-1}Q$ has at least n). This would contradict the hypothesis that a, and thus a^{-1}, have just n negative eigenvalues because of the min-max principle. Since by Problem 8 the ratio of the determinants is up to a positive constant equal to $\langle\chi|a|\chi\rangle - \langle\chi|P(Pa^{-1}P)^{-1}P|\chi\rangle$, the proposition follows. □

3.6.24 Corollary

Let P be an n-dimensional projection. If $H \geq 0$, and PHP is invertible on $P\mathcal{H}$, then

$$\langle\chi|V - t|\chi\rangle \geq \langle\chi|VP(PHP)^{-1}PV|\chi\rangle.$$

If H has n negative eigenvalues, but is otherwise positive, then this equation is still true provided that $PHP > 0$.

Frequently one has intuitive feelings about what would constitute a good approximation to t. These beliefs can be tested with

3.6.25 Kohn's Variational Principle

Let V_t be a comparison potential for which it is possible to calculate $t_t \equiv V_t - V_t(H_0 + V_t - E)^{-1}V_t \equiv V_t\Omega_t$. This differs from the exact t as follows:

$$\begin{aligned} t(k) &= t_t(k) + \Omega_t^*(k)(V - V_t)\Omega_t(k) \\ &\quad -\Omega_t^*(k)(V - V_t)(H - k^2)^{-1}(V - V_t)\Omega_t(k). \end{aligned}$$

3.6.26 Remarks

1. The operator identity (3.6.25) is easy to verify. Its advantage is that the first correction can be calculated when the problem has been solved with V_t, and only the second involves the resolvent of H. Since the second term is quadratic in $(V - V_t)$, there is hope that a good choice of V_t makes it small.
2. If it is known that H is positive aside from n bound states, then (3.6.23) can be used in the last term for $k = 0$, to produce an upper bound for the scattering length. If $V_t = 0$, $\Omega_t = \mathbf{1}$, then it agrees with (3.6.24), which shows that (3.6.24) can be improved on with a superior choice of V_t. If $V \geq 0$, then $0 \leq H_0 \leq H$, so $1/H \leq 1/H_0$, from which we also obtain a lower bound.

3.6.27 Example

Let $|\rangle = 1$ be the vector (of L^∞, not L^2) of a plane wave with $\mathbf{k} = \mathbf{0}$, and let V be such that all inner products of the form

$$b_n = \Big\langle \Big| \underbrace{V \frac{1}{H_0} V \frac{1}{H_0} \cdots V}_{n} \Big| \Big\rangle$$

exist and $H > 0$. Substituting $|\chi\rangle = |\rangle$, $P = H_0^{-1} V|\rangle\langle|VH_0^{-2}V|\rangle^{-1}\langle|VH_0^{-1}$ into (3.6.24), we discover that $\langle|t(0)|\rangle \le b_1 - b_2^2/(b_2+b_3)$. If V is approximated with the separable potential $V_t \equiv V|\rangle\langle|V|\rangle^{-1}\langle|V$, so that $(V - V_t)|\rangle = 0$, then from (3.6.2),

$$\begin{aligned}
\Omega_t(0) &= 1 - (b_1+b_2)^{-1} \frac{1}{H_0} V|\rangle\langle|V, \\
\langle|t_t|\rangle &= \frac{b_1^2}{b_1+b_2}, \\
(V_t - V)\Omega_t(0)|\rangle &= \frac{b_1}{b_1+b_2} V \frac{1}{H_0} V|\rangle - \frac{b_2}{b_1+b_2} V|\rangle, \\
\langle|\Omega_t^*(0)(V - V_t)\Omega_t(0)|\rangle &= \frac{b_1^2 b_3 - b_1 b_2^2}{(b_1+b_2)^2}.
\end{aligned}$$

In Problem 7 it is shown that the upper bound

$$\begin{aligned}
\langle|t|\rangle &\le \langle|t_t|\rangle + \langle|\Omega_t^*(0)(V - V_t)\Omega_t(0)|\rangle \\
&\quad - |\langle|\Omega_t^*(0)(V - V_t)\frac{1}{H_0} V|\rangle|^2 \langle|V \frac{1}{H_0} V + V \frac{1}{H_0} V \frac{1}{H_0} V|\rangle^{-1} \\
&= \frac{b_1^2}{b_1+b_2} - \frac{(b_2^2 - b_3 b_1)b_1}{(b_1+b_2)^2} - \frac{(b_1 b_3 - b_2^2)^2}{(b_2+b_3)(b_1+b_2)^2}
\end{aligned}$$

is valid provided that $H > 0$. If V is also > 0, then there is a complementary lower bound (cf. (3.6.26; 2),

$$\begin{aligned}
&\frac{b_1}{b_1+b_2} - \frac{(b_2^2 - b_3 b_1)b_1 + b_2^3 - 2b_1 b_2 b_3 + b_1^2 b_4}{(b_1+b_2)^2} \\
&= \langle|t_t|\rangle + \langle|\Omega_t^*(0)(V - V_t)\Omega(0)|\rangle \\
&\quad - \langle|\Omega_t^*(0)(V - V_t)H_0^{-1}(V - V_t)\Omega_t(0)|\rangle \le \langle|t_t|\rangle.
\end{aligned}$$

These inequalities hold as well for potentials that do not lend themselves easily to analytical or numerical methods. If V is specialized to the radial case, say $V = \alpha$ if $r < 1$ and otherwise 0, then $\langle|t|\rangle$ can be calculated as $1 - \alpha^{-1/2}\tanh(\alpha^{1/2})$ (Problem 5), which allows the calculation of all the b_n and thereby the bounds. At the radius of convergence $\alpha = \pi^2/4$ of the Born approximation the accuracy is still measured in $\frac{0}{00}$, and they are acceptable well beyond that point (see Figure 3.9).

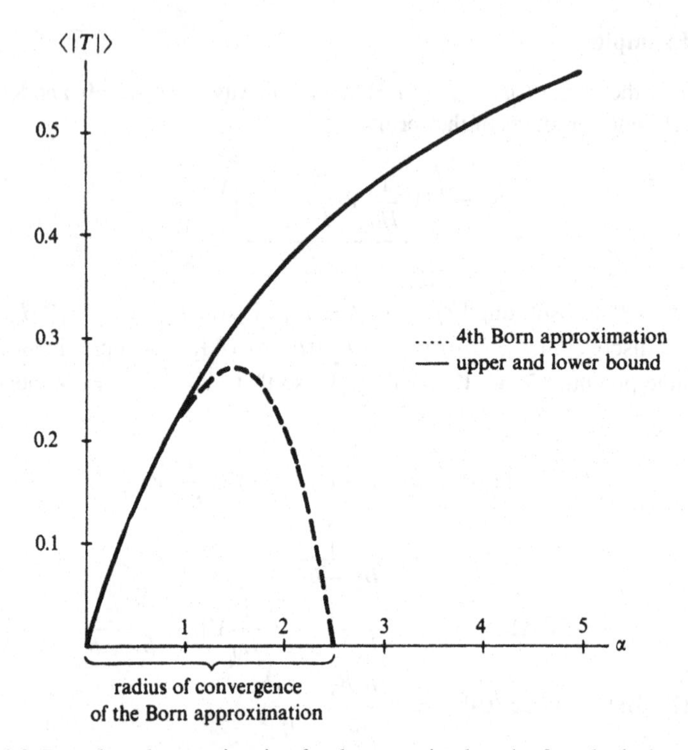

Figure 3.9. Bounds and approximation for the scattering length of a spherical square-well potential.

3.6.28 Problems

1. Show that $V(t) \to V \Rightarrow V = s\text{-}\lim_{\varepsilon \downarrow 0} \varepsilon \int_0^\infty \exp(-\varepsilon t) V(t) dt$.

2. Show that $w\text{-}\lim_{R\to\infty} \int_{-\infty}^0 dt\, \tau_t^0(\Omega_-^* \chi_R \Omega_- - \chi_R) = 0$, if $\| \int_{-\infty}^0 dt (\exp(-iHt)\Omega_- - \exp(-iH_0 t)\varphi\| < \infty$ for all φ, on a dense set $(\chi_R = \Theta(R^2 - |\mathbf{x}|^2))$. Use the result to show that (iii) $\Leftrightarrow$ (iv) in (3.6.13).

3. Let $s\text{-}\lim_{t\to\pm\infty} \mathbf{p}_t = \mathbf{p}_\pm$ (in the sense of (2.5.7; 3)). Show that $s\text{-}\lim_{t\to\pm\infty} \mathbf{x}(t)/|\mathbf{x}(t)| = \pm\mathbf{p}_\pm/|\mathbf{p}_\pm|$, and conclude that $\lim_{t\to\pm\infty} \Delta(\mathbf{x}(t)/|\mathbf{x}(t)|) = \Delta(\mathbf{p}_\pm/|\mathbf{p}_\pm|)$.

4. Suppose that as $r \to \infty$, the potential V goes as $\lambda r^{-2-\varepsilon}$, $\varepsilon > 0$. Show that if $k \neq 0$ and λ is within the radius of convergence of the Born approximation, then $\sigma_t < \infty$. (Since σ_t can be infinite only because of the sum over l, and the Born approximation becomes exact for large l, the statement actually holds for all larger λ as well. If $V = \lambda/r^2$, $\delta_l \sim \sqrt{l^2 + \lambda} - l \sim \lambda/l$ and $\sum_l (2l+1)\sin^2\delta_l \sim \sum_l 1/l$ diverges logarithmically.)

5. Calculate the scattering length for the potential $V(r) = \lambda\Theta(1-r)$. (Write $\psi(\mathbf{x}) = (u_l(r)/r)Y_l^m$, $u(0) = 0$.)

6. Calculate the normalization factors of the scattering amplitude in (3.6.6; 3).

7. Derive the upper bound of (3.6.27).

8. Let Q be the projection onto $P\mathcal{H} \oplus \psi$, $\psi \perp P\mathcal{H}$. Show that Det $QbQ = (\langle\psi|b|\psi\rangle - \langle\psi|bP(PbP)^{-1}Pb\psi\rangle)$Det PbP, and use this fact to fill in the gap in the proof of (3.6.23).

9. Calculate the generator $\delta(E, L)$ for the classical scattering transformation (I: 3.4.10; 2) for the potential γ/r^2, and compare with the phase-shift $\delta_l(E)$. What is the delay time D?

3.6.29 Solutions

1. If $\delta > 0$, there exists a τ such that $\|(V - V(t)\psi\| \leq \delta$ when $t \geq \tau$, so $\varepsilon \int_0^\infty \exp(-\varepsilon t) \times V(t)dt = \varepsilon \int_0^\tau \exp(-\varepsilon t)V(t)dt + \varepsilon \int_\tau^\infty \exp(-\varepsilon t)(V(t) - V)dt + \varepsilon \int_\tau^\infty \exp(-\varepsilon t)V\,dt$, and the first integral $\to 0$ as $\varepsilon \downarrow 0$, the third equals V, and the second is bounded by $\limsup_{\varepsilon\downarrow 0} \varepsilon\| \int \cdots \psi\| \leq \delta$.

2. Let $\psi_t = \exp(-iHt)\Omega_-\varphi$ and $\varphi_t = \exp(-iH_0t)\varphi$. It must first be verified that the expectation value with the state φ is integrable in time: $|\langle\psi_t|\chi_R\psi_t\rangle - \langle\varphi_t|\chi_R\varphi_t\rangle| \leq |\langle(\psi_t - \varphi_t)|\chi_R\varphi_t\rangle| + |\langle\psi_t|\chi_R(\varphi_t - \psi_t)\rangle| \leq 2\|\psi_t - \varphi_t\|$, so by assumption the integral $\int_{-\infty}^0 dt$ is bounded, and indeed uniformly in R. By the dominated convergence lemma, we may then interchange the integration $\int_{-\infty}^0 dt$ and the limit $R \to \infty$, and the latter yields zero, since $\chi_R \to \mathbf{1} = \Omega_-^*\Omega_-$.

   Derivation of the formula for D. The equation $\Omega_+S = \Omega_-$ and $\tau_t^0 S = S$ can be used to rewrite D_R, with $D = \lim_{R\to\infty} D_R$, as

$$D_R = \int_0^\infty dt\, \tau_t^0[S^{-1}\chi_R S - \chi_R] + \int_{-\infty}^0 dt\, \tau_t^0[\Omega_-^*\chi_R\Omega_- - \chi_R] + S^{-1}\int_0^\infty dt\, \tau_t^0[\Omega_+^*\chi_R\Omega_+ - \chi_R]S.$$

As a consequence of the previous result, the last two summands approach 0 weakly as $R \to \infty$. For the first integral we use the Fourier transform $\tilde{S}(t)$ of $S(E)$, the part of S on the energy shell, and write

$$S = \int dt\,\tilde{S}(t)\exp(itH_0), \qquad [\tilde{S}(t), H_0] = 0.$$

The last integral then becomes

$$\begin{aligned} S^{-1}&\int_0^\infty dt \int_{-\infty}^\infty dt'[\tau_t^0\chi_R\exp(it'H_0)\tilde{S}(t') - \tilde{S}(t')\tau_t^0\exp(it'H_0)\chi_R] \\ &= S^{-1}\int_0^\infty dt \int_{-\infty}^\infty dt'[\tau_t^0\chi_R\tilde{S}(t') - \tilde{S}(t')\tau_{t+t'}^0\chi_R]\exp(it'H_0) \\ &= S^{-1}\int_0^\infty dt \int_{-\infty}^\infty dt'[\tau_t^0\chi_R, \tilde{S}('t)]\exp(it'H_0) \\ &+ S^{-1}\int_{-\infty}^\infty dt' \int_0^{t'} dt\,\tilde{S}(t')\tau_t^0\chi_R\exp(it'H_0). \end{aligned}$$

In the limit as $R \to \infty$, χ_R approaches $\mathbf{1}$ strongly, so the first term goes to 0 and the last approaches

$$S^{-1}\int_{-\infty}^\infty dt'\,t'\tilde{S}(t')\exp(it'H_0) = -iS^{-1}\int_0^\infty dE\,\delta(H_0 - E)\frac{\partial S(E)}{\partial E}.$$

3. By assumption,

$$\frac{\mathbf{x}(t)}{t} = \frac{\mathbf{x}(0)}{t} + \frac{1}{t}\int_0^t dt' \mathbf{p}(t')$$

converges strongly to $\mathbf{p}_\pm$, since $s\text{-}\lim_{t\to\infty} a/t = 0$ for every self-adjoint operator a, and by Problem 1 the second term is $\mathbf{p}_\pm$. Hence the bounded functions $\mathbf{x}(t)/|\mathbf{x}(t)|$ converge strongly, which implies the convergence of the mean-square deviation of bounded functions: $a_n \to a \Rightarrow \langle|a_n^2|\rangle = \|a_n|\rangle\|^2 \to \|a|\rangle\|^2 = \langle|a^2|\rangle$.

4. By (3.6.20) we know that $|f| < c|f^{(0)}|$ in the circle of convergence, so it suffices to show that the total cross-section is finite in the Born approximation. This is

$$\langle \exp(i\mathbf{k}\cdot\mathbf{x})|V\delta(H_0 - k^2)V|\exp(i\mathbf{k}\cdot\mathbf{x})\rangle = \int \frac{d^2p}{(2\pi)^3}|\tilde{V}(\mathbf{p}-\mathbf{k})|^2\delta(|\mathbf{p}|^2 - k^2).$$

Since $\tilde{V} \sim |\mathbf{p}-\mathbf{k}|^{-1+\varepsilon}$, the angular part of the integral is finite (cf. (3.4.13; 2)).

5. The solution of the Schrödinger equation

$$\left(-\frac{\partial^2}{\partial r^2} + V(r)\right)u(r) = 0 : u(r) = \begin{cases} (\sqrt{\lambda}\cosh\sqrt{\lambda})^{-1}\sinh\sqrt{\lambda}r & r \le 1 \\ r + \frac{\tanh\sqrt{\lambda}}{\sqrt{\lambda}} - 1 & r \ge 1 \end{cases}$$

has to be compared with $\lim_{k\to 0}(1/k)\sin(kr + \delta(k))$ when $r > 1$. The result is that $a \equiv -\lim_{k\to 0}\delta(k)/k = 1 - \tanh(\sqrt{\lambda})/\sqrt{\lambda}$.

6. We have

$$\begin{aligned} -2\pi i \int_0^\infty \frac{p'^2 dp'}{(2\pi)^3}\delta(p'^2 - p^2)\langle n'|t|n\rangle &= \frac{2pi}{4\pi}\frac{\langle n'|t|n\rangle}{4\pi} \\ &= \sum_l \langle n'|P_l|n\rangle(\exp(2i\delta_l) - 1). \end{aligned}$$

By the addition theorem for the spherical harmonics,

$$\langle n'|P_l|n\rangle = \sum_m Y_l^{-m}(n')Y_l^m(n) = \frac{2l+1}{4\pi}P_l(\cos\theta),$$

so

$$f = \frac{\langle n'|t|n\rangle}{-4\pi} = \sum_l \frac{2l+1}{k}\exp(i\delta_l)\sin\delta_l P_l(\cos\theta).$$

7. $H^{-1} \ge P(PHP)^{-1}P$ and $P = H_0^{-1}|\rangle\langle|VH_0^{-1}(\langle|VH_0^{-2}V|\rangle)^{-1}$ together imply $\langle|\Omega_t^*(0)(V - V_t)H^{-1}(V - V_t)\Omega_t(0)|\rangle \ge |\Omega_t^*(0)(V - V_t)H_0^{-1}V|\rangle|^2\langle|VH_0^{-1}(H_0 + V)H_0^{-1}V|\rangle^{-1}$.

8. $(\text{Det } QbQ)^{-1/2} = \pi^{-(n+1)/2}\int\prod_{i=0}^n dx_i \exp(-\sum_{i,j=0}^n x_i x_j b_{ij})$, $n = \text{Dim } P$, $b_{00} = \langle\psi|b|\psi\rangle$. With the integration variables $\bar{x}_k = x_k + x_0 b_{0j} c_{jk}$, $k = 1, \ldots, n$, $c = (PbP)^{-1}$,

$$\sum_{i,j=0}^n x_i x_j b_{ij} = \sum_{k,l=1}^n \bar{x}_k \bar{x}_l b_{kl} + x_0^2(b_{00} - b_{0k}c_{kl}b_{l0}).$$

The relationship between the determinants results from integration over x_0 and the $\bar{x}_k$. If ψ is not orthogonal to $P\mathcal{H}$, the ratio is simply changed by a positive factor, since $\operatorname{Det} M^t bM = \operatorname{Det} b(\operatorname{Det} M)^2$. The proposition then follows with $b = a^{-1}$ and $\psi = a\psi$.

9. $\delta(E, L) = L - \sqrt{L^2 + 2\gamma}$, $(2/\pi)\delta_l = l + \frac{1}{2} - \sqrt{(l + \frac{1}{2})^2 + 2\gamma} \Rightarrow D = 0$.

4

Atomic Systems

4.1 The Hydrogen Atom

The hydrogen atom is so simple that a complete mathematical analysis can be made. This analysis was a watershed of atomic physics.

The quantum-mechanical treatment of the problem of two particles interacting through a $1/r$ potential follows the outlines of the classical theory (I: §4.2). It starts with the Hamiltonian

$$H = \frac{|\mathbf{p}_1|^2}{2m_1} + \frac{|\mathbf{p}_2|^2}{2m_2} + \frac{\alpha}{|\mathbf{x}_1 - \mathbf{x}_2|}, \quad \alpha = e_1 e_2, \tag{4.1}$$

which acts a priori on $\mathcal{H} = \mathcal{H}_1 \otimes \mathcal{H}_2$, where $\mathcal{H}_i$ is the Hilbert space of the i-th particle. The system can be decomposed into two independent parts by the

4.1.1 Separation into Center-of-Mass and Relative Coordinates

The unitary transformation

$$(\mathbf{x}_1, \mathbf{x}_2; \mathbf{p}_1, \mathbf{p}_2) \rightarrow (\mathbf{x}_{cm}, \mathbf{x}; \mathbf{p}_{cm}, \mathbf{p}),$$

$$\mathbf{x}_{cm} = \frac{m_1\mathbf{x}_1 + m_2\mathbf{x}_2}{m_1 + m_2}, \quad \mathbf{x} = \mathbf{x}_1 - \mathbf{x}_2; \quad \mathbf{p}_{cm} = \mathbf{p}_1 + \mathbf{p}_2;$$

$$\mathbf{p} = \frac{\mathbf{p}_1 m_2 - \mathbf{p}_2 m_1}{m_1 + m_2},$$

changes H into $H = H_{cm} + H_r$, where

$$H_{cm} = \frac{|\mathbf{p}_{cm}|^2}{2M}, \quad H_r = \frac{|\mathbf{p}|^2}{2m} + \frac{\alpha}{|\mathbf{x}|}, \quad M = m_1 + m_2, \quad m = \frac{m_1 m_2}{m_1 + m_2}.$$

4.1.2 Remarks

1. The Hilbert space can also be written as $\mathcal{H} = \mathcal{H}_{cm} \otimes \mathcal{H}_r$, where the operators H_{cm} and H_r act nontrivially only on the factors $\mathcal{H}_{cm}$ and respectively $\mathcal{H}_r$.
2. The question of self-adjointness is answered (3.3.4; 1) and (3.4.25; 2): Since $/1r$ is compact relative to $|\mathbf{p}|^2$, H_r is self-adjoint on $D(|\mathbf{p}|^2)$, and $\sigma_{\text{ess}}(H_r) = \mathbb{R}^+$.
3. The operator H_{cm} generates the free motion of the center of mass. The invariance group is similar to that of the classical situation: The ten generators of the Galilean group, $H_{cm}, \mathbf{p}_{cm}, \mathbf{k}_{cm} \equiv \mathbf{p}_{cm}t - \mathbf{x}_{cm}M$, and $\mathbf{L}_{cm} = [\mathbf{x}_{cm} \times \mathbf{p}_{cm}]$, do not form a Lie algebra, since $[p_l, k_j] = i\,\delta_{lj} \cdot M$ is not a linear combination of them. If M is considered as an additional element of the algebra of observables, then there is an 11-dimensional Lie algebra $\mathcal{A}$. The center $\mathcal{A}' \cap \mathcal{A}''$, which consists of the functions of M alone, creates a superselection rule (see (2.3.6; 7)), unless M is represented as a multiple of $\mathbf{1}$. As in (I: 4.1.10; 3), the Galilean group is the factor group by the center, and $\mathcal{A}$ produces only a ray representation of it. This happens because $\mathbf{p}_{cm}$ and $\mathbf{k}_{cm}$ of course generate the transformations $\mathbf{x}_{cm} \to \mathbf{x}_{cm} + \mathbf{a} + \mathbf{v}t$ and $\mathbf{p}_{cm} \to \mathbf{p}_{cm} + M\mathbf{v}$ on $\mathbb{R}^6$, and, according to (3.1.6; 6), the unitary operators $W(z)$ give only a ray representation of $\mathbb{R}^6$. This means that although

$$\exp(i\mathbf{v} \cdot \mathbf{k}_{cm}) = \exp\left(\frac{-itM|\mathbf{v}|^2}{2}\right) \exp(-i\mathbf{x}_{cm} \cdot \mathbf{v}M) \exp(i\mathbf{p}_{cm} \cdot \mathbf{v}t)$$

causes the transformation $\mathbf{x}_{cm} \to \mathbf{x}_{cm} + \mathbf{v}t$, $\mathbf{p}_{cm} \to \mathbf{p}_{cm} + M\mathbf{v}$, the wave-function gains an additional phase factor $\exp(-itM|\mathbf{v}|^2/2)$, which is, however, not observable, since only relative phases can be measured.

The Hamiltonian H_{cm} was discussed in detail in (3.3.3), so we turn to H_r. The first fact we know is that $\sigma(H_{cm}) = \sigma_{\text{ess}}(H_r) = \mathbb{R}^+$, and the question arises of whether $\sigma_p(H_r) \subset \mathbb{R}^-$ (recall (3.4.14; 4)). This is answered by the

4.1.3 Virial Theorem

If $(H_r - E)\psi = 0$, $\psi \in \mathcal{H}_r$, then

$$E = \frac{1}{2}\left\langle \psi \middle| \frac{\alpha}{r}\psi \right\rangle = -\langle \psi | H_0 \psi \rangle.$$

Proof: The dilatation U such that $U^{-1}(\beta)(\mathbf{x}, \mathbf{p})U(\beta) = (\exp(-\beta)\mathbf{x}, \exp(\beta)\mathbf{p})$, $\beta \in \mathbb{R}$, which was used in (3.3.8; 2), acts on H_r by $U^{-1}(\beta)H_rU(\beta) = \exp(-2\beta)H_0 + \exp(-\beta)\alpha/r$, where $H_0 = |\mathbf{p}|^2/2m$. The equations

$$\left\langle \left(\exp(2\beta)H_0 + \exp(\beta)\frac{\alpha}{r} - E\right)\psi \middle| U(\beta)\psi \right\rangle = 0 = \left\langle \left(H_0 + \frac{\alpha}{r} - E\right)\psi \middle| U(\beta)\psi \right\rangle$$

can be combined so that

$$-\left\langle\left(\frac{1-\exp(2\beta)}{\beta}H_0+\frac{1-\exp(\beta)}{\beta}\frac{\alpha}{r}\right)\psi\middle|U(\beta)\psi\right\rangle=0\quad\text{for all }\beta\in\mathbb{R}\backslash\{0\}.$$

As $\beta\to 0$, the left side converges to $(2H_0+\alpha/r)\psi$ and the right to ψ. Since the convergence is in the strong sense, this proves (4.1.3). □

4.1.4 Corollary

Since $H_0\geq 0$, if $\alpha<0$, then all the eigenvalues of H_r are negative, and if $\alpha\geq 0$, then it has no eigenvalues.

4.1.5 Remarks

1. The usual argument, which runs that $0=\langle\psi|i[H_r,\mathbf{x}\cdot\mathbf{p}]\psi\rangle=\langle\psi|2H_0+\alpha/r|\psi\rangle$ is not quite conclusive, since $D(\mathbf{x}\cdot\mathbf{p})\not\supset D(H_r)$, and a priori it applies only to $\psi\in D(H_r)\cap D(\mathbf{x}\cdot\mathbf{p})$.
2. The analytic perturbation theory of §3.5 works without modification for the negative eigenvalues. Thus an alternative argument using (3.5.17; 2) would run: On dimensional grounds $E(\alpha)=m\alpha^2c$ for some numerical constant c, and hence $\alpha\,\partial E/\partial\alpha=\langle\alpha/r\rangle=2E$.
3. The action of the dilatation also shows that if $\alpha<0$, then there must be infinitely many eigenvalues accumulating at the point 0; given any $\psi\in D(H_r)$, there exists $\tau_0\in\mathbb{R}^+$ such that

$$\begin{aligned}&\langle U(\tau_0)\psi|H_rU(\tau_0)\psi\rangle\\&=\exp(-2\tau_0)\left\langle\psi\middle|\frac{|\mathbf{p}|^2}{2m}\psi\right\rangle+\alpha\exp(-\tau_0)\left\langle\psi\middle|\frac{1}{r}\psi\right\rangle<0.\end{aligned}$$

   If ψ is compactly supported away from zero, then there exists a sequence $\tau_0<\tau_1<\tau_2\ldots$, for which the supports of $U(\tau_i)\psi$ are disjoint, so H_r is a diagonal matrix on the subspace spanned by the vectors $U(\tau_i)\psi$. The claim then follows from the min-max principle (3.5.19).

The next topic will be $\sigma_p(H_r)$, and the last part of this section will be devoted to $\sigma_{\text{ess}}(H_r)$. Both discussions will make use of

4.1.6 Constants of the Motion

The commutant $\{H_r\}'$ contains the vectors $\mathbf{L}$ and $\mathbf{F}=\frac{1}{2}[\mathbf{p}\times\mathbf{L}-\mathbf{L}\times\mathbf{p}]+m\alpha\mathbf{x}/r$. They are related by

(i) $[H,L_m]=0$, $[L_m,F_l]=i\varepsilon_{mls}F_s$,

(ii) $[H,F_m]=0$, $[F_m,F_l]=-2imH_r\varepsilon_{mls}L_s$,

(iii) $\mathbf{L}\cdot\mathbf{F}=\mathbf{F}\cdot\mathbf{L}=0$,

(iv) $|\mathbf{F}|^2=2mH_r(|\mathbf{L}|^2+1)+m^2\alpha^2$.

Proof:

(i) This is because H acts as a scalar and $\mathbf{F}$ as a vector under rotations.

(ii) This is somewhat more subtle; the example of Problem (3.1.17; 4) leaves doubts as to how to proceed. The equation $[H, F_m] = 0$ has to hold on domains that are invariant under finite transformations to show that $[\exp(iHt), \exp(iFs)] = 0$ (recall Definition (3.1.7)). Actually, by Problem 1 it suffices to show that $(d/dt)\exp(iHt)F\exp(-iHt) = 0$, which will be done in Problem 2. The calculation of commutators involving $\mathbf{F}$ is often easier when it is written as $\mathbf{F} = (i/2)[\mathbf{p}, |\mathbf{L}|^2] + m\alpha\mathbf{x}/r$.

(iii) It is clear that $\mathbf{L}\cdot\mathbf{x} = \mathbf{L}\cdot\mathbf{p} = 0$, since there are no products of noncommuting observables. With (i) this implies (iii).

(iv) This requires some calculations, done in Problem 3. □

4.1.7 Corollary

Special combinations of $\mathbf{L}$ and $\mathbf{F}$, namely $A_k = (L_k + F_k/\sqrt{-2mH_r})P/2$ and $B_k = (L_k - F_k/\sqrt{2mH_r})P/2$, where $P \equiv \Theta(-H_r)$ is the projection onto the functions of the negative spectrum of H_r, satisfy the commutation relations of two independent angular momenta:

$$[A_k, A_j] = i\varepsilon_{kjm}A_m, \quad [B_k, B_j] = i\varepsilon_{kjm}B_m, \quad [A_k, B_j] = 0.$$

They are, however, not independent, as

$$|\mathbf{A}|^2 = |\mathbf{B}|^2 = -P\left(\frac{1}{4} + \frac{m\alpha^2}{8H_r}\right). \tag{4.2}$$

According to the discussion of §3.2, $|\mathbf{A}|^2$ and $|\mathbf{B}|^2$ can have only the eigenvalues $\beta(\beta+1)$, $\beta = 0, \frac{1}{2}, 1, \frac{3}{2}, \ldots$, and by (4.2) their eigenvalues are the same. Each eigenvector belongs to a $(2\beta+1)^2$-fold degenerate **supermultiplet**, the members of which have the same eigenvalue for $|\mathbf{A}|^2$ and $|\mathbf{B}|^2$, but differ in their eigenvalues for A_z and B_z. They are eigenvectors of H_r, and by (4.2) its eigenvalues obey **Balmer's formula**

$$E_n = -\frac{m\alpha^2}{2n^2}, \qquad n = 2\beta + 1 = 1, 2, 3 \ldots. \tag{4.3}$$

4.1.8 Remarks

1. The operators $\mathbf{A}$ and $\mathbf{B}$ each satisfy the Lie algebra of O(3), which is identical to that of SU(2). As there is no reason that only the representations of O(3) should arise, the values of β may be either integral or half-odd integral.
2. Of course, $\mathbf{L}$ generates the algebra O(3) and has eigenvalues $l(l+1)$, l integral. If the eigenvalues of $|\mathbf{F}|^2$ in (4.1.6(iv)) are expressed in terms of n and l,

$$|\mathbf{F}|^2|\rangle = m^2\alpha^2\left(1 - \frac{l^2 + l + 1}{n^2}\right)|\rangle,$$

then it is apparent that $l \leq n-1$. Hence for a given n, the values l can assume are $0, 1, \ldots, n-1$.

3. Balmer's formula shows why

$$K \equiv \frac{1}{\sqrt{r}}(H_0 - z)^{-1}\frac{1}{\sqrt{r}}, \qquad z \notin \mathbb{R}^+,$$

belongs to C_p (2.3.21) for p integral only if $p \geq 4$: As we say in the proof of (3.5.36), if $z \in \mathbb{R}^-$, then the eigenvalues λ_n of K are the values of $1/|\alpha|$ for which $H_0 - |\alpha|/r$ has the eigenvalue z. That being so, (4.3) makes $-z\lambda_n^2 = m/(2n^2)$, and the n-th eigenvalue has degeneracy n^2. But then

$$\|K\|_p^p = \sum_{n=1}^{\infty} n^2 \lambda_n^p$$

for p integral is finite only for $p \geq 4$.

4.1.9 Construction of the Eigenvectors of H_r

Since there is no difference in the algebraic situation, we can proceed as with the eigenvectors of the angular momentum (3.2.9). In each supermultiplet there is a state $|\rangle$ of greatest A_3 and B, so

$$A_+|\rangle = B_+|\rangle = 0 \Leftrightarrow F_+|\rangle = L_+|\rangle = 0, \quad F_\pm \equiv F_1 \pm i F_2, \text{ etc.} \tag{4.4}$$

The other states are then obtained by applications of $A_-^p B_-^q$, $0 \leq p, q \leq n-1$. Since F_+ is constructed with x_+, p_+,[1] and observables that commute with $|\mathbf{L}|^2$, it raises l by 1 (cf. (3.2.10)); thus if $F_+|\rangle = 0$, then $|\rangle$ already has the greatest angular momentum possible in the supermultiplet. If an eigenvector $|n, l, m\rangle$ is specified by its eigenvalues for H_r, $|\mathbf{L}|^2$, and L_3, then the original basis state is $|\rangle \equiv |n, n-1, n-1\rangle$.

4.1.10 Eigenfunctions in the x-Representation

If we write

$$F_+ = \frac{i}{2}(p_+|\mathbf{L}|^2 - |\mathbf{L}|^2 p_+) + \frac{m\alpha x_+}{r}$$

and calculate the action of p_+ and x_+ on the eigenvectors (up to a constant—see Problem 4),

$$ip_+ f(r) Y_l^l = \left(\frac{d}{dr} - \frac{l}{r}\right) f(r) Y_{l+1}^{l+1}$$

$$\frac{x_+}{r} f(r) Y_l^l = f(r) Y_{l+1}^{l+1}$$

[1] These should be distinguished from $\mathbf{x}_\pm$, $\mathbf{p}_\pm$, etc. of (3.4.6).

then Equation (4.4) becomes

$$F_+|n, n-1, n-1\rangle = \left(-n\left(\frac{d}{dr} - \frac{n-1}{r}\right) + m\alpha\right)|n, n-1, n-1\rangle = 0.$$

The solution is

$$|n, n-1, n-1\rangle = cr^{n-1}\exp(m\alpha r/n)Y_{n-1}^{n-1}(\theta, \varphi),$$

which is in $L^2((0,\infty), r^2dr) \otimes L^2(S^2)$ when $\alpha < 0$, and is the original basis vector of the supermultiplet.

4.1.11 Remarks

1. The vector $|n, n-1, n-1\rangle$ has the maximal angular momentum and is related to a classical circular orbit, whereas $|n, 0, 0\rangle$ corresponds to a classical trajectory through the origin. However, in quantum theory the latter vector is a spherically symmetric wave-function, since it distinguishes no direction.
2. The wave-function $|n, n-1, n-1\rangle$ decreases in r as $\exp(-r/nr_b)$, where $r_b = 1/|m\alpha| = 0.529 \times 10^{-10}$m., but its maximum is attained at $r = n(n-1)r_b$. This is in accordance with the rough estimate (1.2.3) and the virial theorem, which requires that $\langle 1/r\rangle \sim n^{-2}$.
3. A calculation of the expectation values yields

$$\langle n, n-1, m|r|n, n-1, m\rangle = r_b n\left(n + \tfrac{1}{2}\right),$$
$$\langle n, n-1, m|r^2|n, n-1, m\rangle = r_b^2 n^2\left(n + \tfrac{1}{2}\right)(n+1),$$

   which reveals that the relative mean-square deviation $\Delta r/\langle r\rangle = 1/\sqrt{2n+1}$ vanishes in the limit $n \to \infty$. The state $|n, n-1, n-1\rangle$ gets concentrated ever more strongly on a circle in the $x - y$-plane.
4. States of smaller l, and hence greater eccentricity, can be produced by successive applications of F_- to $|n, n-1, n-1\rangle$ as in (4.1.9), and an example is computed in Problem 5.

We conclude the discussion of σ_p here and turn to σ_{ess}. We begin with the basic fact about this part of the spectrum.

4.1.12 The Absence of Singular Spectrum

$$\sigma_{\text{sing}}(H_r) = \emptyset, \quad so\ \sigma_{ac}(H_r) = \sigma_{\text{ess}}(H_r) = \mathbb{R}^+.$$

Proof: If the dilatation used in the virial theorem (4.1.3) is continued into the complex plane, then $U(\tau)$, $\tau \in \mathbb{C}$, becomes unbounded, but is still defined on the dense set D of analytic vectors (see (2.4.21; 5)). The action of U on H_r can be continued analytically, and

$$\langle\varphi|(H_r - z)^{-1}\psi\rangle = \left\langle U(\tau^*)\varphi\right|\left(\exp(2\tau)H_0 + \exp(\tau)\frac{\alpha}{r} - z\right)^{-1} U(\tau)\psi\Big\rangle,$$

for

$$\tau \in \mathbb{C}, \varphi, \psi \in D.$$

Since multiplication by the complex number e^τ does not affect the relative compactness of α/r, we conclude that

$$\sigma_{\text{ess}}\left(\exp(2\tau)H_0 + \exp(\tau)\frac{\alpha}{r}\right) = \sigma_{\text{ess}}(\exp(2\tau)H_0) = \exp(2\tau)\mathbb{R}^+.$$

The matrix elements of the resolvent in states $\varphi, \psi \in D$ can thus be continued analytically from $\mathbb{R}^+$ into the complex plane as far as $e^{2\tau}\mathbb{R}^+$, the rotated positive axis (see Figure 4.1):

This implies the absence of σ_{sing}, since

$$\begin{aligned}&\langle\varphi|(\Theta(H_r - a) - \Theta(H_r - b))\psi\rangle \\ &= \lim_{\varepsilon\to 0}\int_a^b \frac{dz}{2\pi i}\langle\varphi|((H_r - z - i\varepsilon)^{-1} - (H_r - z + i\varepsilon)^{-1})\psi\rangle,\end{aligned}$$

$a, b > 0$, is the integral of an analytical function and thus approaches 0 like $const \cdot |a - b|$ as $a \to b$. Since D is dense, σ_{ess} contains no part concentrated on a set of Lebesgue measure 0. □

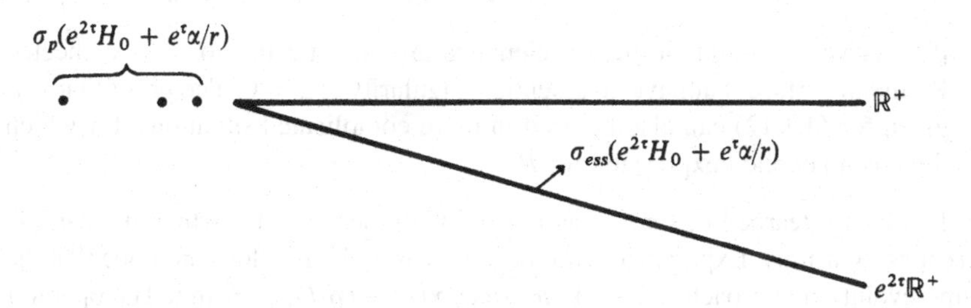

Figure 4.1. Spectrum of the dilated Hamiltonian.

4.1.13 Remarks

1. Only the matrix elements with φ and $\psi \in D$ can be continued beyond $\mathbb{R}^+$; the analytic function $\mathbb{C} \to \mathcal{B}(\mathcal{H}) : z \to (H_r - z)^{-1}$ is not continuable past $\mathbb{R}^+$, as the resolvent becomes unbounded there.
2. As regards the eigenvalues of finite multiplicity of $H(\tau) \equiv \exp(2\tau)H_0 + \exp(\tau)\alpha/r$, they are also analytic in τ; the difference $H(\tau) - H(\tau - \delta)$ is bounded relative to $H(\tau)$ for δ small enough, so the perturbation theory of §3.5 is applicable. This implies that the eigenvalues $E(\tau)$ of $H(\tau)$ are independent of τ (as long as they persist), since if τ is real, $H(\tau)$ and H_r are unitarily equivalent and therefore have identical spectra. But analytic functions that agree on $\mathbb{R}$ are equal everywhere.

3. The eigenfunctions $|\tau\rangle = U(\tau)|0\rangle$ of $H(\tau)$ are likewise analytic in τ, which implies that $(d/d\tau)|\tau\rangle - id|\tau\rangle \in L^2$ if $U(\tau) = \exp(-id\tau)$, so $|\tau\rangle \in D(d)$. This justifies the formal proof (4.1.5; 1) of the virial theorem a posteriori.
4. In this simple case the integral kernel of the resolvent $R_z = (H_r - z)^{-1}$ is known explicitly. In the x-representation it involves Whittaker functions [9], [10]:

$$R_z(\mathbf{x}_1, \mathbf{x}_2) = \frac{\Gamma(1-i\nu)}{4\pi|\mathbf{x}_1 - \mathbf{x}_2|}[W'_{i\nu;1/2}(-i\sqrt{z}(r_1 + r_2 + |\mathbf{x}_1 - \mathbf{x}_2|))$$
$$\times M_{i\nu;1/2}(-i\sqrt{z}(r_1 + r_2 - |\mathbf{x}_1 - \mathbf{x}_2|))$$
$$- W_{i\nu;1/2}(-i\sqrt{z}(r_1 + r_2 + |\mathbf{x}_1 - \mathbf{x}_2|))$$
$$\times M'_{i\nu;1/2}(-i\sqrt{z}(r_1 + r_2 - |\mathbf{x}_1 - \mathbf{x}_2|))],$$
$$\nu = \frac{\alpha}{4\sqrt{z}}, \quad m = \frac{1}{2}.$$

In the p-representation this becomes

$$R_z(\mathbf{p}_1, \mathbf{p}_2) = \frac{\delta^3(\mathbf{p}_1 - \mathbf{p}_2)}{|\mathbf{p}_1|^2 - z} - \frac{\alpha}{8\pi^2 z}\int_0^1 d\rho\rho^{i\nu}\frac{(1-\rho^2)\rho^2}{\left[|\mathbf{p}_1 - \mathbf{p}_2|^2 - \frac{(1-\rho)^2}{\rho}\frac{(z-|\mathbf{p}_1|^2)(z-|\mathbf{p}_2|^2)}{4z}\right]^2}$$

[9]. As we see, suitable matrix elements are analytic in z on a two-sheeted Riemann surface, but have an essential singularity at $z = 0$. The proof that was given for (4.1.12) can also be used in more complicated situations, for which there is no explicit expression for R_z.

In §3.4 we learned that σ_{ac} is associated with the states for which the particle escapes to infinity. Experience with the classical problem allows no hope that the time-evolution approaches that of H_0, since $\mathbf{x}(t) - t\mathbf{p}(t)/m \sim \ln t$. The operator $\exp(iH_r t)\exp(-iH_0 t)$ is not much good, either; it converges weakly to 0 as $t \to \infty$. On the other hand, there again exist some useful

4.1.14 Asymptotic Constants of the Motion

Let P be the projection onto the vectors of $\sigma_{ac}(H_r)$. Then $(P\mathbf{p}P, P(\mathbf{x}/r)P, P(1/r)P) \in \mathcal{A}$ and s-$\lim_{t\to\pm\infty}(P\mathbf{p}P, P(\mathbf{x}/r)P, P(1/r)P) = (\mathbf{p}_\pm, \pm\mathbf{p}_\pm/|\mathbf{p}_\pm|, 0)$ ($\mathbf{p}_\pm$ in the sense of (3.4.6)).

Proof:

(i) Convergence of the Momenta. The claim will be proved only for $\alpha > 0$; the proof is more involved if $\alpha < 0$, and the reader is referred to [11]. In the repulsive case the radial component of the momentum p_r increases monotonically and is bounded, so it is suitable as a Liapunov function. As in classical mechanics (I: 5.3.8), when H_r is expressed in terms of p_r, it becomes

$$H_r = \frac{1}{2m}\left(p_r^* p_r + \frac{|\mathbf{L}|^2}{r^2}\right) + \frac{1}{r}.$$

(details in Problem 6, with $\alpha = 1$). If $L^2(\mathbb{R}^3, d^3x)$ is mapped unitarily to $L^2(S^2, d\Omega)\otimes L^2([0,\infty), dr)$ by $\psi \to u/r$, then p_r becomes identical to the operator p_r of (3.3.5; 4). Let $p_r(t)$ as usual mean $\exp(iH_r t)p_r\exp(-iH_r t)$, so that by (3.3.5; 4(a)),

$$\left\langle Y_l^m \frac{u(r)}{r}\right|\left.\frac{d}{dt}p_r Y_l^m \frac{u(r)}{r}\right\rangle = \int_0^\infty dr\, u^*(r)u(r)\left[\frac{1}{r^2} + \frac{l(l+1)}{r^3}\right] + \frac{1}{2}|u'(0)|^2.$$

Since $\|p_r\psi\|^2 \le \langle\psi|H_r\psi\rangle$, this means that r^{-2} is integrable in time, as is to be expected from $r \sim t$:

$$\int_0^T dt\left\|\frac{1}{r(t)}\psi\right\|^2 \le \langle\psi|p_r(T) - p_r(0)|\psi\rangle \le 2\|\psi\|\langle\psi|H_r\rangle^{1/2}. \qquad \text{(a)}$$

Since this bound is independent of T, given any $\psi \in D(H_r)$ and $\varepsilon > 0$, there exists a T such that

$$\int_T^\infty dt\left\|\frac{1}{r(t)}\psi\right\|^2 < \varepsilon. \qquad \text{(b)}$$

In order to avoid some technical difficulties caused by the presence of unbounded operators, let us examine $R\mathbf{p}R$, where $R = (H_r + c)^{-1}$, $c \in \mathbb{R}^+$. Now,

$$\frac{d}{dt}Rp_i(t)R = R\frac{x_i(t)}{r(t)^3}R$$

so

$$\begin{aligned}\left\|\int_T^\infty dt\frac{d}{dt}Rp_i(t)R\psi\right\| &= \sup_{\|\varphi\|=1}\left|\int_T^\infty dt\left\langle R\varphi\left|\frac{x_i(t)}{r(t)^3}R\psi\right.\right\rangle\right| \\ &\le \sup_{\|\varphi\|=1}\int_T^\infty dt\left|\left\langle R\varphi\left|\frac{1}{r(t)^2}R\psi\right.\right\rangle\right| \\ &\le \sup_{\|\varphi\|=1}\left[\int_0^\infty dt\left\|\frac{1}{r(t)}R\varphi\right\|^2\right]^{1/2}\left[\int_T^\infty dt\left\|\frac{1}{r(t)}R\psi\right\|^2\right]^{1/2} \\ &\le \text{const}\left[\int_T^\infty dt\left\|\frac{1}{r(t)}R\psi\right\|^2\right]^{1/2},\end{aligned}$$

where we have used (a). On account of (b), as $T \to \infty$ this becomes arbitrarily small, which implies the strong convergence of $P\mathbf{p}(t)P$. The operator R maps the Hilbert space into $D(H_r)$, which is a domain of essential self-adjointness for $\mathbf{p}(t)$ and $\mathbf{p}_\pm$. Therefore $\mathbf{p}$ converges strongly to $\mathbf{p}_\pm$ on a domain of essential self-adjointness, and consequently strongly in the sense of (2.5.8; 3) (see Problem 8).

(ii) Convergence of $\mathbf{x}/r$ and $1/r$. The first of these has already been handled in Problem (3.6.32; 3), and the second follows from $\mathbf{x}(t)/t \to \pm\mathbf{p}_\pm/m$. □

4.1.15 Remarks

1. Proposition (4.1.14) implies that all finite functions of these operators converge to the same functions of the asymptotic constants (cf. (2.5.7; 3)). As usual there are domain questions for unbounded operators. The statement $a_n\psi \to a\psi$ for all $\psi \in D$, a dense set, does not suffice for $a_n \to a$. To see this, suppose a and b are two different self-adjoint extensions of a densely defined operator c, and set $a_n = b$ for all n, $D = D(c)$. Since $a_{|D(c)} = b_{|D(c)}$, convergence is trivial, but finite functions a and b may genuinely differ. Only if D is a domain of essential self-adjointness for all a_n and for a can one conclude that $a_n \to a$ (Problem 8).
2. In order that the Møller operators still exist, H_0 might be modified somehow depending on the p_i but so as to describe the correct asymptotic motion. The trouble is that any such H_0 depends on time explicitly, so U_t^0 is not simply $\exp(-itH_0(t))$, and the time-evolution is not being compared with a one-parameter groups automorphisms. One possibility is $U_t^0 = \exp(-i(t|\mathbf{p}|^2/2m + m\alpha(\ln t)/|\mathbf{p}|))$, as $\exp(iH_rt)U_t^0$ in fact converges to the desired wave operator. Since the modified H_0 still commutes with $\mathbf{p}$, this also shows that the $p_i(t)$ converge. We are, however, less interested in pure existence theorems than in explicit computations, so this line of reasoning will not be pursued further.

4.1.16 Corollaries

1. Since H is constant, $PHP = |\mathbf{p}_+|^2/(2m) = |\mathbf{p}_-|^2/(2m)$.
2. $P\mathbf{F}P = \frac{1}{2}[\mathbf{p}_\pm \times \mathbf{L} - \mathbf{L} \times \mathbf{p}_\pm] \pm \frac{m\alpha\mathbf{p}_\pm}{|\mathbf{p}_\pm|} = \frac{i}{2}[\mathbf{p}_\pm, |\mathbf{L}|^2] \pm \frac{m\alpha\mathbf{p}_\pm}{|\mathbf{p}_\pm|}$.

The lesson is that the state of affairs concerning asymptotic constants is just as in the classical theory (I: 4.2.21; 1). In order to find the analogue of (I: 4.2.20) in quantum theory, it is only necessary to take some care with the noncommutativity of the various observables. This being done, one finds the

4.1.17 Connection between $\mathbf{p}_+$ and $\mathbf{p}_-$

$$
\begin{aligned}
p_k^+ &= p_k^- \frac{|\mathbf{L}|^2 + i\eta - \eta^2}{|\mathbf{L}|^2 - i\eta + \eta^2} + [\mathbf{p}_- \times \mathbf{L} - \mathbf{L} \times \mathbf{p}_-]_k \frac{\eta}{|\mathbf{L}|^2 - i\eta + \eta^2} \\
&= p_k^- \frac{|\mathbf{L}|^2(1 + i\eta) + i\eta - \eta^2}{|\mathbf{L}|^2 - i\eta + \eta^2} - i\eta|\mathbf{L}|^2 p_k^- \frac{1}{|\mathbf{L}|^2 - i\eta + \eta^2}, \\
\eta(H_r) &\equiv \frac{Pm\alpha}{\sqrt{2mH_r}}.
\end{aligned}
$$

Proof:

$$\begin{aligned}
2\varepsilon_{ijk}F_jL_k &= \varepsilon_{ijk}[\varepsilon_{jmn}(p_m^{\pm}L_n - L_m p_n^{\pm}) \pm 2\eta p_j^{\pm}]L_k \\
&= -p_i^{\pm}L^2 + p_k^{\pm}L_iL_k - L_k p_i^{\pm}L_k \pm 2\eta(F_i \mp \eta p_i^{\pm}) \pm 2i\eta p_i^{\pm} \\
&= -2p_i^{\pm}L^2 + 2\eta F_i - 2\eta^2 p_i^{\pm} \pm 2i\eta p_i^{\pm}.
\end{aligned}$$

Substitution for **F** from (4.1.16; 2) then proves the claim. □

4.1.18 Remarks

1. The first part of the formula is the analogue of (I: 4.2.20), and follows since $\mathbf{p}_+$ is related to $\mathbf{p}_-$ by reflection through **F** (see I: 4.2.18; 1)).
2. The connection between $\mathbf{p}_+$ and $\mathbf{p}_-$ in the spectral representation of $|\mathbf{L}|^2$, L_3, and $H_r P = |\mathbf{p}_\pm|^2/2m$, can be read off from the second part of the formula:

4.1.19 The Scattering Transformation

$$S = \frac{\Gamma\left(\frac{1}{2} + \sqrt{|\mathbf{L}|^2 + \frac{1}{4}} + i\eta(H_r)\right)}{\Gamma\left(\frac{1}{2} + \sqrt{|\mathbf{L}|^2 + \frac{1}{4}} - i\eta(H_r)\right)}$$

connects the momenta $\mathbf{p}_+$ *and* $\mathbf{p}_-$ *by*

$$\mathbf{p}_+ = S^{-1}\mathbf{p}_- S.$$

Proof: Since S commutes with **L**, it can be written in the tensor-product representation of (3.6.10; 1), with the eigenvectors $|l, m\rangle \in L^2(S^2, d\Omega)$ of $|\mathbf{L}|^2$ and L_3, as

$$S|l, m\rangle \otimes \varphi(k) = \exp(2i\delta_l(k))|l, m\rangle \otimes \varphi(k).$$

The component p_3^+ commutes with L_3 and H_r and, like x_3 of (3.2.10), changes l by 1. If we take the matrix element $\langle l+1, m| \cdot |l, m\rangle$ of the second form of (4.1.17), then we find that

$$\frac{\langle l+1, m|p_3^+|l, m\rangle}{\langle l+1, m|p_3^-|l, m\rangle} = \frac{l+1-i\eta}{l+1+i\eta} = \exp(2i(\delta_l - \delta_{l+1})).$$

Proposition (4.1.19) follows from this recursively, if we set $\exp(2i\delta_0(k)) = \Gamma(1 + i\eta(k))/\Gamma(1 - i\eta(k))$ by convention. □

4.1.20 Remarks

1. As discussed in (I: 4.2.21), S is not determined uniquely by the conditions $S^{-1}\mathbf{p}_- S = \mathbf{p}_+$ and $S^{-1}\mathbf{L}S = \mathbf{L}$; the unitary elements of the commutant of the

algebra $\mathcal{A}$ formed from $\mathbf{p}_-$ and $\mathbf{L}$ remain unspecified. Note that $\{\mathcal{A}\}'$ consists of functions of the energy, so this arbitrariness just corresponds to a choice of $\delta_0(k)$. This amounts to an overall phase and has no effect on the scattering cross-section for $\mathbf{k} \neq \mathbf{k}'$. On the other hand, the time-delay is now infinite instead of $\partial\delta(k)/\partial k^2$, because $m\mathbf{x} - \mathbf{p}t \sim \ln t$ as $t \to \infty$.

2. As $\ell \to \infty$, the phase-shift δ_l diverges as $\sum_l (l+1)^{-1}$. This is why we were not able to normalize δ_∞ to zero, but chose instead to fix δ_0. It thus happens that δ_0 violates the rule for scattering phase-shift δ and scattering length a valid for short-range potentials, that $V > 0 \Rightarrow \delta < 0 \Rightarrow a > 0$ and $V < 0 \Rightarrow \delta > 0 \Rightarrow a < 0$ (see (3.6.5) and (3.6.23; 5)): If δ_∞ were 0, then δ_l would be negative for positive α and positive for negative α.
3. The scattering transformation S does not commute with all the constants of the motion (cf. (I: 4.21.4)). For instance $S^{-1}\mathbf{F}S = \mathbf{F} - 2\eta\mathbf{p}_+$.
4. The scattering matrix has been calculated in the Heisenberg representation. In the interaction representation (see (3.4.21) and (3.4.22; 6)) S_{00} is related to it through $\Omega_-^* S \Omega_- = S_{00}$ and $\Omega_-^* H_r \Omega_- = H_0$. In other words S_{00} is obtained from S by the replacement of H_r with H_0, or $|\mathbf{p}_-|^2$ with $|\mathbf{p}|^2$.
5. Although Proposition (3.6.7) about the poles of S was derived only under restrictive assumptions on V, it remains true that S contains the information of Balmer's formula: $\exp(2i\,\delta_l(k))$ has poles at $k = -im\,\alpha/(l+n_r)$, $n_r = 1, 2, \ldots$. They are in the upper half-plane if $\alpha < 0$, in which case the values of $k^2/2m$ are precisely the energies of the bound states.

Now that the phase-shifts δ_l are known, let us recall that the definition in (3.4.22; 5) and calculate (Problem 7) the

4.1.21 Scattering Amplitude

$$\begin{aligned} f(k; \mathbf{n}', \mathbf{n}) &= \sum_{l=0}^{\infty} \frac{2l+1}{2ik} P_l(\cos\theta) \left[\frac{\Gamma(1+l+i\eta)}{\Gamma(1+l-i\eta)} - 1 \right] \\ &= \frac{i\eta}{2k} \left(\frac{4}{|\mathbf{n}-\mathbf{n}'|^2} \right)^{1+i\eta} \frac{\Gamma(1+i\eta)}{\Gamma(1-i\eta)} - \frac{1}{2ik}\,\delta^2(\mathbf{n}-\mathbf{n}'). \end{aligned}$$

4.1.22 Remarks

1. The sum over l converges on the dense set of finite linear combinations of Y_l^m, for example, but is singular for $\mathbf{n} = \mathbf{n}'$.
2. The first contribution to the scattering amplitude

$$f(k; \mathbf{n}', \mathbf{n}) \sim (\sin\theta/2)^{-2-2i\eta}$$

is a well-defined distribution for all $\mathbf{n}$ and $\mathbf{n}' \in S^2$, and represents the unitary operator S as an integral operator with a kernel. This fact is lost in the Born approximation; the $i\eta$ disappears from the exponent, and f becomes nonintegrably singular. As a whole, f remains singular even after subtraction of the delta function in the forward direction; it is a distribution rather than an ordinary function.

3. The cross-section is the same as classically (I: 4.2.22),

$$\sigma(k; \mathbf{n}, \mathbf{n}') = |f(k; \mathbf{n}, \mathbf{n}')|^2 = \frac{\alpha^2}{16(k^2/2m)^2} \sin^{-4}\frac{\theta}{2}.$$

4.1.23 Problems

1. Show that if a is self-adjoint and b is essentially self-adjoint on a domain D invariant under all $\exp(iat), t \in \mathbb{R}$, then a and $\bar{b} \equiv$ the closure of b in the sense of (3.1.7) commute if $\exp(iat)b\exp(-iat)\psi = b\psi$ for all $\psi \in D$ and $t \in \mathbb{R}$.

2. Show that
   (i) $D(F) \supset D(H_r) \cap D_{\text{fin}}(L^2)$, where D_{fin} consists of the finite linear combinations of the Y_l^m;
   (ii) $\mathbf{F}$ is essentially self-adjoint on $D(H_r) \cap D_{\text{fin}}(L^2)$;
   (iii) For all φ and $\psi \in D(H_r) \cap D_{\text{fin}}(L^2)$, $d/dt\langle\varphi|\exp(iH_r t)F\exp(-iH_r t)|\psi\rangle = 0$. Use Problem 1 to verify that $[H, \mathbf{F}] = \mathbf{0}$.

3. Verify (4.1.6(ii)) and (4.1.6(iv)).

4. Compute the action of $p_\pm$ and p_3 on $|l, m\rangle$.

5. Compute $|2, 0, 0\rangle = F_3|2, 1, 0\rangle$ in the x-representation.

6. Express $|\mathbf{p}|^2$ in terms of $p_r = (1/r)(\mathbf{x} \cdot \mathbf{p}) - i/r$, r, and $|\mathbf{L}|^2$.

7. Calculate the sum over l in (4.1.21).

8. Show that if $a_n\psi \to a\psi$ for all $\psi \in D$, a domain of essential self-adjointness for all a_n and a, then $a_n \to a$ in the sense of (2.5.8; 3).

4.1.24 Solutions

1. $D(\bar{b}) = \{\psi :$ there exists a sequence $\psi_n \in D$ such that $\psi_n \to \psi$, and $b_n\psi$ converges$\}$.

$$b\exp(iat)\psi = \exp(iat)b\psi \qquad \text{for all } \psi \in D \Rightarrow$$
$$\bar{b}\exp(iat)\psi = \exp(iat)\bar{b}\psi \qquad \text{for all } \psi \in D(\bar{b}),$$

since $b\exp(iat)\psi_n = \exp(iat)b\psi_n$.

$$b\psi_n \to \bar{b}\psi,\ \exp(iat)b\psi_n \to \exp(iat)\bar{b}\psi,\ \exp(iat)\psi_n \to \exp(iat)\psi,$$

and consequently $b\exp(iat)\psi_n$ converges, which implies that $\exp(iat)\psi \in D(\bar{b})$, and

$\bar{b}\exp(iat)\psi = \exp(iat)\bar{b}\psi$. Let $\mathcal{C}$ be the finite linear combinations of $\exp(iat)$. Then $\bar{b}c\psi = c\bar{b}\psi$ for all $c \in \mathcal{C}$ and $\psi \in D(\bar{b})$. Furthermore, let $U = (b-i)(\bar{b}+i)^{-1}$. Every vector $\varphi \in \mathcal{H}$ is equal to $(\bar{b}+i)\psi$ for some $\psi \in D(\bar{b})$; thus $U\varphi = (\bar{b}-i)\psi$. $c\varphi = c(\bar{b}+i)\psi = (\bar{b}+i)c\psi$, $Uc\varphi = (\bar{b}-i)c\psi = c(\bar{b}-i)\psi = cU\varphi$, so $Uc = cU$ for all $c \in \mathcal{C}$. Therefore every bounded function of $\bar{b}$ commutes with $\mathcal{C}$.

2. (i) $\|F_j\psi\| \le c_1\| |\mathbf{L}|^2 p_j\psi\| + c_2\|p_j\psi\| + c_3\|\psi\|$ for some $c_i \in \mathbb{R}^+$. Since p_j changes the angular momentum by 1, $\| |\mathbf{L}|^2 p_j\psi\| \le c_4\|p_j\psi\| \le c_5\langle\psi|PH_r\psi\rangle^{1/2}$, and so $\|F_j\psi\| < \infty$ for all $\psi \in D(H_r)\cap D_{\text{fin}}(|\mathbf{L}|^2)$.
   (ii) $D(H_r)\cap D_{\text{fin}}(|\mathbf{L}|^2)$ contains the set $\{x_1^{g_1}x_2^{g_2}x_3^{g_3}\exp(-|\mathbf{x}|^2/2),\ g_i = 0, 1, 2, \ldots,$ and finite linear combinations$\}$, on which $\mathbf{F}$ is essentially self-adjoint.
   (iii)
$$\langle\varphi(t)|i[H, F_k]\psi(t)\rangle = \langle\varphi(t)|i\alpha\left[\frac{x_k}{r^3}, |\mathbf{L}|^2\right] + \alpha i\left[\frac{|\mathbf{L}|^2}{2r^2}, \frac{x_k}{r}\right]|\psi(t)\rangle = 0,$$
   since $[|\mathbf{L}|^2, r] = 0$ and $D(|\mathbf{L}|^2/r^2) \supset D(H_r)\cap D_{\text{fin}}(|\mathbf{L}|^2)$. As a consequence, the matrix elements of $\mathbf{F}(t)$ and $\mathbf{F}(0)$ are the same with the vectors of
$$D(H_r)\cap D_{\text{fin}}(|\mathbf{L}|^2).$$
   Therefore the unique self-adjoint extensions are the same, which is the criterion for commutativity by Problem 1.

3. The commutation relations used below follow from $[\mathbf{p}, f(\mathbf{x})] = -i\nabla f(\mathbf{x})$, $[\mathbf{x}, f(\mathbf{p})] = i\nabla f(\mathbf{p})$, and the identities $[ab, c] = a[b, c] + [a, c]b$ and
$$[ab, cd] = ac[b, d] + a[b, c]d + c[a, d]b + [a, c]db:$$
$$|\mathbf{F}|^2 = \left([\mathbf{p}\times\mathbf{L}] - i\mathbf{p} + \tfrac{m\alpha\mathbf{x}}{r}\right)\cdot\left([\mathbf{p}\times\mathbf{L}] - i\mathbf{p} + \tfrac{m\alpha\mathbf{x}}{r}\right)$$
(because $[\mathbf{p}\times\mathbf{L}] = -[\mathbf{L}\times\mathbf{p}] + 2i\mathbf{p}$).
$$\mathbf{p}\cdot[\mathbf{p}\times\mathbf{L}] = [\mathbf{p}\times\mathbf{p}]\cdot\mathbf{L} = 0,\ [\mathbf{p}\times\mathbf{L}]\cdot\mathbf{p} = (-[\mathbf{L}\times\mathbf{p}] + 2i\mathbf{p})\cdot\mathbf{p} = 2i|\mathbf{p}|^2.$$
$$[\mathbf{p}\times\mathbf{L}]\cdot\mathbf{x} = (-[\mathbf{L}\times\mathbf{p}] + 2i\mathbf{p})\cdot\mathbf{x} = |\mathbf{L}|^2 + 2i\mathbf{p}\cdot\mathbf{x},$$
$$\mathbf{x}\cdot[\mathbf{p}\times\mathbf{L}] = [\mathbf{x}\times\mathbf{p}]\cdot\mathbf{L} = |\mathbf{L}|^2.$$
$$|[\mathbf{p}\times\mathbf{L}]|^2 = |\mathbf{p}|^2|\mathbf{L}|^2 - (\mathbf{p}\cdot\mathbf{L})^2 = |\mathbf{p}|^2|\mathbf{L}|^2.$$
From these equations, $|\mathbf{F}|^2 = |\mathbf{p}|^2|\mathbf{L}|^2 + 2|\mathbf{p}|^2 + m\alpha|\mathbf{L}|^2/r + 2im\alpha(\mathbf{p}\cdot\mathbf{x})/r - |\mathbf{p}|^2 - im\alpha(\mathbf{p}\cdot\mathbf{x})/r + m\alpha|\mathbf{L}|^2/r - im\alpha(\mathbf{x}/r\cdot\mathbf{p}) + m^2\alpha^2$. Since $[\mathbf{x}, \mathbf{p}\cdot\mathbf{x}] = i\mathbf{x}$ and $[\mathbf{p}\cdot\mathbf{x}, 1/r] = i/r$, the final result is that $|\mathbf{F}|^2 = |\mathbf{p}|^2|\mathbf{L}|^2 + |\mathbf{p}|^2 + 2m\alpha|\mathbf{L}|^2/r + 2m\alpha/r + m^2\alpha^2 = m^2\alpha^2 + 2mH_r(|\mathbf{L}|^2 + 1)$. The commutation relations for the components of $\mathbf{F}$ follow from the formula $\mathbf{F} = |\mathbf{p}|^2\mathbf{x} - (\mathbf{p}\cdot\mathbf{x})\mathbf{p} + 2m\alpha\mathbf{x}/r$ and the formulas given above, as well as $[\mathbf{x}, |\mathbf{p}|^2] = 2i\mathbf{p}$, $[\mathbf{p}\cdot\mathbf{x}, |\mathbf{p}|^2] = 2i|\mathbf{p}|^2$, $[\mathbf{p}, 1/r] = i\mathbf{x}/r^3$, $[|\mathbf{p}|^2, 1/r] = i\mathbf{x}/r^3\cdot\mathbf{p} + i\mathbf{p}\cdot\mathbf{x}/r^3$, and $\mathbf{x}\cdot\mathbf{p} - \mathbf{p}\cdot\mathbf{x} = 3i$, which can be verified directly.

4. In the x-representation $\partial/\partial z = \cos\theta(\partial/\partial r) - (\sin\theta/r)(\partial/\partial\theta)$. Because $\cos\theta Y_l^l = cY_{l+1}^l$ and $\sin\theta Y_l^l = lcY_{l+1}^l$, and because of the analogous facts that $\partial/\partial x$ and $\partial/\partial y$,

we find that $ip_3|n,l,l\rangle = c(\partial/\partial r - l/r)|n,l+1,l\rangle$ and

$$ip_+|n,l,l\rangle = c'\left(\frac{\partial}{\partial r} - \frac{l}{r}\right)|n,l+1,l+1\rangle.$$

(The vector $ip_-|n,l,l\rangle$ is a linear combination of $(\partial/\partial r - l/r)|n,l+1,l-1\rangle$ and $(\partial/\partial r + (l+1)/r)|n,l-1,l-1\rangle$.) (See also (3.2.14).)

5. $|2,0,0\rangle = (m\alpha)^{3/2}(1-(m\alpha r/2))\exp(-m\alpha r/2)/\sqrt{8\pi}$.

6. First note that owing to the facts listed in Solution 3, p_r^* is formally equal to p_r, and

$$|\mathbf{L}|^2 = x_k p_i x_t p_s \varepsilon_{ikl}\varepsilon_{stl} = |\mathbf{x}|^2|\mathbf{p}|^2 + i(\mathbf{x}\cdot\mathbf{p}) - (\mathbf{x}\cdot\mathbf{p})(\mathbf{x}\cdot\mathbf{p}),$$

so $|\mathbf{p}|^2 = p_r^2 + |\mathbf{L}|^2/r^2$. If we now map $L^2(\mathbb{R}^3, d^3x)$ unitarily to $L^2(S^2, d\Omega) \otimes L^2(\mathbb{R}^+, dr)$ by $\psi \to u/r$, then p_r becomes the Hermitian operator $-i\,d/dr$ of Example (3.3.5; 4). It fails, however, to be self-adjoint; $p_r^* \supset p_r$, because $D(p_r) = \{\psi \in L^2 : \psi$ is absolutely continuous, $\psi' \in L^2$, and $\psi(0) = 0\} \subset D(p_r^*) = \{\psi \in L^2 : \psi$ is absolutely continuous, and $\psi' \in L^2\}$. A more precise statement is that $|\mathbf{p}|^2 = p_r^* p_r + |\mathbf{L}|^2/r^2$.

7. If $t = \cos\Theta$, then

$$(\sin\Theta/2)^{-2-2i\eta} = \sum_{l=0}^{\infty} 2^{i\eta}(2l+1)P_t(t)\int_{-1}^{1}(1-z)^{-i\eta-1}P_l(z)dz.$$

By recourse to Rodrigues's formula we find that

$$\begin{aligned}\int_{-1}^{1}\cdots &= \frac{1}{2^l l!}\int_{-1}^{1}(1-z)^{-i\eta-1}\frac{d^l}{dz^l}(z^2-1)^l dz\\ &= \frac{1}{2^l l!}(1+i\eta)\cdots(l+i\eta)\times\int_{-1}^{1}(1-z)^{-i\eta-l-1}(1-z)^l(1+z)^l dz\\ &= \frac{1}{2^l l!}\frac{\Gamma(l+1+i\eta)}{\Gamma(1+i\eta)}2^{-i\eta+l}\times\frac{\Gamma(-i\eta)\Gamma(l+1)}{\Gamma(l+1-i\eta)}\\ &= \frac{-1}{i\eta}\frac{\Gamma(1-i\eta)}{\Gamma(1+i\eta)}\frac{\Gamma(l+1+i\eta)}{\Gamma(l+1-i\eta)}2^{-i\eta}\end{aligned}$$

and $(\sin\Theta/2)^{-2-2i\eta} = \sum(1/-i\eta)(2l+1)\exp(i\delta_l)P_l(\cos\Theta)$, provided that δ_0 is chosen as 0.

8. Let $\psi = (a+i)\varphi$, $\varphi \in D$. $((a_n+i)^{-1} - (a+i)^{-1})\psi = (a_n+i)^{-1}(a-a_n)\varphi \to 0$, since $\|(a_n+i)^{-1}\| \le 1$, and analogously if $i \to -i$. However, $(a \pm i)D$ is dense in $\mathcal{H}$, so $(a_n \pm i)^{-1}\psi \to (a \pm i)^{-1}\psi$ for all $\psi \in \mathcal{H}$.

4.2 The Hydrogen Atom in an External Field

Experiments subjecting atoms to constant electric and magnetic fields were indispensable to the understanding of atomic spectra. The effect of weak fields is seemingly just a moderate shift in the energy levels, but in fact the underlying mathematical problem is drastically changed.

The fields applied in laboratory experiments are usually weak in comparison with atomic fields, and appear to have only slight influence on atomic structure. In the other extreme, with the high magnetic fields **B** prevailing on neutron stars, the radius $(eB)^{-1/2}$ of the lowest magnetic orbital (cf. (3.3.5; 3)) can be smaller than the Bohr radius, and the atom contracts around the magnetic lines of force. In very strong electric fields autoionization occurs, and we shall see that even an arbitrarily small electric field destroys the point spectrum of an atom. It is amusing that this problem was one of the first successes of the perturbation theory developed in §3.5, despite its not being applicable in the absence of a point spectrum. One of our goals will be to find the sense in which perturbation theory is still asymptotically valid.

We give the Hamiltonian (4.1) the perturbations of (3.3.5; 1) or (3.3.5; 3), thereby obtaining the

4.2.1 Hamiltonians for the Stark and Zeeman Effects

$$H_E = \frac{|\mathbf{p}|^2}{2m} + \frac{\alpha}{|\mathbf{x}|} + eEx_3 \equiv H_0 + \lambda H',$$

$$H_0 = \frac{|\mathbf{p}|^2}{2m} + \frac{\alpha}{|\mathbf{x}|}, \qquad H' = x_3, \qquad \lambda = eE,$$

$$H_B = \frac{1}{2m}\left(\left(p_1 + \frac{eB}{2}x_2\right)^2 + \left(p_2 - \frac{eB}{2}x_1\right)^2 + p_3^2\right) + \frac{\alpha}{|\mathbf{x}|} \equiv H_0 + \lambda H',$$

$$H_0 = \frac{|\mathbf{p}|^2}{2m} + \frac{\alpha}{|\mathbf{x}|} - \frac{eB}{2m}\hat{L}_3, \quad H' = x_1^2 + x_2^2, \quad \lambda = \frac{e^2B^2}{8m},$$

$\hat{\mathbf{L}} = [\mathbf{x} \times \mathbf{p}]$ is the canonical angular momentum of (3.3.5; 3).

Since H' is not bounded relative to H_0, the question of the self-adjointness of H_E and H_B must be confronted. Serious difficulties are not to be expected, because quantum mechanics mollifies a $1/r$ singularity, and once something has been done about the singularity at $r = 0$, a classical electron in these potentials would evolve in a reasonable way and would not reach any boundaries in a finite time. Roughly speaking, it could be argued that if there is a $c \in \mathbb{R}^+$ such that $|d/dt(|\mathbf{p}|^2+|\mathbf{x}|^2)| \leq c(|\mathbf{p}|^2 + |\mathbf{x}|^2)$, then $|\mathbf{p}(t)|^2 + |\mathbf{x}(t)|^2 \leq \exp(ct)(|\mathbf{p}(0)|^2 + |\mathbf{x}(0)|^2)$, so neither the momentum nor the position coordinate could get unboundedly large in a finite time. The condition that $|\dot{N}| \leq cN$ is equivalent to $\pm\dot{N} + cN \geq 0$, and this argument can be made precise with a lemma on

4.2.2 Self-adjointness on the Domain of Operators Bounded Exponentially in Time

Let H be Hermitian and $N \geq 1$ self-adjoint with $D(N) \subset D(H)$, and suppose there exists a $c \in \mathbb{R}^+$ such that $\langle\psi|(\pm i[H, N] + cN)\psi\rangle > 0$ for all $\psi \in D(N)$. Then H is essentially self-adjoint on $D(N)$.

Proof: Recalling (2.5.10; 1) we shall show that given any $\gamma \in \mathbb{R}^+$, if $\langle\varphi|(H \pm i\gamma)\psi\rangle = 0$ for all $\psi \in D(N)$, then $\varphi = 0$. Specifically, that fact would imply that

$$0 = 2\,\mathrm{Im}\langle\varphi|(H \pm i\gamma)N^{-1}\varphi\rangle = \pm 2\gamma\langle\varphi|N^{-1}\varphi\rangle - \langle N^{-1}\varphi|i[H, N]N^{-1}\varphi\rangle,$$

which is consistent with the assumptions of the lemma and with $\gamma > c/2$ only if $\varphi = 0$. As remarked in (2.5.12; 2), the conclusion then holds for all $\gamma \in \mathbb{R}^+$. □

4.2.3 Application

Now let $N_{E,B} = H_{E,B} + \omega^2|\mathbf{x}|^2$, $\omega \in \mathbb{R}$ and sufficiently large. They are self-adjoint operators on $D(|\mathbf{p}|^2 + |\mathbf{x}|^2)$, because the other terms in $H_{E,B}$ are bounded relative to these. Then note that

$$\pm i[H_{E,B}, N_{E,B}] = \pm i\left[\frac{|\mathbf{p}|^2}{2m}, |\mathbf{x}|^2\right] = \mp\frac{1}{m}(\mathbf{p}\cdot\mathbf{x} + \mathbf{x}\cdot\mathbf{p}) < cN_{E,B}.$$

This leads to the conclusion that H_E and H_B are essentially self-adjoint on $D(|\mathbf{p}|^2 + |\mathbf{x}|^2)$.

The Hamiltonians of (4.2.1) thus determine the time-evolution uniquely; however, there is such a vast difference between $\lambda = 0$ and $\lambda < 0$ that the perturbation theory developed in §3.5 is deprived of its foundations. Moreover, at large distances α/r is insignificant compared with the external field, which therefore controls the action. Suitable bases for comparison are thus free fall (3.3.5; 1) for H_E and motion in a repulsive harmonic force for H_B with $\lambda < 0$.

4.2.4 Existence of the Møller Operators

Let $H_E(\alpha)$ and $H_B(\alpha)$ be as in (4.2.1). Then

$$\Omega_\pm = s\text{-}\lim_{t\to\pm\infty} \exp(iH_{E,B}(\alpha)t)\exp(iH_{E,B}(0)t)$$

exist for H_E if $\lambda \neq 0$ and for H_B if $\lambda < 0$.

4.2.5 Remarks

1. As was discussed in §4.1, the Møller operators do not exist if $\lambda = 0$. The external fields make the time-limit more tractable, because although $1/r$ is not integrable in time when $\mathbf{x} \to \mathbf{x} + \mathbf{p}t$, it is if $\mathbf{x} \to \mathbf{x} + \mathbf{p}t + \mathbf{g}t^2$ or $\mathbf{x} \to \mathbf{x}\cosh(t) + \mathbf{p}\sinh(t)$.
2. It follows from $\Omega^* H(\alpha)\Omega = H(0)$ that $H(\alpha)$ has the same spectrum when restricted to the range of Ω as $H(0)$. This shows that $\sigma_{ac}(H_E) = \mathbb{R}$ if $\lambda \neq 0$ and $\sigma_{ac}(H_B) = \mathbb{R}$ if $\lambda < 0$. The unboundedness below is easy to see using trial functions supported far away from the origin in regions where the potential is very negative. It is clear from this why the analytic perturbation theory §3.5 is impossible.
3. Completeness of the Moller operators $\Omega_\pm$ would imply that $\sigma_p = \sigma_{\mathrm{sing}} = \emptyset$. Though this is the case, the proof is more difficult; the interested reader is referred to [12].

Proof: As in (3.4.11), we begin by taking the time-derivative of $\Omega(t)$, though we shall require only the rough estimate

$$\int_{-\infty}^{\infty} dt \| \exp(iHt)V\exp(-iH_0t)\psi\| = \int_{-\infty}^{\infty} dt\|V\exp(-iH_0t)\psi\|$$
$$\leq \left(\int_{-\infty}^{\infty}\frac{dt}{1+t^2}\right)^{1/2}\left(\int_{-\infty}^{\infty} dt(1+t^2)\|V\exp(-iH_0t)\psi\|^2\right)^{1/2}.$$

Since we are showing that a sequence of bounded operators converges strongly, we may restrict to a total set, $\{\psi = \exp(-|\mathbf{x}-\bar{\mathbf{x}}|^2/2b^2),\ \bar{\mathbf{x}} \in \mathbb{R}^3,\ b \in \mathbb{R}^+\}$. In units where $2m = 1$ and $Ee = 2g$, both $H_E(0)$ and

$$\exp(2itgx_3)\exp(-it^2p_3g)\exp(-it|\mathbf{p}|^2)$$

produce the time-evolution $(x_1, x_2, x_3; p_1, p_2, p_3) \to (x_1 + p_1t, x_2 + p_2t, x_3 + p_3t + gt^2; p_1, p_2, p_3 + 2gt)$; therefore their difference is only a multiple of $\mathbf{1}$. By Solution (3.3.20; 2),

$$(\exp(-it|\mathbf{p}|^2)\psi)(x) = \left[\exp\frac{-|\mathbf{x}-\bar{\mathbf{x}}|^2}{2(b^2+t^2b^{-2})}\right](1+t^2b^{-4})^{-3/4},$$

while $\exp(-it^2p_3g)$ displaces x_3 by gt^2, and $\exp(2itgx_3)$ drops out because it commutes with V. Thus it remains to show that

$$\int_0^{\infty} dt(1+t^2)(1+t^2b^{-4})^{-3/2}\int\frac{d^3x}{r^2}\exp\frac{-|\mathbf{x}-\bar{\mathbf{x}}(t)|^2}{(b^2+t^2b^{-2})} < \infty,$$

$$(\bar{x}_1(t), \bar{x}_2(t), \bar{x}_3(t)) = (\bar{x}_1, \bar{x}_2, \bar{x}_3 + gt^2),$$

which follows from a simple change of variables (Problem 1). The proof for H_B is very similar; it is only necessary to take the harmonic motion in (3.3.20; 2) with an imaginary frequency. This makes $\bar{\mathbf{x}}(t) = \mathbf{x}\cosh(\nu t)$, and the convergence is even easier. □

The foregoing results show immediately that the resolvent

$$(H_0 + \lambda H' - z)^{-1}, \quad z \in \mathbb{C}\backslash\mathbb{R},$$

is not analytic in λ at $\lambda = 0$, where H has been divided into H_0 and H' as in (4.2.1). Perturbation theory will thus fail to converge as an expansion in the external field. It is reasonable to wonder, however, whether the perturbation-theoretic formulas still have some meaning or become pure nonsense. Despite the lack of analyticity, we at least have

4.2.6 Strong Continuity in λ

The function $\lambda \to (H_0 + \lambda H' - z)^{-1}$, for $z \in \mathbb{C}\backslash\mathbb{R}$ and H_0 and H' as in (4.2.1), *is a continuous mapping on $\mathcal{B}(\mathcal{H})$ in the strong topology.*

Proof: The resolvent equation

$$(H_0 - z)^{-1} - (H_0 + \lambda H' - z)^{-1} = \lambda(H_0 + \lambda H' - z)^{-1} H'(H_0 - z)^{-1}$$

obviously holds on $(H_0 - z)D(|\mathbf{p}|^2 + |\mathbf{x}|^2)$, since $D(|\mathbf{p}|^2 + |\mathbf{x}|^2) \subset D(H')$. The operator H_0 is essentially self-adjoint on $D(|\mathbf{p}|^2 + |\mathbf{x}|^2)$, which means that $(H_0 - z)D(|\mathbf{p}|^2 + |\mathbf{x}|^2)$ is dense, because its closure is $(H_0 - z)D(H_0)$, which is all of $\mathcal{H}$ by (2.5.5). Since the resolvent is bounded by $|\mathrm{Im}\, z|^{-1}$ in norm, uniformly in λ, the strong continuity in λ follows from the strong continuity on a dense set. □

Proposition (4.2.6) implies that as $\lambda \to 0$ any bounded, continuous function of $H_0 + \lambda H'$ converges strongly to the same function of H_0. On the same abstract level we can in fact state the following

4.2.7 Continuity Properties of the Spectrum

(i) *For all $z \in \mathrm{Sp}(H_0)$ there exists a $z(\lambda) \in \mathrm{Sp}(H_0 + \lambda H')$ such that*

$$\lim_{\lambda \to 0} z(\lambda) = z.$$

(ii) *For all a and $b \in \mathbb{R}$ such that $a < b$ and $a, b \notin \sigma_p(H_0)$, the projections $P_{(a,b)}(H_0 + \lambda H')$ converge strongly to $P_{(a,b)}(H_0)$.*

4.2.8 Remarks

1. Proposition (i) means that the spectrum of the limiting operator cannot suddenly get larger. Example (3.5.10; 1) shows that it is possible for it to contract suddenly from $\mathbb{R}$ to $\{0\}$. If H' were an analytic perturbation, then norm continuity of the resolvent in λ can be used to exclude this possibility.
2. If the interval (a, b) contains only one eigenvalue of H_0, then (ii) implies that $P_{(a,b)}(H_0 + \lambda H')$ converges to the projection onto the eigenspace of the eigenvalue.
3. The requirement that $a \notin \sigma_p(H_0)$ is a necessary one; suppose as in (3.5.11; 1), that $H_0 = 0$ and $H' = x$, as operators on $L^2((-\infty, \infty), dx)$, and $\chi_{(0,1)}$ be the characteristic function of $(0, 1)$. Then

$$\chi_{(0,1)}(\lambda x) = P_{(0,1)}(H_0 + \lambda H') \to P_{(0,\infty)}(x),$$

but $\chi_{(0,1)}(0) = 0$.

Proof:

(i) We shall show, equivalently, that $(a, b) \cap \mathrm{Sp}(H_0 + \lambda H')$ being empty for all sufficiently small λ implies that $(a, b) \cap \mathrm{Sp}(H_0)$ is empty. By the spectral theorem the latter statement is equivalent to

$$\left\| \left(H_0 - \frac{a+b}{2} + i\frac{b-a}{2} \right)^{-1} \right\| \leq \frac{\sqrt{2}}{(b-a)}.$$

By assumption,

$$\left\| \left(H_0 + \lambda H' - \frac{a+b}{2} + i\frac{b-a}{2} \right)^{-1} \right\| \leq \frac{\sqrt{2}}{(b-a)}$$

for sufficiently small λ. Since the operator norm is strongly lower semicontinuous ($\|\cdot\| = \sup_{\|\psi\|=1} \|\cdot\psi\|$), $R_\lambda \to R$, implies that $\|R\| \leq \liminf \|R_\lambda\|$, from which the proposition follows.

(ii) To generalize from convergence of continuous functions to that of characteristic functions, recall that there exist continuous functions f_n and g_n, $0 \leq f_n \leq \chi_{(a,b)}$ and $g_n \geq \chi_{[a,b]}$, such that $f_n \uparrow \chi_{(a,b)}$ and $g_n \downarrow \chi_{[a,b]}$ pointwise. Hence, by Problem 2, $f_n(H_0) \to \chi_{(a,b)}(H_0)$ and $g_n(H_0) \to \chi_{[a,b]}(H_0)$. Since a and $b \notin \sigma_p(H_0)$, $P_{(a,b)}(H_0) = P_{[a,b]}(H_0)$, so $\chi_{(a,b)}(H_0) = \chi_{[a,b]}(H_0)$. This implies that for all ψ and ε there exist continuous functions $f \leq \chi_{(a,b)} \leq \chi_{[a,b]} \leq g$ such that $\|(f(H_0) - g(H_0))\psi\| \leq \varepsilon$. Consequently,

$$\begin{aligned}
&\|(P_{(a,b)}(H_0 + \lambda H') - P_{(a,b)}(H_0))\psi\| \\
&\quad \leq \|(P_{(a,b)}(H_0 + \lambda H') - f(H_0 + \lambda H'))\psi\| \\
&\qquad + \|(f(H_0 + \lambda H') - f(H_0))\psi\| + \|(P_{(a,b)}(H_0) - f(H_0))\psi\| \\
&\quad \leq \|(g(H_0 + \lambda H') - f(H_0 + \lambda H'))\psi\| \\
&\qquad + \|(f(H_0 + \lambda H') - f(H_0))\psi\| + \|(g(H_0) - f(H_0))\psi\| \\
&\quad \leq \|(g(H_0 + \lambda H') - g(H_0))\psi\| + 2\|(f(H_0 + \lambda H') - f(H_0))\psi\| \\
&\qquad + 2\|(g(H_0) - f(H_0))\psi\|
\end{aligned}$$

is arbitrarily small. □

In the case we have been interested in, H_0 has a point spectrum, but Proposition (4.2.7) does not guarantee that the point spectrum persists when λ is changed from 0. The point spectrum cannot disappear without a trace, however; instead, there is a sort of

4.2.9 Spectral Concentration

Let E_0 be an isolated eigenvalue of H_0 of finite multiplicity and P_0 be the associated projection. Suppose that $P_0 H' P_0$ exists and has eigenvalues E'_j with projections P_j, $\sum_j P_j = P_0$. Then for all $\varepsilon > 0$ and η, $1 < \eta < 2$,

$$s\text{-}\lim_{\lambda \to 0} P_{(E_0 + \lambda E'_j - \varepsilon\lambda^\eta, E_0 + \lambda E'_j + \varepsilon\lambda^\eta)}(H_0 + \lambda H') = P_j.$$

4.2.10 Remarks

1. In the cases we have examined (4.2.1), the exponential fall-off of the eigenvectors H_0 makes them belong to $D(H')$, so the finiteness of $P_0 H' P_0$ is clear.

2. Proposition (4.2.9) states that to order λ^η, $\eta < 2$, the spectrum shrinks down around the eigenvalues predicted by first-order perturbation theory. The proposition is easily generalized to higher order.

Proof: Let ψ_j be one of the vectors spanning the range of P_j, so $H_0\psi_j = E_0\psi_j$ and $P_0 H'\psi_j = P_0 E'_j\psi_j$. Then the ψ constructed by perturbation theory (3.5.16) is undeniably an eigenvector of $H_0 + \lambda H'$ to $O(\lambda^2)$ (cf. (3.5.17; 3)):

$$\begin{aligned}\|(H_0 + \lambda H' - E_0 - \lambda E'_j)(\psi_j - \lambda(H_0 - E_0)^{-1}(H' - E'_j)\psi_j)\|^2 \\ = \lambda^4 \|(H' - E'_j)(H_0 - E_0)^{-1}(H' - E'_j)\psi_j\|^2.\end{aligned}$$

(Recall that $\lambda E'_j$ was incorporated into H_0 in §3.5.) Now, if μ_j is the probability measure associated with the vector $(\mathbf{1} - \lambda(H_0 - E_0)^{-1}(H' - E'_j))\psi_j \equiv \psi_j(\lambda)$, the operator $H_0 + \lambda H'$, and the interval

$$I_j(\lambda) = (E_0 + \lambda E'_j - \varepsilon\lambda^\eta, E_0 + \lambda E'_j + \varepsilon\lambda^\eta),$$

then we get the estimate

$$\begin{aligned}\lambda^4 \|(H' - E'_j)(H_0 - E_0)^{-1}(H' - E'_j)\psi_j\|^2 = \int_{-\infty}^{\infty} d\mu_j(h)(h - E_0 - \lambda E'_j)^2 \\ \geq \varepsilon^2\lambda^{2\eta} \int_{h\notin I_j(\lambda)} d\mu_j(h) = \varepsilon^2\lambda^{2\eta}\|(\mathbf{1} - P_{I_j(\lambda)}(H_0 + \lambda H'))\psi_j(\lambda)\|^2.\end{aligned}$$

Because $\psi_j(\lambda) \to \psi_j$, $\eta < 2$, it follows that $(\mathbf{1} - P_{I_j(\lambda)}(H_0 + \lambda H'))\psi_j \to 0$. Since the vectors ψ_j span the range of P_j, this implies the norm convergence $P_{I_j(\lambda)}(H_0 + \lambda H')P_j \Rightarrow P_j$. By (4.2.7(ii)), once $I_j(\lambda_0)$ no longer contains anything but eigenvalue $E_0 + \lambda E'_j$, the projections $P_{I_j(\lambda_0)}(H_0 + \lambda H')$ converge strongly to $P_{I_j(\lambda_0)}(H_0) = P_j$. Therefore, if $\lambda < \lambda_0$, then

$$P_{I_j(\lambda)}(H_0 + \lambda H') = P_{I_j(\lambda)}(H_0 + \lambda H')P_{I_j(\lambda_0)}(H_0 + \lambda H') \to P_j. \qquad \square$$

At first sight, Theorem (4.2.9) appears without physical significance. For instance, in the trivial example of (4.2.8; 3), perturbation theory does not work, since 0 is an infinitely degenerate eigenvalue of H_0; with $E_0 = E_j = 0$, $\eta < 1$, Theorem (4.2.9) is still valid, yet nothing distinguishes the spectral value of 0 of the operator λx. The experimentally detectable consequences of (4.2.9) are brought to light by consideration of the

4.2.11 Indeterminacy Relation of Time and Energy

The probability that an initial state ψ is again measured at a later time t is $|\langle\psi|\exp(-iHt)\psi\rangle|^2$. For this reason, $\tau(\psi) \equiv \frac{1}{2}\int_{-\infty}^{\infty} dt|\langle\psi|\exp(-iHt)\psi\rangle|^2$ is referred to as the lifetime of ψ. If $\langle\psi|P_{(E_0-\varepsilon,E_0+\varepsilon)}(H)|\psi\rangle \geq 1 - \delta$ then $\tau(\psi) \geq (1-\delta)^2/8\pi\varepsilon$.

Proof: As in the proof of (3.4.11), it follows from Parseval's equation that $\tau(\psi) = \int (d\omega/4\pi)|\langle\psi|\delta(H-\omega)\psi\rangle|^2$, and then by the Cauchy-Schwarz inequality,

$$(1-\delta)^2 \leq \left(\int_{E_0-\varepsilon}^{E_0+\varepsilon} d\omega \langle\psi|\delta(H-\omega)\psi\rangle\right)^2$$
$$\leq \int_{E_0-\varepsilon}^{E_0+\varepsilon} d\omega' \int_{E_0-\varepsilon}^{E_0+\varepsilon} d\omega \langle\psi|\delta(H-\omega)\psi\rangle^2 = 8\pi\varepsilon\tau(\psi). \; \square$$

If a perturbed operator is strongly but not norm continuous, then an eigenvalue E_0 may disappear into a continuum that springs into existence. However, even if this happens, for small λ the state ψ_0 has a long lifetime:

4.2.12 The Lifetimes of Eigenstates that have Disappeared into the Continuum

With the assumptions of (4.2.9), *for all* $\varepsilon > 0$ *there exists a* $\lambda_0 > 0$ *such that* $\tau(\psi_j) > 1/8\pi\varepsilon\lambda^\eta$ *for all* λ, $0 < \lambda < \lambda_0$.

Proof: Let $\psi_I = P_{I(\lambda)}\psi_j$. The strong convergence of the operator $H(\lambda)$ implies the existence of a λ_0 such that $\|\psi_I - \psi_j\| = \|(P_{I(\lambda)} - P_j)\psi_j\| < \varepsilon/2$ for all λ, $0 < \lambda < \lambda_0$. Therefore

$$\|\psi_I\|^2 = \int_{E-\varepsilon\lambda^\eta}^{E+\varepsilon\lambda^\eta} d\omega \langle\psi_j|\delta(H-\omega)\psi_j\rangle \geq 1-\varepsilon,$$

so this proposition follows from (4.2.11). $\square$

Now that the mathematical state of affairs is understood, let us return to the physical problem and examine $H_B = H_0 + \lambda H'$. The situation is only half as bad as it might be, since the point spectrum is preserved for the physical values $\lambda \geq 0$; this follows immediately from the min-max principle, since the term linear in B, which was built into the H_0 of (4.2.1), is simultaneously diagonable with $\hat{L}_3$ (see (3.3.20; 4)), and $e^2B^2(x_1^2 + x_2^2)/8m$ is a positive perturbation. The number of eigenvalues of H_B under a given energy E is therefore at most the same as the number of eigenvalues $E^{(0)}_{n,l,l_3} \equiv -(m\alpha^2/2n^2) - (eB/2m)l_3$ of H_0 under E. This argument leads straightaway to

4.2.13 Bounds for the Eigenvalues of H_B

The lowest eigenvalue E_{l_3} *of* H_B *where* l_3 *is a given eigenvalue of* $\hat{L}_3$ *satisfies*

$$E^{(0)}_{l_3+1,l_3,l_3} \leq E_{l_3} \leq E^{(0)}_{l_3+1,l_3,l_3} + \frac{e^2B^2}{8m}\langle l_3+1, l_3, l_3, |x_1^2 + x_2^2| l_3+1, l_3, l_3\rangle,$$

$$H_0|n, l, l_3\rangle = E^{(0)}_{n,l,l_3}|n, l, l_3\rangle.$$

4.2.14 Remarks

1. These bounds show that the divergence of perturbation theory does not diminish the usefulness of the linear formula for small B. It can in fact be shown that the perturbation series is Borel summable [3].
2. The term α/r is compact relative to the rest of H_B (Problem 6), so the essential spectrum of H_B begins at $eB/2m > 0$, as in (3.3.5; 3).
3. At this stage, (4.2.11) applies only to particles without spin. The presence of spin adds a term $B\mu S_3$ to H_B, where for an electron the spin magnetic moment μ is $2 \cdot [1.0011596] \cdot e/2m$. The new term is simultaneously diagonable with H_0; as long as the relativistic spin-orbit coupling is left out, the difference is a simple additive constant.

To finish the section off, we discuss the Stark effect in greater detail. As we saw that if $E \neq 0$, then $\sigma_{ac}(H_E) = \mathbb{R}$ and $\sigma_p(H_E) = \sigma_s(H_E) = \emptyset$, one might well wonder how so many physicists have made successful careers measuring and calculating the eigenvalues of H_E. The underlying reasons are some nice

4.2.15 Stark-Effect Delicacies for Mathematical Connoisseurs

(i) When $E \neq 0$, suitable expectation values of the resolvent $(H_E - z)^{-1}$ have a branch cut along $\mathbb{R}$, and the poles of the resolvent when $E = 0$ move onto the second sheet when the field E is switched on.

(ii) The imaginary part of the position of the pole associated with the ground state goes as $\exp(-\alpha^3/6eE)$ as $E \to 0$. The small imaginary part shows up as a long lifetime (4.2.12) and as a sharp resonance in the scattering matrix (4.2.4), and hence as a long time-delay (3.6.13).

(iii) Perturbation theory leads to the correct asymptotic power-series for the positions of the poles, all coefficients being integrals of real functions and hence real. The imaginary part is invisible in perturbation theory, since it goes to zero faster than any power of the applied field.

(iv) Since any reasonable procedure for resumming the perturbation series will lead to something real, it cannot give the exact position of the pole. However, if one begins with a complex electric field E, Borel sums the series, and then lets Im E tend to zero, the complex poles can be found exactly.

The proofs of these mathematical facts can be found in [13]. The physics underlying the complex poles is the quantum-mechanical tunneling effect, by which an electron has some probability of reaching a position with large $-x_3$. If the field is not too large, then the time taken for the tunneling is so long that the effect can be neglected in any conceivable experiment.

Let us next ascertain the energies at which the spectrum is asymptotically concentrated. We shall not come up with any bounds for the (nonexistent) eigenvalues.

4.2.16 First-Order Perturbation Theory

The first step is to diagonalize H' in the degeneracy spaces of H_0. From (3.2.10) and the conservation of $\hat{L}_3$ we get

$$\langle n, l, l_3|x_3|n', l', l_3'\rangle = \delta_{l_3,l_3'}\delta_{l,l'\pm 1}\langle n, l, l_3|x_3|n', l \mp 1, l_3\rangle,$$

so in the simplest cases H' looks as follows, represented as a matrix:

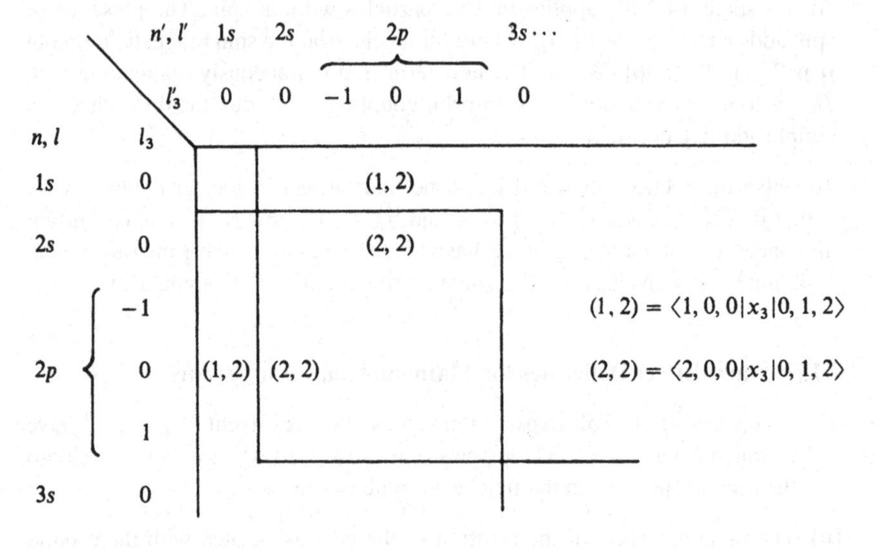

This is diagonalized with the combinations

$$|\pm\rangle = (1/\sqrt{2})(|2, 0, 0\rangle \pm |2, 1, 0\rangle) :$$

		1s	+	2p −1	−	2p 1
1s						
+			(2, 2)			
2p	−1					
−					−(2, 2)	
2p	1					

It is apparent that to first order E, the values E_{n,l,l_3} are unchanged for $n = 1$ and shifted by $\pm eE\langle 2, 0, 0|x_3|2, 1, 0\rangle$ for $n = 2$, $l_3 = 0$ (Problem 5).

4.2.17 Remarks

1. Unsophisticated feelings are vindicated in that, as with the Zeeman effect, the first-order correction to the energy is just the field strength times the dipole moment.
2. This result seems to contradict a general theorem: *The expectation value of the dipole moment is zero in eigenstates of an operator that commutes with the parity P.* Proof: $H|\rangle = E|\rangle \Rightarrow P|\rangle = \pm|\rangle$, so $\langle|x_3|\rangle = -\langle|Px_3P|\rangle = -\langle|x_3|\rangle$. The explanation is that the conditions of the theorem are incomplete; *it must also be assumed that all the eigenvectors with the same E have the same parity*, which is not true in the Stark effect.
3. Relativistically, the $S_{1/2}-P_{3/2}$ degeneracy is removed in the Dirac equation and, moreover, the Lamb shift separates the $P_{1/2}$ and $S_{1/2}$ levels. Then the theorem of Remark 2 applies, and, strictly speaking, there is no linear Stark effect.
4. Since the more precise formula for the splitting of the energies is

$$\frac{\varepsilon_{2s}+\varepsilon_{2p}}{2} \pm \sqrt{\left(\frac{\varepsilon_{2s}-\varepsilon_{2p}}{2}\right)^2 + (eE\langle z\rangle)^2},$$

where the ε's are the eigenvalues for $E = 0$, and since $\varepsilon_{2s} - \varepsilon_{2p}$ is very small, the splitting soon becomes virtually linear in E.

4.2.18 Problems

1. Show that $\int_0^\infty dt(1+t^2)(1+t^2b^{-4})^{-3/2}\int(d^3x/r^2)\exp(-|\mathbf{x}-\bar{\mathbf{x}}(t)|^2/(b^2+t^2b^{-2})) < \infty$, $\bar{\mathbf{x}}(t) = (\bar{x}_1, \bar{x}_2, \bar{x}_3 + gt^2)$.
2. Show that if $f_n(x) \downarrow f(x) \geq 0$, then $f_n \to f$ strongly as a multiplication operator.
3. Calculate $\langle n, l, l_3|x_1^2 + x_2^2|n, l, l_3\rangle$ for $n = 1$ and 2.
4. Find an example of projection operators P_n converging strongly to P for which $\dim P_n = \infty$ but $\dim P < \infty$.
5. Calculate $\langle 2, 0, 0, |x_3|2, 1, 0\rangle$.
6. Let $a_{|M}$ be the restriction of an operator a to the part of the Hilbert space on which $|\hat{L}_3| \leq M$. Show that $H_{B|M}(\alpha)$ is compact relative to $H_{B|M}$ $(\alpha = 0)$.

4.2.19 Solutions

1. First write r^{-2} as $\int_0^\infty ds\exp(-sr^2)$. The resulting Gaussian integrations over x_1, x_2, and x_3 are easy to do by completing the squares in the exponents. To indicate how the calculation then proceeds, consider, for simplicity, $b = 1$ – the calculation for $b \neq 1$ is similar. Then

$$\int_0^\infty \frac{dt}{\sqrt{1+t^2}}\int_0^\infty ds\int_{\mathbb{R}^3} d^3x\exp\left(-sr^2 - \frac{x_1^2+x_2^2+(x_3+gt^2)^2}{1+t^2}\right)$$
$$= \text{const}\int_0^\infty \frac{dt}{\sqrt{1+t^2}}\int_0^\infty \frac{ds}{(s+(1+t^2)^{-1})^{3/2}}$$

$$\times \exp\left(-\frac{g^2 t^4}{1+t^2}\left(1 - \frac{1}{1+s(1+t^2)}\right)\right),$$

where the square in the exponent was completed in the last step. The two remaining integrations are easy to estimate if the domain of integration is divided into $0 \le s, t \le 1$; $1 \le s, t < \infty$; and the rest.

2. By Lebesgue's dominated convergence theorem, $\int d\mu(x)|\psi(x)|^2(f_n(x) - f(x))^2 \to 0$.

3. In units where $m\alpha = 1$,

$$|1, 0, 0\rangle = \frac{1}{\sqrt{\pi}} \exp(-r),$$

$$|2, 0, 0\rangle = \frac{1}{\sqrt{8\pi}}\left(1 - \frac{r}{2}\right) \exp(-r/2),$$

$$|2, 1, \pm 1\rangle = \frac{1}{8\sqrt{\pi}} r \exp(-r/2) \sin\theta \exp(\pm i\varphi),$$

$$|2, 1, 0\rangle = \frac{1}{4\sqrt{2\pi}} r \exp(-r/2) \cos\theta,$$

(cf. (4.1.23; 5)), and the corresponding expectation values are 2, 28, 24, and 12.

4. Let P_n be the operator in $\mathcal{B}(l^2)$, represented diagonally with entries

$$(\underbrace{0, 0, \ldots, 0}_{n}, 1, 1, \ldots).$$

Then $\operatorname{Tr} P_n = \infty$, but $P_n \to 0$.

5. $\langle 2, 0, 0|x_3|2, 1, 0\rangle = -\frac{1}{2}$.

6. We shall show that the graph norm of H_B provides a finer topology on the subspace when $\alpha = 0$, $B > 0$ than when $\alpha = 0$, $B = 0$. The rest of the argument is like the one for the relative compactness when $B = 0$. First note that if $2m = 1$,

$$\begin{aligned} &a\|(|\mathbf{p}|^2 + \lambda(x_1^2 + x_2^2) - w\hat{L}_3)\psi\| + b\|\psi\| \\ &\quad \ge a\|(|\mathbf{p}|^2 + \lambda(x_1^2 + x_2^2))\psi\| + (b - M)\|\psi\|, \end{aligned}$$

and by use of the commutation relations,

$$\begin{aligned} \langle\psi|(|\mathbf{p}|^2 + \lambda(x_1^2 + x_2^2))^2\psi\rangle = \langle\psi||\mathbf{p}|^4 &+ 2\lambda(|\mathbf{p}|(x_1^2 + x_2^2)|\mathbf{p}| - 2) \\ &+ \lambda^2(x_1^2 + x_2^2)^2|\psi\rangle \ge \langle\psi||\mathbf{p}|^4\psi\rangle - 4\lambda\|\psi\|^2, \end{aligned}$$

so

$$a\|(|\mathbf{p}|^2 + \lambda(x_1^2 + x_2^2) - w\hat{L}_3)\psi\| + b\|\psi\| \ge a\||\mathbf{p}|^2\psi\| + (b - M - 2\lambda)\|\psi\|.$$

Since the norms $a\|H_B\psi\| + b\|\psi\|$ are equivalent for all a and $b > 0$, the proposition follows.

4.3 Helium-like Atoms

Although the Schrödinger equation for helium-like atoms is not exactly soluble, it is possible to make statements about it with arbitrarily good accuracy. For that reason it has been a touchstone of quantum mechanics.

The explanation of the spectrum of the helium atom was one of the early successes of the new quantum theory, since the old quantum theory, which was nothing more than classical mechanics bolstered with ad hoc quantum assumptions, was unable to cast much light on the problem. Even today, the set of problems connected with helium must be reckoned among the brilliant successes of mathematical physics. While Schrödinger's equation cannot be solved for helium in terms of familiar functions, it is not only possible to formulate valid general statements about the spectrum of the Hamiltonian, but, indeed, the art of inequalities is so far advanced that rather exact bounds are available for the eigenvalues.

When dealing with two or more electrons, one must bring the exclusion principle into play. However, its importance will be limited in this situation, because of the additional spin degree of freedom. Any orbital can be occupied by two electrons, so long as their spins are antiparallel (a singlet state). Forces that do not affect the spin cause no transitions between states of parallel (triplet) and antiparallel spin, so the spin part of the problem can be dealt with separately. The orbital wavefunctions may be either symmetric or antisymmetric, and associated with either singlet or triplet spin vectors.

In addition to helium, we shall also be interested in the ions H^-, Li^+, Be^{++}, etc., the Hamiltonians of which are the same except that they have different values of the perturbation parameter. Moreover, elementary particle physics has created the possibility of replacing one e^- with a μ^- or other negatively charged particle, i.e., of varying the mass. We began the discussion of hydrogen by introducing center-of-mass and relative coordinates. The mass of the nucleus made its appearance in the latter part of the problem only through the reduced mass, and the problem was otherwise the same as the limit where the nuclear mass was set to infinity. As section 4.6 will be devoted to the study of the nuclear motion, let us immediately pass to this limit. It will remain to be determined how valid the results of this section are for systems like $e^-\mu^+e^-$ or $e^-e^+e^-$.

Once the usual list of necessary remarks has been checked off, things will progress rather rapidly to more detailed and less trivial matters. We start with the

4.3.1 Hamiltonian of an Atom with Two Electrons

$$H = \frac{1}{2m}\left(|\mathbf{p}_1|^2 + |\mathbf{p}_2|^2\right) - Ze^2\left(\frac{1}{|\mathbf{x}_1|} + \frac{1}{|\mathbf{x}_2|}\right) + e^2\frac{1}{|\mathbf{x}_1 - \mathbf{x}_2|}$$

can be put into normal form with a dilatation $\mathbf{p} \to Zme^2\mathbf{p}$, $\mathbf{x} \to (Zme^2)^{-1}\mathbf{x}$ and separation of the factors:

$$H(\alpha) = H(0) + \alpha H' \equiv Z^{-2}e^{-4}m^{-1}H = \frac{1}{2}(|\mathbf{p}_1|^2 + |\mathbf{p}_2|^2)$$

$$-\frac{1}{|\mathbf{x}_1|} - \frac{1}{|\mathbf{x}_2|} + \frac{\alpha}{|\mathbf{x}_1 - \mathbf{x}_2|}, \qquad \alpha = \frac{1}{Z}.$$

4.3.2 Remarks

The perturbation parameter α is not a continuous variable in reality, but it can assume many different values $1, \frac{1}{2}, \frac{1}{3}, \frac{1}{4}$, etc., corresponding to H^-, He, Li^+, Be^{++}, etc.

Since the potential energy is ε-bounded relative to the kinetic energy (see (3.4.2; 2)), we know the

4.3.3 Domain of Self-Adjointness

$$D(H) = (D(|\mathbf{p}|^2) \otimes \mathbb{C}^2) \wedge (D(|\mathbf{p}|^2) \otimes \mathbb{C}^2)$$
$$\subset (L^2(\mathbb{R}^3) \otimes \mathbb{C}^2) \wedge (L^2(\mathbb{R}^3) \otimes \mathbb{C}^2).$$

4.3.4 Gloss

The spin acts on the two-dimensional Hilbert space $\mathbb{C}^2$, so the Hilbert space appropriate for an electron with spin is $L^2(\mathbb{R}^3) \otimes \mathbb{C}^2$. As was mentioned in (3.1.16), a system of two electrons is associated with the antisymmetric tensor product $\wedge$ (recall (I: 2.4.38)) of two such spaces.

We now turn to the task of locating the continuous spectrum of $H(\alpha)$. It turns out to be quite easy, because the perturbation is positive when $\alpha > 0$ and could at most move the spectrum upward.

4.3.5 The Beginning of the Essential Spectrum

$$\sigma_{\text{ess}}(H(\alpha)) = \sigma_{\text{ess}}(H(0)) = [-\frac{1}{2}, \infty).$$

Proof: As remarked in (2.3.18; 5), for all $E \in [-\frac{1}{2}, \infty)$ we must find an orthogonal sequence ψ_n with norms bounded away from zero such that $(H - E)\psi_n \to 0$. Let φ_1 be the ground-state wave-function $|1, 0, 0\rangle$ of (4.1.10), let $R > 0$, and let $\chi_n(r)$ be a sequence of functions supported in $(2^n R, 2^{n+1} R)$ and such that $(|\mathbf{p}|^2/2 - E - \frac{1}{2})\chi_n \to 0$. (For instance, take $\chi_n(r) \sim \exp(ikr)/r, k^2/2 = E + \frac{1}{2}$, cut off outside $(2^n R, 2^{n+1} R)$ and smoothed out at the ends.) Then the sequence $\psi_n \equiv \varphi_1(\mathbf{x}_1)\chi_n(\mathbf{x}_2)$ is as required, since

$$\left\| \frac{1}{|\mathbf{x}_1 - \mathbf{x}_2|} \psi_n \right\| \sim (2^n R)^{-1}. \qquad \square$$

4.3.6 Remarks

1. The physical significance of the continuum starting at $-\frac{1}{2}$ is that one electron stays put in the ground state while the other runs off to infinity.
2. Mathematically speaking, we see that the potential energy may fail to be compact relative to the kinetic energy (it moves the essential spectrum) even when it is relatively ε-bounded.

The next topic is the point spectrum. It is clear that H is semibounded, since $H' > 0$, so $\sigma_p(H) \subset [-1, \infty)$. It will also be shown that if $\alpha < 1$, then there are infinitely many eigenvalues. This is to be expected on physical grounds, since an electron at a large distance would not see a fully screened nuclear charge, and it is known that an arbitrary weak $1/r$ potential has infinitely many bound states. To prove it, it is necessary to find another infinite set of orthogonal trial functions, with which H can be written as a diagonal matrix with eigenvalues less than $-\frac{1}{2}$, the bottom of the essential spectrum of hydrogen. We construct them by putting one electron in the ground state φ_1 of $|\mathbf{p}_1|^2/2 - 1/r_1$ and pulling the other one far away:

$$\langle \varphi_1(\mathbf{x}_1) \otimes \psi(\mathbf{x}_2)|H|\varphi_1(\mathbf{x}_1) \otimes \psi(\mathbf{x}_2)\rangle = -\frac{1}{2} + \langle \psi(\mathbf{x}_2)|\frac{|\mathbf{p}_2|^2}{2} - \frac{1}{r_2}|\psi(\mathbf{x}_2)\rangle$$
$$+\alpha\langle \varphi_1(\mathbf{x}_1) \otimes \psi(\mathbf{x}_2)|\frac{1}{r_{12}}|\varphi_1(\mathbf{x}_1) \otimes \psi(\mathbf{x}_2)\rangle.$$

The repulsion of the first electron shows up for the other one as an effective potential, which we expect to fall off as α/r_2 at large distances. Indeed, in Problem 5 it is calculated as

$$\langle \varphi_1(\mathbf{x}_1)|\frac{\alpha}{r_{12}}|\varphi_1(\mathbf{x}_1)\rangle = \frac{\alpha}{r_2} - \alpha \exp(-2r_2)\left(1 + \frac{1}{r_2}\right). \tag{4.5}$$

Consequently,

$$\langle H\rangle = -\frac{1}{2} + \langle \psi(\mathbf{x}_2)|\frac{|\mathbf{p}_2|^2}{2} - \frac{1-\alpha}{r_2} - \alpha\left(1 + \frac{1}{r_2}\right)\exp(-2r_2)|\psi(\mathbf{x}_2)\rangle.$$

Given disjointly supported functions ψ_j, if $k \neq j$, then $\langle \varphi \otimes \psi_j|H|\varphi \otimes \psi_k\rangle = 0$. By dilating and translating, we can arrange that

$$\langle \psi_j|\frac{|\mathbf{p}|^2}{2} - \frac{1-\alpha}{r} - \alpha\left(1 + \frac{1}{r}\right)\exp(-2r)|\psi_j\rangle < 0,$$

so $\langle \varphi \otimes \psi_j|H|\varphi \otimes \psi_k\rangle = \varepsilon_k\delta_{jk}$, $\varepsilon_k < -\frac{1}{2}$. This proves

4.3.7 The Infinitude of the Point Spectrum

If $\alpha < 1$, then $H(\alpha)$ of (4.3.1) *has infinitely many eigenvalues below $-\frac{1}{2}$, the bottom of its essential spectrum.*

4.3.8 Remark

The exclusion principle was not mentioned, because it still makes no difference. If the two spin states are denoted $\uparrow$ and $\downarrow$, then the antisymmetric state $(\uparrow \varphi(1) \otimes \downarrow \psi(2) - \downarrow \psi(1) \otimes \uparrow \varphi(2))/\sqrt{2}$ leads to the same expectation value.

The virial theorem (4.1.3) made use only of the effect of dilation on the kinetic and potential energies. The existence of more electrons does not change this, so we likewise have a

4.3.9 Virial Theorem

If $(H(\alpha) - E)\psi = 0$, *then* $E = -\langle\psi|T|\psi\rangle = -\frac{1}{2}\langle\psi|1/r_1 + 1/r_2 - \alpha/r_{12}|\psi\rangle$.

4.3.10 Corollary

$H(\alpha)$ *has no eigenvalues* $E \geq 0$.

4.3.11 Remarks

1. One might guess that there is a point spectrum only for $E < -\frac{1}{2}$. We shall, however, discover eigenvalues embedded in the essential spectrum between $-\frac{1}{2}$ and 0 — in fact infinitely many if $\alpha < 1$. They correspond to states whose decay is prevented by conservation laws for various quantum numbers.
2. We shall later rule out the existence of singular spectrum, so σ consists of σ_p between -1 and $-\frac{1}{2}$, both σ_p and σ_{ac} between $-\frac{1}{2}$ and 0, the only σ_{ac} above 0.

This delineates the rough features of the spectrum. Let us next take up some finer details; since the eigenvalues of $H(\alpha)$ are analytic in α, we can start with $H(0)$ and track them as α is switched on.

4.3.12 The Point Spectrum of $H(0)$

Let $\varphi_{n,l,m,s}$ *be the eigenfunctions* $|n, l, m\rangle$ *of* (4.1.10) *times spin eigenfunctions* $(s = \pm\frac{1}{2})$. *Then*

$$\varphi_{n_1,l_1,m_1,s_1^{(1)}}\varphi_{n_2,l_2,m_2,s_2^{(2)}} - \varphi_{n_2,l_2,m_2,s_2^{(1)}}\varphi_{n_1,l_1,m_1,s_1^{(2)}}$$

is an eigenfunction of $H(0)$ *with eigenvalue* $-(n_1^{-2} + n_2^{-1})/2$. *It is* $4n_1^2n_2^2$*-fold degenerate if* $n_1 \neq n_2$, *and* $2n_1^2(2n_1^2 - 1)$*-fold degenerate if* $n_1 = n_2$.

4.3.13 Remarks

1. All states with $n_1 > 1$ and $n_2 > 1$ have energies $\geq -\frac{1}{4}$, and hence live in the continuum beginning at $-\frac{1}{2}$.

2. The operator $H(0)$ possesses a copious commutant,

$$\{H(0)\}' \supset \{\mathbf{L}_1, \mathbf{F}_1, \sigma_1, \mathbf{L}_2, \mathbf{F}_2, \sigma_2\}.$$

   These constants divide the spectrum and keep the discrete states from decaying into the continuum.
3. Parity (3.2.6) was not listed separately among the constants, since it can be expressed with the angular momentum as in (3.2.18; 1). As states evolve according to $H(0)$, the parties of the individual electrons, $P_i \equiv (-1)^{l_i}, l_i = \sqrt{L_i^2 + \frac{1}{4}} - \frac{1}{2}$, are separately conserved. The total parity $P \equiv P_1 P_2 = (-1)^{l_1+l_2}$ is not necessarily $(-1)^l$, however, since $\mathbf{L} \equiv \mathbf{L}_1 + \mathbf{L}_2$ can have any l such that $|l_1 - l_2| \le l \le l_1 + l_2$. There thus exist

4.3.14 States of Natural and Unnatural Parity

If $P = (-1)^l$, then the state is said to have **natural parity**, and if $P = (-1)^{l+1}$, then it has **unnatural parity**.

4.3.15 Example

If n_1 or n_2 equals 1, then l equals l_2 or l_1, and the resultant state has natural parity, $(-1)^{l_2} = (-1)^l$ or $(-1)^{l_1} = (-1)^l$. Hence the isolated point spectrum has natural parity. The first state with unnatural parity has $l_1 = l_2 = l = 1, P = +$. The wave-function is a component of $(\mathbf{x}_1 \wedge \mathbf{x}_2) f(r_1, r_2)$, and if $n_1 = n_2 = 2$, the energy is $-\frac{1}{4}$. In the subspace of unnatural parity, the continuum begins at $E = -\frac{1}{8}$, for $n_1 = 2, n_2 = \infty$.

4.3.16 Constants of the Motion when $\alpha \neq 0$

If $\alpha \neq 0$, then in addition to H, the quantities $\mathbf{L}$, P, σ_1, and σ_2 are conserved.

4.3.17 Physical Consequences of Conservation of Parity

1. Parity must now be listed separately, since it is independent of $\mathbf{L}$. The Hilbert space decomposes into subspaces of natural and unnatural parity, which are not mixed by $H(\alpha)$. Hence discrete states of unnatural parity continue to exist within the continuum of natural parity. Just as in (4.3.5) the beginning of the continuum of unnatural parity at $-\frac{1}{8}$ is not affected when α is switched on. Since the eigenvalue of unnatural parity at $-\frac{1}{4}$ varies continuously with α, it remains isolated from the continuum of this part of the Hilbert space for α sufficiently small.
2. Eigenstates of $H(0)$ with natural parity and energies $E > -\frac{1}{2}$ when $\alpha > 0$ are not prevented from decaying into states with one electron in the ground

state and the other running off to infinity. This is observed as the Auger effect.

3. States of unnatural parity are prevented from decaying to states with one electron in the ground state and the other running to infinity, since the final state would have natural parity. Conservation of parity likewise prevents their creation by direct collisions of electrons with atoms. In reality they are not absolutely stable, since they can decay by the interactions neglected in $H(\alpha)$, for example by electromagnetic radiation. The possible transitions are significantly slower than the Auger transitions.
4. Scattering theory reveals that there are many more constants in the absolutely continuous part of the spectrum, namely all the constants of $H(0)$, such as P_1 and P_2, after being transformed with the Møller operators.

We have seen that the perturbation with $\alpha \neq 0$ has broken the immense symmetry group of $H(0)$ and has separated the highly degenerate unperturbed states. Since H' is a positive perturbation, the eigenvalues $E_i(\alpha)$ are increasing functions of α. The way they depend on it is roughly as depicted in the figure below.

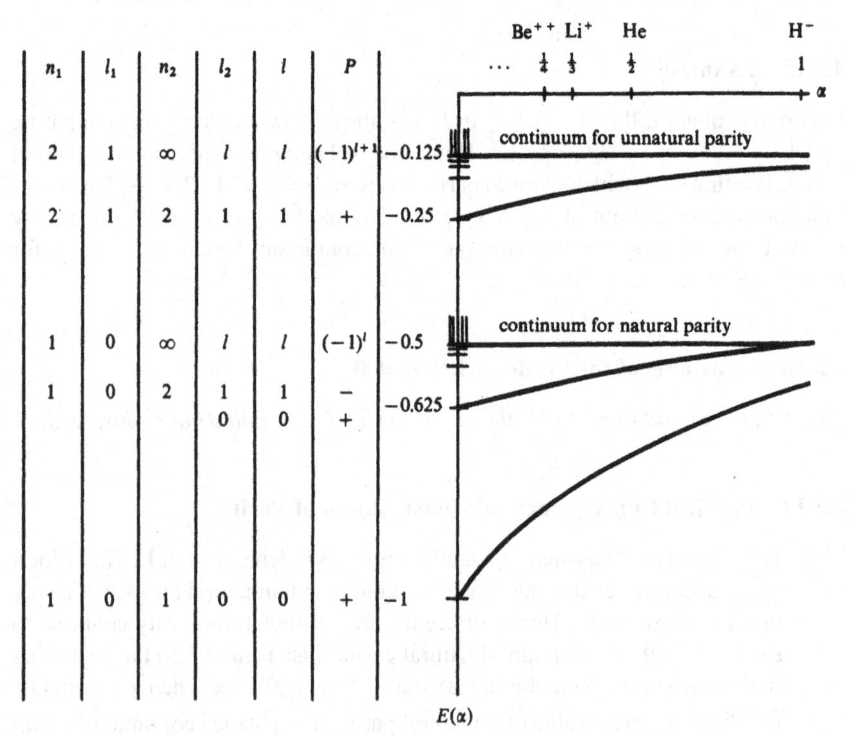

To locate the eigenvalues as functions of α more exactly, recall that not only do they increase monotonically, but that, moreover, the sum of the first n of them, $n = 1, 2, 3, \ldots$, is also concave in α by (3.5.23). We can even state a more refined proposition on the

4.3.18 Concavity of the Eigenvalues

Let $E(\alpha)$ be the sum of the n lowest eigenvalues in a subspace of definite quantum numbers. Then $-(-E(\alpha))^{1/2}$ is concave in α.

Proof: Write $H = \frac{1}{2}(|\mathbf{p}_1|^2 + |\mathbf{p}_2|^2) - \alpha_0(1/r_1 + 1/r_2) + \alpha/r_{12}$. Then a dilatation argument (cf. (4.3.1)) shows that $E(\alpha, \alpha_0)$ is of the form $\alpha_0^2 f(\alpha/\alpha_0)$, which is concave not just in α alone, but in (α, α_0). The condition $f'' \le 0$ generalizes to $E_{,\alpha\alpha} E_{,\alpha_0\alpha_0} - (E_{,\alpha\alpha_0})^2 \ge 0$, which implies that $2ff'' \ge (f')^2$, so

$$\frac{\partial^2}{\partial\alpha^2}\left(-\sqrt{-f(\alpha)}\right) \le 0. \qquad \square$$

4.3.19 Remarks

1. Eigenvalues can cross at finitely many points, at which f may not be differentiable. In such cases, it can be approximated arbitrarily well with C^∞ functions, which justifies the proof.
2. If m is not set equal to 1, then the theorem states that E is concave in the three variables $(1/m, \alpha, \alpha_0)$. The dedicated reader may check that this provides no new information.

4.3.20 Corollaries

1. Linear bounds can be improved by parabolic bounds. For instance, by (3.5.31; 1), if $(H_0 - E_1(0))|0\rangle = 0$, then

$$E_1(0) + \alpha\langle 0|(H')^{-1}|0\rangle^{-1} \le E_1(\alpha) \le E_1(0) + \alpha\langle 0|H'|0\rangle,$$

the lower bound holding provided that

$$\alpha \le \alpha_0 \equiv (E_2(0) - E_1(0))\langle 0|(H')^{-1}|0\rangle.$$

Since $-(-E_1(\alpha))^{1/2}$ is now known to be concave, the linear bounds $g(\alpha) \le g(0) + \alpha g'(0)$, $g = -\sqrt{-f}$, imply:

$$E_1(0)\left(1 + \frac{\alpha}{\alpha_0}\left(\sqrt{\frac{E_2(0)}{E_1(0)}} - 1\right)\right)^2 \le E_1(\alpha) \le E_1(0)\left(1 + \frac{\alpha}{2}\frac{\langle 0|H'|0\rangle}{E_1(0)}\right)^2.$$

2. If $\alpha_1 < \alpha < \alpha_2$, then

$$\frac{-\sqrt{-f(\alpha)} + \sqrt{-f(\alpha_1)}}{\alpha - \alpha_1} \ge \frac{f'(\alpha)}{2\sqrt{-f(\alpha)}} \ge \frac{-\sqrt{-f(\alpha_2)} + \sqrt{-f(\alpha)}}{\alpha_2 - \alpha},$$

so, if $|\alpha\rangle$ is the eigenvector such that $(H_0 + \alpha H' - E_1(\alpha))|\alpha\rangle = 0$, then we obtain bounds on f', and hence on the expectation value of H':

$$\frac{2}{\alpha - \alpha_1}\left(\sqrt{E_1(\alpha)E_1(\alpha_1)} - |E_1(\alpha)|\right) \ge f'(\alpha) = \left\langle \alpha\left|\frac{1}{r_{12}}\right|\alpha\right\rangle$$

$$\geq \frac{2}{\alpha_2 - \alpha}\left(|E_1(\alpha) - \sqrt{E_1(\alpha)E_1(\alpha_2)}\right).$$

4.3.21 Applications

1. The ground state of parahelium ($(1s)^2$): The vector $|0\rangle$ has the form $(\uparrow\downarrow - \downarrow\uparrow)\varphi_1(r_1)\varphi_1(r_2)/\sqrt{2}$, where $\varphi_1(r) = 2\exp(-r)$, and $E_1(0) = -1$, $E_2(0) = -\frac{5}{8}$. Problem 3 is to calculate that $\langle 0|1/r_{12}|0\rangle = \frac{5}{8}$ and $\langle 0|r_{12}|0\rangle = \frac{35}{16}$. Hence $\alpha_0 = \frac{105}{128}$, and

$$\min\left\{-\frac{5}{8}, -\left(1 - \alpha\frac{128}{105}\left(1 - \sqrt{\frac{5}{8}}\right)\right)^2 = -(1 - \alpha \cdot 0.2553)^2\right\}$$

$$\leq E_1(\alpha) \leq -\left(1 - \frac{5\alpha}{16}\right)^2$$

$$= -(1 - \alpha \cdot 0.3152)^2.$$

2. The ground state of orthohelium ($(1s)(2s)$);

$$|0\rangle = \uparrow\uparrow (\varphi_1(r_1)\varphi_2(r_2) - \varphi_2(r_1)\varphi_1(r_2))/\sqrt{2},$$

where $\varphi_2(r) = \exp(-r/2)(1 - r/2)/\sqrt{2}$, $E_1(0) = -\frac{5}{8}$, $E_2(0) = -\frac{5}{9}$.

$$\langle 0|\frac{1}{r_{12}}|0\rangle = \frac{17}{81} - \frac{16}{729},$$

$$\langle 0|r_{12}|0\rangle = \frac{25}{4} - \frac{11}{324} + \frac{2^{12}}{3^9} - \frac{5^2 2^7}{3^9}, \qquad \alpha_0 = 0.4348$$

(see Figures 4.2 and 4.3), and

$$\min\left\{-\frac{5}{9}, -\frac{5}{8}\left(1 - \frac{\alpha}{0.4348}\left(1 - \sqrt{\frac{8}{9}}\right)\right)^2 = -\frac{5}{8}(1 - \alpha \cdot 0.1315)^2\right\}$$

$$\leq E_1(\alpha) \leq -\frac{5}{8}(1 - \alpha \cdot 0.1503)^2.$$

3. The lowest state $L = 1$, i.e., $(1s)(2p)$: $|0\rangle = (\uparrow\downarrow \mp \downarrow\uparrow)(\varphi_1(r_1)\varphi_2(\mathbf{x}_2) \pm \varphi_2(\mathbf{x}_1)\varphi_1(r_2))/2$, $\varphi_2(\mathbf{x}) = Y_1^0(\theta) r \exp(-r/2)/\sqrt{4!}$, once again $E_1 = -\frac{5}{8}$, $E_2 = -\frac{5}{9}$, $\langle 0|1/r_{12}|0\rangle = \frac{59}{243} \pm \frac{112}{6561}$,

$$\langle 0|r_{12}|0\rangle = 5.2449 \mp 0.1366, \quad \alpha = (0.35471, 0.37372),$$

$$\min\left\{-\frac{5}{9}, -\frac{5}{8}\left(1 - \alpha \cdot \frac{0.16123}{0.15303}\right)^2\right\} \leq E(\alpha) \leq -\frac{5}{8}\left(1 - \alpha \cdot \frac{0.2091}{0.1792}\right)^2.$$

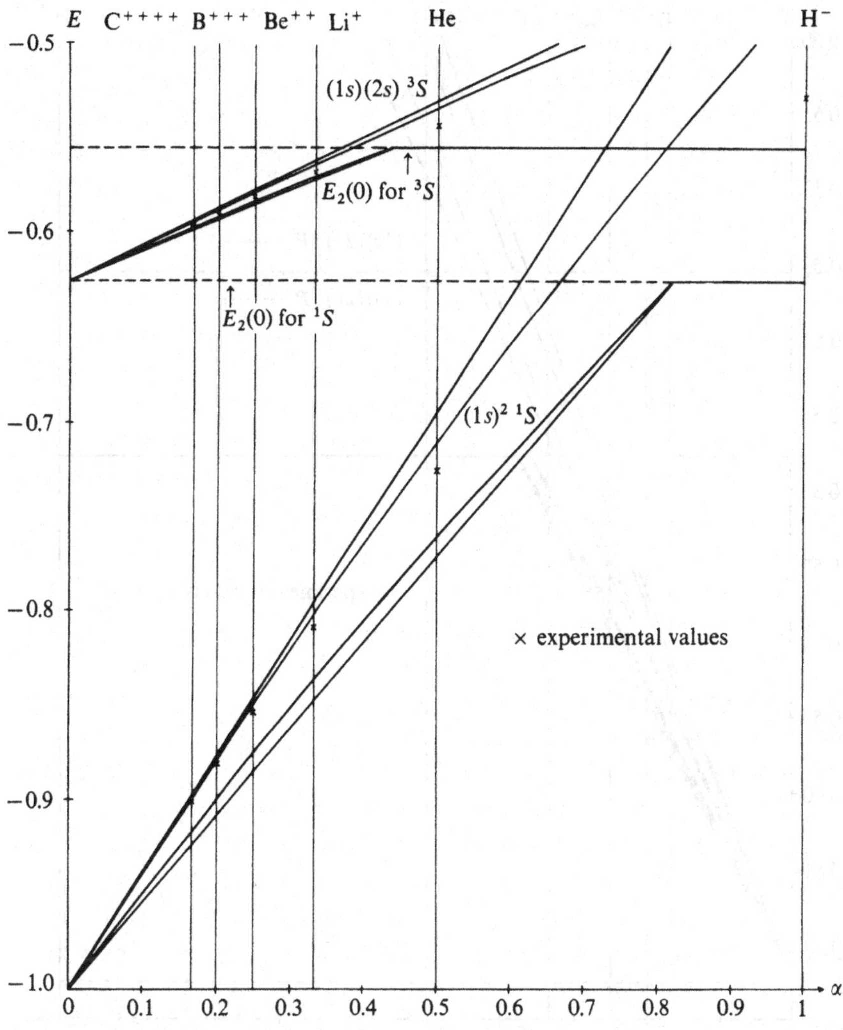

Figure 4.2. Linear and parabolic bounds for $(1s)^2\ {}^1S$ and $(1s)(2s)\ {}^3S$.

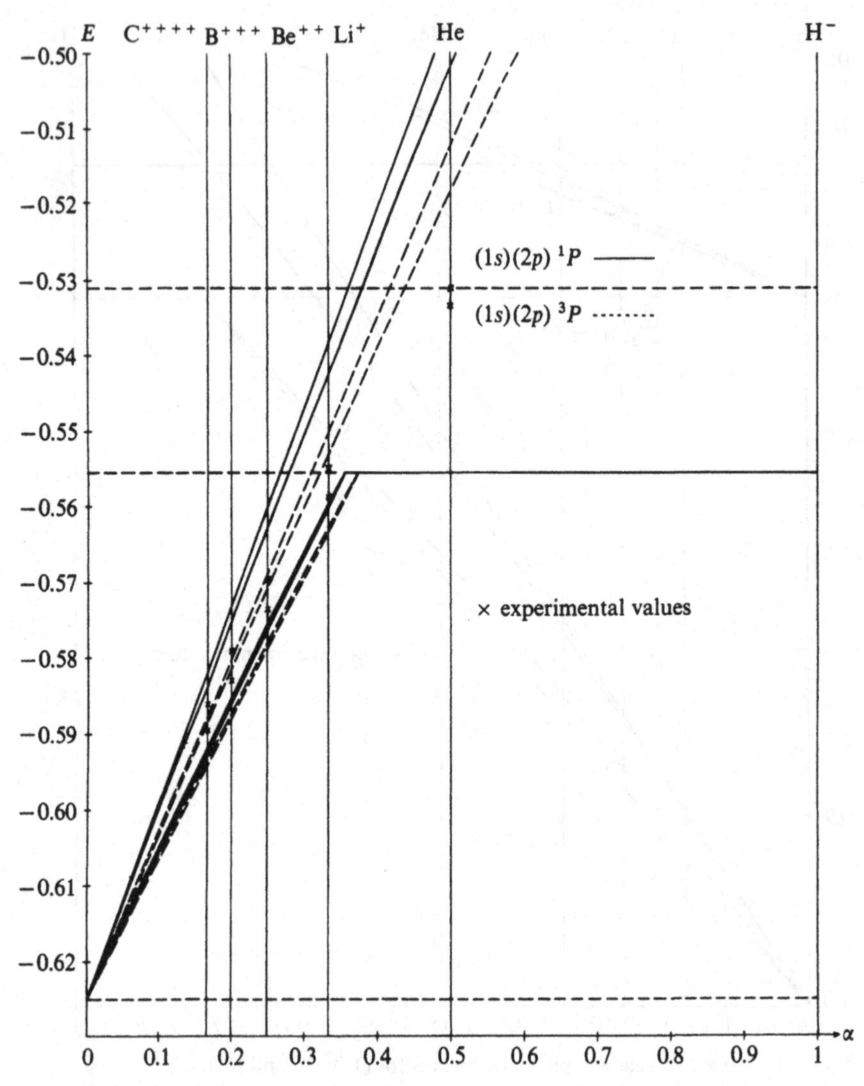

Figure 4.3. Linear and parabolic bounds for $(1s)(2p)\ ^3P$ and $(1s)(2p)\ ^1P$.

4.3.22 Remarks

1. The rate of change at $\alpha = 0$ is exactly $\langle 0|H'|0\rangle$, which shows that $(1s)(2s)$ is the more energetically favorable state with $n = 2$. This is plausible, since the s orbitals are more densely concentrated near the nucleus, and are thus shielded less by the other electron. By first-order perturbation theory the $(1s)(2p)$ state with spin 1 is favored (Hund's rule), since the exclusion principle then makes the electrons avoid each other and feel less Coulombic repulsion. Our bounds, however, are not yet precise enough to prove this tendency for strictly positive α.
2. It is easy to see with $\langle n, l|r|n, l\rangle = (3n^2 - l(l+1))/2$ that the inequalities $\langle r\rangle\langle 1/r\rangle \geq 1$ and $\langle r_{12}\rangle \leq \langle r_1 + r_2\rangle$ are not at all weak.
3. The main drawback of the results that have been presented is that the lower bounds break down when $E(\alpha) > E_2$. The discussion in (3.5.32; 1) shows that only the use of many-dimensional projections can remedy this.

4.3.23 Ground States with Two-Dimensional Projections

1. Consider the two lowest parastates with $\alpha = 0$,

$$|(1s)^2\rangle = \tfrac{1}{\sqrt{2}}(\uparrow\downarrow - \downarrow\uparrow) \otimes \varphi_1(1)\varphi_1(2),$$
$$|(1s)(2s)\rangle = \tfrac{1}{2}(\uparrow\downarrow - \downarrow\uparrow) \otimes (\varphi_1(1)\varphi_2(2) + \varphi_2(1)\varphi_1(2)),$$

and calculate the matrices

$$\begin{pmatrix} \langle (1s)^2|r_{12}|(1s)^2\rangle & \langle (1s)^2|r_{12}|(1s)(2s)\rangle \\ \langle (1s)(2s)|r_{12}|(1s)^2\rangle & \langle (1s)(2s)|r_{12}|(1s)(2s)\rangle \end{pmatrix}$$
$$= \begin{pmatrix} 2.1875 & -0.6371 \\ -0.6371 & 0.1706 \end{pmatrix} \equiv M_L^{-1},$$

$$\begin{pmatrix} \langle (1s)^2|\frac{1}{r_{12}}|(1s)^2\rangle & \langle (1s)^2|\frac{1}{r_{12}}|(1s)(2s)\rangle \\ \langle (1s)(2s)|\frac{1}{r_{12}}|(1s)^2\rangle & \langle (1s)(2s)|\frac{1}{r_{12}}|(1s)(2s)\rangle \end{pmatrix}$$
$$= \begin{pmatrix} 0.625 & 0.1263 \\ 0.1263 & 0.2318 \end{pmatrix} \equiv M_U.$$

The matrix

$$\begin{pmatrix} E_1(0) & 0 \\ 0 & E_2(0) \end{pmatrix} + \alpha M$$

has eigenvalues

$$E_{1,2} = \frac{\varepsilon_1 + \varepsilon_2}{2} + \alpha\,\frac{M_{11} + M_{22}}{2} \mp\sqrt{\left(\frac{\varepsilon_1 - \varepsilon_2}{2} + \alpha\,\frac{M_{11} - M_{22}}{2}\right)^2 + \alpha^2 M_{12}^2}$$

($\varepsilon_1 = E_1(0)$, $\varepsilon_2 = E_2(0)$), which, by use of M_L and M_U, are respectively lower bounds for the first two states, provided that they lie below $E_3(0)$, and upper bounds.

2. Problem 4 gives the analogous calculation for the other states looked at in (4.3.21), with the 2×2 matrices of H' and $(H')^{-1}$ for the states $(1s)(2s)$ and $(1s)(3s)$ and, respectively, $(1s)(2p)$ and $(1s)(3p)$.

4.3.24 Upper Bounds Using Two-Parameter Trial Functions

To comply with the desire of the electron for more freedom of movement, it is reasonable to use the functions $\exp(-\gamma r_1 - \beta r_2) \pm \exp(-\beta r_1 - \gamma r_2)$ and $Y_1(1) r_1 \exp(-\gamma r_1 - \beta r_2) \pm Y_1(2) r_2 \exp(-\beta r_1 - \gamma r_2)$ as trial functions for the ground states with $L = 0, 1$ and $S = 0, 1$, and to minimize the expectation value of H as β and γ vary. The table below lists the optimal parameters as functions of α. It makes it clear that the outer electrons come nearer to escaping when the nuclear charge is smaller.

Variational Calculations

state $^S(L)$	$(1s)(1s)\,^1S$		$(1s)(2s)\,^3S$		$(1s)(2p)\,^3P$		$(1s)(2s)\,^1P$	
α	β	γ	β	γ	β	γ	β	γ
1.	0.283	1.039	0.000		0.000		0.001	1.001
0.75	0.452	1.070	0.094	0.982	0.129	0.999	0.123	1.001
0.5	0.588	1.085	0.161	0.984	0.272	0.997	0.240	1.002
0.3333	0.695	1.097	0.202	0.979	0.361	0.994	0.322	1.003
0.25	0.754	1.108	0.222	0.975	0.400	0.993	0.366	1.003
0.2	0.769	1.082	0.232	0.970	0.424	0.992	0.392	1.001

4.3.25 Remarks

1. Figures 4.4a, b, and c show the greater detail in the pictures that one obtains with the use of two-dimensional projections and the variational ansatz (4.3.24). The plot shows the square root of the energy; it is apparent that the experimental points lie close to the straight lines, and hence that Proposition (4.3.18) is a nearly optimal concavity property.
2. The parabolic upper bounds (4.3.20) can also be obtained with this variational argument, using $\exp(-r Z_{\text{eff}})$ in place of the ground state of H_0 and optimizing Z_{eff}. For the ^1S state this leads to $Z_{\text{eff}} = 1 - 5\alpha/16$, which reflects the partial screening of the nuclear charge.
3. Our bounds are good enough to separate the states ^3S, ^3P, and ^1P, which are degenerate when $\alpha = 0$, and to prove that Hund's rule orders them correctly.
4. With more numerical effort and trial functions having several parameters quite accurate upper bounds are obtainable. Temple's inequality (3.5.30; 2) then pro-

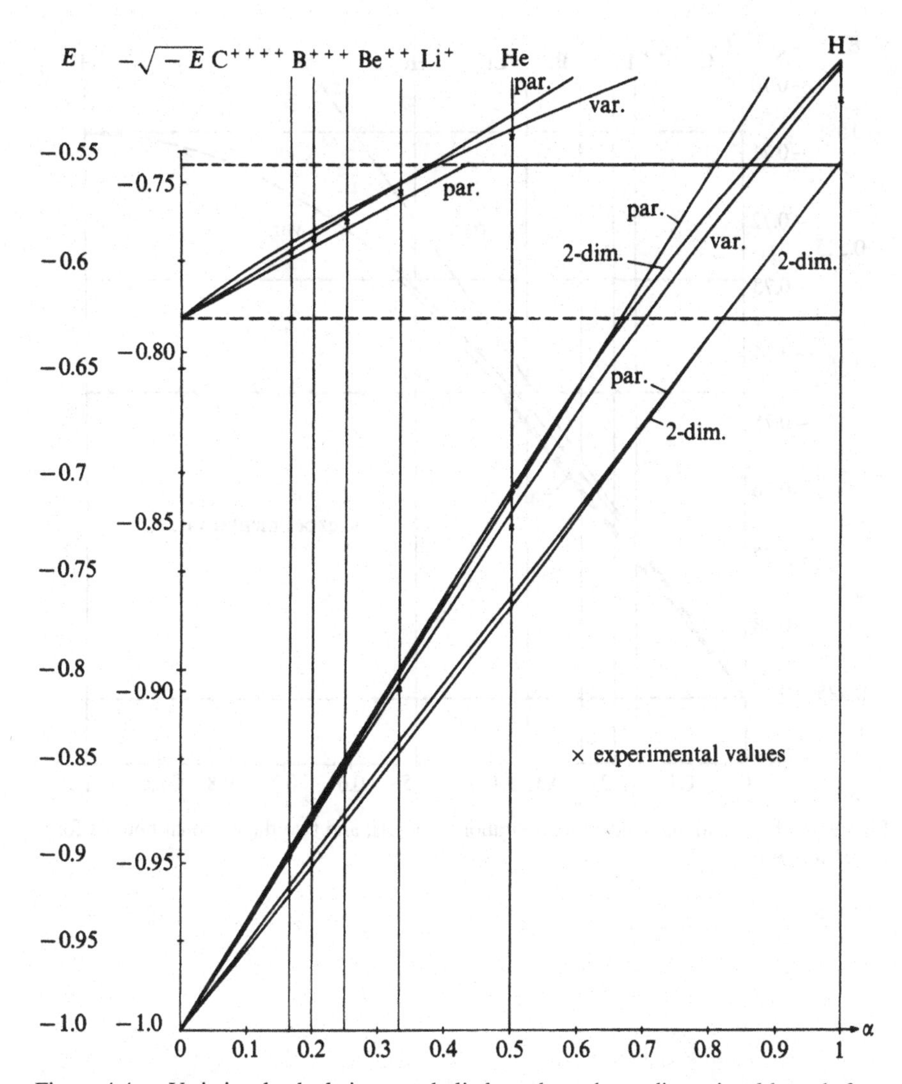

Figure 4.4. a. Variational calculation, parabolic bounds, and two-dimensional bounds for $(1s)^2\ {}^1S$ and $(1s)(2s)\ {}^3S$.

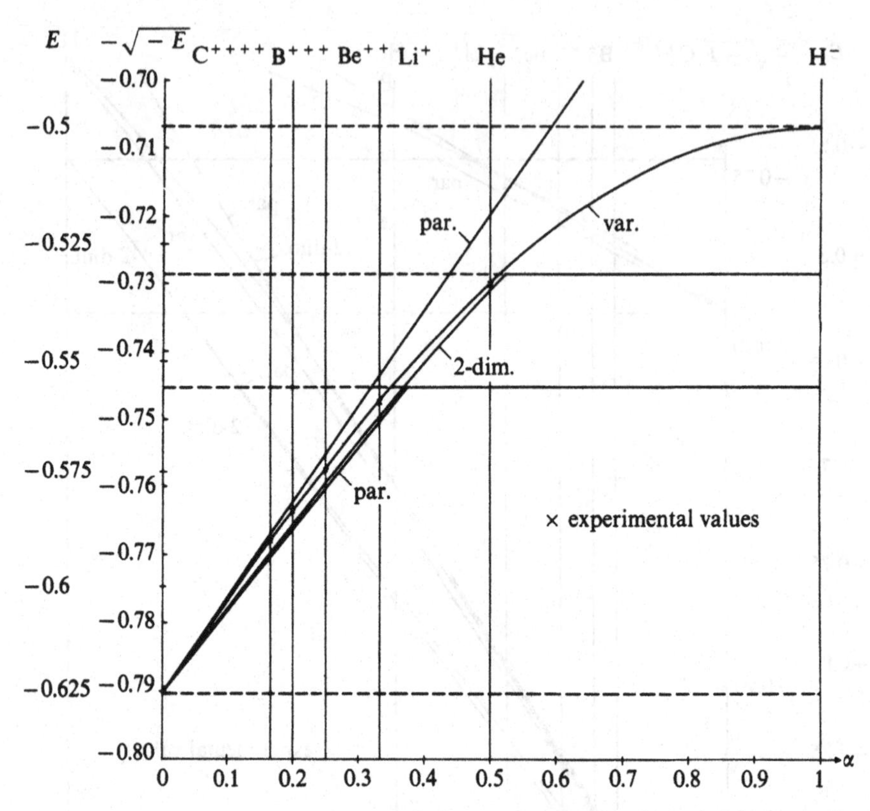

Figure 4.4. b. Variational calculation, parabolic bounds, and two-dimensional bounds for $(1s)(2p)\ {}^3P$.

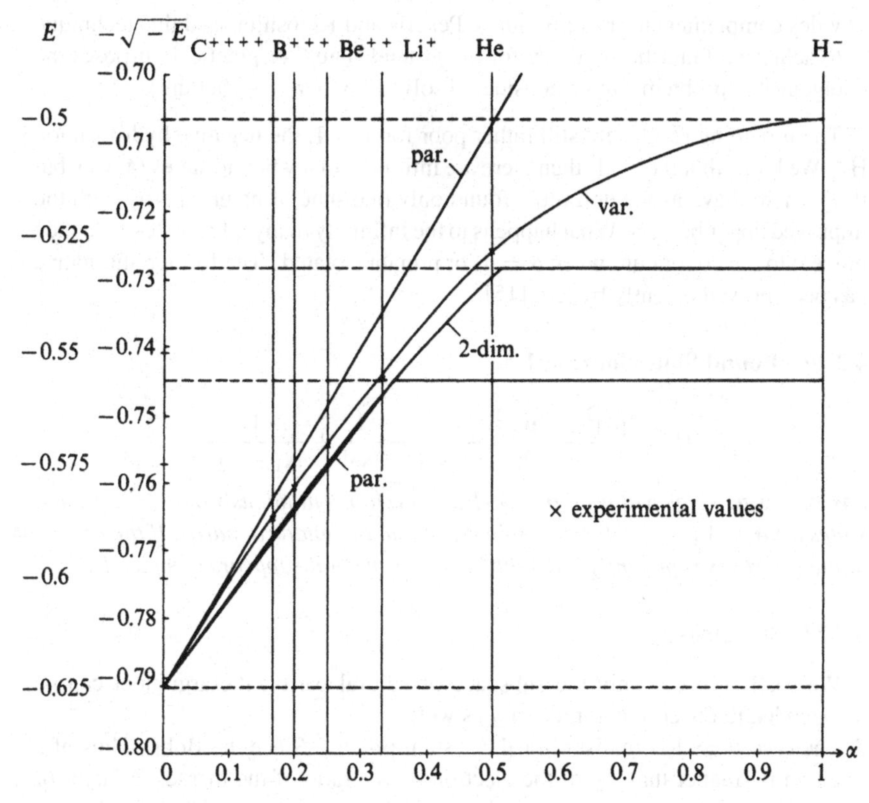

Figure 4.4. c. Variational calculation, parabolic bounds, and two-dimensional bounds for $(1s)(2p)\ ^1P$.

vides complementary lower bounds. Pekeris and Kinoshita used this technique to achieve a fantastic accuracy for the ground state. For practical purposes the eigenvalue problem can be considered solved up to $\alpha = \frac{1}{2}$ (helium).

The foregoing results are still rather poor for $\alpha = 1$, the negative hydrogen ion H^-. We know that if $\alpha < 1$, then there are infinitely many bound states (4.3.7), but if $\alpha = 1$ we have as yet definitely found only one state of natural parity with the improved upper bounds. What happens to the infinitely many other states—do they move into the continuum when $\alpha = 1$, or remain isolated? The former alternative has been proved recently by Hill [15].

4.3.26 Bound States for $\alpha = 1$

$$H = \frac{|\mathbf{p}_1|^2}{2} + \frac{|\mathbf{p}_2|^2}{2\mu} - \frac{1}{|\mathbf{x}_1|} - \frac{1}{|\mathbf{x}_2|} + \frac{1}{|\mathbf{x}_1 - \mathbf{x}_2|}$$

has exactly two eigenvalues if $\mu = 1$. They develop continuously out of the ground states for $\alpha < 1$ in the subspaces of natural and of unnatural parity. If the ratio μ of the masses is sufficiently different from 1, *then σ_P disappears completely.*

4.3.27 Remarks

1. We shall consider only the subspace of natural parity; the argument can be extended to cover unnatural parity as well.
2. The case $\mu \neq 1$ is realistic for the system $p\mu^- e^-$. Since the Bohr radius of a muon is smaller than r_B of the electron by the ratio of the masses, a factor of 207, one would guess that it completely shields the proton, and that the electron is no longer bound. We shall see that the failure to bind happens for much less extreme mass ratios; yet it is difficult to find the exact value of μ at which σ_p disappears.
3. The strategy of the proof is to find an operator $H_L \leq H$ for which it can be shown that there is no (or only one) eigenvalue below the continuum at $-\frac{1}{2}$. More effort is required to show that there are no eigenvalues in the continuum of the same quantum number, and that part of the proof will not be given here. It necessitates verifying that such states are unstable under the addition of H'; see [3] and (4.4.8; 3).

Proof:

(i) $\mu > \pi$. Let us do the easy part first and show that there exist no bound states if $\mu > \pi$. The ground state $H(0)$ has energy $-(1+\mu)/2$, and the continuum begins at $-\mu/2$. If particle 2—let us call it the muon—is excited, then the energy of the state becomes at least $-\frac{1}{2}(1+\mu/4)$, which is in the continuum as soon as $\mu > \frac{4}{3}$. If P_0 is the projection onto the ground state $2\mu^{3/2}/\sqrt{4\pi}\exp(-\mu r_2)$ of the muon, and $P = \mathbf{1} \otimes P_0$, then clearly

$$\frac{|\mathbf{p}_2|^2}{2\mu} - \frac{1}{r_2} \geq -\frac{\mu}{2} P - \frac{\mu}{8}(1-P).$$

As in (3.5.29) we now use the inequality

$$|\mathbf{x}_1 - \mathbf{x}_2|^{-1} \geq P(P|\mathbf{x}_1 - \mathbf{x}_2|P)^{-1}P \equiv V_L(r_1)P,$$

from which it is easy to calculate

$$\begin{aligned} V_L(r_1) &= \left[\left(\frac{\mu^3}{\pi} \right) \int d^3x_2 \exp(-2\mu r_2) |\mathbf{x}_1 - \mathbf{x}_2| \right]^{-1} \\ &= \mu \left[\mu r_1 + \frac{1}{\mu r_1} - \exp(-2\mu r_1) \left(\frac{1}{2} + \frac{1}{\mu r_1} \right) \right]^{-1} \end{aligned} \tag{4.6}$$

(Problem 3). Since

$$\frac{|\mathbf{p}_1|^2}{2} - \frac{1}{|\mathbf{x}_1|} \geq -\frac{1}{2},$$

we finally get

$$H \geq \left(-\frac{\mu}{2} + \frac{|\mathbf{p}_1|^2}{2} - \frac{1}{r_1} + V_L(r_1) \right) P + \left(-\frac{\mu}{8} - \frac{1}{2} \right) (\mathbf{1} - P). \tag{4.7}$$

To show that $H \geq -\mu/2$ it must be verified that $|\mathbf{p}|^2/2 - 1/r + V_L(r) \geq 0$. First note that

$$\begin{aligned} &-\frac{1}{r} + \left[r + \frac{1}{\mu^2 r} - \exp(-2\mu r) \left(\frac{1}{2\mu} + \frac{1}{\mu^2 r} \right) \right]^{-1} \\ &= -\frac{1}{r} \left[1 + \frac{r^2\mu^2}{1 - \exp(-2\mu r)\left(1 + \frac{\mu r}{2}\right)} \right]^{-1} \geq -\frac{1}{r(1 + r^2\mu^2)}. \end{aligned}$$

The Bargmann bound (3.5.35; 1) shows that there is no bound state if $2\int_0^\infty dr/(1 + r^2\mu^2) = (\pi/\mu) < 1$, that is, if the mass of the muon is greater than π times the mass of the electron.

(ii) $\mu = 1$. In this case the P used above, projecting onto wave-functions of the form $f(\mathbf{x}_1)\exp(-\mu r_2)$ does not do the trick, since $\exp(-r_1) f(\mathbf{x}_2)$ has just as low energy. Hence for two electrons it is preferable to use the Hilbert space $\mathcal{H}_2^{\pm}$ of functions of the form

$$\begin{aligned} \bar{f}_{\pm}(\mathbf{x}_1, \mathbf{x}_2) &= (\exp(-r_1) f(\mathbf{x}_2) \pm f(\mathbf{x}_1)\exp(-r_2)) \\ &\otimes \left(\frac{1}{\sqrt{2}} (\uparrow\downarrow - \downarrow\uparrow) \right), \qquad f \in L^2(\mathbf{R}^3). \end{aligned} \tag{4.8}$$

As usual the arrows stand for the spin functions, and we may assume that f is orthogonal to $\exp(-r)$ in $\mathcal{H}_2^-$. The spaces $\mathcal{H}_2^{\pm}$ are invariant under operators of the form

$$\bar{h} = P_0(1) \otimes h(2) + h(1) \otimes P_0(2) \tag{4.9}$$

(the argument of the operator indicates which factor it acts on). The space $L^2(\mathbf{R}^3)$ is mapped into $\mathcal{H}_2^{\pm}$ and $\mathcal{B}(L^2(\mathbf{R}^3))$ is mapped into $\mathcal{B}(\mathcal{H}_2^{\pm})$ by (4.8) and (4.9) in such a way that

$$\langle \bar{f} | h \bar{g} \rangle = 2\langle (1 \pm P_0) f | h (1 \pm P_0) g \rangle. \tag{4.10}$$

The procedure used in the first part of the proof can now be mimicked; first project onto the union of the image spaces of the projections $\mathcal{P}_1$ and $\mathcal{P}_2$:

$$\mathcal{P}_1 = r_{12}{}^{1/2} P_0(1) V_L(x_2) r_{12}{}^{1/2}, \quad \mathcal{P}_2 = r_{12}{}^{1/2} P_0(2) V_L(x_1) r_{12}{}^{1/2},$$

where $r_{12} \equiv |\mathbf{x}_1 - \mathbf{x}_2|$ and V_L is as in (4.6) with $\mu = 1$. Following the notation of (2.2.35) the projection onto the union will be called $\mathcal{P}_1 \vee \mathcal{P}_2$. As we learned there,

$$\mathcal{P}_2 \geq \mathcal{P}_2 \mathcal{P}_1 \mathcal{P}_2 \geq \mathcal{P}_2 \mathcal{P}_1 \mathcal{P}_2 \mathcal{P}_1 \mathcal{P}_2 \geq \cdots \geq \mathcal{P}_1 \wedge \mathcal{P}_2,$$

which can be rewritten for $\mathcal{P}_1 \vee \mathcal{P}_2 = \mathbf{1} - (\mathbf{1} - \mathcal{P}_1) \wedge (\mathbf{1} - \mathcal{P}_2)$, though it involves lengthy expressions. Fortunately, for our purposes the following inequalities are sufficient:

$$\mathcal{P}_1 \wedge \mathcal{P}_2 \leq \frac{1}{2}(\mathcal{P}_1 \mathcal{P}_2 \mathcal{P}_1 + \mathcal{P}_2 \mathcal{P}_1 \mathcal{P}_2) \Leftrightarrow$$

$$\mathcal{P}_1 \vee \mathcal{P}_2 \geq \mathcal{P}_1 + \mathcal{P}_2 - \mathcal{P}_1 \mathcal{P}_2 - \mathcal{P}_2 \mathcal{P}_1 + \frac{1}{2}(\mathcal{P}_1 \mathcal{P}_2 \mathcal{P}_1 + \mathcal{P}_2 \mathcal{P}_1 \mathcal{P}_2). \tag{4.11}$$

This can be simplified with the observation that although $\mathcal{P}_1$ and $\mathcal{P}_2$ do not commute, they have a common eigenfunction $\in L^2(\mathbf{R}^6)$, i.e., when normalized,

$$\chi(\mathbf{x}_1, \mathbf{x}_2) = \frac{1}{\pi}\sqrt{\frac{16 r_{12}}{35}} \exp(-r_1 - r_2), \qquad \mathcal{P}_i \chi = \chi.$$

This function has eigenvalue -1 for the operator $-\mathcal{P}_1 \mathcal{P}_2 - \mathcal{P}_2 \mathcal{P}_1 + (\mathcal{P}_1 \mathcal{P}_2 \mathcal{P}_1 + \mathcal{P}_2 \mathcal{P}_1 \mathcal{P}_2)/2$. Since it can be shown (Problem 2) that in $\mathcal{H}_2^+$ the other eigenvalues of the operator are positive, (4.11) leads to

$$\mathcal{P}_1 \vee \mathcal{P}_2 \geq \mathcal{P}_1 + \mathcal{P}_2 - |\chi\rangle\langle\chi| \tag{4.12}$$

(from here on we work only in $\mathcal{H}_2^+$). The projection method (3.5.31) then shows that

$$\frac{1}{r_{12}} \geq r_{12}^{-1/2} \mathcal{P}_1 \vee \mathcal{P}_2 r_{12}^{-1/2} \geq P_0(1) V_L(2) + P_0(2) V_L(1) - \tfrac{16}{35} P_0(1) P_0(2). \tag{4.13}$$

The next thing to show is that there are no additional negative eigenvalues of

$$H_0 + r_{12}^{-1}, \qquad H_0 = \frac{1}{2} + \frac{1}{2}(|\mathbf{p}_1|^2 + |\mathbf{p}_2|^2) - r_1^{-1} - r_2^{-1}.$$

The operator H_0 leaves the spaces $\mathcal{H}_2^{\pm}$ invariant and acts on them as

$$H_0 \bar{f}_{\pm} = \overline{\left(\frac{1}{2}|\mathbf{p}|^2 - \frac{1}{r}\right) f_{\pm}} = \overline{\left(\frac{1}{2}|\mathbf{p}|^2 - \frac{1}{r} + \frac{\tfrac{1}{4} P_0}{0}\right) f_{\pm}}.$$

On the orthogonal complement of $\mathcal{H}_2^\pm$ no electron is in the ground state, so $H_0 > \frac{1}{4}$. If $P^\pm$ are the projections onto $\mathcal{H}_2^\pm$, then (4.13) allows us to write

$$H_0 + r_{12}^{-1} \geq P^\pm \overline{\left(\frac{1}{2}|\mathbf{p}|^2 - \frac{1}{r} + V_L(r) + \begin{matrix} P_0 \left(\frac{1}{4} - \frac{8}{35} \right) \\ 0 \end{matrix} \right)} + \frac{1}{4}(1 - P^\pm).$$

Since $\frac{1}{4} > \frac{8}{35}$, we have to verify that $h \equiv |\mathbf{p}|^2/2 - 1/r + V_L(r)$ has only one bound state, whereas $(\mathbf{1} - P_0)h(\mathbf{1} - P_0)$ has none. The first statement can be proved analytically by noting, as in (i), that $h \geq |\mathbf{p}|^2/2 - 2/r(1+r)^2$, but making a more detailed calculation. As in the proof of (3.5.36) we find the number of values $\lambda \leq 1$ for which there is a $\psi \in L^2$ satisfying

$$\frac{|\mathbf{p}|^2}{2}\psi = \frac{1}{2}\left(\frac{-1}{r}\frac{\partial^2}{\partial r^2} r + \frac{l(l+1)}{r^2} \right)\psi = \frac{2\lambda}{r(1+r)^2}\psi.$$

With the change of variables $r = z/(1-z)$, $\psi = r^l w(z)$, this becomes the hypergeometric equation $z(1-z)w'' + 2(l+1-z)w' + 4\lambda w = 0$, the well-known properties of which include the requirement that $\lambda = (n+2l+1)(n+2l+2)/4$, $n = 0, 1, 2, \ldots$; $\lambda = \frac{1}{2} < 1$ only if $n = l = 0$. Hence there is at most one bound state. The proof for $\mathcal{H}_2^-$ takes a longer discussion, and can be read in [15]. □

4.3.28 Remarks

1. The proof for H^- can be generalized for finite but large nuclear masses. It is certainly not valid for all nuclear masses, as pe^-p has many bound states.
2. If $Z > 1$ ($\alpha < 1$), then a muon in the ground state should effectively reduce the nuclear charge at larger distances by 1, and the electronic spectrum should be a Balmer spectrum with $Z - 1$ in place of Z, i.e., $\sim (1-\alpha)^2$. To show this mathematically, write

$$\begin{aligned} &\frac{1}{2}\left(|\mathbf{p}_1|^2 + \frac{|\mathbf{p}_2|^2}{\mu} \right) - \frac{1}{r_1} - \frac{1}{r_2} + \frac{\alpha}{r_{12}} \\ &\geq P\left(-\frac{\mu}{2} + \frac{|\mathbf{p}_1|^2}{2\mu_1} + |\mathbf{p}_1|^2 \frac{\mu_1 - 1}{2\mu_1} - \frac{1-\alpha}{r_1} \right. \\ &\qquad \left. + \alpha\left(V_L(r_1) - \frac{1}{r_1} \right) \right) + \left(-\frac{\mu}{8} - \frac{1}{2} \right)(\mathbf{1} - P) \end{aligned}$$

and choose μ_1 large enough that $|\mathbf{p}|^2/2\mu_1 + \alpha(V_L(r) - 1/r)$ just remains ≥ 0. The value $\alpha\mu_1/\mu = 1/\pi$ of part (i) was improved to $\alpha\mu_1/\mu = \frac{1}{2}$ in part (ii), so

$$H \geq \left(-\frac{\mu}{2} + \frac{|\mathbf{p}_1|^2}{2}\left(1 - \frac{2\alpha}{\mu} \right) + \frac{1-\alpha}{r_1} \right) P + \left(-\frac{\mu}{8} - \frac{1}{2} \right)(\mathbf{1} - P).$$

Hence, for the state of the electron with principle quantum number n,

$$-\frac{1}{2}\left(\mu + \frac{(1-\alpha)^2}{n^2}\left(1-\frac{2\alpha}{\mu}\right)^{-1}\right) \leq E_n \leq -\frac{1}{2}\left(\mu + \frac{(1-\alpha)^2}{n^2}\right).$$

The upper bound results from the use of trial functions of the form $\chi \equiv \varphi(\mathbf{x}_1)\exp(-\mu r_2)$. Since

$$\int d^3x_2 |\mathbf{x}_2 - \mathbf{x}_1|^{-1} \exp(-2\mu r_2) \Big/ \int d^2x_2 \exp(-2\mu r_2)$$

stays $\leq 1/r_1$,

$$\langle \chi | H\chi \rangle \leq -\frac{\mu}{2}\langle \varphi | h\varphi \rangle, \qquad h \geq \frac{|\mathbf{p}|^2}{2} - \frac{1-\alpha}{r}.$$

The right side then comes from the min-max principle. If $\mu \sim 200$, the eigenvalues are estimated in this way to within a few percent.

For practical purposes the problem of finding the eigenvalues of H can be considered solved; the next interesting question is what the eigenfunctions are like. Proposition (3.5.33) produces narrow bounds in the L^2 sense for good trial functions, but we wish to answer qualitative questions about the electron-cloud of the two-electron problem, and hope to do so with methods that generalize for complex atoms. These questions concern the limits as $r \to \infty$ and $r \to 0$, and are not only mathematically accessible, but also of interest to chemists and nuclear physicists.

4.3.29 The Asymptotic Behavior of the Electron Density

For $r > r_0$, a sufficiently large constant, the one-electron density of the ground state, $\rho(\mathbf{x}_1) = \int d^3x_2 |\psi(\mathbf{x}_1, \mathbf{x}_2)|^2$, satisfies **Hoffmann–Ostenhof and Morgan's inequality**,

$$c_- r^{(1-\alpha)/\sqrt{2\varepsilon_1}-1} \exp(-\sqrt{2\varepsilon_1}r) \leq \sqrt{\rho(r)} \leq c_+ r^{(1-\alpha)/\sqrt{2\varepsilon_1}-1} \exp(-\sqrt{2\varepsilon_1}r),$$

where $\varepsilon_1 = -E_1 - \frac{1}{2}$ and $r_0 < r < \infty$, $0 < c_- < c_+ < \infty$.

4.3.30 Remark

The basic mathematical reason that the densities for isolated eigenvalues fall off exponentially is as follows: The operator $\exp(i\mathbf{s}\cdot(\mathbf{x}_1 + \mathbf{x}_2))$ generates the transformation $\mathbf{x}_i \to \mathbf{x}_i$, $\mathbf{p}_i \to \mathbf{p}_i + \mathbf{s}$, $H \to H + \mathbf{s}\cdot(\mathbf{p}_1 + \mathbf{p}_2)/m + |\mathbf{s}|^2/m \equiv H_\mathbf{s}$. Since the new term is bounded relative to $|\mathbf{p}_1|^2 + |\mathbf{p}_2|^2$, the family of operators $H_\mathbf{s}$ is analytic in the sense of (3.5.11). The eigenvectors $\varphi_\mathbf{s}$ are connected by $\exp(i\mathbf{s}\cdot(\mathbf{x}_1 + \mathbf{x}_2))$, so

$$\int d^3x_1 d^3x_2 |\exp(i\mathbf{s}\cdot(\mathbf{x}_1 + \mathbf{x}_2))\varphi(\mathbf{x}_1, \mathbf{x}_2)|^2 < \infty$$

for all $\mathbf{s} \in U$, some complex neighborhood of the origin. Theorem (4.3.29) shows in detail how the ionization energy determines the functional behavior at large r.

The proof proceeds via two lemmas which are of some independent interest. We first generalize the fact used in the proof of (3.5.28) that the kinetic energy of $\sqrt{\rho}$ is dominated by the actual kinetic energy.

4.3.31 The Schrödinger Inequality

$$\left(-\frac{\Delta}{2} - \frac{1}{r} + \alpha V_L(r) + \varepsilon_1\right)\sqrt{\rho} \leq 0,$$

where V_L is as in (4.6), *with* $\mu = 1$.

Proof: The Schrödinger equation

$$\left(\frac{1}{2}(|\mathbf{p}_1|^2 + |\mathbf{p}_2|^2) - \frac{1}{r_1} - \frac{1}{r_2} + \frac{\alpha}{r_{12}}\right)\psi = E_1\psi$$

implies that ($r = |x_1|$)

$$E_1\rho(r) = -\frac{1}{r}\rho(r) - \frac{1}{2}\int d^3x_2\psi^*(\mathbf{x}_1, \mathbf{x}_2)\overset{\leftrightarrow}{\Delta}_1\psi(\mathbf{x}_1, \mathbf{x}_2)$$
$$+\int d^3x_2\psi^*(\mathbf{x}_1, \mathbf{x}_2)\left(-\frac{\Delta_2}{2} - \frac{1}{r_2} + \frac{\alpha}{r_{12}}\right)\psi(\mathbf{x}_1, \mathbf{x}_2),$$

where $f\overset{\leftrightarrow}{\Delta}g \equiv (f\Delta g + g\Delta f)/2$. The Cauchy–Schwarz inequality can be used in the equation $\Delta_1\sqrt{\rho} = \nabla_1((\nabla_1\rho)/2\sqrt{\rho})$ to show that, in the sense of distributions,

$$-\sqrt{\rho}\Delta_1\sqrt{\rho} \leq -\int \psi^*\overset{\leftrightarrow}{\Delta}_1\psi d^3x^2.$$

Since $\Delta_1\rho/2 = |\nabla_1\sqrt{\rho}|^2 + \sqrt{\rho}\Delta_1\sqrt{\rho}$, $-\sqrt{\rho}\Delta_1\sqrt{\rho} = -\int\psi^*\overset{\leftrightarrow}{\Delta}_1\psi d^3x_2 - \int|\nabla_1\psi|^2 d^3x_2 + |\nabla_1\rho|^2/4\rho$. The last two terms are then ≤ 0 by Cauchy–Schwarz applied to $|\nabla_1\rho|^2 = 4|\int\psi^*\nabla_1\psi d^3x_2|^2$. Moreover, if $-\frac{1}{2} + \alpha V_L(r_1) \leq -\frac{1}{8}$, then the operator inequality

$$\frac{1}{2}|\mathbf{p}_2|^2 - \frac{1}{r_2} + \frac{\alpha}{r_{12}} \geq \frac{1}{2}|\mathbf{p}_2|^2 - \frac{1}{r_2} + \alpha P_0 V_L(r_1) \geq -\frac{1}{2} + \alpha V_L(r_1),$$

holds for the last term (see (3.5.29)); otherwise a two-dimensional projection is called for. This implies that

$$\left(-E_1 - \frac{1}{r} - \frac{1}{2} + \alpha V_L(r)\right)\rho - \frac{1}{2}\sqrt{\rho}\Delta\sqrt{\rho} \leq 0. \qquad \square$$

4.3.32 Monotony of the Ground-State Wave-Function in the Potential and Sources

Suppose that f, g, V, and W are nonnegative on a domain Ω, that f and g continuous on $\bar{\Omega}$, and that $V \leq W$, $A \leq B$. If $\Delta f \leq Vf + A$ and $\Delta g \geq Wg + B$ on Ω and $f \geq g$ on $\partial\Omega$, then $f \geq g$ on all of Ω.

4.3.33 Gloss

It is assumed that f and g are continuous, so Δf and Δg exist at least in the sense of distributions. If Ω is not a bounded region, then it must be assumed that f and g vanish at infinity. Although the source terms A and B do not appear in the usual Schrödinger equation, they will be needed below.

Proof: Let $D \equiv \{x \in \Omega : g > f\}$; on D we have $\Delta(g-f) \geq Wg - Vf + B - A \geq 0$. In one dimension a function of positive curvature, i.e., a convex function, attains its maximum at one of the end points. Likewise, a subharmonic function F on $\mathbb{R}^n$, i.e., a function for which $\Delta F \geq 0$, attains its maximum on the boundary. Since continuity makes $g = f$ on ∂D, g cannot exceed f on D, and therefore D is empty. □

Proof of (4.3.29):

(i) The upper bound. We know that $V_L > 1/r - 1/r^3$, and by (4.3.31),

$$\Delta\sqrt{\rho} \geq 2\left(\varepsilon_1 - \frac{1-\alpha}{r} - \frac{\alpha}{r^3}\right)\sqrt{\rho}.$$

If

$$j = \frac{1-\alpha}{\sqrt{2\varepsilon_1}} \quad \text{and} \quad f = \frac{1}{r}\exp(-\sqrt{2\varepsilon_1}r)(r^j + \beta r^{j-1}),$$

then

$$\Delta f = \left[2\varepsilon_1 - 2\frac{1-\alpha}{r} + \frac{[j(j-1) + 2\beta\sqrt{2\varepsilon_1}]r^{-2} + (j-1)(j-2)\beta r^{-3}}{1+\beta/r}\right] f.$$

Choosing $\beta < 0$ such that if $r > r_1$, then

$$[j(j-1) + 2\beta\sqrt{2\varepsilon_1}]r^{-2} + (j-1)(j-2)\beta r^{-3} < \left(-\frac{\alpha}{r_3} - \frac{\beta\alpha}{r^4}\right)\cdot 2,$$

we can identify the function $\sqrt{\rho}$ as the g of the subharmonic comparison lemma (4.3.32). In addition, we have to take $r_0 = \max\{-\beta, r_1, r_2\}$, where r_2 is the largest root of $\varepsilon_1 = (1-\alpha)/r_2 + \alpha/r_2^3$. Then the assumptions of

(4.3.32) are satisfied on $\Omega = \{r : r > r_0\}$, if we take

$$c_+ \geq \frac{r_0 \rho(r_0) \exp(\sqrt{2\varepsilon_1} r_0)}{r_0^j (1 - |\beta|/r_0)}.$$

(ii) The lower bound. Let $\varphi(\mathbf{x})$ be the ground state of $|\mathbf{p}|^2/2 - 1/r$ and ψ be the ground state of $H(\alpha)$ (4.3.1). Since φ and ψ are nonnegative, the Cauchy–Schwarz inequality implies that

$$0 \leq f(\mathbf{x}_1) \equiv \int d^3x_2 \varphi(\mathbf{x}_2) \psi(\mathbf{x}_1, \mathbf{x}_2) \leq \sqrt{\rho(\mathbf{x}_1)}.$$

Now,

$$\begin{aligned} 0 &= \int d^3x_2 \varphi(\mathbf{x}_2)(H - E_1)\psi(\mathbf{x}_1, \mathbf{x}_2) \\ &= \left(-\frac{\Delta}{2} - \frac{1}{r_1} + \varepsilon_1\right) f + \alpha \int d^3x_2 \frac{\varphi(\mathbf{x}_2)\psi(\mathbf{x}_1, \mathbf{x}_2)}{|\mathbf{x}_1 - \mathbf{x}_2|}. \end{aligned}$$

If r_1 is large, then we may estimate

$$\begin{aligned} \int d^3x_2 \frac{\varphi(\mathbf{x}_2)\psi(\mathbf{x}_1, \mathbf{x}_2)}{|\mathbf{x}_1 - \mathbf{x}_2|} &= \int_{r_2 \leq \sqrt{r_1}} + \int_{r_2 \geq \sqrt{r_1}} \\ &\leq \frac{f(r_1)}{r_1 - \sqrt{r_1}} + c \exp(-\sqrt{r_1})\sqrt{\rho}, \end{aligned}$$

since $\varphi \sim \exp(-r)$ and ρ remains bounded. Then with the upper bound for $\sqrt{\rho}$ we find that

$$\frac{1}{2}\Delta f \leq \left(\varepsilon_1 - \frac{1}{r} + \frac{\alpha}{r - \sqrt{r}}\right) f + c' \exp\left(-\sqrt{2\varepsilon_1 - \delta}\, r - \sqrt{r}\right),$$

and the assumptions of (4.3.32) can be satisfied by taking $g = c'' \exp(-\sqrt{2\varepsilon_1} r) r^{j-1}(1 + \beta r^{-1/2})$; a somewhat longer argument [24] is needed to convince one that c'' and β can be chosen appropriately. Since $\varphi > 0$, $f(x_1) = 0$ only if $\psi(\mathbf{x}_1, \mathbf{x}_2) = 0$ for all $\mathbf{x}_2$. In particular, it is not possible for f to vanish for all $|\mathbf{x}_1| \in I = (a, \infty)$, as ψ would then be 0 on $I \times I$ and therefore on the domain of analyticity containing $I \times I$. Thus for all $|\mathbf{x}_1|$ sufficiently large, $f(\mathbf{x}_1) > 0$, and we obtain a lower bound of the same asymptotic form as the upper bound. □□

The next topic is the electron density near the nucleus. The equation $\rho(0) = (1/2\pi)\langle dV/dr\rangle$ follows from (3.3.5; 4(b)) for a particle in a central potential V, and a generalization of this equation would be worthwhile. The probability that an electron remains at the nucleus is subject to a focusing of the electronic waves at the origin, which is not easy to understand from the particle point of view. For a convex potential like $V = r^2$, dV/dr increases with r, so $\rho(0)$ is greater for the more highly excited states. Classically, a lessened $\rho(0)$ is to be expected for $V = r^2$, since the particle flies through the origin with a greater speed. If the potential is concave, like $-1/r$, then the ordering of $\rho(0)$ accords with classical intuition.

4.3.34 Bounds for $\rho(0)$

$$\frac{1}{2\pi}|E_1|(E_2 - E_1) + \alpha\rho_{12}(0) \le \rho(0) \le \frac{1}{2\pi}\langle\psi_1|r_1^{-2}|\psi_r\rangle,$$

where

$$\rho(0) = \int d^3x_1 d^3x_2 |\psi_1(\mathbf{x}_1, \mathbf{x}_2)|^2 \delta^3(\mathbf{x}_1),$$

$$\rho_{12}(0) = \int d^3x_1 d^3x_2 |\psi_1(\mathbf{x}_1, \mathbf{x}_2)|^2 \delta^3(\mathbf{x}_1 - \mathbf{x}_2),$$

$$H\psi_1 = E_1\psi_1.$$

Proof: The upper bound (which holds for all eigenvectors). If $u(\mathbf{x}_1, \mathbf{x}_2) = r_1\psi_1(\mathbf{x}_1, \mathbf{x}_2)$, the Schrödinger equation becomes

$$-\frac{\partial^2}{\partial r_1^2}u + Wu = 0,$$

$$W = r_1^{-2}|\mathbf{L}_1|^2 - \Delta_2 - \frac{2}{r_1} + \frac{2\alpha}{r_{12}} - \frac{2}{r_2} - 2E.$$

Since

$$\psi_1(0, \mathbf{x}_2) = \frac{\partial}{\partial r_1}u(\mathbf{x}_1, \mathbf{x}_2)|_{r_1=0},$$

$$\rho(0) = -\frac{1}{2\pi}\int d^3x_1 d^3x_2 \frac{\partial u}{\partial r_1}\frac{\partial^2 u}{\partial r_1^2}r_1^{-2}.$$

With the integration by parts in (3.3.5; 4),

$$\rho(0) = -\frac{1}{2\pi}\int d^3x_1 d^3x_2 \frac{\partial u}{\partial r_1} r_1^{-2} W u$$

$$= (2\pi)^{-1}\left(\langle\psi_1|r_1^{-2} - \frac{|\mathbf{L}_1|^2}{r_1^3}|\psi_1\rangle - \alpha\langle\psi_1|\frac{\mathbf{x}_1\cdot(\mathbf{x}_1 - \mathbf{x}_2)}{r_1|\mathbf{x}_1 - \mathbf{x}_2|^3}|\psi_1\rangle\right).$$

Since $|\psi_1|^2$ is symmetric in $\mathbf{x}_1$ and $\mathbf{x}_2$, the latter term contributes negatively:

$$\langle\psi_1|\frac{r_1^2 - \mathbf{x}_1\cdot\mathbf{x}_2}{r_1 r_{12}^3}|\psi_1\rangle = \frac{1}{2}\langle\psi_1|r_{12}^{-3}\left(\frac{1}{r_1} + \frac{1}{r_2}\right)(r_1 r_2 - \mathbf{x}_1\cdot\mathbf{x}_2)|\psi_1\rangle \ge 0.$$

The lower bound. This part of the proof requires a lemma:

4.3.35 A Bound on the Mean-Square Deviation

$$(\Delta a)^2 \le \frac{\langle 1|[a,[H,a]]|1\rangle}{2(E_2 - E_1)},$$

where $H|1\rangle = E_1|1\rangle$, *and it is assumed that the two lowest eigenvalues* E_1 *and* E_2 *are* $\le \inf \sigma_{\text{ess}}$, *that* $|1\rangle \in D(a)$, $a^* = a$ *and* Δa *is calculated with* $|1\rangle$.

Proof:

$$\begin{aligned}\langle 1|a(Ha - aH)|1\rangle &= -\langle 1|(Ha - aH)a|1\rangle \\ &= \langle 1|a(\mathbf{1} - |1\rangle\langle 1|)(Ha - aH)|1\rangle \\ &\ge (E_2 - E_1)(\langle 1|a^2|1\rangle - \langle 1|a|1\rangle^2).\end{aligned} \quad \square$$

If we now set $a = \mathbf{p}_1$ in the lemma, then after partial integration

$$\begin{aligned}|\Delta \mathbf{p}_1|^2 = \langle |\mathbf{p}_1|^2\rangle = |E_1| &\le \frac{\langle |\Delta_1(-1/r_1 + \alpha/r_{12})|\rangle}{2(E_2 - E_1)} \\ &= \frac{2\pi}{E_2 - E_1}[\rho(0) - \alpha\rho_{12}(0)].\end{aligned} \quad \square\square$$

4.3.36 Remarks

1. We calculated only the contribution from $\mathbf{L} = 0$ to $\rho(0)$ in (3.3.5; 4(b)), but since $|\psi_l|^2 \sim r^{2l}$ for reasonable potentials, the term $|\mathbf{L}|^2/r^3$ contributes nothing in the case of a single particle.
2. If $\bar{\rho}$ denotes the spherical average of $\rho(r)$ at the nucleus, then its derivative satisfies the **cusp condition**

$$\left.\frac{d\bar{\rho}}{dr}\right|_{r=0} = -2\rho(0). \tag{4.14}$$

This can be seen by multiplying the equation $u'' = Wu$ by ψ and integrating over Ω and $\mathbf{x}_2$:

$$\frac{d\bar{\rho}}{dr} = \frac{r}{4\pi}\int \left(\psi W \psi - \psi \frac{\partial^2}{\partial r_1^2}\psi\right) d\Omega\, dx_2.$$

In the limit $r \to 0$, the only contribution is $-2r_1^{-1}$ from W, yielding (4.14). Hence we have the bounds

$$-\frac{1}{\pi}\langle \psi_1|r^{-2}|\psi_1\rangle \le \left.\frac{d\bar{\rho}}{dr}\right|_{r=0} \le -\frac{1}{\pi}|E_1|(E_2 - E_1) - 2\alpha\rho_{12}(0).$$

This section closes with a discussion of more general expectation values of r^ν. The first facts of interest are the

4.3.37 Monotonic Properties of $\langle a^\nu \rangle$

If a is positive operator, then (i) for $\nu \geq 1$ the functions $\mathbb{R} \to \bar{\mathbb{R}} : \nu \to \langle a^\nu \rangle^{1/\nu}$ increase monotonically; (ii) the functions $\nu \to -\ln\langle a^\nu \rangle$ are concave as long as $\langle a^\nu \rangle < \infty$.

Proof: (i) follows immediately from Jensen's inequality when the expectation values are written in a spectral representation. For (ii) calculate $\partial^2/\partial r^2$. □

4.3.38 Remarks

1. If the only available estimates for $\langle a^\nu \rangle$ with certain values of ν are not very satisfactory, then (4.3.37) can help out by using better estimates for other values of ν and interpolating.
2. Often the calculation of expectation values of an operator can be reduced to finding accurate bounds for some energy; for instance, if $E(\beta)$ is the eigenvalue of $H + \beta b$, then $\langle b \rangle = \partial E(\beta)/\partial\beta$. Since $E(\beta)$ is concave, bounds on $\partial E/\partial\beta$ can be obtained from bounds on $E(\beta)$. In this way our knowledge of $E(\alpha)$ gives us the expectation value of $1/r_{12}$, and thereby, with the virial theorem, of $\langle |\mathbf{p}_1|^2 \rangle = \langle |\mathbf{p}_2|^2 \rangle = |E|$ and $\langle 1/r_1 \rangle = \langle 1/r_2 \rangle$.

The line of reasoning indicated in Remark 2 will not be pursued further, as it soon leads to extensive computations. Instead, we shall only cite some general inequalities for expectation values involving r and $\mathbf{p}$. Since they are simply variants of the indeterminacy relations, and the more detailed form of the interaction does not enter, numerically they leave much to be desired in our special cases.

4.3.39 Lower Bounds for $\langle r^\nu \rangle^{1/\nu}$

$\nu = 2$: Since $|\mathbf{p}|^2/2 + r^2\omega^2/2 \geq 3\omega/2$, it follows that if $\omega = \frac{2}{3}\langle |\mathbf{p}|^2 \rangle$ then $\langle r^2 \rangle \geq 9/4\langle |\mathbf{p}|^2 \rangle$ and $\langle r^2 \rangle^{1/2} \geq \frac{3}{2}\langle |\mathbf{p}|^2 \rangle^{-1/2}$.
$\nu = 1$: $|\mathbf{p}|^2/2 + gr \geq (g/2)^{2/3} \cdot c$, where c is the first zero of the Airy function $Ai(x)$, $c = 2.338\ldots$. Then $\langle r \rangle \geq 1.2446\langle |\mathbf{p}|^2 \rangle^{-1/2}$.
$\nu = -1$: $|\mathbf{p}|^2/2 - \alpha/r \geq -\alpha^2/2 : \langle 1/r \rangle^{-1} \geq \langle |\mathbf{p}|^2 \rangle^{-1/2}$.
$\nu = -2$: By (2.5.22; 6), $|\mathbf{p}|^2/2 - 1/8r^2 \geq 0$, so $\langle 1/r^2 \rangle^{-1/2} \geq \frac{1}{2}\langle |\mathbf{p}|^2 \rangle^{-1/2}$.

4.3.40 Upper Bounds for $\langle r^\nu \rangle^{1/\nu}$

If we take $\mathbf{a} = \mathbf{x}_1 r_1^{q-1}$ in (4.3.35), and note that in a rotationally invariant state $\langle \mathbf{a} \rangle = 0$ and $[\mathbf{a}, [H, \mathbf{a}]] = |\nabla \mathbf{a}|^2 = r^{2q-2}(2 + q^2)$, then $\langle r^{2q} \rangle \leq (2 + q^2)\langle r^{2q-2} \rangle (2(E_2 - E_1))^{-1}$. Combining this with earlier results we get some useful inequalities for integral values of ν, $-2 \leq \nu \leq 2$:
$\nu = 2$: With $q = 1$ in the formula just derived,

$$\langle r^2 \rangle^{1/2} \leq \sqrt{3/2}(E_2 - E_1)^{1/2}.$$

$\nu = 1$: With $q = \frac{1}{2}$, $\langle r \rangle \leq \frac{9}{8}\langle 1/r \rangle (E_2 - E_1)^{-1} \leq \frac{9}{8}\langle |\mathbf{p}|^2 \rangle^{1/2}/(E_2 - E_1)$. As an imprecise inequality is used twice, the result is very weak.

$\nu = -1$: In the case of interest the virial theorem implies that $\langle 1/r_1 \rangle \geq |E_1|$, $\langle 1/r_1 \rangle^{-1} \leq |E_1|^{-1}$ (in atomic units).

$\nu = -2$: With $q = 0$ in the formula $\langle r^{-2} \rangle^{-1/2} \leq (E_2 - E_1)^{-1/2}$.

The thrust of these inequalities is that an average of r cannot get too small without increasing the kinetic energy greatly, and it cannot get too large without reducing the spacing of the eigenvalues. If E_1 and E_2 are taken as known for helium, then with the virial theorem the following bounds are found for $\langle r^\nu \rangle^{1/\nu}$ in atomic units, some of which can be improved with (4.3.38; 1):

ν	lower bound	upper bound
2	0.88	1.408
1	0.73	2.535
−1	0.587	0.689
−2	0.293	1.150

By Corollary (4.3.20; 2), $\langle 1/r \rangle^{-1} \simeq 0.61$. These inequalities are too general to be more precise; much greater numerical effort would be called for.

Since $\langle r^{-2} \rangle$ is important for calculating $\rho(0)$, we shall improve the result for $\nu = -2$ with the aid of (2.2.33; 4). It follows from $i\langle a^* b - b^* a \rangle \leq \langle a^* a \rangle + \langle b^* b \rangle$ with $a = p_r + ic/r$ and $b, c \in \mathbb{R}$ that

$$\langle r^{-2} \rangle c(1-c) + \langle r^{-1} \rangle 2bc - b^2 - \langle |\mathbf{p}|^2 \rangle \leq 0.$$

If this is optimized in b and c, then

$$\langle r^{-2} \rangle^2 - 4\langle |\mathbf{p}|^2 \rangle \langle r^{-2} \rangle + 4\langle r^{-1} \rangle^2 \langle |\mathbf{p}|^2 \rangle \leq 0 \Rightarrow$$

$$\langle r^{-2} \rangle \lesseqgtr 2\langle |\mathbf{p}|^2 \rangle \left(1 \pm \sqrt{1 - \langle r^{-1} \rangle^2 / \langle |\mathbf{p}|^2 \rangle}\right).$$

Then the virial theorem finally produces

4.3.41 Bounds for $\langle r^{-2} \rangle$ given E_1

Given two electrons and $\alpha > 0$,

$$\langle r^{-2} \rangle \lesseqgtr 2|E_1| \left(1 \pm \sqrt{1 - |E_1|}\right).$$

4.3.42 Problems

1. Show that the operator with integral kernel $\varphi(\mathbf{x}) V_L(\mathbf{x}) |\mathbf{x} - \mathbf{x}'| V_L(\mathbf{x}') \varphi(\mathbf{x}')$ has only one positive eigenvalue, and the rest of its spectrum is negative. (Hint: dominate $|\mathbf{x} - \mathbf{y}|$ as the integral kernel of a quadratic form by a form with kernel $f(\mathbf{x}) f(\mathbf{y})$.
2. Show that $r_{12}^{-1/2}(\mathcal{P}_1\mathcal{P}_2 + \mathcal{P}_2\mathcal{P}_1) r_{12}^{-1/2}$ has only one positive eigenvalue on the space of symmetric functions, and conclude that $r_{12}^{-1} \geq r_{12}^{-1/2}(\mathcal{P}_1 + \mathcal{P}_2 - |\chi\rangle\langle\chi|) r_{12}^{-1/2}$.

3. Calculate $\langle 0|1/r_{12}|0\rangle$ and $\langle 0|r_{12}|0\rangle$ in (4.3.20; 1). Note that the normalization of $\varphi_1(r)$ above did not contain the angular factor $1/\sqrt{4\pi}$.
4. Calculate the 2×2 matrices M_L^{-1} and M_L for $(1s2p)$ and $(1s3p)$ (see (4.3.23; 2)).
5. Calculate the screened potential of (4.5).
6. Show that for Coulomb systems $\langle V^2\rangle = 3E^2 + \langle T^2\rangle$, where T and V are respectively the kinetic and the potential energy, and expectation values are taken with eigenvectors of H.

4.3.43 Solutions

1. Let $f(\mathbf{x},\mathbf{y}) = 8\pi\int d^3q(\exp(i\mathbf{q}\cdot\mathbf{x})-1)(\exp(-i\mathbf{q}\cdot\mathbf{y})-1)/|\mathbf{q}|^4 - |\mathbf{x}| - |\mathbf{y}| + |\mathbf{x}-\mathbf{y}|$. Then note that $f(\mathbf{x},\mathbf{y}) = 0$, since $\Delta_{\mathbf{x}} f(\mathbf{x},\mathbf{y}) = \Delta_{\mathbf{y}} f(\mathbf{x},\mathbf{y}) = f(0,\mathbf{y}) = f(\mathbf{x},0) = 0$ and $f(\lambda\mathbf{x},\lambda\mathbf{y}) = \lambda f(\mathbf{x},\mathbf{y})$. The integral is evidently the kernel of a positive form. Consequently,
$$\begin{aligned}|\mathbf{x}-\mathbf{y}| &\le |\mathbf{x}|+|\mathbf{y}|\\ &= \frac{1}{2}(1+|\mathbf{x}|)(1+|\mathbf{y}|) - \frac{1}{2}(1-|\mathbf{x}|)(1-|\mathbf{y}|)\\ &\le \frac{1}{2}(1+|\mathbf{x}|)(1+|\mathbf{y}|)\end{aligned}$$
as a form inequality. The proposition then follows from the min-max principle applied to the operators obtained by multiplying on both sides by φV_L.
2. The operator $P_0(1)V_L(\mathbf{x}_2)r_{12}P_0(2)V_L(\mathbf{x}_1)+P_0(2)V_L(\mathbf{x}_1)r_{12}P_0(1)V_L(\mathbf{x}_2) = K^{(2)}$ has the expectation values
$$\langle\Phi|K^{(2)}|\Phi\rangle = 2\int d^3x_2 d^3x_2' f(\mathbf{x}_2)K^{(1)}(x_2,x_2')f(x_2'),$$
where
$$f(\mathbf{x}_2) = \int d^3x_1\varphi_1(\mathbf{x}_1)\Phi(\mathbf{x}_1,\mathbf{x}_2),$$
and
$$K^{(1)}(\mathbf{x}_2,\mathbf{x}_2') = \varphi_1(\mathbf{x}_2)V_L(\mathbf{x}_2)|\mathbf{x}_2-\mathbf{x}_2'|V_L(\mathbf{x}_2')\varphi_1(\mathbf{x}_2'),$$
provided that $\Phi(\mathbf{x}_1,\mathbf{x}_2) = \Phi(\mathbf{x}_2,\mathbf{x}_1)$. By problem 1, $K^{(1)}$ and thus $K^{(2)}$ each have only one positive eigenvalue, and the same must be true of $\mathcal{P}_1\mathcal{P}_2+\mathcal{P}_2\mathcal{P}_1$. Since we know that the positive eigenvalue is 2 and that $|\chi\rangle$ is the associated eigenvector, $\mathcal{P}_1\mathcal{P}_2+\mathcal{P}_2\mathcal{P}_1 < 2|\chi\rangle\langle\chi|$. The proposition then follows from $\mathcal{P}_1\mathcal{P}_2\mathcal{P}_1+\mathcal{P}_2\mathcal{P}_1\mathcal{P}_2 \ge 2|\chi\rangle\langle\chi|$.
3. $\langle 0|\frac{1}{r_{12}}|0\rangle = 16\int_0^\infty dr_1\, r_1^2\exp(-2r_1)\int_0^\infty dr_2\, r_2^2\exp(-2r_2)$
$$\times\int_{-1}^{1}\frac{dz}{2}(r_1^2+r_2^2-2r_1r_2z)^{-1/2}$$
$$= 16\int_0^\infty dr_1\, r_1\exp(-2r_1)\int_0^\infty dr_2\, r_2\exp(-2r_2)\cdot\begin{cases} r_2, & \text{when } r_1 > r_2\\ r_1, & \text{when } r_2 > r_1.\end{cases}$$
For $\langle 0|r_{12}|0\rangle$ the z-integral is $\int_{-1}^{1}dz(r_1^2+r_2^2-2r_1r_2z)^{1/2}/2$ instead of $\int_{-1}^{1}dz(r_1^2+r_2^2-r_1r_2z)^{-1/2}/2$.

4. $M_L^{-1} =$

$$\frac{1}{2}\begin{bmatrix} \langle(1s2p) \pm (2p1s)|r_{12}|(1s2p) \pm (2p1s)\rangle & \langle(1s2p) \pm (2p1s)|r_{12}|(1s3p) \pm (3p1s)\rangle \\ \langle(1s3p) \pm (3p1s)|r_{12}|(1s2p) \pm (2p1s)\rangle & \langle(1s3p) \pm (3p1s)|r_{12}|(1s3p) \pm (3p1s)\rangle \end{bmatrix}.$$

Symmetric:

$$M^{-1} = \begin{bmatrix} 5.11 & -1.77 \\ -1.77 & 12.58 \end{bmatrix}, \qquad M = \begin{bmatrix} 0.21 & 0.02 \\ 0.02 & 0.08 \end{bmatrix},$$

$$E_{1,2} = -\frac{5.17}{144} \pm \frac{\alpha}{2}(0.206 + 0.08) \mp \sqrt{\left(-\frac{5}{144} + \frac{\alpha}{2}(0.206 - 0.08)\right)^2 + \alpha^2 0.029^2},$$

$$E_{1,2} = -0.590 + \alpha 0.145 \mp \sqrt{(-0.035 + \alpha 0.061)^2 + \alpha^2 0.00084}.$$

Antisymmetric:

$$M^{-1} = \begin{bmatrix} 5.38 & -1.65 \\ -1.65 & 12.64 \end{bmatrix}, \qquad M = \begin{bmatrix} 0.194 & 0.025 \\ 0.025 & 0.082 \end{bmatrix},$$

$$E_{1,2} = -0.590 + \alpha 0.138 \mp \sqrt{(-0.035 + \alpha 0.056)^2 + \alpha^2 0.00064}.$$

5. $\langle \varphi_1(\mathbf{x}_1)|\frac{1}{r_{12}}|\varphi_1(\mathbf{x}_1)\rangle = \frac{4}{r_2}\int_0^{r_2} \exp(-2r_1)r_1^2 dr_1 + 4\int_{r_2}^{\infty} \exp(-2r_1)r_1\, dr_1.$

6. It follows from $V = H - T$ that $\Delta V = \Delta T$ if the deviation is calculated in an eigenstate of H. The proposition then follows from the virial theorem.

4.4 Scattering Theory of Simple Atoms

After the point spectrum of two particles in a Coulomb field has been analyzed, we are ready to study the continuous spectrum. Its physical significance can be understood by seeing how scattering theory applies in this case.

The scattering of several particles is a set of complicated diffraction problems in $\mathbb{R}^{3n}$, with potentials falling off in some directions but not in others. Due to the lack of simplicity the subject has long been used as a proving ground for any conceivable assumption or approximation, the validity of which was wholly obscure. Our first job will be to make physical sense of the continuous spectrum. We have seen that the wave-functions of the bound states fall off exponentially, making the particles localized near the nucleus. One would expect that in the other states one or more particles are asymptotically free. The first step in proving this is to show that there is no singular spectrum, which is associated with particles that wander arbitrarily far away but keep returning on occasion. The only known realistic examples of nontrivial singular spectrum are some clever models due to D. B. Pearson in which the potential consists of an array of barriers as for a band model, but for which the allowed bands are Cantor sets, and the energy spectrum is $\mathbb{R}^+$. The potentials either extend to infinity or fall off very slowly, and one would not expect any singular spectrum in scattering from an atom. In fact, it does not exist for Coulomb systems [22].

We next wish to exclude the possibility, mentioned in (3.4.10; 1), of waves entering the region of interaction, never to come out again. The relative compactness of the potentials, which was a key fact in the proof of asymptotic completeness for one-particle scattering, is now lacking. L. Faddeev was the first to figure out how to group compact parts of the resolvent. For energies less than the ionization energy it is straightforward to prove existence and completeness of the Møller operators, yet above the ionization threshold there is still no simple way to verify asymptotic completeness.

A physicist's work has only begun with the proof of existence of the wave-operators; actual calculations of what can be measured are needed. What accuracy can be guaranteed for the numerical results of various computations? The experience of volume II, §3.4, prepares us for the worst, as the interference effects in diffraction problems get so complicated as to make a mockery of mankind's calculating skills. In consequence it is all the more welcome to learn that the persistent efforts of L. Spruch, R. Blankenbecler, R. Sugar, and many others in recent decades have led to operator inequalities that bound the scattering parameters in many limiting cases with amazing precision.

Let us start by bettering our understanding of σ_{ess} and discovering the fate of the discrete states of H_0 that are embedded in the continuum. Our earlier analysis made ready use of the compactness of the $V(H_0-z)^{-1}$ occurring in $(H-z)^{-1} = (H_0-z)^{-1}(\mathbf{1}+V(H_0-z)^{-1})^{-1}$, but, as already mentioned, this fact breaks down for several particles interacting through pair potentials. We therefore start by finding a representation of the resolvent in which the effect of the interaction on the spectrum can be deduced with arguments about compactness. We shall restrict ourselves to the 3-body problem, though the method clearly generalizes to n particles.

If the potentials are relatively bounded, the resolvent can always be expanded in a norm-convergent series.

$$
\begin{aligned}
(H(\alpha)-z)^{-1} &= \sum_{n=0}^{\infty}\sum_{\gamma_1=1}^{3}\cdots\sum_{\gamma_n=1}^{3}(-1)^n R_0 v_{\gamma_1} R_0 v_{\gamma_2}\cdots R_0 v_{\gamma_n} R_0,\\
R_0 &\equiv (T-z)^{-1},\quad v_1 = -\frac{1}{r_1},\quad v_2 = -\frac{1}{r_2},\quad v_3 = \frac{\alpha}{r_{12}},\\
T &= \frac{1}{2}(|\mathbf{p}_1|^2+|\mathbf{p}_2|^2),
\end{aligned}
\tag{4.15}
$$

with the proviso that $d(z, \mathrm{Sp}(T)) = d(z, \mathbb{R}^+) = |\mathrm{Im}\, z|$ when $\mathrm{Re}\, z > 0$ or $|z|$ when $\mathrm{Re}\, z \le 0$ is large enough for

$$
\left\|\frac{1}{r}R_0\psi\right\| \le \varepsilon\|\psi\| + a\|R_0\psi\| \le (\varepsilon + a[d(z,\mathbb{R}^+)]^{-1})\|\psi\|.
$$

With only one electron all the summands other than the first, R_0, were compact. Since any norm limit of compact operators is compact, we concluded that the essential spectrum remains unchanged. The situation at hand is different, since the operators act on tensor products, and even though each contribution is compact on one factor when $n > 0$, it is not compact on both. In order to understand how the resolvent factorizes, write R_0 as the norm-convergent integral

$$R_0 = \int_C \frac{d\xi/2\pi i}{(\frac{1}{2}|\mathbf{p}_1|^2 - \xi)(\frac{1}{2}|\mathbf{p}_2|^2 - z + \xi)}, \tag{4.16}$$

where the contour of complex integration is $C = (\mathbb{R}^+ - i\varepsilon) \cup (-i\varepsilon, i\varepsilon) \cup (\mathbb{R}^+ + i\varepsilon)$, $0 < \varepsilon < |\mathrm{Im}\, z|$. Then

$$\begin{aligned} R_0 \frac{1}{r_1} R_0 = \int_{C\times C} & \frac{d\xi_1\, d\xi_2}{(2\pi i)^2} \frac{1}{(\frac{1}{2}|\mathbf{p}_1|^2 - \xi_1)} \frac{1}{r_1} \frac{1}{(\frac{1}{2}|\mathbf{p}_1|^2 - \xi_2)} \\ & \otimes \frac{1}{(\frac{1}{2}|\mathbf{p}_2|^2 - z + \xi_1)} \frac{1}{(\frac{1}{2}|\mathbf{p}_2|^2 - z + \xi_2)}, \end{aligned} \tag{4.17}$$

and the integrand is of the form compact $\otimes$ bounded. Yet if there are two different potentials, both factors become compact:

$$\begin{aligned} R_0 \frac{1}{r_1} R_0 \frac{1}{r_2} R_0 = \int & \frac{d\xi_1\, d\xi_2\, d\xi_3}{(2\pi i)^3} \frac{1}{(\frac{1}{2}|\mathbf{p}_1|^2 - \xi_1)} \frac{1}{r_1} \frac{1}{(\frac{1}{2}|\mathbf{p}_1|^2 - \xi_2)} \frac{1}{(\frac{1}{2}|\mathbf{p}_1|^2 - \xi_3)} \\ & \otimes \frac{1}{(\frac{1}{2}|\mathbf{p}_2|^2 - z + \xi_1)} \frac{1}{(\frac{1}{2}|\mathbf{p}_2|^2 - z + \xi_2)} \frac{1}{r_2} \frac{1}{(\frac{1}{2}|\mathbf{p}_2|^2 - z + \xi_3)}. \end{aligned} \tag{4.18}$$

There is also the case if r_1 or r_2 is replaced with r_{12}. Hence, while the contributions to (4.15) having all γ_i equal are not compact, all the others are. There is a graphical shorthand for the terms in (4.15), whereby electrons are drawn as lines and interactions $1/r_\gamma$ are drawn as wavy lines connecting electronic lines if $\gamma = (12)$ and extending outward from electronic lines if $\gamma = 1$ or 2:

R_0 $\qquad$ $R_0 \frac{1}{r_{12}} R_0$

$R_0 \frac{1}{r_1} R_0$  $\qquad$ $R_0 \frac{1}{r_2} R_0 \frac{1}{r_{12}} R_0$

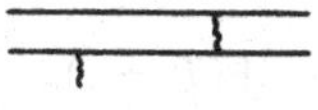

etc. The graphs of the products in (4.15) are the graphs of the factors joined together, so the set of graphs contains the whole algebraic structure of the operators. The noncompact operators are the disconnected graphs, which means that the electronic lines are either not connected to one another or not connected to the outside. The connected graphs, that is, the compact operators, are an ideal of the algebra. Thus, with the notation the graphical representation of the resolvent equation is

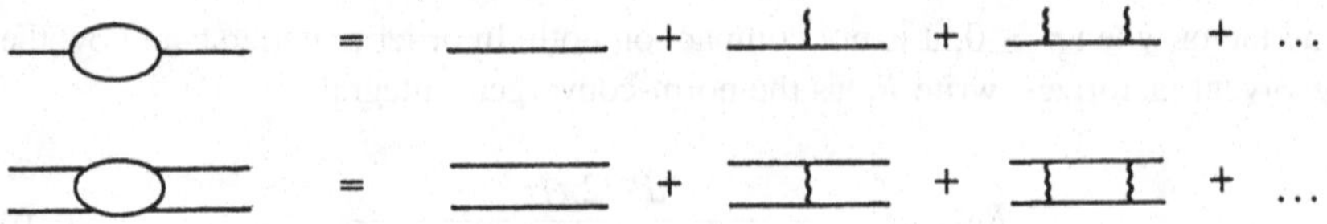

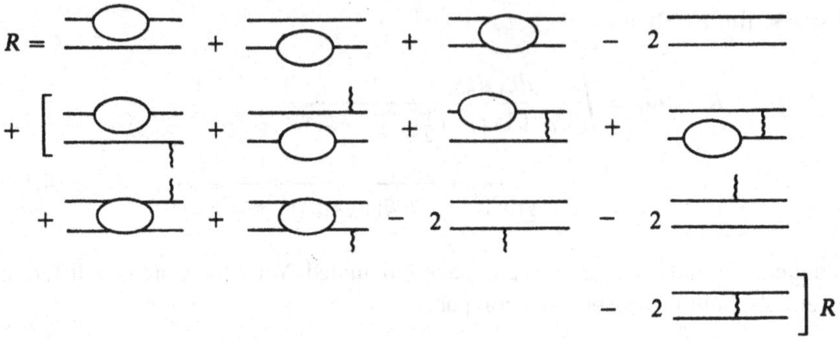

The translation of this as formulas is known as the

4.4.1 Weinberg–Van Winter Equation

$$R = D + JR,$$

$$D = \left(T - \frac{1}{r_1} - z\right)^{-1} + \left(T - \frac{1}{r_2} - z\right)^{-1} + \left(T + \frac{\alpha}{r_{12}} - z\right)^{-1} - 2(T-z)^{-1},$$

$$J = \left(T - \frac{1}{r_1} - z\right)^{-1}\left(-\frac{1}{r_2} + \frac{\alpha}{r_{12}}\right) + \left(T - \frac{1}{r_2} - z\right)^{-1}\left(-\frac{1}{r_1} + \frac{\alpha}{r_{12}}\right)$$

$$- \left(T + \frac{\alpha}{r_{12}} - z\right)^{-1}\left(\frac{1}{r_1} + \frac{1}{r_2}\right) - 2(T-z)^{-1}\left(-\frac{1}{r_1} - \frac{1}{r_2} + \frac{\alpha}{r_{12}}\right)$$

$$= (T-z)^{-1}\frac{1}{r_1}\left(T - \frac{1}{r_1} - z\right)^{-1}\left(-\frac{1}{r_2} + \frac{\alpha}{r_{12}}\right)$$

$$+(T-z)^{-1}\frac{1}{r_2}\left(T - \frac{1}{r_2} - z\right)^{-1}\left(-\frac{1}{r_1} + \frac{\alpha}{r_{12}}\right)$$

$$-(T-z)^{-1}\frac{\alpha}{r_{12}}\left(T + \frac{\alpha}{r_{12}} - z\right)^{-1}\left(\frac{1}{r_1} + \frac{1}{r_2}\right).$$

4.4.2 Consequences

As shown above, J is compact for $d(z, \mathbb{R})$ sufficiently large. Moreover, its compactness extends by analyticity to the whole region $\mathbb{C}\backslash\mathbb{R}^+ \bigcup_{n=1}^{\infty}\{-1/2n^2\}$ (Problem 3). This means that $(\mathbf{1} - J(z))^{-1}$ has only isolated poles of finite multiplicity in the region of analyticity, and the only additional singularities of $R(z) = (\mathbf{1} - J(z))^{-1} D(z)$ are those of $D(z)$. This reasoning leads to a result due to Hunziker, Van Winter, and Zhislin, generalizing (4.3.5) for not necessarily positive α.

4.4.3 The HVZ Theorem

$$\sigma_{\text{ess}}(H(\alpha)) = \sigma_{\text{ess}}\left(T - \frac{1}{r_1}\right) \cup \sigma_{\text{ess}}\left(T - \frac{1}{r_2}\right) \cup \sigma_{\text{ess}}\left(T + \frac{\alpha}{r_{12}}\right).$$

4.4.4 Remark

This result casts more light on Remark (4.3.6; 1). The essential spectrum describes particles that escape to infinity, so σ_{ess} begins at the various ionization energies.

In order to eliminate the possibility of σ_{sing}, we look at the $H(\alpha)$ dilated with a complex parameter τ as in (4.1.12):

$$H_\alpha(\tau) \equiv U(\tau)H(\alpha)U^{-1}(\tau) = \exp(2\tau)T + \exp(\tau)V,$$
$$(H_\alpha(\tau) - z)^{-1} = (\mathbf{1} - J(\tau))^{-1} D(\tau),$$

$$D(\tau) = \left(\exp(2\tau)T - \frac{\exp(\tau)}{r_1} - z\right)^{-1} + \left(\exp(2\tau)T - \frac{\exp(\tau)}{r_2} - z\right)^{-1} + \left(\exp(2\tau)T + \frac{\alpha\exp(\tau)}{r_{12}} - z\right)^{-1} - 2(\exp(2\tau)T - z)^{-1},$$

$$\begin{aligned}\exp(2\tau)J(\tau) = {} & (T - \exp(-2\tau)z)^{-1}\frac{1}{r_1}\left(T - \frac{\exp(-\tau)}{r_1} - \exp(2-\tau)z\right)^{-1} \\ & \times\left(-\frac{1}{r_2} + \frac{\alpha}{r_{12}}\right) + (T - \exp(-2\tau)z)^{-1}\frac{1}{r_2} \\ & \times\left(T - \frac{\exp(-\tau)}{r_2} - \exp(-2\tau)z\right)^{-1}\left(-\frac{1}{r_1} + \frac{\alpha}{r_{12}}\right) \\ & + (T - \exp(-2\tau)z)^{-1}\frac{\alpha}{r_{12}}\left(T + \frac{\alpha\exp(-\tau)}{r_{12}} - \exp(-2\tau)z\right) \\ & \times\left(\frac{1}{r_1} + \frac{1}{r_2}\right). \end{aligned} \tag{4.19}$$

The operator $J(\tau)$ is again compact, so $(\mathbf{1} - J(\tau))^{-1}$ affects only the point spectrum of $H_\alpha(\tau)$. The essential spectrum originates from D, which, however, contains only the sorts of expressions encountered in (4.1.12). Since

$$\sigma(A \otimes \mathbf{1} + \mathbf{1} \otimes B) = \sigma(A) + \sigma(B)$$

(see [3], Section XIII.9) and

$$\sigma\left(\exp(2\tau)\frac{|\mathbf{p}|^2}{2} - \frac{\exp(\tau)}{r}\right) = \bigcup_{n\geq 1}\left\{\frac{-1}{2n^2}\right\} \cup \exp(2\tau)\mathbb{R}^+$$

(see (4.1.13; 2)),

$$\sigma\left(\exp(2\tau)\frac{1}{2}(|\mathbf{p}_1|^2 + |\mathbf{p}_2|^2) - \frac{\exp(\tau)}{r_1}\right)$$
$$= \bigcup_{n\geq 1}\left\{\frac{-1}{2n^2} + \exp(2\tau)\mathbb{R}^+\right\} \cup \exp(2\tau)\mathbb{R}^+.$$

This means that the continuum starting at each energy $-1/2n^2$ gets swung out into the complex plane by the dilatation (see Figure 4.5):

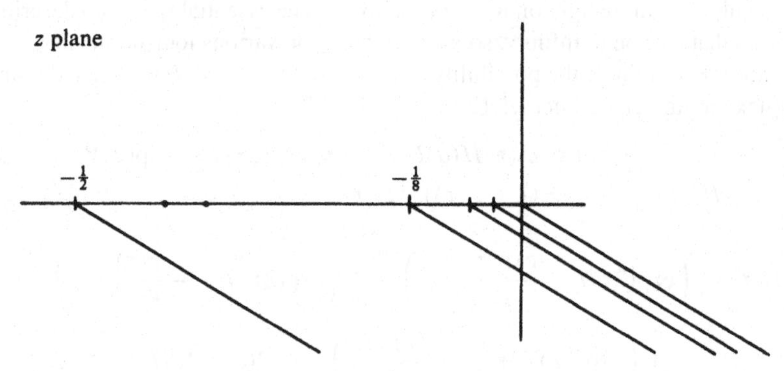

Figure 4.5. $\mathrm{Sp}(\exp(2\tau)\frac{1}{2}(|\mathbf{p}_1|^2 + |\mathbf{p}_2|^2) - \exp(\tau)/r_1)$ when $\mathrm{Im}\exp(2\tau) < 0$.

As a consequence, the matrix elements of the resolvent in the dense set D of entire vectors with respect to $U(\tau)$ can be continued analytically past the real axis, since it intersects σ_{ess} only at $\bigcup_n\{-1/2n^2\} \cup \{0\}$. As in (4.1.12), we conclude the

4.4.5 Absence of Singular Spectrum

$$\sigma_{\mathrm{sing}}(H_\alpha) = \emptyset, \quad \textit{so } \sigma_{\mathrm{ess}}(H_\alpha) = \sigma_{ac}(H_\alpha).$$

Because $\langle U^{-1}(\tau^*)\varphi|(H_\alpha - a)^{-1}U^{-1}(\tau)\psi\rangle = \langle\varphi|(H_\alpha(\tau) - z)^{-1}\psi\rangle$, the spectrum of $H_\alpha(\tau)$ reflects the singularities that occur when matrix elements of the resolvent are continued analytically. The next topic to investigate is how the eigenvalues in the continuum of H_0 move onto the second sheet of z when the perturbation α/r_{12}, is switched on after this analytic continuation. We start by noting the

4.4.6 Spectrum of $H_0(\tau)$

$$\sigma_P(H_0(\tau)) = \sigma_p(H_0(0)) = \bigcup_{n,m\geq 1}\left\{-\frac{1}{2n^2} - \frac{1}{2m^2}\right\},$$

$$\sigma_{ac}(H_0(\tau)) = \bigcup_{n\geq 1}\left\{-\frac{1}{2n^2} + \exp(2\tau)\mathbb{R}^+\right\} \cup \{\exp(2\tau)\mathbb{R}^+\}.$$

As to the eigenvalues of the non-Hermitian operator $H_\alpha(\tau)$, if Im $\exp(2\tau) < 0$, then some of them may have negative imaginary parts. The argument of (4.1.13), by which the eigenvalues were analytic in τ and invariant for real τ, and therefore constant, works only assuming that they remain isolated. But σ_{ess} sweeps over the eigenvalues with negative imaginary parts as Im $\tau \to 0$, invalidating the argument. In fact, the eigenvalues $\bigcup_{n,m>2}\{-1/2n^2 - 1/2m^2\}$ of $H_0(\tau)$ move to complex positions as α/r_{12} is turned on with τ fixed, unless their movement is prevented by some selection rule. Since the continuum of $H_0(\tau)$ swings out of the way, the eigenvalues are isolated for Im $\tau < 0$ and $\alpha = 0$, and lowest-order perturbation theory (3.5.16) can be used to compute their displacement as α/r_{12} is turned on (the derivation of the perturbation-theoretic formula (3.5.16) did not require the operator to be Hermitian). As expected, the position of an eigenvalue is independent of τ, which merely specifies which complex half-plane the limit to real values is to be taken from.

4.4.7 Perturbation Theory of the Complex Eigenvalues

Let $|\tau\rangle = U(\tau)|0\rangle$ in the notation of (3.5.16), *taking $|\tau\rangle$ and $E(0)$ as the eigenvector and eigenvalue of $H_0(\tau)$. Then the eigenvalue of $H_\alpha(\tau)$ is*

$$\begin{aligned}E(\alpha) &= E(0) + \alpha\langle\tau|\frac{\exp(\tau)}{r_{12}}|\tau\rangle - \alpha^2\langle\tau|\frac{\exp(\tau)}{r_{12}} P_\perp(\tau)(H_0(\tau)\\ &\quad -E(0))^{-1}P_\perp(\tau)\frac{\exp(\tau)}{r_{12}}|\tau\rangle + o(\alpha^2)\\ &= E(0) + \alpha\langle 0|\frac{1}{r_{12}}|0\rangle - \alpha^2\lim_{\varepsilon\downarrow 0}\langle 0|\frac{1}{r_{12}} P_\perp(\tau)(H_0(0)\\ &\quad -E(0) - i\varepsilon)^{-1}P_\perp(\tau)\frac{1}{r_{12}}|0\rangle + o(\alpha^2).\end{aligned}$$

4.4.8 Remarks

1. To $o(\alpha^2)$, the imaginary part of $E(\alpha)$ is

$$\text{Im } E(\alpha) = -\pi\alpha^2\langle 0|\frac{1}{r_{12}} P_\perp\delta(H_0 - E(0))P_\perp\frac{1}{r_{12}}|0\rangle,$$

a formula known as **Fermi's golden rule**. Although this is certainly an eigenvalue of $H_\alpha(\tau)$ for sufficiently large Im τ, the Hermitian operator $H_\alpha(0)$, of course,

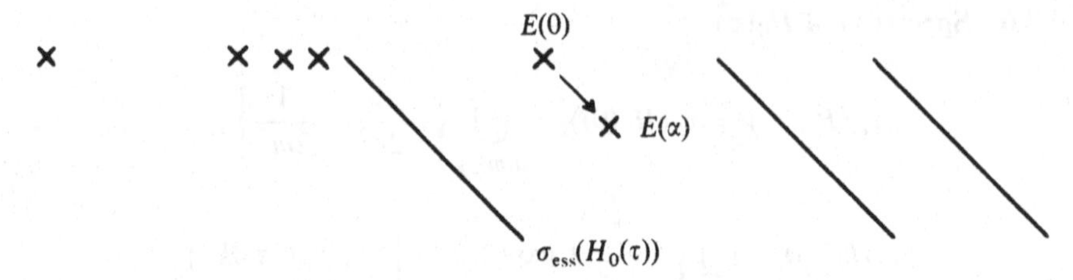

Figure 4.6. The motion of the eigenvalues of $H_\alpha(\tau)$ for $\alpha > 0$.

has no complex eigenvalues. They only appear when the matrix elements of the resolvent are continued analytically (see Figure 4.6).

2. The interpretation of the poles as resonances is based on the lifetime (4.2.11) as well as on the scattering operator $V - V(H - E)^{-1}V$. If α is small, they are the dominant contribution to the resolvent at $E = \text{Re}(E(\alpha))$, and in that case $\text{Im}\, E(\alpha)$ determines the probability of decay.
3. For the bound states of unnatural parity, $H_0(0)$ has no continuum in the subspace of equal quantum numbers, so $P_\perp \delta(H(0) - E_0) P_\perp$ vanishes and thus so does $\text{Im}\, E(\alpha)$. However, for the other states, $\text{Im}\, E(\alpha) < 0$; this answers the question that arose earlier about $\sigma_p(H(\alpha))$ embedded in the continuum. For sufficiently small α, it develops continuously out of the eigenvalues $-1/2n^2 - 1/2m^2$ of $H(0)$, which have $\text{Im}\, E(\alpha) = 0$.

The next problem is to show that the vectors associated with $\sigma_{ac}(H(\alpha))$ are scattering states, in order to be able eventually to calculate the relevant scattering parameters with accuracies good enough for comparison with experimental data. This is a formidable goal, so we shall content ourselves with the simplest nontrivial situation, where an electron e^- scatters from an atom consisting of a μ^- and a p. We consider only energies less than 1 KeV, so the muon remains in the ground state. If the energy is great enough to excite the muon to higher bound states, then the calculation is quite similar in principle, but much more complicated to write out. If the muon can be ionized, then the ordinary Møller operators do not exist, and recourse must be had to comparison with the kind of modified time-evolution of (4.1.15; 2). The Coulomb field is screened only provided that the muon remains bounded. Only then is the range of the interaction short enough for the scattering theory of §3.4 to work. As in (3.4.16) we shall study (for sufficiently large μ)

4.4.9 Channel Hamiltonian

$$H = \frac{|\mathbf{p}_1|^2}{2\mu} + \frac{|\mathbf{p}_2|^2}{2} - \frac{1}{r_1} - \frac{1}{r_2} + \frac{1}{r_{12}},$$

$$H_1 = \frac{|\mathbf{p}_1|^2}{2\mu} - \frac{1}{r_1} + \frac{|\mathbf{p}_2|^2}{2}, \qquad I_1 = -\frac{1}{r_2} + \frac{1}{r_{12}},$$

$$P_1(E) \equiv \Theta(E - H_1) = |\varphi\rangle\langle\varphi| \otimes \Theta\left(E + \frac{\mu}{2} - \frac{|\mathbf{p}_2|^2}{2}\right) \quad \text{for } E < -\frac{\mu}{8},$$

$Q_1(E) \equiv \Theta(E - H)$,

$\varphi(r_1) = 2\mu^{3/2}/\sqrt{4\pi}\exp(-\mu r_1)$ is the ground state of $\dfrac{|\mathbf{p}_1|^2}{2\mu} - \dfrac{1}{r_1}$.

This notation allows the principal result to be stated:

4.4.10 The One-Channel Møller Operators

If $E < -\mu/8$, then the Møller operators

$$\Omega_{1\pm} \equiv s\text{-}\lim_{t\to\pm\infty} Q_1(E)\exp(iHt)\exp(-iH_1t)P_1(E)$$

and

$$\Omega_{1\pm}^{*} \equiv s\text{-}\lim_{t\to\pm\infty} P_1(E)\exp(iH_1t)\exp(-iHt)Q_1(E)$$

exist and are complete,

$$\Omega_{1\pm}^{*}\Omega_{1\pm} = P_1(E), \qquad \Omega_{1\pm}\Omega_{1\pm}^{*} = Q_1(E).$$

4.4.11 Remarks

1. It is crucial that the effective interaction between e^- and μ^-p falls off faster than $1/r$; if there were a nucleus of charge $Z > 1$ instead of p, it would be necessary to choose
$$H_1 = \frac{|\mathbf{p}_1|^2}{2\mu} + \frac{|\mathbf{p}_2|^2}{2} - \frac{Z}{r_1} - \frac{Z-1}{r_2}.$$
2. The scattering of e^+ from μ^-p is the same, with some changes of sign.
3. If the electronic spins are parallel, the scattering of e^- from a hydrogen atom is complicated by the Pauli principle: Since H_1 is not invariant under permutation of the two electrons, $\exp(iHt)\exp(-iH_1t)$ does not send the subspace of antisymmetric states to itself. It is necessary instead to take the limit of $\exp(iHt)\exp(-iH_1t)|\psi_1\rangle \pm \exp(iHt)\exp(-iH_2t)|\psi_2\rangle$, which is equivalent to an antisymmetrization of the scattering amplitude.
4. If the electronic spins are antiparallel, then H has a bound state, which must be known precisely to calculate the scattering length; if there is an eigenvalue at zero, the scattering length becomes infinite.
5. The S-matrix in the interaction representation, $S_{11} = \Omega_{1+}^{*}\Omega_{1-}$, satisfies the unitary condition $S_{11}^{*}S_{11} = S_{11}S_{11}^{*} = P_1(E)$.

Proof of (4.4.10): As in the proof of (3.4.11), we check whether

$$Q_1(E)\exp(iHt)I_1\exp(-iH_1t)P_1(E)$$

and

$$P_1(E)\exp(iH_1t)I_1\exp(-iHt)Q_1(E)$$

are integrable in time. The conditions of (3.4.11) for H and H_1 have to be met after multiplication by Q_1 and, respectively, P_1, so the supremum needs to be taken only over

$$\omega \in I \equiv \left(-\frac{\mu}{2}+\delta, -\frac{\mu}{8}-\delta\right), \qquad \delta \downarrow 0.$$

(i) $\sup_{\omega\in I}\|\sqrt{I_1}\delta(H_1-\omega)\sqrt{I_1}\|$. For this part it is intuitively clear that for $\omega < -\mu/8$

$$\delta\left(\frac{|\mathbf{p}_1|^2}{2\mu}-\frac{1}{r_1}+\frac{|\mathbf{p}_2|^2}{2}-\omega\right) = |\varphi\rangle\langle\varphi| \otimes \delta\left(\frac{|\mathbf{p}_2|^2}{2}-\omega-\frac{\mu}{2}\right);$$

and this is derived formally in Problem 1. With $k^2/2 = \omega + \mu/2$, we find

$$\begin{aligned}
&\operatorname{Tr}(\sqrt{I_1}\delta(H_1-\omega)\sqrt{I_1})^n \\
&= \int d^3x_1\cdots d^3x_n u(\mathbf{x}_1)g_k(\mathbf{x}_1-\mathbf{x}_2)u(\mathbf{x}_2)g_k(\mathbf{x}_2-\mathbf{x}_3)\cdots g_k(\mathbf{x}_n-\mathbf{x}_1),
\end{aligned}$$

where

$$g_k(\mathbf{x}) = \frac{1}{4\pi^2 r}\sin kr$$

and

$$\begin{aligned}
u(\mathbf{x}) &= \int d^3x'|\varphi(\mathbf{x}')|^2|I_1(\mathbf{x},\mathbf{x}')| \\
&= \int d^3x'|\varphi(\mathbf{x}')|^2\left|\frac{1}{|\mathbf{x}|}-\frac{1}{|\mathbf{x}-\mathbf{x}'|}\right|.
\end{aligned}$$

The potential u is larger than the potential

$$V(\mathbf{x}) = \int d^3x'|\varphi(\mathbf{x}')|^2\left(\frac{1}{|\mathbf{x}-\mathbf{x}'|}-\frac{1}{|\mathbf{x}|}\right) = -\left(\mu+\frac{1}{r}\right)\exp(-2\mu r) \tag{4.20}$$

(see (4.5)), which is the potential due to the proton shielded by the muon-cloud. However, the dipole expansion $1/|\mathbf{x}-\mathbf{x}'| = 1/r + (\mathbf{x}\cdot\mathbf{x}')/r^3 + O(r^{-3})$ shows that $u \sim r^{-2}$ as $r \to \infty$. We thus have a one-body situation of the kind treated in (3.4.11). Since u has only a $1/r$ singularity at the origin, (3.4.14; 1) shows that indeed

$$\sup_{\omega\in I}\|\sqrt{I_1}\delta(H_1-\omega)\sqrt{I_1}\| < \infty.$$

(ii) $\sup_{\omega\in I}\|\sqrt{I_1}\delta(H-\omega)\sqrt{I_1}\|$.

$$\delta(H-\omega) = \frac{1}{2\pi i}\lim_{\varepsilon\downarrow 0}\left(\frac{1}{H-\omega-i\varepsilon}-\frac{1}{H-\omega+i\varepsilon}\right).$$

The resolvent still has the same structure as in (4.15) in the trivial generalization to $\mu > 1$, so

$$\sqrt{I_1}(H-z)^{-1}\sqrt{I_1} = (1-\sqrt{I_1}J(z)(\sqrt{I_1})^{-1})^{-1}\sqrt{I_1}D(z)\sqrt{I_1},$$

and it is again a question of showing the boundedness of $\sqrt{I_1}D(z)\sqrt{I_1}$ and the compactness of $\sqrt{I_1}J(z)/\sqrt{I_1}$. We already know this if z is not real, and it only remains to see what happens for $z \in [-\mu/2+\delta\pm i\varepsilon, -\mu/8-\delta\pm i\varepsilon] = I$, $\delta > 0$, $\varepsilon \downarrow 0$. The only singular part of $D(z)$ is

$$\left(\frac{|\mathbf{p}_1|^2}{2\mu}+\frac{1}{r_1}+\frac{|\mathbf{p}_2|^2}{2}-z\right)^{-1} = |\varphi\rangle\langle\varphi|\otimes\left(\frac{|\mathbf{p}_2|^2}{2}-z-\frac{\mu}{2}\right)^{-1}$$
$$+(\mathbf{1}-|\varphi\rangle\langle\varphi|\otimes\mathbf{1})\left(\frac{|\mathbf{p}_1|^2}{2\mu}-\frac{1}{r_1}+\frac{|\mathbf{p}_2|^2}{2}-z\right)^{-1},$$

and in fact only the first term in this expression. If the muon is excited, its energy becomes $\geq -\mu/8$, and the second term is uniformly bounded by $1/\delta$ for all $z \in I$. In part (i) we find that

$$\sqrt{I_1}|\varphi\rangle\langle\varphi|\otimes\left(\frac{|\mathbf{p}_2|^2}{2\mu}-\omega-\frac{\mu}{2}\pm i\varepsilon\right)^{-1}\sqrt{I_1}$$

is bounded (the proof also works for the term with $\delta(|\mathbf{p}|^2-z)$ replaced by $(|\mathbf{p}|^2-z)^{-1}$). Similarly, the only term in question in $\sqrt{I_1}J/\sqrt{I_1}$ is

$$\sqrt{I_1}(T-z)^{-1}\frac{1}{r_1}\left(T-\frac{1}{r_1}-z\right)^{-1}\sqrt{I_1},$$

and in fact again only the contribution from the muon in the ground state. To prove the compactness of

$$\sqrt{I_1(\mathbf{x}_1,\mathbf{x}_2)}(T-z)^{-1}\frac{1}{r_1}|\varphi(\mathbf{x}_1)\rangle\langle\varphi(\mathbf{x}_1)|$$
$$\otimes\left(\frac{|\mathbf{p}_2|^2}{2}-z-\frac{\mu}{2}\right)^{-1}\sqrt{I_1(\mathbf{x}_1,\mathbf{x}_2)}$$

we argue as in (3.4.13). It is rather easy (Problem 2) to convince oneself that $K = \sqrt{I_1(\mathbf{x}_1,\mathbf{x}_2)}(T-z)^{-1}(1/r_1)|\varphi(\mathbf{x}_1)\rangle\langle\varphi(\mathbf{x}_1)|\delta(|\mathbf{p}_2|^2-\omega)\sqrt{I_1(\mathbf{x}_1,\mathbf{x}_2)})$ belongs to $\mathcal{C}_4$ and is Hölder continuous in ω. This guarantees the compactness of the operator in question for $z = x+iy$, $-\mu/2 < x < 0$, as $y \downarrow 0$ (Problem 2). The points z_i at which $J(z)$ has eigenvalue 1 are eigenvalues of H with finite multiplicity. The associated eigenfunctions fall off exponentially (4.3.29), so $\sqrt{I_1}\psi$ is in $\mathcal{H}$. The operator $\sqrt{I_1}J(z)/\sqrt{I_1}$ also has eigenvalue 1 iff z equals one of the z_i. Since we know that there is no eigenvalue of H in the interval I, $(1-\sqrt{I_1}J(z)/\sqrt{I_1})^{-1}$ is bounded uniformly on I, and thus so is

$$\sup_{\omega\in I}\|\sqrt{I_1}\delta(H-\omega)\sqrt{I_1}\| < \infty. \qquad \square$$

Once it is known that the Møller operators exist, and hence that the S-matrix is well defined, it becomes possible to calculate the scattering amplitude of e^- from μ^-p. This is a matter of finding the expectation value of $I_1 - I_1(H-E)^{-1}I_1$ with $E = -\mu/2 + k^2/2$ in the (unnormalizable) state $\varphi(\mathbf{x}_1)\exp(i\mathbf{k}\cdot\mathbf{x}_2)$. Physical intuition says that the large mass of the muon prevents an electron of low energy from having much influence on it. As a consequence, the electron should simply feel the familiar effective potential $V_t = 1 \otimes V$, where V is in (4.20). This feeling can at least be tested by using V_t as the comparison potential in Kohn's variational principle (3.6.25) and substituting for Ω_t; setting $k = 0$ for this purpose, we get a

4.4.12 Variational Principle for the Scattering Length

The scattering length is $1/4\pi$ times the expectation value of

$$T = T_t + T_t V_t^{-1}(I_1 - V_t)V_t^{-1}T_t - T_t V_t^{-1}(I_1 - V_t) \times \left(H + \frac{\mu}{2}\right)^{-1}(I_1 - V_t)V_t^{-1}T_t,$$

$$T_t = V_t - V_t\left(H_t + \frac{\mu}{2}\right)^{-1} V_t,$$

$$H_t = H + V_t - I_1 = \frac{|\mathbf{p}_1|^2}{2\mu} - \frac{1}{r_1} + \frac{|\mathbf{p}_2|^2}{2} + V_t(\mathbf{x}_2)$$

in the state $\varphi(\mathbf{x}_1) \otimes \mathbf{1}$.

4.4.13 Calculation of the Three Contributions to T

(i) T_t. This amounts to the scattering of an electron by a short-range potential V_t. There is no interaction between the muon and the electron, so everything factorizes. Since V_t is radially symmetric, T_t could be evaluated numerically on a computer. However, it is hardly worth the trouble, since with the actual value $\mu = 207$, V_t is such a small perturbation that the first Born approximation a_B has an accuracy measured in $\frac{0}{00}$. To see this, refer to (3.6.20) and calculate the quantity defined there,

$$\|K\|^2 \le \|v\|_R^2 \equiv \int d^3x\, d^3x' v(\mathbf{x})v(\mathbf{x}')(4\pi|\mathbf{x}-\mathbf{x}'|)^{-2}$$
$$= \int \frac{d^3k}{(2\pi)^3}\left(\frac{\tilde{v}(\mathbf{k})}{4\pi}\right)^2 \frac{2\pi^2}{|\mathbf{k}|}$$
$$= \frac{1}{2}\int_0^\infty dk^2 \frac{[|\mathbf{k}|^2 + 8\mu^2]^2}{[|\mathbf{k}|^2 + 4\mu^2]^4} = \frac{1}{4\mu^2}\frac{7}{6},$$

since the Fourier transform of $|\varphi(\mathbf{x})^2|$ is $(1 + |\mathbf{k}|^2/4\mu^2)^{-2}$, so

$$\frac{\tilde{v}(\mathbf{k})}{4\pi} = \frac{-1}{|\mathbf{k}|^2}\left(1 - \frac{(4\mu)^2}{(|\mathbf{k}|^2 + 4\mu^2)^2}\right).$$

The scattering length a_t is thus equal to $a_B(1 \pm \|K\|/(1-\|K\|)) = a_B(1 \pm 0.002)$, and a calculation of a_B yields

$$a_B = \int \frac{d^3x}{4\pi} v(\mathbf{x}) = \frac{\tilde{v}(\mathbf{0})}{4\pi} = -\frac{1}{2\mu^2} = -\frac{1}{2}(206.8)^{-2} \cdot \text{(Bohr radius)}$$
$$= -0.619 \times 10^{-13}\text{cm}.$$

(ii) The term linear $I_1 - V_t$ vanishes: Clearly,

$$T_t\varphi(\mathbf{x}_1) \otimes \exp(i\mathbf{k}\cdot\mathbf{x}_2)$$
$$= \varphi(\mathbf{x}_1) \otimes \left(V_t - V_t\left(\frac{|\mathbf{p}_2|^2}{2} - |\mathbf{k}|^2\right)^{-1} V_t\right)\exp(i\mathbf{k}\cdot\mathbf{x}_2),$$

and $\int d^3x_1 |\varphi(\mathbf{x}_1)|^2 (I_1(\mathbf{x}_1,\mathbf{x}_2) - V_t(\mathbf{x}_2))$ is zero by definition. The V_t we have chosen is optimal in the sense that the difference between $\langle T\rangle$ and $\langle T_t\rangle$ is quadratic in $I_1 - V_t$.

(iii) To estimate the effect of $(H+\mu/2)^{-1}$ in the last term, recall from §4.3 that H has no bound state, so $H + \mu/2 \geq 0$. Indeed, H dominated a certain one-particle Hamiltonian with no bound states:

$$H + \frac{\mu}{2} \geq \left[\frac{|\mathbf{p}_2|^2}{2} - \frac{1}{r_2} + V_L(r_2)\right] P + \left(\frac{3\mu}{8} - \frac{1}{2\gamma} + \frac{1-\gamma}{2}|\mathbf{p}_2|^2\right)$$
$$\times(\mathbf{1} - P) = H_L,$$
$$P = |\varphi\rangle\langle\varphi| \otimes \mathbf{1},$$

where we used $|\mathbf{p}|^2/2 - 1/r \geq -1/2\gamma + (1-\gamma)|\mathbf{p}|^2/2$ for $0 < \gamma < 1$ in the last term. Let us finally choose γ to optimize the bound. If $a \geq b > 0$, then $a^{-1} \leq b^{-1}$ (2.2.38; 11), so we arrive at the upper bound

$$\left(H + \frac{\mu}{2}\right)^{-1} \leq H_L^{-1} = P\left(\frac{|\mathbf{p}_2|^2}{2} - \frac{1}{r_2} + V_L(r_2)\right)^{-1}$$
$$+(\mathbf{1} - P)\left(\frac{1-\gamma}{2}|\mathbf{p}_2|^2 - \frac{1}{2\gamma} + \frac{3\mu}{8}\right)^{-1}.$$

Zero would be a trivial lower bound for $(H + \mu/2)^{-1}$, but it is not good enough to assess how much difference the correction to T_t makes. For this purpose we can use the projection methods of (3.5.29), and take P' in $(H + \mu/2)^{-1} \geq P'(P'(H+\mu/2)P')^{-1} \cdot P'$ as the projection $|\varphi'\rangle\langle\varphi'| \otimes 1$ with some φ'. Then

$$P'\left(\frac{|\mathbf{p}_2|^2}{2} - \frac{1}{r_1} + \frac{1}{r_{12}}\right)P' \leq P'\frac{|\mathbf{p}_2|^2}{2},$$

since

$$\int \frac{d^3x' \rho(\mathbf{x}')}{|\mathbf{x} - \mathbf{x}'|} \leq \frac{1}{r}$$

for any spherically symmetric ρ. Let $P'(|\mathbf{p}_1|^2/2\mu - 1/r_1)P'$ be $\varepsilon_1 P'$. The not exactly known resolvent is thus finally bounded by the following one-particle operators:

$$\frac{2P'}{|\mathbf{p}_2|^2 + \kappa'^2} \leq \frac{1}{H + \mu/2} \leq \frac{P}{|\mathbf{p}_2|^2/2 - 1/r_2 + V_L(r_2)} + \frac{2}{1-\gamma}\frac{\mathbf{1} - P}{|\mathbf{p}_2|^2 + \kappa^2},$$

$$\kappa'^2 = \mu + 2\varepsilon_1, \qquad \kappa^2 = \frac{2}{1-\gamma}\left[\frac{3\mu}{8} - \frac{1}{2\gamma}\right].$$

Since V_t has been chosen so that $P(I_1 - V_t)P = 0$, only the last term on the right contributes to T, and so the bounds on the scattering length are

$$\begin{aligned}\langle\psi|T_t|\psi\rangle - \|R(I_1 - V_t)V_t^{-1}T_t\psi\|^2 &\leq \langle\psi|T|\psi\rangle \\ &\leq \langle\psi|T_t|\psi\rangle - \|P'R'(I_1 - V_t)V_t^{-1}T_t\psi\|^2,\end{aligned}$$

$$R = \left[\frac{2/(\gamma-1)}{|\mathbf{p}_2|^2 + \kappa^2}\right]^{1/2}, \qquad R' = \left[\frac{2}{|\mathbf{p}_2|^2 + \kappa'^2}\right]^{1/2}, \qquad \psi = \varphi \otimes 1.$$

The term

$$V_t^{-1}T_t\psi = \varphi(\mathbf{x}_1) \otimes (1 + \tfrac{1}{2}|\mathbf{p}_2|^{-2}V_t(\mathbf{x}_2))^{-1} \cdot 1 \equiv \varphi(\mathbf{x}_1) \otimes \psi_t(\mathbf{x}_2)$$

factorizes as (ground state of the muon) $\otimes$ (scattering wave-function of the electron). The second factor can be evaluated by numerical integration of the radial Schrödinger equation, but once more it is hardly worth the trouble, since the Born approximation $\psi_t \simeq 1$ is quite good. Specifically, the norm of $|\mathbf{p}|^{-2}V_t/2$ in L^∞ is

$$\|\tfrac{1}{2}|\mathbf{p}|^{-2}V_t\|_\infty = \tfrac{1}{8\pi}\sup_{\mathbf{x}}\int\tfrac{d^3x' V_t(\mathbf{x}')}{|\mathbf{x}-\mathbf{x}'|} = \tfrac{3}{8\mu},$$

so $\|\psi_t - 1\|_\infty \leq \|\frac{1}{2}|\mathbf{p}|^{-2}V_t\|_\infty = 0.001$. As a consequence, ψ_t can be replaced with 1 in the following integrals to an accuracy on the order of $\frac{0}{00}$, after which they are all elementary by Fourier transformation:

$$\begin{aligned}\|R(I_1 - V_t)\psi_t\|^2 &= \int d^3x_1\, d^3x_2\, d^3x_2' \frac{\exp(-\kappa|\mathbf{x}_2 - \mathbf{x}_2'|)}{2\pi(1-\gamma)|\mathbf{x}_2 - \mathbf{x}_2'|}\psi_t(\mathbf{x}_2) \\ &\times \left(-\frac{1}{|\mathbf{x}_2|} + \frac{1}{|\mathbf{x}_2 - \mathbf{x}_1|} - V_t(\mathbf{x}_2)\right) \\ &\left(-\frac{1}{|\mathbf{x}_2'|} + \frac{1}{|\mathbf{x}_2' - \mathbf{x}_1|} - V_t(\mathbf{x}_2')\right)\psi_t(\mathbf{x}_2')|\varphi(\mathbf{x}_1)|^2 \\ &\overset{\psi_t \to 1}{\cong} \int \frac{d^3k}{(2\pi)^3}\frac{\kappa^2}{|\mathbf{k}|^2 + \kappa^2}\left[2 - \frac{2}{(1 + |\mathbf{k}|^2/4\mu^2)^2}\right. \\ &\left. - \left(1 - \frac{1}{(1 + |\mathbf{k}|^2/4\mu^2)^2}\right)^2\right]\frac{(4\pi)^2}{|\mathbf{k}|^4}\frac{2}{1-\gamma}\end{aligned}$$

$$= \frac{2}{1-\gamma}\int \frac{d^3k}{(2\pi)^3}\frac{\kappa^2}{|\mathbf{k}|^2+\kappa^2}\frac{(4\pi)^2}{|\mathbf{k}|^4}\left[1-\frac{1}{(1+|\mathbf{k}|^2/4\mu^2)^4}\right].$$

The corresponding contribution to the upper bound is

$$\int d^3x_1 d^3x_1' d^3x_2 d^3x_2' \varphi'(\mathbf{x}_1)\varphi(\mathbf{x}_1)\left(-\frac{1}{|\mathbf{x}_2|}+\frac{1}{|\mathbf{x}_1-\mathbf{x}_2|}-V_t(\mathbf{x}_2)\right)$$

$$\cdot\frac{\exp(-\kappa'|\mathbf{x}_2-\mathbf{x}_2'|)}{2\pi|\mathbf{x}_2-\mathbf{x}_2'|}\left(-\frac{1}{|\mathbf{x}_2'|}+\frac{1}{|\mathbf{x}_1'-\mathbf{x}_2'|}-V_t(\mathbf{x}_2')\right)\varphi'(\mathbf{x}_1')\varphi(\mathbf{x}_1').$$

If φ' is now chosen as $(1+\alpha r)\exp(-\beta r)$ and the bounds are optimized in α, β and γ, then there result the following

4.4.14 Bounds for the Scattering Length of $e^- - (p\mu)$

The scattering length $a = \langle\psi|T\psi\rangle/4\pi$ satisfies

$$-0.835\times 10^{-13}\text{cm} = a_t - \frac{4}{\sqrt{3}}\mu^{-5/2}\left(1-\frac{2}{\sqrt{3\mu}}\right)^{-1}$$
$$\leq a \leq -0.62\times 10^{-13}\text{cm}.$$

4.4.15 Remarks

1. Although muonium μ^-p is about $r_b/\mu \sim 2.4\times 10^{-11}$ cm. across, the scattering length is smaller by a factor μ^{-1}: The kinetic energy prevents the wave-function from reacting much to a short-range potential. This explains why muonium diffuses through matter without significant interaction with the electrons.
2. The deviation from a_t can be interpreted as a virtual excitation of the muon. The lower bound shows that the ratio between the correction to a_t and a_t goes to 0 as $\mu^{-1/2}$ when $\mu\to\infty$. It is still on the order of 10% for the muon.
3. If $k > 0$,then $(H-E)^{-1}$ is no longer positive, and the resolvent must be analyzed more carefully.

4.4.16 Problems

1. Prove that $\delta(|\mathbf{p}_1|^2/2\mu - 1/r_1 + |\mathbf{p}_2|^2/2-\omega) = |\varphi\rangle\langle\varphi|\otimes\delta(|\mathbf{p}_2|^2/2-\omega-\mu/2)$ for $\omega < 3\mu/8$ (see the proof of (4.4.10)).

2. Show that

$$K = \sqrt{I_1}\left(\frac{|\mathbf{p}_1|^2+|\mathbf{p}_2|^2-z}{2}\right)^{-1}\frac{1}{r_1}|\varphi(\mathbf{x}_1)\rangle\langle\varphi(\mathbf{x}_1)|\otimes\delta(|\mathbf{p}_2|^2-\omega)\sqrt{I_1}$$

belongs to C_4 if $z<0$.

3. Show that if $I(z)$ is an analytic family of operators on a connected, open region G and $I(z)$ is compact in a neighborhood of $z_0\in G$, then $I(z)$ is compact for all $z\in G$. (Hint:

the series expansion

$$\sum_{n=0}^{\infty} \frac{(z-z_0)^n}{n!} I^{(n)}(z_0)$$

is convergent in norm on a sufficiently small neighborhood of z_0. Show that all $I^{(n)}(z_0)$ are compact and use the method of overlapping discs described, for example, in [17].

4.4.17 Solutions

1. Use $\delta(a \otimes \mathbf{1} + \mathbf{1} \otimes b) = \int_{-\infty}^{\infty} d\alpha\, \delta(a-\alpha) \otimes \delta(\alpha + b)$:

$$\int_{-\infty}^{\infty} d\alpha\, \delta\left(\frac{|\mathbf{p}_1|^2}{2\mu} - \frac{1}{r} - \alpha\right) \otimes \delta\left(\frac{|\mathbf{p}_2|^2}{2} - \omega + \alpha\right) = \int_{-\infty}^{\omega} d\alpha\, \delta \otimes \delta$$

$$= |\varphi\rangle\langle\varphi| \otimes \delta\left(\frac{|\mathbf{p}_2|^2}{2} - \omega - \frac{\mu}{2}\right).$$

2. The operator is of the form

$$\sqrt{I_1(\mathbf{x}_1, \mathbf{x}_2)}|\psi(\mathbf{x}_1)\rangle\langle\varphi(\mathbf{x}_1)|\delta(|\mathbf{p}_2|^2 - \omega)\sqrt{I_1(\mathbf{x}_1, \mathbf{x}_2)},$$

where $\psi(\mathbf{x}_1) = ((|\mathbf{p}_1|^2 + \omega)/2 - z)^{-1}(1/r_1)\varphi(\mathbf{x}_1)$. If $z < 0$, then this function is as good as φ, since $\hat{u}(\mathbf{x}) = \int d^3x' |\psi(\mathbf{x}')|^2 |I_1(\mathbf{x}, \mathbf{x}')|$ has the same $1/r^2$ behavior as $u(\mathbf{x})$. Hence we conclude from (3.4.13; 2) that $\mathrm{Tr}(KK^*)^2 < \infty$.

3. $I^{(n)}(z_0) = (n!/2\pi i) \int_{C_0} (I(z)/(z-z_0)^{n+1}) dz$ is the norm-limit of Riemann sums of compact operators and therefore compact, provided that C_0 is a circle in the region of compactness of z_0. However, the power series converges on the largest circle about z_0 completely contained in the domain of analyticity G. Hence compactness can be continued to the whole region G in analogy with analytic continuation through overlapping discs.

4.5 Complex Atoms

The motion of the inner electrons in a large atom can be approximated to within a few percent by motion in an averaged field.

The first genuinely many-body problem that we shall discuss in much depth is that of an infinitely heavy atomic nucleus of charge Z and N electrons. The methods will be the same as those developed for helium, though the Pauli exclusion principle will now emerge as the decisive fact. Insofar as the interelectronic repulsion can be neglected, these ideas lead to the familiar shell structure of atoms, which is so evident in the periodic table of the elements. To date it has not been possible to determine whether electronic shells are also predicted by the Schrödinger equation including the effect of the electrons on one another. A derivation would require extraordinarily tight upper and lower bounds for the energy eigenvalues. It is easy to get upper bounds with variational procedures, elementary calculations providing about 10% accuracy and more elaborate ones attaining about 1%. The methods that have been presented for lower bounds break down, however, as the ground state is pushed well into the continuum of H_0 by the interelectronic repulsion, though lower bounds with an accuracy of a few percent are obtainable with another strategy. Unfortunately, the accuracy is with respect to the total energy, which goes as $Z^2 N^{1/3}$eV, amounting to MeV for large atoms, while the energy differences of importance for the shell structure and, for that matter, all of chemistry are only a few eV. The required precision is far too great, so we shall have to make do with the qualitative traits of the spectrum, which are due to various conserved quantities.

We follow the steps of §4.3.

4.5.1 The Normal Form H_N of the Hamiltonian

$$H = \frac{1}{2m} \sum_{i=1}^{N} |\mathbf{p}_i|^2 - Ze^2 \sum_{i=1}^{N} \frac{1}{|\mathbf{x}_i|} + e^2 \sum_{i>j} \frac{1}{|\mathbf{x}_i - \mathbf{x}_j|}$$

can be transformed to $Z^2 e^4 m H_N$, where

$$H_N(\alpha) = \sum_{i=1}^{N} \left(\frac{1}{2} |\mathbf{p}_i|^2 - \frac{1}{|\mathbf{x}_i|} \right) + \alpha \sum_{i>j} \frac{1}{|\mathbf{x}_i - \mathbf{x}_j|} = H_0 + \alpha H', \quad \alpha = \frac{1}{Z},$$

by a dilatation of the coordinates.

4.5.2 Remarks

1. The potential energy is once again ε-bounded relative to the kinetic energy, and H is self-adjoint on the domain of the kinetic energy and bounded from below.
2. The analysis of the resolvent in (4.4.1) can be adapted for the many-electron Hamiltonian. All the connected graphs in the expansion of the resolvent are compact, and the noncompact parts correspond to distributions of the nucleus and electrons into groups within which the particles interact, but which are not bound together. The clusters determine where the essential spectrum of the overall Hamiltonian begins, and if $\alpha > 0$, Theorem (4.4.3) generalizes:

4.5.3 The HVZ Theorem

$$\sigma_{\text{ess}}(H_N(\alpha)) = \cup_{M=1}^{N-1} \text{Sp}(H_M(\alpha)).$$

The point spectrum also resembles that of (4.3.9) and (4.3.18); the effect of dilatations on the Hamiltonian implies the

4.5.4 Concavity of the Ground-State Energy

The functions $E_1(\alpha)$ and even $-\sqrt{-E_1(\alpha)}$ are concave in α.

4.5.5 The Virial Theorem

If $(H(\alpha) - E)\psi = 0$, then

$$E = -\langle\psi|\sum_{i=1}^{N}\frac{|\mathbf{p}_i|^2}{2}|\psi\rangle = \frac{1}{2}\langle\psi| -\sum_{i=1}^{N}\frac{1}{|\mathbf{x}_i|} + \alpha\sum_{i>j}\frac{1}{|\mathbf{x}_i - \mathbf{x}_j|}|\psi\rangle.$$

4.5.6 Remarks

1. As before, it follows that there are no eigenvalues $E \geq 0$.
2. If $E(\alpha)$ is known, then (4.5.5) together with

$$\frac{\partial E}{\partial \alpha} = \langle\psi|\sum_{i>j}\frac{1}{|\mathbf{x}_i - \mathbf{x}_j|}|\psi\rangle$$

allows the expectation values of $1/|\mathbf{x}_i|$ to be determined, too.

It is furthermore true that positive ions and atoms have infinitely many bound states. Because of the symmetry requirements on the wave-function the proof is more difficult than for helium.

4.5.7 The Infinitude of the Point Spectrum

If $\alpha < 1/(N-1)$, then $H_N(\alpha)$ has an infinite point spectrum.

Proof: We follow the time-honored recipe of the proof of (4.3.7), but this time the exclusion principle comes into play. To take care of the symmetry, assume a trial function

$$\Psi_{n,\tau}(\mathbf{x}_1, \ldots, \mathbf{x}_N) = \mathcal{N}\sum_{j=1}^{N}(-1)^j\varphi_n(r_j\tau)\chi(\mathbf{x}_1, \ldots, \hat{\mathbf{x}}_j, \ldots, \mathbf{x}_N),$$

$$\tau \in \mathbb{R}^+, \quad r_j \equiv |\mathbf{x}_j|,$$

where χ is the antisymmetrized, normalized ground state of H_{N-1}, with energy E_{N-1}. The symbol $\hat{\mathbf{x}}_j$ is used to indicate the absence of the coordinate $\mathbf{x}_j$, which stands for both the spatial and the spin coordinates of the j-th particle. For the present we assume only that $\varphi \in C^\infty$ and $\varphi_n(r) \neq 0$ only for $n < r < n+1$, $n = 1, 2, \ldots$. The $\mathcal{N}$ is a normalization constant. Since H is real, we may suppose that φ and χ are real:

$$\mathcal{N}^2\sum_{j=1}^{N}(-1)^j\sum_{k=1}^{N}(-1)^k\int d^3x_1\cdots d^3x_N\varphi_n(\tau r_j)\varphi_n(\tau r_k)\chi(\mathbf{x}_1\cdots\hat{\mathbf{x}}_j\cdots\mathbf{x}_N)$$
$$\times\chi(\mathbf{x}_1\cdots\hat{\mathbf{x}}_k\cdots\mathbf{x}_N) = 1.$$

The mixed terms like

$$\int d^3x_1\cdots d^3x_N\varphi_n(\tau r_1)\varphi_n(\tau r_2)\chi(\mathbf{x}_2\cdots\mathbf{x}_N)\chi(\mathbf{x}_1, \mathbf{x}_3, \ldots, \mathbf{x}_N)$$

cause trouble. However, we know that the functions χ fall off exponentially, as the argument leading to (4.7) does not depend on the number of particles: consider the group generated by $\exp(isr_1)$; by analytic continuation, for some $\delta > 0$, $\exp(sr_1)\chi(x_1, \ldots, x_{N-1}) \in \mathcal{H}$ for all s, $|s| < \delta$. It then follows that the mixed term is $O(\exp(-2\tau\delta))$ as $\tau \to \infty$, because $\| \exp(-sr)\varphi_n(r\tau)\| \leq \exp(-ns\tau)$, and so $\mathcal{N}^{-2} = N\tau^{-3} + O(\exp(-2\tau\delta))$. If $\langle\psi|H_N\psi\rangle$ is calculated next, there are some more mixed terms:

$$O(\exp(-2\tau\delta)) + \langle\varphi|\frac{|\mathbf{p}|^2}{2} - \frac{1}{r} + \alpha V(r)|\varphi\rangle + E_{N-1}(\alpha),$$

$$V(r_1) = (N-1)\int d^3x_2 \cdots d^3x_N \frac{1}{r_{12}}|\chi(\mathbf{x}_2 \cdots \mathbf{x}_N)|^2.$$

Because of the exponential fall-off, $V(r) = (N-1)/r + O(\exp(-2r\delta))$ as $r \to \infty$. All told, if τ is sufficiently large and $\alpha < 1/(N-1)$, then

$$\begin{aligned}\langle\Psi_{n,\tau}|H_N|\Psi_{n,\tau}\rangle &= E_{N-1} + \tau^{-2}\langle\varphi_n|\frac{|\mathbf{p}|^2}{2}\varphi_n\rangle - (1-\alpha(N-1)) \\ &\quad \times \langle\varphi_n|\frac{1}{r}|\varphi_n\rangle\tau^{-1} + O(\exp(-2\tau\delta)) < E_{N-1}.\end{aligned}$$

Although the functions Ψ_n are not orthogonal, it is still true that $\langle\Psi_n|\Psi_{n'}\rangle = \delta_{n,n'} + O(\exp(-2\tau\delta))$. They can thus be orthogonalized in the limit $\tau \to \infty$ without affecting the inequality $\langle H_N\rangle \leq E_{N-1}$. The off-diagonal matrix elements $\langle\Psi_{n,\tau}|H_N|\Psi_{n',\tau}\rangle$ are all likewise $O(\exp(-2\tau\delta))$. If τ is sufficiently large, then for any $n \in \mathbb{Z}^+$ there is an n-dimensional subspace on which all eigenvalues of H_N lie below E_{N-1}, which proves (4.5.7). □

4.5.8 Bound States Within the Continuum

The continuous spectrum between E_{N-1} and 0 usually has embedded eigenvalues, of states whose decay is forbidden by conservation laws. Consider as an example lithium, $N = 3$. The ground state is the $(1s)^2(2s)$ configuration ${}^2S^+$. The lowest state with spin $\frac{3}{2}$ is $(1s)(2s)(2p)$, i.e., ${}^4P^-$, which is well within the continuum of the ${}^2S^+$ states $(1s)^2(\infty s)$. Moreover, the states of unnatural parity ${}^4P^+$ of the $(1s)(2p)^2$ configuration are within the continuum of the ${}^2S^+$ states. (The notation is ${}^{(2S+1)}L^P$.)

It would take us too far afield to go into great detail about the host of energy levels. To gain an overview of the dependence of the energy on N, it is most convenient to investigate the ground states of atoms with full shells. They are milestones in the periodic table, and their theoretical analysis is relatively straightforward.

4.5.9 Upper Bounds for $E_1(\alpha)$ Using Eigenfunctions of $H(0)$

The most primitive estimate is the expectation value of H in the ground state of $H(0)$, which is easy to find; the Balmer levels are simply filled up with the available electrons in accordance with the exclusion principle. If the states with quantum numbers (n, l, l_3, s) are enumerated with the index j, then the wave-function of the N noninteracting particles is

$$\Psi(\mathbf{x}_1 \cdots \mathbf{x}_N) = \frac{1}{\sqrt{N!}} \sum_p (-1)^p \prod_{j=1}^N \psi_j(\mathbf{x}_{p_j}), \tag{4.21}$$

where (p_j) is a permutation of $1 \cdots N$. Since the levels of energy $-1/2n^2$ are $2n^2$-fold degenerate, the eigenvalue of $H(0)$ with this eigenfunction and filled n-shells for $n \geq n_0$ is

$$E = -\frac{1}{2} \sum_{n=1}^{n_0} \frac{2n^2}{n^2} = -n_0,$$

where

$$N = 2 \sum_{n=1}^{n_0} n^2 = \frac{2n_0^3 + 3n_0^2 + n_0}{3}$$

(see Problem (4.5.19)). The calculation of $\langle \psi | H | \psi \rangle$ requires in addition the evaluation

$$\begin{aligned} \langle \Psi | H' | \Psi \rangle &= \frac{N(N-1)}{2} \langle \Psi | \frac{1}{r_{12}} | \Psi \rangle \\ &= \frac{1}{2} \int \frac{d^3x_1 d^3x_2}{|\mathbf{x}_1 - \mathbf{x}_2|} \sum_{j,j'} (|\psi_j(\mathbf{x}_1)|^2 |\psi_{j'}(\mathbf{x}_2)|^2 \\ &\quad - \psi_j(\mathbf{x}_1)\psi_{j'}(\mathbf{x}_2)\psi_{j'}^*(\mathbf{x}_1)\psi_j^*(\mathbf{x}_2)). \end{aligned} \tag{4.22}$$

The latter part of this, known as the **exchange term**, only occurs if ψ_j and $\psi_{j'}$ have the same spin, since the scalar product includes the scalar product in spin space as a factor. The exchange term is ≤ 0 for all j and j':

$$\int \frac{d^3x_1 d^3x_2}{|\mathbf{x}_1 - \mathbf{x}_2|} \rho^*(\mathbf{x}_1)\rho(\mathbf{x}_2) \geq 0$$

holds because the Fourier transform $1/2\pi^2 k^2$ of $1/|\mathbf{x}|$ is positive.

The exchange term was calculated for the simplest states in §4.3 and was about 10% as large as the first term. If it is ignored when $j \neq j'$, then at any rate there results an upper bound. The remaining term simplifies for closed L-shells when it is noted that

$$\frac{1}{|\mathbf{x}_1 - \mathbf{x}_2|} = \frac{1}{r_2} \sum_{l=0}^{\infty} \sum_{m=-l}^{l} \left(\frac{r_1}{r_2}\right)^l \frac{4\pi}{2l+1} Y_l^m(\Omega_1) Y_l^{-m}(\Omega_2) \tag{4.23}$$

for $r_2 > r_1$, and otherwise $r_1 \leftrightarrow r_2$. Since $\sum_{m=-l}^{l} |Y_l^m(\Omega)|^2$ is independent of the angles, it is clear that the first contribution to the sum $\sum_{j,j'}$ is spherically symmetric, and when the angular integration is done, the term with $l \neq 0$ in the decomposition (4.23) cancels out:

$$\langle\Psi|H'|\Psi\rangle \leq \frac{1}{2}\int d^3x_1 d^3x_2 \sum_{j,j'} |\psi_j(\mathbf{x}_1)|^2 |\psi_{j'}(\mathbf{x}_2)|^2 \times \left(\frac{\Theta(r_1-r_2)}{r_1}+\frac{\Theta(r_2-r_1)}{r_2}\right)$$

$$\leq \frac{1}{2}\sum_{j\neq j'} \min\left\{\langle\psi_j|\frac{1}{r}|\psi_j\rangle, \langle\psi_{j'}|\frac{1}{r}\psi_{j'}\rangle\right\}. \tag{4.24}$$

Because of the virial theorem, $\langle\psi_j|1/r|\psi_j\rangle = 1/n^2$, $j=(n,l,l_3,s)$, so

$$\langle\Psi|H'|\Psi\rangle \leq \frac{1}{2}\sum_{n,n'=1}^{n_0} 2n^2 2n'^2 \min\left\{\frac{1}{n^2},\frac{1}{n'^2}\right\} = \sum_{n=1}^{n_0} 4\sum_{n'=1}^{n} n'^2$$

$$= \frac{1}{3}[n_0^4 + 4n_0^3 + 5n_0^2 + 2n_0]$$

(see Problem (4.5.19)). The net result is that

$$\langle\Psi|H|\Psi\rangle \leq -n_0 + \frac{\alpha}{3}[n_0^4 + 4n_0^3 + 5n_0^2 + 2n_0].$$

As $N \to \infty$, $n_0 = (3N/2)^{1/3} + O(1)$, and if $\alpha = O(1/N)$, then

$$E \leq -N^{1/3}\left(\frac{3}{2}\right)^{1/3}\left(1-\frac{\alpha N}{2}\right) + O(1). \tag{4.25}$$

Because of (4.5.4), Equation (4.24) can be bettered with a parabolic bound, as before, and since $H' > 0$, we obtain

4.5.10 Rough Bounds for E_N for Neutral Atoms

If $\alpha N = N/Z = 1$, then to $O(N^{-1/3})$ the lowest eigenvalue E_N of H_N is

$$-\left(\frac{3}{2}\right)^{1/3} = -1.145 \leq \frac{E_N}{N^{1/3}} \leq -\left(\frac{3}{2}\right)^{1/3}\left(1-\frac{1}{4}\right)^2 = -0.6439.$$

4.5.11 Remarks

1. This eigenvalue E_N is $Z^{-2}e^{-4}m^{-1}$ times the eigenvalue of the H of (4.5.1), so the dependence on N and Z anticipated in (1.2.11) is verified.
2. An estimate $O(N^{-1/3})$ is useless for actual atoms, since the corrections neglected in (4.25) with the replacement of $\sum_{n'=1}^{n}(n')^2$ by $n^3/3$ are still about 50% for $n_0 = 10$, which makes $N = 770$.
3. The estimate (4.24) is not very good for electrons within the same shell. For instance, the right side is 1 for $(1s)^2$, while in reality the left side is only $\frac{5}{8}$ (see (4.3.21; 1)). However, it is pretty accurate for electrons in different shells; for

$(1s)(2s)$ it gives 0.25 instead of 0.2318. In all the error is not so bad for large atoms, since it is the interaction between shells that causes E_N to be $\sim N^{1/3}$.

4. If α is arbitrary, then the upper bound is $-\left(\frac{3}{2}\right)^{1/3} N^{1/3}(1-\alpha N/4)^2$, which has the maximum $\left(\frac{6}{7}\right)^{7/3} \alpha^{-1/3}$ when $N\alpha = \frac{4}{7}$. Thus a more favorable trial function when $N > 4Z/7$ is the one having all additional electrons at infinity with energy 0; with it, $E/N^{1/3} \leq -\left(\frac{6}{7}\right)^{7/3} = -0.6978$. According to Thomas-Fermi theory (see volume IV) the correct asymptotic value is -0.77.

4.5.12 $E_1(\alpha)$ Estimated with Trial Functions Having Two Parameters

In the effort to improve eigenvalue bounds, we recall that the parabolic bonds corresponded to the use of trial functions $\exp(id\tau)\Psi$ with an optimized dilatation parameter τ. This takes the partial screening of the field of the nucleus into account, which is clearly most significant for the external electrons. It would thus make more sense to stretch the different ψ_j with different τ_j. Although this would ruin the orthogonality of the ψ_j, which is essential for the calculation, it is still compatible with orthogonality to dilate the ψ's with different l independently. In order to have more flexibility in the choice of trial functions, let us use eigenfunctions of a Hamiltonian containing an additional $1/r^2$ potential, since the eigenvalues are known in that case. If the angular momentum is l, we shall take

$$H_l = \frac{p_r^* p_r}{2} + \frac{(l+\delta_l)(1+\delta_l+1)}{r^2} - \frac{\tau_l}{r},$$

which has eigenvalues

$$E_{n_r,l} = -\frac{\tau_l^2}{2}(n_r + l + \delta_l + 1)^{-2}.$$

Later, τ_l and δ_l will be optimized. The expectation values $1/r$ and $1/r^2$, and therefore of $|\mathbf{p}|^2$ can be calculated by taking derivatives of E by τ and, respectively, δ: with $n = n_r + l + 1$,

$$\left\langle \frac{1}{r} \right\rangle = \tau_l (n+\delta_l)^{-2},$$

$$\langle |\mathbf{p}|^2 \rangle = \tau_l^2 (n+\delta_l)^{-3} \left(n - \frac{\delta_l (l+\frac{1}{2})}{\delta_l + l + \frac{1}{2}} \right).$$

To fill the shells in order, the sets of quantum numbers (n, l, l_3, s) will be enumerated with an index j. Each l-shell must be filled with $2l+1$ or $2(2l+1)$ electrons to be spherically symmetric. In that case,

$$\langle H \rangle = \sum_l \sum_n \nu(n,l) \left\{ \frac{\tau_l^2}{2}(n+\delta_l)^{-3} \left[n - \frac{\delta_l (l+\frac{1}{2})}{\delta_l + l + \frac{1}{2}} \right] \right.$$

$$-\frac{\tau_l}{(n+\delta_l)^2}(1-\alpha N(n,l))\Bigg\},$$

$$N(n,l)=\sum_{j'<j}\nu(n(j),l(j')),$$

where $\nu(n,l)$ is the occupation number, and δ_l is allowed to vary only as long as $n+\delta_l$ is monotonic in j. The optimal value of $\langle H\rangle$ as a function of τ is attained when

$$\tau_l=\sum_n \nu(n,l)\frac{1-\alpha N(n,l)}{(n+\delta_l)^2}\left[\sum_{n'}\frac{\nu(n',l)}{(n+\delta_l)^3}\left(n'-\frac{\delta_l(l+\frac{1}{2})}{\delta_l+l+\frac{1}{2}}\right)\right]^{-1},$$

i.e.,

$$\langle H\rangle=-\frac{1}{2}\sum_l\left\{\sum_n \nu(n,l)\frac{1-\alpha N(n,l)}{(n+\delta_l)^2}\right\}^2\left\{\sum_{n'}\frac{\nu(n',l)}{(n+\delta_l)^3}\left(n'-\frac{\delta_l(l+\frac{1}{2})}{\delta_l+l+\frac{1}{2}}\right)\right\}^{-1} \tag{4.26}$$

The optimization by δ_l has to be done on a computer, and improves the result of our earlier ansatz that $\delta_l=0$ by several percent. The table given below contains calculations of energies for several typical atoms, along with the screening constants and the δ_l. Note that because of (4.24), Formula (4.26) is still a strict upper bound even for a spherical configuration plus one more electron.

Z	configuration of non-filled n-shells	δ_0 / τ_0	δ_1 / τ_1	δ_2 / τ_2	δ_3 / τ_3	δ_4 / τ_4	δ_5 / τ_5	$EN^{-7/3}$ by (4.26)	$EN^{-7/3}$ by Hartree-Fock
		−0.03	−0.39						
10	$2s^2p^33p^3$			–	–	–	–	−0.524	−0.594
		0.86	0.28						
		−0.01	0	−0.82					
20	$3d^54d^5$				–	–	–	−0.562	−0.620
		0.94	0.69	0.17					
		−0.01	−0.11	0	0				
40	$3p^3d^{10}4f^{14}$					–	–	−0.591	−0.646
		0.96	0.72	0.48	0.19				
		−0.02	−0.09	−0.19	0	0	0		
60	$4d^5f^75g^96h^{11}$							−0.607	−0.648
		0.94	0.78	0.53	0.40	0.27	0.10		
		−0.02	−0.17	−0.59	−0.86	0			
80	$4s^1p^6d^{10}f^{14}5d^5f^7g^9$						–	−0.613	−0.656
		0.94	0.73	0.41	0.20	0.06			

Remarkably, this simple analytical ansatz arrives nearly within 10% of the correct values. The road to greater precision is rocky, but the trip is made easier by the reasonable assumption of a

4.5.13 Self-Consistent Field

Hartree and Fock discovered one of the favorite methods for inventing good trial functions. The first step is to take the infimum of $\langle\Psi|H\Psi\rangle$ when Ψ is a product or determinantal trial function (4.21). The question immediately arises of whether the infimum is actually a minimum, i.e., whether there are minimizing ψ_j. If so, they satisfy the appropriate variational equations,

$$(H_\Psi\varphi_i)(\mathbf{x}) = \sum_{j=1}^{N} e_{ij}\varphi_j(\mathbf{x}), \qquad i = 1,\dots N,$$

$$(H_\Psi\varphi)(\mathbf{x}) = \left\{-\frac{\Delta}{2} - \frac{1}{|\mathbf{x}|} + \alpha U_\Psi(\mathbf{x})\right\}\varphi(\mathbf{x}) - \alpha(K_\psi\varphi)(\mathbf{x}),$$

$$U_\Psi(\mathbf{x}) = \sum_{i=1}^{N}\int d^3x' \frac{|\varphi_i(\mathbf{x}')|^2}{|\mathbf{x}-\mathbf{x}|},$$

$$(K_\psi\varphi)(\mathbf{x}) = \sum_{i=1}^{N}\varphi_i(\mathbf{x})\int d^3x'\varphi(\mathbf{x}')\varphi_i^*(\mathbf{x}')\frac{1}{|\mathbf{x}-\mathbf{x}'|};$$

where e_{ij} are the Lagrange multipliers due to the orthonormalization. Since H_Ψ is self-adjoint, e_{ij} is a Hermitian matrix, and can thus be put into diagonal form $e_{ij} = \delta_{ij}e_i$. Hence we look for the N lowest eigenvalues of H_Ψ. If $\alpha \leq 1/N$ (positive ions and neutral atoms), then, as has recently been proved in [18], minimizing solutions exist, but an analytic solution of this system of nonlinear equations is hardly to be hoped for. An iterative procedure on a computer is called for, and even if this also fails to find exact solutions, $\langle\psi|H\psi\rangle$ is at least still an upper bound for the lowest eigenvalue. The procedure is unfortunately very cumbersome, especially because of the exchange terms $K\varphi$. The upper bounds obtained are also shown in the table for comparison. It is assumed that these values are within a few percent of the exact values. Granting that accuracy, the product ansatz of (4.21) is not at all bad, which suggests looking for lower bounds by using effective one-particle potentials approaching the Coulombic repulsion of the electrons from below. The principal theoretical shortcoming so far is the lack of a good lower bound; great refinement of the upper bound does not reduce the theoretical error much. The methods used in §4.3 are of no value because of the huge number of levels of $H(0)$. Some help is provided by

4.5.14 Lower Bound for Interactions of Positive Type

Let $V(x)$ be a potential of positive type (i.e., with a Fourier transform $\tilde{V}(k) \geq 0$, $V(0) < \infty$), and suppose $\Phi(\mathbf{x}) \in L^1(\mathbb{R}^3) \cap L^\infty(\mathbb{R}^3)$, so that $\tilde{\Phi}(\mathbf{k})$ exists. Then

$$\sum_{n>m} V(\mathbf{x}_n - \mathbf{x}_m) \geq \sum_{n=1}^{N}\Phi(\mathbf{x}_n) - \frac{1}{2}\int\frac{d^3k}{(2\pi)^3}\left[\frac{|\tilde{\Phi}(\mathbf{k})|^2}{\tilde{V}(\mathbf{k})} + N\tilde{V}(k)\right].$$

Proof: It follows from

$$0 \leq \int d^3x d^3x' \left(\sum_{n=1}^{N} \delta(\mathbf{x}-\mathbf{x}_n) - \rho(\mathbf{x})\right) V(\mathbf{x}-\mathbf{x}') \left(\sum_{m=1}^{N} \delta(\mathbf{x}'-\mathbf{x}_m) - \rho(\mathbf{x}')\right)$$

that

$$\sum_{n>m} V(\mathbf{x}_n - \mathbf{x}_m) \geq \sum_{n=1}^{N} \int d^3x \rho(\mathbf{x}) V(\mathbf{x}-\mathbf{x}_n) - \frac{N}{2} V(0)$$
$$- \frac{1}{2} \int d^3x\, d^3x' \rho(\mathbf{x})\rho(\mathbf{x}') V(\mathbf{x}-\mathbf{x}').$$

Now set $\Phi(\mathbf{x}) = \int d^3x' \rho(\mathbf{x}') V(\mathbf{x}-\mathbf{x}')$; the proposition then follows by Fourier transformation. □

4.5.15 Remarks

1. The significance of (4.5.14) is that a pair interaction is bounded by an effective one-particle potential Φ.
2. The degree to which Φ is arbitrary may seem surprising. A poor choice of Φ, however, renders (4.5.14) a triviality, saying that something positive is greater than something negative.

4.5.16 The Atomic Potential Bounded Below

We have seen that a V of positive type can be estimated from below with arbitrary one-particle potentials, at the cost of a constant. Since the Coulomb potential becomes infinite at 0, it is first necessary to find a smaller function finite at 0 and with a positive Fourier transformation;

$$V(\mathbf{x}) \equiv \frac{1-\exp(-\mu r)}{r} \leq \frac{1}{r}, \quad \tilde{V}(k) = \frac{4\pi\mu^2}{k^2(k^2+\mu^2)}, \tag{4.27}$$

will do. The effective repulsive potential

$$\Phi(\mathbf{x}) = \int d^3x' \frac{n(\mathbf{x}')}{|\mathbf{x}-\mathbf{x}|},$$

where $n(\mathbf{x})$ is the electron density, should be fairly realistic and lead to the best result. Then (transforming back to x-space),

$$H' \geq \sum_{n=1}^{N} \Phi(\mathbf{x}_n) - \frac{1}{2}\left\{\int d^3x\, \Phi(\mathbf{x}) n(\mathbf{x}) + 4\pi\mu^{-2} \int d^3x\, n(\mathbf{x})^2 + N\mu\right\}. \tag{4.28}$$

Since μ appears only in the constants, it can be optimized immediately, so we may set

$$\mu = \left[\frac{8\pi}{N} \int d^3x\, n(\mathbf{x})^2\right]^{1/3}.$$

As an analytically convenient approximation to $n(\mathbf{x})$ we use the semiempirical expression (for $N = Z$),

$$\Phi(\mathbf{x}) = N \frac{r + 2(9/2N)^{1/3}}{[r + (9/2N)^{1/3}]^2}, \quad n(\mathbf{x}) = \frac{N^{2/3}}{4\pi r} \frac{6(9/2N)^{1/3}}{[r + (9/2N)^{1/3}]^4},$$

$$\frac{1}{2}\{\ \} = \frac{2}{5}\left(\frac{2}{9}\right)^{1/3} N^{7/3} + \frac{3}{2}\left(\frac{2}{7}\right)^{1/3} N^{5/3}. \tag{4.29}$$

Although the Schrödinger equation cannot be solved analytically with this potential, the solution of a radial equation poses no difficulty for a computer. The following one-particle levels are found for

$$H_N \equiv Z^{-2}\left[-\frac{\Delta}{2} - \frac{Z}{r} + \Phi_N(\mathbf{x}) - \frac{\{\ \}}{2N}\right]:$$

n	l	H_0	$\hat{H}_{10}$	$\hat{H}_{20}$	$\hat{H}_{40}$
1	0	−0.5	−0.395	−0.412	−0.436
2	0	−0.125	−0.112	−0.084	−0.083
2	1	−0.125	−0.103	−0.077	−0.079
3	0	−0.055	−0.097	−0.053	−0.036
3	1	−0.055	—	−0.051	−0.034
3	2	−0.055	—	−0.051	−0.031
4	0	−0.031	—	−0.051	−0.028
4	1	−0.031	—	—	−0.028
4	2	−0.031	—	—	−0.028

After these levels have been filled up, the resultant lower bounds compare with the Hartree-Fock upper bounds as follows:

N	$\leq$	$EZ^{-2}N^{-1/3}$	$\leq$
10	−0.761		−0.594
20	−0.730		−0.620
40	−0.715		−0.646
60	−0.712		−0.648
80	−0.711		−0.656

(4.30)

4.5.17 Remarks

1. This method works better for larger N. Indeed, one of the most important results of Thomas–Fermi theory (see volume IV) is that the product ansatz for the wave-functions in an averaged field becomes exact as $N \to \infty$.
2. The potential (4.29) guessed here is still not the best possible. If $n(\mathbf{x}) = c\exp(-1.56r)$, then the lower bound one gets is −0.698 for $Z = N = 36$.
3. In volume IV, (4.28) will be recognized as a special case of a family of inequalities of Thomas–Fermi theory. In this case all the inequalities are, however, numerically about equally accurate.

4. Unlike the upper bounds, these lower bounds work just as well for the individual excited states; the n-th eigenvalue of the lower-bound Hamiltonian lies below the true n-th eigenvalue.
5. The lower bounds depend critically on the form of Ψ as an antisymmetrized product, whereas the upper bounds can be improved significantly by the use of linear combinations of determinants.
6. The bounds that have been derived reveal that the asymptotic value -0.77 is approached from above, but that the exact value is still not very close to -0.77 when $N = 80$.
7. The relativistic effects are comparable to the theoretical errors for heavy elements. The experimental values lie within the bounds found here, after relativistic corrections.

As to the properties of the electron density $\rho(\mathbf{x})$, there is an immediate generalization of the upper bounds of §4.3 as $r \to \infty$ and $r = 0$ to the case of N electrons. For the purposes of a qualitative discussion it suffices to have

4.5.18 Bounds for $\langle r^\nu \rangle$

In §1.2 it was explained why the average value of r should go as $N^{-1/3}$. This does not mean that heavier atoms are smaller than lighter ones. What is perceived as the size of an atom is the diameter of the outermost electronic orbital, whereas the mean value of r is dominated by the dense interior electron-cloud. In order to see whether the conjecture of §1.2 is a rigorous consequence of quantum mechanics, let us find some bounds for $\langle r^\nu \rangle^{1/\nu}$. If $\nu = -1$, then the virial theorem yields

$$\left\langle \frac{1}{r} \right\rangle^{-1} \leq \frac{NZ}{2|E_N|},$$

so if $N = Z$ then (4.25),

$$\left\langle \frac{1}{r} \right\rangle^{-1} \leq N^{-1/3}(0.6978)^{-1}\frac{1}{2}(1 + O(N^{-1/3})).$$

On the other hand, for fermions with spin $\frac{1}{2}$,

$$\sum_{i=1}^{N} \left(\frac{|\mathbf{p}_i|^2}{2} - \beta \frac{1}{r_i} \right) \geq -\beta^2 N^{1/3} \left(\frac{3}{2} \right)^{1/3} (1 + O(N^{-1/3})), \qquad \beta > 0,$$

so

$$N \left\langle \frac{1}{r} \right\rangle \leq \frac{1}{\beta} \left\langle \sum_i \frac{|\mathbf{p}_i|^2}{2} \right\rangle + \beta \left(\frac{3N}{2} \right)^{1/3} (1 + O(N^{-1/3})).$$

If

$$\beta^2 = \left\langle \sum_i \frac{|\mathbf{p}_i|^2}{2} \right\rangle \left(\frac{3N}{2} \right)^{-1/3}$$

and

$$\sum_i \left\langle \frac{|\mathbf{p}_i|^2}{2} \right\rangle = |E_N| \leq \left(\frac{3}{2}\right)^{1/3} N^{7/3}(1 + O(N^{-1/3})),$$

then

$$\left\langle \frac{1}{r} \right\rangle^{-1} \geq N^{-1/3} \left(\frac{3}{2}\right)^{-1/3} \frac{1}{2}(1 + O(N^{-1/3})).$$

The result, to $O(N^{-1/3})$, is that $0.436 \leq N^{1/3}\langle 1/r\rangle^{-1} \leq 0.716$. The asymptotic Thomas–Fermi value is 0.556. If $\nu = 2$, then we obtain a lower bound by filling up the harmonic-oscillator levels:

$$\frac{1}{2}\sum_{i=1}^{N}(|\mathbf{p}_i|^2 + \omega^2|\mathbf{x}_i|^2) \geq \omega N^{4/3}\frac{3^{4/3}}{4}(1 + O(N^{-1/3})),$$

so

$$\left\langle \sum_{i=1}^{N} |\mathbf{x}_i|^2 \right\rangle \geq \frac{(3N)^{8/3}}{16\langle \sum |\mathbf{p}_i|^2\rangle} \geq \frac{6^{1/3}9N^{1/3}}{32} \Rightarrow N^{1/3}\langle r^2\rangle^{1/2} \geq 0.71 \text{ to } O(N^{-1/3}).$$

These rough numbers provide only an overview; for particular atoms one can do much better with the more accurate values of (4.30) for E.

4.5.19 Problems

Calculate $\sum_{n=1}^{n_0} n^\nu$ for $\nu = 1, 2$ and 3.

4.5.20 Solutions

As a consequence of the binomial theorem,

$$(n_0 + 1)^{\nu+1} - 1 = (\nu + 1)\sum_n n^\nu + \binom{\nu+1}{2}\sum_n n^{\nu-1} + \cdots + \binom{\nu+1}{\nu}\sum_n n + n_0,$$

from which the individual sums can be determined recursively. The results are:

$$\nu = 1 : \tfrac{1}{2}(n_0^2 + n_0),$$
$$\nu = 2 : \tfrac{1}{6}(2n_0^3 + 3n_0^2 + n_0),$$
$$\nu = 3 : \tfrac{1}{4}(n_0^4 + 2n_0^3 + n_0^2).$$

4.6 Nuclear Motion and Simple Molecules

The large masses of atomic nuclei make them move so slowly within atoms and molecules that to a higher degree of approximation they can be treated as static centers of force.

In the previous sections atomic nuclei were considered as fixed centers of force, but the validity of this approximation remains to be determined. The question is of central importance in molecular theory, which, as we shall see shortly, is based on the Born-Oppenheimer approximation (4.6.8), in which the nuclei are at first regarded as fixed while the electrons move in the field of the static force centers. The energy of the system of electrons then serves as the potential in which the nuclei move. The intuition behind this approximation is that the light electrons move much more rapidly than the heavy nuclei, so from the standpoint of the electrons the potential is nearly static. Plausible as this may sound, it does not release us from the obligation to investigate whether this conceptual division of the action actually follows from an analysis of the Schrödinger equation of the whole atomic or molecular system.

4.6.1 Separation of the Center-of-Mass of an Atom

We shall initially continue to investigate the case of an atom with N electrons. Let $(\mathbf{r}_0, \mathbf{k}_0)$ and $(\mathbf{r}_1, \mathbf{k}_1, \ldots, \mathbf{r}_N, \mathbf{k}_N)$ be the positions and momenta of the nucleus of mass M and, respectively, the electrons of mass m. (The symbols $\mathbf{x}_i$ are reserved for the relative coordinates below.) The kinetic energy is

$$T = \frac{|\mathbf{k}_0|^2}{2M} + \sum_{i=1}^{N} \frac{|\mathbf{k}_i|^2}{2m}.$$

In everything before now the limit $1/M \to 0$ was taken. Yet it is not possible to carry out any sort of perturbative expansion in $1/M$, at least directly. If $1/M = 0$, then the states are infinitely degenerate, and if $1/M < 0$, then T is not even positive definite. Let us introduce center-of-mass and relative coordinates $\mathbf{x}_0$ and $\mathbf{x}_1, \ldots, \mathbf{x}_N$:

$$\mathbf{x}_0 = \left(M\mathbf{r}_0 + m \sum_{i=1}^{N} \mathbf{r}_i \right) (M + Nm)^{-1}, \quad \mathbf{x}_i = \mathbf{r}_i - \mathbf{r}_0, \quad i = 1, \ldots, N.$$

In order to calculate the momenta conjugate to these coordinates, write T in terms of the velocities:

$$\begin{aligned} 2T &= M|\dot{\mathbf{r}}|^2 + m \sum_{i=1}^{N} |\dot{\mathbf{r}}_i|^2 \\ &= (M + Nm)|\dot{\mathbf{x}}_0|^2 + m \sum_{i=1}^{N} |\dot{\mathbf{x}}_i|^2 - \frac{m^2}{M + Nm} \left| \sum_{i=1}^{N} \dot{\mathbf{x}}_i \right|^2. \end{aligned}$$

The momenta are thus

$$\mathbf{p}_0 = \frac{\partial T}{\partial \dot{\mathbf{x}}_0} = (M + Nm)\dot{\mathbf{x}}_0, \quad \mathbf{p}_i = \frac{\partial T}{\partial \dot{\mathbf{x}}_i} = m \left(\dot{\mathbf{x}}_i - \sum_{j=1}^{N} \dot{\mathbf{x}}_j \frac{m}{M + Nm} \right).$$

When substituted in above, this yields the

4.6.2 Kinetic Energy in Center-of-Mass and Relative Coordinates

$$T = \frac{|\mathbf{p}_0|^2}{2(M+Nm)} + \frac{M+m}{2mM} \sum_{i=1}^{N} |\mathbf{p}_i|^2 + \frac{1}{M} \sum_{i>j>0} \mathbf{p}_i \cdot \mathbf{p}_j .$$

4.6.3 Remarks

1. We see there are three terms, the kinetic energy of the center of mass with the total mass; the kinetic energy of the electrons, with reduced masses depending on the nuclear mass; and finally a correction on the order of $1/M$, known as the **Hughes–Eckart term**. Since it is obviously bounded relative to the second term, nothing prevents it from being handled with analytic perturbation theory. Note, however, that it can be either positive or negative.
2. Since the Hughes–Eckart term is not compact relative to T, it is a reasonable question whether it influences the essential spectrum. We have seen that without the correction the essential spectrum of H_N begins at the lowest point of the spectrum of H_{N-1}. This is physically reasonable, and is interpreted as the threshold of ionization. Expressed differently, we have proved that without the term on the order of $1/M$, $\inf \sigma_{\text{ess}}(T + V_N) = \inf \operatorname{Sp}(T + V_{N-1})$, where V_{N-1} is the potential neglecting the last particle. But this state of affairs should not be affected by the presence of the Hughes–Eckart term. Indeed, the compactness of the individual terms is not destroyed by a relatively bounded perturbation.

4.6.4 Estimate of the Effect of a Finite Nuclear Mass on the Energy Eigenvalues

Since the center-of-mass motion separates off, we consider only the relative energy, which can also be written

$$H_r = \sum_{i=1}^{N} \frac{|\mathbf{p}_i|^2}{2m} + \frac{|\sum_{i=1}^{N} \mathbf{p}_i|^2}{2M} + V.$$

Because mass has the dimensions of energy in units where $\hbar = e = 1$, and no other constants encumbered with dimensions appear, the ground-state energy must be of the form

$$E = mf\left(\frac{m}{M}\right) < 0.$$

The coefficient of $1/M$ in H_r is positive, so f increases monotonically. Since E must be concave in $(1/m, 1/M)$, it follows that

$$\frac{\partial^2 E}{\partial(1/m)^2} \frac{\partial^2 E}{\partial(1/M)^2} - \left(\frac{\partial^2 E}{\partial(1/m)\partial(1/M)}\right)^2 \geq 0,$$

so $f'' \le 2(f')^2/f$, and $-1/f$ is concave. Since f is negative, this is a stronger concavity property than $f'' < 0$.

$$-\frac{1}{f(m/M)} \le -\frac{1}{f(0)} + \frac{f'(0)}{f(0)^2}\frac{m}{M},$$

so

$$f\left(\frac{m}{M}\right) \le \frac{f(0)}{1-(f'(0)/f(0))(m/M)}. \tag{4.31}$$

We use the inequality $|\sum_{i=1}^{N} \mathbf{p}_i|^2 \le N\sum_{i=1}^{N} |\mathbf{p}_i|^2$ to bound $f'(0)$: If the expectation value of H is calculated in the ground state for $m/M = 0$, then $\langle \sum_i |\mathbf{p}_i|^2 \rangle = 2m\,E(m/M = 0) = 2m|f(0)|$. A priori we have

$$f(0) \le f\left(\frac{m}{M}\right) \le f(0)\left(1 - N\frac{m}{M}\right), \tag{4.32}$$

from which it follows that $0 \le f'(0) \le N|f(0)|$. With the aid of (4.31), Inequality (4.32) is refined to

$$f(0) \le f\left(\frac{m}{M}\right) \le \frac{f(0)}{1+Nm/M}. \tag{4.33}$$

4.6.5 Remarks

1. If $N = 1$, this method improves the upper bound $f(0)(1 - m/M)$ to $f(0)/(1 + m/M)$, which is the exact result for the correction due to the reduced mass.
2. If $N = 2$ and $Z = 1$, then the upper bound is good enough to prove the binding of $e^-\mu^+e^-$ but not of $e^-e^+e^-$, i.e. $m = M$. In that case $f(0)$ is the energy of H^-, -0.528, and it would be necessary to have

$$\frac{-0.528}{1+2(m/M)} < \frac{-0.5}{1+(m/M)},$$

   i.e., $m/M < 0.06$. It requires highly sophisticated trial functions to show the existence of a bound state of $e^-e^+e^-$, as recently observed.
3. For neutral atoms, $M \ge N$ proton masses, and the correction for nuclear motion is less than 0.1% of the energy for $M = \infty$. This means that the ratio of the nuclear velocity to the electronic velocities is $O(m/M)$.
4. Whereas (4.32) says that big N may destroy binding, the refinement (4.33) only admits a strong reduction of the binding with increasing N. Actually in bosonic matter the energy density for $M = \infty$ goes like $-N^{5/3}$ but for $M < \infty$ only like $-N^{7/5}$. For $N = 10^{30}$ this makes a reduction by 10^8!

We now take up the problem with $\mathcal{N}$ nuclei, the coordinates of which will be written as capital letters. We first write down the

4.6.6 Molecular Hamiltonian

$$
\begin{aligned}
H &= \sum_{i=1}^{N} \frac{|\mathbf{p}_i|^2}{2m} + \sum_{k=1}^{\mathcal{N}} \frac{|\mathbf{P}_k|^2}{2M_k} - \sum_{i=1}^{N}\sum_{k=1}^{\mathcal{N}} \frac{\alpha Z_k}{|\mathbf{x}_i - \mathbf{X}_k|} \\
&\quad + \alpha \sum_{i<j} \frac{1}{|\mathbf{x}_i - \mathbf{x}_j|} + \alpha \sum_{k<l} \frac{Z_k Z_l}{|\mathbf{X}_k - \mathbf{X}_l|} \\
&\equiv H_\infty + \sum_{k=1}^{\mathcal{N}} \frac{|\mathbf{P}_k|^2}{2M_k}.
\end{aligned}
$$

4.6.7 Remarks

1. There are no new difficulties in verifying self-adjointness and semiboundedness.
2. The general conclusions that were drawn about the Coulomb potential from its behavior under dilatations are still valid.
3. The motion of the center of mass will henceforth be considered as separated off.

The Hamiltonian (4.6.6) describes a fairly intractable many-body problem. In order to be able to frame detailed propositions about it, we shall rely on the extreme ratio of the masses, $m/M_k < 0.001$, to break the motion into that of fast electrons and slow nuclei. The expectation is that the nuclei can be considered static for the motion of the electrons, so we next investigate the accuracy of this description.

4.6.8 The Born–Oppenheimer Approximation

The first step is to find the energy eigenvalues $E_n(X)$, $X = (\mathbf{X}_1, \dots, \mathbf{X}_{\mathcal{N}})$, of H_∞. These become the potentials for the motion of the nuclei, so the next step is to find the eigenvalues E_{n1} of

$$
H_{nuc} \equiv \sum_{k=1}^{\mathcal{N}} \frac{|\mathbf{P}_k|^2}{2M_k} + E_n(X).
$$

4.6.9 The Accuracy of the Born–Oppenheimer Approximation

Let E_1 be the lowest eigenvalue of H, $\Psi_X(x)$ the ground state of H_∞, and $\Phi(X)$ that of H_{nuc}. Then, defining $x = (\mathbf{x}_1, \dots, \mathbf{x}_N)$,

$$
E_{11} \le E_1 \le E_{11} + \sum_k \frac{1}{2M_k} \int d^{3N}x\, d^{3\mathcal{N}}X |\Phi(X) \nabla_{\mathbf{X}_k} \Psi_X(x)|^2.
$$

Proof: The lower bound follows from the operator inequalities

$$
H \ge E_1(X) + \sum_{k=1}^{\mathcal{N}} \frac{|\mathbf{P}_k|^2}{2M_k} \ge E_{11}.
$$

The upper bound. If $\Psi_X(x)\Phi(X)$ is used as a trial function, then clearly the expectation value of H_∞ is $E_1(X)$ and, on the one hand, H_{nuc} produces E_{11}. On the other hand, since the electron wave-functions depend on the nuclear coordinates, the operator $\sum_{k=1}^{\mathcal{N}} |\mathbf{P}_k|^2/2M_k$ also acts on $\Psi_X(x)$;

$$P_k \Psi_X(x)\Phi(X) = -i\left[\Psi_X(x)\frac{\partial \Phi(X)}{\partial X_k} + \Phi(X)\frac{\partial \Psi_X(x)}{\partial X_k}\right].$$

When squared and combined with $\langle E_1(X)\rangle$ the first term yielded E_{11}, while the second term is the correction in (4.6.9). Because of the normalization of $\Psi_X(x)$ the mixed terms drop out when integrated by $d^{3N}x$:

$$\int d^{3N}x \left(\Psi_X^*(x)\frac{\partial \Psi_X(x)}{\partial \mathbf{X}_k} + \frac{\partial \Psi_X^*(x)}{\partial \mathbf{X}_k}\Psi_X(x)\right) = \frac{\partial}{\partial \mathbf{X}_k}\int d^{3N}x |\Psi_X(x)|^2 = 0.$$

□

4.6.10 Remarks

1. The only dependence on M_k in the integral is that of Φ. Provided that this quantity remains finite as $M_k \to \infty$, the difference between the bounds is $O(\max_k\{1/M_k\})$.
2. The vibrational energy of the nuclei is on the order of $(E(X)''/M_k)^{1/2} = O(M_k^{-1/2})$. The Born–Oppenheimer approximation is thus accurate enough that it makes sense to calculate this energy.
3. Rotational energy is inversely proportional to the moment of inertia, so for the nuclei it is $O(M_k^{-1})$. This is comparable to the error of the approximation, and it will thus not be possible to make any firm statements about it.

We shall now survey some general properties of $E(X)$.

4.6.11 Lower Bound: The Energy of the United Atom

Disregarding the nuclear repulsion, $E(X)$ has its infimum when $\mathbf{X}_i = \mathbf{X}_k$ for all i and k.

Proof:

$$H_\infty - \alpha \sum_{k>l} \frac{Z_k Z_l}{|\mathbf{X}_k - \mathbf{X}_l|} = \sum_{i=1}^{N} \frac{|\mathbf{p}_i|^2}{2m} - \sum_{i=1}^{N}\sum_{k=1}^{\mathcal{N}} \frac{\alpha Z_k}{|\mathbf{x}_i - \mathbf{X}_k|} + \alpha \sum_{i<j} \frac{1}{|\mathbf{x}_i - \mathbf{x}_j|}$$

$$= \sum_{k=1}^{\mathcal{N}} \frac{Z_k}{Z}\left\{\sum_{i=1}^{N}\left(\frac{|\mathbf{p}_i|^2}{2m} - \frac{Z\alpha}{|\mathbf{x}_i - \mathbf{X}_k|}\right) + \sum_{i>j} \frac{\alpha}{|\mathbf{x}_i - \mathbf{x}_j|}\right\},$$

$$Z = \sum_{k=1}^{\mathcal{N}} Z_k,$$

and the expression in the curly brackets { } is precisely the Hamiltonian of an atom of charge Z. If its lowest eigenvalues is denoted $E(N, Z)$, then

$$E(X) \geq \alpha \sum_{k>l} \frac{Z_k Z_l}{|\mathbf{X}_k - \mathbf{X}_l|} + E(N, Z). \qquad \square$$

4.6.12 Remark

Proposition (4.6.11) means that the electrons would prefer the nuclei to be all bunched together. To understand how molecules are formed it remains to be determined where this attraction balances the Coulombic repulsion of the nuclei.

4.6.13 The Effects of the Dilation Group

Let $\mathbf{X}_i = R\bar{\mathbf{X}}_i$. *Consider the coordinates* $\mathbf{X}_i$ *as fixed, so that the molecule preserves its shape as it expands or contracts as* R *varies. If* $H_\infty = T + V$, *where* $T = \sum_i |\mathbf{p}_i|^2/2m$, *then the expectation values with the electronic wave-function* $|\rangle$ *such that* $H_\infty|\rangle = E(R)|\rangle$ *satisfy*

$$\langle T \rangle = -E(R) - R\frac{\partial E(R)}{\partial R}$$

$$\langle V \rangle = 2E(R) + R\frac{\partial E(R)}{\partial R}.$$

Proof: The operator H_∞ is put into the form

$$H_\infty = \frac{1}{mR^2}\left(\frac{1}{2}\sum_i |\mathbf{p}_i|^2 + mR\alpha V\right),$$

by a dilatation $\mathbf{x}_i \to R\mathbf{x}_i$, $\mathbf{p}_i \to R^{-1}\mathbf{p}_i$, where V depends on $\mathbf{x}_i$ and $\bar{\mathbf{X}}_i$, but not on R. Hence $E(R)$ is of the form $(1/mR^2)f(\alpha mR)$, where we will not bother to indicate the dependence on $\bar{\mathbf{X}}_i$. By the Feynman–Hellmann Theorem (3.5.17; 2), $\langle V \rangle = \alpha(\partial/\partial\alpha)E$ and $\langle T \rangle = E - \langle V \rangle = m\,\partial E/\partial(1/m)$, which shows (4.6.13). $\square$

4.6.14 Remarks

1. At the equilibrium position, where $\partial E/\partial R = 0$, the virial theorem for the electrons as usual states that $\langle V \rangle = 2E = -2\langle T \rangle$. If $\partial E/\partial R \neq 0$, then the kinetic energy is less than $|E|$ when $\partial E/\partial R$ is greater than 0, and vice versa. This agrees with physical intuition, according to which if R is too small the kinetic energy of the electrons is too large.
2. A false argument for the formation of molecules is sometimes advanced, that their greater volume allows some savings in the kinetic energy of the electrons. It is certainly true that $\langle T \rangle < |\langle V \rangle|/2$ as R is decreased from infinity to the region where $\partial E/\partial R > 0$. However, $\langle T \rangle = |E|$ at the equilibrium position,

and is thus greater than in an atom if $|E|$ is to be greater than the energy of the isolated atoms.

3. Of course, the virial theorem also holds for H as a whole. If $\langle\!\langle\ \rangle\!\rangle$ denotes the expectation value in the ground state E_1 of H, and T_k denotes the kinetic energy of the nuclei, then

$$|E_1| = \langle\!\langle T \rangle\!\rangle + \langle\!\langle T_k \rangle\!\rangle < |E_{11}| = \langle T \rangle,$$

so

$$\langle T \rangle - \langle\!\langle T \rangle\!\rangle > \langle\!\langle T_k \rangle\!\rangle.$$

This shows that the expectation value calculated in the Born–Oppenheimer approximation is within $O(M_k^{-1})$ of the exact expectation value.

4. Since V is bounded relative to T, isolated eigenvalues are analytic in α. Therefore f is analytic, so $E(R)$ is analytic in R for $R \neq 0$, provided that the eigenvalues remain isolated.

4.6.15 Upper Bound to $E_1(R)$

Let R_0 be the equilibrium position, i.e., $\partial E_1/\partial R|_{R=R_0} = 0$. Then for all $R > 0$,

$$E_1(R) \leq E_1(R_0)\frac{R_0^2}{R^2}\left(1 + 2\frac{R - R_0}{R_0}\right).$$

Proof: In the proof of (4.6.13) it was shown that $E_1 = (1/mR^2)f(\alpha m R)$, where f was some concave function by (3.5.21). Therefore $R^2E_1(R)$ is concave in R, and is always less than its tangent (see Figure 4.7):

$$R^2E(R) \leq R_1^2E(R_1) + (R - R_1)(2R_1E(R_1) + R_1^2E'(R_1))$$

for all R and R_1. Proposition (4.6.15) follows with $R_1 = R_0$. □

4.6.16 Application to Diatomic Molecules

If there are two nuclei, R may be identified with $|\mathbf{X}_1 - \mathbf{X}_2|$, and the upper bound for the nuclear motion is

$$H_k \leq \frac{|\mathbf{P}_{cm}|^2}{2M_{cm}} + \frac{|\mathbf{P}|^2}{2M} + E(R_0)\frac{R_0^2}{R^2}\left(1 + 2\frac{R - R_0}{R}\right),$$

where $M_{cm} = M_1 + M_2$, $M = M_1M_2/(M_1 + M_2)$, and $\mathbf{P}$ is the momentum conjugate to $\mathbf{X}_1 - \mathbf{X}_2$. Since the potential is a superposition of $1/R$ and $1/R^2$ potentials, the Schrödinger equation can be solved analytically (Problem (4.6.25)), producing the general inequality

$$E(R_0) \leq E_{11} \leq \frac{E(R_0}{(\sqrt{1+x} + \sqrt{x})^2} \quad \text{where } x = \frac{1}{4R_0^2M|E(R_0)|}. \tag{4.34}$$

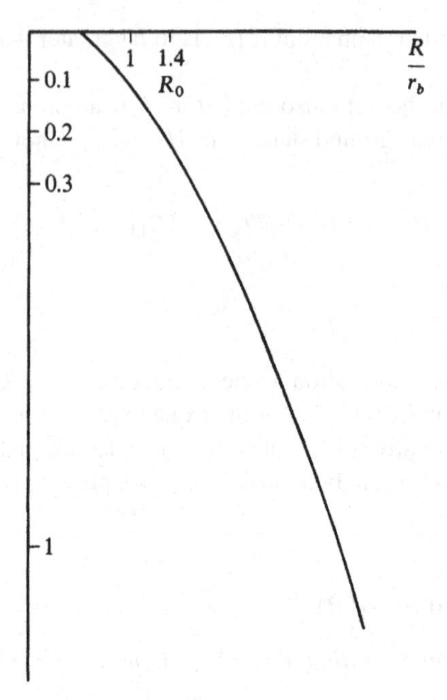

Figure 4.7. $E \cdot R^2$ in atomic units for H_2.

4.6.17 Remarks

1. Since $|E_0(R)| \sim m$ and $R_0 \sim m^{-1}$, we see explicitly that $E_{11} - E(R_0) = O((m/M)^{1/2})$.
2. Inequality (4.34) is too general to be numerically accurate in special cases. For instance, for H_2^+ and H_2 it states that the zero-point energies of vibration, $E_{11} - E(R_0)$, are less than 0.24 and, respectively, 0.49 eV, whereas the observed values are roughly 0.14 and 0.26 eV.

As was shown in (4.6.11), $E_1(X)$ is always greater than the sum of the ground-state energy of the united atom and the Coulomb repulsion of the nuclei. The amount by which the ground state of a diatomic molecule can exceed this lower bound can be estimated by a

4.6.18 Bound to $E_1(X)$ in Terms of the Electron Density

Let $\rho(\mathbf{x})$ be the electron density of the ground state of

$$H_{N,Z} = \sum_{i=1}^{N} \frac{|\mathbf{p}_i|^2}{2m} - Z \sum_{i=1}^{N} \frac{1}{r_i} + \sum_{i<j} \frac{1}{|\mathbf{x}_i - \mathbf{x}_j|} \qquad Z = Z_1 + Z_2,$$

and let $E(N, Z)$ be the ground-state energy. For a diatomic molecule,

$$E(N, Z) \leq E_1(R) - \frac{Z_1 Z_2}{R} \leq E(N, Z) + Z \int_{r \leq R/2} d^3x\, \rho(\mathbf{x}) \left(\frac{1}{r} - \frac{2}{R} \right),$$

$$R = |\mathbf{X}_1 - \mathbf{X}_2|, \qquad d^3x\, \rho(\mathbf{x}) = N$$

Proof: Consider H_∞ as a function $H_{\mathbf{R}}$ of the vector $\mathbf{R} = \mathbf{X}_1 - \mathbf{X}_2$ and average over the angle. This affects only the potential

$$-\sum_{i=1}^{N} \left(\frac{Z_1}{|\mathbf{x}_i - \mathbf{R}/2|} + \frac{Z_2}{|\mathbf{x}_i + \mathbf{R}/2|} \right),$$

which is turned into the potential

$$-\sum_{i=1}^{N} \frac{Z}{r_i} + Z \sum_{i=1}^{N} \left(\frac{1}{r_i} - \frac{2}{R} \right) \Theta \left(\frac{R}{2} - r_i \right), \qquad R = |\mathbf{R}|,\ r_i = |\mathbf{x}_i|,$$

of a spherical shell of radius $R/2$ and charge Z. By the min-max principle, the expectation value of

$$\int \frac{d\Omega}{4\pi} H_{\mathbf{R}} = H_{N,Z} + \frac{Z_1 Z_2}{R} + Z \sum_{i=1}^{N} \left(\frac{1}{r_i} - \frac{2}{R} \right) \Theta \left(\frac{R}{2} - r_i \right)$$

in the ground state of $H_{N,Z}$ is an upper bound to $E_1(R)$. □

4.6.19 Remark

If $\rho_m = \sup_x \rho(x)$, we see

$$E(N, Z) \le E_1(R) - \frac{Z_1 Z_2}{R} \le E(N, Z) + \frac{Z\pi}{6} R^2 \rho_m.$$

Thus, after the Coulombic term $Z_1 Z_2 / R$ is subtracted off, the energy $E(R)$ approaches $E(N, Z)$ with a horizontal tangent as $R \to 0$. This also follows from the fact proved in [19] with a somewhat more difficult argument, that, although $E(R)$ is not analytic at $R = 0$, it is at least C^2 there.

The rough outlines of the general features of the important observables are now known. We shall conclude by working out some of the finer details for especially simple molecules.

4.6.20 The Properties of $E_1(R)$ for H_2^+

If $\mathcal{N} = 2$, $N = 1$, $Z_1 = Z_2$, and $\mathbf{X}_1 - \mathbf{X}_2 = \mathbf{R}$, then H_∞ of (4.6.6) is unitarily equivalent to $Z_1^2 m H$, where

$$H = \frac{|\mathbf{p}|^2}{2} - \frac{\alpha}{|\mathbf{x} - \mathbf{R}/2|} - \frac{\alpha}{|\mathbf{x} + \mathbf{R}/2|} + \frac{\alpha}{R}.$$

If $E_1(R)$ is the lowest eigenvalue of H, then

(i) *$E_1(R) - \alpha/R$ increases monotonically in R; and*

(ii) *$R^2 E_1(R) - R\alpha$ is concave and decreasing in R.*

Proof:

(ii) This follows from (4.6.13). Since $H - \alpha/R$ decreases as a function of α,

$$\frac{\partial}{\partial R}\left(R^2\left(E_1 - \frac{\alpha}{R}\right)\right) - \frac{\alpha}{R}\frac{\partial}{\partial \alpha}\tilde{f} < 0,$$

where $\tilde{f} = R^2$ times the lowest eigenvalue of $H - \alpha/R$.

(i) This is less trivial, and requires a variant of (4.3.32) for not necessarily positive potentials: □

4.6.21 Monotony of the Schrödinger Wave-Function in the Potential

Let Ω be an open set $\subset \mathbb{R}^n$, and

(i) $f, g \in C^0(\bar{\Omega})$, $f, g > 0$,

(ii) $f(\mathbf{x})$, $g(\mathbf{x}) \to 0$ *as* $|\mathbf{x}| \to \infty$,

(iii) $f(\mathbf{x}) \geq g(\mathbf{x})$ *for all* $\mathbf{x} \in \partial\Omega$,

(iv) Δf, $\Delta g \in L^1(\Omega)$,

(v) *suppose that* $V(\mathbf{x}) < W(\mathbf{x})$ *for all* $\mathbf{x} \in \Omega$ *and that, in the sense of distributions,* $-\Delta f + Vf \geq 0$ *and* $-\Delta g + Wg \leq 0$.

Then $f(\mathbf{x}) \geq g(\mathbf{x})$ *for all* $\mathbf{x} \in \bar{\Omega}$.

Proof: To avoid some complications the proof will be sketched for sufficiently nice g, f, and Ω. Let $D = \{\mathbf{x} \in \Omega : g(\mathbf{x}) > f(\mathbf{x})\}$ as in the proof of (4.3.32). Assumptions (i), (ii), and (iii) imply that $g = f$ on ∂D. Because of (i) and (iv), with Green's theorem,

$$0 < \int_D (W - V) f g \, d^3x \leq \int_D (f\Delta g - g\Delta f) d^3x = \int_{\partial D} dS \, f \frac{\partial}{\partial n}(g - f),$$

where $\partial/\partial n$ is the derivative in the direction of the outward normal to ∂D. Since $g_{|\partial D} = f_{|\partial D}$ and $g > f$ on D, the difference $g - f$ cannot increase in the outward direction, so we conclude that $D = \emptyset$. □

The proof of (4.6.20) can now be completed. The operator $H - \alpha/R$ is unitarily equivalent to

$$h = -\frac{\Delta}{2} - \frac{\alpha}{r} - \frac{\alpha}{[(x-R)^2 + y^2 + z^2]^{1/2}}.$$

By the Feynman–Hellmann theorem,

$$\frac{\partial e_1}{\partial R} = \alpha\langle\psi|(R - x)[x - R)^2 + y^2 + z^2]^{-3/2}\psi\rangle$$

$$= \alpha \int_{-\infty}^{\infty} \int_{-\infty}^{\infty} dy\, dz \int_{R}^{\infty} dx (x - R)[(x - R)^2 + y^2 + z^2]^{-3/2}$$
$$\times [\psi^2(2R - x, y, z) - \psi^2(x, y, z)],$$

where ψ is the ground-state eigenfunction of h and $e_1 = E_1 - \alpha/R$ is its eigenvalue. But now $\psi(2R - x, y, z) \geq \psi(x, y, z)$ for all $x > R$, since the assumptions of (4.6.21) are satisfied with $\Omega = \{(x, y, z) : x > R\}$, $f = \psi(2R - x, y, z)$, and $g = \psi(x, y, z)$: As in (3.5.26), ψ is nonnegative, and it can in fact be proved strictly positive [3]. The functions f and g are equal on $\partial\Omega = \{(x, y, z) : x = R\}$, and we take

$$W(x) \equiv -\frac{\alpha}{r} - \frac{\alpha}{[(x - R)^2 + y^2 + z^2]^{1/2}} - E_1(R) + \frac{\alpha}{R},$$

and

$$V \equiv \frac{-\alpha}{[(x - 2R)^2 + y^2 + z^2]^{1/2}} - \frac{\alpha}{[(x - R)^2 + y^2 + z^2]^{1/2}} - E_1(R) + \frac{\alpha}{R}.$$

Clearly, $W > V$ for all $x > R$. Therefore $\partial e_1/\partial R \geq 0$. □□

4.6.22 Remarks

1. With (4.6.13) it now follows that $\langle V \rangle - 2E_1 \geq -\alpha/R$ for all R.
2. If H is of the form of (4.6.20), then

$$\frac{\partial}{\partial R}\left(E_1 - \frac{\alpha}{R}\right)$$
$$= \frac{\alpha}{4}\langle\psi| \left(\frac{R - 2x}{[(x - R/2)^2 + y^2 + z^2]^{3/2}} + \frac{R + 2x}{[(x + R/2)^2 + y^2 + z^2]^{3/2}}\right)$$
$$\times |\psi\rangle \geq 0$$

   means that the electron prefers to be between the nuclei.
3. It is certainly not true that all eigenstates are monotonic in the internuclear separation. For example, as $R \to \infty$ the eigenvector with eigenvalue $e_2(R)$ that becomes the $2p$ state when $R = 0$ is asymptotically

$$\exp\left(-\left|x - \frac{R}{2}\right|\right) - \exp\left(-\left|x + \frac{R}{2}\right|\right),$$

   up to normalization. The corresponding eigenvalue is that of the ground state of the hydrogen atom, and thus the same as $e_2(0)$. Since $e_2(R)$ is not constant, it is definitely not monotonic in R.

4.6.23 Bounds for $E_1(R)$ for H_2^+

The most convenient methods are the Rayleigh–Ritz variational principle and Temple's inequality (3.5.30; 2). An accuracy of 0.1% is attainable with the trial

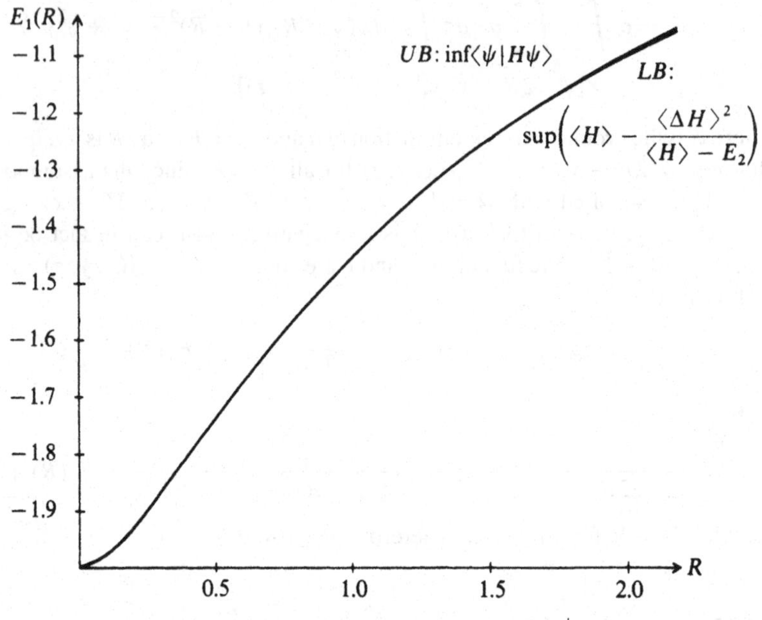

Figure 4.8. Bounds to $E_1(R) - \frac{\alpha}{R}$ for H_2^+.

functions

$$\psi = \left(1 + \frac{\beta R^2 \nu^2}{4}\right) \exp\left(-\frac{\alpha R \mu}{2}\right),$$

$$(\mu, \nu) = \frac{1}{2}\left(\left|\mathbf{x} - \frac{\mathbf{R}}{2}\right| \pm \left|\mathbf{x} + \frac{\mathbf{R}}{2}\right|\right),$$

by adjusting the parameters α and β (see the table and Figure 4.8).

$E_{1LB} \le E_1(R) - \frac{\alpha}{R} \le E_{1UB}$ for H_2^+

R	Temple			Rayleigh–Ritz		
	E_{1LB}	α	β	E_{1UB}	α	β
0.2	−1.929	1.91	0.64	−1.929	1.94	0.67
0.4	−1.801	1.80	0.61	−1.801	1.84	0.61
0.6	−1.672	1.71	0.58	−1.671	1.75	0.57
0.8	−1.555	1.62	0.55	−1.554	1.67	0.54
1.0	−1.452	1.55	0.52	−1.451	1.59	0.52
1.2	−1.363	1.49	0.50	−1.361	1.53	0.50
1.4	−1.285	1.44	0.49	−1.284	1.48	0.48
1.6	−1.217	1.40	0.48	−1.216	1.43	0.46
1.8	−1.157	1.36	0.47	−1.156	1.39	0.45
2.0	−1.104	1.32	0.47	−1.102	1.35	0.45

4.6.24 Remarks

1. For lack of a better lower bound, the value $-0.5 + 1/R$ was used for the energy $E_2(R)$ of the next higher gerade (even) state.
2. The bounds become inaccurate in the regime of large R. It can be shown [20] that $E_1(R)$ goes asymptotically as $-\frac{1}{2} - \frac{9}{4}R^4$ as $R \to \infty$. The expressions for the gerade and ungerade states are the same to all orders in a formal $1/R$ expansion; however, it is feasible to calculate the gap between them [23].
3. Since the relativistic corrections approach the level of 0.1%, it is not worthwhile to pursue greater accuracy within the framework of the Schrödinger equation.
4. The eigenvalues E_i of H_2^+ can be calculated to arbitrary accuracy as a continued fraction.
5. The increased density of states makes it more difficult to obtain accurate lower bounds when there are more electrons in the molecule. One first needs a rough lower bound for E_2 to get a better one for E_1. It takes a rather more laborious computation to reach the accuracy we have gotten for H_2^+ [21].

4.6.25 Problem

Study the Schrödinger equation with $H = |\mathbf{p}|^2/2 + \alpha/r^2 - \beta/r$.

4.6.26 Solution

Replace $\ell(\ell+1)$ with $\ell(\ell+1) + 2\alpha$ (cf. I: 3.4.24; 6)).

4.6.27 Some Difficult Problems

1. Investigate the three-body Coulomb system with charges $+, -, -$, and masses $m_1, m_2, 1$. For what region of the m_1, m_2-plane does there exist a point spectrum (cf. (4.3.26))? In particular, is there a bound state of e^+ H?
2. Two helium atoms attract with a Van der Waals potential $E_1(R) \sim -1/R^6$ as $R \to \infty$. Find a lower bound to $E_1(R)$ with a flat enough potential minimum to show that two helium atoms do not bind.
3. Find bounds for the imaginary parts of the resonances E of (4.4.8; 1).
4. Bound the scattering cross-section for e^1 – H near the resonances (4.4.8; 1).
5. Prove asymptotic completeness for the scattering of e^1 – H above the ionization energy.
6. Study the monotonic properties of $E_1(R) - Z_1 Z_2/R$ for complicated diatomic molecules.
7. In what sense does the Born–Oppenheimer approximation converge? The operator $H \to H_\infty$ as $M_k \to \infty$, but how does $H_\infty(R)$ converge as $R \to \infty$ and to what?
8. The proof of (4.3.29) provides no numerical values for $c_\pm$ or r_0. Find some.
9. The upper bound (4.3.34) for $\rho(0)$ is the exact value if there is only one particle, while the lower bound is too small by a factor $\frac{3}{16}$. With more electrons the upper

bound degrades somewhat and the lower bound gets much worse. Find better lower bounds.

Part II

Quantum Mechanics of Large Systems

1
Systems with Many Particles

1.1 Equilibrium and Irreversibility

Macroscopic bodies act in an irreversible and deterministic manner in contrast with the reversible and indeterministic character of the underlying laws of quantum physics. How can the apparent contradiction be understood?

We have learned to describe systems of finitely many particles with an algebra $\mathcal{A}$ of observables, and information about the systems with a state w on the algebra (cf. (III: 2.2.32)). As our main goal is the study of everyday matter, our framework will be that of nonrelativistic quantum theory. For the purposes of contrast, or of aiding intuition, we shall also have occasion to call upon classical mechanics, where states are measures on phase space, and extremal states are point measures. In either framework time-evolution can be represented as an automorphism $a \to a_t$, for $a \in \mathcal{A}$ in the Heisenberg picture. If desired, time-dependence can alternatively, in the Schrödinger picture, be put upon the state: $w \to w_t$ such that $w_t(a) = w(a_t)$. If the algebra is Abelian (classical mechanics), then the point of an extremal state moves along a classical trajectory in phase-space.

In our earlier experience, systems of N particle are so complex for large N that it becomes impossible to reach precise, quantitative conclusions. It turns out, however, that the theoretical analysis again simplifies in the limit $N \to \infty$. Many properties become independent of the exact number of particles and other detailed characteristics of the physical system, somewhat in analogy to what happens in the central limit theorem of probability theory. This may seem peculiar at first; we have always had $\mathcal{A} = \mathcal{B}(\mathcal{H})$, $\mathcal{H}$ a separable Hilbert space, and time-evolution was given by a unitary group on $\mathcal{H}$. What, then, appears so special about a many-particle system? Just that the information contained in a pure state about a many-particle system is so overwhelming that it would be too ambitious to employ the whole of $\mathcal{B}(\mathcal{H})$ for the observables. Actual measurements could never be made on more than a few observables, so $\mathcal{B}(\mathcal{H})$ has to be cut down to size. For instance, suppose that a device is only equipped to observe one particle at a time, and is unable to detect correlations between particles. Then, rather than taking the entire tensor product of the individual particles as the algebra of observables, it is reasonable to regard $\mathcal{A}$ as a single factor. Accordingly, many states differing on $\mathcal{B}(\mathcal{H})$ reduce to the same state when restricted to $\mathcal{A}$. (The classical situation is similar; the restriction of

$$w(\mathbf{q}_1, \ldots, \mathbf{p}_N)$$

is

$$w(\mathbf{q}_1, \mathbf{p}_1) = \int d^3q_2 \ldots d^3q_N d^3p_2 \ldots d^3p_N w(\mathbf{q}_1, \ldots, \mathbf{p}_N),$$

so whole cylindrical regions of phase-space reduce to a single restricted state.) As a consequence large portions of the space of states on $\mathcal{B}(\mathcal{H})$ are quite similar from the point of view of the reduced algebra $\mathcal{A}$. If, in the Schrödinger picture, the state w_t travels throughout the space of states, then its restriction takes on a certain value with a very high probability, unless prevented by some constants of the motion. This most probable state is called the **equilibrium state** over $\mathcal{A}$.

The irreversible tendency toward equilibrium has always aroused wonder, especially as the basic equations of dynamics are invariant under reversal of the motion (III: 3.3.17). We have even seen in classical mechanics that the trajectory of any point on a compact energy surface returns arbitrarily close to its initial position (I: 2.6.13). In quantum theory the Hamiltonian H of a system confined to a finite volume has purely discrete spectrum. If ε_j and $|j\rangle$ denote the eigenvalues and eigenvectors of H, then the time-dependence of an observable a is given by

$$w_t(a) = \sum_{j,k} c_j^* c_k \exp(it(\varepsilon_j - \varepsilon_k))\langle j|a|k\rangle,$$

where the state w is represented by the vector $\sum_j c_j|j\rangle$. The state $w_t(a)$ is now an almost-periodic function of t; if the sum is finite, and the ε_j are rationally dependent, then it is actually strictly periodic. At any rate, to arbitrarily good accuracy, $w_t(a)$ again becomes nearly $w(a)$ after some sufficiently long delay. The trouble is that the recurrence times are so unimaginably long that they have no physical relevance. Suppose, for instance, that there are N distinct energy differences ω_j. The recurrence time can then be estimated as follows. The factors $\exp(i\omega_j t)$ can be pictured as N clocks with hands moving at N different rates. The question is how long it takes for a certain configuration of clock faces to reappear to within some angular accuracy $\Delta\varphi$. The configuration in the space of angles has measure $(\Delta\varphi/2\pi)^N$, so the recurrence time is on the order of $(\Delta\varphi/2\pi)^{-N}/\omega$, where the reciprocal angular velocity $1/\omega$ is an average of the $1/\omega_j$. Even for just $N = 10$, $1/\omega = 1$ sec., and $(\Delta\varphi/2\pi) = 1/100$, so that w_t returns to w to within

1% accuracy, the recurrence time is 10^{20} sec., which is much longer than the age of the universe.

The approach to equilibrium is connected to a loss of information; to be more precise, information does not get lost, but only less accessible. We have seen that when the wave-packet of a free particle spreads (III:3.3.3), Δx grows linearly with time, although the state remains pure and thus has maximal information content. The observable with least deviation from the mean is, however, not $x(t)$ but $x(0) \equiv x(t) - pt$.

This behavior can be seen even in classical motion if a minimal spread of the support of the probability distribution function in phase space is hypothesized to account for quantum effects. If, say, the initial probability density $\rho(p, q)$ is concentrated on a part of the energy shell $\{(q; p)|p_1 \leq p \leq p_2\}$ and is not pointlike, and it moves freely on a torus, then it eventually fills the energy shell densely with a "fuzzy" distribution. Faster particles overtake the slower ones, as bicycles racing in a stadium start packed closely together but later draw apart and eventually spread around the whole track (Figure 1.1).

The ergodic hypothesis has figured importantly in the history of statistical mechanics; it is the assumption that the trajectory of almost every point winds densely around the energy shell in phase space, so that the time average can be replaced with the average over the energy shell. On the one hand this requires more than is necessary, since it suffices to fill a sufficiently typical part of the energy shell, the average on which equals the average on the whole shell for the reduced algebra of observables. On the other hand, although macroscopic measurements last much longer than the collision time, they last much less than the recurrence time, so one does not wait for the whole energy shell to be sampled. We shall discuss examples in which the equilibrium state is actually attained by the state in a reasonable time after reduction to one particle.

A pictorial description of the situation is as follows. The information about a subsystem (i.e., the opposite of the entropy, to be defined later) as a function on the space of states of the total system consists mainly of a plain with few hills and still fewer mountains. The larger the total system, the further apart the prominences. Even if a path begins on a peak, it soon descends to the plain, and there is only the slightest probability that it will ascend another mountain in any conceivable time. The time of descent to the plain and the recurrence time are of completely different orders of magnitude. It takes only the time corresponding physically to a few collisions to descend to a level near that of the plain, whereas the other mountains lie in the unfathomable distance. This means that equilibrium is reached long before the immense recurrence time required to wind throughout the space of states; generally, a path soon reaches states that can not be distinguished from equilibrium because of the limits of our measuring abilities. Of course, there is still the question of how one happened, at the beginning, to be at the top of the mountain, but that brings up the one of how the current state of the universe came about and is outside the scope of this book.

Another puzzle is the apparent causal behavior that classical thermodynamics prescribes for macroscopic bodies. According to the arguments that have been

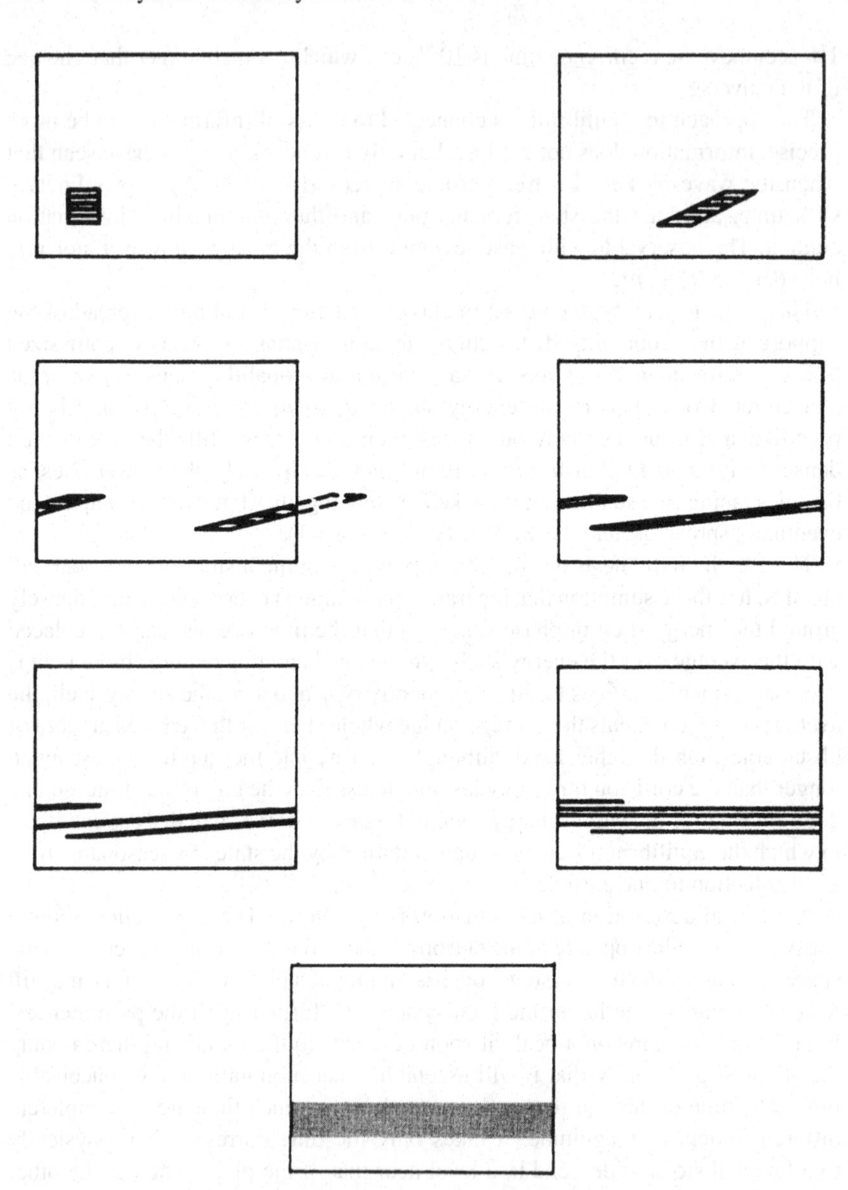

Figure 1.1. The motion of the density in phase space for a free particle on a torus

advanced, one would rather suspect that the fluctuations of the observables are increased by the loss of information. This is actually true for microscopic variables like the positions and momenta of individual particles. However, if only the so-called macroscopic observables are considered, that is, roughly what was accessible to the more primitive experimental arts of an earlier epoch, then deterministic features arise. Their origin is simply that statistically independent quantities are being averaged: if $a = (1/N)\sum_{j=1}^{N} a_j$, where $w(a_i a_j) = w(a_i)w(a_j)$ for $i \neq j$, then

$$(\Delta a)^2 = \frac{1}{N^2}\Big[w\Big(\sum_{j,k}(a_j a_k)\Big) - \sum_{j,k} w(a_j)w(a_k)\Big] = \frac{1}{N^2}\sum_{j=1}^{N}(\Delta a_j)^2.$$

Thus $\Delta a \sim N^{-1/2}$, and for sufficiently large N the deviations from the average are negligible. We shall learn that in the quantum-theoretical formalism such an a approaches a multiple of the identity operator as $N \to \infty$. The limiting coefficient depends on the representation of the algebra.

Let us verify the phenomena described above in two explicitly soluble models. Of necessity they will lack some of the complications arising in reality, but they exhibit the important features. They are embryonic forms of systems of fermions and bosons.

1.1.1 The Chain of Spins

Let the algebra of observables of the total system be generated by $\boldsymbol{\sigma}_j$, $j = 1, \ldots, N$, where each $\boldsymbol{\sigma}_j$ is a copy of the usual Pauli matrices $\boldsymbol{\sigma}$. Instead of Cartesian components we use $\sigma \equiv \sigma^z$ and $\sigma^{\pm} \equiv (\sigma^x \pm i\sigma^y)/2$, which satisfy the commutation relations

$$\begin{aligned} \left[\sigma_j, \sigma_k^{\pm}\right] &= \pm\delta_{jk} 2\sigma_k^{\pm}, \\ \left[\sigma_j^{+}, \sigma_k^{-}\right] &= \delta_{jk}\sigma_k. \end{aligned} \tag{1.1}$$

The chain is closed by the identification of $\boldsymbol{\sigma}_{j+N}$ with $\boldsymbol{\sigma}_j$, and the Hamiltonian that determines the time-evolution will be assumed to be of the form

$$H = B\sum_{j=1}^{N}\mu_j\sigma_j + \sum_{n=1}^{N-1}\sum_{j=1}^{N}\sigma_j\sigma_{j+n}\varepsilon(n). \tag{1.2}$$

The physical meaning of this is that the spins are coupled with magnetic moments μ_j to an external magnetic field B, and in addition there is an Ising-like spin-spin interaction with the nth neighbor. The strength $\varepsilon(n)$ of this interaction is a function that can be specified later, and the periodicity allows us to assume $\varepsilon(n) = 0$ for $n > N/2$. If the contributions to H are denoted as in

$$H \equiv H_0 + \sum_n H_n, \tag{1.3}$$

then the H_k commute with one another and with the σ_j. They are therefore constant in time, and the time-evolution of σ^+ and $\sigma^- = (\sigma^+)^*$ can be calculated easily from the relationship

$$f(\sigma)\sigma^+ = \sigma^+ f(\sigma + 2), \tag{1.4}$$

which follows from (1.1.2). We find

$$\begin{aligned}\sigma_k^+(t) = (\sigma_k^-(t))^* &= \sigma_k^+(0)\exp\{2it[B\mu_k + \sum_n \varepsilon(n)(\sigma_{k+n} + \sigma_{k-n})]\} \\ &= \sigma_k^+(0)\exp(2itB\mu_k)\prod_n(\cos 2t\varepsilon(n) + i\sigma_{k+n}\sin 2t\varepsilon(n))(\cos 2t\varepsilon(n) \\ &\quad + i\sigma_{k-n}\sin 2t\varepsilon(n)),\end{aligned} \tag{1.5}$$

where $a(t) = \exp(iHt)a\exp(-iHt)$.

The time-evolution consists of Larmor precession in the external field and a kind of diffusion along the chain due to the spin-spin interaction. Suppose that the state at $t = 0$ is pure and has the form of a product, where the spins have a 3-component s and σ_k^+ has phase α_k:

$$\langle\sigma_k(0)\rangle = s, \quad \langle\sigma_k^+(0)\rangle = \frac{1}{2}\sqrt{1-s^2}\exp(i\alpha_k), \quad \langle\prod_j \sigma_j\rangle = \prod_j\langle\sigma_j\rangle. \tag{1.6}$$

Then

$$\begin{aligned}\langle\sigma_k^+(t)\rangle &= \frac{1}{2}\sqrt{1-s^2}\exp\{i(\alpha_k + 2tB\mu_k)\}f^2(t), \\ f(t) &= \prod_{n=1}^{N/2}(\cos 2t\varepsilon(n) + is\sin 2t\varepsilon(n)).\end{aligned} \tag{1.7}$$

If N is finite, then f is almost periodic, and if $N = \infty$, then $f(t)$ will generally tend to zero as $t \to \infty$ (supposing that $\varepsilon(n)$ tends to zero in such a way that the infinite product makes sense). To make this more explicit, let us consider the special case $s = 0$ and $\varepsilon(n) = 2^{-n-1}$. If $N = \infty$, then f satisfies the equation

$$f(t) = \prod_{n=1}^{\infty}\cos 2^{-n}t = \frac{f(2t)}{\cos t}. \tag{1.8}$$

Since f is an entire function, this functional equation and the condition $f(0) = 1$ determine f uniquely – differentiate (1.8) to get the Taylor series of f. Since the function $(\sin t)/t$ satisfies (1.8), it equals f. Hence, as $N \to \infty$, the expectation value of $\sigma^\pm$ approaches zero. For finite N it follows from (1.8) that

$$f_N(t) = \prod_{n=1}^{N/2}\cos 2^{-n}t = \frac{\sin t}{t}\left[\frac{\sin t2^{-N/2}}{t2^{-N/2}}\right]^{-1}. \tag{1.9}$$

Therefore, as discussed earlier, the recurrence time $2^{N/2}/\pi$ grows exponentially with N, while the time it takes to reach equilibrium is independent of N.

To summarize, we have ascertained that for $N \to \infty$ the initially pure state of the algebra reduced to one spin tends as $t \to \infty$ to $\langle\sigma\rangle = s$, $\langle\sigma^\pm\rangle = 0$, which corresponds to a mixture:

$$\langle \sigma \rangle = \mathrm{Tr}(\rho\sigma), \quad \rho = \frac{\exp(-\eta\sigma)}{\mathrm{Tr}\exp(-\eta\sigma)}, \quad \tanh\eta = s. \tag{1.10}$$

Even though the expectation values of the $\sigma_k^{\pm}$ go to zero, their fluctuations remain nonzero, since $\sigma_k^{+}\sigma_k^{-} = (1+\sigma_k)/2$ is constant. The average magnetization

$$\mathbf{M}_N(t) = \frac{1}{N}\sum_k \sigma_k(t) \tag{1.11}$$

works differently. In the state (1.6) of our example, $\langle M_N^z \rangle = s$, whereas $\langle M_N^{\pm} \rangle$ is $O(N^{-1/2})$, provided either that the initial phases are disordered or that the $\sigma_k^{\pm}$ get out of phase after a while because the μ_k differ. The latter situation can in fact be undone by a sudden reversal of B, in the spin-echo effect. If $N = \infty$, the diffusion caused by suitable $\varepsilon(n)$ is irreversible, and $lim_{t\to\infty}\langle M_{\infty}^{\pm}(t)\rangle = 0$. At $t = 0$ the fluctuations are $O(N^{-1/2})$ and remain at this magnitude for all time: If $\sigma_k^{+}(t)\sigma_{k'}^{-}(t)$ is calculated by multiplying together two expressions of the form (1.1.6), then it should be recalled that $\sigma^2 = 1$. However, if the function $\varepsilon(n)$ falls off sufficiently rapidly with n, then the σ^2 terms make little difference for large $k-k'$, and the argument given earlier for the deviations of statistically independent quantities remains valid.

1.1.2 Chain of Oscillators

Now represent the total system by positions and momenta $q_1, \ldots, q_N, p_1, \ldots, p_N$, such that $[q_j, p_k] = i\delta_{jk}$, and let the time-evolution be determined by

$$H = \sum_{j=1}^{N} \frac{1}{2}(p_j^2 + (q_j - q_{j+1})^2). \tag{1.12}$$

This Hamiltonian contains interactions only between nearest neighbors, and the chain can be closed by the condition of periodicity $q_{j+N} = q_j$, $p_{j+N} = p_j$. The masses and force constants have been set to 1, which amounts to measuring the time in units of the natural period of oscillation. The equations of motion are

$$j = p_j, \quad \dot{p}_j = q_{j+1} + q_{j-1} - 2q_j. \tag{1.13}$$

With a periodic extension of the variables, $\xi_1, \ldots, \xi_{2N}$, such that

$$\xi_{2n} = p_n, \quad \xi_{2n+1} = q_{n+1} - q_n, \tag{1.14}$$

they are put into the form

$$\dot{\xi}_j = \xi_{j+1} - \xi_{j-1}. \tag{1.15}$$

The variables ξ_n satisfy

$$\xi_{n+2N} = \xi_n, \quad \sum_n \xi_{2n+1} = 0.$$

Recall that the Bessel functions satisfy the recursion formula $\dot{J}_n = (J_{n-1} - J_{n+1})/2$; as a consequence we see that the solution of the initial-value problem is

$$\xi_n(t) = \sum_{k=-\infty}^{\infty} \xi_k(0) J_{k-n}(2t). \tag{1.16}$$

1.1.3 Remarks

1. Since $|J_\nu(z)| \sim |z/\nu|^{|\nu|}$ as $|\nu| \to \infty$, the sum over k in (1.1.18) converges for, say, bounded $\{\xi_k(0)\}$.
2. If $N < \infty$, then (1.16) still holds provided that $\xi_{k+2N}(0) = \xi_k(0)$.
3. Since the equations of motion are linear, the classical and quantum time-automorphisms are identical.
4. There are still N constants of motion with the variables ξ:

$$I_k = \sum_{j=1}^{2N} \xi_j \xi_{j+k}, \qquad k = 1, \ldots, N.$$

With the auxiliary condition that $\sum_n \xi_{2n+1} = 0$, only $N - 1$ of the constants are independent, and we find that $\sum_n I_{2n+1} = 0$. If $N = \infty$, then I_k remains significant classically, provided that $\{\xi_k\} \in l^2$.

In order to have a useful framework for discussing the questions that will arise as in these two examples, it is convenient for technical reasons to make use of the Weyl algebra (cf.(III,§ 3.1)). With one particle, the Weyl algebra consists of the operators $W(r + is) = \exp(i(pr + qs))$, $r, s \in \mathbb{R}$, along with their linear combinations and norm-limits. A state on the Weyl algebra is uniquely characterized by the function $E(r, s) \equiv \langle \exp(i(pr + qs)) \rangle$. We shall only concern ourselves with coherent states (III:3.1.13), which are of the form $W(z')|u\rangle$, where $|u\rangle$ is a Gaussian function, the width of which determines the ratio between Δp and Δq. Since

$$\langle u|W(r + is)|u\rangle = \exp[-\frac{1}{4}(\omega r^2 + \frac{s^2}{\omega})],$$

it follows that

$$(\Delta p)^2 = -\frac{d^2}{dr^2} \ln E_{|r,s=0} = \frac{\omega}{2}, \qquad (\Delta q)^2 = -\frac{d^2}{ds^2} \ln E_{|r,s=0} = \frac{1}{2\omega}.$$

The expectation value in the more general state $W(z')|u\rangle$ can be calculated according to (III:3.1.2;1) as

$$\begin{aligned} \langle W(z')u|W(z)|W(z')u\rangle &= \langle u|W(-z')W(z)W(z')|u\rangle \\ &= \langle u|W(z)|u\rangle \exp[\frac{i}{2}\mathrm{Im}(z^* z' - z^{*'} z)] \\ &= \exp[-\frac{1}{4}(\omega r^2 + \frac{s^2}{\omega}) + i(rs' - r's)]. \end{aligned} \tag{1.17}$$

Thus, the quantities Δp and Δq are the same as with $|u\rangle$, but the expectation values of p and q are now s' and $-r'$.

Let us return to the issue of how the restriction of the many-particle state to a subsystem evolves in time. The operators $\exp[i(r\xi_0(t)+s\xi_1(t))]$, which describe the momentum of a single particle and its position relative to its neighbor, are useful to this end. Since $[\xi_0(t),\xi_1(t)] = i$, they form a Weyl system. A state characterized by

$$\Big\langle \exp\Big[i\sum_{n=-\infty}^{\infty}(\xi_{2n}r_n+\xi_{2n+1}s_n)\Big]\Big\rangle = \exp\Big[-\frac{1}{4}\sum_{n=-\infty}^{\infty}\Big(\omega r_n^2+\frac{s_n^2}{\omega}\Big) + i(r_n s_n' - r_n' s_n)\Big] \tag{1.18}$$

can be regarded as the generalization of (1.1.20).

1.1.4 Remarks

1. The exponent on the left is a linear combination of p_k and q_n, as appropriate for a Weyl system for several particles, yet the variables ξ_{2n} and ξ_{2n+1} are not pairs of canonically conjugate variables, since $[\xi_{2n},\xi_{2n-1}]\neq 0$. Thus (1.1.21) is not simply the tensor product of coherent states of a tensor product of Weyl systems.
2. The significance of (1.18) is once again that the variables ξ_{2n} (resp. ξ_{2n+1}) all have deviation ω and expectation values s_n' (resp. $1/\omega$ and $-r_n'$).

With (1.18), the desired state on the one-particle system turns out to be

$$\begin{aligned} E(r,s) &\equiv \langle \exp(i(r\xi_0(t)+s\xi_1(t)))\rangle \\ &= \Big\langle \exp\Big(i\sum_{n=-\infty}^{\infty}\big[\xi_{2n}(0)(rJ_{2n}+sJ_{2n-1})+\xi_{2n+1}(0)(rJ_{2n+1}+sJ_{2n})\big]\Big)\Big\rangle \\ &= \exp\sum_{n=-\infty}^{\infty}\Big\{-\frac{1}{4}\Big(\omega(rJ_{2n}+sJ_{2n-1})^2+(rJ_{2n+1}+sJ_{2n})^2\frac{1}{\omega}\Big) \\ &\quad + is_n'(rJ_{2n}+sJ_{2n-1}) - ir_n'(rJ_{2n+1}+sJ_{2n})\Big\}. \end{aligned} \tag{1.19}$$

The sums can be evaluated by recourse to the formulas

$$\begin{aligned} \sum_{n=-\infty}^{\infty} J_{2n}(2t)J_{2n+j}(2t) &= \frac{1}{2}(\delta_{0j}+J_j(4t)), \quad j\in\mathcal{Z} \\ \sum_{n=-\infty}^{\infty} J_{2n+1}(2t)J_{2n+1+j}(2t) &= \frac{1}{2}(\delta_{0j}-J_j(4t)), \end{aligned} \tag{1.20}$$

which are derived in Problem 2. As $t\to\infty$, only the terms with $j=0$ remain. Moreover, it can be seen from the integral representations and the Riemann–Lebesgue lemma that the contributions linear in the J_k go to zero as $t \to \infty$. In all, we get

$$\lim_{t\to\infty} E(r,s) = \exp\left[-\frac{1}{4}\left(\omega + \frac{1}{\omega}\right)(r^2+s^2)\right]. \tag{1.21}$$

1.1.5 Remarks

1. The limiting state corresponds to the mixture $E = \operatorname{Tr} \rho W(z)$, $\rho = \exp[-\eta(p_1^2 + q_1^2)]/\operatorname{Tr}\exp[-\eta(p_1^2+q_1^2)]$, $\coth\eta = (\omega + 1/\omega)/2$ (Problem 3). As $\omega \to 1$, that is, for minimal mean-square deviation, $\eta \to \infty$, and the state becomes pure. With larger mean-square deviations, $\omega \neq 1$, $(\omega+1/\omega)/2 > 1$, the limiting state is a mixture.
2. Whereas at $t = 0$ the ratio of Δp to Δq is ω^2, they become equal as $t \to \infty$, i.e., their ratio, 1, becomes the one defined by H. This corresponds to equal amounts of kinetic and potential energy.
3. The reason that the existence of the constants (1.1.3; 4) does not prevent the onset of equilibrium is again the choice of the initial state. Of course, equilibrium can not occur if the system starts off in an eigenstate of a normal mode of oscillation.

These few remarks will serve as our first orientation to irreversible phenomena. We have already studied an example of an irreversible phenomenon in volume II, the emission of light. It is always important to take the limit $N \to \infty$ before $t \to \infty$, as in a finite volume the light returns to the point of emission, and the behavior is almost periodic rather than irreversible. The next section will deal with how the energy is affected by the first limiting process.

1.1.6 Problems

1. Calculate the entropy $S(t) = -\operatorname{Tr}\rho(t)\ln\rho(t)$ for one spin, where f is given by (1.1.9).
2. Calculate $\sum_{n=-\infty}^{\infty} J_{2n}(x)J_{2n+j}(x)$ and $\sum_{n=-\infty}^{\infty} J_{2n+1}(x)J_{2n+1+j}(x)$.
3. Show that the density matrix ρ has the property stated in (1.1.5;1).

1.1.7 Solutions

1. Since $\operatorname{Tr}\rho(t) = 1$, the density matrix is of the form $\rho(t) = \frac{1}{2} + (t)\cdot\boldsymbol{\sigma}$. Let $c(t) = |(t)|$, which $\leq \frac{1}{2}$. The eigenvalues of $\rho(t)$ are $\frac{1}{2} \pm c(t)$, so

$$S(t) = -\left[\frac{1+c(t)}{2}\ln\frac{1+c(t)}{2} + \frac{1-c(t)}{2}\ln\frac{1-c(t)}{2}\right].$$

Because $\operatorname{Tr}\sigma_i\sigma_j = 2\delta_{ij}$, we find $(t) = \frac{1}{2}\langle\boldsymbol{\sigma}\rangle$, and therefore $c(t) = (s^2 + (1-s^2)f^4(t))^{1/2}$. Observe that f is not monotonic, and hence that S does not increase monotonically from 0 to its equilibrium value,

$$-\left[\frac{1+s}{2}\ln\left(\frac{1+s}{2}\right) + \frac{1-s}{2}\ln\left(\frac{1-s}{2}\right)\right].$$

2.

$$\exp\left[\frac{z}{2}\left(t-\frac{1}{t}\right)\right]=\sum_{n=-\infty}^{\infty} t^n J_n(z).$$

Putting $z = x + y$ yields,

$$\sum_{j=-\infty}^{\infty} t^j J_j(x+y)=\exp\left[\frac{x}{2}\left(t-\frac{1}{t}\right)\right]\exp\left[\frac{y}{2}\left(t-\frac{1}{t}\right)\right]$$

$$=\Big(\sum_k t^k J_k(x)\Big)\Big(\sum_t t^t J_t(y)\Big)=\sum_{j=-\infty}^{\infty} t^j \sum_{n=-\infty}^{\infty} J_n(x)J_{j-n}(y),$$

so $J_j(x+y)=\sum_{n=-\infty}^{\infty} J_n(x)J_{j-n}(y)$, which is the addition theorem of Schläfli and Neumann. Putting $y = -x$ and changing j to $-j$ then yields $\sum_n J_n(x)J_{n+j}(x)=\delta_{j0}$, and with $y = x$, there results

$$\sum_n J_n(x)J_{-j-n}(x)=\sum_n(-1)^{n+j}J_n(x)J_{n+j}(x)=J_{-j}(2x)=(-1)^j J_j(2x),$$

from which formulas (1.20) follow.

3.

$$\operatorname{Tr}\exp[-\eta(p_1^2+q_1^2)]=\sum_{n=0}^{\infty}\exp[-\eta(1+2n)]$$

and

$$\langle p^2+q^2\rangle=\Big(-\frac{\partial^2}{\partial r^2}-\frac{\partial^2}{\partial s^2}\Big)E(r,s)=-\frac{\frac{\partial}{\partial\eta}\operatorname{Tr}\exp[-\eta(p^2+q^2)]}{\operatorname{Tr}\exp[-\eta(p^2+q^2)]}$$

lead to the result.

1.2 The Limit of an Infinite Number of Particles

The first issues to confront for large systems are what happens to macroscopic properties like energy and volume as $N \to \infty$.

The models examined in § 1.1 were only caricatures of reality. We shall now determine the physical properties of large bodies. The first question is how the volume V has to vary as $N \to \infty$, in order to ensure that the potential and kinetic energies will be comparable in magnitude and that the interaction between the particles is correctly accounted for. In particular, when are E and V normal, extensive quantities proportional to N? In order to fix our ideas, we shall pay particular attention to certain special cases, large atoms and macroscopic or cosmic objects. The dominant force is then electrostatic, except that in cosmic matter gravity also has a decisive effect. Heuristic arguments will sometimes be adduced in this section for guidance in finding which quantities have limits as $N \to \infty$ in these systems. The mathematical derivation of these results will be supplied in § 4.

1.2.1 Free Particles

We begin with a consideration of noninteracting particles confined to a box of side R. The energy consists of the quantum-mechanical zero-point energy plus a thermal component proportional to the temperature T. As we are only interested in the dependence on N for large N, we set $\hbar = k = m = 1$. As explained in (III:1.11) the zero-point energy of a system of fermions is $\sim (\Delta p)^2 \sim (\Delta x)^{-2}$, where Δx is about $RN^{-1/3}$, since the volume available per fermion is only R^3/N. We arrive at

$$E = \frac{N^{5/3}}{2R^2} + \frac{3}{2}NT. \tag{1.22}$$

If the two contributions are to remain comparable as $N \to \infty$, and if T goes as N^t for some power t, then R must be $\sim N^{1/3-t/2}$, and EN^{-1-t} will tend to a limiting value. The type of interaction will determine the value of t at which the limit is nontrivial and thus of physical interest. For this to happen the kinetic and potential energies have to remain of the same order of magnitude.

Bosons do not have the solitary temperament, so Δx may be set equal to R. The energy is then on the order of

$$E = \frac{N}{2R^2} + \frac{3}{2}NT. \tag{1.23}$$

If the two contributions are to have the same dependence on N and we make $T \sim N^t$, then $R \sim N^{-t/2}$ and $E \sim N^{t+1}$. If it is insisted that T remain constant and $R \sim N^{1/3}$, then $E \sim N$, but the zero-point energy drops below the thermal energy. The exact calculation for free bosons in fact reveals that, with a fixed particle density and below a critical temperature, a certain fraction $\lambda(T) > 0$ of the particles are to be found in the ground state with $E_0 \sim N^{1/3}$, and thus N may be replaced with $(1 - \lambda(T))N$. This makes this usual limit also nontrivial.

1.2.2 Large Atoms

The Hamiltonian of a large atom (with $e^2 = 1$) is

$$H = \sum_{i=1}^{N}\left(\frac{|\mathbf{p}_i|^2}{2} - Z|\mathbf{x}_i|^{-1}\right) + \sum_{i>j}|\mathbf{x}_i - \mathbf{x}_j|^{-1}, \tag{1.24}$$

which can, if one wishes, be confined in a box. Recall that in volume III we figured out that if $T = 0$ and $Z = N$, the energy is about $N^{5/3}/2R - N^2e^2/R$, which has a minimum about $-\frac{1}{2}N^{7/3}$ for $R \sim N^{-1/3}$. Therefore, in the limit $N \to \infty$ we should expect to set $t = \frac{4}{3}$. In § 4.1 it will not only be proved that these limits converge, but even that the Thomas-Fermi theory becomes exact in that limit. The problem can thus be solved in the limit $N \to \infty$, though the solution is not suitable for a direct numerical comparison of theory and experiment. Since there are corrections of about $N^{-1/3}$, 10% accuracy can not be expected for $N \lesssim 10^3$. On the other hand, relativistic effects become significant when $N \sim 10^2$. The kinetic energy is then $\sim N^{4/3}/R$ and if $Ze^2 > 1$ the energy is no longer bounded below. Hence the picture that emerges of a large atom is only an idealization, but at least one with many instructive aspects.

Systems of bosons depend on N in a different way. They all settle into the ground state, and with $Z \sim N$ the radius goes as N^{-1} and the energy as N^3. The limits of EN^{-3} and $N^3\rho(xN)$ would be expected to exist, where ρ is the one-particle density distribution. For thermal effects to remain significant, T must be chosen $\sim N^2$. This problem is mostly of academic interest, and the convergence of the quantities mentioned above has been proved only for the ground state.

1.2.3 Jellium

Like an atom, jellium consists of particles repelling one another with a Coulomb force and immersed in the field of an external charge distribution. The difference is that the charge distribution is not concentrated at a point, but rather homogeneously spread with density ξ, through a box Λ (Λ will also sometimes denote the volume of Λ). It can be regarded as a model of highly compressed matter, with the homogeneous background charge coming from fast-moving electrons, and the particles with explicit coordinates being the nuclei. It is nevertheless often used to describe electrons in a metal, although it is rather far-fetched to speak of the assemblage of ions as a homogeneous background. The Hamiltonian is

$$H = \sum_{i=1}^{N} \frac{|\mathbf{p}_i|^2}{2} + \sum_{i>j} |\mathbf{x}_i - \mathbf{x}_j|^{-1} - \sum_{i=1}^{N} U(\mathbf{x}_i) + \frac{\xi}{2} \int_\Lambda d^3x U(\mathbf{x}), \tag{1.25}$$

where $U(\mathbf{x}) = \xi \int_\Lambda d^3x'/|\mathbf{x}-\mathbf{x}'|$. For the system to be neutral, $\xi \int_\Lambda d^3x = N$. The electrostatic energy of the background has been added in so that the potential energy will remain bounded below by $N(RN^{-1/3})^{-1}$, where R is the linear dimension of Λ. The proof of this relies on the well-known fact of electrostatics that the Coulomb repulsion of two homogeneously charged spheres is less than or equal to that of two point charges at their centers – the inequality occurs when they overlap. Now imagine blowing the charged particles up to homogeneously charged spheres of radius a, and let

$$\left(\frac{4\pi a^3}{3}\right)^{-2} \int\limits_{\substack{|\mathbf{x}-\mathbf{x}_i|\le a \\ |\mathbf{x}'-\mathbf{x}_j|\le a}} \frac{d^3x d^3x'}{|\mathbf{x}-\mathbf{x}'|} = U_{ij}(a),$$

$$\left(\frac{4\pi a^3}{3}\right)^{-1} \int\limits_{|\mathbf{x}-\mathbf{x}_i|\le a} d^3x U(\mathbf{x}) = U_i(a). \tag{1.26}$$

Then H may be written in the form

$$H = \sum_{i=1}^{N} \frac{|\mathbf{p}_i|^2}{2} + \overbrace{\frac{1}{2}\sum_{i,j=1}^{N} U_{ij}(a) - \sum_{i=1}^{N} U_i(a) + \frac{\xi}{2}\int d^3x U(\mathbf{x})}^{\alpha}$$
$$+ \overbrace{\sum_{i=1}^{N}(U_i(a) - U(\mathbf{x}_i))}^{\beta} - \overbrace{\frac{1}{2}\sum_i U_{ii}(a)}^{\gamma} + \overbrace{\sum_{i<j}(|\mathbf{x}_i - \mathbf{x}_j|^{-1} - U_{ij}(a))}^{\delta}. \quad (1.27)$$

Contribution α is positive, since it is of the form

$$\int \frac{dxdx'}{|\mathbf{x} - \mathbf{x}'|}\rho(\mathbf{x})\rho(\mathbf{x}'),$$

and $1/|\mathbf{x}|$ has a positive Fourier transform. It is easy to show (Problem 1) that $\beta \geq -(2\pi/5)\xi a^2 N$, equality holding provided that all the spheres lie within Λ, and $\gamma = (N/2)(6/5a)$, the self-energy of homogeneously charged spheres. As discussed earlier, $\delta \geq 0$. The lower bound $-N((2\pi/5)\xi a^2 + (3/5a))$ is optimized at $a = (3/4\pi\xi)^{1/3} \equiv r_s$, which is precisely the radius at which the sum of the volumes of the spheres equals that of Λ. This computation leads to the

1.2.4 Lower Bound for the Energy

$$H \geq \sum_{i=1}^{N} \frac{|\mathbf{p}_i|^2}{2} - \frac{9}{10}\frac{N}{r_s}.$$

1.2.5 Remarks

1. Nothing has yet been assumed about the shape of Λ or the statistics of the particles. In particular, if Λ is spherical, then by Problem 2,

$$-\sum_{i=1}^{N} U(\mathbf{x}_i) + \frac{\xi}{2}\int_{\Lambda} d^3x U(\mathbf{x}) \leq \frac{N}{2R^3}\sum_{i=1}^{N} |\mathbf{x}_i|^2 - \frac{9}{10}\frac{N^2}{R},$$

where equality holds if $x_i \in \Lambda$ for all i.

2. Despite its great generality, the numerical accuracy of the bound (1.2.4) is surprisingly good. If $\mathbf{x}_i$ are the sites of a simple, face-centered, or body-centered cubic lattice, computer studies have been made of the limit as $N \to \infty$ of the potential energy over Nr_s^{-1}, yielding respectively the values – 0.880, – 0.895, and – 0.896 [3].

Lower bounds for H depending on the particle statistics may be derived from (1.2.4). The energy of free fermions is, as seen earlier, $\sim N^{5/3}/R^2 \sim Nr_s^{-1}$, and with the aid of the more precise proportionality factor,

$$H \geq N(1.1r_s^{-2} - 0.9r_s^{-1}) \geq -\frac{0.81}{4.4}N \quad \text{for} \quad \text{all} \quad r_s \in \mathbb{R}^+ \tag{1.28}$$

for spin-$\frac{1}{2}$ particles. Even if the volume and consequently r_s are treated as variables, the resultant lower bound is $\sim N$. We shall discover later that with no more than first-order perturbation theory we can obtain an upper bound not much different from (1.2.12): the Pauli exclusion principle makes the electrons stay at a distance r_s apart, and this correlation imitates the energetically favorable configurations of (1.2.5; 2). Since the minimizing radius r_s does not depend on N, in this model $E \sim N$ and $R \sim N^{1/3}$, so the exponent t of (1.2.1) equals zero.

A very different picture emerges of bosons. With the kinetic energy (1.2.3) we find, ignoring precise coefficients, that

$$H \geq \frac{N^{1/3}}{r_s^2} - \frac{N}{r_s}. \tag{1.29}$$

The minimizing r_s is $\sim N^{-2/3}$, and so $E \sim N^{5/3}$

1.2.6 Remarks

1. If the background charge is concentrated at discrete points of a lattice, then trial functions can be thought up that show $E < -cN^{5/3}$, and thus in this case the energy in fact goes as $N^{5/3}$ [2].
2. So far only the electrostatic energy has been accommodated in the background, and minimized according to the density ξ. If the background consists of electrons, then its zero-point energy must also be calculated. In a jellium of deuterium atoms, which are bosons, the energy turns out to be $\sim N$: The background density prevents them from collapsing, and for fixed r_s (1.2.13) is on the order of N.

1.2.7 Real Matter

Real matter consists of positive and negative point-particles interacting with a Coulomb force, so

$$H = \sum_{i=1}^{N} \frac{|\mathbf{p}_i|^2}{2m_i} + \sum_{i>j} \frac{e_i e_j}{|\mathbf{x}_i - \mathbf{x}_j|} \tag{1.30}$$

for particles confined to a box of volume $\Lambda \sim R^3$. We shall often particularize to the situation wherein all negative particles are identical with $m = |e| = 1$ and all positive particles are identical with mass M and charge Z. Provided that Z is not so large that relativistic effects become significant, (1.30) gives a reasonably accurate description of ordinary matter. We therefore expect to find that $E \sim -N$ for $R \sim N^{1/3}$.

The proof of this fact, known as the "stability of matter", has to be deferred to § 4.3. At this point we shall make do with several

1.2.8 Remarks

1. Roughly speaking, the difficulty is that the double sum for the potential energy contains $\sim N^2$ terms, so many cancellations are needed for the result to be only $\sim N$. If, as in the gravitating system to be described shortly (1.2.9), all the contributions are of like sign, then cancellations certainly do not occur. Similarly, if the total charge $Q \equiv \sum_i e_i$ is $\sim N^{2/3+\varepsilon}$ and the system is restricted to a region of linear dimension $R \sim N^{1/3}$, the energy fails to be extensive. The electrostatic energy Q^2/R is $\leq N$ only if $Q \leq N^{2/3}$.
2. Even requiring that $Q = 0$ will not guarantee that $|E| \sim N$ if all the particles are bosons. To prove this, rewrite (1.30) (with $M = Z = 1$) as

$$H = \sum_{i=1}^{N^-} \frac{|\mathbf{p}_i|^2}{2} + \sum_{\alpha=1}^{N^+} \frac{|\mathbf{p}_\alpha^+|^2}{2} + \sum_{i>j} |\mathbf{x}_i^- - \mathbf{x}_j^-|^{-1} + \sum_{\alpha>\beta} |\mathbf{x}_\alpha^+ - \mathbf{x}_\beta^+|^{-1}$$
$$- \sum_{i,\alpha} |\mathbf{x}_i^- - \mathbf{x}_\alpha^+|^{-1}, \tag{1.31}$$

where $N^+ = N^-$ for a neutral system. Now take the expectation value in a state with $\Psi^+ \otimes \Psi^-$, where $\Psi^\pm$ are the trial functions that led to $E \sim -N^{7/5}$ for Bose-jellium. Although the particles are correlated, the charge density is homogeneous, as for instance

$$\langle \Psi^+ | - \sum_{i,\alpha} |\mathbf{x}_i^- - \mathbf{x}_\alpha^+|^{-1} | \Psi^+ \rangle = -\xi \sum_i \int_\Lambda \frac{d^3x}{|\mathbf{x}_i^- - \mathbf{x}|}.$$

The last term in (1.31) is therefore equivalent to $-\sum_i U(\mathbf{x}_i^-) - \sum_\alpha U(\mathbf{x}_\alpha^+) + 2(\xi/2) \int d^3x U(\mathbf{x})$, and there results the sum of the energies of the positive and negative Bose-jellia. The expectation value is consequently about $-N^{7/5}$, which is an upper bound to the energy by the min-max principle (III:3.5.19). This "instability," which corresponds to the ground-state energy being nonextensive and the spatial contraction of many-particle aggregates of charged bosons, does not imply that individual atoms consisting of oppositely charged bosons would be unstable. A single, nonrelativistic atom of He^4 with its electrons subjected to Bose statistics (but with their original mass and charge) would have the same ground-state energy as real He^4, since the two-particle ground-state wave-function is symmetric in the spatial coordinates. The lesson here is that experience with two-electron molecules is not a trustworthy guide to the problem of the stability of matter: Since the Pauli exclusion principle makes no difference, the two electrons might just as well be bosons, but a system of many bosons would be unstable, whereas a many-fermion system is stable.
3. Since He^3 is just as stable as He^4, stability is not a matter of the type of statistics of one of the kinds of charge-carrier. Moreover, the relevant energy is always

measured in Rydbergs, using the electronic mass, so matter should remain stable even in the limit of infinite nuclear masses.

4. It could be argued heuristically that the potential energy should go as $-N^{4/3}R^{-1}$, since each charge sees an opposite charge at a distance $RN^{-1/3}$, while charges further away should be screened. If this is added to the kinetic energy $N^{5/3}R^{-2}$ of fermions or NR^{-2} of bosons, the minimum is respectively $\sim -N$ at $R \sim N^{1/3}$ or $\sim -N^{5/3}$ at $R \sim N^{-1/3}$. This is actually the true dependence on N if the nuclei have infinite masses whereas for finite nuclear masses the energy goes with $N^{7/5}$.
5. In relativistic dynamics the kinetic energy is $\sim |\mathbf{p}| \sim 1/\Delta x$, so the system is softer. The heuristic arguments would evaluate the total energy of bosons as $\sim N/R - e^2 N^{4/3}/R$, which is unbounded below when N is sufficiently large. Whereas nonrelativistic energies are always semibounded for any fixed N, it may happen that the relativistic energy goes to $-\infty$ for sufficiently large, but still finite, values of N.
6. The instability of a Coulomb system of bosons has nothing to do with the long range of the $1/r$ potential, but comes from its short-range features. If the singularity is chopped off by changing the potential to $V(x) = (1 - \exp(-\mu r))/r$, the system of bosons also becomes stable: Since the Fourier transform of V is

$$\widetilde{V}(\mathbf{k}) = \frac{4\pi\mu^2}{|\mathbf{k}|^2(|\mathbf{k}|^2 + \mu^2)} > 0,$$

with $|e_i| = e$, we find that

$$V \equiv \sum_{i>j} e_i e_j V(\mathbf{x}_i - \mathbf{x}_j) = \frac{1}{2}\int \frac{d^3k}{(2\pi)^3}\widetilde{V}(\mathbf{k})\Big|\sum_j \exp(i\mathbf{k}\cdot\mathbf{x}_j)e_j\Big|^2$$

$$-\frac{1}{2}\sum_{i=1}^{N} e_i^2 V(\mathbf{0}) > -\frac{N}{2}e^2 V(0) = -\frac{N}{2}e^2\mu,$$

so H is bounded below by $-cN$. It could be argued that nuclei have a form factor, and that if μ is taken as the reciprocal of the nuclear radius, then V would be a more realistic potential than $1/r$. This would lead to a simple proof of stability, but it misses the real point. Since the Rydberg, which is measured in electronvolts (eV), is determined by the mass of the electron, it is the kinetic energy of the electrons rather than the size of the nuclei that matters most for stability. The lower bound from the size of the nuclei alone would be $\sim -N$ MeV.

1.2.9 Cosmic Bodies

The $1/r$ potentials in an object with gravitationally interacting particles are all attractive, so the situation is drastically different. The ground state of the

Hamiltonian

$$H_G = \sum_{i=1}^{N} \frac{|\mathbf{p}_i|^2}{2} - \kappa \sum_{i>j} |\mathbf{x}_i - \mathbf{x}_j|^{-1} \tag{1.32}$$

goes as $-N^{7/3}$ for fermions. By the now familiar argument, $E \sim N^{5/3}/R^2 - N^2/R$, which has its minimum value $\sim -N^{-7/3}$ for $R \sim N^{-1/3}$. This can easily be translated into an exact upper bound by the use of trial functions localized in $\mathbb{R}^3$. Lower bounds are harder to come by, since energetically more favorable possibilities have to be ruled out. In this case there is an easier way: Write

$$H_G = \sum_{i=1}^{N} \sum_{j \neq i} \left(\frac{|\mathbf{p}_j|^2}{2(N-1)} - \frac{\kappa}{2} |\mathbf{x}_i - \mathbf{x}_j|^{-1} \right) \equiv \sum_{i=1}^{N} h_i, \tag{1.33}$$

so that each h_i is the Hamiltonian of an atom with electrons having no Coulomb repulsion. Particle number i stands for the atomic nucleus, as it has no kinetic energy, and the others are electrons, with mass $N-1$ and potential $-|\mathbf{x}_i - \mathbf{x}_j|^{-1}/2$. According to (III:4.5.9) it follows that $h_i \geq -cN^{4/3}$, and indeed the result is a

1.2.10 Bound for the Energy of Gravitating Fermions

$$H_G > -cN^{7/3}, \qquad c = O(1).$$

1.2.11 Remarks

1. Fermi statistics were not fully taken into account, since we have only antisymmetrized with respect to $N-1$ particles when filling the energy levels. Since complete antisymmetrization restricts the set of admissible functions further, (1.2.10) is at any rate a lower bound.
2. The limit as $N \to \infty$ in this case exists with the scaling behavior $t = 4/3$ of (1.2.1), as in (1.2.2). This does not mean that the limit with $t = 4/3$ fails to exist for ordinary matter, but only that it is trivial. The potential energy goes to zero and the particles remain free.
3. If the particles are bosons, then they can all be put into the ground state, and $E \sim -N^3$. The radius of the ground state then goes as N^{-1}.
4. The Hamiltonian (1.32) was appropriate for the discussion of electrically neutral particles; if they are instead charged, then κ must be replaced with $\kappa - e_i e_j$. If we bear normal matter in mind, the gravitational force comes from the protonic mass, and in units where the mass of the proton is 1, $\kappa/e^2 \sim 10^{-36}$. Inequality (1.2.10) then *a fortiori* provides a lower bound, since

$$\frac{1}{2}\sum_i \frac{|\mathbf{p}_i|^2}{2} + \sum_{i>j} \frac{e_i e_j}{|\mathbf{x}_i - \mathbf{x}_j|} + \frac{1}{2}\sum_i \frac{|\mathbf{p}_i|^2}{2} - \sum_{i>j} \frac{\kappa}{|\mathbf{x}_i - \mathbf{x}_j|}$$
$$\geq -2c_e e^4 N - 2c\kappa^2 N^{7/3},$$

c_e being the constant for charged particles. The number of particles determines which N-dependence dominates. Gravity begins to win out when $N \sim (e^2/\kappa)^{3/2} \sim 10^{54}$, which is about the mass of Jupiter, and the energies of larger heavenly bodies are controlled mainly by gravitation. A concrete consequence is that the atoms get squashed and turn into a plasma of nuclei and electrons. This inequality provides a more rigorous foundation for the heuristic considerations of (II:4.5.1).

We shall see in § 4.2 that the system (1.32) can be solved in the limit $N \to \infty$, as the Thomas-Fermi theory becomes exact. Thomas-Fermi theory provides an idealization of stars, various corrections again being needed to make it realistic. In particular, if $N \sim 10^{54}$ relativistic effects become important. As with atoms with $Z > 137$, the Hamiltonian is unbounded below, which leads to a catastrophe. Nonetheless, Thomas-Fermi theory reflects the thermodynamic properties of stars rather well.

This section concludes with Table 1.1 displaying the many possibilities:

Table 1.1. The N-dependence of the kinetic energy K and the potential energy V when N is large.

			K	V	$R_{\min}$	$E(R_{\min})$
Nonrelativistic	electric	Bose	N/R^2	$-N^{4/3}/R$	$N^{-1/3}$	$-N^{5/3}$
		Fermi	$N^{5/3}/R^2$	$-N^{4/3}/R$	$N^{1/3}$	$-N$
	gravitational	Bose	N/R^2	$-N^2/R$	N^{-1}	$-N^3$
		Fermi	$N^{5/3}/R^2$	$-N^2/R$	$N^{-1/3}$	$-N^{7/3}$
Relativistic	electric	Bose	N/R	$-N^{4/3}/R$	0	$-\infty$
		Fermi	$N^{4/3}/R$	$-N^{4/3}/R$	0	$-\infty$
		†			or ∞	or 0
	gravitational	Bose	N/R	$-N^2/R$	0	$-\infty$
		Fermi	$N^{4/3}/R$	$-N^2/R$	0	$-\infty$

† If $R_{\min}$ tends to $+\infty$ more rapidly than $N^{1/3}$, then the kinetic energy per particle, $N^{1/3}/R$, becomes arbitrarily small, eventually $\ll m$, and the system is nonrelativistic. Hence $R_{\min}$ certainly can not increase faster than $N^{1/3}$. Which energy breaks the stalemate depends on the strength of the charge. If $Z < 137$, the kinetic energy wins out, and if $Z > 137$, the potential energy wins out.

1.2.12 Problems

1. Calculate the β and γ of (1.2.3).
2. Verify (1.2.5;1).

1.2.13 Solutions

1.

$$\gamma : \int\limits_{\substack{|x|\le a\\|x'|\le a}} \frac{d^3x d^3x'}{|\mathbf{x}-\mathbf{x}'|} =$$

$$= \int r^2 dr d\Omega r'^2 dr' d\Omega' \sum_{n,m} \Big[\frac{r^n}{r'^{n+1}} \Theta(r'-r) + \frac{r'^n}{r^{n+1}} \Theta(r-r') \Big] \frac{4\pi}{2n+1}$$

$$\cdot\, Y_n^m(\Omega) Y_n^{m*}(\Omega') = \int_0^a \int_0^a r^2 dr r'^2 dr' \Big(\frac{\Theta(r'-r)}{r'} + \frac{\Theta(r-r')}{r} \Big) (4\pi)^2$$

$$= \frac{2a^5}{15} (4\pi)^2.$$

$$\beta : \int\limits_{\substack{|\mathbf{x}|\le a\\ \mathbf{x}'\in\Lambda}} d^3x d^3x' \Big(\frac{1}{|\mathbf{x}-\mathbf{x}'|} - \frac{1}{|\mathbf{x}'|} \Big) = \int\limits_{|\mathbf{x}|,|\mathbf{x}'|\le a} \cdots + \int\limits_{\substack{|\mathbf{x}|\le a\\ |\mathbf{x}'|\ge a}} \cdots$$

The second integral equals 0, as can be seen by expanding $|\mathbf{x}-\mathbf{x}'|^{-1}$ in spherical harmonics. The first integral equals $-(2\pi a^2/5)(4\pi a^3/3)$ if $\{\mathbf{x}' : |\mathbf{x}'| \le a\} \subset \Lambda$, and is otherwise greater than or equal to this.

2. $U(\mathbf{x}_i) \le -(3N/2R) + (N/2R)(|\mathbf{x}_i|^2/R^2)$, equality holding for $|x_i| < R$. The self-energy of the background charge is $3N^2/5R$.

1.3 Arbitrary Numbers of Particles in Fock Space

The properties of large systems should not depend on the exact number of particles, so it is convenient to use a representation with a variable number of particles.

We are used to dealing with atomic systems on $\mathcal{H}_n$, the n-particle Hilbert space. As it is impossible to count the particles in a large system, it is convenient to regard the number N of particles as an observable capable of assuming various values. Accordingly, we shall study **Fock space**

$$\mathcal{H}_F = \bigoplus_{n=0}^{\infty} \mathcal{H}_n, \qquad N_{|\mathcal{H}_n} = n, \tag{1.34}$$

as the foundation for later analysis. The space $\mathcal{H}_0$ is one-dimensional and spanned by the **vacuum vector** $|0\rangle$. If the particles under consideration are either all bosons or all fermions, then $\mathcal{H}_n$ is either the n-fold symmetric or totally antisymmetric tensor product of $\mathcal{H}_1 = L^2(\mathbb{R}^3, d^3x)$ with itself, which will be denoted $\mathcal{H}_1 \circledS \mathcal{H}_1 \circledS \cdots \circledS \mathcal{H}_1$ or $\mathcal{H}_1 \mathbf{w} \mathcal{H}_1 \mathbf{w} \cdots \mathbf{w} \mathcal{H}_1$. If f_j, j=1,2,..., is a complete orthonormal set of functions on $\mathcal{H}_1$, then the vectors $|f_{j_1} \circledS f_{j_2} \circledS \cdots \circledS f_{j_n}\rangle$ or respectively $|f_{j_1} \mathbf{w} f_{j_2} \mathbf{w} \cdots \mathbf{w} f_{j_n}\rangle$ are a basis for $\mathcal{H}_n$. In the latter case all the j_k are to be taken different. For bosons the same f's can be collected together and written as $|f_{j_1}^{n_1}, \ldots, f_{j_n}^{n_k}\rangle$, with $\sum_k n_k = N$. The C^* algebra generated on the individual $\mathcal{H}_n$ of the boson Fock space by the symmetrized Weyl operators

$$\sum_{\pi} \exp\Big[i \sum_j (r_{\pi_j} x_j + s_{\pi_j} p_j)\Big],$$

where $(\pi_1, \ldots, \pi_n)$ is a permutation of $(1, \ldots, n)$, will be called the **Weyl algebra**, and is represented reducibly on $\mathcal{H}_F$ – all bounded functions of N alone belong to the commutant of the representation.

The irreducible **field algebra** on $\mathcal{H}_F$ turns out to be invaluable for the many-body problem:

1.3.1 Definition

Let $|f_1, f_2, \ldots\rangle \equiv |f_1 Ⓢ f_2 \ldots\rangle$, and define the **creation and annihilation operators** $a^*(f)$ and $a(f)$ by linear extension

$$\begin{aligned}
a(f_m)|f_{j_1}^{n_1}, \ldots, f_{j_k}^{n_k}\rangle &= \delta_{mj_1}\sqrt{n_1}|f_{j_1}^{n_1-1}, f_{j_2}^{n_2}, \ldots, f_{j_k}^{n_k}\rangle \\
&\quad + \delta_{mj_2}\sqrt{n_2}|f_{j_1}^{n_1}, f_{j_2}^{n_2-1}, \ldots, f_{j_k}^{n_k}\rangle + \ldots \\
&\quad + \delta_{mj_k}\sqrt{n_k}|f_{j_1}^{n_1}, f_{j_2}^{n_2}, \ldots, f_{j_k}^{n_k-1}\rangle \qquad \text{(for bosons)}, \\
a(f_m)|f_{j_1} \wedge \cdots \wedge f_{j_n}\rangle &= \delta_{mj_1}|f_{j_2} \wedge \cdots \wedge f_{j_n}\rangle - \delta_{mj_2}|f_{j_1} \wedge f_{j_3} \wedge \cdots \wedge f_{j_n}\rangle \\
&\quad + \cdots + (-1)^{n+1}|f_{j_1}, \ldots, f_{j_{n-1}}\rangle \qquad \text{(for fermions)}, \\
a^*(f_m)|f_{j_1}^{n_1}, \ldots, f_{j_k}^{n_k}\rangle &= \delta_{mj_1}\sqrt{n_1+1}|f_{j_1}^{n_1+1}, f_{j_2}^{n_2}, \ldots, f_{j_k}^{n_k}\rangle \\
&\quad + \delta_{mj_2}\sqrt{n_2+1}|f_{j_1}^{n_1}, f_{j_2}^{n_2+1}, \ldots, f_{j_k}^{n_k}\rangle + \ldots \\
&\quad + \delta_{mj_k}\sqrt{n_k+1}|f_{j_1}^{n_1}, \ldots, f_{j_k}^{n_k+1}\rangle \\
&\quad + \Big(1 - \sum_{l=1}^{k} \delta_{mj_l}\Big)|f_m f_{j_1}^{n_1}, \ldots, f_{j_k}^{n_k}\rangle \qquad \text{(for bosons)}, \\
a^*(f_m)|f_{j_1} \wedge \cdots \wedge f_{j_n}\rangle &= |f_m \wedge f_{j_1} \wedge \cdots \wedge f_{j_n}\rangle \qquad \text{(for fermions)},
\end{aligned}$$

and $a(\alpha f + \beta g) = \alpha a(f) + \beta a(g)$ for f and $g \in \mathcal{H}_1$.

1.3.2 Remarks

1. The prototypes of the a's for bosons are the a and a^* of a harmonic oscillator (III:3.3.5;2), and for fermions they are the matrices $\sigma^{\pm}$ of (1.1.1). The formal analogy is not just superficial; the operators $a(f)$ show up when one quantizes coupled oscillators and then passes to a continuous limit, in the procedure known as **field quantization**, or **second quantization**.

2. Formally, the a's satisfy the commutation or anticommutation relations (with $(f|g)$ being the scalar product on $\mathcal{H}_1$):

$$\left.\begin{aligned}\left[a(f), a^*(g)\right] &= (f|g)\\ \left[a(f), a(g)\right] &= 0\end{aligned}\right\} \text{ for bosons}$$

$$\left.\begin{aligned}a(f)a^*(g) + a^*(g)a(f) \equiv \left[a(f), a^*(g)\right]_+ &= (f|g),\\ \left[a(f), a(g)\right]_+ &= 0\end{aligned}\right\} \text{ for fermions}$$

Conversely, (1.3.1) can be derived from the commutation relations and $a(f)|0\rangle = 0$. The commutation relations are invariant under unitary transformations of the f_j, so (1.3.1) is independent of the choice of the basis. In the spirit of the GNS construction, vector states may be identified with operators:

$$|f_{j_1}^{n_1}, \dots, f_{j_k}^{n_k}\rangle = (n_1! \dots n_k!)^{-1/2} a^*(f_{j_1})^{n_1} \dots a^*(f_{j_k})^{n_k} |0\rangle,$$

or

$$|f_{j_1} \wedge \cdots \wedge f_{j_k}\rangle = a^*(f_{j_1}) \dots a^*(f_{j_k})|0\rangle.$$

3. As in (III:3.1.10;2) the commutation relations reveal that the operators $a(f)$ are unbounded. To get a C^* algebra, it is necessary to use the bounded operators $\exp[i(\alpha a(f) + \alpha^* a^*(f))]$; the algebra they generate is called $\mathcal{A}_B$.
4. The anticommutation relations for fermion fields are the same as those of $\sigma^\pm$, for which reason their $a(f)$ are bounded: $\|a(f)\Psi\|^2 + \|a^*(f)\Psi\|^2 = \langle\Psi|(a^*(f)a(f) + a(f)a^*(f))|\Psi\rangle = (f|f)\|\gamma w\|^2$, so $\|a(f)\| \le \|f\|$. Because $\langle 0|a(f)a^*(f)|0\rangle = \|f\|^2$, this means $\|a(f)\| = \|a^*(f)\| = \|f\|$. The operators $a(f)$ generate a C^* algebra $\mathcal{A}_F$, which is the normclosure of the polynomials in a and a^*.
5. It follows from Remark 4 that the mapping $f \to a^*(f)$ is an isometric homomorphism of the Banach-space structure of $\mathcal{H}_1$, to that of $\mathcal{A}_F$. (The mapping $f \to a(f)$ is continuous but antilinear, that is, $a(\lambda f + \mu g) = \lambda^* a(f) + \mu^* a(g)$.) For every unitary transformation $U \in \mathcal{B}(\mathcal{H}_1)$ there is a linear transformation $a(f) \to a(Uf)$, which can be extended to an automorphism u:

$$\begin{aligned}&u(a(f_1) \dots a(f_k)a^*(g_1) \dots a^*(g_j))\\ &\quad = a(Uf_1) \dots a(Uf_k)a^*(Ug_1) \dots a^*(Ug_j).\end{aligned} \tag{1.35}$$

In particular, for every strongly continuous unitary group $U(t)$ there is a norm-continuous group of automorphisms u_t on $\mathcal{A}_F$ (i.e., the mapping $t \to u_t(a)$ from $\mathbb{R}$ to $\mathcal{B}(\mathcal{H}_F)$ is continuous in norm for all a). Therein lies a difference from the Weyl algebra, for which, although the free time-evolution $\exp[i(rp + sx)] \to \exp[i(rp + s(x + pt))]$ is strongly continuous in t, it is not continuous in norm. The time-evolution on $\mathcal{A}_B$ is also not continuous in norm, so the property of continuity can not be expressed without reference to a representation. In this regard the field algebra of fermions is much the nicer, owing ultimately to its being modeled on the matrices

$$\begin{pmatrix} 0 & 1 \\ 0 & 0 \end{pmatrix}, \quad \begin{pmatrix} 0 & 0 \\ 1 & 0 \end{pmatrix}.$$

Fermion fields will consequently be preferred when investigating more problematic cases.

6. The algebras $\mathcal{A}_F$ and $\mathcal{A}_B$ may be thought of as constructed from local algebras $\mathcal{A}_\Lambda$, containing only those $a(f)$ and $a^*(f)$ for which supp $f \subset \Lambda$. Clearly, $\mathcal{A}_\Lambda \subset \mathcal{A}_{\Lambda'}$, when $\Lambda \subset \Lambda'$. Since $\mathcal{H}_1$ is the norm-closure of $\bigcup_{\Lambda\subset\mathbb{R}^3} L^2(\Lambda, d^3x)$, $\mathcal{A}_F$ equals the norm-closure of $\bigcup_{\Lambda\subset\mathbb{R}^3} \mathcal{A}_\Lambda$.
7. It is common for annihilation operators to be introduced at single points, for which formally $[a(\mathbf{x}), a^*(\mathbf{x}')] = \delta^3(\mathbf{x}-\mathbf{x}'), a(f^*) = \int d^3x a(\mathbf{x}) f(\mathbf{x}), a^*(f) = \int d^3x' a^*(\mathbf{x}') f(\mathbf{x}')$. Although $a(\mathbf{x})$ is densely defined as an operator, it is not closeable, so $a^*(\mathbf{x})$ exists only in the sense of a quadratic form and not as an operator (Problem 8). The object $a^*(\mathbf{x})$ is called an operator-valued distribution.
8. Since a annihilates a particle and a^* creates one, the spaces $\mathcal{H}_n$ are not invariant subspaces of Fock space. It can in fact be shown that $\mathcal{A}_F$ and $\mathcal{A}_B$ are irreducibly represented on $\mathcal{H}_F$ (Problem 1). The algebra $\mathcal{A}_F$ is said to be **quasilocal.**

Remark (1.3.2;5) implies that such things as translations and free time-evolution correspond to norm-continuous one-parameter groups of automorphisms on $\mathcal{A}_F$. The question arises as to whether they can be presented as strongly continuous, one-parameter unitary groups on $\mathcal{H}_F$. If the representation called for is just like the GNS representation of (III: 2.3.9) with the vacuum $|0\rangle$ as a cyclic, and also invariant, vector, then the answer is yes (however, see Problems 6 and 7):

1.3.3 The Unitary Representability of the Automorphism

Let u_g be a group of automorphisms of a C^* algebra $\mathcal{A}$, w be an invariant state (i.e., $w(u_g(a)) = w(a)$ for all g), and π_w be the representation constructed with w. Then the group of automorphisms has a unique unitary representation U_g on the Hilbert space $\mathcal{H}_F$, such that

$$\pi_w(u_g(a)) = U_g \pi_w(a) U_g^{-1}, \quad U_g\Omega = \Omega, \tag{1.36}$$

where Ω is the cyclic vector.

Proof: If we let $U_g\pi_w(a)\Omega = \pi_w(u_g(a))\Omega$, then the U_g thereby defined satisfies the stated requirements. It is unique, since if there existed another $\tilde{U}_g$, with the same properties, then it would follow that $(\tilde{U}_g U_g^{-1} - \mathbf{1})\Omega = 0$, $\tilde{U}_g U_g^{-1} \in \pi(\mathcal{A})'$. Now, because Ω is cyclic for $\pi(\mathcal{A})$, it separates $\pi(a)'$, and therefore $\tilde{U}_g U_g^{-1} = \mathbf{1}$ (cf.Problem 5). (**Separating** means that for $a' \in \pi(\mathcal{A})', a'|\Omega\rangle = 0$ implies $a' = 0$.) □

1.3.4 Remarks

1. If the group is topological and the realization as a group of automorphisms is weakly continuous, then U_g is strongly continuous,
$$\|(U_g - \mathbf{1})\pi_w(a)\Omega\|^2 = 2w(a^*a) - w(a^*u_g(a)) - w(u_g(a^*)a) \to 0$$
as g approaches the identity.
2. Our representation of $\mathcal{A}_F$ (1.3.1) is a π_w such that $w(a) = \langle 0|a|0\rangle$ for $a \in \mathcal{A}_F$. Therefore Ω is the vacuum vector $|0\rangle$, and is invariant under the transformations brought up in (1.3.2; 5). It follows that the Euclidean group and free time-evolution can be represented by strongly continuous unitary groups of operators on Fock space. They consequently have self-adjoint generators (Problem 2), which are, however, not bounded. Even the operators U_g do not belong to $\mathcal{A}_F$. To prove this fact we shall make use of

1.3.5 Definition

The C^* algebra obtained by closing the even polynomials in a and a^* in norm is denoted $\mathcal{A}_G$. The norm-closure of the polynomials having the same number of a's as a^*'s in each summand is $\mathcal{A}_E$.

1.3.6 Remarks

1. $\mathcal{A}_F \supset \mathcal{A}_G \supset \mathcal{A}_E$. In the Fock representation, $\mathcal{A}_E = \{N\}' \cap \mathcal{A}_F$.
2. Because $[ab, c] = a[b, c]_+ - [a, c]_+ b = a[b, c] + [a, c]b$, if $d \in \mathcal{A}_{\Lambda G}$ and $c \in \mathcal{A}_{\bar{\Lambda}}$, $\bar{\Lambda} \cap \Lambda = \emptyset$, then $[d, c] = 0$.

1.3.7 Asymptotic Commutativity

Let $V(t) \in \mathcal{B}(L^2(\mathbb{R}^3))$ be a one-parameter, unitary group of operators with absolutely continuous spectrum, such that $V(t) \rightharpoonup 0$ as $t \to \infty$, and let $u_t(a(f)) \equiv a(V(t)f)$. Then $\lim_{t\to\infty} \|[a, u_t(b)]\| = 0$ for all $a \in \mathcal{A}_G$ and $b \in \mathcal{A}_F$; this state of affairs is described by saying that $\mathcal{A}_G$ is **asymptotically Abelian** *with respect to u_t.*

Proof: First note that $\|[a(f), u_t(a^*(g))]_+\| = \|[a^*(f), u_t(a(g))]_+\| = |(V(t)g|f)| \to 0$ as $t \rightharpoonup \infty$. If d is an even polynomial and c is any polynomial in $a(f)$ and $a^*(g)$, then with Remark (1.3.6;2) it follows that the commutator vanishes asymptotically. Because the algebraic operations are continuous in norm, this extends to $\mathcal{A}_G$ and $\mathcal{A}_F$. □

1.3.8 Corollaries

1. *Since the generators of the spatial translation group and the free time-evolution have purely continuous spectrum, for them* $V(t) \rightharpoonup 0$*, and the appropriate commutators involving them go to zero.*
2. *The corresponding one-parameter groups of unitary operators on Fock space,* $U_t \in \mathcal{B}(\mathcal{H}_B)$*, can not belong to* $\mathcal{A}_F$*. Since every* U_t *commutes with* N*, it must belong to* $\mathcal{A}_E$*, and hence* $\|[U_t, u_{t'}(a)]\| < \varepsilon$ *for all* $\varepsilon \in \mathbb{R}^+$*,* $a \in \mathcal{A}_F$*, and sufficiently large* t'*. Note that* $\|U_t U_{t'} a U_{t'}^{-1} - U_{t'} a U_{t'}^{-1} U_t\| = \|U_t a U_t^{-1} - a\|$ *which obviously can not be arbitrarily small for all* t*. It is even true that* $\mathcal{A}_F \cap \prod_t U_t = U_0$*.*
3. Since $\mathcal{A}_F$ is irreducible, $\mathcal{A}_F'' = \mathcal{B}(\mathcal{H}_B)$ (III:2.3.4), so U_t is certainly attainable as the strong limit of elements of $\mathcal{A}_F$, or even $\mathcal{A}_E$.

1.3.9 Remarks

1. Since commuting observables are jointly diagonable, and hence can be measured simultaneously, if V is a group of translations, this implies that measurements separated by a large spatial distance do not interfere with each other. The local character of the algebra is important for this, and it does not apply to the Weyl operators, as $\exp[i(rp+sx)]$ and $\lim_{a\to\infty} \exp[i(r'p+s'(x+a))]$ do not commute. Even the bicommutant $\mathcal{A}_F''$ in the Fock representation is not asymptotically Abelian – for instance, the generators of the Euclidean group belong to the strong closure of $\mathcal{A}_F$ and are constant with respect to the free time-evolution but do not commute. Therefore $\mathcal{A}_F''$ is not asymptotically Abelian with respect to free time-evolution.
2. The point of (1.3.7) for the time-evolution is that as time passes the disturbance due to a measurement diffuses so widely that local observables are not affected at much later times. This does not apply to the observables x and p, as p and $x + pt$ fail to commute even at large t. Observe that we have as yet proved commutativity only for free time-evolution; the question of whether it also holds for more realistic time-evolutions remains open.
3. This phenomenon does not occur for compact groups like the rotations; for them U is a sum of finite-dimensional representations, for which it is impossible that $U \rightharpoonup 0$.

1.3.10 Global Observables

The particle-number operator N was defined in (1.3.1). It is unbounded and thus $\notin \mathcal{B}(\mathcal{H}_F)$, which $\supset \mathcal{A}_F$. Its domain of self-adjointness is

$$D_N = \{\psi_0 \oplus \psi_1 \oplus \cdots \oplus \psi_n \oplus \cdots \in \mathcal{H}_F : \sum_{n=1}^{\infty} n^2 \|\psi_n\|^2 < \infty\}.$$

Moreover, unitary gauge transformations $U(\alpha) = \exp(iN\alpha) \in \mathcal{B}(\mathcal{H}_F)$ also do not belong to $\mathcal{A}_F$, but can be attained as strong limits of elements of $\mathcal{A}_E$. In the

Fock representation,

$$U(\alpha) = \text{s-}\lim_{M\to\infty} \exp\Big(i\alpha \sum_{j=1}^{M} a^*(f_j)a(f_j)\Big),$$

where $\{f_j\}$ is an orthonormal basis. Although $U(\alpha)$ does not depend on the basis, it can only be defined in certain representations.

1.3.11 Remark

Since N is conserved in all of the systems treated here, it is not physically possible to measure the relative phase of states of different N. This means that N creates a superselection rule in the sense of (III: 2.3.6; 7), and the algebra of observables should, properly speaking, be $\{N\}' = \mathcal{A}_E''$. The representation of this algebra on $\mathcal{H}_F$ is reducible, as its commutant is $\{N\}'' \neq \{\lambda \cdot \mathbf{1}\}$.

1.3.12 Observables at a Point

One frequently considers the particle density and current at a point,

$$\rho(\mathbf{x}) = a^*(\mathbf{x})a(\mathbf{x}) = \sum_{j,k} a^*(f_j)a(f_k) f_j^*(\mathbf{x}) f_k(\mathbf{x}),$$

$$\begin{aligned}\mathbf{j}(\mathbf{x}) &= -\frac{1}{2mi}(a^*(\mathbf{x})\nabla a(\mathbf{x}) - (\nabla a^*(\mathbf{x}))a(\mathbf{x})) \\ &= \sum_{j,k} a^*(f_j)a(f_k)\Big(\frac{1}{2mi}(f_j^*(\mathbf{x})\nabla f_k(\mathbf{x}) - (\nabla f_j^*(\mathbf{x}))f_k(\mathbf{x}))\Big).\end{aligned}$$

The f_k in these formulas must be chosen as an orthonormal basis of C^1 functions, in which case these observables are densely defined as quadratic forms. They are not, however, closeable: Their restrictions to $\mathcal{H}_1$ are the quadratic forms of

$$\psi^*(\mathbf{x})\psi(\mathbf{x}) \quad \text{and} \quad \frac{1}{2mi}(\psi^*(\mathbf{x})\nabla\psi(\mathbf{x}) - (\nabla\psi^*(\mathbf{x}))\psi(\mathbf{x})),$$

the former of which is recognizable as the prototype of this phenomenon as encountered in (III: 2.5.17; 3). Matrix elements with, say, $\rho(\mathbf{x})$ may be understood as distributional limits of matrix elements of the bounded operators $a^*(f)a(f)$ as $f \to \delta^3(\mathbf{x})$. Similarly, the continuity equation $\dot\rho + \nabla \cdot \mathbf{j} = 0$ holds at least for matrix elements if, evolving freely in time, $i\dot f = -\Delta f/2m$.

1.3.13 Problems

1. Show that the representations of $\mathcal{A}_F$ and $\mathcal{A}_B$ on $\mathcal{H}_F$ are irreducible.
2. Construct the generators of free time-evolution and of translation.
3. Find dense domains of definition for the quadratic forms $\rho(\mathbf{x})$ and $\mathbf{j}(\mathbf{x})$.

4. Define the number of particles in the volume V, $N_V = \int_V d^3x\rho(\mathbf{x})$, as an unbounded, self-adjoint operator.

5. For $\mathcal{A} \subset \mathcal{B}(\mathcal{H})$ and $\Omega \in \mathcal{H}$, show that Ω is cyclic for $\mathcal{A}$ iff Ω separates $\mathcal{A}'$.

6. The mapping $a \to b : b(f) = a(f) + L(f)$ is an automorphism α_L of the Bose algebra whenever L is a linear, but not necessarily continuous, functional. Show that α_L is unitarily implementable on $\mathcal{H}_F$, i.e., there exists a $U_L \in \mathcal{B}(\mathcal{H}_F)$ such that $\mathbf{1} = U_L^* U_L = U_L U_L^*$ and $U_L a(f) U_L^{-1} = b(f)$, iff L is continuous, which means that it can be written as $L(f) = (\rho|f)$ for some $\rho \in \mathcal{H}_1$.

7. Let $b(f) = a(\Phi f) + a^*(\bar{\Psi} f)$, $\Phi, \Psi \in \mathcal{B}(\mathcal{H}_1)$, Φ invertible. Show
(i) that $a \to b$ is an automorphism of the Bose (resp. Fermi) field algebra if
$$\Phi\Phi^* \mp \Psi\Psi^* = 1 = \Phi^*\Phi \mp (\Psi^*\Psi)^t,$$
$$\Phi\Psi^t \mp \Psi\Phi^t = 0 = (\Psi^*\Phi)^t \mp \Psi^*\Phi,$$
where $\bar{\Psi} = \Psi^{*t}$; and
(ii) that it can be represented as a unitary operator on $\mathcal{H}_F$ iff $\Phi^{-1}\Psi \in \mathcal{C}_2(\mathcal{H}_1)$.

8. Show that although the $a(\mathbf{x})$ of (1.3.2; 7) is densely defined, it is not closeable, and the domain of definition of its adjoint $a^*(\mathbf{x})$ contains only the zero vector.

1.3.14 Solutions

1. Let b be an operator such that $[b, a(f)] = [b, a^*(f)] = 0$ for all $f \in \mathcal{H}_F$. From the commutation relations of (1.3.2; 2) and $a(f)|0\rangle = 0$, it follows that $\langle 0|a(f_1)\dots a(f_m)ba^*(g_1)$ $\dots a^*(g_n)|0\rangle = \langle 0|b|0\rangle \cdot \langle 0|a(f_1)\dots a^*(g_n)|0\rangle$, which implies that $\langle x|bx\rangle = \langle 0|b|0\rangle \|x\|^2$ on a dense set, and therefore $b = \langle 0|b|0\rangle \cdot \mathbf{1}$.

2. With Theorem (1.3.5) and the fact that the $\mathcal{H}_n$ are invariant, by reasoning as in (1.3.10) we find that the two generators are
$$\text{s-}\lim_{M\to\infty} \sum_{i,j}^{M} \int \nabla f_j(\mathbf{x}) \cdot \nabla f_i^*(\mathbf{x}) a^*(f_j) a(f_i) d^3x$$
and
$$\text{s-}\lim_{M\to\infty} i \sum_{k,j}^{M} \int \nabla f_j^*(\mathbf{x}) f_k(\mathbf{x}) a(f_k) d^3x,$$
where the strong limit is defined as in (III: 2.5.7; 3). Formally, these can be written as $\int d^3x \nabla a^*(\mathbf{x}) \cdot \nabla a(\mathbf{x})$ and $i \int d^3x a^*(\mathbf{x}) \overleftrightarrow{\nabla} a(\mathbf{x})$.

3. For $\rho(\mathbf{x})$, linear combinations of $\prod_j a^*(f_j)|0\rangle$ with continuous f_j. For $\mathbf{j}(\mathbf{x})$, the f_k have to be continuously differentiable.

4. $N_V = \sum_{j,k} a^*(f_j) a(f_k) \int_V d^3x f_j^*(\mathbf{x}) f_k(\mathbf{x})$, $0 \le N_V \le N$, is a Hermitian operator on D_N (1.3.10), and hence the domain of its Friedrichs extension contains D_N.

5. "If": Let P be the projection onto the orthogonal complement of $\{a|\Omega\rangle\}$ for $a \in \mathcal{A}$. Then $P \in \mathcal{A}'$ and $P|\Omega\rangle = 0$, so $P = 0$.

"Only if": Let $a' \in \mathcal{A}'$, $a'|\Omega\rangle = 0$. Then $a'a|\Omega\rangle = 0$ for all $a \in \mathcal{A}$, which implies that $a' = 0$ on a dense set, so $a' = 0$.

6. The mapping $a \to b$ is unitarily implementable on $\mathcal{H}_F$ iff there exists a vector $|0_b\rangle \in \mathcal{H}_F$ such that $b(f)|0_b\rangle = 0$ for all $f \in \mathcal{H}_1$. It is clear that the existence of U implies that of $|0_b\rangle = U|0\rangle$. On the other hand, the mapping

$$\prod_{i=1}^{n} a_i^*|0\rangle \to \prod_{i=1}^{n} b_i^*|0_b\rangle,$$

where $a_i = a(f_i), b_i = b(f_i)$ and $\{f_i\}$ is an orthonormal basis, defines a unitary operator U, since this set of vectors is total. (Every vector is cyclic for an irreducible representation.) If L is not continuous, then ker L is dense in $\mathcal{H}_1$, and therefore $a(f)|0_b\rangle = 0$ for a dense set of f's. This implies that $|0_b\rangle = |0\rangle$ and thus that $L \equiv 0$, which is continuous. Therefore $|0_b\rangle \notin \mathcal{H}_F$. If, however, $L(f) = (g|f)$, $g \in \mathcal{H}_1$, it is possible to choose $f_1 = g/\|g\|$. Because $a\exp[-a^*\|g\|] = \exp[-a^*\|g\|](a - \|g\|)$, the vector $|0_b\rangle = \exp[-a_1^*\|g\|]|0\rangle$ formally satisfies $b_k|0_b\rangle = (a_k + \delta_{k1}\|g\|)|0_b\rangle = 0$. It is also normalizable provided that

$$\infty > \langle 0|\exp\Big[-\|g\|a_1\Big]\exp\Big[-\|g\|a_1^*\Big]|0\rangle = \sum_{n=0}^{\infty} \frac{1}{(n!)^2}\|g\|^{2n} n! = \exp\|g\|^2,$$

so $\langle 0_b|0_b\rangle < \infty$ if $\|g\|^2 < \infty$.

7. (i) In matrix notation, for $b = \Phi a + \Psi a^*$, (i) must hold: $1 = [b, b^*]_\mp = \Phi\Phi^* \mp \Psi\Psi^*$, and $0 = [b, b]_\mp = \Phi\Psi^t \mp \Psi\Phi^t$. Written as block matrices, this becomes

$$\begin{pmatrix} \Phi & \Psi \\ \Psi^{*t} & \Phi^{*t} \end{pmatrix} \begin{pmatrix} \Phi^* & \mp\Psi^t \\ \mp\Psi^* & \Phi^t \end{pmatrix} = \mathbf{1}.$$

For invertibility it is necessary that

$$\begin{pmatrix} \Phi^* & \mp\Psi^t \\ \mp\Psi^* & \Phi^t \end{pmatrix} \begin{pmatrix} \Phi & \Psi \\ \Psi^{*t} & \Phi^{*t} \end{pmatrix} = \mathbf{1}.$$

which produces the second line of the conditions.

(ii) The Fock vacuum $|0_b\rangle$ satisfies $0 = (\Phi^{-1}b)_k|0_b\rangle = (a_k + M_{kl}a_l^*)|0_b\rangle$, where $M = \Phi^{-1}\Psi$. Because $[a, a^*Ma^*] = 2Ma^*$, it can be written formally as $|0_b\rangle = c\exp[-a^*Ma^*/2]|0\rangle$. (Observe that by (i), $M = M^t$ (resp. $M = -M^t$).) To determine the normalization constant c, we shall calculate

$$\langle 0|\exp\Big[-\frac{1}{2}aNa\Big]\exp\Big[-\frac{1}{2}a^*Ma^*\Big]|0\rangle$$

when $M = \pm M^t$, $N = \pm N^t$, $[M, N^*] = 0$ and M and N are for the moment real. They can then simultaneously be put into the normal forms

$$\begin{pmatrix} n_1 & & & \\ & n_2 & & \\ & & n_3 & \\ & & & \ddots \end{pmatrix}, \quad \begin{pmatrix} m_1 & & & \\ & m_2 & & \\ & & m_3 & \\ & & & \ddots \end{pmatrix}$$

and respectively

$$\begin{pmatrix} & n_1 & & & \\ -n_1 & & & & \\ & & & n_2 & \\ & & -n_2 & & \\ & & & & \ddots \end{pmatrix}, \quad \begin{pmatrix} & m_1 & & & \\ -m_1 & & & & \\ & & & m_2 & \\ & & -m_2 & & \\ & & & & \ddots \end{pmatrix}$$

with real, orthogonal transformations. The transformations preserve the commutation relations of the field operators, so we may use this basis to calculate

$$\langle 0| \exp\Big[-\frac{n_1}{2}a_1^2\Big] \exp\Big[-\frac{m_1}{2}a_1^{*2}\Big]|0\rangle = \sum_{n=1}^{\infty} \frac{(n_1 m_1)^n}{4^n (n!)^2}(2n)! = (1-n_1 m_1)^{-1/2}$$

and, respectively, for fermions,

$$\langle 0| \exp\Big[-n_1 a_2 a_1\Big] \exp\Big[-m_1 a_1^* a_2^*\Big]|0\rangle = 1 + n_1 m_1.$$

Therefore,

$$\begin{aligned}\langle 0| \exp\Big[-\frac{1}{2}aNa\Big] \exp\Big[-\frac{1}{2}a^* M a^*\Big]|0\rangle &= \prod_i (1-n_i m_i)^{-1/2} \\ &= \left(\operatorname{Det}\begin{pmatrix} 1 & M \\ N & 1 \end{pmatrix}\right)^{-1/2}\end{aligned}$$

and, respectively,

$$\prod_i (1+n_i m_i) = \left(\operatorname{Det}\begin{pmatrix} 1 & M \\ N & 1 \end{pmatrix}\right)^{1/2}.$$

This can be continued analytically to complex matrix elements, and, in particular, in our case,

$$|c|^2 \left(\operatorname{Det}\begin{pmatrix} 1 & M \\ M^* & 1 \end{pmatrix}\right)^{\mp 1/2} = 1.$$

The determinant is finite for $M \in \mathcal{C}_2$. Observe that in the case of bosons, $\Phi^*\Phi \geq 1$, and so $\Phi = V(\Phi^*\Phi)^{1/2}$ is always invertible. The result for fermions is valid for M acting on either even or odd dimensional spaces.

8. The dense domain of definition of $a(\mathbf{x})$ consists of vectors with continuous, bounded f's. For example, for fermions,

$$\begin{aligned}a(\mathbf{x})|f_{j_1} \wedge \cdots \wedge f_{j_n}\rangle = f_{j_1}(x)|f_{j_2} \wedge \cdots \wedge f_{j_n}\rangle - f_{j_2}(\mathbf{x})|f_{j_1} \wedge f_{j_3} \wedge \cdots \wedge f_{j_n}\rangle + \cdots \\ + (-1)^{n+1} f_{j_n}(\mathbf{x})|f_{j_1} \wedge \cdots \wedge f_{j_{n-1}}\rangle.\end{aligned}$$

The operator $a(\mathbf{x})$ is not closeable. Suppose that $f_\lambda(\mathbf{x}') = \exp[-|\mathbf{x}-\mathbf{x}'|^2\lambda]$; then $|f_\lambda\rangle \to 0$ as $\lambda \to \infty$, but $a(\mathbf{x})|f_\lambda\rangle = |0\rangle \not\to 0$. Formally, $a^*(\mathbf{x})$ creates a particle with wave-function $f(\mathbf{x}') = \delta^3(\mathbf{x}-\mathbf{x}')$. Since this is not normalizable, $a^*(\mathbf{x})$ makes every vector $|f_{j_k} \mathbf{w} \cdots \mathbf{w} f_{j_n}\rangle$ infinitely long.

1.4 Representations with $N = \infty$

Systems of N particles are represented on a Hilbert space that is the tensor product of N Hilbert spaces for single particles. The infinite tensor product opens the door to the new mathematical features of field theory.

The scalar product on an N-fold tensor product of spaces $\mathcal{H}_1$, was defined multiplicatively by

$$\langle x|x\rangle = \prod_{i=1}^{N}(x_i|x_i), \quad |x\rangle = |x_1) \otimes |x_2) \ldots |x_n), \quad x_i \in \mathcal{H}_1. \tag{1.37}$$

If $N \to \infty$, the vectors $|x\rangle$ that can be used in this formula are initially only those for which the infinite product converges. The product might well converge to 0 even though $(x_i|x_i) > 0$ for all i. In order to form the quotient space with respect to the zero vectors, it will first be necessary to form the equivalence class not only of vectors with some factor zero but also containing the vectors for which the product

$$\prod_{i=1}^{\infty}(x_i|x_i)$$

converges to zero. On the quotient space, (1.37) defines a separating norm, so the space can be completed to a Hilbert space $\mathcal{H}$, with the linear structure defined in the usual way.

This does not yet, however, suffice to define the scalar product of different vectors $|x\rangle$ and $|y\rangle$. Though only vectors such that $(x_i|x_i) = (y_i|y_i) = 1$ for all i need to be considered, there are still two possibilities, namely

$$\prod_{i=1}^{\infty}|(x_i|y_i)| \to c > 0, \quad \text{and} \quad \prod_{i=1}^{\infty}|(x_i|y_i)| \to 0,$$

$\prod_{i=1}^{\infty}(x_i|y_i) \to 0$ as well, and the vectors may be considered orthogonal. Possibility (I), on the other hand, does not guarantee that $\prod_i(x_i|y_i)$ converges. If $(x_j|y_j) = \exp(i\varphi_j)|(x_j|y_j)|$, then their product is said to converge if not only $\prod_i|(x_i|y_i)|$ but also $\sum_i|\varphi_i|$ converges. One now encounters the convention that vectors may be deemed orthogonal whenever $\sum_i|\varphi_i| \to \infty$ (case (I_b)). Let us thus agree on a

1.4.1 Definition of the Scalar Product

$$\langle x|y\rangle = c \text{ providedthat } \textstyle\prod_i(x_i|y_i) \to c \neq 0, \text{ (case(Ia));}$$

$$\langle x|y\rangle = 0 \text{ providedthat } \textstyle\prod_i(x_i|y_i) \to 0, \quad \text{(case(II), orinthe divergentsense(Ib)).}$$

1.4.2 Remarks

1. It is easy to see that the scalar product thus defined on $\mathcal{H}$ obeys all the rules of the game.
2. The space $\mathcal{H}_1$ has been assumed separable, yet even if $\mathcal{H}_1 = \mathcal{C}^2$, the larger space $\mathcal{H}$ is nonseparable. Let $|\mathbf{n})\in\mathcal{C}^2$ be defined such that $(\mathbf{n}|\mathbf{n}) = 1$, $(\mathbf{n}|\boldsymbol{\sigma}|\mathbf{n}) = \mathbf{n}\in\mathbb{R}^3$, $|\mathbf{n}|^2 = 1$, and $|\mathbf{n}\rangle = |\mathbf{n})\otimes|\mathbf{n})\otimes\ldots$. Then $\langle\mathbf{n}|\mathbf{n}'\rangle = 1$ if $\mathbf{n} = \mathbf{n}'$ and is otherwise 0, showing that there is an uncountable orthonormal system of vectors.
3. Possibilities (Ia) and (I) create equivalence relations between vectors, because the convergence of $\prod_i(x_i|y_i)$ and $\prod_i(y_i|z_i)$ implies that of $\prod_i(x_i|z_i)$, and, likewise, that of $\prod_i|(x_i|y_i)|$ and $\prod_i|(y_i|z_i)|$ implies that of $\prod_i|(x_i|z_i)|$ (Problem 2). It is accordingly necessary to distinguish between strong (Ia) and weak (I) equivalence classes:

$$(\mathrm{Ia}):\prod_i{}'(x_i|y_i)\to c\neq 0,\quad (\mathrm{I}):\prod_i{}'|(x_i|y_i)|\to c>0.$$

The symbol $\prod'$ means that any finite number of factors 0 are to be left out. The equivalence classes span linear subspaces, so $\mathcal{H}$ can be decomposed into (uncountably) many weak equivalent classes, for which vectors of different classes are orthogonal. Each weak equivalence class can be further decomposed into mutually orthogonal strong equivalence classes. Since the latter differ only by phase factors within a given weak equivalence class, they contain the same physical information.

1.4.3 Representations of $\mathcal{A}$ on Infinite Tensor Products

For the reasons stated in § 1.1 and § 1.3 we shall be interested in the algebra generated by the operators $\mathcal{B}(\mathcal{H}_i)$. More precisely, let $\mathcal{A}$ be the algebra generated by $\mathcal{B}(\mathcal{H}_1)\otimes\mathbf{1}\otimes\mathbf{1}\ldots, \mathbf{1}\otimes\mathcal{B}(\mathcal{H}_2)\otimes\mathbf{1}\ldots$, etc., and let $\mathcal{A}''$ be its strong (= weak) closure. The first thing to notice is that an element a of $\mathcal{A}$ sends no vector of $\mathcal{H}$ out of its strong equivalence class; since other than a finite number of entries there is always an infinite $\mathbf{1}\otimes\mathbf{1}\otimes\mathbf{1}\ldots$, nothing alters the convergence of $\prod_{i=1}^{\infty}(x_i|y_i)$. The representation of $\mathcal{A}$ on $\mathcal{H}$ is consequently reducible to a high degree; every strong equivalence class is an invariant subspace. The formation of the weak closure changes nothing, since $\langle x|a_n y\rangle = 0$ for $|x\rangle$ and $|y\rangle$ in different equivalence classes, and if $a_n \rightharpoonup a$, then clearly $\langle x|ay\rangle = 0$. Thus every strong equivalence class provides a representation of $\mathcal{A}$ and of $\mathcal{A}''$, and it is a peculiarity of the infinite tensor product that these representations are inequivalent so long as they arise from different weak equivalence classes.

1.4.4 Example

Return to the simple case of (1.4.2; 2), and define $\boldsymbol{\sigma}_j\cdot\mathbf{n} = \sigma_j$ and $\sigma_j^{\pm}$ in analogy with (1.1.2) such that $\sigma_j^-|\mathbf{n}) = |-\mathbf{n})$, $\sigma_j^+|-\mathbf{n}) = |\mathbf{n})$, $\sigma_j^+|\mathbf{n}) = \sigma_j^-|-\mathbf{n}) = 0$. Let

$\mathcal{A}$ be the algebra generated by σ_j and $\sigma_j^{\pm}$, $j = 1, 2, \ldots$, let $\pi_{\mathbf{n}}$, be its representation on the strong equivalence class of $|\mathbf{n}\rangle$, and define $\mathcal{A}_{\mathbf{n}} \equiv \pi_{\mathbf{n}}(\mathcal{A})$. The representation is constructed like the Fock representation, the operators $\pi_{\mathbf{n}}(\sigma_j^{\pm})$ corresponding to creation and annihilation operators and $|\mathbf{n}\rangle$ to the vacuum: $\pi_{\mathbf{n}}(\sigma_j^{+})|\mathbf{n}\rangle = 0$ for all j. The vectors $\pi_{\mathbf{n}}(\sigma_{j_1}^{-}, \ldots, \sigma_{j_k}^{-})|\mathbf{n}\rangle$ are total for the (strong) equivalence class, and the representation $\mathcal{A}_n$ is irreducible (likewise for $\mathcal{A}_n''$ *a fortiori*).

1.4.5 Remarks

1. These representations of the σ's are always equivalent on finite tensor products; the Hilbert space constructed with the GNS procedure contains every vector $|\mathbf{n}'\rangle$, in contrast to the infinite case, where the σ's never send vectors out of equivalence classes, which, however, contain no vectors $|\mathbf{n}'\rangle$ with $\mathbf{n}' \neq \mathbf{n}$.
2. The mean magnetization

$$\mathbf{s} = \lim_{N\to\infty} \sum_{j=1}^{N} \frac{1}{N} \pi_{\mathbf{n}}(\boldsymbol{\sigma}_j)$$

exists as a strong limit, so $\mathbf{s} \in \mathcal{A}_n''$. As $N \to \infty$ the commutator of this observable with any element of the algebra goes to zero in the norm topology, so $\mathbf{s}$ is in the center of $\mathcal{A}_n''$. In any irreducible representation, $\mathbf{s}$ must be a multiple of the identity, and is thus the same as $\mathbf{n}$, its expectation value in the state $|\mathbf{n}\rangle$. If $\mathbf{n} \neq \mathbf{n}'$, then $\pi_{\mathbf{n}}$ and $\pi_{\mathbf{n}'}$ are inequivalent: If there existed a unitary transformation U mapping the equivalence classes of $\mathbf{n}$ and $\mathbf{n}'$ onto each other and such that $U\pi_{\mathbf{n}}(\boldsymbol{\sigma}_j)U^{-1} = \pi_{\mathbf{n}'}(\boldsymbol{\sigma}_j)$, then this could be extended to a transformation of the strong closures $\mathcal{A}_{\mathbf{n}}''$ and $\mathcal{A}_{n'}''$, and when applied to $\mathbf{s}$ it would imply that $U\mathbf{n}U^{-1} = \mathbf{n}'$. This is impossible, since two different multiples of the identity can not be unitarily related.
3. On the space $\mathcal{H}$ there exists a unitary transformation sending $|\mathbf{n}\rangle$ to $|\mathbf{n}'\rangle$. Let $n_j' = M_{jk}n_k$, $MM^t = 1$; then the transformation $|\mathbf{n}) \to |M\mathbf{n})$ (on every factor of $|\mathbf{n}\rangle$) is clearly the unitary transformation that brings this about. Upon restriction to an equivalence class, its action is

$$U\pi_{\mathbf{n}}(\sigma_j)U^{-1} = \pi_{\mathbf{n}}(\sigma_k)M_{kj},$$

in contrast to the previous U, and so it creates an isomorphism between $\pi_{\mathbf{n}}(\mathcal{A})$ and $\pi_{\mathbf{n}'}(\mathcal{A})$.
4. Within a given representation the rotation

$$\pi_{\mathbf{n}}(\sigma_j) \to \pi_{\mathbf{n}}(\sigma_k)M_{kj}$$

represents an automorphism of the C^* algebra generated by the σ's, and as such it preserves norms. Yet it can not be extended continuously to the weak closure. If there were such an extension, then $n_j \cdot \mathbf{1} \to n_k M_{kj} \cdot \mathbf{1}$, but $\lambda \cdot \mathbf{1}$ is invariant under every automorphism. Consequently, in the representation space of $\pi_{\mathbf{n}}$ there exists no unitary transformation $U^{-1}\pi_{\mathbf{n}}(\sigma_j)U = M_{jk}\pi_{\mathbf{n}}(\sigma_k)$, as it would extend to $\pi_{\mathbf{n}}(\mathcal{A})''$. Formally, it would turn $|\mathbf{n}\rangle$ into $|\mathbf{n}'\rangle$, but there is no vector $|\mathbf{n}'\rangle$ in the representation space of $\pi_{\mathbf{n}}$ (cf. Problems (1.3.13; 6) and (1.3.13; 7)).

5. Let $M(t)$ be a one-parameter group of rotations on $\mathbb{R}^3$ – for definiteness about the 3-axis – and let $U(t)$ be its representation on $\mathcal{H}$ as discussed in Remark 3. On a formal level, $\sum_{j=1}\sigma_j^3$ could be regarded as the generator of the group. The unitary operators $U(t)$ map the equivalence class of $|\mathbf{n}\rangle$ into itself only if $\mathbf{n}$ points in the 3-direction, and in that case the restriction of $U(t)$ to this equivalence class belongs to $\mathcal{A}_{\mathbf{n}}''$. Although it is not possible to define $\sum_{j=1}^{\infty}\sigma_j^3$, densely, $\sum_{j=1}^{\infty}(\sigma_j^3 - \mathbf{1})$ is essentially self-adjoint in the representation $\pi_{\mathbf{n}}$ on the dense set specified in (1.4.4) and is the generator of the rotations about the 3-axis. In other representations there is no workable definition of this operator, as all its matrix elements are infinite. It is natural to ask at this point what the generator of $U(t)$ looks like. It turns out, though, that $U(t)$ has no generator: By Stone's theorem (III: 2.4.22) the existence of a generator is equivalent to strong continuity of $U(t)$, but $U(t)$ is not even weakly continuous, for if $\mathbf{n}$ does not point in the 3-direction, then $\langle\mathbf{n}|U(t)|\mathbf{n}\rangle = 1$ if $t = 0$ and is otherwise 0. It is true that the mapping $t \to U(t)$ is weakly measurable, but the generalization of Stone's theorem for weakly measurable groups works only on separable Hilbert spaces.
6. "Local" rotations of m spins are generated by $\sum_{j=1}^{m}\sigma_j^3$, and always exist.

The representations of the σ's on the individual strong equivalence classes studied until now have all been irreducible, and correspond to GNS constructions using a pure state (cf. (III: 2.3.10; 5)). We shall also see in (2.1.6; 5) that mixed states likewise correspond to vectors in a larger Hilbert space on which the algebra is represented reducibly. That space is the tensor product of the irreducible representation space with another Hilbert space. The key fact to bear in mind when constructing such representations of the σ's is that the infinite tensor product is no longer associative,. for instance $C^4 \otimes C^4 \otimes C^4 \otimes \cdots = (C^2 \otimes C^2) \otimes (C^2 \otimes C^2) \otimes (C^2 \otimes C^2) \otimes \cdots \neq C^2 \otimes C^2 \otimes C^2 \otimes \cdots$: The vector

$$\frac{1}{\sqrt{2}}\left[\begin{pmatrix}1\\0\end{pmatrix}\otimes\begin{pmatrix}0\\1\end{pmatrix}+\begin{pmatrix}0\\1\end{pmatrix}\otimes\begin{pmatrix}1\\0\end{pmatrix}\right]$$
$$\otimes\frac{1}{\sqrt{2}}\left[\begin{pmatrix}1\\0\end{pmatrix}\otimes\begin{pmatrix}0\\1\end{pmatrix}+\begin{pmatrix}0\\1\end{pmatrix}\otimes\begin{pmatrix}1\\0\end{pmatrix}\right]\otimes\ldots$$

on the left has no counterpart on the right. For this reason we shall not simply take the tensor product of the space examined in Example 1.4.4 with another Hilbert space, but shall instead proceed as follows.

1.4.6 Thermal Representations

If there is only one spin, i.e., $\mathcal{A}$ is generated by $\mathbf{1}$ and $\boldsymbol{\sigma}$, then the GNS representation using the state given in (1.6) becomes a reducible representation on $C^4 : \pi(\mathcal{A}) =$

$\mathcal{B}(\mathcal{C}^2) \otimes \mathbf{1}$, $\pi(\sigma) = \sigma \otimes \mathbf{1}$, $\pi(\mathcal{A})' = \mathbf{1} \otimes \mathcal{B}(\mathcal{C}^2)$, $Z = \pi(\mathcal{A}) \cap \pi(\mathcal{A})' = \{\alpha \cdot \mathbf{1}\}$,

$$\Omega = \begin{pmatrix}1\\0\end{pmatrix} \otimes \begin{pmatrix}1\\0\end{pmatrix} \sqrt{\tfrac{1+s}{2}} + \begin{pmatrix}0\\1\end{pmatrix} \otimes \begin{pmatrix}0\\1\end{pmatrix} \sqrt{\tfrac{1-s}{2}}, \quad 0 < s < 1,$$

$$\langle \sigma \rangle = \langle \Omega | \sigma \Omega \rangle = (0, 0, s).$$

Despite being reducible ($\mathcal{A}' \neq \{\alpha \cdot \mathbf{1}\}$), this representation is a factor (its center is $Z = \{\alpha \cdot \mathbf{1}\}$). Accordingly, when passing to infinitely many spins we consider the representation on $\mathcal{C}^4 \otimes \mathcal{C}^4 \otimes \mathcal{C}^4 \otimes \cdots$ constructed with $\Omega \otimes \Omega \otimes \Omega \otimes \cdots$. We find, analogously, that

$$\begin{aligned}
\pi(\mathcal{A}) &= (\mathcal{B}(\mathcal{C}^2) \otimes \mathbf{1}) \otimes (\mathcal{B}(\mathcal{C}^2) \otimes \mathbf{1}) \otimes \ldots, \\
\pi(\mathcal{A})' &= (\mathbf{1} \otimes \mathcal{B}(\mathcal{C}^2)) \otimes (\mathbf{1} \otimes \mathcal{B}(\mathcal{C}^2)) \otimes \cdots + \text{weak limits} \\
\pi(\mathcal{A})'' &\quad \text{is the weak closure of } \mathcal{A}, \text{ and } Z = \{\alpha \cdot \mathbf{1}\},
\end{aligned}$$

which is a reducible factor representation.

1.4.7 Remarks

1. This representation is not equivalent to any of those found in (1.4.5); as mentioned above, the vector $\Omega \otimes \Omega \otimes \Omega \otimes \ldots$ has no counterpart in the earlier representations π_n, since the corresponding functional in π_n would then be strongly continuous. The state defined by $\Omega \otimes \Omega \otimes \Omega \otimes \cdots$ on $\mathcal{A}$.
$$\langle (\sigma_{j_1} \cdot \mathbf{n}_1)(\sigma_{j_2} \cdot \mathbf{n}_2) \ldots (\sigma_{j_k} \cdot \mathbf{n}_k) \rangle = s^k n_1^z n_2^z \ldots n_k^z$$
is a (norm) continuous linear functional, and therefore extensible to the whole C^* algebra generated by $\mathcal{A}$, but it still need not be strongly continuous in a representation: For instance, in the representation using π_n,
$$P_N = \prod_{i=N}^{2N} \frac{1 + \sigma_i \cdot \mathbf{n}}{2}$$
converges strongly to $\mathbf{1}$, but $\langle P_N \rangle = ((1 + sn^z)/2)^N \to 0 \neq 1$. Recall that a refinement of the topology on the range space or a coarsening of the topology on the domain space may destroy the continuity of a mapping.
2. The fact that with only one spin, $\langle \sigma \rangle = \operatorname{Tr} \sigma \exp(-\eta\sigma_3)/\operatorname{Tr}\exp(-\eta\sigma_3)$, might mislead one into thinking that for infinitely many spins, in the notation of (1.1.1),
$$\langle \cdot \rangle = \operatorname{Tr} \cdot \rho, \quad \rho = \frac{\exp(-\eta \sum_j \sigma_j)}{\operatorname{Tr}\exp(-\eta \sum_k \sigma_k)}.$$
What goes wrong is that
$$\frac{\exp(-\eta \sum_{j=1}^N \sigma_j)}{\operatorname{Tr}\exp(-\eta \sum_{j=1}^N \sigma_j)} \Rightarrow 0 \quad \text{as} \quad N \to \infty.$$
3. In the thermal representation (1.4.6) it is of course possible to write $\langle \cdot \rangle = \operatorname{Tr} \cdot P_\Omega$, where P_Ω is the projection onto the cyclic vector, but $P_\Omega \notin \mathcal{A}''$.

1.4.8 Decomposition of the Representations

Because of the analogy between $\sigma^{\pm}$ and the operators a and a^* for fermions, the phenomena we have discussed are also characteristic of systems of infinitely many fermions. It is not so important that the σ's commute whereas the a's anticommute; the distinction can be gotten around with the right transformation. For a system of bosons the individual factors of the tensor product are already infinite-dimensional, which causes additional complications. In either case there are a great number of inequivalent representations; the uniqueness theorem (III: 3.1.5) for finite systems does not hold any more. Thus it would be desirable to find a point of view that organizes them somehow. The concept of a factor was introduced in (III: 2.3.4), as an algebra with a trivial center, $Z = \{\alpha \cdot \mathbf{1}\}$. On a finite-dimensional space it amounts to a direct sum of equivalent irreducible representations. The first step in any decomposition is to collect the equivalent irreducible representations together in factors and then write the whole representation as a sum of various factors. In the finite-dimensional case this appears as shown in Figures 1.2 and 1.3.

It will be observed that the projections onto the space $\mathcal{H}_{ik}$ of the irreducible representations belong to $\pi(\mathcal{A})'$ and the projections onto the spaces $\mathcal{H}_i$ of the factors belong to the center. Both $\pi(\mathcal{A})$ and $\pi(\mathcal{A})'$ map $\mathcal{H}_i$ into itself. The elements of the center become multiples of the identity when projected onto $\mathcal{H}_i$; they can assume different values only on different $\mathcal{H}_i$. The decomposition into factors is thus uniquely fixed by Z and consequently by $\pi(\mathcal{A})$. The further decomposition into irreducible representations is not likewise fixed; some arbitrariness is connected with the spaces $\mathcal{H}_{ik}$. If, for example, $\mathcal{H}_1 = \mathcal{H}_{11} \otimes C^n = \mathcal{H}_{11} \otimes e_1 \oplus \cdots \oplus \mathcal{H}_{11} \otimes e_2 \oplus \cdots \mathcal{H}_{11} \otimes e_n$, then the choice of the basis $\{e_i\}$ for C^n remains free, since the

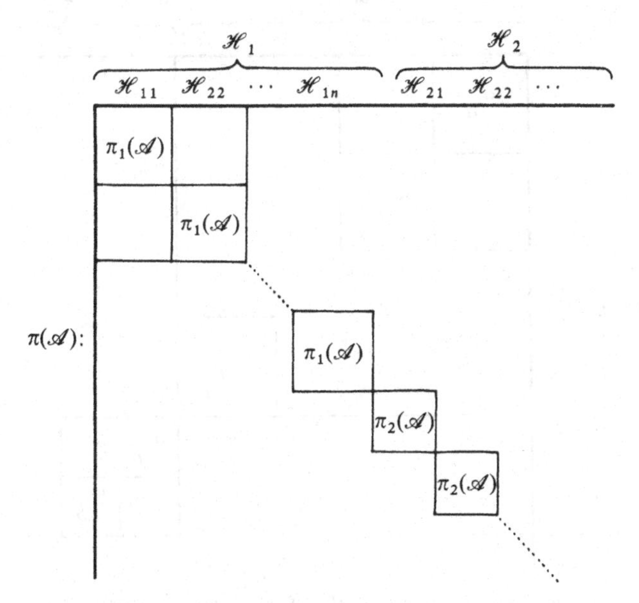

Figure 1.2. The representation of $\mathcal{A}$ in matrix form.

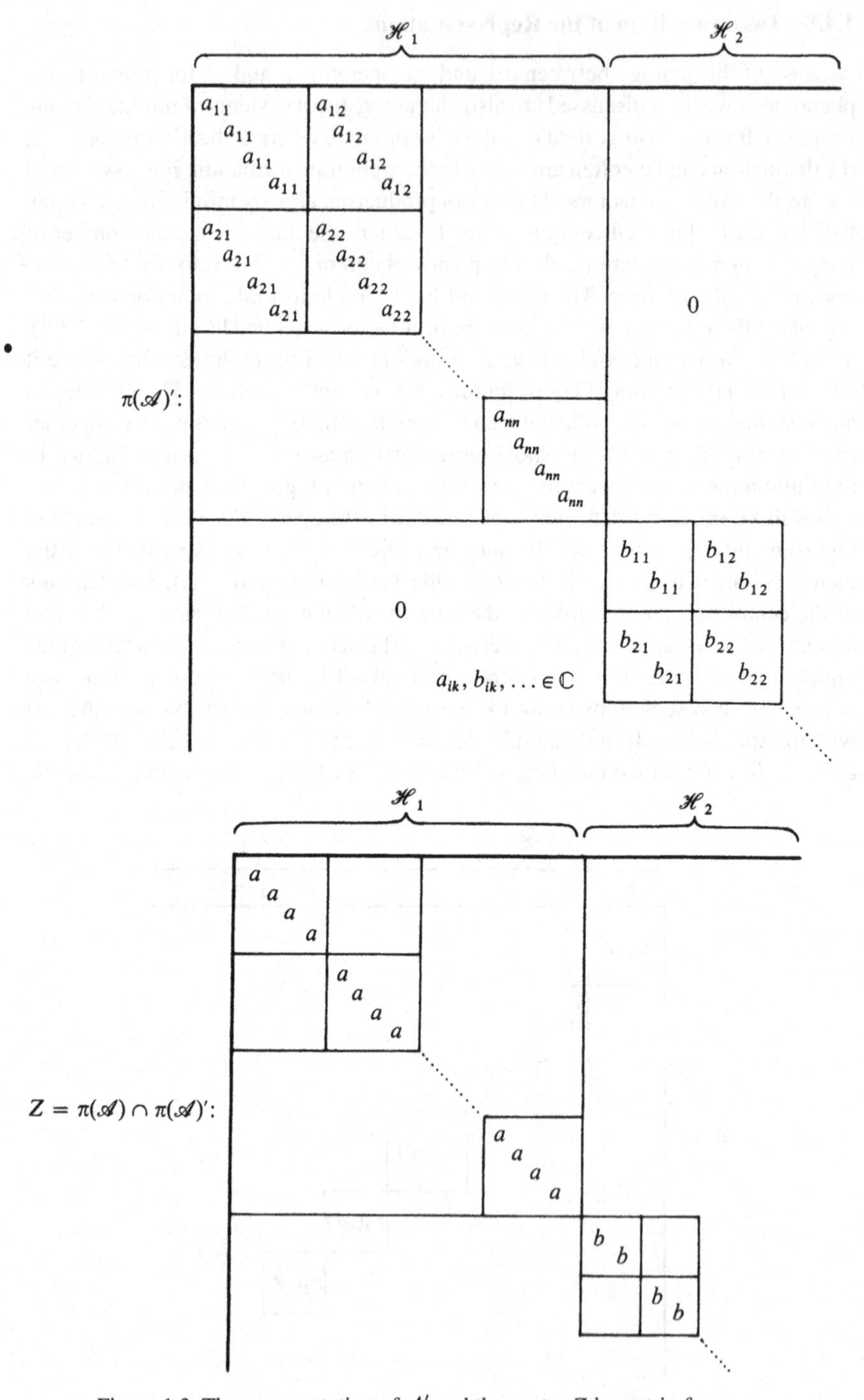

Figure 1.3. The representation of $\mathcal{A}'$ and the center Z in matrix form.

space is the same for every choice of orthogonal basis. Different bases correspond to the different maximally Abelian subalgebras of $\pi(\mathcal{A})'$ that they diagonalize.

The passage to an infinite dimension requires the generalization of sums to integrals. The spectral theorem (III: 2.3.11) states that a Hermitian operator $a \in \mathcal{B}(\mathcal{H})$ may be represented as a multiplication operator on some space $L^2(d\mu, \mathrm{Sp}(a))$. If there is degeneracy, then a spectral value $\alpha \in \mathrm{Sp}(a)$ is associated not with a single complex number but with a many-dimensional Hilbert space $\mathcal{H}_\alpha$. If $v(\alpha)$ denotes the component of $v \in \mathcal{H}$ in $\mathcal{H}_\alpha$, then the scalar product on $\mathcal{H}$ can be written as

$$\langle v|w\rangle = \int d\mu(\alpha)\langle v(\alpha)|w(\alpha)\rangle.$$

The action of a on v is $(av)(\alpha) = \alpha v(\alpha)$. The center $Z = \pi(\mathcal{A}) \cap \pi(\mathcal{A})'$ is a commutative algebra, and its elements may be simultaneously diagonalized, and so any $z \in Z$ may be written as $(zv)(\alpha) = f(\alpha)v(\alpha)$, where f assigns a complex number to α. Any element a of $\mathcal{A}$ can then be represented by $[\pi(a)v](\alpha) = \pi_\alpha(a)v(\alpha)$, $\pi_\alpha(a) \in \mathcal{B}(\mathcal{H}_\alpha)$, and $b \in \pi(\mathcal{A})' \Rightarrow (bv)(\alpha) = b(\alpha)v(\alpha)$, $b(\alpha) \in \mathcal{B}(\mathcal{H}_\alpha)$, $[b(\alpha), \pi_\alpha(a)] = 0$ for all $a \in \mathcal{A}$. In a finite number of dimensions every $\mathcal{H}_\alpha$ can be written $\mathcal{H}_\alpha = \mathcal{H}_\alpha^{(1)} \otimes \mathcal{H}_\alpha^{(2)}$, $\pi_\alpha(\mathcal{A}) = \mathcal{B}(\mathcal{H}_\alpha^{(1)}) \otimes \mathbf{1}_{\mathcal{H}_\alpha^{(2)}}$, and $b(\alpha)$ is of the form $\mathbf{1}_{\mathcal{H}_\alpha^{(1)}} \otimes b$, $b \in \mathcal{B}(\mathcal{H}_\alpha^{(2)})$. This is as far as the finite-dimensional analogy goes; it will not be possible to write every factor π_α in the form $\mathcal{B}(\mathcal{H}) \otimes \mathbf{1}$.

1.4.9 Classification of Factors

We pause now to take stock of the factors, which will function as basic building blocks. The possibility that comes to mind first for a preliminary, rough classification is to define a trace. In (III: 2.3.19) the trace was defined as a mapping from $\mathcal{A}_+$, the positive operators, to $\bar{\mathbb{R}}^+$, and it was extended to a linear mapping from the trace class $\mathcal{C}_1(\mathcal{H})$ to $\mathcal{C}$. The trace is discontinuous in all topologies weaker than the trace topology given by $\|\cdot\|_1$. It may even occur that the only element of an algebra $\mathcal{A}$ in the trace class is the zero operator, as for example with the factor $\mathcal{B}(\mathcal{H}) \otimes \mathbf{1}$, where $\mathbf{1}$ is the identity on an infinite-dimensional space. In this case there is plainly the possibility of defining a trace by $\Phi(a \otimes \mathbf{1}) = \mathrm{Tr}_1\, a$, which has all the necessary properties. This observation suggests an abstract

1.4.10 Definition of the Trace

Let $\mathcal{A}_+$ be the positive cone of a strongly closed algebra $\mathcal{A}$, i.e., a von Neumann algebra. A **trace** is a mapping $\Phi : \mathcal{A}_+ \to \bar{\mathbb{R}}^+$ with the following properties.

(i) $\Phi(\lambda_1 a_1 + \lambda_2 a_2) = \lambda_1 \Phi(a_1) + \lambda_2 \Phi(a_2)$ for $a_i \in \mathcal{A}_+$ and $\lambda_i \in \mathbb{R}^+$;

(ii) $\Phi(a) = \Phi(uau^{-1})$ for all $a \in \mathcal{A}_+$ and all unitary $u \in \mathcal{A}$.

The trace Φ is said to be

1. **faithful**, if $\Phi(a) = 0$ and $a \in \mathcal{A}_+ \Leftrightarrow a = 0$;
2. **finite**, if $\Phi(a) < \infty$ for all $a \in \mathcal{A}_+$;
3. **semifinite**, if for all $a \in \mathcal{A}_+$ there exists a nonzero $b < a$ such that $\Phi(b) < \infty$;

4. **normal**, if for every increasing filter (see (III:2.2.21)) $F \subset \mathcal{A}_+$ with supremum s, $\Phi(s) = \sup_{a \in F} \Phi(a)$.

1.4.11 Examples

1. $\Phi(a) = 0$ for all $a \in \mathcal{A}_+$. The trace is unfaithful, finite, and normal.
2. $\Phi(0) = 0$, $\Phi(a) = \infty$ for all $a \neq 0$. The trace is faithful, not semifinite, and normal (purely infinite).
3. Let $\mathcal{A}$ be the $n \times n$ matrices and $\Phi(a) = \operatorname{Tr} a$. The trace is faithful, finite, and normal.
4. $\mathcal{A} = \mathcal{B}(\mathcal{H})$, $\mathcal{H}$ infinite-dimensional, and $\Phi(a) = \operatorname{Tr}(a)$. The trace is faithful, semifinite, and normal.
5. $\mathcal{A} = \mathcal{B}(\mathcal{H}_1) \oplus \mathcal{B}(\mathcal{H}_2)$, $\Phi(a \oplus b) = \alpha \operatorname{Tr} a + \beta \operatorname{Tr} b$, α and $\beta \in \mathbb{R}^+$. The trace is faithful only if α and β are nonzero and finite only if the $\mathcal{H}_i$ are finite-dimensional. In all cases it is semifinite and normal. (Note that although Φ is invariant under unitary transformations belonging to $\mathcal{A}$ for $\alpha \neq \beta$, it is not invariant under all unitary transformations in $\mathcal{B}(\mathcal{H}_1 \oplus \mathcal{H}_2)$.
6. Let $\mathcal{A}$ be the algebra of multiplication operators $L^\infty(\mathbb{R}, d\mu)$ on $L^2(\mathbb{R}, d\mu)$, and $\Phi(a) = \int d\mu(x) a(x) \rho(x)$ for some non-negative, measurable ρ. If $\rho > 0$ a.e., then Φ is faithful; if $\rho \in L^1(\mathbb{R}, d\mu)$, then Φ is finite; and if $\rho < \infty$ a.e., then Φ is semifinite. In all cases the trace is normal.
7. Let $\mathcal{A}$ be the algebra of multiplication operators l^∞ on l^2, and $\Phi(a) = \lim_{i \to \infty} a_i$ when the limit exists, and otherwise let the trace be defined by linear extension with the Hahn–Banach theorem. The trace is finite and neither faithful nor normal: If $F = \{(a_i)$, where $a_i = 1$ for finitely many i and otherwise $= 0\}$, then $s = (a_i = 1)$, and $\Phi(s) = 1$, but $\Phi(a) = 0$ for all $a \in F$.

1.4.12 Remarks

1. Property (ii) may be replaced with (ii)′: $\Phi(aa^*) = \Phi(a^*a)$ for all $a \in \mathcal{A}$ (Problem 3).
2. It can be shown in general that $\{a \in \mathcal{A}_+ : \Phi(a) < \infty\}$ consists of the positive elements of a two-sided self-adjoint ideal $\mathcal{M}_\Phi$ onto which Φ can be extended as a linear form (also denoted Φ). It is discontinuous in every topology that is strictly coarser than the one defined by the norm $\|a\|_\Phi = \Phi((a^*a)^{1/2})$. All continuous linear functionals on $\mathcal{M}_\Phi$ with this topology are of the form $a \to \Phi(ab)$, $a \in \mathcal{M}_\Phi$, $b \in \mathcal{A}$ (Problem 4), and nonzero for $b \neq 0$.
3. Property (ii) implies for $a \in \mathcal{M}_\Phi$ and any unitary $u \in \mathcal{A}$ that $\Phi(ua) = \Phi(au)$. Moreover, since every element of $\mathcal{A}$ is a linear combination of unitary operators, $\Phi(ab) = \Phi(ba)$, $a \in \mathcal{M}_\Phi$, $b \in \mathcal{A}$.
4. The requirement of normality originates in the theory of integration, where monotonic convergence can be permuted with integration. The trace can consequently be regarded as a generalization of the integral to noncommutative integrands.
5. If Φ is normal, then $\mathcal{A}$ may be written as $\mathcal{A} = \mathcal{A}_1 \oplus \mathcal{A}_2 \oplus \mathcal{A}_3$, where $\Phi_{|\mathcal{A}_3}$ is faithful and semifinite, $\Phi_{|\mathcal{A}_1} = 0$, and $\Phi_{\mathcal{A}_2}$ is purely infinite (Problem 5). As

we shall be interested solely in normal traces and shall ignore the trivial cases of Examples 1 and 2, we may confine our attention to faithful, semifinite traces.

The ordering of operators induces an ordering of traces, whereby $\Phi \leq \Psi$ shall mean $\Phi(a) \leq \Psi(a)$ for all $a \in \mathcal{A}_+$. For the ordering of the trace there is a theorem on

1.4.13 The Form of a Dominating Trace

Let Φ and Ψ be normal, semifinite traces on a von Neumann algebra $\mathcal{A}$. Then $\Phi \leq \Psi$ iff there exists $b \in \mathcal{A} \cap \mathcal{A}'$, $0 < b \leq 1$, such that $\Phi(a) = \Psi(ab)$ for all a.

Proof: Let $\mathcal{M}_\Psi$ be the ideal on which $\Psi < \infty$, given the norm $\|a\| = \Psi((aa^*)^{1/2})$. The mapping $a \to \Phi(a)$ is then a continuous linear form on $\mathcal{M}_\Psi$, and by Remark (1.4.12; 2) it is $\Psi(ab)$ for some $b \in \mathcal{A}$. To prove that $b \in \mathcal{A}'$, observe that for all $a \in \mathcal{M}_\Phi$ and $c \in \mathcal{A}$, $0 = \Phi(ac - ca) = \Psi(acb - cab) = \Psi(a[c, b])$, so, according to (1.4.12; 2), $[c, b] = 0$. □

1.4.14 Corollary

Any two faithful, normal, semifinite traces on the same factor are proportional.
More specifically, if Φ_1 and Φ_2 are two such traces, then $\Phi_1 < \Phi_1 + \Phi_2$ and $\Phi_2 < \Phi_1 + \Phi_2$. Since the center of the factor consists of multiples of the identity, $\Phi_i = \lambda(\Phi_1 + \Phi_2)$, $0 < \lambda_i < 1$, so $\Phi_1 = \lambda_1 \lambda_2^{-1} \Phi_2$.

Because the trace is essentially unique on any factor, it may be asked whether the trace of a projection is an integer c, which would allow a reasonable definition of the dimension of the subspaces onto which they project.

1.4.15 The Types of Factors

Factors of Type I
The range of the trace of the projections of factors of type I is $c \in \mathcal{Z}^+$, and they are of the form $\mathcal{B}(H) \otimes \mathbf{1}$, with $\mathcal{H}$ separable, i.e., a sum of identical copies of an irreducible algebra of operators. The trace is given by $\Phi(a \otimes \mathbf{1}) = c \operatorname{Tr} a$, and if the dimension of $\mathcal{H}$ is n, then it is finite for $n < \infty$ and not finite but only semifinite for $n = \infty$. This creates a distinction between subtypes I_n and I_∞.

Factors of Type II
On Factors of Type II there is a semifinite, normal, faithful trace the range of which when applied to the projections is either $[0, 1]$ or $\mathbb{R}^+$. Depending on whether the trace is finite or only semifinite, one distinguishes between subtypes II_1 and II_∞. An example of type II_1 is the algebra of infinitely many spins (1.1.2) represented with the GNS construction using the state $\Phi : \Phi(\mathbf{1}) = 1$, $\Phi(\prod \sigma_j) = 0$ ((1.4.7;1) with $s = 0$). This state has the properties of a trace; commutativity (1.4.11(ii)) holds trivially, and this representation is a factor. Since the factor is obviously not isomorphic to anything of the form $\mathcal{B}(\mathcal{H}_n) \otimes \mathbf{1}$, $n < \infty$, and the trace is finite, it

must be of type II_1. It is reducible but not of type I, since it can not be written as a direct sum of identical irreducible algebras. Type II_∞ factors are of the form type $\mathrm{I}_\infty \otimes$ type II_1, where the trace is defined multiplicatively on the tensor product.

Factors of Type III

They have no normal, faithful, semifinite trace. The infinite spin algebra (1.1.2) again provides an example, this time with the GNS representation using the state (1.10) with $s \neq 0$, in other words (1.4.7).

1.4.16 Remarks

1. The type with the properties familiar from finite matrices is I, while types II and III are less intuitive. All three types occur in the GNS representation of the spin algebra with a state of the form (1.10), I_∞ with $s = 1$, II_1 with $s = 0$, and III with $0 < s < 1$. To the malicious delight of many mathematicians the initial impression that type III is the rule for infinite systems has panned out with the passage of time. Types I and II turn out to be peripheral possibilities.
2. It was ascertained in (III: 2.3.6; 5) that factor representations with maximally Abelian subalgebras are irreducible. As a result, representations of types II and III have no in $\mathcal{B}(\mathcal{H})$ maximally Abelian subalgebras.
3. If a factor includes an irreducible subrepresentation, then a semifinite, normal trace can be defined on it, mapping the projections to a discrete set of values, and it must therefore be of type I. It was remarked in (III: 2.3.10; 5) that the GNS construction yields an irreducible representation iff the state it builds on is pure. This means that no vector in the Hilbert space of a representation of type II or III corresponds to a pure state on the algebra.
4. Any operator a of an algebra of type III is of course bounded, so $\mathrm{Tr}\,\rho a$ is well defined for any $\rho \in \mathcal{C}_1(\mathcal{H})$, only ρ can not come from the algebra, which contains no element of a trace class (other than 0).

Let us end the section by recapitulating the physical significance of the new mathematical phenomena that make an appearance in infinite systems.

1. **Inequivalent Representations**
   Since vectors that differ globally are always orthogonal, globally different situations lead to inequivalent representations. Within a given representation different elements of the algebra produce vectors that differ only locally.
2. **Non-normal States**
   Expectation values with a vector of a different, inequivalent representation constitute a state on the algebra, but one that fails to be strongly continuous with respect to the original representation, and hence it is not normal. They are representations of different global circumstances, and thus assign different values to global observables like densities, which are only defined with strong limits.
3. **Factors**
   Whereas $\pi(\mathcal{A})$ describes microscopic observables, $\pi(\mathcal{A})''$ covers macroscopic observables as well. Factors associate certain numerical values to the

global observables lying in the center $\pi(\mathcal{A})'' \cap \pi(\mathcal{A})'$ – factors are the macroscopically pure states. In factors, Khinchin's ergodic theorem applies to them, stating that these global quantities exhibit no fluctuation. Even if vectors of a factor are pure with respect to this subalgebra, they may produce mixed states. The ground state is associated with type I, finite temperature with type III, and infinite temperature with type II.

4. **Unitary Representation of the Time-Evolution**
If the algebra changes globally as time passes, then a representation may change at any moment into an inequivalent representation, and it is not possible to represent the time-evolution with a group of unitary transformations within the representation. Yet if the representation is based on a time-invariant state, then the other vectors of the representation differ only locally, and thus do not change in time, from the global point of view. This establishes the possibility of a unitary time-evolution.

1.4.17 Problems

1. Show that with vectors $|x^{(1)}\rangle, \ldots, |x^{(n)}\rangle$ and $\lambda_1, \ldots, \lambda_n \in C$, Definition (1.4.1) implies that $\sum_{i,k} \lambda_i^* \lambda_k \langle x^{(i)} | x^{(k)} \rangle \geq 0$. (Hint: it suffices to show this for the case where the $|x^{(i)}\rangle$ are strongly equivalent. Prove that $\sum_{i,k} \lambda_i^* \lambda_k \prod_{j=1}^N (x_j^{(i)} | x_j^{(k)}) \geq 0$ for any N and take the limit $N \to \infty$.)

2. (i) Show that $|x\rangle$ and $|y\rangle$ are strongly equivalent iff $\sum_i |1 - (x_i | y_i)| < \infty$ and weakly equivalent iff $\sum_i |1 - |(x_i | y_i)|| < \infty$.
   (ii) Conclude from (i) that the $\underset{\text{strong}}{\sim}$ of $|x\rangle \underset{\text{strong}}{\sim} |y\rangle$ has all the properties of an equivalence relation, namely reflexivity, symmetry, and transitivity. (Hint: use the inequality $|1 - (x|z)| \leq 4[|1 - (x|y)| + |1 - (y|z)|]$, which holds for unit vectors. This 4 is a generous constant.)
   (iii) Show that $|x\rangle \underset{\text{weak}}{\sim} |y'\rangle$ iff there exists a sequence $\{\varphi_j\}$ such that $|x\rangle \underset{\text{strong}}{\sim} |y'\rangle$, $|y'\rangle \equiv \exp(i\varphi_1)|y_1\rangle \otimes \exp(i\varphi_2)|y_2\rangle \otimes \ldots$.
   (iv) Show that $\underset{\text{weak}}{\sim}$ is also an equivalence relation.

3. Show that condition (ii) of the definition of the trace (1.4.10), i.e., $\Phi(a) = \Phi(UaU^{-1})$, may be replaced with: $\Phi(a^*a) = \Phi(aa^*)$ for all a in a von Neumann algebra $\mathcal{A}$.

4. Show that for a faithful, normal, semifinite trace Φ, all continuous linear forms on $a \in \mathcal{M}_\Phi$ may be written as $a \to \Phi(ab)$ for some $b \in \mathcal{A}$. (Hint: use the inequality $|\Phi(ab)| \leq \Phi(|ab|) \leq \|b\| \Phi(|a|)$.)

5. Show that with any normal trace Φ, $\mathcal{A}$ can be written $\mathcal{A} = \mathcal{A}_1 \oplus \mathcal{A}_2 \oplus \mathcal{A}_3$, where $\Phi_{|\mathcal{A}_1} \equiv 0$, $\Phi_{|\mathcal{A}_2}$ is faithful and semifinite, and $\Phi_{|\mathcal{A}_3}$ is purely infinite. (Use the following corollaries of von Neumann's density theorem (III: 2.3.24; 4):
   (I) Let $\mathcal{M} \subset \mathcal{A}$ be a strongly closed, two-sided ideal. Then $\mathcal{M}$ contains a projection operator P such that $P \in \mathcal{A} \cap \mathcal{A}'$ and $P \geq Q$ for all projection operators $Q \in \mathcal{M}$.
   (II) Let $\mathcal{N}$ be a two-sided ideal and suppose a is in the positive part of the weak closure of $\mathcal{N}$. Then there exists an increasing filter $\subset \mathcal{N}^+$ having a for its supremum.

1.4.18 Solutions

1. The $n \times n$ matrix

$$\begin{pmatrix} (x_j^{(1)}|x_j^{(1)}) \dots (x_j^{(1)}|x_j^{(n)}) \\ \vdots \\ (x_j^{(n)}|x_j^{(1)}) \dots ((x_j^{(n)}|x_j^{(n)}) \end{pmatrix}$$

is Hermitian and non-negative, and is thus a sum of projections, i.e., matrices of the form

$$\begin{pmatrix} h_1^* h_1 \dots h_1^* h_n \\ \vdots \\ h_n^* h_1 \dots h_n^* h_n \end{pmatrix}.$$

This implies

$$(x_j^{(i)}|x_j^{(k)}) = \sum_{l_j=1}^{n} h_i^{l_j *} h_k^{l_j},$$

and

$$\sum_{i,k} \lambda_i^* \lambda_k \prod_{j=1}^{N} (x_j^{(i)}|x_j^{(k)}) = \sum_{i,k} \lambda_i^* \lambda_k \sum_{l_1,\dots,l_N} h_i^{l_1 *} h_k^{l_1} \dots h_i^{l_N *} h_k^{l_N} \geq 0,$$

since

$$\sum_{i,k} \lambda_i^* \lambda_k h_i^{l_1 *} \dots h_k^{l_N} = \left| \sum_k \lambda_k h_k^{l_1} \dots h_k^{l_N} \right|^2 \geq 0.$$

2. (i) follows from the theory of infinite products [12].
   (ii) To prove the inequality, choose a basis for the subspace spanned by $|x)$, $|y)$, and $|z)$ such that they correspond to the vectors $(\alpha, \beta, 0)$, $(1, 0, 0)$ and $(\gamma, \delta, \varepsilon)$, where $|\alpha|^2 + |\beta|^2 = |\gamma|^2 + |\delta|^2 + |\varepsilon|^2 = 1$. Then $(x|y) = \alpha^*$, $(y|z) = \gamma$, $(x|z) = \alpha^*\gamma + \beta^*\delta$. $|1 - \alpha^*\gamma - \beta^*\delta| \leq |1 - \alpha^*\gamma| + |\beta||\delta| \leq 2|1-\alpha| + 2|1-\gamma| + (1-|\alpha|^2)^{1/2}(1-|\gamma|^2)^{1/2} \leq 2(|1-\alpha|^{1/2} + |1-\gamma|^{1/2})^2 \leq 4[|1-\alpha| + |1-\gamma|]$. The reflexivity and symmetry of the equivalence relation are trivial, and transitivity follows from (i) together with the inequality.
   (iii) $\Rightarrow$: Choose $\varphi_j = -\arg(x_j|y_j)$.
   $\Leftarrow$: This is trivial.
   (iv) follows from (ii) and (iii).

3. (ii) $\Rightarrow$ (ii′): With a polar decomposition, $a = V|a|$, where $a^*a = |a|^2 = V^*V|a|^2$, $aa^* = V|a|^2V^*$. Let $\mathcal{M}_\Phi$ be the trace-class ideal: $a \in \mathcal{M}_\Phi \Rightarrow a^*a \in \mathcal{M}_\Phi$ and $aa^* \in \mathcal{M}_\Phi \Rightarrow Va^*a \in \mathcal{M}_\Phi$, since $V = \text{w-}\lim_{\varepsilon \downarrow 0} a(|a|^2 + \varepsilon)^{-1/2} \in \mathcal{A}$, which, with Remark (1.4.13; 3) implies $\Phi(V^*Va^*a) = \Phi(Va^*aV^*)$.
   (ii′) $\Rightarrow$(ii): Let $a \geq 0$. $\Phi(UaU^{-1}) = \Phi(Ua^{1/2}a^{1/2}U^*) = \Phi(a^{1/2}U^*Ua^{1/2}) = \Phi(a)$, and every operator is a linear combination of positive operators.

4. To prove the inequality, let a and b be non-negative. $\Phi(ab) = \Phi(a^{1/2}ba^{1/2}) \leq \|b\|\Phi(a)$, since for any a and b, $a^{1/2}ba^{1/2} \leq a^{1/2}\|b\|a^{1/2}$. Thus $|\Phi(ab)|^2 \leq \Phi(|a^*||b|)\Phi(|a||b^*|)$ and is consequently $\leq \||b|\|\Phi(|a^*|)\||b^*|\|\Phi(|a|)^2$, in which the Cauchy-Schwarz inequality $|\Phi(ab)|^2 \leq \Phi(aa^*)\Phi(bb^*)$ (see (III: 2.2.20; 1)) was used

in the form $|\Phi(ab)|^2 = |\Phi(U|a|V|b|)|^2$ (with the polar decompositions $a = U|a|$ and $b = V|b| = \|b\|^2\Phi(|a|)^2 = |\Phi(|b|^{1/2}U|a|^{1/2}|a|^{1/2}V|b|^{1/2})|^2 \leq \Phi(|b^{1/2}U|a|^{1/2} \times |a|^{1/2}U^*|b|^{1/2})\cdot\Phi(|b|^{1/2}V^*|a|^{1/2}|a|^{1/2}V|b|^{1/2}) = \Phi(|b|U|a|U^*)\times\Phi(V|b|V^*|a|) = \Phi(|b||a^*|)\Phi(|b^*||a)$. Now let $ab = W|ab|$; then $\Phi(|ab|) = \Phi(W^*ab) \leq \|bW^*\| \times \Phi(|a|) \leq \|b\|\Phi(|a|)$. The first part of the inequality follows from $|\Phi(ab)| = |\Phi(ab \cdot \mathbf{1})| \leq \|\mathbf{1}\|\Phi(|ab|) = \Phi(|ab|)$.

It is a corollary of the inequality that the norm of the mapping $a \to \Phi(ab)$ is $\|b\|$. This allows $\mathcal{A}$ to be identified with a closed subspace of $\mathcal{M}_\Phi^*$. To see that $\mathcal{A} = \mathcal{M}_\Phi^*$, first suppose $a \in \mathcal{M}_\Phi^+$. Then the mapping $\mathcal{A} \to \mathcal{C} : b \to \Phi(ab)$ is normal, entailing ultraweakly continuous (see (2.1.4)), which implies that for any $a \in \mathcal{M}_\Phi$, $b \to \Phi(ab)$ is ultraweakly continuous. Because of the inequality again, the norm of this mapping is $\Phi(|a|)$, which implies that $\mathcal{M}_\Phi$ can be imbedded isometrically and isomorphically in the predual $\mathcal{A}_*$, i.e., the space of ultraweakly continuous linear functions. Thus $\mathcal{M}_\Phi \subset \mathcal{A}_*$. We shall see in (2.1.3) and (2.1.4) that $\mathcal{C}_1 = \mathcal{B}(\mathcal{H})_*$ and $\mathcal{C}_1^* = \mathcal{B}(\mathcal{H})$. Since $\mathcal{A}$ is ultraweakly closed, $\mathcal{A}_\perp = \mathcal{C}_1/\mathcal{C}_1^\perp$ with $\mathcal{C}_1^\perp = \{\rho \in \mathcal{C}_1 : \operatorname{Tr} \rho a = 0 \text{ for all } a \in \mathcal{A}\}$, so $\mathcal{A} = (\mathcal{C}_1/\mathcal{C}_1^\perp)^*$. Therefore $\mathcal{M}_\Phi^* \subset (\mathcal{A}_*)^* = \mathcal{A}$, which implies $\mathcal{M}_\Phi^* = \mathcal{A}$.

Remark: $\mathcal{M}_\Phi$ is dense in $\mathcal{A}_*$ but not in general closed.

5. For more about types I and II, see Chapter I, § 3 of [4]. The set $\{a \in \mathcal{A}^+ : \Phi(a) = 0\}$ is the positive part of a two-sided ideal $\mathcal{N}$. Let $\mathcal{M}$ be the trace class, let $\bar{\mathcal{N}}$ and $\bar{\mathcal{M}}$ be the strong closures of $\mathcal{N}$ and $\mathcal{M}$, and P_1 and P_2 be respectively the largest projections they contain (see Corollary I). The Hilbert space $\mathcal{H}$ can be decomposed as $\mathcal{H}_1 \oplus \mathcal{H}_2 \oplus \mathcal{H}_3$, where $\mathcal{H}_1 \equiv P_1\mathcal{H}$, $\mathcal{H}_2 \equiv (P_2 - P_1)\mathcal{H}$, $\mathcal{H}_3 \equiv (1 - P_2)\mathcal{H}$ in which case $\mathcal{A} = \mathcal{A}_1 \oplus \mathcal{A}_2 \oplus \mathcal{A}_3$, where $\mathcal{A}_i \equiv \mathcal{A}_{|\mathcal{H}_i}$, since P_1 and P_2 belong to $\mathcal{A} \cap \mathcal{A}'$.

It is obvious that $\Phi_{|\mathcal{A}_1} = 0$. To see that $\Phi_{|\mathcal{A}_2}$ is semifinite, apply Corollary II: Let $a \in \bar{\mathcal{M}}^+ \backslash \mathcal{M}^+$; then there exists an operator $b \in \mathcal{M}^+$, $b \leq a$, such that $\Phi(b) > 0$. The remaining claims are trivial.

2 Thermostatics

2.1 The Ordering of the States

The heuristic concepts of purer and more chaotic states can be made mathematically precise with reference to a lattice structure of the classes of equivalent density matrices.

States are by definition (III: 2.2.18) normed, positive linear functionals on an algebra $\mathcal{A}$ of observables. If the dimension of the underlying space is finite, $\mathcal{A} = \mathcal{B}(\mathbb{C}^n)$, then all linear functionals are of the form $\mathcal{A} \ni a \to \operatorname{Tr} \rho a \equiv (\rho|a)$, $\rho \in \mathcal{B}(\mathbb{C}^n)$, and $\mathcal{B}(\mathbb{C}^n)$ is its own dual space. The inequality of (1.4.18;4),

$$|(\rho|a)| \leq \|a\| \|\rho\|_1, \qquad \|\rho\|_1 = \operatorname{Tr}(\rho^* \rho)^{1/2} \tag{2.1}$$

then holds, and is optimal in the sense that

$$\sup_{\|\rho\|_1=1} |(\rho|a)| = \|a\|, \qquad \sup_{\|a\|=1} |(\rho|a)| = \|\rho\|_1. \tag{2.2}$$

If the dimension of $\mathcal{H}$ is infinite, the inequality applies initially to the operators of finite rank (cf. (III: 2.3.21)), denoted $\mathcal{E}$ or $\mathcal{E}_1$, depending on whether the norm $\| \, \|$ or $\| \, \|_1$ is used. In these topologies continuous linear functionals are of the form

$$\mathcal{E} \ni a \to \operatorname{Tr} \rho a \quad \text{with} \quad \|\rho\|_1 < \infty$$

or

$$\mathcal{E}_1 \ni a \to \operatorname{Tr} \rho a \quad \text{with} \quad \|\rho\| < \infty.$$

The linearity and continuity of the functionals thus defined are obvious, and it can be seen as follows that all functionals with these properties are of that form.

By what was said earlier, a linear functional on $\mathcal{E}$ determines the restriction of an operator ρ to any finite-dimensional subspace. To guarantee that $|(\rho|a)| \leq c\|a\|$ for all $a \in \mathcal{E}$ or $|(\rho|a)| \leq c\|a\|_1$ for all $a \in \mathcal{E}_1$, by (2.1) it is necessary to ensure that $\|\rho\|_1$ or respectively $\|\rho\|$ is bounded. If the spaces $\mathcal{E}$ and $\mathcal{E}_1$ are now completed, becoming the Banach spaces $\mathcal{C}$ and $\mathcal{C}_1$ of (III: 2.3.21), then their dual spaces are unaffected – the dual spaces of a space and of a dense subspace are the same. The state of affairs is analogous to that of l^0, l^1, and l^∞, the spaces of sequences (x_i) satisfying respectively $\lim_{i\to\infty} x_i = 0$, $\sum_i |x_i| < \infty$, and $\sup_i |x_i| < \infty$:

2.1.1 Duality for the Subspaces of $\mathcal{B}(\mathcal{H})$

$\mathcal{C}^* = \mathcal{C}_1$, $\mathcal{C}_1^* = \mathcal{B}(\mathcal{H})$, where $\mathcal{C}$ and $\mathcal{B}(\mathcal{H})$ are given the norm $\|\,\|$, and $\mathcal{C}_1$ the norm $\|\,\|_1$. These norms on $\mathcal{C}_1$ and $\mathcal{B}(\mathcal{H})$ produce the strong topology on the dual spaces, as can be seen from a comparison of (2.1) with (III: 2.2).

The Banach space $\mathcal{C}$ is thus not reflexive, so $\mathcal{B}(\mathcal{H})^*$ is strictly larger than $\mathcal{C}_1$. If a Banach space $\mathcal{E}$ is nonreflexive, then the same is true of $\mathcal{E}^*$, $\mathcal{E}^{**}$, etc.: Let $a \in \mathcal{E}^{**}$ but $a \notin \mathcal{E}$. The functional $w : e + \lambda a \to \lambda$ defined on $\{E + \lambda a\}$ can be extended continuously to $\mathcal{E}^{**}$ by the Hahn – Banach theorem. Therefore, $w \in \mathcal{E}^{***}$, but $w_{|\mathcal{E}} = 0$. Hence $\mathcal{C}_1$ and $\mathcal{B}(\mathcal{H})$ are also not reflexive; $\mathcal{B}(\mathcal{H})^*$ is strictly larger than $\mathcal{C}_1$. All trace-class operators provide linear functionals on the bounded operators by $a \to \operatorname{Tr} \rho a$, and these linear functionals are even continuous if $\mathcal{B}(\mathcal{H})$ is equipped with a weaker topology than the one from $\|\,\|$: If the neighborhood basis is defined by

$$U_{\rho,\varepsilon}(a) = \{a' \in \mathcal{B}(H) : |\operatorname{Tr} \rho(a - a')| < \varepsilon\}, \tag{2.3}$$

and ρ ranges only over $\mathcal{E}$, then this is the weak topology. If ρ is allowed to range over $\mathcal{C}_1$, then it is known as the ultraweak topology, and is genuinely finer than the weak topology but coarser than the $\|\,\|$-topology. The linear functionals $a \to \operatorname{Tr} \rho a$ for $\rho \in \mathcal{C}_1$ are, however, obviously continuous if $\mathcal{B}(\mathcal{H})$ has the ultraweak topology. These functionals have in addition the property of normality (III: 2.2.21): the order of taking weakly continuous linear functionals and suprema over bounded sets can be interchanged, since by Vigier's theorem (III: 2.3.24; 11) the supremum is the limit of a strongly, and therefore also weakly, convergent sequence. Since the weak and ultra-weak topologies are equivalent on bounded sets, normality carries over to ultraweakly continuous, linear functionals. A somewhat deeper theorem ([4], I, § 4, Theorem I) states that these include all normal linear functionals on $\mathcal{B}(\mathcal{H})$. We summarize by stating the

2.1.2 Characterization of Normal States

The following properties are equivalent for a state w on $\mathcal{B}(\mathcal{H})$:

(i) w is normal (III: 2.2.21);

(ii) w is given by a **density matrix** ρ such that $w(a) = \operatorname{Tr} \rho a$, $\rho \geq 0$, $\operatorname{Tr} \rho = 1$;

(iii) w is ultraweakly continuous.

2.1.3 Remarks

1. The density matrices form a norm-closed, convex subset of the unit sphere of $\mathcal{C}_1$, the trace-class operators with the trace norm $\| \|_1$.
2. If the system is classical, then instead of $\mathcal{B}(\mathcal{H})$ there is an Abelian von Neumann algebra, and we are familiar with the normal traces in the guise of probability measures. Specifically, on the L^∞ functions on phase space they are of the form $\rho(p,q)d\Omega$, $d\Omega$ being Liouville measure (I: 3.1.2; 3), $\rho \in L^1$, $\rho \geq 0$, $\int d\Omega\rho = 1$. Yet it may be that $|\rho| = \sup_{p,q} |\rho(p,q)| \not< 1$: Suppose that χ_A is the characteristic function of a set A such that $\Omega(A) \equiv \int d\Omega_{\chi_A} < 1$; then an example is furnished by $\rho = \chi_A/\Omega(A)$.
3. All states over $\mathcal{B}(\mathcal{H})$ constructed with a vector of $\mathcal{H}$ are pure, normal, and even weakly continuous – the density matrix for them is a one-dimensional projection. Conversely, any one-dimensional projection yields a pure state on $\mathcal{B}(\mathcal{H})$.
4. The spectrum of a density matrix is discrete, as it is in the trace class (and hence compact). The sum of the eigenvalues ρ_i is 1.
5. The density matrix can be thought of as a combination of the vectors that diagonalize it, or as a pure state on a larger Hilbert space $\mathcal{H}_g \equiv \mathcal{H}\otimes\mathcal{H}$, in which $\mathcal{B}(\mathcal{H})$ is imbedded as $\mathcal{B}(\mathcal{H})\otimes\mathbf{1}$. The vector of $\mathcal{H}_g$ corresponding to $\rho = \sum_j |j\rangle\langle j|\rho_j$ is $\sum_j |j\rangle \otimes |j\sqrt{\rho_j}$; (cf.(1.4.6)). If $\mathcal{H}$ is separable, then the weak topology on $\mathcal{H}_g$ induces the ultraweak topology of $\mathcal{B}(\mathcal{H})$ on $\mathcal{B}(\mathcal{H}) \otimes \mathbf{1}$.
6. The normal states are weak*- dense in the positive unit sphere of $\mathcal{B}(\mathcal{H})^*$ (see (III: 2.1.19)), but are a proper subset rather than the whole of it. Hence they are not also weak*- compact.

Traces offer many advantages for doing calculations, owing to the commutativity property (1.4.12; 3). Inequalities for ordinary numbers often extend to traces, even when noncommutativity prevents them from extending directly to operators. Some of these inequalities will be used frequently later, and so are listed below. It will always be assumed that whatever the trace is taken of belongs to the trace class, though many of them have the generalization that if the lesser side of an inequality becomes infinite, then so does the greater side. For greater flexibility general forms are presented, while the name attached refers to the original version. The symbol Tr will always mean the trace on $\mathcal{B}(\mathcal{H})$. These inequalities apply trivially to factors of type I, and many also apply to type II.

2.1.4 Basic Inequalities

1. Peierls's Inequality. Let k be a convex function from $\mathbb{R}$ to $\mathbb{R}^+$ and $\{|i\rangle\}$ be a not necessarily complete, orthonormal set. Then

$$\operatorname{Tr} k(a) = \sup_{\{|i\rangle|\}} \sum_i k(\langle i|a|i\rangle).$$

2. Convexity. Let k be a convex function from $\mathbb{R}$ to $\mathbb{R}$ and $0 \leq \alpha \leq 1$. Then

$$\operatorname{Tr} k(\alpha a + (1-\alpha)b) \leq \alpha \operatorname{Tr} k(a) + (1-\alpha) \operatorname{Tr} k(b).$$

3. The Peierls–Bogoliubov Inequality. Let k be a strictly monotonically increasing, convex, differentiable function $\mathbb{R} \to \mathbb{R}$ (and thus the inverse function k^{-1} exists), and suppose k/k' is convex. Then

$$k^{-1}(\operatorname{Tr} k(\alpha a + (1-\alpha)b)) \leq \alpha k^{-1}(\operatorname{Tr} k(a)) + (1-\alpha)k^{-1}(\operatorname{Tr} k(b)).$$

4. Monotony. If m is a monotonically increasing function $\mathbb{R} \to \mathbb{R}$,

$$a \geq b \Rightarrow \operatorname{Tr} m(a) \geq \operatorname{Tr} a(b).$$

5. Klein's Inequality. Let f, g, and h be functions $\mathbb{R} \to \mathbb{R}$ such that for all $\alpha \in \operatorname{Sp} a$, $\beta \in \operatorname{Sp} b$, and $c_k \in \mathbb{R}$,

$$\sum_k c_k f_k(\alpha) g_k(\beta) h_k(\alpha) \geq 0.$$

Then

$$\operatorname{Tr} \sum_k c_k f_k(a) g_k(b) h_k(a) \geq 0.$$

6. Hölder's Inequality. Suppose that k_1 and k_2 are convex, strictly monotonic functions $\mathbb{R} \to \mathbb{R}$, the mapping $(\alpha, \beta) \to k_1^{-1}(\alpha) k_2^{-1}(\beta)$ is concave, and $\mathcal{H}$ has dimension $N < \infty$. Then

$$\left| \frac{1}{N} \operatorname{Tr} ab \right| \leq k_1^{-1} \left(\operatorname{Tr} \frac{1}{N} k_1(|a|) \right) k_2^{-1} \left(\operatorname{Tr} \frac{1}{N} k_2(|b|) \right).$$

7. The Cauchy–Schwarz Inequality. $|\operatorname{Tr}(ab)^2| \leq \operatorname{Tr} a^* a b b^*$.
8. Lieb's Theorem. Let a and b be non-negative, $a, b, c \in \mathcal{B}(\mathcal{H})$, and $0 \leq \alpha \leq 1$. Then the functions $a \to \operatorname{Tr} \exp(c + \ln a)$ and $(a, b) \to \operatorname{Tr} a^\alpha c b^{1-\alpha} c^*$ are concave.

Proof:

1. By the spectral theorem and Jensen's inequality, for any unit vector $|i\rangle$, $\langle i|k(a)|i\rangle \geq k(\langle i|a|i\rangle)$, and therefore $\sum_i \langle i|k(a)|i\rangle \geq \sum_i k(\langle i|a|i\rangle)$. Equality holds if the $|i\rangle$ are eigenvectors of a. It suffices to take the supremum over finite sets $\{|i\rangle\}$.

2. Let $|i\rangle$ be the eigenvectors of $\alpha a + (1-\alpha)b$. By Peierls's inequality,

$$\begin{aligned}
\operatorname{Tr} k(\alpha a + (1-\alpha)b) &= \sum_i k(\alpha\langle i|a|i\rangle + (1-\alpha)\langle i|b|i\rangle) \\
&\leq \alpha \sum_i k(\langle i|a|i\rangle) + (1-\alpha) \sum_i k(\langle i|b|i\rangle) \\
&\leq \alpha \operatorname{Tr} k(a) + (1-\alpha) \operatorname{Tr} k(b).
\end{aligned}$$

Note that the inequality $k(\alpha a + (1-\alpha)b) \leq \alpha k(a) + (1-\alpha)k(b)$ can be false in the sense of operator ordering.

3. If k/k' is convex, then for sequences of numbers $\{\beta_i\}$ and $\{\gamma_i\}$,

$$k^{-1}\left(\sum_i k(\beta_i\alpha + \gamma_i(1-\alpha))\right)$$

$$\leq \alpha k^{-1}\left(\sum_i k(\beta_i)\right) + (1-\alpha)k^{-1}\left(\sum_i k(\gamma_i)\right)$$

by Problem 2. Hence, as with Inequality 2,

$$\begin{aligned} k^{-1}(\operatorname{Tr} k(\alpha a + (1-\alpha)b)) &= k^{-1}\left(\sum_i k(\alpha\langle i|a|i\rangle + (1-\alpha)\langle i|b|i\rangle)\right) \\ &\leq \alpha k^{-1}\left(\sum_i k(\langle i|a|i\rangle)\right) \\ &\quad +(1-\alpha)k^{-1}\left(\sum_i k(\langle i|b|i\rangle)\right) \\ &\leq \alpha k^{-1}(\operatorname{Tr} k(a)) + (1-\alpha)k^{-1}(\operatorname{Tr} k(b)), \end{aligned}$$

using Inequality 1 again.

4. If $a \geq b$, then the min-max principle implies for their ordered eigenvalues that $a_i \geq b_i$, so $\sum_i m(a_i) \geq \sum_i m(b_i)$. Once again, the inequality $m(a) \geq m(b)$ may fail for operators.

5. Let a_i and b_i be the eigenvalues of a and b, and c_{ij} be the scalar product of the eigenvectors of a with those of b. Then

$$\operatorname{Tr} \sum_k c_k f_k(a) g_k(b) h_k(a) = \sum_{i,j} |c_{ij}|^2 \sum_k c_k f_k(a_i) g_k(b_j) h_k(a_i) \geq 0.$$

6. Let a_i and b_i be the ordered eigenvalues of $|a|$ and $|b|$, and let $|i\rangle$ denote the eigenvectors of a. By the min-max principle (III: 3.5.19),

$$\begin{aligned} \operatorname{Tr} ab = \sum_{i,j} \langle i|a|j\rangle\langle j|b|i\rangle &\leq \sum_i (a_i - a_{i+1}) \sum_{k=1}^{i} \langle k\|b\|k\rangle \\ &\leq \sum_i (a_i - a_{i+1}) \sum_{k=1}^{i} b_k = \sum_i a_i b_i. \end{aligned}$$

The inequality

$$\frac{1}{N}\sum_{i=1}^{N} k_1^{-1}(\alpha_i) k_2^{-1}(\beta_i) \leq k_1^{-1}\left(\frac{1}{N}\sum_{i=1}^{N}\alpha_i\right) k_2^{-1}\left(\frac{1}{N}\sum_{i=1}^{N}\beta_i\right),$$

$$\text{for} \quad \alpha_i \equiv k_1(a_i) \quad \text{and} \quad \beta_i \equiv k_2(b_i),$$

is just the assumption of concavity.

7. By the Cauchy–Schwarz inequality (III: 2.2.20; 1) for states,

$$|\operatorname{Tr} abab|^2 \le \operatorname{Tr} abb^*a^* \operatorname{Tr} b^*a^*ab = (\operatorname{Tr} a^*abb^*)^2.$$

The order of the operations is important; it is not true in general that $\operatorname{Tr}(ab)^2 \le \operatorname{Tr} a^*ab^*b$.

8. The proof of this rather deep proposition in the noncommutative case is too laborious to be repeated here—see [5]. □

2.1.5 Corollaries

1. For any orthonormal system $\{|i\rangle\}$,

$$\beta F(H) \equiv -\ln \operatorname{Tr} \exp(-\beta H) \le -\ln \sum_i \exp(-\beta\langle i|H|i\rangle).$$

2. The function $H \to \operatorname{Tr} \exp(-\beta H)$ is convex.
3. In fact, even $H \to \ln \operatorname{Tr} \exp(-\beta H)$ is convex, so $F(H)$ is concave. By recourse to $(\partial/\partial\alpha) \operatorname{Tr} f(H + \alpha V)|_{\alpha=0} = \operatorname{Tr} V f'(H)$, and the fact that F is majorized by any tangent, one finds that

$$F(H_0) + \langle V\rangle_H \le F(H_0 + V) \le F(H_0) + \langle V\rangle_{H_0},$$

where $\langle a\rangle_H = \operatorname{Tr} a \exp(-\beta H)/\operatorname{Tr} \exp(-\beta H)$.
4. $H_1 \ge H_2 \Rightarrow F(H_1) \ge F(H_2)$.
5. If k is convex, then $\operatorname{Tr}(k(a) - k(b) - (a-b)k'(b)) \ge 0$, so $\operatorname{Tr}(a \ln a - a \ln b - (a-b)) \ge 0$, too. If $f_1(\alpha) = \int_0^\alpha d\alpha' g(\alpha')$ and $f_2(\beta) = \int_0^\beta d\beta' g^{-1}(\beta')$, then by Young's inequality, $\alpha\beta \le f_1(\alpha) + f_2(\beta)$, and therefore $\operatorname{Tr} ab \le \operatorname{Tr} f_1(a) + \operatorname{Tr} f_2(b)$. In particular, if p and q are ≥ 1 and related by $1/p + 1/q = 1$, and a and b are nonnegative, then $\operatorname{Tr} ab \le (1/p) \operatorname{Tr} a^p + (1/q) \operatorname{Tr} b^q$.
6. With $k_1(\alpha) = \alpha^p$, $k_2(\beta) = \beta^q$, Corollary 5 can be improved to $\operatorname{Tr} ab \le (\operatorname{Tr} |a|^p)^{1/p} (\operatorname{Tr} |b|^q)^{1/q}$; since this no longer involves N, it also holds when $N = \infty$. By iteration,

$$\left\| \prod_{i=1}^n a_i \right\|_p \le \prod_{i=1}^n \|a_i\|_{p_i},$$

$$\|a\|_p = (\operatorname{Tr} |a|^p)^{1/p}, \quad \text{where} \quad \sum_i \frac{1}{p_i} = \frac{1}{p}, \quad p, p_i \ge 1.$$

As $p \to \infty$, $\|a\|_p \to \|a\|$, so $|\operatorname{Tr} ab| \le \|a\| \operatorname{Tr} |b|$; the trace class is a two-sided ideal of $\mathcal{B}(\mathcal{H})$ (cf. (III: 2.3.20; 3)).
7. If a and b are Hermitian, then $\operatorname{Tr}(ab)^2 \le \operatorname{Tr} a^2b^2$, $a = a^*$, $b^{-1} = b^*$: $|\operatorname{Tr}(ab)^2| \le \operatorname{Tr} a^2$. By iterating this, $|\operatorname{Tr}(ab)^{2p}| \le \operatorname{Tr}(abb^*a^*)^{2p-1} = \operatorname{Tr}(|a|^2|b^2|)^{2p-1} \le \ldots \le \operatorname{Tr} |a|^{2p}|b|^{2p}$. Because of the Trotter product formula $\exp(a + b) = \text{s-}\lim_{n\to\infty}(\exp(a/n)\exp(b/n))^n$ (see (III: 2.4.9)), $|\operatorname{Tr} \exp(\alpha a + \beta b)| \le \operatorname{Tr} |\exp(\alpha a)||\exp(\beta b)|$, for $\alpha, \beta \in \mathbb{C}$, and initially for Hermitian operators of finite rank. It then extends to $\exp(\alpha a + \beta b) \in \mathcal{B}_1(\mathcal{H})$, $\exp(\alpha a) \in \mathcal{B}_1(\mathcal{H})$, $\exp(\beta b) \in \mathcal{B}(\mathcal{H})$ and thereby yields a generalization of Corollary 3 known as the Golden-Thompson-Symanzik inequality [6], $\exp(-\beta\langle V\rangle_{H_0}) \le \operatorname{Tr} \exp[-\beta(H_0 + V - F(H_0))] \le \langle \exp(-\beta V)\rangle_{H_0}$.

8. The function $(a, b) \to \lim_{\alpha \downarrow 0} \operatorname{Tr}(1/\alpha)(a - a^{1-\alpha} b^{\alpha}) = \operatorname{Tr} a(\ln a - \ln b)$ is convex.

Our next task is to give the density matrices an ordering that indicates which of two ρ's corresponds to the more chaotic state. The ordering must of course be independent of the basis, and so it can depend only on the eigenvalues ρ_i. If the eigenvalues are thought of as ordered by their magnitudes, then pure states are associated with sequences $(1, 0, 0, \ldots)$, i.e., with the greatest possible first eigenvalue. Because $\sum_{i=1}^{\infty} \rho_i = 1$, two density matrices can not be strictly ordered by the natural ordering of Hermitian operators. However, by the min-max principle (III: 3.5.19),

$$\rho(n) \equiv \sum_{i=1}^{n} \rho_i = \sup_{\mathcal{H}_n} \operatorname{Tr}_{\mathcal{H}_n} \rho,$$

which permits the following

2.1.6 Definition of the Ordering of the Density Matrices

A density matrix $\tilde{\rho}$ is said to be **more mixed**, or **more chaotic**, than ρ if $\tilde{\rho}(n) \leq \rho(n)$ for all n. In symbols, $\tilde{\rho} \succeq \rho$ (or $\rho \preceq \tilde{\rho}$).

2.1.7 Remarks

1. This clearly defines a preordering of the density matrices, i.e., $\rho \preceq \rho$; and if $\rho \preceq \tilde{\rho}$ and $\tilde{\rho} \preceq \tilde{\tilde{\rho}}$, then $\rho \preceq \tilde{\tilde{\rho}}$. If two density matrices are equivalent, that is, $\rho \preceq \tilde{\rho}$ and $\tilde{\rho} \preceq \rho$, then $\rho_i = \tilde{\rho}_i$, and so they are related by $\tilde{\rho} = V \rho V^*$. If the space is finite-dimensional, then V can be chosen unitary, and otherwise it is only an isometric mapping $(\operatorname{Ker} \rho)^{\perp} \to (\operatorname{Ker} \tilde{\rho})^{\perp}$; if $\operatorname{Dim} \operatorname{Ker} \rho \neq \operatorname{Dim} \operatorname{Ker} \tilde{\rho}$, then it has no unitary extension.
2. If the equivalent density matrices are classed together, then (2.1.6) gives the classes a lattice structure, characterized by the sequences of numbers $\{\rho(n)\}$. The sequence $\{\min(\rho(n), \tilde{\rho}(n))\}$ yields the equivalence class of the purest states more mixed than either ρ or $\tilde{\rho}$. The concave hull of $\max(\rho(n), \tilde{\rho}(n))$ with respect to n characterizes the most mixed states purer than either ρ or $\tilde{\rho}$. The sequences thus defined are positive, increasing, and concave in n, and tend to 1 as $n \to \infty$ (or equal 1 when $n = \operatorname{Dim} \mathcal{H}$). Their successive differences are therefore decreasing sequences of positive numbers summing to 1, which correspond to an equivalence class of density matrices. The lattice contains a class of purest elements, namely the extremal states. If the dimension of $\mathcal{H}$ is finite, then there is also a most mixed state with $\rho = \mathbf{1}/\operatorname{Dim} \mathcal{H}$, but if it is infinite, there is none.
3. The ordering and convexity are compatible on the space of states in the sense that if $\rho \preceq \mu$ and $\rho \preceq \nu$ then $\rho \preceq \alpha\mu + (1 - \alpha)\nu$ for $0 \leq \alpha \leq 1$:

$$\sup_{\mathcal{H}_n} \operatorname{Tr}_{\mathcal{H}_n} (\alpha\mu + (1 - \alpha)\nu) \leq \alpha \sup_{\mathcal{H}_n} \operatorname{Tr}_{\mathcal{H}_n} \mu + (1 - \alpha) \sup_{\mathcal{H}_n} \operatorname{Tr}_{\mathcal{H}_n} \nu \leq \rho(n).$$

4. Since the operators $\rho(n)$ are suprema of the weakly continuous functions $\mathrm{Tr}_{\mathcal{H}_n} \rho$, they are weakly lower semicontinuous. Moreover, it will be shown later (2.4.19; 1) that sequences of density matrices converging weakly to a density matrix are convergent even in the trace norm. Hence the maps $\rho \to \rho(n)$ are actually weakly continuous, and a limit belongs to the same mixing class.
5. The ordering of the density matrices is not total – for instance

$$\begin{pmatrix} \frac{1}{2} & & \\ & \frac{1}{2} & \\ & & 0 \end{pmatrix} \quad \text{and} \quad \begin{pmatrix} \frac{3}{4} & & \\ & \frac{1}{8} & \\ & & \frac{1}{8} \end{pmatrix}$$

are not related by it.

2.1.8 Examples

1. In the Schrödinger picture the time-evolution of a system is given by $\rho \to \rho_t \equiv U(t)\rho U^{-1}(t)$, which shows that density matrices remain in their equivalence classes.
2. The time-average $(1/T)\int_0^T dt \rho_t$ is more mixed than the original density matrices. This operation involves combinations and weak limits, which can only make density matrices more chaotic.
3. If the time-evolution of a density matrix is a linear transformation of the eigenvalues, $\rho_i(t) = M_{ik}(t)\rho_k(0)$, then for $\mathrm{Tr}\,\rho = 1$ and $\rho \geq 0$ it must be true that $\sum_i M_{ik} = 1$ for all k, and $M_{ik} \geq 0$ for all i and k. If, for finite dimension N, it is also required that the chaotic state $\rho_i = 1/N$ be stationary for all i, then, moreover, $\sum_k M_{ik} = 1$ for all i. The matrix M is then said to be **doubly stochastic**. Such matrices clearly form a convex set, and are consequently convex combinations of the extremal elements by the Krein-Milman theorem. The extremal elements have entries $M_{ik} = 0$ or 1, and so $1 = \sum_i M_{ik} = \sum_k M_{ik}$ implies that each row and each column has exactly one 1; this makes them permutation matrices, mapping any ρ to an equivalent ρ. Therefore, $\rho(t) \succeq \rho(0)$, as $\rho(t)$ is a convex combination of ρ's equivalent to $\rho(0)$. This kind of time-evolution thus increases the mixing. Its differential version $\dot{\rho}(t) = \dot{M}\rho(0)$ is a **master equation** $\dot{\rho}_i = \sum_k W_{ik}(\rho_k - \rho_i)$, where W satisfies $\sum_i W_{ik} = \sum_i W_{ki}$.
4. If an observable has one-dimensional projections P_i, then the state is immediately converted to $\tilde{\rho} \equiv \sum_i P_i \rho P_i$ when the observable is measured. Once it is perceived that the kth eigenvalue has been measured, ρ becomes P_k. The first stage of the measurement increases the mixing of the state, $\tilde{\rho} \succeq \rho$. This follows from the min-max principle: If $P_i|i\rangle = |i\rangle$, then

$$\tilde{\rho}(n) = \sum_{i=1}^{n} \langle i|\rho|i\rangle \leq \rho(n) = \sup_{\mathcal{H}_n} \mathrm{Tr}_{\mathcal{H}_n} \rho.$$

The second stage makes the state pure. This can be interpreted in that the interaction with the measuring apparatus extracts information, which unmixes the state upon transmission to the human mind.

5. The "coarse-grained" density matrix $\tilde{\rho} \equiv \sum_i P_i \lambda_i$, $\lambda_i = \operatorname{Tr} \rho P_i$, is more mixed than $\sum_i P_i \rho P_i$ by Problem 1, and *a fortiori* $\tilde{\rho} \succeq \rho$.
6. Suppose that the function k is convex from $\mathbb{R}^+$ to $\mathbb{R}^+$ and $k(0) = 0$; then clearly the smaller eigenvalues are suppressed to a greater degree in $k(\rho)$. In fact, $\rho \succeq k(\rho)/\operatorname{Tr} k(\rho)$ by Problem 3, and the resulting states are purer. In particular, if $k(x) = x^{\beta'/\beta}$, $\beta' > \beta$, then $\exp(-\beta H)/\operatorname{Tr}\exp(-\beta H) \succeq \exp(-\beta' H)/\operatorname{Tr}\exp(-\beta' H)$. The physical significance is that the mixing of the canonical density matrices is greater at higher temperatures.

We have seen that convex combinations of $U\rho U^{-1}$ and weak limits increase the mixing of ρ. This exhausts the possibilities:

2.1.9 Theorem

$\tilde{\rho} \succeq \rho$ iff $\tilde{\rho}$ is in the weakly closed convex hull of $\{U\rho U^{-1}\}$.

2.1.10 Remark

The weak closure of $\{a \in \mathcal{B}^+(\mathcal{H}), \|a\| = 1\}$ is $\{a \in \mathcal{B}^+(\mathcal{H}), \|a\| \leq 1\}$, and density matrices may converge weakly to zero. This means that the set of density matrices is not weakly closed, which causes technical difficulties in the proof, which is put off to Problem 4 for that reason.

2.1.11 Corollary

If $\tilde{\rho} \succeq \rho$, then for any convex function k, $\operatorname{Tr} k(\tilde{\rho}) \leq \operatorname{Tr} k(\rho)$.

Proof: If $\tilde{\rho} = \sum_i c_i U_i \rho U_i^{-1}$, $0 \leq c_i \leq 1$, $\sum_i c_i = 1$, and the sum is finite, then by the convexity inequality (2.1.4; 2), $\operatorname{Tr} k(\tilde{\rho}) \leq \sum_i c_i \operatorname{Tr} k(U_i \rho U_i^{-1}) = \operatorname{Tr} k(\rho)$. Moreover, $\rho \to \operatorname{Tr} k(\tilde{\rho})$ is weakly lower semicontinuous, so the limiting case of an infinite sum is likewise bounded by $\operatorname{Tr} k(\rho)$. □

Corollary (2.1.11) gives rise to the possibility of defining mappings of the density matrices to the real numbers, monotonic with respect to the ordering $\succeq$, and so enables the degree of disorder to be measured. For instance, if $k(\rho) = \rho^2$, then $\operatorname{Tr} k(\rho)$ can equal 1 only for pure states, and is otherwise smaller. The next section will discuss some other properties distinguished by the function $-k(\rho) = -\rho \ln \rho$ used to define the entropy. For now, note that the converse of (2.1.11) is also true:

2.1.12 Theorem

$\tilde{\rho} \succeq \rho$ iff for every convex function k, $\operatorname{Tr} k(\tilde{\rho}) \leq \operatorname{Tr} k(\rho)$.

Proof: Because of (2.1.11), we need only show that if $\tilde{\rho} \not\succeq \rho$, then there exists a convex function k such that $\operatorname{Tr} k(\tilde{\rho}) \geq \operatorname{Tr} k(\rho)$. Let m be the first integer such that

$\tilde{\rho}_1+\tilde{\rho}_2+\cdots+\tilde{\rho}_m > \rho_1+\rho_2+\cdots+\rho_m$, and let $k(x) = (x-\rho_m)$, when $x \geq \rho_m$, and otherwise 0. Then $k(\rho_1) = \rho_1 - \rho_m, \ldots, k(\rho_m) = \rho_m - \rho_m = 0 = k(\rho_{m+1}) = k(\rho_{m+2}) = \cdots$. By assumption, $\tilde{\rho}_1 + \tilde{\rho}_2 + \cdots + \tilde{\rho}_{m-1} \leq \rho_1 + \rho_2 + \cdots + \rho_{m-1}$, so $\tilde{\rho}_m > \rho_m$, which implies $k(\tilde{\rho}_i) = \tilde{\rho}_i - \rho_m > 0$ for all $i \leq m$. Therefore, $\operatorname{Tr} k(\rho) = \rho_1 + \rho_2 + \cdots + \rho_m - m\rho_m < \tilde{\rho}_1 + \tilde{\rho}_2 + \cdots + \tilde{\rho}_m - m\rho_m \leq \operatorname{Tr} k(\tilde{\rho})$.

Since expectation values in mixed states are averages of different spectral values, they do not reach the extremes of the spectrum so easily. This observation creates a new way to define the ordering relationship. □

2.1.13 Theorem

(i) $\tilde{\rho} \succeq \rho \Leftrightarrow \sup\limits_{\substack{U \\ U^*=U^{-1}}} \operatorname{Tr} U\tilde{\rho}U^{-1}a \leq \sup\limits_{\substack{U \\ U^*=U^{-1}}} \operatorname{Tr} U\rho U^{-1}a$ for all $a \in \mathcal{B}^+(\mathcal{H})$,

(ii) $\tilde{\rho} \succeq \rho \Leftrightarrow \inf\limits_{\substack{U \\ U^*=U^{-1}}} \operatorname{Tr} U\tilde{\rho}U^{-1}a \geq \inf\limits_{\substack{U \\ U^*=U^{-1}}} \operatorname{Tr} U\rho U^{-1}a$for all$a \in \mathcal{B}^+(\mathcal{H})$.

Proof: See Problem 5. □

2.1.14 Corollary

Let $(\Delta_\rho a)^2 \equiv \operatorname{Tr} \rho a^2 - (\operatorname{Tr} \rho a)^2 = \inf_\lambda \operatorname{Tr} \rho(a-\lambda)^2$. *Then* $\tilde{\rho} \succeq \rho$ *implies that*

$$\inf_U \Delta_{U\tilde{\rho}U^{-1}}a \geq \inf_U \Delta_{U\rho U^{-1}}a \text{ for all } a.$$

This means that if one is interested in the least deviation Δa of a within the equivalence classes of ρ and $\tilde{\rho}$, then it is smaller for the state that is less mixed.

The various aspects of the relationship can be summarized as follows:

2.1.15 Conditions for Density Matrices to be Compared

The ordering relationship $\tilde{\rho} \succeq \rho$ is equivalent to each of the following:

(i) $\tilde{\rho}(n) \leq \rho(n)$ for all n;

(ii) $\tilde{\rho} = \text{w-}\lim_\alpha \sum_i c_{i\alpha} U_{i\alpha} \rho U_{i\alpha}^{-1}$, $c_{i\alpha} > 0$, $\sum_i c_{i\alpha} = 1$, $U_{i\alpha}^{-1} = U_{i\alpha}^*$;

(iii) $\operatorname{Tr} k(\tilde{\rho}) \geq \operatorname{Tr} k(\rho)$ for every concave function k;

(iv) $\begin{smallmatrix}\sup \\ \inf \\ U\end{smallmatrix} \operatorname{Tr} U\tilde{\rho}U^{-1}a \begin{smallmatrix}\leq\sup \\ \leq\inf \\ U\end{smallmatrix} \operatorname{Tr} U\rho U^{-1}a$, $a \in \mathcal{B}^+(\mathcal{H})$, $U^{-1} = U^*$.

2.1.16 Problems

1. Let P_i be pairwise orthogonal projections of dimensions $n_i < \infty$ and $\sum_i P_i = = \mathbf{1}$. Show that $\sum_i (1/n_i) P_i \operatorname{Tr} P_i \rho \succeq \sum_i P_i \rho P_i$.

2. Let $k(x) > 0$, $k' > 0$, $k'' > 0$, k/k' convex. Show that the mapping $(\beta_1, \ldots, \beta_n) \to k^{-1}(\sum_{i=1}^n k(\beta_i))$ of $\mathbb{R}^n$ to $\mathbb{R}$ is convex. (Hint: note that: (i) A mapping $(\beta_1, \ldots, \beta_n)$

is convex if $\chi''(0) \geq 0$, where χ is the function $\chi(t) = f(\beta_1 + u_1 t, \ldots, \beta_n + u_n t)$ and $(u_1, \ldots, u_n)$ and $(\beta_1, \ldots, \beta_n)$ are arbitrary. (ii) If the function $K(\delta)/\delta$ increases monotonically, then $K(\sum_i \delta_i) \geq \sum_i K(\delta_i)$, $\delta_i > 0$.)

3. Let k be a convex, monotonically increasing function, $k(x) \geq 0$ for $x \geq 0$, and $k(0) = 0$. Show that $\rho \succeq k(\rho)/\operatorname{Tr} k(\rho)$.

4. Show that $\tilde{\rho} \succeq \rho \Leftrightarrow \tilde{\rho} \in \overline{\operatorname{Conv}\{U\rho U^{-1}\}}^{\text{weak}}$.

   (i) Let $\mathcal{K}(\rho) = \{a \geq 0$: a is compact, and $\alpha_1 + \cdots + \alpha_n \leq \rho(n)$ for all n, where α_i are the eigenvalues in increasing order $\}$. Show that $\mathcal{K}(\rho)$ is convex and weakly compact.

   (ii) Let $\mathcal{E}(\rho) = \{a \in \mathcal{K}(\rho) : \alpha_1 = \rho_1, \ldots, \alpha_n = \rho_n, \alpha_{n+1} = \ldots = 0$ or $\alpha_i = \rho_i$ for all $i\}$. Show that $\mathcal{E}(\rho)$ contains the extremal points of $\mathcal{K}(\rho)$.

   (iii) Show that $\mathcal{E}(\rho) \subset \overline{\{U\rho U^{-1}\}}^{\text{weak}}$.

   (iv) Finish the proof by applying the Krein–Milman theorem: Every compact, convex set equals the closure of the convex hull of its extremal points.

5. Prove Theorem (2.1.13).

2.1.17 Solutions

1. Let $d\mu(U)$ be the invariant measure on the compact group $U(n)$, normalized to 1. For all $a \in \mathcal{B}(\mathbb{C}^n)$, $\mathbf{1}_{|\mathbb{C}^n}(1/n)\operatorname{Tr} a = \int d\mu U a U^{-1}$, since the right side is invariant under all U and hence proportional to $\mathbf{1}_{|\mathbb{C}^n}$, and $\operatorname{Tr}\int d\mu U a U^{-1} = \operatorname{Tr} a$. Similarly,

$$\frac{1}{n} P \operatorname{Tr} P\rho + (1 - P)\rho(1 - P) = \int d\mu_P U_P \rho U_P^{-1}$$
$$= \int d\mu_P U_P (P\rho P + (1 - P)\rho(1 - P)) U_P^{-1},$$

if the operators U_P vary over the unitary transformations of $\mathcal{H}$ equaling $\mathbf{1}$ on $(1 - P)\mathcal{H}$. Therefore, $(1/n)P \operatorname{Tr} P\rho + (1 - P)\rho(1 - P) \succeq P\rho P + (1 - P)\rho(1 - P)$, which proves the claim by iteration.

2. (i) is trivial, and (ii) follows from

$$\textstyle\sum_i \delta_i \geq \delta_k \Rightarrow \delta_k K\left(\sum_i \delta_i\right) \geq \left(\sum_i \delta_i\right) K(\delta_k),$$
$$\left(\textstyle\sum_k \delta_k\right) K\left(\sum_i \delta_i\right) \geq \left(\sum_i \delta_i\right) \sum_k K(\delta_k).$$

Now let $\chi(t) \equiv k^{-1}(\sum k(\beta_i + u_i t))$. The function $\chi(t)$ is convex iff $\chi''(t) \geq 0$. $[k'(\chi)]^3 \chi'' = [k'(\chi)]^2 [\sum_i u_i^2 k''(\beta_i)] - k''(\chi)[\sum_i u_i k'(\beta_i)]^2$ (where $\chi \equiv \chi(0)$, $\chi'' \equiv \chi''(0)$), so it remains to show that $[k'(\chi)]^2 \sum_i u_i^2 k''(\beta_i) \geq k''(\chi) \times$ $\times [\sum_i u_i k'(\beta_i)]^2$. By the Cauchy–Schwarz inequality, $[\sum_i u_i k'(\beta_i)]^2 =$ $= [\sum_i u_i \sqrt{k''(\beta_i)} \sqrt{k'(\beta_i)^2 / k''(\beta_i)}]^2 \leq [\sum_i u_i^2 k''(\beta_i)][\sum_i k'(\beta_i)^2 / k''(\beta_i)]$, and the desired inequality is certainly satisfied if $\psi(\chi) \equiv k'(\chi)^2 / k''(\chi) \geq$

$\geq \sum_i k'(\beta_i)^2/k''(\beta_i) = \sum \psi(\beta_i)$. By (ii), this is the case if $K(\delta)/\delta$ increases monotonically, where K is defined by $\delta_i = k(\beta_i)$, $K(\delta_i) = \psi(\beta_i)$. Finally, $K(\delta)/\delta$ increases monotonically $\Leftrightarrow k'^2/kk''$ increases monotonically $\Leftrightarrow k/k'$ is convex.

3. If $0 \leq x \leq y$, then $x = (x/y)y + (1 - (x/y))0$, and hence $k(x) \leq (x/y)k(y)$, $yk(x) \leq xk(y)$. Consequently

$$\sum_{i=1}^{m} k(\rho_i)\left(\sum_{j=1}^{m} \rho_j + \sum_{j=m+1}^{\infty} \rho_j\right) \geq \sum_{j=1}^{m} \rho_j \left(\sum_{j=1}^{m} k(\rho_i) + \sum_{i=m+1}^{\infty} k(\rho_i)\right),$$

i.e.,

$$[k(\rho_1) + \cdots + k(\rho_m)]\left(\sum_{i=1}^{\infty} k(\rho_i)\right)^{-1} \geq [\rho_1 + \cdots + \rho_m]\left(\sum_{i=1}^{\infty} \rho_i\right)^{-1}.$$

Remark: If k is concave, then $\rho \succeq k(\rho)/\operatorname{Tr} k(\rho)$.

4. (i) By (2.1.7; 3) the set $\mathcal{K}(\rho)$ is convex. Moreover, $\alpha_1 + \cdots + \alpha_n = \sup_{\mathcal{H}_\mathbf{n}} \operatorname{Tr}_{\mathcal{H}_\mathbf{n}} a$ is weakly lower semicontinuous in a, so $\mathcal{K}(\rho)$ is weakly closed and, since $\|a\| = \alpha_1 \leq \rho_1 = \|\rho\| \leq \operatorname{Tr} \rho = 1$, also weakly compact.
   (ii) By considering all the possibilities, one realizes that it is possible to write any $a \in \mathcal{K}(\rho)$ as $\alpha\rho_1 + (1-\alpha)\rho_2$, $0 < \alpha < 1$, with $\rho_i \in \mathcal{K}(\rho)$, unless $a \in \mathcal{E}(\rho)$.
   (iii) Let $a = \sum_{i=1}^n \rho_i |1, i\rangle\langle 1, i|$, $\rho = \sum \rho_i |2, i\rangle\langle 2, i|$, where $\{|1, i\rangle\}$ and $\{|2, i\rangle\}$ are two orthonormal systems. Let $U|2, i\rangle = |1, i\rangle$, $U_l|1, n + i\rangle = |1, n + l - i\rangle$ for $1 \leq i \leq l - 1$, $U_t|1, i\rangle = |1, i\rangle$ otherwise. $a = \text{s-}\lim_{l\to\infty} U_l U \rho U^{-1} U_t^{-1}$.
   (iv) By the Krein-Milman theorem, $\mathcal{K}(\rho) = \overline{\operatorname{Conv} \mathcal{E}(\sigma)}^{\text{weak}} = \overline{\operatorname{Conv} \{U\rho U^{-1}\}}^{\text{weak}}$ (by (iii)), and $\tilde{\rho} \in \mathcal{K}(\rho)$, if $\tilde{\rho} \succeq \rho$.

5. By a replacement of a with $a + \|a\|$ if necessary, a may be assumed positive. Then $\operatorname{Tr} \rho a = \sup_n \operatorname{Tr} \rho^{(n)} a$, where the $\rho^{(n)}$ have the eigenvalues $\rho_1, \rho_2, \ldots, \rho_n, 0, 0, \ldots$. The changes of the orders of operation in what follows are justified for the $\rho^{(n)}$, and the suprema can also be interchanged:
   (i)$\Rightarrow$: Let $\alpha_1 \geq \alpha_2 \geq \cdots$ be the decreasing sequence of eigenvalues of a and α_∞ be the upper boundary of $\sigma_{\text{ess}}(a)$ (to be understood in the sense analogous to (III: 3.5.19)). If $\rho = \sum \rho_i |i\rangle\langle i|$, then

$$\begin{aligned}\operatorname{Tr} \rho a &= \sum \rho_i \langle i|a|i\rangle = (\rho_1 - \rho_2)\langle 1|a|1\rangle + (\rho_2 - \rho_3)[\langle 1|a|1\rangle + \langle 2|a|2\rangle] + \cdots \\ &\leq (\rho_1 - \rho_2)\alpha_1 + (\rho_2 - \rho_3)(\alpha_1 + \alpha_2) + \cdots = \sum \rho_i \alpha_i,\end{aligned}$$

   and $\sup \operatorname{Tr} U\rho U^{-1} a = \sum \rho_i \alpha_i$.

$$\begin{aligned}\sum_i \tilde{\rho}_i \alpha_i &= \tilde{\rho}_1(\alpha_1 - \alpha_2) + (\tilde{\rho}_1 + \tilde{\rho}_2)(\alpha_2 - \alpha_3) + \cdots + \alpha_\infty \\ &\leq \rho_1(\alpha_1 - \alpha_2) + (\rho_1 + \rho_2)(\alpha_2 - \alpha_3) + \cdots + \alpha_\infty = \sum \rho_i \alpha_i.\end{aligned}$$

   $\Leftarrow$:Choose an n-dimensional projection for a and use the min-max principal. The proof of (ii) is similar.

2.2 The Properties of Entropy

The information about a system in a mixed state is incomplete. The entropy is a measure of how far from maximal the information is.

In statistical physics, entropy is not an observable in the sense of an operator on Hilbert space, but rather a property of the state of the system, measuring the lack of our knowledge as expressed in the specification of the state. This section will consider what sorts of conditions single out a particular measure of this lack of knowledge and will see what conclusions can be drawn from it.

A primary requirement would be monotony with respect to the ordering introduced in the preceding section (we consider only normal states). In other words, a density matrix that is more mixed should have more entropy, which we denote $S : \tilde{\rho} \succeq \rho \Rightarrow S(\tilde{\rho}) \geq S(\rho)$. This leaves many possibilities open for the definition; every monotonic function of the trace of a concave function of ρ would satisfy this requirement (cf. (2.1.11)). A further reasonable requirement is the additivity of the entropies of independent systems. If their combination is represented on the tensor product of their Hilbert spaces, this means

$$S(\rho' \otimes \rho'') = S(\rho') + S(\rho''). \tag{2.4}$$

The two requirements together do not yet quite determine S uniquely. The whole one-parameter family of

2.2.1 α-Entropies

$$S_\alpha(\rho) = \frac{1}{1-\alpha} \ln \operatorname{Tr} \rho^\alpha, \quad \alpha \in \mathbb{R}^+ \setminus \{1\},$$

satisfy the general

2.2.2 Properties of Entropy

(i) $0 \leq S_\alpha(\rho) \leq \ln \dim \mathcal{H}$;

(ii) $\tilde{\rho} \succeq \rho \Rightarrow S_\alpha(\tilde{\rho}) \geq S_\alpha(\rho)$;

(iii) $S_\alpha(\rho' \otimes \rho'') = S_\alpha(\rho') + S_\alpha(\rho'')$;

(iv) If $\rho = P/\dim P$, $P = P^2 = P^*$, then $S_\alpha(\rho) = \ln \dim P$.

(In particular, $S_\alpha(\rho) = 0$ iff ρ is a pure state, and $S_\alpha(\rho) = \ln \dim \mathcal{H}$ iff ρ is the chaotic state $1/\dim \mathcal{H}$.)

Proof:

(i) If $\alpha > 1$, then $\sum_i \rho_i^\alpha \leq (\sum_i \rho_i)^\alpha = 1$, and if $\alpha < 1$, then $\sum_i \rho_i = 1 \leq (\sum_i \rho_i^\alpha)^{1/\alpha}$. This shows the left side of the inequality, and the right follows from (iv) and (ii).

(ii) The function ρ^α is concave for $\alpha < 1$ and convex for $\alpha > 1$. The logarithm is monotonic, and the $1-\alpha$ accounts for the sign (see (2.1.18(iii)).

(iii) $\mathrm{Tr}(\rho' \otimes \rho'')^\alpha = \mathrm{Tr}[(\rho')^\alpha \otimes (\rho'')^\alpha] = \mathrm{Tr}(\rho')^\alpha \cdot \mathrm{Tr}(\rho'')^\alpha$.

(iv) If $n = \dim P$, then $S_\alpha(\rho) = (1/(1-\alpha)) \ln(nn^{-\alpha})$.

The entropy can be fixed uniquely by a more stringent assumption of additivity (2.4), with which monotony emerges as a consequence rather than a separate axiom:
□

2.2.3 Characterization of the von Neumann Entropy

The only entropy satisfying the following conditions is $S(\rho) = -\mathrm{Tr}\,\rho \ln \rho$

(i) *$S(\rho)$ is a continuous function of the eigenvalues of ρ ;*

(ii) $S\begin{pmatrix} 1/2 & 0 \\ 0 & 1/2 \end{pmatrix} = \ln 2;$

(iii) *If*

$$\mathcal{H} = \bigoplus_{n=1}^{N} \mathcal{H}_n, \quad \rho = \bigoplus_{n=1}^{N} p_n \rho_n, \quad \sum_n p_n = 1,$$
$$0 \le p_i \le 1, \quad \mathrm{Tr}\,\rho_n = 1,$$

then, regardless of the dimension of $\mathcal{H}_n$, $S(\rho) = \sum_{n=1}^N p_n S(\rho_n) + S(p)$, where p is the diagonal matrix on $\mathbb{C}^n$ having eigenvalues p_n.

2.2.4 Remarks

1. Since the representation should make no difference, S can only depend on the eigenvalues. It certainly does not seem unreasonable to demand continuity.
2. Condition (ii) is a normalization.
3. If all the $\mathcal{H}_n$ in condition (iii) have the same dimension and all ρ_n are equal, then $\mathcal{H} = \mathcal{H}_1 \otimes \mathbb{C}^n$, and (iii) reduces to (2.4). This generalization of (2.4), which makes possible an inductive proof, has the following interpretation: Suppose a system consists of two subsystems, one described by $\mathbb{C}^n$ and the other having several variants according to the position of the state vector of the first in $\mathbb{C}^n$. Then the entropy of the total system is just the sum of the entropy of the first subsystem and those of the second, averaged according to their probabilities.
4. The formula $S = -\mathrm{Tr}\,\rho \ln \rho$ can be justified in the spirit of Boltzmann as follows. Let the state corresponding to ρ be realized as a vector of a reducible representation of the algebra $\mathcal{A}$ of observables consisting of N identical representations. The ensemble described by ρ can be thought of as having been subjected to a sequence of N measurements, where ρ_i is N_i/N, N_i being the number of times the eigenvector e_i has been measured. The Hilbert space is $\mathcal{H} = \bigoplus_{j=1}^N \mathcal{H}_j$, where the spaces $\mathcal{H}_j$ are all identical and are spanned by $\{e_i\}$.

The observables are represented as a direct sum of N identical representations. With the use of doubled indices, this can be written as $\mathcal{H}_j = \bigoplus_{i=1}^r e_{i,j}$. A ρ of rank r and with $\rho_i = N_i/N$, $i = 1, \ldots, r$, is represented by the vector

$$\frac{1}{\sqrt{N}}(e_{1,1} + e_{1,2} + \cdots + e_{1,N_1} + e_{2,N_1+1} + \cdots + e_{2,N_1+N_2} + \cdots \\ + e_{r,N_1+N_2+\cdots+N_{r-1}+1} + \cdots + e_{r,N})$$

of $\mathcal{H}$. If the e_i are chosen from other spaces $\mathcal{H}_j$, the same state results, and there are clearly $W \equiv N!/\prod_i N_i!$ different vectors for the same ρ. If the numbers N_i are large enough, then $\ln W \cong N \ln N - \sum_i N_i \ln N_i = N \sum_i \rho_i \ln \rho_i$, so $(1/N) \ln W \to -\operatorname{Tr} \rho \ln \rho$. Assuming that every vector of $\mathcal{H}$ is assigned the same probability, S turns out to be roughly the logarithm of the probability of the configuration, and there is an identification: the most mixed state = the state of greatest entropy = the most probable state.

5. $S(\rho) = \lim_{\alpha\to 1} S_\alpha(\rho)$, yet if the dimension is infinite, then $S(\rho)$ may become $+\infty$. However, Properties (2.2.2) remain valid in this limit, and apply to S as well.
6. A particular consequence of (2.2.2(ii)) is that $S(\alpha\rho + (1-\alpha)U\rho U^{-1}) \geq S(\rho)$. More generally, (2.1.4; 2) implies that the mapping $\rho \to S(\rho)$ is concave: $S(\alpha\rho_1 + (1-\alpha)\rho_2) \geq \alpha S(\rho_1) + (1-\alpha)S(\rho_2)$. This means that the entropy of a mixed state is greater than the constituent entropies weighted as in the mixing. If $\rho = \sum_n p_n \rho_n$, $0 \leq p_n \leq 1$, $\sum_n p_n = 1$, then the inequalities

$$\sum_n p_n S(\rho_n) \leq S(\rho) \leq \sum_n p_n S(\rho_n) + \sum_n p_n \ln \frac{1}{p_n}$$

necessarily follow (Problem 4). They are optimal in the sense that equality holds on the left if all ρ_n are equal, and on the right if all ρ_n have disjoint support, by (2.2.3(iii)).
7. Although by (2.2.2(iv)) all the S_α are the same with the chaotic state, with the canonical state $\rho = \exp(-\beta(H - F(\beta)))$, $\operatorname{Tr}\exp(-\beta H) = \exp(-\beta F(\beta))$, they are different (Problem 6).

Proof of (2.2.3): We write $S(\rho_1, \rho_2, \ldots)$ for $S(\rho)$.

(a) Let $\mathcal{H} = \mathbb{C}^1$. Then $S(1) = 0$, because on $\mathbb{C}^2$, $S(\rho_1, \rho_2) = \rho_1 S(1) + \rho_2 S(1) + S(\rho_1, \rho_2)$.

(b) Let $\mathcal{H} = \mathbb{C}^n$, $f(n) \equiv S(1/n, 1/n, \ldots, 1/n)$, and let $n = m_1 m_2$. We write $\mathbb{C}^n = \mathbb{C}^{m_1} \oplus \mathbb{C}^{m_1} \oplus \cdots \oplus \mathbb{C}^{m_1}$ and use (iii) with

$$N = m_2, \qquad p_i = m_2^{-1}, \qquad \rho_i = \begin{pmatrix} 1/m_1 & & \\ & \ddots & \\ & & 1/m_1 \end{pmatrix},$$

$$f(m_1 m_2) = m_2 \frac{1}{m_2} f(m_1) + f(m_2) = f(m_1) + f(m_2).$$

The solution of this equation is $f(n) = C \ln n$ and the normalization (ii) makes $C = 1$. Other solutions are excluded by the continuity requirement [5].

(c)

$$f(m) = S\left(\overbrace{\frac{1}{m}, \frac{1}{m}, \ldots, \frac{1}{m}}^{n}, \overbrace{\frac{1}{m}, \ldots, \frac{1}{m}}^{m-n}\right)$$

$$= \frac{n}{m} f(n) + \frac{m-n}{m} F(m-n) + S\left(\frac{n}{m}, \frac{m-n}{m}\right),$$

so by step (b),

$$S\left(\frac{n}{m}, 1 - \frac{n}{m}\right) = -\frac{n}{m} \ln \frac{n}{m} - \left(1 - \frac{n}{m}\right) \ln \left(1 - \frac{n}{m}\right).$$

This holds initially only for integers n and m, and then by continuity holds generally, $S(\rho_1, \rho_2) = -\sum_{i=1}^{2} \rho_i \ln \rho_i$.

(d) The rest of the proof proceeds inductively: with $\mathbb{C}^{n+1} = \mathbb{C}^n \oplus \mathbb{C}$, $p_1 = 1 - \rho_n$, $p_2 = \rho_n$,

$$\begin{aligned} S(\rho_1, \rho_2, \ldots, \rho_{n-1}, \rho_n) &= (1 - \rho_n) S\left(\frac{\rho_1}{1-\rho_n}, \ldots, \frac{\rho_{n-1}}{1-\rho_n}\right) + \rho_n S(1) \\ &\quad + S(1 - \rho_n, \rho_n) \\ &= -\sum_{i=1}^{n-1} \rho_i \ln \frac{\rho_i}{1-\rho_n} - \rho_n \ln \rho_n \\ &\quad - (1-\rho_n) \ln(1-\rho_n) = -\sum_{i=1}^{n} \rho_i \ln \rho_i. \end{aligned}$$

2.2.5 The Classical Entropy

For a classical density $p_{\mathrm{cl}}(\mathbf{x}, \mathbf{p})$ on phase space the entropy would be defined as $-\int d\Omega \rho_{\mathrm{cl}} \ln \rho_{\mathrm{cl}}$. This is not *a priori* positive-definite; for instance $\rho_{\mathrm{cl}} = \chi(A/\Omega(A))$ as in (2.1.3; 2) leads to $-\int d\Omega \chi \ln \chi = \ln \Omega(A)$, which is negative if $\Omega(A) < 1$. It is easy to see that this entropy also depends on the measure of volume in phase space.

There are many ways to associate a density ρ_{cl} with a density matrix ρ or vice versa. The most useful such expressions are obtained with a method of A. Wehrl, in which for a given density matrix ρ one calculates expectation values in coherent states, and, conversely, a classical density is used to mix coherent states. The coherent states $W(\mathbf{z})|u\rangle \equiv |\mathbf{x}\rangle$ of (III: 3.1.13) can be generalized for functions u that are even and normalized, but not necessarily Gaussian. The state $|\mathbf{z}\rangle$ has the wavefunction $\exp(i\mathbf{k}\cdot\mathbf{x})u(\mathbf{x}-\mathbf{q})$ if $\mathbf{z} = \mathbf{q}+i\mathbf{k} \in \mathbb{C}^{dN}$, which is the phase space for

N particles in a physical space of dimension d. It is easy to check that $\mathbf{z} = \langle \mathbf{z}|\mathbf{x}|\mathbf{z}\rangle + i\langle \mathbf{z}|\mathbf{p}|\mathbf{z}\rangle$ still holds and that the states are complete, $\int d^{2Nd}z(2\pi)^{-Nd}|\mathbf{z}\rangle\langle\mathbf{z}| = 1$.

2.2.6 The Density Matrix and the Phase-Space Density

If to an N-particle density matrix ρ we associate the phase-space density $\rho_{\mathrm{cl}}(\mathbf{z}) = \langle \mathbf{z}|\rho|\mathbf{z}\rangle$, and to a classical density $f(\mathbf{z})$ on phase space we associate the density matrix $\rho_{\mathrm{qu}} = \int d\Omega_z f(\mathbf{z})|\mathbf{z}\rangle\langle\mathbf{z}|$, $d\Omega_z^N = (2\pi)^{-Nd}d^{2N}z$, then

$$\rho \geq 0, \quad \mathrm{Tr}\,\rho = 1 \Rightarrow 0 \leq \rho_{\mathrm{cl}}(\mathbf{z}) \leq 1, \quad \int d\Omega_z^N \rho_{\mathrm{cl}}(\mathbf{z}) = 1,$$

$$f \geq 0, \quad \int d\Omega_z f(z) = 1 \Rightarrow 0 \leq \rho_{\mathrm{qu}} \leq 1, \quad \mathrm{Tr}\,\rho_{\mathrm{qu}} = 1. \tag{2.5}$$

Proof: Positivity is trivial, and the connection between the trace and the phase-space integral follows from the n-dimensional version of a formula of (III:3.1.14;1):

$$\mathbf{1} = \int d\Omega_z^N |\mathbf{z}\rangle\langle\mathbf{z}| \Rightarrow \mathrm{Tr}\,a = \sum_i \langle i|a|i\rangle = \sum_i \int d\Omega_z^N \langle i|\mathbf{z}\rangle\langle\mathbf{z}|a|i\rangle$$
$$= \int d\Omega_z^N \langle\mathbf{z}|a|\mathbf{z}\rangle.$$

Conversely, $\mathrm{Tr}\int d\Omega_z f(\mathbf{z})|\mathbf{z}\rangle\langle\mathbf{z}| = \sum_i \int d\Omega_z^N f(\mathbf{z})|\langle\mathbf{z}|i\rangle|^2 = \int d\Omega_z f(\mathbf{z})$, since $\langle\mathbf{z}|\mathbf{z}\rangle = 1$. The denominator $(2\pi)^{2N}$ in $d\Omega_z^N$ reveals that the phase-space volume is measured in units of h rather than $\hbar = h/2\pi = 1$. □

2.2.7 Inequalities for the Classical and Quantum-Mechanical Entropies

(i) $S(\rho) \leq -\int d\Omega_z^N \rho_{\mathrm{cl}}(\mathbf{z}) \ln \rho_{\mathrm{cl}}(z) \equiv S_{\mathrm{cl}}(\rho)$;

(ii) $-\int d\Omega_z^N f(\mathbf{z}) \ln f(\mathbf{z}) \leq S(\rho_{\mathrm{qu}})$.

2.2.8 Remarks

1. Inequality (i) implies that the ρ_{cl} of (2.2.7) always has more entropy than $S(\rho)$. This classical entropy is therefore always positive; the density ρ_{cl} defined in (2.2.7) can never be so concentrated as to make the classical entropy negative, and indeed $\rho_{\mathrm{cl}} \leq 1$.
2. It can also be shown that this classical entropy equals 1 if ρ is extremal, and otherwise it is greater than 1 [32].
3. If a quantum-mechanical density is associated with a classical density f by mixing the coherent states with f, then Inequality (ii) states that the quantum-mechanical entropy is greater than the classical entropy. The latter may even tend to $-\infty$, for instance if f tends to a delta function.

4. Inequality (ii) shows that the continuous analogue of the last inequality of (2.2.4;6) is false: $S(|\mathbf{z}\rangle\langle\mathbf{z}|) = 0$, and in this case the inequality goes in the other direction, with the replacements $p_n \to f(\mathbf{z})$, $\sum_n \to \int d\Omega_z^N$:

$$-\int d\Omega_z^N f(\mathbf{z}) \ln f(\mathbf{z}) + \int d\Omega_z^N f(\mathbf{z}) S(|\mathbf{z}\rangle\langle\mathbf{z}|) \leq S\left(\int d\Omega_z^N f(\mathbf{z})|\mathbf{z}\rangle\langle\mathbf{z}|\right).$$

5. If the particles are identical, states must be either symmetrized or antisymmetrized according to the statistics. For bosons this is accomplished most easily with the aid of the creation operator

$$a_{\mathbf{z}}^* \equiv a^*(\exp[i\mathbf{k}\cdot\mathbf{x}]u(\mathbf{q}-\mathbf{x})), \quad |\mathbf{z}_1, \ldots, \mathbf{z}_N\rangle = a_{\mathbf{z}_1}^* \ldots a_{\mathbf{z}}^*|0\rangle,$$

with which

$$\mathbf{1} = \sum_{n=0}^{\infty} \frac{(2\pi)^{-nd}}{n!} \int d^{2d}\mathbf{z}_1 \cdots d^{2d}\mathbf{z}_n |\mathbf{z}_1, \ldots, \mathbf{z}_n\rangle\langle\mathbf{z}_1, \ldots, \mathbf{z}_n|.$$

So, with identical bosons, when the trace is taken the volume of the classical phase space has to be divided by $n!$. The states are not yet normalized to norm 1.

$$\langle\mathbf{z}_1', \ldots, \mathbf{z}_N'|\mathbf{z}_1, \ldots, \mathbf{z}_N\rangle = \sum_P (\pm 1)^P \prod_{i=1}^N \langle\mathbf{z}_i'|\mathbf{z}_{P_i}\rangle \equiv {}^{\text{Per}}_{\text{Det}}(\langle\mathbf{z}_i'|\mathbf{z}_k\rangle),$$

where $P_1, \ldots, P_n$ is a permutation of $1, \ldots, n$, because the coherent states are not orthogonal:

$$\langle\mathbf{z}'|\mathbf{z}\rangle = \int d^d x \exp[i x \cdot (\mathbf{k}-\mathbf{k}')] u^*(\mathbf{x}-\mathbf{q}') u(\mathbf{x}-\mathbf{q}).$$

These determinants and permanents crop up along with $d\Omega_z^N$ in the calculations of expectation values, making them more laborious.

6. Since these inequalities are valid for coherent states with a great degree of arbitrariness in u, they can be optimized by varying u.

Inequalities (2.2.7) will follow from a lemma of Berezin on the

2.2.9 Relationship Between the Trace and the Phase-Space Integral

Let K be a convex function and suppose $a^* = a$. Then

(i) $\operatorname{Tr} K(a) \geq \int d\Omega_z^N K(\langle\mathbf{z}|a|\mathbf{z}\rangle)$;

(ii) $\int d\Omega_z^N K(f(\mathbf{z})) \geq \operatorname{Tr} K(a)$, *where* $a = \int d\Omega_z^N f(\mathbf{z})|\mathbf{z}\rangle\langle\mathbf{z}|$, $K(a) \in C^1$, *and* f *is a measurable function* $\mathbb{C}^N \to \mathbb{R}$.

Proof:

(i) As noted in the proof of Peierls's inequality, $\langle |K(a)|\rangle \geq K(\langle |a|\rangle)$ for expectation values in an arbitrary vector, so

$$\operatorname{Tr} K(a) = \int d\Omega_z^N \langle \mathbf{z}|K(a)|\mathbf{z}\rangle \geq \int d\Omega_z^N K(\langle \mathbf{z}|a|\mathbf{z}\rangle).$$

(ii) If $|j\rangle$ denotes an eigenfunction of a, then

$$\begin{aligned}\operatorname{Tr} K(a) &= \sum_j K(\langle j|a|j\rangle) = \sum_j K\left(\int d\Omega_z^N f(\mathbf{z})|\langle \mathbf{z}|j\rangle|^2\right)\\ &\leq \sum_j \int d\Omega_z^N |\langle \mathbf{z}|j\rangle|^2 K(f(\mathbf{z}))\\ &= \int d\Omega_z^N K(f(\mathbf{z})).\end{aligned}$$

□

Proof of (2.2.9): The function $x \ln x$ is convex, and for the concave function $-x \ln x$ the inequalities for convex functions are reversed. □

The additivity of the entropy when $\rho = \rho_1 \otimes \rho_2$ generalizes to an inequality when ρ is not in the form of a product. To cover general ρ_1 and ρ_2 requires the

2.2.10 Definition of Partial Traces

Let $\mathcal{H} = \mathcal{H}_1 \otimes \mathcal{H}_2$. The **partial traces** Tr_1 and Tr_2 are defined by $\operatorname{Tr}_{1,2} a = \sum_j \langle j|a|j\rangle \in \mathcal{B}(\mathcal{H}_{2,1})$ for any $a \in \mathcal{C}_1(\mathcal{H})$, where $\{|j\rangle\}$ is any complete orthonormal set in $\mathcal{H}_{1,2}$.

A consequence of this is the

2.2.11 Subadditivity of the Entropy

Let $\rho_{1,2} = \operatorname{Tr}_{2,1} \rho$. *Then* $S(\rho) \leq S(\rho_1) + S(\rho_2)$, = *holds only for* $\rho = \rho_1 \otimes \rho_2$.

2.2.12 Remarks

1. If $\rho = \rho_1 \otimes \rho_2$, then $\rho_{1,2} = \operatorname{Tr}_{2,1} \rho$ and by (2.2.2(iii)) equality holds in (2.2.11).
2. The partial traces reproduce the reduced density matrices used in § 1.1. At that time we noticed that the reduction entailed a loss of information. Inequality (2.2.11) indicates that there is less information in ρ_1 and ρ_2 than in the original ρ. ρ is a state ω over $\mathcal{B}(\mathcal{H}_1 \otimes \mathcal{H}_2) = \mathcal{B}(\mathcal{H}_1) \otimes \mathcal{B}(\mathcal{H}_2)$ and $\rho_{1,2}$ are the density matrices of the restrictions $\omega_{|\mathcal{B}(\mathcal{H}_{1,2})}$ of ω to the such algebras $\mathcal{B}(\mathcal{H}_1) \otimes \mathbf{1}$ resp. $\mathbf{1} \otimes \mathcal{B}(\mathcal{H}_1)$. This terminology will frequently used in the sequel.
3. If $\alpha \neq 1$, then the α-entropies S_α (2.2.1) are not subadditive (Problem 2). It is consequently not necessarily true that $\rho_1 \otimes \rho_2 \succeq \rho$.
4. Subadditivity allows axiom (iii) of (2.2.3) to be replaced [7] with

   (iii (a)) $S(\rho) = S(V^* \rho V)$ for all isometrics V; and (iii (b)) $S(\rho) \leq S(\rho_1) + S(\rho_2)$, equality holding iff $\rho = \rho_1 \otimes \rho_2$.

Proof: By Klein's inequality (2.1.4; 5), $\mathrm{Tr}\, a \ln a - \mathrm{Tr}\, a \ln b \geq \mathrm{Tr}(a - b)$. Put $a = \rho$ and $b = \rho_1 \otimes \rho_2$ and note that $\ln \rho_1 \otimes \rho_2 = \ln \rho_1 \otimes \mathbf{1} + \mathbf{1} \otimes \ln \rho_2$. □

2.2.13 Corollary

Consider a sequence of ever larger systems on the tensor product $\mathcal{H}^n$, $n = 1, 2, 3, \ldots$. Suppose that the density matrices ρ_n are compatible so that when reduced to a subsystem they always become the density matrix of the smaller system: $\rho_m = \mathrm{Tr}_{n-m}\, \rho_n$, $m \leq n$. If $\sigma_n = -(1/n) \mathrm{Tr}\, \rho_n \ln \rho_n$, then $n\sigma_n \leq m\sigma_m + (n-m)\sigma_{n-m}$. In particular, $\sigma_{2n} \leq \sigma_n$, and hence the limits $\lim_{n\to\infty} \sigma_n = \inf_n \sigma_n$ must exist and be ≥ 0. Although the entropy itself does not tend to a limit as the size of the system gets arbitrarily large, the specific entropy does.

It will be asked by how far (2.2.11) misses equality. More precisely, it might be supposed that the entropy of a united system is always greater than that of any single one of its parts. Surprisingly, this is not necessarily so with quantum statistics; ρ could be a pure state, thus having entropy zero, while the ρ_i correspond to mixtures. This is the case that arose in the discussion of the time-evolution in § 1.1; the additional information contained in ρ has to do with the correlations between the subsystems. The correlations are precisely pinned down in

2.2.14 Lemma

Let ρ be pure; then ρ_1 and ρ_2 have the same spectrum with the same multiplicities, except possibly for an eigenvalue at 0.

Proof: See Problem 3. □

2.2.15 Corollary

If ρ is pure, then $S(\rho_1) = S(\rho_2)$. Our information about the subsystems is correlated, so they possess the same amount of disorder.

In this case, $S(\rho) = S(\rho_1) - S(\rho_2)$; more generally there is a

2.2.16 Triangle Inequality

$$|S(\rho_1) - S(\rho_2)| \leq S(\rho) \leq S(\rho_1) + S(\rho_2).$$

(Lieb and Araki [8]).

2.2.17 Remarks

1. This inequality has no classical analogy; a counter example is provided by a ρ with $S(\rho) < 0$ but $S(\rho_1) = S(\rho_2)$.

2. Even if the entropy of a subsystem can be greater than that of the whole system, the triangle inequality reveals that it can not exceed the sum of the total entropy and the entropy of the complementary subsystem.
3. Astonishingly, the classical entropy (2.2.9) of a quantum-mechanical density matrix is monotonic; it is always larger for the whole than for a part: $S_{\mathrm{cl}}(\rho) \geq S_{\mathrm{cl}}(\rho_1)$. (For the proof see Problem 5.)

Proof: According to Remark (2.1.3; 5), ρ may be regarded as a pure state ρ_{123} on a large Hilbert space $\mathcal{H}_1 \otimes \mathcal{H}_2 \otimes \mathcal{H}_3$, for which $\rho = \mathrm{Tr}_3\, \rho_{123}$. Let $\rho_3 = \mathrm{Tr}_{12}\, \rho_{123}$, $\rho_{23} = \mathrm{Tr}_1\, \rho_{123}$; then by Corollary (2.2.15), $S(\rho) = S(\rho_3)$, $S(\rho_1) = S(\rho_{23})$. Because of subadditivity, $S(\rho_1) = S(\rho_{23}) \leq S(\rho_3) + S(\rho_2) = S(\rho) + S(\rho_2)$, and along with the same thing with 1 and 2 interchanged, this yields the left inequality of (2.2.16). □

An ideal measurement leaves the system in a pure state, reducing the entropy to 0. For this reason, $S(\rho)$ may be regarded as a measure of the amount of information to be gained by an ideal measurement. The difference $S(\rho) - S(\rho_1)$ specifies how much more information a measurement of the total system can yield than a measurement of a subsystem. Inequality (2.2.16) bounds this relative information gain by $S(\rho_2)$:

$$|S(\rho) - S(\rho_1)| \leq S(\rho_2).$$

With quantum statistics the difference can be either positive or negative. If ρ is pure, so that the greatest possible information about the total system is available, but ρ_1 is a mixture, then more information can be obtained by measuring the subsystem. On the other hand, there are some inequalities for this entropy difference that are analogous to those of the classical entropy:

2.2.18 Inequalities for the Entropy Difference

Let ρ_{123} be given on $\mathcal{H}_1 \otimes \mathcal{H}_2 \otimes \mathcal{H}_3$, and $\rho_{12} = \mathrm{Tr}_3\, \rho_{123}$, $\rho_1 = \mathrm{Tr}_2\, \rho_{12}$, etc. Then (Lieb and Ruskai [8])

(i) $S(\rho_{12}) - S(\rho_1)$ *is concave in* ρ_{12}*;*

(ii) $S(\rho_{13}) - S(\rho_1) + S(\rho_{23}) - S(\rho_2) \geq 0$ *; and*

(iii) $S(\rho_{123}) - S(\rho_2) \leq S(\rho_{12}) - S(\rho_2) + S(\rho_{32}) - S(\rho_2)$.

2.2.19 Remarks

1. Proposition (i) implies that mixing increases the relative information gain. In particular, the relative information gain is a monotonic function in ρ_{12} with the ordering introduced in (2.1.6).
2. If Roman numerals are used to denote the systems corresponding to the Hilbert spaces $\mathcal{H}_i$, then Inequality (ii) implies that more information can be obtained

by measuring I ∪ III and II ∪ III than I and II. If $\mathcal{H}_2$ is one-dimensional, so $S(\rho_2) = 0$ and $S(\rho_{23}) = S(\rho_3)$, then this proposition reduces to (2.2.18).
3. Inequality (iii) is subadditivity for the entropy difference. The information content of I ∪ II and III ∪ II relative to II is greater than that of I ∪ II ∪ III relative to II.

Proof: These general inequalities for density matrices reflect mixing properties of the entropy like those used in phenomenological thermodynamics, and thereby provide a deeper foundation for those classical rules. Their proof (2.2.25;7) is greatly aided by the properties of the

2.2.20 Relative Entropy

$$S(\sigma|\rho) \equiv \operatorname{Tr} \rho(\ln \rho - \ln \sigma), \quad \rho, \sigma \geq 0, \quad \operatorname{Tr} \rho = \operatorname{Tr} \sigma = 1$$

For states Φ, Ψ over a C^*-algebra $\mathcal{A}$ this generalizes (see (3.2.17;6)) to $S(\Phi, \Psi) = \sup \int_0^\infty \frac{dt}{t}[\frac{\Phi(1)}{1+t} - \Psi(y^*(t)y(t) - \frac{1}{t}\Phi(x(t)x^*(t))]$ where $x, y : R^+ \to \mathcal{A}$ $x(t) + y(t) = 1$ and the sup is taken over all step function supported away from 0. $S(\sigma|\rho)$ enjoys the properties

(i) $S(\sigma|\rho) \geq 0$;

(ii) *the function* $(\Phi, \Psi) \to S(\Phi, \Psi)$ *is convex and lower semicontinuous*;

(iii) $S(\sigma \otimes \tau|\rho \otimes \tau) = S(\sigma|\rho)$ *for any density matrix* τ; *and*

(iv) $S(\Phi_{|\mathcal{B}}|\Psi_{|\mathcal{B}}) \leq S(\Phi|\Psi)$ *if* $\Phi_{|\mathcal{B}}$, $\Psi_{|\mathcal{B}}$ *are the restrictions to a subalgebra* $\mathcal{B} \in \mathcal{A}$.

Proof of the Properties of the Relative Entropy

(i) This was shown in the proof of subadditivity (2.2.12).

(ii) S is the sup of linear functional of continuous functions.

(iii)

$$\begin{aligned} S(\sigma \otimes \tau|\rho \otimes \tau) &= \operatorname{Tr}_{12} \rho \otimes \tau[(\ln \rho) \otimes \mathbf{1} + \mathbf{1} \otimes \ln \tau \\ &\quad -(\ln \sigma) \otimes \mathbf{1} - \mathbf{1} \otimes \ln \tau] \\ &= \operatorname{Tr}_1 \rho(\ln \rho - \ln \sigma) \operatorname{Tr}_2 \tau = S(\sigma|\rho). \end{aligned}$$

(iv) For the subalgebra $\mathcal{B}$ the sup is taken over a smaller set.

2.2.21 Remarks

1. If σ is the canonical density matrix $\sigma = \exp(-\beta H)/\exp(-\beta F)$, and the free energy is $F = -\beta^{-1} \ln \operatorname{Tr} \exp(-\beta H)$, then $S(\sigma|\rho) = \beta(\operatorname{Tr} \rho H - F) - S(\rho)$. If a free energy $F(\rho) \equiv \operatorname{Tr} \rho H - \beta^{-1} S(\rho)$ is ascribed to ρ, then $S(\sigma|\rho) = F(\rho) - F$. The relative entropy $S(\sigma|\rho)$ measures the difference from the canonical free

energy $F(\sigma) = F$, which always lies lower because of (i). $S(\sigma, \rho)$ is strictly convex in ρ so the minimizing ρ is unique.

2. By Property (ii), mixing and passing to limits bring the free energy closer to the canonical free energy.
3. $S(0|\rho)$ can be written as $\lim_{\lambda\to 0}[S((1-\lambda)\sigma + \lambda\rho) - (1-\lambda)S(\sigma) - \lambda S(\rho)]$ and is thus the mixing entropy if ρ is infinitesimally mixed to σ.
4. (iv) implies that if a subsystem is weakly coupled, $H_{12} \cong H_1 \otimes \mathbf{1}_2 + \mathbf{1}_1 \otimes H_2$, i.e., $\exp(-\beta(H_1 - F_1)) \cong \mathrm{Tr}_1 \exp(-\beta(H_{12} - F_{12}))$, then its difference from its canonical free energy is always less than that of the whole system. The analogous argument for the entropy only leads to $S(\rho_1) \leq S(\rho) + \ln d_2$, which already follows from (2.2.4;3).
5. (i) can actually be sharpened to $S(\rho|\sigma) \geq \frac{1}{2}\|\rho - \sigma\|^2$.

A final matter to investigate is how sensitive S is to small changes in ρ.

2.2.22 Theorem

The mapping $C_1^+ \to \mathbb{R}^+ : \rho \to S(\rho)$ is lower semicontinuous in the trace-norm topology

2.2.23 Remarks

1. The set C_1 is topologized with the trace norm $\| \|_1$. If a sequence $\{\rho_N\}$ converges in this topology to ρ, then $S(\rho)$ is at most $\lim_{N\to\infty} S(\rho_N)$. However, we shall see in (2.4.18; 1) that for density matrices all topologies between the trace topology and the weak topology are equivalent.
2. Continuity does not occur, because in every $\| \|_1$-neighborhood of ρ there are density matrices with arbitrarily much entropy. This follows directly from concavity,

$$S\left(\frac{1}{N}\rho_n + \left(1 - \frac{1}{N}\right)\rho\right) \geq \frac{1}{N}S(\rho_N) + \left(1 - \frac{1}{N}\right)S(\rho).$$

Let $S(\rho) = 0$, and $S(\rho_N) = N^2$; then $S((1/N)\rho_N + (1-1/N)\rho) \geq N$, although

$$\left\| \frac{1}{N}\rho_N + \left(1 - \frac{1}{N}\right)\rho - \rho \right\|_1 \leq \frac{2}{N},$$

so the density matrices converge to ρ. The two terms in the expression $(1/N)\rho_N + (1-1/N)\rho$, however, can not be comparable in the sense of (2.1.6); that would contradict (2.1.7; 4), by which the limit of a sequence of equivalent density matrices can not be purer than the elements of the sequence.

3. The mappings $C_1^+ \to \mathbb{R}^+ : \rho \to S_\alpha(\rho), \alpha > 1$ are continuous (see below).
4. By lower semicontinuity the sets $S_n \equiv \{\rho : S(\rho) \leq n\}$ are closed, and by Remark 2 they are nowhere dense. This means that the set $\bigcup_n S_n$ of ρ's of finite entropy is of the first category, the topological analogue of a null set. In this sense the entropy is almost always $+\infty$.

Proof: Because $\operatorname{Tr}\rho^\alpha = \|\rho\|_\alpha^\alpha \leq \|\rho\|^{\alpha-1} \cdot \|\rho\|_1$, the mapping of $\mathcal{C}_1$ to $\mathbb{R}^+ : \rho \to S_\alpha(\rho)$ is continuous. As the supremum of a set of continuous functions, $S(\rho) = \sup_{\alpha>1} S_\alpha(\rho)$ is lower semicontinuous. □

The failure of $S(\rho)$ to be continuous does not diminish its usefulness. The density matrices ρ of very large S have their eigenvalues ρ_i spread so far apart that the average of the energy diverges.

2.2.24 The Continuity of the Entropy at Finite Energy

Suppose that $H \geq 0$ *and* $\operatorname{Tr}\exp(-\beta H) < \infty$ *for some* $\beta > 0$. *If the density matrices having* $\operatorname{Tr}\rho H < \infty$ *are topologized with the norm* $\|\rho\|_H = \operatorname{Tr}\rho(1+H)$, *then* $S(\rho)$ *is a continuous mapping* $\mathcal{C}_H \to \mathbb{R}^+$, *where* $\mathcal{C}_H = \{\rho \in \mathcal{C}_1, \|\rho\|_H < \infty\}$.
Proof: According to Remark (2.2.21; 1), $S(\rho) = \beta(\operatorname{Tr}(\rho H) - F) - S(\sigma|\rho)$, where $\sigma = \exp(-\beta H)/\exp(-\beta F)$. The function $\operatorname{Tr}\rho H$ is continuous in the $\|\|_H$-topology, and $-S(\sigma|\rho)$ is upper semicontinuous, because the $\|\|_H$-topology is finer than the trace topology. Since $S(\rho)$ is lower semicontinuous in the trace-norm topology, it is also lower semicontinuous in $\|\|_H$, and hence continuous in $\|\|_H$. □

2.2.25 Problems

1. Generalize the entropy of a state over a C^*-algebra $\mathcal{A}$ by
$$S_\omega(\mathcal{A}) = \sup_{\sum_i \lambda_i\omega_i=\omega} \sum_i \lambda_i S(\omega|\omega_i)$$
where ω_i are states and $\lambda_i \in R^+, \sum_i \lambda_i = 1$. Show that if the states can be described by density matrices this reduces to (2.2.3). For a subalgebra $\mathcal{B} \in \mathcal{A}$ define $H_\omega(\mathcal{B}) = \sup_{\sum\lambda_i\omega_i=\omega} \sum_i \lambda_i S(\omega_{|\mathcal{B}}|\omega_{i|\mathcal{B}})$ and show that this is dominated by $S_{\omega|\mathcal{B}}(\mathcal{B})$ and is monotonic $\mathcal{B}_1 \in \mathcal{B}_2 \Rightarrow H_\omega(\mathcal{B}_1) \leq H_\omega(\mathcal{B}_2)$ which $S_{\omega|\mathcal{B}}$ is not.
2. For $\alpha \neq 1$, show that the α-entropies S_α of (2.2.1) are not subadditive.
3. Prove Lemma (2.2.14).
4. Show that $S(\sum_i \lambda_i\rho_i) \leq \sum_i \lambda_i S(\rho_i) - \sum_i \lambda_i \ln\lambda_i$, $\lambda_i > 0$, $\sum_i \lambda_i = 1$.
5. Show that $S_{\text{cl}}(\rho_1) \leq S_{\text{cl}}(\rho)$ if $\mathcal{H} = \mathcal{H}_1 \otimes \mathcal{H}_2$, where $\mathcal{H}_i$ are one-particle Hilbert spaces, particles 1 and 2 are distinguishable, $\rho_1 = \operatorname{Tr}_2 \rho$, and $S_{\text{cl}}(\rho)$ is defined as in (2.2.5).
6. Calculate $S_\alpha(\exp(-\beta[H - F(\beta)])$, where $\exp(-\beta F(\beta)) = \operatorname{Tr}\exp(-\beta H)$.
7. Prove (2.2.20). Hint: use $S(\rho_1 \otimes \mathbf{1}|\rho_{12}) = S(\rho_1) - S(\rho_{12})$ thus $S(\rho_{23}) + S(\rho_{12}) - S(\rho_{123}) - S(\rho_2) = S(\mathbf{1} \otimes \rho_{23}|\rho_{123}) - S(\mathbf{1} \otimes \rho_2|\rho_{12})$.
8. Prove the formula for the identity operator in (2.2.8;5).

2.2.26 Solutions

1. For density matrices $\sum_i \lambda_i S(\rho|\rho_i) = S(\rho) - \sum_i \lambda_i S(\rho_i)$ and the sup is attained if the ρ_i are the 1-dimensional spectral projectors of ρ for which $S(\rho_i) = 0$. $H_\omega(\mathcal{B})$ and $S_{\omega_{|\mathcal{B}}}(\mathcal{B})$

look formally the same but for the latter the sup is taken over $\sum_i \lambda_i \omega_{i|\mathcal{B}} = \omega_{|\mathcal{B}}$. This is less stringent than $\sum_i \lambda_i \omega_i = \omega$ and thus a larger set $\Rightarrow H_\omega(\mathcal{B}) \leq S_{\omega|\mathcal{B}}(\mathcal{B})$. Clearly H inherits the monotonicity from $S(\omega|\omega_i)$.

2. Let $\mathcal{H} = \mathbb{C}^2 \otimes \mathbb{C}^2$ and $\rho = (\rho_{ik,jl})$, where $\rho_{ik,jl} = \delta_{ij}\delta_{kl} r_{ik}$; $r_{11} = pq + \varepsilon$, $r_{12} = p(1-q) - \varepsilon$, $r_{21} = (1-p)q - \varepsilon$, $r_{22} = (1-p)(1-q) + \varepsilon$ with $0 < p, q < 1$, $p, q \neq \frac{1}{2}$. Since ρ is diagonal, this allows $S_\alpha(\rho)$ to be read off with no further ado: If $\varepsilon = 0$, then $\rho = \rho_1 \otimes \rho_2$,

$$\rho_1 = \mathrm{Tr}_2\, \rho = \begin{pmatrix} p & 0 \\ 0 & 1-p \end{pmatrix}, \qquad \rho_2 = \mathrm{Tr}_1\, \rho = \begin{pmatrix} q & 0 \\ 0 & 1-q \end{pmatrix}.$$

If $S_\alpha(\rho)$ were $\leq S_\alpha(\rho_1) + S_\alpha(\rho_2)$, then the function $g(\varepsilon) \equiv (pq+\varepsilon)^\alpha + (p(1-q) - \varepsilon)^\alpha + ((1-p)q - \varepsilon)^\alpha + ((1-p)(1-q) + \varepsilon)^\alpha$ would have an extremum at $\varepsilon = 0$, but $g'(0) \neq 0$ if $\alpha \neq 1$.

3. Let $|x\rangle \in \mathcal{H}_1 \otimes \mathcal{H}_2 |x\rangle = \sum_{i,k} c_{ik} |i\rangle_1 \otimes |k\rangle_2$, where $\{|i\rangle_1\}$ and $\{|k\rangle_2\}$ are orthonormal sets in $\mathcal{H}_1$ and $\mathcal{H}_2$ respectively, and $\rho = |x\rangle\langle x|$.

$$\begin{aligned} \mathrm{Tr}_2\, |x\rangle\langle x| &= \mathrm{Tr}_2 \sum_{ijkl} c_{ik} c^*_{jl} |i\rangle_{11}\langle j| \otimes |k\rangle_{22}\langle l| \\ &= \sum_{ijkl} c_{ik} c^*_{jl} |i\rangle_{11}\langle j| \delta_{kl} = \sum_{ijk} c_{ik} c^*_{jk} |i\rangle_{11}\langle j|, \end{aligned}$$

which implies that the positive eigenvalues of $\mathrm{Tr}_2\, |x\rangle\langle x|$ are the same as those of the matrix CC^*, where $C = (c_{ij})$. A similar argument shows that the positive eigenvalues of $\mathrm{Tr}_1\, |x\rangle\langle x|$ are the same as those of C^*C and thus of CC^*.

4. Let $\lambda_i \rho_i = a_i$; then the proposition is equivalent to $S(\sum_i a_i) \leq \sum_i S(a_i)$ for all $a_i \in \mathcal{C}_1^+$. Since $\ln x$ is monotonic as an operator function (III: 2.2.38., 11), if $a_k \geq 0$, then $\ln a_i \leq \ln(\sum_j a_j)$, which implies $a_i^{1/2}(\ln a_i) a_i^{1/2} \leq a_i^{1/2}(\ln \sum_j a_j) a_i^{1/2}$, and therefore $\sum_i \mathrm{Tr}(a_i \ln a_i) \leq \mathrm{Tr}[(\sum_i a_i) \ln(\sum_i a_i)]$.

5. $\rho_1(\mathbf{z}_1) \equiv \langle \mathbf{z}_1 | \rho_1 | \mathbf{z}_1 \rangle = \sum_i \langle \mathbf{z}_1 \otimes e_i | \rho | \mathbf{z}_1 \otimes e_i \rangle \geq \langle \mathbf{z}_1 \otimes \mathbf{z}_2 | \rho | \mathbf{z}_1 \otimes \mathbf{z}_2 \rangle \equiv \rho(\mathbf{z}_1, \mathbf{z}_2)$, since $\{e_i\}$ may be chosen to be an arbitrary basis. Therefore

$$S_{\mathrm{cl}}(\rho) - S_{\mathrm{cl}}(\rho_1) = \int d\Omega_z^2\, \rho(\mathbf{z}_1, \mathbf{z}_2) \ln\left(\frac{\rho(\mathbf{z}_1)}{\rho(\mathbf{z}_1, \mathbf{z}_2)}\right) \geq 0.$$

6.

$$S_\alpha = \frac{1}{1-\alpha} \ln \mathrm{Tr} \exp(-\alpha\beta(H - F(\beta))) = \frac{\alpha\beta}{\alpha - 1}[F(\alpha\beta) - F(\beta)].$$

As $\alpha \to 1$, $S \to \partial F(\beta)/\partial\beta$.

7. That $\mathbf{1} \otimes \sigma$ is not normalized does not disturb convexity and monotonicity of $S(\mathbf{1} \otimes \sigma | \rho)$ since $\Psi(1) = 1$ was not used. (i) follows from convexity and (iii) from monotonicity. As to (ii) the left hand side is concave in ρ_{123} according too (i). Thus it assumes its minimum at the extremal points, i.e. pure ρ_{123}. There ((2.2.14) tell us $S(\rho_1) = S(\rho_{23})$ and $S(\rho_2) = S(\rho_{13})$ and the minimum becomes zero.

8. The right side of the equation clearly leaves the number of particles invariant. Hence the formula is shown by

$$\langle f_1, \ldots, f_N | \int \frac{dz_1 \cdots dz_N}{N!(2\pi)^{2N}} |\mathbf{z}_1, \ldots, \mathbf{z}_N\rangle\langle \mathbf{z}_1, \ldots, \mathbf{z}_N | g_1, \ldots, g_N\rangle$$

$$= \sum_{P,Q} (\pm 1)^{P+Q} \int \frac{dz_1 \cdots dz_N}{N!(2\pi)^{dN}} \prod_{i=1}^{N} \langle f_{P_i} | \mathbf{z}_i \rangle \langle \mathbf{z}_i | g_{Q_i} \rangle$$

$$= \sum_{P,Q} (\pm 1)^{P+Q} \prod_i \langle f_{P_i} | g_{Q_i} \rangle \frac{1}{N!}$$

$$= \sum_{P'} (\pm 1)^{P'} \prod_i \langle f_i | g_{P'_i} \rangle = \langle f_1, \ldots, f_N | g_1, \ldots, g_N \rangle .$$

2.3 The Microcanonical Ensemble

Insight into the fundamental thermodynamic laws is gained by investigating the chaotic state below the energy surface.

Two trains of thought are usually followed to justify regarding the equilibrium state as predominating for macroscopic systems. Like Boltzmann, one can investigate the time-evolution of a system and show that most states tend to equilibrium. Alternatively, one can follow Gibbs and examine an ensemble of identical copies of the system and identify states of scanty information with equilibrium states. The set of problems connected with the first procedure is the subject of the next chapter, while in this section we shall study systems for which the only information concerns the energy. If it is known that the energy does not exceed some maximum value E_m then, as remarked in (2.1.7;2), the most mixed state containing no further information corresponds to the

2.3.1 Microcanonical Density Matrix

$$\rho = \Theta(E_m - H) / \operatorname{Tr} \Theta(E_m - H), \qquad \Theta(x) = \begin{cases} 1 & \text{for} \quad x \geq 0 \\ 0 & \text{for} \quad x < 0 \end{cases},$$

where $E_m \geq \varepsilon_1 \equiv$ the lowest eigenvalue of H. Its

2.3.2 Entropy and Average Energy

Are

$$S = \ln \operatorname{Tr} \Theta(E_m - H), \quad E = \exp(-S) \operatorname{Tr} H \Theta(E_m - H).$$

2.3.3 Remarks

1. The discontinuous function Θ of a self-adjoint operator is defined with the spectral representation of the operator.
2. It is assumed that H is bounded below and that $\sigma_{\text{ess}}(H)$ is empty, so the traces in (2.3.2) are finite.

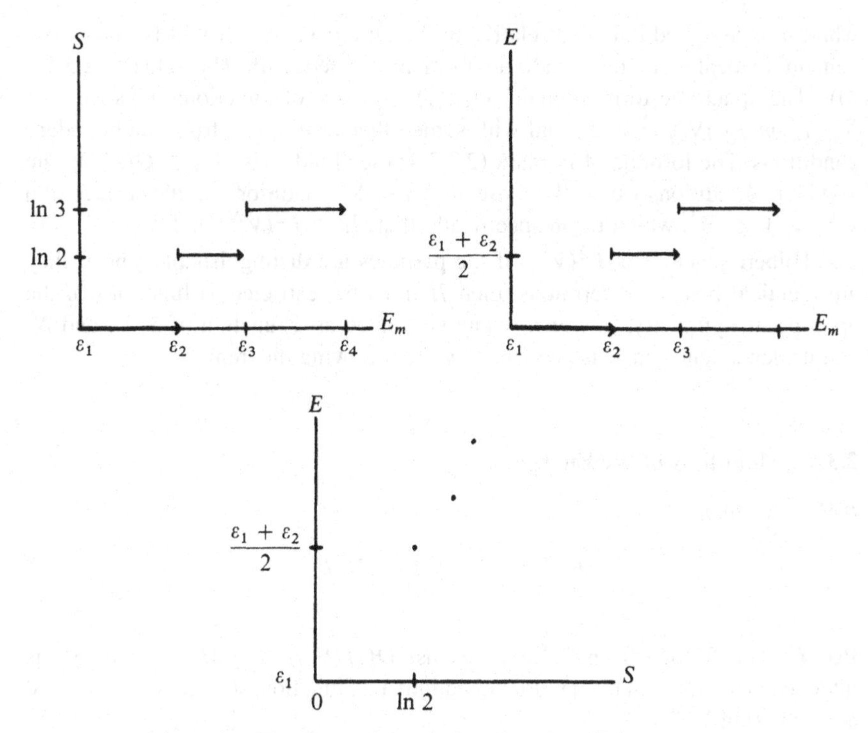

Figure 2.1. The thermodynamic functions for a finite system.

3. The entropy S is a discontinuous function of E_m and has no well-defined inverse. On the other hand, E may be constructed as a function of S, as shown in Figure 2.1. The function $E(S)$ increases monotonically.
4. By the min-max principle, $E(S)$ is also given by $E(S) = \exp(-S) \times \inf_{\mathcal{H}_{\mathbf{n}}} \mathrm{Tr}_{\mathcal{H}_{\mathbf{n}}} H$, where $\mathcal{H}_{\mathbf{n}}$ is an n-dimensional subspace of $D(H)$ and $n = \exp(S)$. It is consequently a concave function of all parameters on which the dependence of H is concave.
5. By Property (2.2.2(iv)), all α-entropies S_α lead to the same S (2.3.2), which can be identified as the entropy of phenomenological thermodynamics.
6. It will be seen shortly that in the systems under consideration here the density of states increases so rapidly with the energy that in the limit of an infinite system, any density matrix $\rho \sim \Theta(E - H) - \Theta(E(1 - \varepsilon) - H)$ yields the same entropy density for all $\varepsilon > 0$.

The further properties of $E(S)$ follow from the special form of the Hamiltonian,

$$H_N = \sum_{i=1}^{N} \frac{|\mathbf{p}_i|^2}{2m_i} + \sum_{i>j} v(\mathbf{x}_i - \mathbf{x}_j),$$

where v is assumed bounded relative to the kinetic energy. It will be most convenient to deal with the quadratic form associated with H_N (cf.(III: 2.5.17; 2)). The quadratic-form domain $Q(H_N)$ consists of functions ψ such that $\sum_i (1/2m_i) \int |\nabla_i \psi|^2 < \infty$ and with some other restrictions from the boundary conditions. The formula of Remark (2.3.3;4) then holds with $\mathcal{H}_\mathbf{n} \subset Q(H_N)$. The boundary conditions we shall choose are Dirichlet conditions on the surface of a volume $V \subset \mathbb{R}^3$, which mean specifically that: $\mathcal{H} \subset L^2(V^N)$ and $\psi|_{\partial(V^N)} = 0$. The Hilbert space $\mathcal{H}$ is $L^2(V^N)$ if the particles are distinguishable, and if they are identical bosons or fermions, then $\mathcal{H}$ must be restricted to functions of the appropriate symmetry. The energy can be treated as a function of S, V, and N, and its dependence on V is described by the following theorem.

2.3.4 Monotony of the Energy

If $V' \supset V$, then

$$E(S, V', N) \leq E(S, V, N).$$

Proof: This follows from (2.3.3;4) because $Q(H(V')) \supset Q(H(V))$, where $\supset$ is intended in the sense of the natural imbedding, i.e., functions ψ such that $\psi|_{\partial V} = 0$ are set to 0 in $V' \setminus V$. □

Subadditivity generalizes this monotony when particles in separated volumes do not repel one another.

2.3.5 Subadditivity of the Energy

If $V_1 \cap V_2 = \emptyset$ and $v(\mathbf{x}_i - \mathbf{x}_j) \leq 0$ for all $\mathbf{x}_i \in V_1$, $\mathbf{x}_j \in V_2$, then

$$E(S_1 + S_2, V_1 \cup V_2, N_1 + N_2) \leq E_1(S_1, V_1, N_1) + E(S_2, V_2, N_2).$$

Proof: This again follows from (2.3.3;4), since the right side results from taking the infimum over a subspace of $Q(H)$, which consists of tensor products of $\exp(S_1)$ vectors, for which N_1 particles lie within the volume V_1, with $\exp(S_2)$ vectors having N_2 particles within V_2. The tensor products have to be symmetrized or antisymmetrized if there are Bose or Fermi statistics. However, since symmetrization does not affect the expectation values of (2.3.5) when the functions have disjoint supports, (2.3.5) is independent of the statistics.

The existence of $\lim_{V \to \infty} E/V$ can be derived from the subadditivity, though it is rather difficult to go beyond the restriction $v \leq 0$. This problem will have to be investigated later for each of the systems discussed in § 1.2, and for now convergence will simply be assumed. The condition is satisfied trivially for free particles ($v = 0$). To draw conclusions like those of (2.2.13), assume that V is a cube, the volume of which will also be fearlessly denoted $V \in \mathbb{R}^+$. If eight cubes are packed together as a single cube of double the side, then (2.3.5) implies

$$E(8S, 8V, 8N) \leq 8E(S, V, N). \tag{2.6}$$

Assuming in addition that there exists $A \in \mathbb{R}^+$ such that

$$H_N \geq -AN \quad \text{for all} \quad N \in \mathbb{Z}^+, \tag{2.7}$$

the limit

$$\lim_{\mathbb{Z}^+ \ni \nu \to \infty} 8^{-\nu} E(8^\nu S, 8^\nu V, 8^\nu N) = \inf_\nu 8^{-\nu} E(8^\nu S, 8^\nu V, 8^\nu N)$$

exists. This allows the passage to an infinite system, for which the energy, entropy, and particle densities are defined by $E/V = \varepsilon$, $S/V = \sigma$, and $N/V = \rho$. □

2.3.6 The Thermodynamic Limit of the Energy Density

$$\varepsilon(\sigma, \rho) = \inf_{\mathbb{Z}^+ \ni \nu} 8^{-\nu} \rho E(8^\nu \sigma \rho^{-1}, 8^\nu \rho^{-1}, 8^\nu).$$

2.3.7 Remarks

1. Equation (2.7) guarantees that $\varepsilon > -\infty$, so the infimum always exists; but (2.3.6) is only of interest when there is a well-defined limit, for only then is it certain that the thermodynamic properties do not depend on the exact number of particles. Even if the limit exists, as in the case of (2.6), it does not guarantee that the resulting ε is nontrivial. If, say, the particles can be distinguished (which does not invalidate the general conclusions), then classically,

$$\exp(S) = \int_{V^N} d^{3N}x \int d^{3N}p \Theta\left(E_m - \sum_{i=1}^{N} |\mathbf{p}_i|^2\right) = \pi^{3N/2} \frac{E_m^{3N/2} V^N}{(3N/2)!},$$

and

$$E = \frac{E_m}{1 + 2/3N}.$$

Therefore, as $N \to \infty$,

$$\frac{E}{V} = \frac{3}{2\pi e} \frac{\rho^{5/3}}{N^{2/3}} \exp\left(\frac{2}{3}\sigma\rho^{-1}\right) \to 0.$$

The familiar result obtains only with the replacement $\exp(S) \to (1/N!) \exp(S)$ to account for the particles being identical. A later calculation of $\varepsilon(\sigma, \rho)$ will reveal that (2.3.6) is then not without content.

2. Though the result has been derived only for cubes, the limit clearly exists for other shapes if they are not too different from cubes.
3. The effect of dilatations on the kinetic energy of free particles (cf. (III: 3.3.20;8) and (III: 4.1.3)) implies, moreover, that

$$E(S, VN) = \exp(2\tau)E(S, \exp(-3\tau)V, N).$$

Hence the one-parameter family of limits

$$\lim_{\nu\to\infty} 8^{-\nu(1-2\tau)} E(8^{\nu} S, 8^{\nu(1-3\tau)} V, 8^{\nu} N)$$

exist (cf. (1.2.1)). Ordinarily, the limit is taken with $\tau = 0$, and quantities proportional to N, like E, S, and V, are described as extensive, while N-independent quantities like ε, ρ and σ are called intensive. The existence of some limit is important, for, whatever it may be like, it enables precise propositions to be formulated. In reality systems are large but still finite, but if a quantity converges as $N \to \infty$ the limit may be expected to be attained for practical purposes when, say, $N = 10^{24}$. Indeed, it will be shown in realistic situations that the limit is sometimes attained to $O(N^{-1/6})$, which is sufficient accuracy for macroscopic bodies. There are various ways to interpret the limit $N \to \infty$. As has been done here, the system may be thought of as becoming larger and larger, or, alternatively, the atoms may be imagined smaller and smaller with their number in the fixed volume of the container being increased at the same time.

Since monotony and convexity survive pointwise limits, there are the following

2.3.8 Properties of the Energy Density

For the function $\mathbb{R}^+ \times \mathbb{R}^+ \to \mathbb{R}^+ : \sigma, \rho \to \varepsilon(\sigma, \rho)$,

(i) *ε increases monotonically in σ;*

(ii) *$\rho^{-1}\varepsilon(\alpha\rho, \rho)$ increases monotonically in ρ;*

(iii) *ε is convex in (σ, ρ);*

(iv) *moreover, for free particles, $\varepsilon(\sigma, \rho) = \rho^{5/3} f(\sigma/\rho)$.*

Proof: Property (i) holds as remarked in (2.3.3; 3), and Property (ii) follows from Theorem (2.3.4). From subadditivity (2.3.5),

$$\varepsilon\left(\frac{1}{2}(\sigma_1 + \sigma_2), \frac{1}{2}(\rho_1 + \rho_2)\right) \leq \frac{1}{2}\left(\varepsilon(\sigma_1, \rho_1) + \varepsilon(\sigma_2, \rho_2)\right),$$

which implies (iii), and (iv) follows from (2.3.7; 3). □

2.3.9 Remarks

1. Since $N \in \mathbb{Z}^+$, $S \in \ln \mathbb{Z}^+$, ε is at first defined only on the dense set for which $\sigma\rho^{-1}$ is a power of $(\ln z)/2$, $z \in \mathbb{Z}^+$. It extends continuously to $\mathbb{R}$, because

monotony and concavity with the coefficient $\frac{1}{2}$ imply uniform continuity. There are discontinuous functions that are concave with coefficient $\frac{1}{2}$, such as

$$f(x) = \begin{cases} x, x & \text{rational,} \\ 0, & \text{otherwise,} \end{cases}$$

for which the equation $f(\alpha x) = \alpha f(x)$ holds for all rational α. However, this can not occur if the function is monotonic. The extension then in addition satisfies the inequality

$$\varepsilon(\alpha\sigma_1 + (1-\alpha)\sigma_2, \alpha\rho_1 + (1-\alpha)\rho_2) \leq \alpha\varepsilon(\sigma_1, \rho_1) + (1-\alpha)\varepsilon(\sigma_2, \rho_2)$$
$$\text{for all} \quad \alpha \in \mathbb{R}, \quad 0 \leq \alpha \leq 1.$$

2. Subadditivity (2.3.5) is sufficient but not necessary for Property (iii); (2.3.5) may be violated if the interaction is partially repulsive, which is a necessary assumption or $H_N \geq -AN$ when the particles interact. However, if the potential goes to zero rapidly enough at infinity, the correction to (2.3.5) on any finite region is a surface effect, so the convexity of the energy density is still guaranteed in the thermodynamic limit. On the other hand, the special form (2.3.8) is crucial, and in § 4.2 it will be seen that convexity (2.3.8(iii)) is violated in gravitating systems, although (2.3.5) is valid.
3. Since the limiting function is continuous, Dini's theorem ensures that the monotonic limit (2.3.8) is uniform on compact sets.
4. Let H be defined so that $\inf \varepsilon = 0$. Since ε is convex in σ, unless $\varepsilon \equiv 0$, there exists a σ_0 such that ε is strictly monotonic in σ for all $\sigma > \sigma_0$. There is consequently an inverse function $\sigma(\varepsilon, \rho)$ (see Figure 2.2), which is concave and monotonically increasing in ε.
5. As long as σ is strictly monotonically increasing in ε, the density matrices

$$\rho = \Theta(E_m - H)\exp(-S)$$

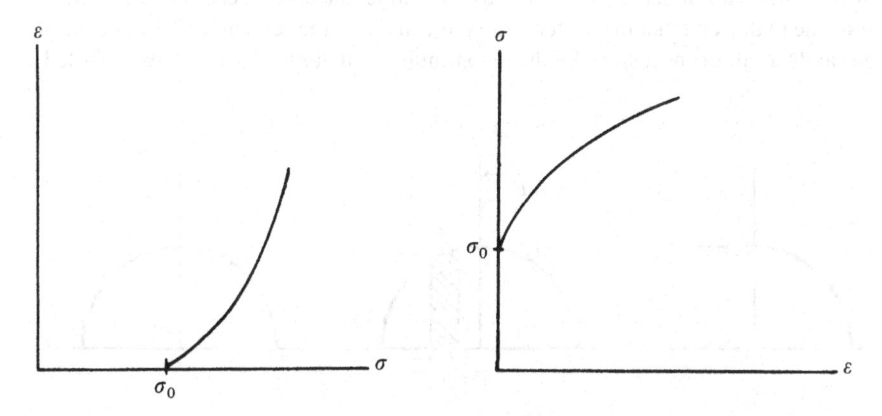

Figure 2.2. The thermodynamic functions for an infinite system.

and

$$\rho_\delta = (\Theta(E_m - H) - \Theta(E_m - V\delta - H))\exp(-S_\delta)$$

yield the same entropy densities in the limit $N \to \infty$:

$$\begin{aligned}\sigma_\delta &= \lim_{V\to\infty} \frac{1}{V} \ln \operatorname{Tr}(\Theta(E_m - H) - \Theta(E_m - V\delta - H)) \\ &= \sigma(\varepsilon, \rho) + \lim_{V\to\infty} \frac{1}{V} \ln(1 - \exp[-V(\sigma(\varepsilon, \rho) - \sigma(\varepsilon - \delta, \rho))]) = \sigma.\end{aligned}$$

This means that as $N \to \infty$ most of the states crowd just under the energy surface with arbitrarily high density.

6. For some systems $\sigma(\varepsilon)$ is constant for ε greater than some ε_1, in which case ρ and ρ_δ may have different entropies. Consider for example N spins in an external field ((1.2) with $\varepsilon = 0$). The density of states $(\partial/\partial E)\exp[S(E)]$ is invariant under $\sigma \to -\sigma$ and thus an even function in E. This makes $\operatorname{Tr}\rho_\delta$ a decreasing function of E_m when $E_m + \delta > 0$, which is impossible for $\operatorname{Tr}\rho$ (see Figure 2.3); Definition (2.3.1) rules negative temperatures out.
7. The number of energy levels below E_m is $\exp(N\sigma/\rho)$, which is immense for macroscopic bodies, $N \sim 10^{24}$. It would never be possible to isolate the energy levels completely – their widths are on the order of (macroscopic time)$^{-1}$, which is much larger than their spacing. Systems will later be idealized as infinite, having continuous energy spectra, which comes closer to reality than does the fiction of a discrete spectrum.

After this first exposure to these ideas, let us consider two systems the interaction between which is so weak that it can be neglected in comparison with other energies. They are to be considered as parts of a larger system with $\mathcal{H} = \mathcal{H}_1 \otimes \mathcal{H}_2$, $H = H_1 + H_2$. The question is how the energy and entropy are shared by the two subsystems. Even though H is a sum, the microcanonical density matrix (2.3.1) is not in the form of a product $\rho = \rho_1 \otimes \rho_2$, and we will have to see how the entropy of this state can nonetheless be additive for independent, macroscopic systems. Assume to this end that the systems are large and that the sequence (2.3.6) converges and has all the necessary kinds of continuity so that $\varepsilon = E/V$ can be regarded

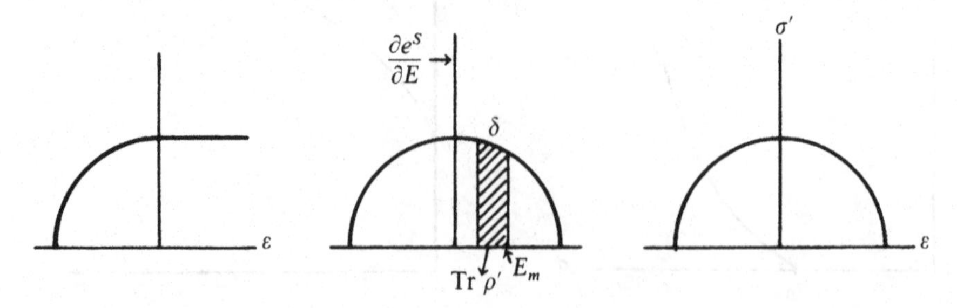

Figure 2.3. Inequivalence of the microcanonical ensembles for spins in a magnetic field.

as a continuous variable for the purposes of integration and differentiation. For the problem at hand and other estimates we shall need

2.3.10 Lemma

Let $\sigma(\varepsilon) \leq 0$ and be concave on [0, 1], and $\sigma(1) = 0$, $-\infty < \sigma(0) < 0$; this implies that a is nondecreasing and that there exists an ε_0, $0 < \varepsilon_0 \leq 1$ such that $\sigma' \equiv \sigma'(\varepsilon_0) > 0$. Then

$$\frac{1 - \exp(-V|\sigma(0)|)}{V|\sigma(0)|} \leq \int_0^\infty d\varepsilon \exp(V\sigma(\varepsilon)) \leq 1 - \varepsilon_0 + \frac{1 - \exp(-V\varepsilon_0\sigma')}{V\sigma'}.$$

Proof: By assumption (see Figure 2.4),

$$(1-\varepsilon)\sigma(0) \leq \sigma(\varepsilon) \leq \begin{cases} 0 & \text{for } \varepsilon_0 \leq \varepsilon \leq 1 \\ -(\varepsilon_0 - \varepsilon)\sigma' & \text{for } 0 \leq \varepsilon \leq \varepsilon_0. \end{cases}$$

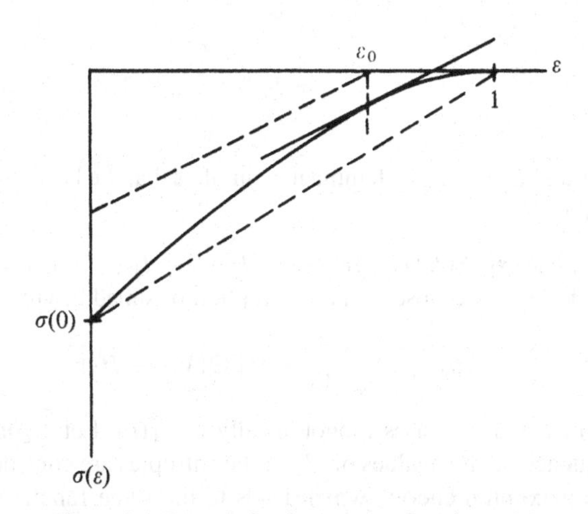

Figure 2.4. Bounds for the concave function $\sigma(\varepsilon)$.

2.3.11 Corollaries

1. If σ is concave but not necessarily negative, then the formula

$$\int_a^b V d\varepsilon \exp(V\sigma(\varepsilon)) = \exp(V\bar{\sigma}) \int_a^b d\varepsilon V \exp(V(\sigma(\varepsilon) - \bar{\sigma}))$$

with $\bar{\sigma} = \max_{a \le \varepsilon \le b} \sigma(\varepsilon)$ can be used instead, since $-\infty < \bar{\sigma} < \infty$ unless $\sigma \equiv \pm\infty$. By an application of the lemma, possibly after subdivision of the region of integration,

$$\lim_{V\to\infty} \frac{1}{V} \ln \int_a^b d\varepsilon V \exp(V(\sigma(\varepsilon) - \bar{\sigma})) = 0.$$

Thus only the maximum value of σ contributes in the infinite limit:

$$\lim_{V\to\infty} \frac{1}{v} \ln \int_a^b V d\varepsilon \exp(V\sigma(\varepsilon)) = \sup_{a\le\varepsilon\le b} \sigma(\varepsilon) = \bar{\sigma}.$$

2. Remark (2.3.9;5) leads one to expect that E_m and E may become equal for large systems. More precisely, if σ is concave in ε, $d\sigma/d\varepsilon > 0$, then $\lim_{V\to\infty}(E - E_m)/V = 0$. This follows because E may be written as

$$E = \exp(-S) \operatorname{Tr} H\Theta(E_m - H) = \int_0^{E_m} dE' E' \frac{\partial}{\partial E'} \operatorname{Tr} \Theta(E' - H) \exp(-S)$$

$$= E_m - \int_0^{E_m} dE' \operatorname{Tr} \Theta(E' - H) \exp(-S).$$

With $\varepsilon_0 = 1$ and $E' = \varepsilon V$ the lemma now implies that the last integral is $O(1)$, whereas $E_m \sim V$.

3. We next calculate $\exp(S(E)) = \operatorname{Tr} \Theta(E - H_1 - H_2)$, $H_i \ge 0$, as $V = V_1 + V_2 \to \infty$ with V_i/V fixed. Because of the assumption of subadditivity,

$$\sigma_{1,V_1}(\varepsilon) \equiv \frac{1}{V_1} \ln \operatorname{Tr}_1 \Theta(V_1\varepsilon - H_1)$$

is concave in ε and increases monotonically to $\sigma_1(\varepsilon)$. Let $E_2[n]$ denote the ordered sequence of eigenvalues of H_2. If the entropies are considered as functions of the maximum energy, which leads to the same function in the limit $V \to \infty$ because of Corollary 2, then n may be identified with $\exp S$, and $E_2(S_2) \equiv E_2[\exp(S_2)]$ becomes the function introduced in (2.3.3;3). With $E = \varepsilon V$,

$$\sigma(\varepsilon) = \lim_{V\to\infty} \frac{1}{V} \ln \operatorname{Tr} \Theta(E - H_1 - H_2)$$

$$= \lim_{V\to\infty} \frac{1}{V} \sum_{n=1}^{\exp(S_2(E))} \exp(S_1(E - E_2[n])).$$

Now regard n as a continuous variable, and interpolate $E_2[n]$ linearly. Since the integrand decreases monotonically, the sum $\sum_{n=1}^{\exp(S_2(E))} \cdots$ lies between

$\int_0^{\exp(S_2(E))} dn \cdots$ and $\int_1^{\exp(S_2(E))+1} dn \cdots$, and the evaluation of the error is unnecessary, since $\exp(S_2(E)) \sim \exp(10^{23})$. With the variables $\sigma_2 = (1/V_2)\ln n$, $\sigma(\varepsilon)$ can be written as

$$\lim_{V\to\infty} \frac{1}{V} \ln \int_0^{\sigma_2(\varepsilon)} V_2 d\sigma_2 \exp\left[V_1 \sigma_{1,V_1}\left(\frac{V}{V_1}\varepsilon - \frac{V_2}{V_1}\varepsilon_{2,V_2}(\sigma_2)\right) + V_2\sigma_2 \right].$$

Now note that $\sigma_2 \to a - b\varepsilon_2(\sigma_2)$ is concave if $b \geq 0$, σ_{1,V_1} is concave and increasing, and that (concave, increasing) ∘ concave=concave. This allows the lemma to be applied, to show

$$\begin{aligned}\sigma(\varepsilon) &= \lim_{V\to\infty} \sup_{0\leq\sigma_2\leq\sigma_2(\varepsilon)} \left[\frac{V_1}{V}\sigma_{1,V_1}\left(\frac{V}{V_1}\varepsilon - \frac{V_2}{V_1}\varepsilon_{2,V_2}(\sigma_2)\right) + \frac{V_2}{V}\sigma_2 \right] \\ &= \sup_{0\leq\sigma_2\leq\sigma_2(\varepsilon)} \left[\frac{V_1}{V}\sigma_1\left(\frac{V}{V_1}\varepsilon - \frac{V_2}{V_1}\varepsilon_2(\sigma_2)\right) + \frac{V_2}{V}\sigma_2 \right].\end{aligned}$$

The interchange of the limit $V \to \infty$ and the supremum is justified because $\varepsilon_{2,V_2}(\sigma_2)$ increases monotonically in σ_2 for all V_2, and since $\sigma_{1,V_1}(\varepsilon)$ likewise increases in ε, it decreases in σ_2, and consequently the first term in the brackets [ ] converges uniformly on compact sets to

$$\sigma_2 \to \sigma_1\left(\frac{V}{V_1}\varepsilon - \frac{V_2}{V_1}\varepsilon_2(\sigma_2)\right).$$

Although the concavity of σ is preserved in the limit $V \to \infty$, strict concavity, which is needed to guarantee that the maximum is attained at only one point, may break down. A lack of strict concavity means that there is a phase transition, and will be examined in detail later. If, however, $\sigma_i(\varepsilon_i)$ are strictly concave and continuously differentiable, then the result of Corollary 3 can be improved upon and the additivity of the entropies demonstrated.

2.3.12 Equilibrium Condition

Let $\sigma_i(\varepsilon_i) = \lim_{V_i\to\infty}(1/V_i)\ln \operatorname{Tr}\Theta(V_i\varepsilon_i - H_i)$ *be strictly concave and continuously differentiable,* $\lim_{\varepsilon\to 0}\sigma'(\varepsilon) = \infty$ *and* $\lim_{V\to\infty} V_i/V \equiv \alpha_i$, $\alpha_1 + \alpha_2 = 1$. *Then*

$$\lim_{V\to\infty} \frac{1}{V} \ln \operatorname{Tr}\Theta(V\varepsilon - H_1 - H_2) \equiv \sigma(\varepsilon) = \alpha_1\sigma_1(\varepsilon_1) + \alpha_2\sigma_2(\varepsilon_2),$$

where ε_i *are determined uniquely by*

$$\alpha_1\varepsilon_1 + \alpha_2\varepsilon_2 = \varepsilon, \qquad \frac{\partial}{\partial\varepsilon_1}\sigma_1(\varepsilon_1) = \frac{\partial}{\partial\varepsilon_2}\sigma_2(\varepsilon_2).$$

2.3.13 Remarks

1. The energy densities can equally well be regarded as functions of the entropy densities, which reformulates the equilibrium condition as

$$\frac{\partial}{\partial \sigma_1}\varepsilon_1(\sigma_1) = \frac{\partial}{\partial \sigma_2}\varepsilon_2(\sigma_2) \quad \text{and} \quad \alpha_1\sigma_1 + \alpha_2\sigma_2 = \sigma.$$

2. Convexity of $\varepsilon(\sigma)$ is equivalent to concavity of $\sigma(\varepsilon)$, which is equivalent to the number of states below E_m not increasing faster than exponentially with the energy. This is not a general property of quantum-mechanical systems, and has to be checked in individual cases. A simple counterexample is the hydrogen atom, for which $E_n \sim -1/n^2$ $\exp(S(E_n)) \sim n^3$, where n is the principal quantum number, and therefore

$$E \sim -\exp\left(-\tfrac{2}{3}S\right), \quad \tfrac{\partial E}{\partial S} \sim \tfrac{2}{3}\exp\left(-\tfrac{2}{3}S\right) > 0,$$
$$\tfrac{\partial^2 E}{\partial S^2} \sim -\tfrac{4}{9}\exp\left(-\tfrac{2}{3}S\right) < 0.$$

   In such cases there may be many solutions of the equilibrium condition (see Figure 2.5).
3. Condition (2.3.12) implies that the energy is apportioned between the two systems so as to maximize the total entropy. From the point of view of $\varepsilon(\sigma)$ this means distributing entropy so as to minimize the total energy. As a consequence, the subadditivity inequality (2.3.5) becomes an equality in the limit $V \to \infty$.
4. If $\varepsilon_i(\sigma) \in \mathbb{C}^2$, then at the minimum, $\varepsilon_1''/\alpha_1 + \varepsilon_2''/\alpha_2 \geq 0$, where $\varepsilon'' = \partial^2\varepsilon/\partial\sigma^2$. Then by Problem 4, at the minimum, $1/\varepsilon'' = \alpha_1/\varepsilon_1'' + \alpha_2/\varepsilon_2''$.

If the total system consists of a system immersed in a thermal reservoir, then the system of interest is not affected by the fine details of the reservoir, but only by $\partial\sigma_2/\partial\varepsilon_2$, which not only determines $\partial\sigma_1/\partial\varepsilon_1$, but also equals $\partial\sigma/\partial\varepsilon$, because

$$\frac{d}{d\varepsilon}\left(\alpha_1\sigma_1(\varepsilon_1(\varepsilon)) + \alpha_2\sigma_2\left(\frac{\varepsilon}{\alpha_2} - \frac{\alpha_1}{\alpha_2}\varepsilon_1(\varepsilon)\right)\right)$$
$$= \sigma_2'(\varepsilon_2(\varepsilon)) + \alpha_1\frac{d\varepsilon_1}{d\varepsilon}(\sigma_1'(\varepsilon_1(\varepsilon)) - \sigma_2'(\varepsilon_2(\varepsilon))),$$

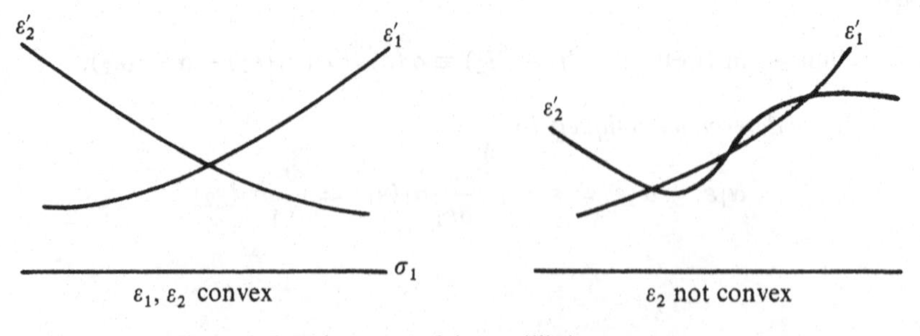

Figure 2.5. Uniqueness of the equilibrium temperature.

where

$$\varepsilon_2(\varepsilon) \equiv \frac{\varepsilon}{\alpha_2} - \frac{\alpha_1}{\alpha_2}\varepsilon_1(\varepsilon),$$

and the latter term vanishes because of (2.3.12). This is the justification for

2.3.14 Definition

The **temperature** is

$$T = \frac{\partial \varepsilon}{\partial \sigma}.$$

2.3.15 Remarks

1. The temperature has the dimension of energy in units where Boltzmann's constant k is set to 1.
2. The temperature is always positive with the microcanonical ρ (2.3.1), but ρ_δ gives the spin system of (2.3.9;6) a negative temperature at $E > 0$.
3. The concavity of σ means that the specific heat at constant volume,

$$V^{-1}C_N \equiv c_V = \frac{d\varepsilon}{dT} = \frac{d\varepsilon}{d\sigma}\left(\frac{dT}{d\sigma}\right)^{-1} = \frac{T}{d^2\varepsilon/d\sigma^2}$$

is positive. In particular, by Remark (2.3.13; 4), the heat capacity (at constant volume) $C_V = V \cdot 1/\varepsilon''$ of the total system is the sum of the heat capacities $V_i \cdot 1/\varepsilon_i''$ of the subsystems. The condition of stability $\varepsilon_1''/\alpha_1 + \varepsilon_2''/\alpha_2 \geq 0$ implies that two systems of negative specific heat can not coexist in equilibrium. Heat transferred from the hotter system to the colder one would make the hot one hotter and the cold one colder. Large temperature fluctuations would arise, making the situation unstable. If only subsystem 1 has negative specific heat, while that of subsystem 2 is positive, then the heat capacities must satisfy $|C_1| > C_2$: The transfer of heat from 1 to 2 would warm subsystem 1 less than 2, so 2 would immediately cool off by transferring heat back to 1, making the temperature equilibrium between the subsystems stable. This means that the temperature of a system of negative specific heat should be taken with a small thermometer, and never with a large thermal reservoir.

Now allow the wall between the subsystems to be slowly movable. The energy as a function of V acts as a potential energy for the wall, just as the electron energy acted as the potential for the atomic nuclei in the Born-Oppenheimer approximation in volume III. Stable equilibrium occurs when the total volume V is apportioned so as to minimize the energy. Let $V_2 = V - V_1$, and look for

$$E(S, V, N_1 + N_2) = \inf_{\substack{0 \leq S_1 \leq S \\ 0 \leq V_1 \leq V}} (E_1(S_1, V_1, N_1) + E_2(S - S_1, V - V_1, N_2)). \quad (2.8)$$

In the cases of interest here, E depends differentiably on V even for finite systems, and $E \to \infty$ if $V \to 0$. Hence the infimum is attained within the interval $0 < V_1 < V_2$, and is determined by the

2.3.16 Equilibrium Condition

For E of (2.8), the equilibrium volume V, satisfies

$$\frac{\partial E_1}{\partial V_1} = \frac{\partial E_2}{\partial V_2}\Big|_{V_2 = V - V_1}.$$

2.3.17 Remarks

1. Because the energy is monotonic (2.3.4), with the boundary conditions $\psi|_{\partial V} = 0$, it follows that $\partial E/\partial V < 0$, and so (2.3.16) definitely has a solution V_1. At that minimum,

$$\frac{\partial E}{\partial V} = \frac{\partial E_1}{\partial V_1} = \frac{\partial E_2}{\partial V_2}\Big|_{V_2 = V - V_1},$$

   and

$$\frac{\partial^2 E_1}{\partial V_1^2} + \frac{\partial^2 E_2}{\partial V_2^2} \geq 0, \quad \left(\frac{\partial^2 E}{\partial V^2}\right)^{-1} = \left(\frac{\partial^2 E_1}{\partial V_1^2}\right)^{-1} + \left(\frac{\partial^2 E_2}{\partial V_2^2}\right)^{-1}.$$

2. With other boundary conditions it may not be true that $\partial E/\partial V < 0$. For example, if a hydrogen atom is confined to a sphere on the surface of which $d\psi|_{\partial V} = 0$, then $E = E_\infty - \alpha V^{-1/3}$, so $\partial E/\partial V > 0$. This kind of boundary condition can be approximately realized physically with a very strong δ' potential. The lesson of this is that it is necessary to verify the hope that in infinite systems the pressure (see (2.3.21)) satisfies $P \equiv -\partial E/\partial V \geq 0$. This does not guarantee that $\partial^2 E/\partial V^2 \geq 0$ even with the boundary condition $\psi|_{\partial V} = 0$, which makes the proof of the convexity of $\varepsilon(\sigma, \rho)$ all the more important for real matter.
3. Since $\partial E/\partial V|_S = -\partial E/\partial S|_V \partial S/\partial V|_E$, another interpretation of (2.3.16) is that the condition $\partial(S_1(E_1, V_1) + S_2(E_2, V - V_1))/\partial V_1 = 0$ determines V_1; that is, the volumes arrange themselves to maximize the total entropy.

   Analogous to (2.3.16) is

2.3.18 Definition

The **pressure** is $P \equiv -\partial E/\partial V$. In the limit $V \to \infty$ it becomes

$$P = -\varepsilon + \rho\frac{\partial \varepsilon}{\partial \rho} + \sigma\frac{\partial \varepsilon}{\partial \sigma} = T\left(\sigma - \varepsilon\frac{\partial \sigma}{\partial \varepsilon} - \rho\frac{\partial \sigma}{\partial \rho}\right).$$

2.3.19 Remarks

1. For realistic systems it can be shown how the pressure defined in (2.3.21) arises from the forces exerted by the system on the wall [9].
2. The equilibrium condition states that the pressures of the two subsystems are equal, with the same value as the total system has.
3. Remark (2.3.17;1) implies for the compressibility

$$\kappa = -\left[V\frac{\partial P}{\partial V}\right]^{-1} \overset{V\to\infty}{\longrightarrow} \left[\rho^2\frac{\partial^2\varepsilon}{\partial\rho^2} + 2\rho\sigma\frac{\partial^2\varepsilon}{\partial\rho\partial\sigma} + \sigma^2\frac{\partial^2\varepsilon}{\partial\sigma^2}\right]^{-1}$$

   that

$$\kappa = \frac{V_1}{V}\kappa_1 + \frac{V_2}{V}\kappa_2.$$

4. For the systems to be stable against displacements of their interface, their volumes and compressibilities must be related by $(\kappa_1 V_1)^{-1} + (\kappa_2 V_2)^{-1} \geq 0$. For reasons like those of (2.3.15;3) it is not possible for two systems of negative compressibility to coexist, because the pressure of one system would increase with its volume and force that of the other one down. If only subsystem 1 has negative compressibility, then a necessary condition for stable equilibrium is $V_1 \geq V_2 \cdot \kappa_2/|\kappa_1|$. The increase of pressure in subsystem 1 when it expands is then less than that of 2 when it contracts. If V_1 is large enough in comparison with V_2, then subsystem 2 undergoes a large relative compression and exerts more pressure back on 1 than 1 exerts on 2. The volumes adjust in the other direction and stable equilibrium is established.

Consider finally what happens to the particle configuration if the subsystems can exchange particles to maximize the entropy. Formally, this means that the Hilbert space is

$$\mathcal{H} = \bigoplus_{N_1=1}^{N} \mathcal{H}_{N_1,V_1} \otimes \mathcal{H}_{N_2,V_2},$$

and the quantity to be calculated is

$$\operatorname{Tr}\Theta(E - H) = \sum_{N_1=0}^{N} \exp(S(N_1))\exp(S(N - N_1)). \tag{2.9}$$

In the limit $V \to \infty$, $N \to \infty$, $V_i/V \to \alpha_i$, $N_i/V_i \to \rho_i$, if S is concave in N, then arguments like those made earlier yield

$$\sigma(\rho) = \sup_{\alpha_1\rho_1+\alpha_2\rho_2=\rho} (\alpha_1\sigma_1(\rho_1) + \alpha_2\sigma_2(\rho_2)). \tag{2.10}$$

If the functions $\sigma_i(\rho_i)$ are nice, we obtain the

2.3.20 Equilibrium Condition

Let $\sigma_i(\rho_i)$ be strictly concave and continuously differentiable. Then $\sigma(\rho) = \alpha_1\sigma_1(\rho_1) + \alpha_2\sigma_2(\rho_2)$, where ρ_i are determined uniquely by the conditions

$$\alpha_1\rho_1 + \alpha_2\rho_2 = \rho \quad \text{and} \quad \frac{\partial\sigma_1}{\partial\rho_1} = \frac{\partial\sigma_2}{\partial\rho_2}.$$

2.3.21 Remarks

1. For a given ε and a given ρ, the six variables $\varepsilon_i, \rho_i, \alpha_i$ satisfy the three equations $\alpha_1\varepsilon_1 + \alpha_2\varepsilon_2 = \varepsilon$, $\alpha_1\rho_1 + \alpha_2\rho_2 = \rho$, $\alpha_1 + \alpha_2 = 1$. The three variations δE, δV, and δN corresponding to the equilibrium conditions are not independent, because $S(E, V, N)$ is of the special form $V\sigma(E/V, N/V)$, and there is one equation too few to fix six variables. Suppose for simplicity that the two subsystems are identical, $\sigma_1 = \sigma_2 = \sigma$; then because of the concavity, the maximum of $\alpha_1\sigma(\varepsilon_1, \rho_1) + \alpha_2\sigma(\varepsilon_2, \rho_2)$ is assumed when $\varepsilon_1 = \varepsilon_2 = \varepsilon$, $\rho_1 = \rho_2 = \rho$, and $\alpha_1 = 1 - \alpha_2$ is not determined by (2.3.20) and can be specified arbitrarily. Equality of the temperatures and the chemical potentials (see (2.3.22)) suffices to guarantee that the pressures are equal. After the onset of equilibrium, the wall allowing the exchange of energy and particles no longer exerts any force, and can be placed anywhere.
2. It is still possible to minimize the energy instead of maximizing the entropy. But this does not furnish a new stability condition, since if $\partial\varepsilon/\partial\sigma > 0$ the concavity of $(\varepsilon, \rho) \to \sigma(\varepsilon, \rho)$ is equivalent to the convexity of $(\sigma, \rho) \to \varepsilon(\sigma, \rho)$ (Problem 2). Besides $c_V > 0$ and $\kappa > 0$, this requires that

$$\frac{\partial^2 E}{\partial S^2}\frac{\partial^2 E}{\partial V^2} > \left(\frac{\partial^2 E}{\partial S \partial V}\right)^2,$$

or, in terms of the adiabatic expansivity

$$\alpha = \frac{1}{V}\frac{\partial V}{\partial T}\Big|_S, \quad \alpha^2 > c_V\kappa/T.$$

This amounts physically to the requirement of stability under a simultaneous change in the entropy and volume, related by

$$\delta S \sim \frac{\partial^2 E}{\partial V^2}, \quad \delta V \sim -\frac{\partial^2 E}{\partial S \partial V}.$$

The equilibrium condition (2.3.20) requires the chemical potentials of the subsystems to be equal, if they are defined as with (2.3.21; 2) by minimizing the energy:

2.3.22 Definition

The **chemical potential** is

$$\mu = \frac{\partial \varepsilon}{\partial \rho} = - \left.\frac{\partial \varepsilon}{\partial \sigma}\right|_{\rho} \left.\frac{\partial \sigma}{\partial \rho}\right|_{\varepsilon} .$$

2.3.23 Remarks

1. The intuitive meaning of the temperature is the amount of energy it would take to raise the system from the quantum number n to $en (e = 2.718 \cdots)$. Analogously, the chemical potential is the energy increase when a particle is added to the system without changing V or S.
2. Although T and P are always positive with the assumptions and boundary conditions that have been postulated, μ can in general have either sign. Because the density of states increases with N, the e^Sth eigenvalue may decrease with N even if $H \geq 0$.

In phenomenological thermodynamics entropy increases if the energy, volume, or particle number increases, according to the relationship $TdS = dE + PdV - \mu dN$. As we have seen, some of these differentials are well defined only in the thermodynamic limit, and are then considered as intensive properties. For future convenience, we collect the

2.3.24 Interrelationships Among the Thermodynamic Properties

$$T = \frac{\partial \varepsilon}{\partial \sigma}, \quad \mu = \frac{\partial \varepsilon}{\partial \rho} = -T \frac{\partial \sigma}{\partial \rho},$$

$$P = -\varepsilon + \sigma \frac{\partial \varepsilon}{\partial \sigma} + \rho \frac{\partial \varepsilon}{\partial \rho} = T \left(\sigma - \varepsilon \frac{\partial \sigma}{\partial \varepsilon} - \rho \frac{\partial \sigma}{\partial \rho} \right),$$

$$c_V = T \left[\frac{\partial^2 \varepsilon}{\partial \sigma^2} \right]^{-1}, \quad \kappa = \left[\sigma^2 \frac{\partial^2 \varepsilon}{\partial \sigma^2} + 2\rho\sigma \frac{\partial^2 \varepsilon}{\partial \rho \partial \sigma} + \rho^2 \frac{\partial^2 \varepsilon}{\partial \rho^2} \right]^{-1} .$$

Gloss: The sense of the partial derivatives is that, of the two variables on which a function has been regarded as depending, the one not written explicitly is to be held fixed. In any doubtful case the fixed argument will be indicated explicitly.

2.3.25 Remark

Without knowledge of the Hamiltonian nothing can be said about the values the thermodynamic functions can assume. In (2.3.9; 6) there was an example in which $\varepsilon(\sigma)$ was even bounded above. If the function $\varepsilon(\sigma)$ is convex and asymptotically linear, then there is a maximum temperature. This is quite possibly the case realized in Nature, and $T_{\max} = 140$ MeV. In a model to be investigated shortly (2.3.27; 2), the function $\varepsilon(\sigma)$ has a kink, so T skips over certain values. It depends on the system whether the minimum entropy σ_0 defined in (2.3.9; 4) equals zero as postulated in the third law of thermodynamics. For instance, with a system consisting of N spins without energy $\otimes$ a system with entropy $N\sigma$, the total entropy divided by N equals $\sigma + \ln 2$, and when $\sigma \to 0$ the total entropy is the $\ln 2$ left over. It is true that the ground state of this system is degenerate, but it is also easy to find examples with nondegenerate ground states for which the third law fails, simply by taking the previous Hamiltonian $\oplus$ a one-dimensional system with a lower energy level. The resulting ground state is simple, but that has no effect on what happens as $N \to \infty$.

It has been seen that the concavity of the function $\sigma(\varepsilon, \rho)$ is at the root of thermodynamic stability. Concavity is jeopardized when σ is maximized with respect to all of its parameters – the supremum of a set of concave functions is not necessarily concave, in contrast to the infimum. However, there is a useful

2.3.26 Lemma on the Envelope of a Set of Concave Functions

If $\sigma(\varepsilon, \alpha)$ is jointly concave in ε and α, then $\bar{\sigma}(\varepsilon) = \sup_\alpha \sigma(\varepsilon, \alpha)$ is concave in ε.

Picture of the Proof

Think of the silhouette of a concave mountain slope and of a mountain with hollows.

Formal Proof if $\sigma(\varepsilon, \alpha) \in C^2$

With this assumption, the maximum is attained at a point $\alpha(\varepsilon), \bar{\sigma}(\varepsilon) = \sigma(\varepsilon, \alpha(\varepsilon))$, and

$$\sigma_{,\alpha}(\varepsilon, \alpha(\varepsilon)) = 0 \Rightarrow \sigma_{,\alpha\varepsilon} + \frac{d\alpha(\varepsilon)}{d\varepsilon}\sigma_{,\alpha\alpha} = 0.$$

Then

$$\frac{d^2\bar{\sigma}}{d\varepsilon^2} = \sigma_{,\varepsilon\varepsilon} + \sigma_{\varepsilon\alpha}\frac{d\alpha(\varepsilon)}{d\varepsilon} = \frac{\sigma_{,\varepsilon\varepsilon}\sigma_{,\alpha\alpha} - (\sigma_{,\varepsilon\alpha})^2}{\sigma_{,\alpha\alpha}}.$$

Since $\sigma_{,\alpha\alpha} \le 0$ and $\sigma_{,\varepsilon\varepsilon}\sigma_{,\alpha\alpha} - (\sigma_{,\varepsilon\alpha})^2 \ge 0$, $d^2\bar{\sigma}/d\varepsilon^2 \le 0$. If $\sigma_{,\alpha\alpha} = 0$, it follows that $\sigma_{,\alpha\varepsilon}(\varepsilon, \alpha(\varepsilon)) = 0$, and therefore $\bar{\sigma}_{,\varepsilon\varepsilon} = \sigma_{,\varepsilon\varepsilon} \le 0$. (For the proof without the assumption that $\sigma(\varepsilon, \alpha) \in C^2$, see Problem 3.)

If the entropy is maximized with respect to parameters in the absence of joint concavity, then thermodynamic stability may be lost, and it will be necessary to reconsider the foregoing assumptions.

2.3.27 Examples

1. Model of a star

Consider N classical particles in a container V and attracting each other pairwise only within some $V_0 \subset V$. Suppose the potentials are constant in V_0 and $\sim N^{-1}$, to ensure that E be extensive.

$$H_N = \sum_{i=1}^{N} |\mathbf{p}_i|^2 - \frac{1}{N}\sum_{i,j=1}^{N} \chi_{V_0}(\mathbf{x}_i)\chi_{V_0}(\mathbf{x}_j),$$

$$\chi_{V_0}(\mathbf{x}) = \begin{cases} 1 & \text{for } \mathbf{x} \in V_0 \\ 0 & \text{otherwise} \end{cases}.$$

With indistinguishable particles, the volume of phase space below the energy surface,

$$\exp(S(E, V, N))$$
$$= \frac{1}{N!}\int d^{3N}p\, d^{3N}x\,\Theta\left(E - \sum_{i=1}^{N}|\mathbf{p}_i|^2 + \frac{1}{N}\sum_{i,j=1}^{N}\chi_{V_0}(\mathbf{x}_i)\chi_{V_0}(\mathbf{x}_j)\right)$$
$$= \frac{\pi^{3N/2}}{N!(3N/2)!}\int_{(\cdots)>0} d^{3N}x\left(E + \frac{1}{N}\sum_{i,j=1}^{N}\chi_{V_0}(\mathbf{x}_i)\chi_{V_0}(\mathbf{x}_j)\right)^{3N/2},$$

can be calculated exactly, because the integrand is piecewise constant. Let N_0 be the number of the $\mathbf{x}_i$ in V_0. Then

$$\exp(S) = \frac{V_0^N \pi^{3N/2}}{(3N/2)!}\sum_{-NE\le N_0^2\le N^2}\left(\frac{V}{V_0}-1\right)^{N-N_0}\frac{(E+N_0^2/N)^{3N/2}}{N_0!(N-N_0)!}$$
$$\equiv \sum_{N_0=1}^{N}\exp(S(E, V, N; N_0)).$$

Only the dependence on E matters, so let $E = \varepsilon \cdot N$, $\rho = N/V = 1$, $N_0/N \equiv \alpha$, $(\max(0, -\varepsilon))^{1/2} \le \alpha \le 1$. Then it remains to evaluate

$$\sigma(\varepsilon) = \sup_{\alpha}\lim_{N\to\infty}\frac{1}{N}S(N\varepsilon, N, N; \alpha N) \equiv \sup_{\alpha}\sigma(\varepsilon, \alpha),$$

and with the help of Stirling's formula,

$$\sigma(\varepsilon, \alpha) = \tfrac{3}{2}\ln(\varepsilon+\alpha^2) - \alpha\ln\alpha - (1-\alpha)\ln(1-\alpha) + F(1-\alpha) + \text{constant},$$
$$F = \ln\left(\tfrac{V}{V_0} - 1\right). \tag{2.11}$$

A calculation of the derivatives yields

$$\sigma_{,\varepsilon} = \frac{3}{2(\varepsilon+\alpha^2)}, \qquad \sigma_{,\alpha} = \frac{3\alpha}{\varepsilon+\alpha^2} + \ln\left(\frac{1}{\alpha}-1\right) - F,$$

$$\sigma_{,\varepsilon\varepsilon} = -\frac{3}{2(\varepsilon+\alpha^2)^2}, \quad \sigma_{,\varepsilon\alpha} = -\frac{3\alpha}{(\varepsilon+\alpha^2)^2}, \quad \sigma_{,\alpha\alpha} = \frac{3\varepsilon - 3\alpha^2}{(\varepsilon+\alpha^2)^2} - \frac{1}{\alpha(1-\alpha)}.$$

The maximum is achieved on the curve

$$\varepsilon(\alpha) = -\alpha^2 + \frac{3\alpha}{F - \ln(1/\alpha - 1)},$$

and the ranges of values of the variables are such that $\varepsilon + \alpha^2 \geq 0$, so only the branch of $F > \ln(1/\alpha - 1)$ comes into consideration. Because

$$\sigma_{,\alpha\alpha} = -\frac{(\varepsilon - \varepsilon_1(\alpha))(\varepsilon - \varepsilon_2(\alpha))}{(\varepsilon + \alpha^2)^2 \alpha(1-\alpha)},$$

$$\varepsilon_{1,2} = \frac{3\alpha}{2}\left(1 - \frac{5}{3}\alpha \pm \sqrt{1-\alpha}\sqrt{1 - 11\alpha/3}\right),$$

$\sigma(\varepsilon, \alpha)$ is concave in α except when $\varepsilon_2 < \varepsilon < \varepsilon_1$. The sign of $d\varepsilon/d\alpha = -\sigma_{,\alpha\alpha}/\sigma_{,\alpha\varepsilon}$, changes in the interval $\varepsilon_2 < \varepsilon < \varepsilon_1$, so three values of α belong to a single ε, and the maximum needed is the greater of the two. Joint concavity requires that

$$\sigma_{,\varepsilon\varepsilon}\sigma_{,\alpha\alpha} - (\sigma_{,\alpha\varepsilon})^2 = \frac{3(\varepsilon - 3\alpha + 4\alpha^2)}{2(\varepsilon + \alpha^2)^3 \alpha(1-\alpha)} \geq 0$$

and implies $\varepsilon \geq 3\alpha - 4\alpha^2$. If $\varepsilon(\alpha)$ lies in this range of values, then the system has positive specific heat, and otherwise not (see Figure 2.6). Indeed,

$$\frac{3}{2}T = \varepsilon + \alpha^2 = \frac{3\alpha}{F - \ln(1/\alpha - 1)}$$

behaves as a function of ε as shown in Figure 2.7. The physical significance is that if energy is removed, the temperature falls until a certain fraction of the particles reside in V_0, which causes the system to start heating back up. If most of the particles are eventually in V_0, then they behave normally again. The system can be thought of as a normal system with

$$\sigma(\varepsilon, \rho) = \rho\left(\frac{3}{2}\ln\varepsilon - \frac{5}{2}\ln\rho\right)$$

put into contact with a peculiar system with

$$\sigma(\varepsilon, \rho) = \rho\left(\frac{3}{2}\ln(\varepsilon + \rho^2) - \frac{5}{2}\ln\rho\right) - F\rho.$$

If the energy is apportioned between them according to

$$\sigma(\varepsilon, \alpha) = \sup_{\varepsilon_1}\left(\frac{3}{2}(\alpha\ln(\varepsilon_1 + \alpha^2) + (1-\alpha)\ln(\varepsilon - \varepsilon_1))\right) - \alpha F$$
$$-\frac{5}{2}(\alpha\ln\alpha + (1-\alpha)\ln(1-\alpha)),$$

then the entropy becomes exactly that of (2.11).

2. Model of a Ferromagnet

This problem is quantum-mechanical, but its analysis soon begins to resemble that of Example 1, for which reason we shall boldly plunge on to the estimates without

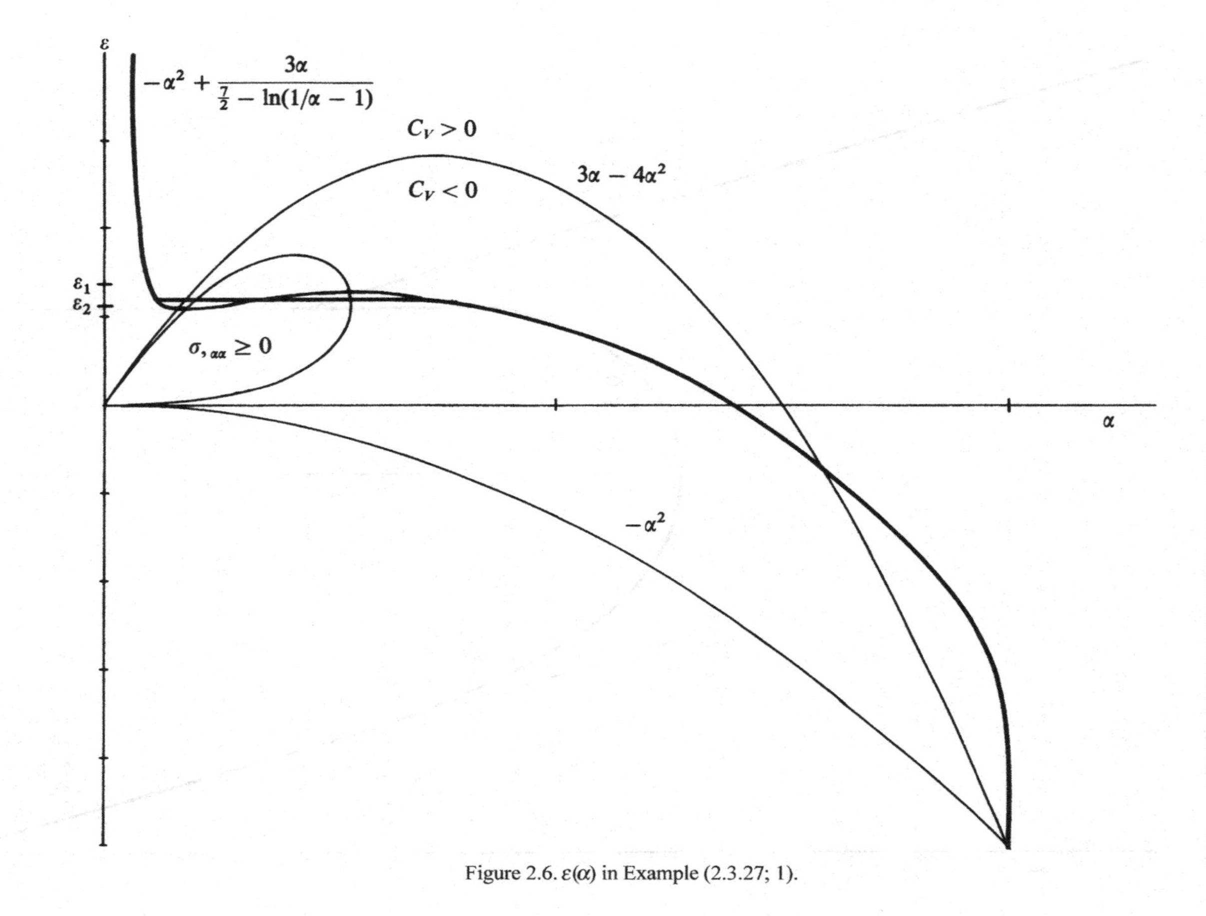

Figure 2.6. $\varepsilon(\alpha)$ in Example (2.3.27; 1).

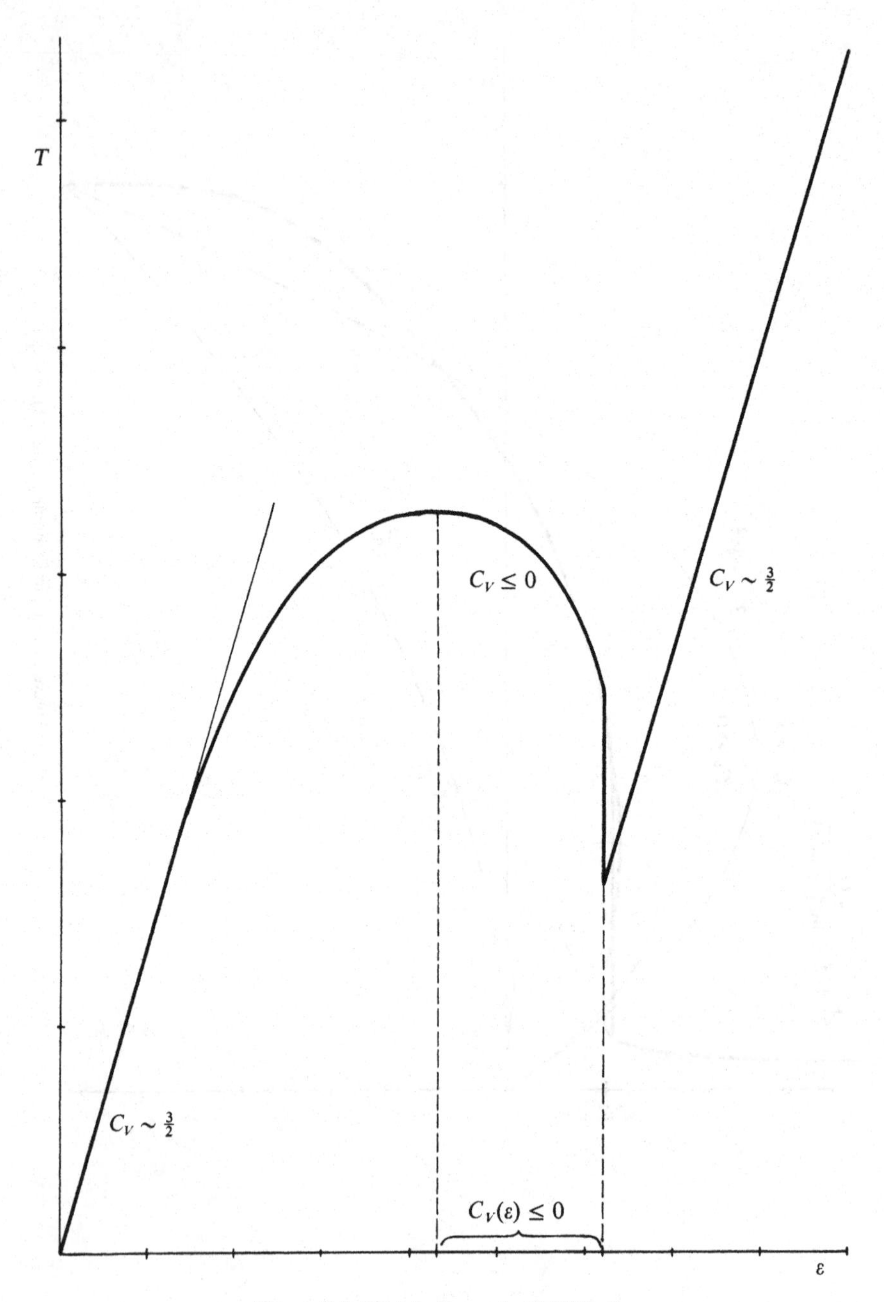

Figure 2.7. $T(\varepsilon)$ in Example (2.3.27; 1).

wasting time about epsilontic details. The Hamiltonian of (1.2) is modified to

$$H = B\sum_{j=1}^{N}\sigma_j^{(z)} - \frac{1}{N}\sum_{i,j=1}^{N}\sigma_i\cdot\sigma_j,$$

which contains a magnetic field in the z-direction and a spin-spin interaction favoring parallel spins. The strength of the interaction is the same for all pairs and must be $\sim 1/N$ for H to be $\sim N$. The mean magnetization $\mathbf{M}_N$ can be introduced as before, $H/N = BM_N^{(z)} - \mathbf{M}_N\cdot\mathbf{M}_N$, and it was shown in (III,§ 3.2) that the two parts of H can be diagonalized simultaneously. If the eigenvalues of $M_N^{(z)}$ are m_z and those of $\mathbf{M}_N\cdot\mathbf{M}_N$ are $m(m+2/N)$, $0\le m\le 1$, $-m\le m_z\le m$, then m_z and m are always multiples of $1/N$ spaced $2/N$ apart. To calculate Tr $\Theta(E-H)$ it is also necessary to find the multiplicities of the eigenvalues: If $m=1$, then all spins must be parallel, and for one of these vectors, $m_z=1$. There are N vectors with $m_z=1-2/N$, corresponding to the N possible ways to flip one spin. One of those possibilities has $m=1$ (apply M^- to the previous vector) and the others must have $m=1-2/N$. The general rule is that of the $\binom{N}{r}$ vectors with $m_z=1-2r/N$, $\binom{N}{r-1}$ of them have $m>1-2r/N$, and the remaining

$$\binom{N}{r}-\binom{N}{r-1}=\frac{N!(N-2r+1)}{r!(N-r+1)!}$$

have $m=1-2r/N$. This means that the number of vectors with the eigenvalues (m,m_z) is

$$\frac{N!(Nm+1)}{((N/2)(1-m))!((N/2)(1+m)+1)!}\sim\sqrt{\frac{2}{\pi(1-m^2)N}}\frac{2m}{m+1}$$
$$\times\exp\left\{N\left[\ln 2-\left(\frac{1+m}{2}\right)\ln(1+m)-\left(\frac{1-m}{2}\right)\ln(1-m)\right]\right\}.$$

The last step used Stirling's formula $x!\sim(x/e)^x\sqrt{2\pi x}$, which is justified only for $m<1$ even when $N\gg 1$, but in the limit being taken the contributions from the boundaries of the summation region are inconsequential. Since the integrand is a continuous function, as $N\to\infty$ the sum $\sum_{m=0}^{1}\sum_{m_z=-m}^{m}\cdots$ can be replaced with the integral $(N/2)^2\int_0^1 dm\int_{-m}^{m}dm_z\cdots$, and with $\varepsilon=E/N$ this leaves

$$\exp(S(\varepsilon)) = N^{3/2}\int_0^1\frac{dm}{m+1}\sqrt{\frac{m^2}{2\pi(1-m^2)}}$$
$$\times\exp\left\{N\left[\ln 2-\left(\frac{1+m}{2}\right)\ln(1+m)-\left(\frac{1-m}{2}\right)\ln(1-m)\right]\right\}$$

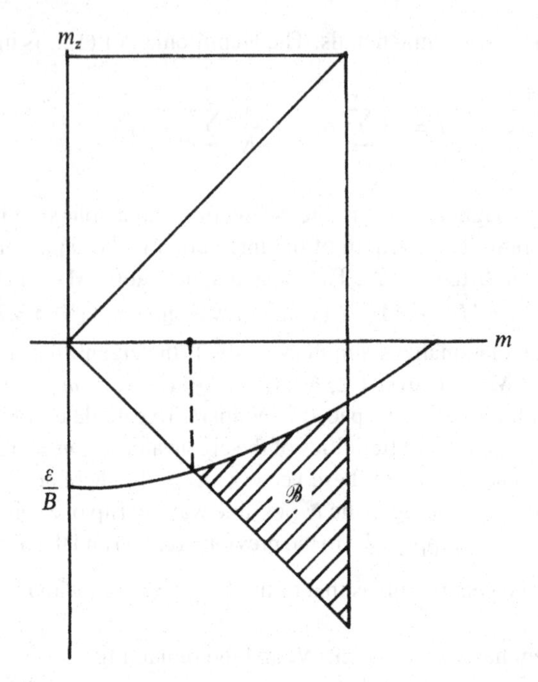

Figure 2.8. The region of integration in the $(m - m_z)$-plane.

$$\times \int_{-m}^{m} dm_z \Theta(\varepsilon + m^2 - Bm_z). \tag{2.12}$$

Therefore the domain $\mathcal{B}$ of integration is $\{(m, m_z) : 0 \leq m \leq 1, -m \leq m_z \leq m\} \cap \{(m, m_z) : m_z \leq (\varepsilon + m^2)/B\}$. The entropy S is obviously even in B, so we may restrict consideration to $B \geq 0$ (see Figure 2.8).

Since the exponential function decreases rapidly with m, the appropriate generalization of Lemma (2.3.10) makes $\sigma = \lim_{N\to\infty} S/N$ sensitive only to $m_0 \equiv \inf_{m,m_z \in \mathcal{B}} m$ (the exponent in (2.3.34) decreases monotonically in m):

$$m_0 = \Theta(-\varepsilon) \left(\sqrt{\frac{B^2}{4} - \varepsilon} - \frac{B}{2} \right),$$

$$\sigma = \ln 2 - \frac{1}{2} \Theta(-\varepsilon) \left[(1+m_0) \ln(1+m_0) + (1-m_0) \ln(1-m_0) \right], \tag{2.13}$$

if $\varepsilon \geq -1 - B$, and is otherwise 0. Since σ is concave but decreasing in m_0, the concavity in ε remains to be verified:

$$T^{-1} = \frac{d\sigma}{d\varepsilon} = \frac{\Theta(-\varepsilon)}{4(B^2/4 - \varepsilon)^{1/2}} \ln \frac{1+m_0}{1-m_0} \geq 0,$$

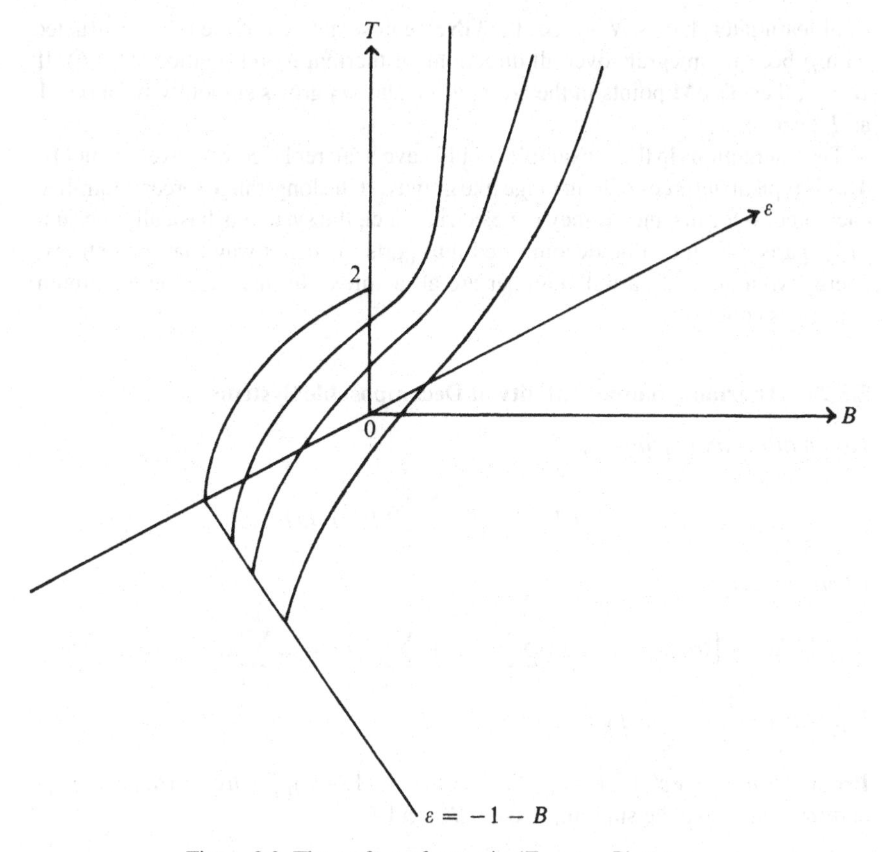

Figure 2.9. The surface of states in $(T - \varepsilon - B)$-space.

$$-\frac{1}{T^2 c_B} = \frac{d^2\sigma}{d\varepsilon^2} = -\frac{\Theta(-\varepsilon)}{8}$$
$$\times\left[-\left(\frac{B^2}{4} - \varepsilon\right)^{-3/2} \ln\frac{1+m_0}{1-m_0} + \frac{2}{(B^2/4 - \varepsilon)(1-m_0^2)}\right]. \quad (2.14)$$

In a lucky break, the positive term in the brackets $[\cdots]$ is always greater than the negative one, and c_B is always positive. If $-1 > -1-B \leq \varepsilon \leq 0$, then T increases continuously from 0 to ∞. The heat capacity c_B increases from 0 to a maximum value and then falls back to 0. If $B = 0$, then T reaches the value 2 for $\varepsilon = 0$, at which c_B has risen to $\frac{3}{2}$. Afterwards, T jumps up to ∞ and c_V falls back to 0 (see Figure 2.9).

Thus if $B = 0$ and $T < 2$, the thermal motion is no longer strong enough to counter the ordering tendency of H, and a spontaneous magnetization m_0 appears. As no direction is preferred, the thermal expectation value $|\operatorname{Tr} \rho\mathbf{M}|$ remains 0. We shall learn later that as $N \to \infty$, the GNS representations of the σ's constructed with ρ become integrals over all directions of thermal representations (1.4.6). If $B > 0$, then Tr $\rho\mathbf{M}$ points in the z-direction, and m_0 grows smoothly from 0 to 1 as T decreases.

The interactions in these examples could have been replaced with average fields. This is typical of forces of long range like gravity. If the long-range forces neutralize each other – for instance if they are electric – then the system is basically the sum of its parts, i.e., it can be decomposed into parts in such a way that the entropy, energy, volume, and particle number are all additive. In that case the maximum entropy is concave.

2.3.28 Thermodynamic Stability of Decomposable Systems

For an arbitrary function σ,

$$\bar{\sigma}(\varepsilon, \rho) \equiv \sup_n \sup_{K_n} \sum_{i=1}^n \alpha_i \sigma(\varepsilon_i, \rho_i),$$

where

$$K_n = \Big\{(\alpha_i), (\varepsilon_i), (\rho_i) : \sum_{i=1}^n \alpha_i = 1, \sum_{i=1}^n \alpha_i \varepsilon_i = \varepsilon, \sum_{i=1}^n \alpha_i \rho_i = \rho\Big\},$$

is jointly concave in its two variables.

Proof: Let $\varepsilon = \gamma\varepsilon' + (1-\gamma)\varepsilon''$, $\rho = \gamma\rho' + (1-\gamma)\rho''$. Divide (α_i) into (α_i') and (α_i''), and take the supremum over $K_{n'}$ and $K_{n''}$:

$$K_{n'} \equiv \Big\{(\alpha_i'), (\varepsilon_i'), (\rho_i') : \sum_{i=1}^{n'} \alpha_i' = \gamma, \sum_{i=1}^{n'} \alpha_i' \varepsilon_i' = \varepsilon', \sum_{i=1}^{n'} \alpha_i' \rho_i' = \rho'\Big\},$$

$$K_{n''} \equiv \Big\{(\alpha_i''), (\varepsilon_i''), (\rho_i'') : \sum_{i=1}^{n''} \alpha_i'' = 1-\gamma, \sum_{i=1}^{n''} \alpha_i'' \varepsilon_i'' = \varepsilon'', \sum_{i=1}^{n''} \alpha_i'' \rho_i'' = \rho''\Big\}.$$

Since this is only a particular division,

$$\begin{aligned}\bar{\sigma}(\varepsilon, \rho) &\geq \sup_{n', n''} \sup_{K_{n'}, K_{n''}} \left(\sum_i \alpha_i' \sigma(\varepsilon_i', \rho_i') + \sum_i \alpha_i'' \sigma(\varepsilon_i'', \rho_i'')\right) \\ &= \gamma\bar{\sigma}(\varepsilon', \rho') + (1-\gamma)\bar{\sigma}(\varepsilon'', \rho'').\end{aligned}$$

2.3.29 Remarks

1. The construction (2.3.28) gives the concave envelope of σ, but nothing guarantees that $\bar{\sigma}$ is strictly concave. If σ is linear, then $\bar{\sigma} = \sigma$, and σ is of the form of Example (2.3.27;1). The convex part of the curve gets bridged by a straight line, as shown in Figure 2.10.

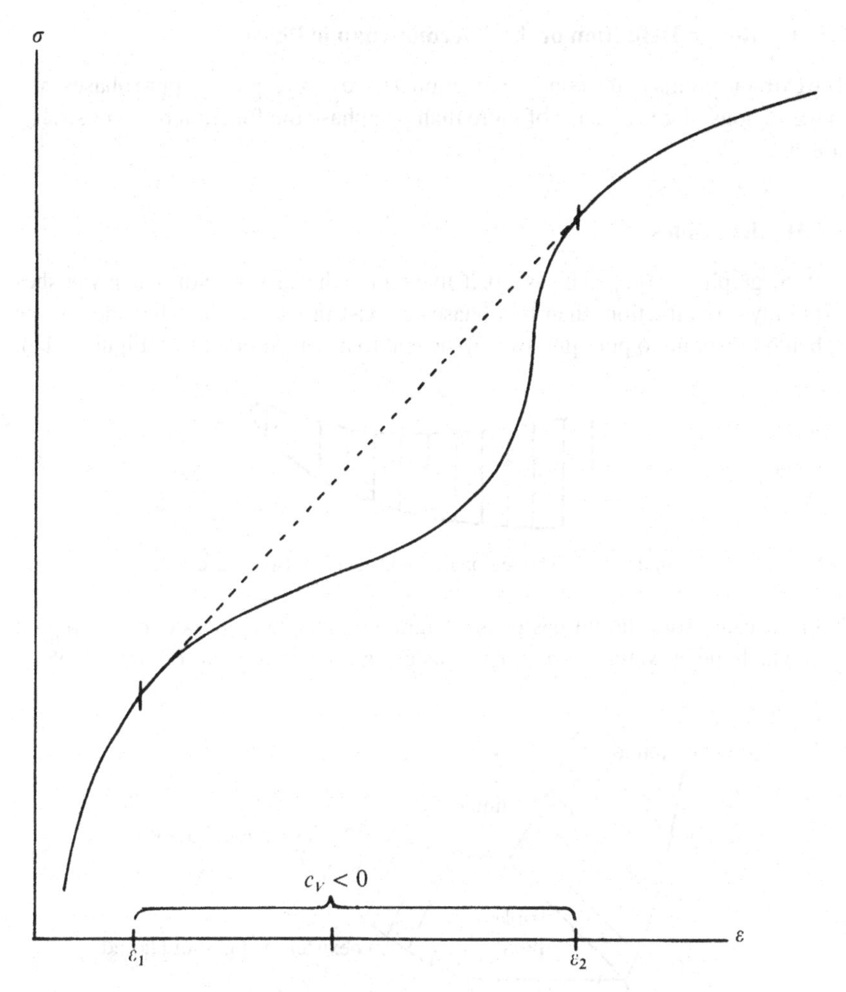

Figure 2.10. The region of negative specific heat.

2. In (2.3.32;1) $\sup_{\alpha_1 \varepsilon_1}$ was an expression of a different form and $\bar{\sigma}(\varepsilon)$ was not concave.

The function $\bar{\sigma}$ is simply $\alpha\sigma(\varepsilon_1) + (1 - \alpha)\sigma(\varepsilon_2)$ in the intervening region where $\varepsilon = \alpha\varepsilon_1 + (1 - \alpha)e_2$ for fixed ε_1 and ε_2. An interpretation is that the system consists of two phases in this region, having energies ε_1 and ε_2 and the temperature remains constant as the total energy varies, while the proportions of the phases present change. This suggests a

2.3.30 Rough Definition of the Thermodynamic Phases

The extreme points of the concave function $\sigma(\varepsilon, \rho)$ correspond to pure phases, and in the regions of coexistence of more than one phase the function σ is not strictly concave.

2.3.31 Examples

1. If the graph of $\sigma(\varepsilon, \rho)$ shows a belt-like region the curvature of which vanishes in only one direction, then two phases coexist in its interior. The sides of the belt correspond to pure phases and the end to a critical point (see Figure 2.11):

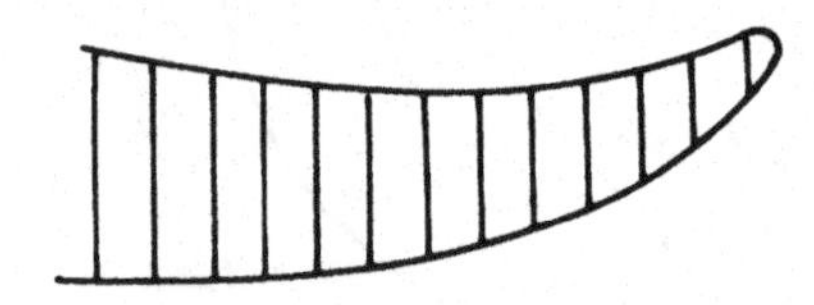

Figure 2.11. The region of coexistence of two phases.

2. In the usual solid-liquid-gas phase diagram, the triple point occurs in a region at which the curvature of $\sigma(\varepsilon, \rho)$ vanishes in both directions (Figure 2.12):

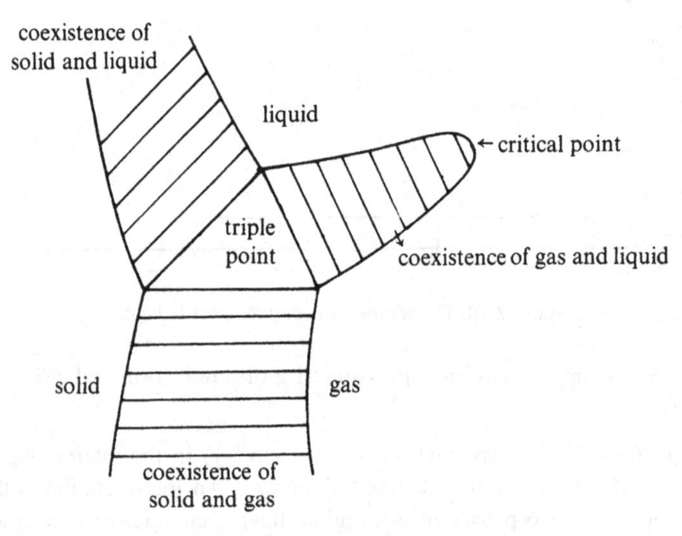

Figure 2.12. Regions of coexistence.

2.3.32 Remarks

1. The sum, in the sense of (2.3.28), of many copies of Example (2.3.27; 1) produces a concave $\bar{\sigma}$, since the convex part lies below the phase-transition line. Some concave pieces of the curve are also bridged over, and are known as metastable phases, which arise in superheated stars and supercooled gases. They have positive specific heats and are locally stable (see Figure 2.13):
2. Gibbs's phenomenological phase rule states that whenever a material has two coexisting phases, there is always a one-parameter family of coexisting phases described by $T(\alpha)$ and $\mu(\alpha)$. Three coexisting phases can only exist at discrete values of (T, μ). This is exactly what went on in (2.3.31; 1) and (2.3.31; 2), where the parts that are flat in one direction are two-dimensional, but is not a consequence of concavity alone; for instance the function $\sigma = -\varepsilon^p \rho^{-q}$, $p > q + 1 > 1$, has a straight line segment only if $\varepsilon = 0$, but is nonetheless concave in (ε, ρ).
3. A quadruple point of a substance would be a flat rectangle in the energy surface. The nonexistence of quadruple points does not follow from concavity, but amounts to the assumption that the flat pieces of the energy surface form a simplex. If they do not form a simplex, then the ratio of the phases in the mixture is not even necessarily determined by ε and ρ:

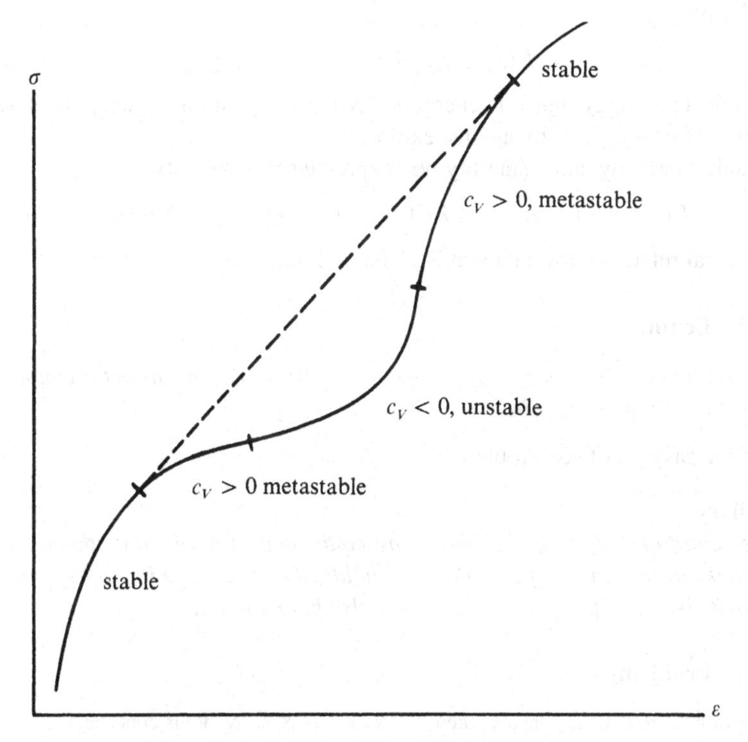

Figure 2.13. Stability of the regions of (2.3.32;1).

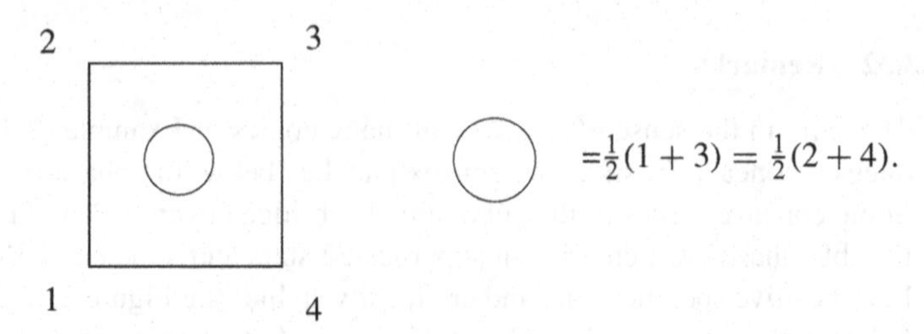

At this point we have no arguments that would show that quadruple points do not occur, and in fact it is easy to construct models with quadruple points by taking the sum of two independent systems each of which has a phase transition. We shall have to take the issue up anew in (3.2.12;2).

2.3.33 Stability Conditions

The first stability condition was the extensivity of E, S, V, N, that is the existence of $\varepsilon = \lim_{\lambda\to\infty} \frac{1}{\lambda} E(\lambda S, \lambda V, \lambda N)$. If it fails the system implodes. Mathematically it means that for $\lambda \to \infty$ ε becomes homogeneous function of first degree, i.e. a function $F : (R^+)^n \to R$ which satisfies

$$F(\lambda z) = \lambda F(z), \quad \lambda \in R^+, z \in (R^+)^n \tag{H}$$

subadditivity

$$F(z_1 + z_2) \leq F(z_1) + F(z_2) \tag{S}$$

means for the energy that it is energetically more favourable to keep the system together. Thus it is stability against explosion.

Finally thermodynamic stability was expressed by convexity

$$F(\lambda z_1 + (1-\lambda)z_2) \leq \lambda F(z_1) + (1-\lambda)F(z_2) \quad \lambda \in [0,1] \tag{C}$$

The logical relation between them is given by Landsbergs

2.3.34 Lemma

For a continuous function $F : (R^+)^n \to R^+$, $F(0) = 0$ each pair of the conditions (H), (S), (C) implies the third.

For the easy proof see Problem 5.

Corollary

If one condition holds the two others are equivalent, if it fails the others cannot hold both. In particular, if subadditivity holds, dynamic stability and extensivity are equivalent. If it fails one of the two is also bound to fail.

2.3.35 Problems

1. Show that if $\sigma(\varepsilon, \rho)$ is concave, then $(E, N, V) \to S(E, N, V)$ is concave.
2. Show that for $\varepsilon_{,\sigma} > 0$, $\sigma(\varepsilon, \rho)$ is concave iff $\varepsilon(\sigma, \rho)$ is convex.

3. Without assuming differentiability, show that if $\sigma(\varepsilon, \alpha)$ is concave, then $\bar{\sigma}(\varepsilon) = \sup_\alpha \sigma(\varepsilon, \alpha)$ is concave.

4. Prove the relationship $V/\varepsilon'' = V_1/\varepsilon_1'' + V_2/\varepsilon_2''$ of (2.3.15;4).

5. Prove (2.3.34).

2.3.36 Solutions

1. For simplicity assume that σ is twice differentiable. Then

$$D^2 S = \frac{1}{V} \begin{vmatrix} \sigma_{,\varepsilon\varepsilon} & \sigma_{,\varepsilon\rho} & -\varepsilon\sigma_{,\varepsilon\varepsilon} - \rho\sigma_{,\varepsilon\rho} \\ \sigma_{,\varepsilon\rho} & \sigma_{,\rho\rho} & -\varepsilon\sigma_{,\varepsilon\rho} - \rho\sigma_{,\rho\rho} \\ -\varepsilon\sigma_{,\varepsilon\varepsilon} - \rho\sigma_{,\varepsilon\rho} & -\varepsilon\sigma_{,\varepsilon\rho} - \rho\sigma_{,\rho\rho} & \varepsilon^2\rho_{,\varepsilon\varepsilon} + 2\varepsilon\rho\sigma_{,\varepsilon\rho} + \rho^2\sigma_{,\rho\rho} \end{vmatrix}.$$

Observe that the concavity of S is equivalent to $D^2 S \leq 0$, which means that $D^2\sigma \leq 0$ and $\det D^2 S \leq 0$. However, $\det D^2 S = 0$ because the mapping $\lambda \to S(\lambda E, \lambda N, \lambda V)$ is affine.

2. The function σ is concave iff the concave hull $\bar{\Gamma} = \{(x, y, z) = \sum \lambda_i (x_i, y_i, z_i), (x_i, y_i, z_i) \in \Gamma, 0 \leq \lambda_i \leq 1, \sum_i \lambda_i = 1\}$ of the graph $\Gamma = \{(x, \varepsilon, \rho) : x = \sigma(\varepsilon, \rho)\}$ lies completely below Γ. However, looked at from the other side, Γ is also the graph of the inverse function $\varepsilon(\sigma, \rho)$, except that "below" becomes "above" and vice versa.

3. Let $\varepsilon = \gamma\varepsilon_1 + (1-\gamma)\varepsilon_2$, and choose $\alpha_{1,2}$ so that $\sup_\alpha \sigma(\varepsilon_i, \alpha) = \sigma(\varepsilon_i, \alpha_j)$, $i = 1, 2$, or at least comes arbitrarily close to equality,

$$\begin{aligned} \sup_\alpha \sigma(\varepsilon, \alpha) &\geq \sigma(\gamma\varepsilon_1 + (1-\gamma)\varepsilon_2, \gamma\alpha_1 + (1-\gamma)\alpha_2) \\ &\geq \gamma\sigma(\varepsilon_1, \alpha_1) + (1-\gamma)\sigma(\varepsilon_2, \alpha_2) \\ &= \gamma\bar{\sigma}(\varepsilon_1) + (1-\gamma)\bar{\sigma}(\varepsilon_2). \end{aligned}$$

4.

$$\varepsilon_1'(\sigma_1) = \varepsilon_2'\left(\frac{V\sigma - V_1\sigma_1}{V_2}\right) \Rightarrow \sigma_1'\varepsilon_1'' = \left(\frac{V}{V_2} - \frac{V_1}{V_2}\sigma_1'\right)\varepsilon_2'' \Rightarrow \sigma_1' = \frac{V}{V_2}\frac{\varepsilon_2''}{\varepsilon_1'' + \varepsilon_2'' V_1/V_2},$$

$$\varepsilon'(\sigma) = \varepsilon_1'(\sigma_1) \Rightarrow \varepsilon'' = \sigma_1'\varepsilon_1'' = \frac{V}{V_2}\varepsilon_2''\frac{\varepsilon_1''}{\varepsilon_1'' + \varepsilon_2'' V_1/V_2} \Rightarrow \frac{V}{\varepsilon''} = \frac{V_1}{\varepsilon_1''} + \frac{V_2}{\varepsilon_2''}.$$

5. $S \wedge H \Rightarrow C : F(\lambda z_1 + (1-\lambda)z_2) \overset{\leq}{\underset{S}{}} F(\lambda z_1) + F(1-\lambda z_2) \overset{=}{\underset{H}{}} \lambda F(z_1) + (1-\lambda)F(z_2)$.
$C \wedge H \Rightarrow S : F(\lambda z_1) + F((1-\lambda)z_2) \overset{=}{\underset{H}{}} \lambda F(z_1) + (1-\lambda)F(z_2)$
$\overset{\geq}{\underset{C}{}} F(\lambda z_1 + (1-\lambda)z_2)$.
$C \wedge S \Rightarrow H : F(\lambda x) \overset{\leq}{\underset{C}{}} \lambda F(x), \lambda \leq 1, F(Nz) \overset{\leq}{\underset{S}{}} NF(z), N \in Z^+$.
$\Rightarrow F(\lambda z) \geq \lambda F(z), \lambda \in 1/Z^+, \Rightarrow NF(x/N) = F(x) \Rightarrow NF(Mx) = F(NMx) = MF(Nx)$
$\Rightarrow F(\lambda z) = \lambda F(z), \lambda \in Q \Rightarrow F(\lambda z) = \lambda F(z), \lambda \in R^+$ by continuity.

2.4 The Canonical Ensemble

The Maxwell–Boltzmann distribution arises from the state of a system in contact with a thermal reservoir. If the system is large, this state is indistinguishable from that of the microcanonical ensemble.

In the preceding section it was shown that the entropy of two large subsystems without interaction is additive. The entropy was always defined with the microcanonical density matrix (2.3.1), but when the density matrix is restricted to a subsystem,

$$\rho_1 = \frac{\mathrm{Tr}_2\, \Theta(E - H_1 - H_2)}{\mathrm{Tr}\, \Theta(E - H_1 - H_2)} \equiv \exp(S(E - H_1))/\,\mathrm{Tr}_1 \exp(S(E - H_1)), \quad (2.15)$$

it appears quite different. It will now be shown that ρ_1 does not depend on the nature of the second system if it is infinitely large (a thermal reservoir). We shall also find out that this so-called canonical density matrix is equivalent to the microcanonical density matrix if the system is large. The convergence of ρ_1 as the second subsystem becomes infinitely large is described by

2.4.1 Lemma

Suppose that the concave, increasing functions $(1/V)S(E) \equiv \sigma_V(E/V)$ *and their derivatives converge uniformly on some neighborhood of* $\varepsilon = E/V$ *to a function* $\sigma(\varepsilon) \in C^1$ *and to* $\sigma'(\varepsilon)$*. Then as* $V \to \infty$,

$$\rho_V \equiv \frac{\exp[V\sigma_V((E - H_1)/V)]}{\mathrm{Tr}\exp[V\sigma_V((E - H_1)/V)]} \to \frac{\exp(-H_1\sigma'(\varepsilon))}{\mathrm{Tr}\exp(-H_1\sigma'(\varepsilon))}$$

in the trace norm, provided that $\exp(-H_1\sigma'(\varepsilon))$ *is of the trace class* $\mathcal{C}_1$.

2.4.2 Remarks

1. As in (2.3.11;2), E and E_m can be identified.
2. *A priori*, $S(E)$ has been defined only for discrete values. We assume that it can be interpolated with a concave, strictly increasing, continuously differentiable function.
3. The facts $\sigma_{\mathrm{ess}}(H) = \emptyset$ and $H \geq 0$ do not suffice to make $\exp(-\beta H) \in \mathcal{C}_1$; $\mathrm{Sp}(H)$ could be $\mathbb{Z}^+$ and the eigenvalues $n \in \mathbb{Z}^+$ could have multiplicity n^n. More assumptions are needed than (2.3.3;2).
4. The significance of the lemma is that temperature is the only property of a reservoir in the infinitely large limit that enters into the reduced density matrix. The reduced density matrix has the canonical form regardless of the structure of the reservoir, when the energy of interaction can be neglected.

Proof of (2.4.1): With $\mathrm{Tr}_1\,\Theta(E_1 - H_1) = \exp(S_1(E_1))$, $\mathrm{Tr}\,\Theta(E - H_1 - H_2) = \int dE_1 \exp(S(E - E_1) + S_1(E_1))S_1'(E_1)$, ρ_V can be written as

$$\rho_V = \frac{\exp\{V[\sigma_V(\varepsilon - (H_1/V)) - \sigma_V(\varepsilon)]\}}{\int dE_1 \exp\{S_1(E_1) + \ln S_1'(E_1) + V[\sigma_V(\varepsilon - (E_1/V)) - \sigma_V(\varepsilon)]\}}.$$

Because of concavity, if $H_1 \geq 0$, then $H_1\sigma_V'(\varepsilon) \leq V[\sigma_V(\varepsilon) - \sigma_V(\varepsilon - (H_1/V))] \leq H_1\sigma_V'(\varepsilon - (H_1/V))$ (see Figure 2.14). The assumption that σ' converges uniformly then makes $V[\sigma_V(\varepsilon - (H_1/V)) - \sigma_V(\varepsilon)]$ converge uniformly to $-H_1\sigma'(\varepsilon)$ on compact sets in $\mathrm{Sp}(H_1)$. Moreover, there exist V' and β such that for all $V > V'$, there is an operator inequality, $\exp[V(\sigma_V(\varepsilon - (H_1/V)) - \sigma_V(\varepsilon))] \leq \exp(-\beta H_1)$. In the spectral representation of H_1, $\exp[V(\sigma_V(\varepsilon - (H_1/V)) - \sigma_V(\varepsilon))] \to \exp(-H_1\sigma'(\varepsilon))$ in the strong topology, by the Lebesgue dominated convergence theorem. If the operator on the right belongs to $\mathcal{C}_1$, then by the dominated convergence theorem again,

$$\begin{aligned}\mathrm{Tr}\exp[-H_1\sigma'(\varepsilon)] &= \int dE_1 \exp[S_1(E_1) + \ln S_1'(E_1) - E_1\sigma'(\varepsilon)] \\ &= \lim_{V\to\infty} \int dE_1 \exp\bigg\{S_1(E_1) + \ln S_1'(E_1) \\ &\qquad + V\left[\sigma_V\left(\varepsilon - \frac{E_1}{V}\right) - \sigma_V(\varepsilon)\right]\bigg\}.\end{aligned}$$

The proof is completed by appealing to the theorem (Problem 1) that strong convergence of density matrices to a density matrix implies convergence in the trace norm.

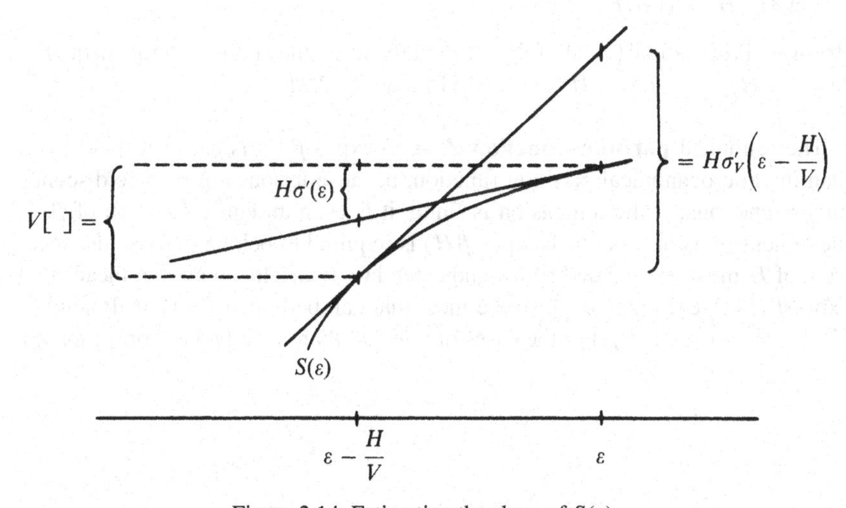

Figure 2.14. Estimating the slope of $S(\varepsilon)$.

2.4.3 Corollaries

1. Since ρ_V converges in the sense of the strong topology of $\mathcal{B}(\mathcal{H})^*$ (cf. (2.1.2)), $\mathrm{Tr}\, \rho_V a \to \mathrm{Tr}\, a \exp[-\beta(H_1 - F)]$ for all $a \in \mathcal{B}(\mathcal{H}_1)$, where $\beta \equiv \sigma'(\varepsilon)$, $\exp(-\beta F) = \mathrm{Tr} \exp(-\beta H_1)$.
2. Because of Theorem (2.2.24), $S(\exp[-\beta(H_1 - F)]) \leq \underline{\lim}_{V \to \infty} S(\rho_V)$.

Recall that the microcanonical state is the most mixed state below E_m. The canonical state instead satisfies

2.4.4 The Maximum Principle for the Canonical Entropy

Let $\rho = \exp(-\beta H)/\mathrm{Tr} \exp(-\beta H)$ *and let* $\bar{\rho}$ *be any density matrix such that* $\mathrm{Tr}\, \bar{\rho} H = \mathrm{Tr}\, \rho H$. *Then* $S(\rho) \geq S(\bar{\rho})$.

2.4.5 Remarks

1. Proposition (2.4.4) states that with a given average energy, the canonical state has the greatest possible entropy. The proposition does not work for all α-entropies S_α, so it can not be improved to the statement that $\rho \succeq \bar{\rho}$.
2. According to inequality (2.1.4; 2), since $x \to -x \ln x$ is strictly concave, S is a strictly concave function on the convex set of density matrices ρ such that $\mathrm{Tr}\, \rho H = E$. This means that the maximum is unique, and there can not even be local maxima elsewhere.
3. Not all $S_\alpha(\rho)$ are equal with the canonical $\rho : S_\alpha = \alpha\beta(F(\alpha\beta) - F(\beta))$.
4. This maximum principle is sometimes invoked as the motivation for the canonical density matrix, without appealing to the microcanonical state.
5. The free energy satisfies the inequality $F(\bar{\rho}) \geq F(\rho)$ without the assumption that $\mathrm{Tr}\, \bar{\rho} H = \mathrm{Tr}\, \rho H$.

Proof: Proposition (2.4.4) follows directly from the more general inequality $S(\rho) - \langle H \rangle_\rho - S(\bar{\rho}) + \langle H \rangle_{\bar{\rho}} \geq 0$, see Remark (2.2.21; 1). □

The canonical **partition function** $Z \equiv \mathrm{Tr} \exp(-\beta H)$ is easier to work with than the microcanonical partition function, because it does not involve discontinuous functions; if the dimension is finite, it is even an entire function of β. If the dimension is infinite, then $\exp(-\beta H)$ is required to belong to $\mathcal{C}_1$ so the spectrum of H must be bounded below and extend to $+\infty$. This, however, means that $\exp(-\beta H) \notin \mathcal{C}_1$, for $\beta < 0$, so the most that can be hoped for is analyticity in $\mathbb{C}^+ \equiv \{x + iy : x > 0\}$. For the cases of interest, there is in fact a proposition on

2.4.6 The Analyticity of the Partition Function of Finite Systems

Let $\exp(-\beta H_0) \in \mathcal{C}_1$, *for all* $\beta > 0$ *and suppose* v *is* ε*-bounded with respect to* H_0 *(cf. (III: 3.4.1)). Then the mapping* $\mathbb{C}\times\mathbb{C}^+ \to \mathbb{C} : (\alpha, \beta) \to \operatorname{Tr}\exp[-\beta(H_0+\alpha v)]$ *is analytic, and* $(\partial/\partial\alpha)\operatorname{Tr}\exp[-\beta(H_0+\alpha v)]_{|\alpha=0} = -\operatorname{Tr}\beta v\exp[-\beta H_0]$.

2.4.7 Remarks

1. Since the operator $H_0 + \alpha v$ is not normal when α is nonreal, the exponential function has to be defined. This can be done as in (2.1.5;7) or by integrating the resolvent,
$$\exp[-\beta(H_0+\alpha v)] = \int_C \frac{dz}{2\pi i}\frac{\exp(-\beta z)}{(H_0+\alpha v - z)},$$
in which the integration contour runs through the region of analyticity (cf. (III:3.5.13)) so that the integral converges in norm.
2. The next task is to make sense of $\operatorname{Tr}\exp[-\beta(H_0 + \alpha v)]$ and show that $\exp[-\beta(H_0 + \alpha v)]$ belongs to $\mathcal{C}_1$ for $(\alpha, \beta) \in \mathbb{C}\times\mathbb{C}^+$. If $\alpha, \beta \in \mathbb{R}\times\mathbb{R}^+$, then this follows from $H_0 + \alpha v \geq H_0/2 - C(\alpha)$, $\exp(-\beta H_0) \in \mathcal{C}_1$, and the observation that if $0 < a < b \in \mathcal{C}_1$, then $a \in \mathcal{C}_1$. If α and β are complex, then Corollary (2.1.5; 7) can be appealed to for $|\operatorname{Tr}\exp(\alpha a + \beta b)| \leq \operatorname{Tr}|\exp(\alpha a)||\exp(\beta b)|$, with $\exp(a)$ and $\exp(b)$ Hermitian, and in particular $|\operatorname{Tr}\exp[-aH_0 - bv + i(cH_0 + dv)] \leq \operatorname{Tr}\exp(-aH_0 - bv)$ for all real a, b, c, and d.
3. The proposition implies that the free energy $F = -T\ln Z$ can have singularities only at the zeros of Z. If $(\alpha, \beta) \in \mathbb{R}\times\mathbb{R}^+$ then $Z > 0$, so F is analytic in a neighborhood of $\mathbb{R}\times\mathbb{R}^+$. In addition, Corollary (2.1.5; 3) states that $-\ln Z$ is concave in $(\beta, \alpha\beta) \in \mathbb{R}\times\mathbb{R}^+$, so F is concave in $(T, \alpha/T)$ (cf. (III: 3.5.22)). The equation $\partial F/\partial\alpha = \langle v\rangle$ generalizes the Feynman–Hellmann formula (III: 3.5.17; 2).

Proof: See Problem 2. □

Since the exponential function is convex, the free energy can be bounded in terms of phase-space integrals by means of (2.2.9), and the upper bound of (2.2.9) can be improved upon with Corollary (2.1.5; 7).

2.4.8 The Connection with the Classical Free Energy

Let
$$H = \sum_{i=1}^{N} |\mathbf{p}_i|^2 + v(\mathbf{x}), \quad \exp(-\beta F) = \operatorname{Tr}\exp(-\beta H) < \infty,$$

and

$$\exp[-\beta F_{\mathrm{cl}}(v)] = \int d^{3N}x \frac{d^{3N}p}{(2\pi)^{3N}} \exp\left[-\beta\left(\sum_{i=1}^{N} |\mathbf{p}_i|^2 + v(\mathbf{x})\right)\right].$$

Then

$$F_{\mathrm{cl}}(v) \leq F \leq \inf_u F_{\mathrm{cl}}(v_u),$$

where

$$v_u(\mathbf{x}) = \int d^{3N}x' v(\mathbf{x}')|u(\mathbf{x}-\mathbf{x}')|^2 + \int d^{3N}x |\nabla u(\mathbf{x})|^2.$$

2.4.9 Remarks

1. The function $v(\mathbf{x})$ contains the interaction between the particles, as well as a possible external field. It must even account for the box confining the system, as the Hilbert space is $L^2(\mathbb{R}^{3N})$.
2. The proposition shows that quantum effects can only increase the free energy, either with a kinetic zero-point energy or a smeared-out effective potential.
3. The particles have been assumed distinguishable; the modifications needed for indistinguishable particles will be discussed below.
4. Countless attempts at expansions in $\hbar$ have been made in the literature, but the results are not conclusive because rigorous bounds on the higher-order contributions have not been obtained.
5. If $\hbar$ is not set to 1, the dimensionless volume in phase space becomes $d^{3N}xd^{3N}ph^{-3N}$, rather than $d^{3N}xd^{3N}p\hbar^{-3N}$.

Proof: The lower bound for F. By Corollary (2.1.5; 7),

$$\begin{aligned}\operatorname{Tr}\exp[-\beta(H_0+v)] &\leq \operatorname{Tr}\exp(-\beta H_0)\exp(-\beta v)\\ &= \int d^{3N}x \langle\mathbf{x}|\exp(-\beta H_0)|\mathbf{x}\rangle \exp(-\beta v(\mathbf{x})),\end{aligned}$$

and it was observed in (III: 3.3.3) that $\exp(-\beta H_0)$ has the integral kernel

$$K(\mathbf{x},\mathbf{x}) = \left(\frac{1}{4\pi\beta}\right)^{3N/2} = \int \frac{d^{3N}p}{(2\pi)^{3N}} \exp\left(-\beta\sum_{k=1}^{N}|\mathbf{p}_k|^2\right).$$

The upper bound for F follows immediately from (2.2.9), for $\langle z\|\mathbf{p}|^2|z\rangle = (\Im z^2) + \int dx|\nabla u|^2$.

2.4.10 Example

The one-dimensional harmonic oscillator: $u(x) = \exp(-bx^2/2)\sqrt[4]{b/\pi}$, $H = p^2 + \omega^2 x^2$,

$$\operatorname{Tr}\exp(-\beta H) = \sum_{n=0}^{\infty} \exp[-\beta\omega(2n+1)] = \frac{\exp(\omega\beta)}{1-\exp(-2\omega\beta)},$$

$$v_u = \omega^2\left(x^2 + \frac{1}{2b}\right) + \frac{b}{2},$$

which has the minimum $\omega^2 x^2 + \omega$ when $b = \omega$. Since

$$\int_{-\infty}^{\infty} \frac{dpdx}{2\pi} \exp[-\beta(p^2 + \omega^2 x^2)] = \frac{1}{2\omega\beta},$$

the bounds (2.4.9) yield the inequalities

$$\frac{\exp(-\alpha/2)}{\alpha} \leq \frac{\exp(-\alpha/2)}{1-\exp(-\alpha)} \leq \frac{1}{\alpha}, \quad \alpha \equiv 2\omega\beta \in \mathbb{R}^+.$$

The interest in the bounds (2.4.8) is mainly academic, since the particles in real physics are either fermions or bosons. In addition to multiplying the volume element of the phase-space integral by $1/N!$, the generalization for indistinguishable particles entails an effective interaction that vanishes as $mT \to \infty$, and is repulsive for fermions and attractive for bosons.

2.4.11 Bounds on F for Indistinguishable Particles

Suppose that

$$H = \frac{1}{2m}\sum_{i=1}^{N} |\mathbf{p}_i|^2 + v(\mathbf{x}_1, \ldots, \mathbf{x}_N),$$

$$\exp[-\beta F_{\text{cl}}(H)] = \frac{1}{(2\pi)^{3N} N!}\int d^{3N}x d^{3N}p \exp[-\beta H(\mathbf{p}_1, \ldots, \mathbf{p}_N, \mathbf{x}_1, \ldots, \mathbf{x}_N)],$$

and that $F_B(H)$ and $F_F(H)$ equal $-T \ln \operatorname{Tr}\exp(-\beta H)$, where the trace is taken over the symmetric (resp. antisymmetric) tensor product of the one-particle spaces. Then

$$F_{\text{cl}}(H) \leq F_F(H) \leq F_{\text{cl}}(h + v_F),$$
$$F_{\text{cl}}(H + v_B) \leq F_B(H) \leq F_{\text{cl}}(h),$$

where the function $h(\mathbf{p}_i, \mathbf{x}_i)$ is the expectation value of H in the symmetrized (resp. antisymmetrized) states of (2.2.8; 5):

$$h(\mathbf{z}_1, \ldots, \mathbf{z}_N) = \frac{\langle \mathbf{z}_1, \ldots, \mathbf{z}_N | H | \mathbf{z}_1, \ldots, \mathbf{z}_N\rangle}{\langle \mathbf{z}_1, \ldots, \mathbf{z}_N | \mathbf{z}_1, \ldots, \mathbf{z}_N\rangle}, \quad \mathbf{z}_i = \mathbf{x}_i + i\mathbf{p}_i.$$

If the coherent states are chosen with $u(\mathbf{x}) = \exp(-mT|\mathbf{x}|^2/2)$, then the effective potentials are

$$v_F = \begin{cases} T \ln 2 \sum_{i \neq k} \exp(-mT|\mathbf{x}_i - \mathbf{x}_k|^2) & \text{if } \sup_j \sum_{i \neq j} \exp\left(\frac{-mT|\mathbf{x}_i - \mathbf{x}_j|^2}{2}\right) \leq \frac{1}{2} \\ \infty & \text{otherwise} \end{cases}$$

and

$$v_B = -T \sum_{i,k} \exp\left(\frac{-mT|\mathbf{x}_i - \mathbf{x}_k|^2}{2}\right).$$

Proof: The lower bounds. For one particle in x-space (see (III: 3.3.3)),

$$\langle \mathbf{x}| \exp\left(\frac{-\beta|\mathbf{p}|^2}{2m}\right) |\mathbf{x}'\rangle = \left(\frac{mT}{2\pi}\right)^{3/2} \exp\left(\frac{-mT|\mathbf{x} - \mathbf{x}'|^2}{2}\right),$$

so in the properly symmetrized or antisymmetrized basis, if there are N particles, then

$$\langle \mathbf{x}_1, \ldots, \mathbf{x}_N| \exp\left(\frac{-\beta \sum_i |\mathbf{p}_i|^2}{2m}\right) |\mathbf{x}_1, \ldots, \mathbf{x}_N\rangle$$
$$= \frac{1}{N!} \left(\frac{mT}{2\pi}\right)^{3N/2} \sum_P (\pm 1)^P \exp\left(\frac{-mT \sum_i |\mathbf{x}_i - \mathbf{x}_{P_i}|^2}{2}\right).$$

The sum over permutations amounts to just a permanent or determinant of the form $\langle \mathbf{z}_1, \ldots, \mathbf{z}_N | \mathbf{z}_1, \ldots, \mathbf{z}_N\rangle$, by (2.2.8; 5). It is therefore ≥ 1 or, respectively, ≤ 1, since the length of a vector is increased or, respectively, decreased when acted upon by a_f^* with $\|f\| = 1$:

$$\|a_f^* |\rangle\|^2 = \langle |a_f a_f^*| \rangle = \langle | \rangle \pm \langle |a_f^* a_f| \rangle \gtreqless \langle | \rangle.$$

For fermions, $\mathrm{Det}(\langle \mathbf{z}_i | \mathbf{z}_k \rangle) \leq 1$, whereas for bosons the permanent has an upper bound from Problem 4, $\mathrm{Per}(\langle \mathbf{z}_i | \mathbf{z}_k \rangle) \leq \exp[\sum_{i,k} |\langle \mathbf{z}_i | \mathbf{z}_k \rangle|]$. The rest of the proof is similar to that of the lower bound of (2.4.10):

$$\mathrm{Tr} \exp[-\beta(H_0 + v)] \leq \mathrm{Tr} \exp(-\beta H_0) \exp(-\beta v)$$
$$= \frac{1}{N!(2\pi)^{3N}} \int d^{3N}x d^{3N}p \exp[-\beta(H_0(\mathbf{p}_1, \ldots, \mathbf{p}_N) + v(\mathbf{x}_1, \ldots, \mathbf{x}_N))]$$
$$\times {}^{\mathrm{Per}}_{\mathrm{Det}} \left(\exp\left(-\frac{m}{2}|\mathbf{x}_i - \mathbf{x}_j|^2 T\right)\right)$$
$$\leq \frac{1}{N!(2\pi)^{3N}} \int d^{3N}x d^{3N}p \exp\Bigg[-\beta\Bigg(H_0(\mathbf{p}_1, \ldots, \mathbf{p}_N) + v(\mathbf{x}_1, \ldots, \mathbf{x}_N)$$
$$-T \begin{Bmatrix} \exp(-\sum_{i,j} (mT/2)|\mathbf{x}_i - \mathbf{x}_j|^2) \\ 0 \end{Bmatrix}\Bigg)\Bigg].$$

The upper bounds. Since the symmetrized and antisymmetrized coherent states are not normalized,

$$\langle \mathbf{z}_1, \ldots, \mathbf{z}_N | \mathbf{z}_1, \ldots, \mathbf{z}_N \rangle \equiv n(z) = {}^{\text{Per}}_{\text{Det}}(\langle \mathbf{z}_i | \mathbf{z}_k \rangle) \gtrless 1,$$

the normalization has to be accounted for in (2.2.9(i)):

$$\operatorname{Tr} k(a) \geq \int d\Omega_z n(z) k \left(\frac{\langle z | a | z \rangle}{n(z)} \right).$$

For bosons the inequality follows now from $n(z) \geq 1$. For fermions, with $u(\mathbf{x}) = \exp(-mT|\mathbf{x}|^2/4)$, it is necessary to estimate $\operatorname{Det}(1 + K)$, where

$$K_{ij} = \begin{cases} \exp(-mT/2)|\mathbf{x}_i - \mathbf{x}_j|^2), & i \neq j \\ 0, & i = j. \end{cases}$$

Since

$$\|K\| \leq \sup_j \sum_{i \neq j} \exp \left(-\frac{mT}{2} |\mathbf{x}_i - \mathbf{x}_j|^2 \right),$$

we find

$$\begin{aligned} \ln \operatorname{Det}(\langle \mathbf{z}_i | \mathbf{z}_j \rangle) &= \ln \operatorname{Det}(1 + K) = \operatorname{Tr} \ln(1 + K) \\ &= \sum_{n=2}^{\infty} \operatorname{Tr} K^n \frac{(-1)^n}{n} \leq \operatorname{Tr} K^2 \sum_{n=0}^{\infty} \frac{\|K\|^n}{n+2} \\ &\leq \ln \frac{1}{1 - \|K\|} \operatorname{Tr} K^2 \\ &\leq \begin{cases} \ln 2 \operatorname{Tr} K^2 & \text{for } \|K\| \leq \frac{1}{2} \\ \infty & \text{otherwise.} \end{cases} \end{aligned}$$

Finally,

$$\operatorname{Tr} K^2 = \sum_{i \neq j} \exp[-mT|\mathbf{x}_i - \mathbf{x}_j|^2].$$

2.4.12 Remarks

1. If $\min_{i,j} |\mathbf{x}_i - \mathbf{x}_j| \equiv b > 0$, then $\|K\| \cong b^{-3} \int_b^\infty dr r^2 \exp(-r^2 mT/2) \cong \exp(-mTb^2/2)$, so v_F can be replaced with a hard-core potential with a radius depending on T and energy $\sim N$.
2. The ranges of the potentials v_B and v_F are approximately the thermal wavelength, i.e., the wavelength of a particle with kinetic energy $3T/2$, so when the particles are about this close together, as in a degenerate quantum gas, the bounds spread wide apart.

In closing, let us study the limit $N \to \infty$ in the framework of the canonical ensemble. Not only the reservoir but also the subsystem will be made infinite at the same time, and we wish to know whether the free energy density F/V tends to a limit φ. This should be the case whenever this limit exists microcanonically. Then the issue is how to recover the microcanonical quantities from knowledge of φ:

2.4.13 Theorem

Suppose that, with $H \geq 0$, $\sigma_V(\varepsilon, \rho) = (1/V) \ln \operatorname{Tr} \Theta(V\varepsilon - H)$ converges uniformly on compact sets to a concave function $\sigma(\varepsilon, \rho)$ and is bounded above by a function $s(\varepsilon, \rho)$ such that $0 = s(\varepsilon_0, \rho) = \lim_{\varepsilon \to \infty} s(\varepsilon, \rho)/\varepsilon$, when V is big enough. Writing as usual $\beta = 1/T$, then

$$\lim_{V \to \infty} \left(-\frac{T}{V} \ln \operatorname{Tr} \exp(-\beta H) \right) = \inf_{\varepsilon} (\varepsilon - T\sigma(\varepsilon, \rho)) \equiv \varphi(T, \rho).$$

2.4.14 Remarks

1. Since σ is concave, it has a right derivative,

$$\sigma' \equiv \lim_{\delta \downarrow 0} (\sigma(\varepsilon + \delta, \rho) - \varepsilon(\varepsilon, \rho)) \frac{1}{\delta}.$$

   The infimum is attained at the point $\varepsilon(T, \rho)$ for which $\sigma'(\varepsilon(T, \rho), \rho) = 1/T$ (see Figure 2.15). If σ' has a discontinuity, jumping over the value $1/T$, then $\varepsilon(T, \rho)$ is the point at which the jump takes place. The usual thermodynamic relationship $\varphi(T, \rho) = \varepsilon(T, \rho) - T\sigma(\varepsilon(T, \rho)\rho)$ holds for the free energy.
2. The function $\beta\varphi$ is a Legendre transform $\mathcal{L}(\sigma)(\beta) = \inf_{\varepsilon}(\beta\varepsilon - \sigma(\varepsilon))$. The transformation $\mathcal{L}$ has the following properties:

   (i) $\mathcal{L} \circ \mathcal{L}$ produces the concave envelope of any function so $\mathcal{L} \circ \mathcal{L} = 1$ on concave functions;

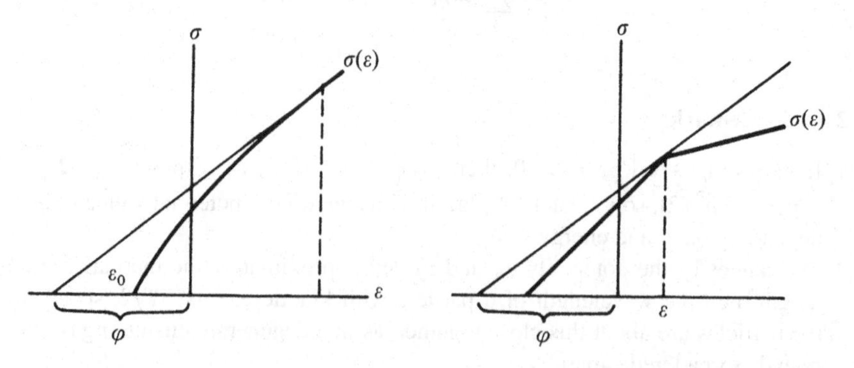

Figure 2.15. The geometric meaning of the free energy.

(ii) $\mathcal{L}$ maps a linear piece of a concave function to the point of a corner and vice versa;

(iii) $\mathcal{L}$ maps the set of strictly concave, continuously differentiable functions into itself. By Property (i),

$$\sigma(\varepsilon) = \inf_{\beta}(\beta\varepsilon - \mathcal{L}(\sigma)(\beta)) = \inf_{T} \frac{\varepsilon - \varphi(T)}{T}.$$

3. If $\sigma(\varepsilon)$ is strictly concave and continuously differentiable, then by Problem 3 the limit $V \to \infty$ and the derivative by β can be taken in either order. The energy and entropy densities calculated with the canonical density matrix are

$$\lim_{V\to\infty} \operatorname{Tr} \frac{H}{V} \frac{\exp(-\beta H)}{\operatorname{Tr}\exp(-\beta H)} = -\lim_{V\to\infty} \frac{\partial}{\partial\beta}\frac{1}{V} \ln \operatorname{Tr}\exp(-\beta H)$$
$$= -T^2 \frac{\partial}{\partial T}\frac{\varphi}{T} = \varphi + T\sigma$$

and

$$\lim_{V\to\infty} \frac{T}{V} S\left(\frac{\exp(-\beta H)}{\operatorname{Tr}\exp(-\beta H)}\right) = \varepsilon - \varphi,$$

which are obviously identical to the microcanonical energy and entropy densities. This fact is known as the **equivalence of the ensembles**.

4. The concavity of σ in ε is a necessary condition for the ensembles to be equivalent, since the specific heat in the canonical ensemble,

$$\frac{\partial\varepsilon}{\partial T} = \frac{\beta^2}{V}\frac{\partial^2}{\partial\beta^2} \ln \operatorname{Tr}\exp(-\beta H)$$

is automatically positive by Corollary (2.1.5;3).

5. The bounding function s is necessary to ensure that

$$\lim_{V\to\infty} \sup_{\varepsilon}(T\sigma_V(\varepsilon,\rho) - \varepsilon) = \sup_{\varepsilon}(T\sigma(\varepsilon,\rho) - \varepsilon);$$

without it, $T\sigma_V(\varepsilon) - \varepsilon = 1 - (1 - \varepsilon/V)^2$ is a counterexample. (The assumption that $H \geq 0$ is a normalization.)

Proof of (2.4.13):

$$\operatorname{Tr}\exp(-\beta H) = \int_0^\infty dE \exp(-\beta E) \frac{\partial}{\partial E} \operatorname{Tr}\Theta(E - H)$$
$$= \beta \int_0^\infty dE \exp[-\beta E + S(E)]$$
$$= \beta V \exp[-\beta V \varphi_V(T,\rho)]$$
$$\times \int_0^\infty d\varepsilon \exp[-\beta V(\varepsilon - T\sigma_V(\varepsilon) - \varphi_V)],$$

where

$$\varphi_V(T, \rho) = \inf_{\varepsilon}(\varepsilon - T\sigma_V(\varepsilon, \rho)).$$

If V is taken large enough, then the infimum lies between 0 and $\varepsilon_0 : \varepsilon_0 - T\sigma(\varepsilon_0, \rho) = 0$. By assumption the functions σ_V converge uniformly on this compact interval, so $\varphi_V(T, \rho) \to \varphi(T, \rho)$. A modification of Lemma (2.3.12) shows that the contribution of the integral to φ is negligible in this limit. This step uses the assumption to ensure that for all $T > 0$ the exponent is dominated by $-\beta E$ for large E, so that the dominated convergence lemma applies.

Several general properties of the Legendre transform of σ can be deduced from those of the microcanonical energy density (2.3.8), and are listed below:

2.4.15 Properties of the Free Energy Density

1. As the infimum of a set of linear functions, $\varphi(T, \rho)$ is concave in T. If $H \geq O$, then $\varphi(T, \rho) \leq 0$, and $\varphi(0, \rho) = 0$.

2. The function $\varphi(T, \rho)$ is convex in ρ, because $f(x, y)$ being convex in (x, y) implies that $\inf_x f(x, y)$ is convex in y (see (2.3.26)).

3. $\rho^{-1}\varphi(T, \rho)$ is an increasing function of ρ, since $\mathrm{Tr}\exp(-\beta H)$ is an increasing function of V when N and β are fixed.

4. $T^{-1}\varphi(T, \rho)$ is a decreasing function of T, since for $H \geq 0$, $\exp(-\beta H)$ is a decreasing function of β.

2.4.16 Remark

Although convexity survives the thermodynamic limit, the analyticity (2.4.7; 3) of F is less hardy. The zeros of Z may approach the real axis as the system is made infinite, causing discontinuities in the derivatives of φ. Example (2.3.27;2) can be modified to a degenerate BCS model, with

$$H = B \sum_{j=1}^{N} \sigma_j^{(z)} - \frac{1}{N} \sum_{i,j=1}^{N} (\boldsymbol{\sigma}_i \cdot \boldsymbol{\sigma}_j - \sigma_i^{(z)} \sigma_j^{(z)}).$$

This Hamiltonian has the eigenvalues $N(Bm_z - m(m + 2/N) + m_z^2)$, and, as in (2.13),

$$\varphi(T, B) = \inf_{0 \leq |m_z| \leq m \leq 1} \left(-m^2 + \left(m_z + \frac{B}{2} \right)^2 - \frac{B^2}{4} - T\sigma(m) \right),$$

$$\sigma(m) = \ln 2 - \frac{1+m}{2} \ln(1+m) - \frac{1-m}{2} \ln(1-m).$$

The infimum with respect to m_z is attained at $\max\{-B/2, -m\}$, assuming $B \geq 0$. If $m_z = -B/2$, then setting the derivative by m to zero leads to the equation

$$m(T) = \tanh\left(\frac{2m(T)}{T}\right).$$

If $m_z = -m$, then instead of this, the minimizing value is $m(T, B) = \tanh(B/T)$. The two different possibilities give critical temperatures

$$T_c(B) \equiv \begin{cases} B/\operatorname{arctanh}(B/2) & \text{if } 0 < B < 2, \\ 0 & \text{if } 2 \leq B. \end{cases}$$

Figure 2.16 depicts $\varphi(T, B)$. The values of m and m_z are continuous at the transition point, but their derivatives are not. The function φ remains continuous along with its first derivatives – the derivatives by m and m_z vanish – but the second derivatives of $\varphi(T, B)$ are discontinuous at $T = T_c(B)$. Such properties as the specific heat display the discontinuity characteristic of a phase transition.

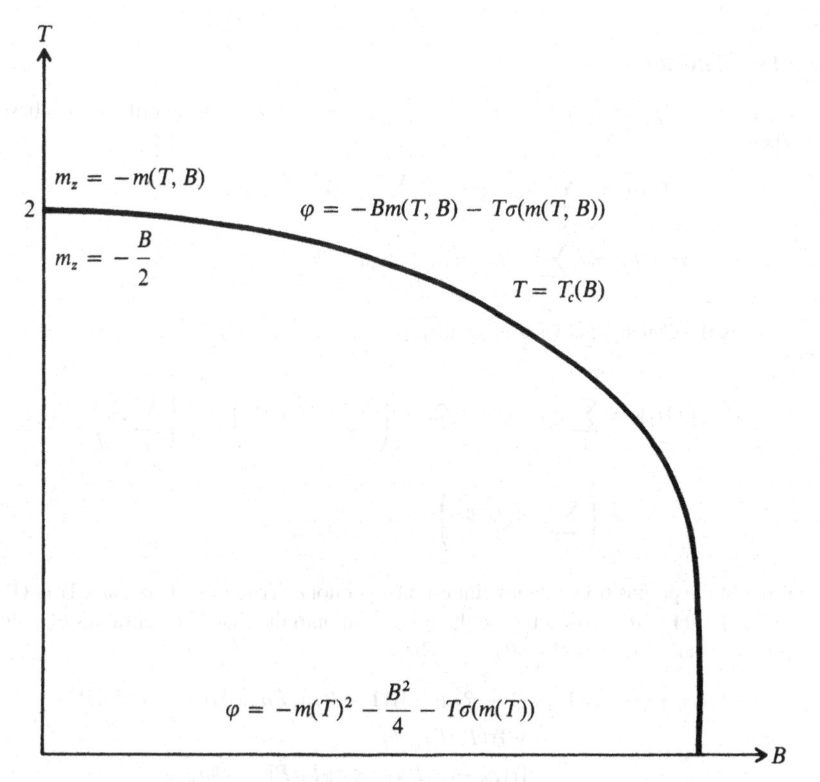

Figure 2.16. The free energy in Example (2.3.27;2).

2.4.17 Problems

1. Let ρ_n and ρ be density matrices for which $\rho_n \rightharpoonup \rho$. Show that $\mathrm{Tr}\,|\rho_n - \rho| \to 0$. (Hint: use the following lemma: *If ρ is a density matrix and Q a projection such that* $\mathrm{Tr}\,\rho Q < \varepsilon$, *then for all $a \in \mathcal{B}(\mathcal{H})$,* $|\mathrm{Tr}\,\rho Q a| < \|a\|\sqrt{\varepsilon}$).

2. Prove (2.4.6) by applying Hartogs's theorem: *If $f(z_1, z_2)$ is separately analytic in z_1 and z_2, then it is jointly analytic.* Also observe that the trace is a continuous mapping $\mathcal{C}_1 \to \mathbb{C}$, where $\mathcal{C}_1$ has the norm $\|\cdot\|_1$.

3. Suppose $\varphi_V(\varepsilon)$ is a sequence of concave functions converging pointwise to $\varphi(\varepsilon)$. Let $\varphi'_{V,r}(\varepsilon)$ and $\varphi'_{V,l}(\varepsilon)$ denote the right and left derivatives of $\varphi_V(\varepsilon)$, and likewise for $\varphi'_r(\varepsilon)$ and $\varphi'_l(\varepsilon)$. Show that for all ε,

$$\varphi'_r(\varepsilon) \leq \lim_{V\to\infty} \inf \varphi'_{V,r}(\varepsilon) \leq \lim_{V\to\infty} \sup \varphi'_{V,l}(\varepsilon) \leq \varphi'_l(\varepsilon),$$

and that if φ_V and φ are differentiable at the point ε, then $\lim \varphi'_V(\varepsilon) = \varphi'(\varepsilon)$.

4. Show that $|\mathrm{Per}\langle \mathbf{z}_i|\mathbf{z}_k\rangle| \leq \exp \sum_{i,k} |\langle \mathbf{z}_i|\mathbf{z}_k\rangle|$.

5. Find a function of x and y that is convex in each variable separately but not jointly convex.

2.4.18 Solutions

1. Lemma: $\rho = \sum c_i |x_i)(x_i|$, where $c_i \geq 0$, $\sum_i c_i = 1$, and $\{x_i\}$ is an orthonormal basis. Then

$$\mathrm{Tr}\,\rho Q = \sum_i c_i (x_i|Qx_i) = \sum_i c_i \|Qx_i\|^2 < \varepsilon,$$

$$|\mathrm{Tr}\,\rho Q a| = |\sum c_i (Qx_i|ax_i)| \leq \|a\| \cdot \sum_i c_i \|Qx_i\| < \|a\|\sqrt{\varepsilon},$$

since by the Cauchy–Schwarz inequality,

$$\sum_i c_i \|Qx_i\| = \sum_i \sqrt{c_i}\|Qx_i\|\sqrt{c_i} \leq \left(\sum_i c_i \|Qx_i\|^2\right)^{1/2} \cdot \left(\sum_i c_i\right)^{1/2} = \left(\sum_i c_i \|Qx_i\|^2\right)^{1/2}.$$

Proof of the proposition: For any finite-rank operator a, $\mathrm{Tr}\,\rho_n a \to \mathrm{Tr}\,\rho a$, and $\mathrm{Tr}\,\rho_n(\mathbf{1} - a) \to \mathrm{Tr}\,\rho(\mathbf{1} - a)$. Now let P be the projection onto the first N eigenvalues of ρ and choose N such that $\mathrm{Tr}\,\rho(\mathbf{1} - P) < \varepsilon$. Then

$$\mathrm{Tr}(\rho_n - \rho)a = \mathrm{Tr}\,\rho_n(\mathbf{1} - P)a + \mathrm{Tr}(\mathbf{1} - P)\rho_n P a + \mathrm{Tr}(\rho_n - \rho)PaP + \mathrm{Tr}(P\rho P - \rho)a.$$

$$\mathrm{Tr}(\rho_n - \rho)PaP < \varepsilon\|PaP\| < \varepsilon\|a\|$$

for sufficiently large n, since all topologies are equivalent on the finite-dimensional space $P\mathcal{B}(H)P$, and $\mathrm{Tr}(\rho_n - \rho)PaP \to 0$. $|\mathrm{Tr}(P\rho P - \rho)a| \le \|a\| \,\mathrm{Tr}(\mathbf{1} - P)\rho < \|a\| \cdot \varepsilon$, $\mathrm{Tr}\,\rho_n(\mathbf{1}-P) \to \mathrm{Tr}\,\rho(\mathbf{1}-P) < \varepsilon$, which implies that for n large enough, $\mathrm{Tr}\,\rho_n(\mathbf{1}-P) < 2\varepsilon$. Hence, by the lemma,

$$|\mathrm{Tr}\,\rho_n(\mathbf{1} - P)a| < \sqrt{2\varepsilon}\|a\|,$$
$$|\mathrm{Tr}(\mathbf{1} - P)\rho_n Pa| = |\mathrm{Tr}\,\rho_n(\mathbf{1} - P)a^* P\| \le \sqrt{2\varepsilon}\|a^* P\| \le \sqrt{2\varepsilon}\|a\|.$$

Consequently,

$$|\mathrm{Tr}(\rho_n - \rho)a| < (2\varepsilon + 2\sqrt{2\varepsilon})\|a\|,$$
$$\mathrm{Tr}\,|\rho_n - \rho| = \sup_{\|a\|\le 1} |\mathrm{Tr}(\rho_n - \rho)a| < 2\varepsilon + 2\sqrt{2\varepsilon}.$$

2.

$$U(\alpha, \beta) \equiv \exp[-\beta(H_0 + \alpha v)] \in \mathcal{C}_1$$

(i) Analyticity (=complex differentiability) in β:

$$\left\| \frac{U(\alpha, \beta + \beta') - U(\alpha, \beta)}{\beta'} + (H_0 + \alpha v)U(\alpha, \beta) \right\|_1$$
$$\le \left\| \left(\frac{U(\alpha, \beta') - 1}{\beta'} + (H_0 + \alpha v) \right) U\left(\alpha, \frac{\beta}{2}\right) \right\| \left\| U\left(\alpha, \frac{\beta}{2}\right) \right\|_1 \to 0$$

as $\beta' \to 0$, since U is a $\|\cdot\|$-convergent integral of $\|\cdot\|$-analytic functions and therefore a $\|\cdot\|$-analytic mapping, $\mathbb{C} \times \mathbb{C}^+ \to \mathcal{B}$.

(ii) Analyticity in α:

$$U(\alpha + \alpha', \beta) - U(\alpha, \beta) = -\beta\alpha' \int_0^1 d\tau U(\alpha + \alpha', \beta(1 - \tau))vU(\alpha, \tau\beta),$$

$$\|U(\alpha + \alpha', \beta(1 - \tau))vU(\alpha, \tau\beta)\|_1 \le \|U(\alpha + \alpha', \beta(1 - \tau))\|$$
$$\times \left\| vU\left(a, \frac{\tau\beta}{2}\right) \right\| \left\| U\left(\alpha, \frac{\tau\beta}{2}\right) \right\|_1 \le \text{constant},$$

when $\frac{1}{2} \le \tau \le 1$. If $0 \le \tau \le \frac{1}{2}$, then the first factor has to be divided up. This shows that the mapping $\mathbb{C} \times \mathbb{C}^+ \to \mathcal{B}_1 : (\alpha, \beta) \to U(\alpha, \beta)$ is analytic, and therefore the mapping $\mathbb{C} \times \mathbb{C}^+ \to \mathbb{C} : (\alpha, \beta) \to \mathrm{Tr}\,U(\alpha, \beta)$ is analytic, because the trace is continuous and linear $\mathcal{B}_1 \to \mathbb{C}$, and thus also analytic.

3. Concavity yields $(1/\varepsilon')(\varphi_V(\varepsilon + \varepsilon') - \varphi_V(\varepsilon)) \le \varphi'_{V,r}(\varepsilon) \le \varphi'_{V,l}(\varepsilon) \le (1/\varepsilon')(\varphi_V(\varepsilon - \varepsilon') - \varphi_V(\varepsilon))$ for all $\varepsilon' > 0$, and the statement follows from this with the limits $\lim_{\varepsilon' \to \infty} \lim_{V \to \infty}$.

4.

$$\mathrm{Per}\langle \mathbf{z}_i | \mathbf{z}_k \rangle \leq \mathrm{Per}|\langle \mathbf{z}_i | \mathbf{z}_k \rangle| = \sum_P \prod_{i=1}^N |\langle \mathbf{z}_i | \mathbf{z}_{P_i} \rangle| \leq \prod_{(i,j)} (1 + |\langle \mathbf{z}_i | \mathbf{z}_j \rangle|)$$

$$\leq \exp\left(\sum_{i,j} |\langle \mathbf{z}_i | \mathbf{z}_j \rangle| \right).$$

5. $f(x, y) = -xy$. The Hessian matrix $\begin{pmatrix} 0 & -1 \\ -1 & 0 \end{pmatrix}$ is not positive.

2.5 The Grand Canonical Ensemble

The thermodynamic functions are easier to calculate explicitly if the constraint of a fixed number N of particles is dropped. It is physically realistic for a system coupled to a reservoir of particles.

This section will investigate the situation of a system with a reservoir with which it can exchange particles as well as heat. As in (2.9), the underlying Hilbert space is taken as

$$\bigoplus_{N_1=0}^{N} \mathcal{H}_{N_1,V_1} \otimes \mathcal{H}_{N-N_1,V_2},$$

and the Hamiltonian is

$$H = \bigoplus_{N_1=0}^{N} (H_1(V_1, N_1) + H_2(V_2, N - N_1)).$$

We consider the limit as $N \to \infty$ and $V_2 \to \infty$, and begin by collecting the immediate generalizations of some of the results of § 2.4. Proofs will not be given, as they entail only slight modifications of the earlier ones.

2.5.1 Convergence of the Reduced Density Matrix

Suppose that the concave, increasing functions

$$\frac{1}{V_2} \mathrm{Tr}_2\, \Theta(E_2 - H_2) = \frac{1}{V_2} S_2(E_2, V_2, N_2) \equiv \sigma_{V_2}\left(\frac{E_2}{V_2}, \frac{N_2}{V_2}\right)$$

and their derivatives converge uniformly on a neighborhood of $\varepsilon = E_2/V_2$ and $\rho = N_2/V_2$ to $\sigma(\varepsilon, \rho)$, $\partial\sigma/\partial\varepsilon$, and $\partial\sigma/\partial\rho$. Then with $V = V_1 + V_2$, $N = N_1 + N_2$,

$$\lim_{V_2 \to \infty} \frac{\mathrm{Tr}_2\, \Theta(E - H)}{\mathrm{Tr}\, \Theta(E - H)} \to \frac{\exp\left[-H_1(V_1, N_1)\frac{\partial\sigma}{\partial\varepsilon} - N_1 \frac{\partial\sigma}{\partial\rho}\right]}{\mathrm{Tr}_1 \exp\left[-H_1(V_1, N_1)\frac{\partial\sigma}{\partial\varepsilon} - N_1 \frac{\partial\sigma}{\partial\rho}\right]} = \rho_{GC}$$

in the trace norm.

2.5.2 Remarks

1. The symbol Tr_2 denotes the trace in the second factor of

$$\bigoplus_{N_1=0}^{N} \mathcal{H}_{N_1,V_1} \otimes \mathcal{H}_{N-N_1,V_2},$$

so in the limit $N \to \infty$, $H_1(N_1, V_1)$ operates on $\sum_{N_1=0}^{\infty} \mathcal{H}_{N_1,V_1}$. This operator on the Hilbert space of an indefinite number of particles is most conveniently written in terms of the field operators (1.3.2).
2. The values of μ for which $\exp[-\beta(H - \mu N)] \in \mathcal{C}_1$ depend on the problem. If, for instance,

$$-\ln \mathrm{Tr}_{|\mathcal{H}_{N_1}} \exp[-\beta H_1(N_1)] > -cN_1,$$

then the trace exists whenever $\Re\beta\mu < -c$.

Many of the results of § 2.4 may be reformulated for the grand canonical ensemble merely be replacing H with $H - \mu N$. An example is

2.5.3 The Principle of Maximum Entropy

Let $\bar{\rho}$ be a density matrix such that $\mathrm{Tr}\, \bar{\rho} H = \mathrm{Tr}\, \rho_{GC} H$, $\mathrm{Tr}\, \bar{\rho} N = \mathrm{Tr}\, \rho_{GC} N$. *Then* $S(\rho_{GC}) \geq S(\bar{\rho})$.

If system 1 is now taken infinitely large, presupposing the extensivity following from $H > -cN$, then T/V times the logarithm of the grand canonical partition function has a limit, which may be identified as the pressure, with reference to (2.3.18).

2.5.4 The Thermodynamic Limit

If the assumptions of (2.4.13) are satisfied, then

$$\begin{aligned}
&\lim_{V\to\infty} \frac{T}{V} \ln \mathrm{Tr} \exp[-\beta(H - \mu N)] \\
&= \lim_{V\to\infty} \frac{T}{V} \ln \sum_{N=0}^{\infty} \exp\left[-\beta V \left(\varphi_V \left(T, \frac{N}{V}\right) - \mu N\right)\right] \\
&= \sup(\mu\rho - \varphi(T, \rho)) = P(T, \mu).
\end{aligned}$$

2.5.5 Remarks

1. The supremum is attained where the right derivative

$$\lim_{\delta \downarrow 0} (\varphi(T, \rho + \delta) - \varphi(T, \rho))\delta^{-1} = \mu,$$

unless μ is on an endpoint of the interval on which $P(T, \mu)$ is defined. This means that with (2.3.24), μ can be identified with

$$\frac{\partial \varepsilon}{\partial \rho}\Big|_\sigma = \frac{\partial \varepsilon}{\partial \rho}\Big|_T - T\frac{\partial \sigma}{\partial \rho}\Big|_T = \frac{\partial \varphi}{\partial \rho}\Big|_T.$$

Because

$$\mu\rho - \varphi = \rho\frac{\partial \varepsilon}{\partial \rho}\Big|_\sigma + \sigma\frac{\partial \varepsilon}{\partial \sigma}\Big|_\rho - \varepsilon = P,$$

the grand canonical partition function turns out to be $\exp(PV/T)$. We shall also speak of P as the pressure when the system is finite, although it does not always agree with the definition as the force per area on the wall.

2. As before, the ensembles are equivalent, on account of the identities

$$\rho = \frac{\partial P}{\partial \mu}\Big|_T, \quad \varepsilon = T\frac{\partial P}{\partial T}\Big|_{\mu/T} - P = \mu\rho - T\frac{\partial \varphi}{\partial T}\Big|_\rho - \mu\rho + \varphi = \varphi + T\sigma,$$
$$T\sigma = \varepsilon - \mu\rho + P.$$

Observe that the grand canonical averages of N/V and H_N/V approach ρ and ε, and that the entropy density of ρ_{GC} equals σ.

2.5.6 Properties of the Pressure

1. The function $(T, \mu) \to P$ is convex, since it is the supremum of convex functions.
2. The pressure increases with μ, since it is the supremum of increasing functions.
3. If $H - \mu N \geq 0$, then $T^{-1}P$ is an increasing function of T, since $\exp[-\beta(H - \mu N)]$ is a decreasing function of β.

The grand canonical ensemble is particularly useful for identical particles, and allows the thermodynamic functions of bosons or fermions interacting with an external field to be evaluated more explicitly. For this purpose, we write the Hamiltonian and the particle number in terms of the field operators (1.3.2) and our orthogonal basis $\{f_m\}$, as

$$\begin{aligned} H &= \sum_{m,n} a_m^* a_n \left[\int d^3x \nabla f_m^*(\mathbf{x}) \cdot \nabla f_n(\mathbf{x}) + f_m^*(\mathbf{x}) f_n(\mathbf{x}) v(\mathbf{x}) \right] \\ &\equiv \sum_{m,n} a_m^* a_n \langle f_m | h | f_n \rangle, \\ N &= \sum_m a_m^* a_m, \end{aligned} \tag{2.16}$$

where $h = |\mathbf{p}|^2 + v(\mathbf{x})$ is the one-particle Hamiltonian, and a_m stands for $a(f_m)$. If h has pure-point spectrum with eigenvalues ε_m, and f_m are taken as the eigenvectors associated with ε_m, then

$$\operatorname{Tr}\exp[-\beta(H-\mu N)] = \operatorname{Tr}\exp\left[-\beta\sum_m a_m^* a_m(\varepsilon_m-\mu)\right]. \tag{2.17}$$

Taking the trace leads to easily computed sums, since a^*a has the eigenvalues 0 and 1 for fermions and $0, 1, 2, \ldots,$ for bosons. In these cases, P_F and P_B become

$$P_F(z) = -P_B(-z) = \frac{T}{V}\sum_m \ln(1+z\exp(-\beta\varepsilon_m)), \tag{2.18}$$

where $z \equiv exp(\beta\mu)$ is known as the **fugacity**. When written in terms of the one-particle Hamiltonian $h = |\mathbf{p}|^2 + V(\mathbf{x})$ and the trace tr on the one-particle space $L^2(\mathbb{R}^3)$,

2.5.7 The Pressure of Fermions or Bosons in an External Field

becomes

$$P_F(T,z) = \frac{T}{V}\operatorname{tr}\ln(1+z\exp(-\beta h)) = -P_B(T,-z).$$

2.5.8 Remarks

1. In the limit $z \to 0$, $P_F(T,z) = P_B(T,z) = z(T/V)\sum_m \exp(-\beta\varepsilon_m)$, which corresponds to very dilute matter, for which both Bose and Fermi statistics become the same (Boltzmann statistics).
2. If $h \geq 0$ and $\exp(-\beta h) \in C_1$, then the singularities of $\exp(P)$ occur where $z = -\exp(\beta\varepsilon_m) < -1$, $m = 0, 1, 2, \ldots$. The function $\exp(P)$ is analytic in z until the singularities are reached, i.e., the power series in z converges. The analytic function $P_F(T,z)$ describes all three kinds of statistics. Fermi statistics correspond to $z = \exp(\mu/T) > 0$, Boltzmann statistics to $z \to 0$, and Bose statistics to $-\exp(\varepsilon_0) < z < 0$ (see Figure 2.17).

   It is easy to calculate expectation values as well as the partition function:

$$\begin{aligned}\langle a_m^* a_{m'}\rangle &\equiv \operatorname{Tr} a_m^* a_{m'} \exp[-\beta(H-\mu N+PV)] \\ &= \frac{\delta_{mm'}}{\exp[\beta(\varepsilon_m-\mu)]\pm 1}.\end{aligned} \tag{2.19}$$

Since every one-particle vector $|f\rangle \in L^2(\mathbb{R}^3)$ can be expanded in eigenvectors of h, and when restricted to $L^2(\mathbb{R}^3)$, $a_f^* a_f$ equals $P_f = |f\rangle\langle f|$, the information about the one-particle observables is contained in the

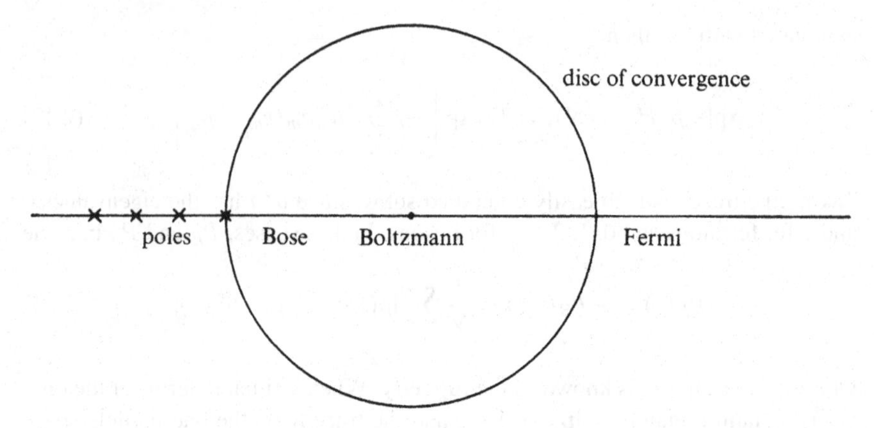

Figure 2.17. Singularities of P in the complex z-plane.

2.5.9 Effective One-Particle Density Matrix

One-particle expectation values are given by $\rho_1 = (\exp[\beta(h-\mu)] \pm 1)^{-1}$ with the formula $\langle a_f^* a_f \rangle = \operatorname{tr} \rho_1 P_f = \langle f|\rho_1|f\rangle$. The density matrix ρ_1 has the properties

$$\operatorname{tr} \rho_1 = \bar{N},$$

and

$$0 \le \rho_1 \le \begin{cases} 1 & \text{for fermions} \\ \bar{N} & \text{for bosons.} \end{cases}$$

2.5.10 Remarks

1. The number $\bar{N}$ is defined by $\langle \sum_m a_m^* a_m \rangle = \operatorname{tr}[\exp(\beta(h-\mu)) \pm 1]^{-1}$. If it is preferred to deal with these more understandable variables of the canonical ensemble, then this can be taken as the equation determining μ.
2. Similarly, $\langle H \rangle = \operatorname{tr} \rho_1 h$, etc.
3. If a reduced density matrix on the one-particle phase space is defined with coherent states (cf. (2.2.6) with (2.2.8; 5)), $\rho(\mathbf{x}, \mathbf{p}) = \langle a_z^* a_z \rangle = \langle z|\rho_1|z\rangle$, then the properties of ρ_1 generalize as

$$\int \frac{d^3x d^3p}{(2\pi)^3} \rho(\mathbf{x}, \mathbf{p}) = \bar{N},$$

and

$$0 \le \rho(\mathbf{x}, \mathbf{p}) \le \begin{cases} 1 & \text{for fermions} \\ \bar{N} & \text{for bosons.} \end{cases}$$

This shows that the exclusion principle of fermions has the effect of reducing the maximum value $\bar{N}$ of $\rho(z)$ allowed in quantum mechanics to 1.

As well as the one-particle observables, global properties like $\langle H\rangle$ and P can be calculated with ρ_1, and even the many-particle entropy can be expressed in terms of ρ_1:

2.5.11 The Effective One-Particle Entropy

$$\begin{aligned} S(\rho_{GC}) &= -\operatorname{Tr}\rho_{GC}\ln\rho_{GC} = \frac{VP}{T} + \beta\langle H-\mu N\rangle \\ &= \operatorname{tr}\left[\pm\ln(1\pm\exp[-\beta(h-\mu)]) + \beta\frac{h-\mu}{\exp[\beta(h-\mu)]\pm 1}\right] \\ &= -\operatorname{tr}[\rho_1\ln\rho_1 \pm (1\mp\rho_1)\ln(1\mp\rho_1)]. \end{aligned}$$

2.5.12 Remarks

1. The part in addition to the normal $-\operatorname{tr}\rho\ln\rho$ in S reveals that the many-particle system has increased disorder. The addition shows up for fermions in the entropy of a spin-$\frac{1}{2}$ density matrix,

$$S\begin{pmatrix}\rho & 0\\ 0 & 1-\rho\end{pmatrix} = -\rho\ln\rho - (1-\rho)\ln(1-\rho),$$

where ρ is the probability for spin-up. For bosons it is the entropy of an oscillator,

$$S\left((1-x)\begin{pmatrix}1 & & & \\ & x & & \\ & & x^2 & \\ & & & \ddots\end{pmatrix}\right) = -\rho\ln\rho + (1+\rho)\ln(1+\rho),$$

where

$$\rho = \sum_{n=0}^{\infty} n\rho_{nn} = \frac{x}{1-x}$$

is the expectation value of the number of bosons.

2. In accordance with the maximum-entropy principle (2.5.3), the one-particle ρ_1 (2.5.11) is the $\rho\in\mathcal{C}_1(L^2(\mathbb{R}^3))$ that maximizes

$$\frac{PV}{T} = -\operatorname{tr}[\rho\ln\rho \pm (1\mp\rho)\ln(1\mp\rho) + \rho\beta(h-\mu)]$$

(Problem 4). Also, on a formal level,

$$\begin{aligned} 0 = -\frac{V}{T}\frac{\delta P}{\delta\rho}\Big|_{\rho=\rho_1} &= \beta(h-\mu) + \ln\rho_1 - \ln(1\mp\rho_1) \\ \Rightarrow \rho_1 &= [\exp[\beta(h-\mu)]\pm 1]^{-1}. \end{aligned}$$

The density matrix ρ_1 describes the distribution of bosons or fermions. Its significance is brought out most clearly in the classical limit.

2.5.13 Classical Bounds for the Pressure of Particles in External Fields

With notation like that of (2.2.6), let

$$h = |\mathbf{p}|^2 + v(\mathbf{x}) = \int d\Omega_z f(\mathbf{z})|\mathbf{z}\rangle\langle\mathbf{z}|, \quad h(\mathbf{z}) = \langle\mathbf{z}|h|\mathbf{z}\rangle,$$

$$\rho(\mathbf{z}) = \mathrm{Tr}\, a_z^* a_z \rho_{GC} = \langle\mathbf{z}|[\exp[\beta(h-\mu)] \mp 1]^{-1}|\mathbf{z}\rangle = \langle\mathbf{z}|\rho_1|\mathbf{z}\rangle,$$

where v is such that all expressions appearing are well depend. Then, with $\mathbf{z} = \mathbf{q} + i\mathbf{p}$, for bosons,

$$-\int d\Omega_z \ln(1 - \exp[-\beta(h(\mathbf{z}) - \mu)]) \leq \beta P(\beta,\mu)V$$
$$\leq -\int d\Omega_z \ln(1 - \exp[-\beta(|\mathbf{p}|^2 + v(\mathbf{q}) - \mu)]),$$
$$\beta P(\beta,\mu)V \leq \int d\Omega_z \ln(1 + \rho(\mathbf{z})),$$

and for fermions,

$$\int d\Omega_z \ln(1 + \exp[-\beta(h(\mathbf{z}) - \mu)]) \leq \beta P(\beta,\mu)V$$
$$\leq \int d\Omega_z \ln(1 + \exp[-\beta(f(\mathbf{z}) - \mu)]),$$
$$-\int d\Omega_z \ln(1 - \rho(\mathbf{z})) \leq \beta P(\beta,\mu)V.$$

In analogy with (2.4.9), one gathers that

$$h(\mathbf{q} + i\mathbf{p}) = |\mathbf{p}|^2 + v_u(\mathbf{q}) + \int |\nabla u(\mathbf{x})|^2 d^3x$$

and

$$f(\mathbf{q} + i\mathbf{p}) = |\mathbf{p}|^2 + v^u(\mathbf{q}) - \int |\nabla u(\mathbf{x})|^2 d^3x,$$

where

$$v_u(\mathbf{q}) = \int v(\mathbf{x})|u(\mathbf{x} - \mathbf{q})|^2 d^3x$$

and

$$v(\mathbf{x}) = \int v^u(\mathbf{q})|u(\mathbf{x} - \mathbf{q})|^2 d^3q,$$

and u is an arbitrary vector of $L^2(\mathbb{R}^3)$ such that $\|u\|_2 = 1$ and $\|\nabla u\|_2 < \infty$.

Proof: *Bosons.* The first two inequalities are the analogues of (2.4.11), where the lower bound relies on (2.2.9) with the convex function $x \to -\ln(1 - \exp(-x))$.

The upper follows from Corollary (2.1.5; 7) if it is borne in mind that $h - \mu$ must be positive, so $\| \exp[-(h-\mu)]\| < 1$, and the series

$$- \ln(1 - \exp[-\beta(h-\mu)]) = \sum_{n=1}^{\infty} \frac{\exp[-n\beta(h-\mu)]}{n}$$

converges in the norm $\| \cdot \|$.It must even converge in the norm $\| \cdot \|_1$, since it was assumed that $-\ln(1 - \exp[-\beta(h-\mu)] \in \mathcal{C}_1$, and the series is monotonic. With recourse again to (2.4.9), each term is bounded by

$$-\int d\Omega_z (1/n) \exp[-n\beta(|\mathbf{p}|^2 + V(\mathbf{q}) - \mu)],$$

which also converges by assumption. Since all terms are positive, $\sum_n$ and $\int d\Omega_z$, can be interchanged. The final inequality follows from the concavity of the function $x \to \ln(1+x)$:

$$-\langle \mathbf{z}| \ln(1 - \exp[-\beta(h-\mu)])|\mathbf{z}\rangle = \langle \mathbf{z}| \ln(1+\rho_1)|\mathbf{z}\rangle \leq \ln(1 + \langle \mathbf{z}|\rho_1|\mathbf{z}\rangle)$$

implies that

$$-\operatorname{tr} \ln(1 - \exp[-\beta(h-\mu)]) \leq \int d\Omega_z \ln(1 + \rho(\mathbf{z})).$$

Fermions. The first two inequalities again come from (2.2.9) with the convex function $x \to \ln(1+\exp(-x))$, and the last one is a consequence of the convexity of $x \to -\ln(1-x)$. □

2.5.14 Remarks

1. If $x > 0$, then $(\exp(x) \pm 1)^{-1}$ is convex, and if $x < 0$, then it is concave. For bosons, $x > 0$, and so
$$\rho(\mathbf{z}) = \langle \mathbf{z}|(\exp[\beta(h-\mu)] - 1)^{-1}|\mathbf{z}\rangle \geq (\exp[\beta(h(\mathbf{z}) - \mu)] - 1)^{-1}.$$
The analogous inequality for fermions is true only if $h - \mu > 0$.
2. In Problem 3 it is shown that $\langle \mathbf{z}|(-\Delta)|\mathbf{z}\rangle = |\mathbf{p}|^2 + K$, $K = \int d^3x |\nabla u(\mathbf{x})|^2$, where $\mathbf{z} = \mathbf{q} + i\mathbf{p}$, and on the other hand, $-\Delta = \int d\Omega_z (|\mathbf{p}|^2 - K)|\mathbf{z}\rangle\langle\mathbf{z}|$. Similarly, $\langle \mathbf{z}|v\mathbf{z}\rangle = \int d^3x |u(\mathbf{x}-\mathbf{q})|^2 v(\mathbf{x}) = v_u(\mathbf{q})$, and $v = \int d\Omega_z v^u(\mathbf{q})|\mathbf{z}\rangle\langle\mathbf{z}|$, if $v(\mathbf{x}) = \int d^3q |u(\mathbf{x}-\mathbf{q})|^2 v^u(\mathbf{q})$. What goes on with the lower bound is thus that the classical Hamiltonian h is increased by the kinetic energy K of u, and the potential is smeared out by convolution with $|u|^2$. With the upper bound the classical Hamiltonian is reduced by K and the potential is unsmeared. If v is of slow enough variation that even for u with small K, $v^u(q)$ is approximately equal to $v_u(\mathbf{q}) = \langle \mathbf{z}|v|\mathbf{z}\rangle$, then the bounds draw close together.
3. In the very dilute limit of (2.5.8; 1) the bounds produce the classical result, if the indistinguishablity of the particles is accounted for by a $1/N!$ in the phase

space:

$$\exp\left(\frac{PV}{T}\right) = \sum_{N=0}^{\infty} \frac{1}{N!} \int d^3x_1 \cdots d^3p_N$$
$$\times \exp[-\beta(|\mathbf{p}_1|^2 + \cdots |\mathbf{p}_N|^2 + v(\mathbf{x}_1) + \cdots + v(\mathbf{x}_N) - N\mu)]$$
$$= \exp[\exp(-\beta(F_{\text{cl}} - \mu))],$$

so by (2.5.4),

$$\frac{PV}{T} = \int d^3x d^3p \exp[-\beta(|\mathbf{p}|^2 + v(\mathbf{x}) - \mu)] = N,$$

which is the ideal gas law. Unless $\exp(\beta(h - \mu)) \gg 1$, the statistics matter. They are built into the bounds, but the indeterminacy relation forces the bounds apart.

4. In the classical limit, in which Inequalities (2.5.13) become equalities, $\rho_1(\mathbf{x}, \mathbf{p}) = (\exp[-\beta(|\mathbf{p}|^2 + V(\mathbf{x}) - \mu)] \pm 1)^{-1}$ is the density on phase space that optimizes

$$\frac{PV}{T} = S(\rho_1) - \beta\langle h - \mu\rangle \equiv -\int d\Omega_z[\rho_1(\mathbf{z}) \ln \rho_1(\mathbf{z})$$
$$\pm (1 \mp \rho_1(\mathbf{z})) \ln(1 \mp \rho_1(\mathbf{z})) + \rho_1(\mathbf{z})\beta(h(\mathbf{z}) - \mu)]$$

(Problem 4).

5. If, more generally, ρ is a density matrix of the many particle system on Fock space, and

$$\rho_1 = \int d\Omega_z d\Omega_{z'} |\mathbf{z}\rangle\langle\mathbf{z}'| \operatorname{Tr}(\rho a_{z'}^* a_z)$$

and

$$\rho(\mathbf{z}) = \operatorname{Tr}(\rho a_z^* a_z) = \langle\mathbf{z}|\rho_1|\mathbf{z}\rangle$$

are the associated one-particle density matrix and density, then it follows from (2.5.3) and (2.5.11) that

$$S(\rho) = -\operatorname{Tr}\rho \ln \rho \le -\operatorname{tr}[\rho_1 \ln \rho_1 \pm (1 \mp \rho_1) \ln(1 \mp \rho_1)]$$
$$\le -\int d\Omega_z[\rho(z) \ln \rho(z) \pm (1 \mp \rho(z)) \ln(1 \mp \rho(z))],$$

where the H in (2.5.3) is taken as the second quantization of $(1/\beta)[\ln(1 \mp \rho_1) - \ln \rho_1]$, and μ is set to 0. The first inequality becomes an equality with ρ_{GC}, which is the density matrix of greatest entropy for a given one-particle density matrix ρ_1. The second inequality follows from (2.2.9), since

$$x \to -[x \ln x \pm (1 \mp x) \ln(1 \mp x)]$$

is concave with the upper signs for $0 < x < 1$ and with the lower signs for $x < 0$.

The extent of the validity of the classical picture will be delineated through a series of examples.

2.5.15 Free Bosons and Fermions in a Box with Soft Walls

With a harmonic potential $v(\mathbf{x}) = \omega^2|\mathbf{x}|^2$, the N-particle Hamiltonian is

$$H = \sum_{i=1}^{N}(|\mathbf{p}_i|^2 + \omega^2|\mathbf{x}_i|^2) = \sum_{i=1}^{N}|\mathbf{p}_i|^2 + \frac{\omega^2}{2N}\sum_{i,j=1}^{N}|\mathbf{x}_i - \mathbf{x}_j|^2 + \frac{\omega^2}{N}|\sum_{i=1}^{N}\mathbf{x}_i|^2,$$

containing harmonic forces between the particles and a harmonic force acting on the center of mass. As before (cf. (2.4.10) and (2.5.14; 1)), let $\mathbf{z} = \mathbf{q} + i\mathbf{k}$ and $u(\mathbf{x}) = \exp(-\omega|\mathbf{x}|^2/2) : h(\mathbf{z}) = |\mathbf{k}|^2 + \omega^2|\mathbf{q}|^2 + 3\omega$, $f(|\mathbf{z}|) = |\mathbf{k}|^2 + \omega^2|\mathbf{q}|^2 - 3\omega$. Because

$$\mp\int\frac{d^3kd^3q}{(2\pi)^3}\ln(1 \mp \exp[-\beta(|\mathbf{k}|^2 + \omega^2|\mathbf{q}|^2 - \mu)])$$
$$= \pm\frac{T^3}{(2\omega)^3}\sum_{\nu=1}^{\infty}(\pm 1)^\nu\frac{\exp(\nu\beta\mu)}{\nu^4},$$

(2.5.17) implies

$$\pm\frac{T^3}{(2\omega)^3}F_4(\pm(\exp[\beta(\mu - 3\omega)]) \leq \ln \operatorname{Tr}_{\substack{B\\F}} \exp[-\beta(H - \mu N)]$$
$$\leq \pm\frac{T^3}{(2\omega)^3}F_4(\pm\exp[\beta(\mu + 3\omega)]),$$

where

$$F_\sigma(x) \equiv \sum_{\nu=1}^{\infty}\frac{x^\nu}{\nu^\sigma}. \tag{2.20}$$

The result can be calculated exactly in this case, since the eigenvalues are $\varepsilon_m = 3\omega + 2\omega(m_1 + m_2 + m_3)$, $\mathbf{m} \in (\mathbb{Z}^+)^3$, and so

$$\mp\sum_m \ln(1 \mp \exp[-\beta(\varepsilon_m - \mu)])$$
$$= \pm\sum_{\nu=1}^{\infty}(\pm 1)^\nu\frac{\exp[\nu\beta(\mu - 3\omega)]}{\nu}[(1 - \exp(-2\beta\omega\nu))]^{-3}.$$

The bounds draw together to this value in the limit $\omega \to 0$. This limit is related to the limit $V \to \infty$, since the average of, for instance, $|\mathbf{x}|^2$ is $\sim T/\omega^2$. Accordingly, we eliminate ω in favor of the effective volume $V = (\pi T)^{3/2}/\omega^3$ and take the limit $V \to \infty$. Then with $z = \exp(\beta\mu)$, (2.20) yields

$$P_{\substack{B\\F}}(T, z) = \pm\frac{T^{5/2}}{8\pi^{3/2}}F_4(\pm z). \tag{2.21}$$

2.5.16 Remarks

1. As $\omega \to 0$, the potential v goes to zero pointwise, and the density (2.5.14;3) on phase space turns into the well-known Bose or Fermi distribution,

$$\rho(\mathbf{x}, \mathbf{p}) = [\exp(\beta(|\mathbf{p}|^2 - \mu)) \pm 1]^{-1}.$$

2. The energy spectrum of this example resembles that of a massless particle in a box $\{\mathbf{x} : |x_i| < L/2\}$, $E = (p_1^2 + p_2^2 + p_3^2)^{1/2}$, $p_i = m_i\pi/L$, $\mathbf{m} \in (\mathbb{Z}^+)^3$. In the limit $L \to \infty$, this E produces the same pressure up to a constant as $(m_1 + m_2 + m_3)\omega$, when ω is identified with π/L. Then

$$P_{\substack{B\\F}}(T, z) = \pm T^4 F_4(\pm z)\pi^{-3}.$$

2.5.17 A Box with Hard Walls

Now suppose that the potential $v_L(\mathbf{x}) \geq 0$ is significantly smaller than $1/L^2$ for $|x_i| < L/2$ but increases exponentially as soon as $|x_i| > L/2$. Since what happens should not depend on the precise form of v_L, only certain bounds will be imposed on v_L. Because of the monotonic property, all the steps up to (2.5.17) and (2.5.18;1) proceed as before.[1]

$$\gamma_-^3\varphi(x)\varphi(y)\varphi(z) \leq v_L(\mathbf{x}) \leq \gamma_+^3\varphi(x)\varphi(y)\varphi(z), \quad 0 < \gamma_- < \gamma_+,$$

$$\varphi(x) = \exp\left(\tfrac{-cL}{2}\right)\cosh(cx),$$

$$\mathcal{N}^2 \int_{-\infty}^{\infty} dx' \exp(-bx'^2)\varphi(x + x') = \exp\left(-\tfrac{c^2}{4b}\right)\varphi(x),$$

so for the other bound,

$$\varphi(x) = \int_{-\infty}^{\infty} dx \exp(-bx'^2)\exp\left(-\frac{c^2}{4b}\right)\varphi(x' + x)\mathcal{N}^2.$$

The x-space portion of the calculation of $\sum_{\nu=1}^{\infty}(-1)^{\nu+1}[\exp(\nu\beta\mu)/\nu] \times \int d\Omega_z \exp(-\beta\nu g(z))$, where $g(z) = f(z)$ or respectively $h(z)$ (cf.(2.5.17) and (2.2.11)), leads to

$$\int_{-\infty}^{\infty} dx \exp(-B_\pm \cosh cx) = \frac{2}{c}\int_1^{\infty} \frac{dv}{\sqrt{v^2 - 1}} \exp(-B_\pm v)$$

$$\overset{B_\pm \to 0}{=} \frac{2}{c}\left(\ln\frac{1}{B_\pm} + O(1)\right)$$

with $B_\pm = \gamma_\pm\beta\exp[\pm c^2/4b - cL/2])$, since it is being evaluated in the limit $V = L^3 \to \infty$. If a sequence $(v_L(\mathbf{x}))_{L\to\infty}$ of wall potentials has bounds of the

[1] From this point until right before (2.5.24), + and − will indicate upper and lower bounds for the potential due to the wall rather than Bose and Fermi statistics.

above-mentioned form with $c(L) = o(L)$ and $\ln(\beta\gamma_\pm(L)) = o(c(L)\cdot L)$, then

$$\frac{2}{cL}\ln\frac{1}{B_\pm} = 1 \mp \frac{c}{2bL} - \frac{2}{cL}\ln\beta\gamma_\pm$$

converges to 1 for both bounds. The p-integral is the same as in (2.5.19), and so, finally,

$$P_{\substack{B\\F}}(T,z) = \pm\frac{T^{5/2}}{8\pi^{3/2}}F_{5/2}(\pm z). \qquad (2.22)$$

2.5.18 Remarks

1. This is the same result as that of summing over all the eigenvalues of a free particle in a box with Dirichlet boundary conditions on the wall (Problem 5). The bounds (2.5.17) show that in very large part it is only the total volume of V that matters, rather than its detailed form.
2. The nature of the wall is expressed by $F_{5/2}$ in (2.5.23) and F_4 in (2.5.21). For lower densities, $z \ll 1$, they coincide, as $F_\sigma = z + O(z^2)$.

2.5.19 The Thermodynamic Functions of Free Particles

All the thermodynamic functions can be obtained from $P(T,z)$, so (2.5.24) will allow the gaps left by (2.3.10) to be filled in, and the functions can be written down explicitly. We shall investigate the limiting cases where $z \to \infty$, $z \to 0$, and $z \to -1$, corresponding to the extremes of Fermi, Boltzmann, and Bose statistics. The limits $z \to \infty, -1$ are what is referred to as a **degenerate gas**. By Problem 1, F has the asymptotic forms

$$-F_{5/2}(-z) \longrightarrow \begin{cases} -\zeta(\frac{5}{2}) + (z+1)\zeta(\frac{3}{2}) & z \to -1 \\ z - z^2 \cdot 2^{-5/2} & z \to 0 \\ 4\left[\frac{2}{5}(\ln z)^{5/2} + \frac{\pi^2}{4}(\ln z)^{1/2}\right]/3\sqrt{\pi} & z \to \infty, \end{cases} \qquad (2.23)$$

where $\zeta(\sigma)$ is the Riemann zeta function,

$$\zeta(\sigma) = \sum_{\nu=1}^{\infty}\frac{1}{\nu^\sigma} = F_\sigma(1), \quad \sigma \in \mathbb{C}, \quad \mathrm{Re}\sigma > 1.$$

The zeta function has an analytic continuation to the punctured complex plane $\{\sigma \in \mathbb{C} | \sigma \neq 1\}$. In the three limits,

$$\frac{P}{T} = \frac{2\varepsilon}{3T} \longrightarrow \begin{cases} T^{3/2}\left[\zeta(\frac{5}{2}) + (z-1)\zeta(\frac{3}{2})\right]/8\pi^{3/2} & \textit{Bose} \\ T^{3/2}z(1 \pm z\cdot 2^{-5/2})/8\pi^{3/2} & \textit{Boltzmann} \\ [8pt]T^{3/2}\left[\frac{2}{5}(\ln z)^{5/2} + \frac{\pi^2}{4}(\ln z)^{1/2}\right]/6\pi^2 & \textit{Fermi}, \end{cases} \qquad (2.24)$$

so, writing

$$\varepsilon_{\substack{B\\F}} = T^2 \frac{\partial}{\partial T} \frac{1}{T} P_{\substack{B\\F}}(T,z) = \frac{3}{2} P_{\substack{B\\F}}(T,z) = \pm \frac{3}{2} T^{5/2} F_{5/2}(\pm z) \frac{1}{8\pi^{3/2}},$$

$$\rho_{\substack{B\\F}} = z \frac{\partial}{\partial z} \frac{1}{T} P_{\substack{B\\F}}(T,z) = \pm T^{3/2} F_{3/2}(\pm z) \frac{1}{8\pi^{3/2}},$$

$$\sigma_{\substack{B\\F}} = \pm \frac{T^{3/2}}{8\pi^{3/2}} \left[\frac{5}{2} F_{5/2}(\pm z) - \ln z F_{3/2}(\pm z) \right], \tag{2.25}$$

to the lowest nonvanishing order,

$$\rho \longrightarrow \begin{cases} z\zeta(\frac{3}{2}) T^{3/2}/8\pi^{3/2} & \textit{Bose} \\ T^{3/2} z(1 \pm z \cdot 2^{-3/2})/8\pi^{3/2} & \textit{Boltzmann} \\ T^{3/2}(\ln z)^{3/2}/6\pi^2 + T^{3/2}(\ln z)^{-1/2}/48 & \textit{Fermi}, \end{cases} \tag{2.26}$$

$$\sigma \longrightarrow \begin{cases} T^{3/2} \frac{5}{2} \zeta(\frac{5}{2})/8\pi^{3/2} & \textit{Bose} \\ T^{3/2}/(8\pi^{3/2} \left(\frac{5}{2} z - \ln z\right)) & \textit{Boltzmann} \\ T^{3/2}(\ln z)^{1/2}/12 & \textit{Fermi}. \end{cases} \tag{2.27}$$

When expressed in terms of the more intuitively appealing variables ρ and T,

$$P = \frac{2}{3}\varepsilon \longrightarrow \begin{cases} \rho T + T^{5/2} \left(\zeta(\frac{5}{2}) - \zeta(\frac{3}{2})\right)/8\pi^{3/2} & \textit{Bose} \\ \rho T \quad \text{it(“ideal gas”)} & \textit{Boltzmann} \\ (6\pi^2 \rho)^{5/3}/15\pi^2 + (6\pi^2\rho)^{1/3} T^2/24 & \textit{Fermi}, \end{cases}$$

$$\sigma \longrightarrow \begin{cases} \frac{5}{2}\zeta(\frac{5}{2}) T^{3/2}/8\pi^{3/2} & \textit{Bose} \\ \left.\begin{array}{l} 5\rho/2 - \rho \ln(\rho 8\pi^{3/2}/T^{3/2}) \\ = \rho \ln(T^{3/2} \exp(5/2)/\rho 8\pi^{3/2}) \end{array}\right\} & \textit{Boltzmann} \\ T(6\pi^2\rho)^{1/3}/12 & \textit{Fermi}. \end{cases} \tag{2.28}$$

2.5.20 Remarks

1. As $z \to 0$, (2.28) gives the classical result (2.3.7; 1) with an additional factor $1/N!$ in the volume of phase space. If V_P denotes the volume available in the one-particle phase space, and the $1/N!$ is incorporated into the general definition, then

$$S \sim \ln \frac{1}{N!} \left(\frac{V_P}{h^3}\right)^N$$

leads to

$$\frac{S}{N} \sim \ln \frac{V_P}{Nh^3}.$$

On the other hand, in configuration space and with units for which $\hbar = m = 1$, (2.5.27) informs us that $S \sim \ln VT^{3/2}/N$. Since $T^{-1/2}$ equals the thermal de

Broglie wavelength λ, with these units, the following rule of thumb applies to the entropy: Entropy per particle=ln{ volume of phase space per particle, as measured in h^3}=ln{ volume of configuration space per particle, as measured in λ^3}.

2. Fermions have a zero-point energy $E_0 = V\varepsilon_0$ left over when $T \to 0$, where $\varepsilon_0 \equiv (6\pi^2\rho)^{5/3}/10\pi^2$, and a zero-point pressure $2\varepsilon_0/3$. Because

$$T = \frac{4(\varepsilon - \varepsilon_0)^{1/2}}{6(\pi^2\rho)^{1/6}},$$

it is also possible to write

$$\sigma = \left(\frac{\varepsilon}{\varepsilon_0} - 1\right)^{1/2} \frac{\rho 2\pi}{\sqrt{10}},$$

showing that the number M of states in the interval $[E_0, E]$ is

$$M \cong \exp\{N\left(\frac{E}{E_0} - 1\right)^{1/2} \frac{2\pi}{\sqrt{10}}\}.$$

For example, in an atomic nucleus the kinetic energy is $E_0 \cong N \cdot 20$ eV, so with a fixed kinetic excitation energy $\delta E = E - E_0$ the number of states in the interval is $\sim \exp 2\sqrt{N}\sqrt{\delta E/20\text{MeV}}$. If $\delta E \sim 1$ MeV, then for $N = 20$ there are about e^2, i.e., 7 or 8, states; whereas if $N \sim 200$, then the number increases to about $e^{6.5} \sim 0.5 \times 10^3$. This is in agreement with the experimental observation that the density of the energy states of heavy nuclei is on the order of $(\text{eV})^{-1}$.

3. If the energy of the ground state is redefined to zero, then z must be less than 1 for bosons – otherwise by (2.19) $n_0 \equiv \langle a_0^* a_0 \rangle = z/(1-z)$ is either infinite or negative. Because $F_{3/2}(z) < \zeta(\frac{3}{2})$ when $0 < z < 1$, it follows from (2.25) that $T > T_c \equiv (8\pi^{3/2}\rho/\zeta(\frac{3}{2}))^{2/3}$. On the other hand, n_0 can be made arbitrarily big by taking z close enough to 1. The difficulty with this is that the two limits $z \to 1$ and $V \to \infty$ have to be taken jointly if the density has been fixed. If $z(V) = 1 - 1/\rho_0 V$ and $T < T_c(\rho)$, then

$$\rho = \rho_0 + \zeta\left(\frac{3}{2}\right)\frac{T^{3/2}}{8\pi^{3/2}},$$

$$P = \frac{2}{3}\varepsilon = \zeta\left(\frac{5}{2}\right)\frac{T^{5/2}}{8\pi^{3/2}} = \lim_{V\to\infty}\frac{T}{V}\ln \operatorname{Tr}\exp\left[-\frac{1}{T}(H_V - \mu_V(T,\rho)N)\right],$$

if

$$\mu = T\Theta(T_c(\rho) - T)\ln(1 - 1/\rho_0 V)$$

$$\sigma = \frac{5}{2}\zeta\left(\frac{5}{2}\right)\frac{T^{3/2}}{8\pi^{3/2}} = \frac{5}{2}(\rho - \rho_0)\zeta(\frac{5}{2})/\zeta(\frac{3}{2}).$$

This shows that a nonzero fraction $\rho_0/\rho = 1 - (T/T_c)^{3/2}$ of the particles reside in the ground state and contribute nothing to the energy, pressure, or entropy (provided H is replaced with $H - E_0$). The number of particles in the first excited state, $n_1 = 1/(z^{-1}\exp(\beta/L^2)-1) \sim L^2$, is rather large, but $n_1/V \to 0$. For similar reasons, the relative mean-square deviation $(\Delta n_i)^2/\langle n_i\rangle^2$ remains positive for n_0 as $V \to \infty$, but goes to zero for the higher states. The specific heat

$$c_V = \frac{\partial E}{\partial T}\Big|_{V,N}$$

is continuous at T_c and $\partial c_V/\partial T$ is discontinuous (Problem 2). If $T = T_c$ then the choice of ρ_0 has to depend on V.

4. The values $\mu = 0$ and $z = 1$ apply to a situation where N is not conserved, such as a gas of photons or phonons (cf.(2.5.22; 2)). It is easy to calculate $\operatorname{Tr}\exp(-\beta H)$ with the H of (2.5.7). The pressure $P = -\varphi$, and

$$\frac{\partial P}{\partial \rho}\Big|_T = -\frac{\partial \varphi}{\partial \rho}\Big|_T = \mu = 0,$$

so the compressibility is infinite. The system behaves much like a gas at the condensation point, the vacuum state, i.e., no particles, being analogous to the condensed state. It therefore has $\varepsilon = \sigma = P = V = 0$, and the system can be compressed into the vacuum. The entropy density σ is then simply the quantity $\Delta s/\Delta v$ of the Clausius–Clapeyron equation which simply assumes the form

$$\frac{\partial P}{\partial T}\Big|_\rho = \sigma.$$

Since $P = -\varphi$, Theorem (2.4.14) implies that this equation holds identically. The quantities $\varepsilon/T \approx \rho \approx \sigma$ depend only on T and correspond to a particle of energy T in each wavelength cube. Consequently, entropy $\cong$ particle number $\cong$ energy/T.

2.5.21 Particles in a Magnetic Field

The Hamiltonian was given in (III: 3.3.5; 3):

$$H = |\mathbf{p} - e\mathbf{A}|^2 = p_3^2 + 2eB(a^* a) + \frac{1}{2}.$$

The boundary conditions are that the wave-function must vanish at $x_3 = 0$ and $x_3 = L$, the 3-axis pointing along B, so the eigenvalues of p_3 are $\pi m/L$, $m = 1, 2, 3, \ldots$. The center $\bar{\mathbf{x}}$ of the orbit is confined to $|\bar{\mathbf{x}}|^2 = (2/eB)(g + \frac{1}{2}) < R^2$ in the plane perpendicular to B, so the geometry is cylindrical. The "wall potential" $\infty \cdot \Theta(|\bar{\mathbf{x}}|^2 - R^2)$ confining the particle is not a multiplication operator by a real-valued function $V(x_1, x_2)$, but rather a function of the operator

$$|\bar{\mathbf{x}}|^2 = \frac{1}{4}\left(x_1^2 + x_2^2\right) + \frac{1}{e^2B^2}\left(p_1^2 + p_2^2\right) + \frac{1}{eB}\left(x_1 p_2 - x_2 p_1\right),$$

representing the sum of a two-dimensional harmonic oscillator in the $(x_1 - x_2)$-plane and the x_3-component of the angular momentum. The construction of such a momentum-dependent wall potential will be left to the ingenuity of the experimentalists. By (III: 3.3.5; 3), $|\bar{\mathbf{x}}|^2$is quantized so that g is a whole number, and a^*a has the eigenvalues $n = 0, 1, 2, \ldots$. As $L \to \infty$, the sum $\sum_{g=0}^{R^2eB/2} \sum_{m=1}^{\infty} \sum_{n=0}^{\infty}$ turns into

$$\int_0^\infty dp_3 \frac{L}{\pi} \sum_{n=0}^{\infty} \frac{R^2 eB}{2} = \frac{VeB}{2\pi^2} \int_0^\infty dp_3 \sum_{n=0}^{\infty},$$

where V denotes the volume of the cylinder. The classical bounds amount to the replacement

$$\sum_{n=0}^{\infty} \to \int_0^\infty dn,$$

in which all magnetic effects are swept away. We have to resort to the exact expression (2.5.9), with which the grand canonical partition functions becomes

$$\begin{aligned} \beta P_{\substack{B\\F}}(z) &= \mp \frac{eB}{2\pi^2} \int_0^\infty dp_3 \sum_{n=0}^{\infty} \ln\left(1 \mp z \exp\left[-\beta(p_3^2 + eB(2n+1))\right]\right) \\ &= \pm \frac{T^{3/2}}{8\pi^{3/2}} \sum_{\nu=1}^{\infty} \frac{(\pm z)^\nu}{\nu^{5/2}} \frac{\nu eB\beta}{\sinh \nu eB\beta}, \end{aligned} \tag{2.29}$$

where the B in $P_{\substack{B\\F}}$ denotes Bose statistics as usual and has nothing to do with the magnetic field B. This reveals right away that, as in (2.3.32; 2), an arbitrarily weak magnetic field ruins the phase transition of the Bose gas, since for any T.

$$\rho_{\substack{B\\F}} = z\frac{\partial}{\partial z}\beta P_{\substack{B\\F}} = \pm \frac{T^{3/2}}{8\pi^{3/2}} \sum_{\nu=1}^{\infty} \frac{(\pm z)^\nu}{\nu^{1/2}} \frac{eB\beta}{\sinh \nu eB\beta}$$

can get arbitrarily big as $z \to \exp(\beta E_0) = \exp(\beta eB)$. This happens because the particles are free to move only parallel to $\mathbf{B}$ and are trapped in orbits in the direction perpendicular to $\mathbf{B}$ even though the radius of the cylinder goes to infinity. The system acts as though confined to a cylinder only the length of which tends to ∞, and in one dimension there is no Bose condensation. If the magnetic energy eB is much less than the thermal energy T, then the next correction to the foregoing result is $\sim B^2$:

$$\beta P_{\substack{B\\F}} \to \pm \frac{T^{3/2}}{8\pi^{3/2}} \left[F_{5/2}(\pm z) - \frac{1}{6}\left(\frac{eB}{T}\right)^2 F_{1/2}(\pm z) \right]. \tag{2.30}$$

If this is used to calculate the magnetization per volume in the limit $B \to 0$ with T fixed,

$$m = \frac{\partial P_{\substack{B\\F}}}{\partial eB} = \frac{1}{V}\langle \sum_{\substack{\text{all}\\ \text{particles}}} (x_1 p_2 - x_2 p_1 - eB(x_1^2 + x_2^2))\rangle$$

$$= \mp \frac{T^{3/2}}{8\pi^{3/2}} \frac{eB}{3T} F_{1/2}(\pm z), \tag{2.31}$$

then with (2.5.19) and the formula $F_{\sigma-n}(z) = (z(d/dz))^n F_\sigma(z)$ (see (2.5.20)), its limits in the three extreme cases of the different statistics are

$$m \longrightarrow \begin{cases} -eB \cdot \infty & \textit{Bose} \\ -eB\rho/3T & \textit{Boltzmann} \\ -eB(6\pi^2\rho)^{1/3}/12\pi^2 & \textit{Fermi}. \end{cases} \tag{2.32}$$

2.5.22 Remarks

1. The negative sign indicates diamagnetism, which is to be expected quantum-mechanically: By Lenz's law the classical orbits rotate in the direction with negative L_z. However, a current appears in the other direction when particles bounce off the wall of the box (see Figure 2.18). With classical statistics the circulating currents cancel out at every point of the interior, leaving only a current circulating along the surface, which is exactly compensated for by the "reflected" current, since the partition function

$$\int d^3x d^3p \exp[-\beta|\mathbf{p} - \mathbf{A}(\mathbf{x})|^2] = \int d^3x d^3p \exp(-\beta|\mathbf{p}|^2)$$

   is completely independent of B. This means that if either ρ is fixed and $T \to \infty$ or T is fixed and $\rho \to 0$, then m tends to 0. Diamagnetism is therefore a characteristically quantum-mechanical effect; if the sum $\sum_{n=0}^{\infty}$ is replaced with an integral $\int_0^\infty dn$, and $2n+1$ becomes $2n$, which is in essence the limit $\hbar \to 0$, then P becomes independent of B (a theorem of Bohr and van Leeuwen).
2. In quantum theory, states with negative L_z are energetically favored (III: 3.3.19; 4), so a quantum gas is diamagnetic. The reason that the magnetization m of a completely degenerate Bose gas tends to ∞ is that P fails to be analytic at $z = 1$, $B = 0$. This topic will shortly be discussed in more detail.

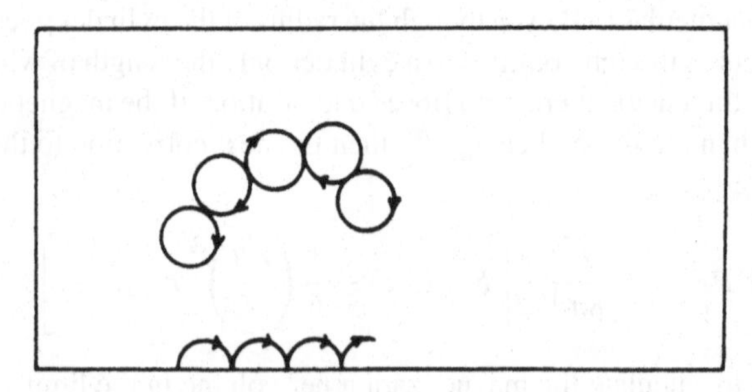

Figure 2.18. Classical trajectories of particles in a box with a magnetic field.

3. Since P depends only on R^2L,

$$R\frac{\partial}{\partial R}P = 2L\frac{\partial}{\partial L}P,$$

i.e., the pressure remains isotropic.

In order to make sense of the limit of degenerate Bose gas, let $\beta\mu = \ln z$, and write

$$\frac{\beta P_B}{V} = \frac{T^{3/2}}{8\pi^{3/2}}\sum_{\nu=1}^{\infty}\frac{\exp[-\beta\nu(eB-\mu)]}{\nu^{5/2}}\frac{2eB\beta\nu}{1-\exp[-2eB\beta\nu]},$$

$$\rho = \frac{T^{3/2}}{8\pi^{3/2}}\sum_{\nu=1}^{\infty}\frac{\exp[-\beta\nu(eB-\mu)]}{\nu^{3/2}}\frac{2eB\beta\nu}{1-\exp[-2eB\beta\nu]},$$

$$m = -\rho + \frac{T^{3/2}}{4\pi^{3/2}}\sum_{\nu=1}^{\infty}\frac{\exp[-\beta\nu(eB-\mu)]}{\nu^{3/2}(1-\exp[-2eB\beta\nu])^2} \times[1-\exp[-2eB\beta\nu](1+2eB\beta\nu)] > -\rho \qquad (2.33)$$

without expanding in B. The convergence of the series for m and ρ in (2.33) (by domination for $B \geq 0$ with μ fixed) implies that

$$\lim_{B\to+0} m(T,\mu,B) = 0$$

for all fixed $T > 0$ and $\mu < 0$. Yet if $B \to 0$ with T fixed and $\mu < 0$ then all the densities ρ are less than $\zeta(\frac{3}{2})T^{3/2}/8\pi^{3/2}$, as in (2.5.20;3). If $T \leq T_c(\rho)$ (see (2.5.33;3)), then the limits $B \to 0$ and $\mu \to 0$ must again be appropriately coordinated. Since for $B > 0$ and for all values $\rho > 0$ and $T > 0$ there exists a unique $\mu(T,\rho,B) < eB$ such that $\lim_{B\to 0}\mu(T,\rho,B) = 0$ for $T \leq T_c(\rho)$), and since the series for $m+\rho$ from (2.33) also converges uniformly in B on an interval containing $\mu = eB$, the limit $B \to 0$ can be taken term by term. This yields

$$\lim_{B\to 0} m(T,\rho,B) = -\rho_0 = -\rho\left[1-\left(\frac{T}{T_c(\rho)}\right)^{3/2}\right],$$

provided that $T \leq T_c(\rho)$ (cf. (2.5.20; 3)). If $T \geq T_c(\rho)$ then the limit is zero as observed earlier.

2.5.23 Remarks

1. The physical interpretation of this result is that in the limit $B \to 0$ only the particles in the ground state contribute to the magnetization. The ground state has $L_z = -1$, so for $B = 0$ the contribution to m is simply the sum of L_z over the particles in a unit volume in the ground state.
2. The notation B is perhaps misleading, since it stands only for the external field and not for that due to the system itself. Actually, the field due to the system has to be taken into account, as it screens B throughout the interior of the system.

2.5.24 Black-Body Radiation in Partial (i.e., Anisotropic) Equilibrium

If the particles are massless, as in (2.5.16; 2) and (2.5.20; 4), and they have a density matrix like ρ_{GC} but containing only states in a certain dilatation-invariant part D of p-space, then we can still write

$$\varphi = T \int_D \frac{d^3 p}{(2\pi)^3} \ln(1 - \exp[-\beta|\mathbf{p}|]) = -cT^4,$$

where the constant c depends on D (but not on T). It is then still true that

$$\varepsilon = 3P = -3\varphi = \frac{3}{4} T\sigma = 3cT^4.$$

A realistic example of this situation is sunlight falling on the earth, for which essentially all the p-vectors come from the direction of the sun. The constant c is reduced by a factor $\sim 10^{-5}$, the solid angle subtended by the sun, in comparison with the isotropic equilibrium value with $D = \mathbb{R}^3$. Once the radiation is made isotropic without changing ε significantly by the time it reaches the earth, T is lowered by a factor of about $10^{5/4}$, from $\sim 6000°$ K to $\sim 300°$ K. At the same time, $\sigma = 4\varepsilon/3T$ is increased by this factor of 20. This just means that by energy conservation we make 20 room temperature photons out of one photon from the sun. It is consistent with an increase in the total entropy that this physical process creates highly ordered structures with little entropy; their decrease of entropy is nothing compared with the gigantic increase of the radiation entropy. About 10^{20} photons per cm^2 arrive from the sun each minute, and this times 20 is the entropy increase of sunlight per cm^2 and min. In an hour this comes to roughly the total entropy of a cubic centimeter of matter for each square centimeter of ground, so, for example, a newly planted forest could grow to a height of 10 meters over a summer without violating the second law of thermodynamics. The sun thus expends entropy as well as energy. Although isotropic black-body radiation at 300° K would be just as energetic, the energy would be unusable for the creation of life (as would be the case as the universe subsided into heat death).

The grand canonical ensemble determines the expectation values of field operators as well as the thermodynamic functions. Equation (2.19) showed how to calculate quadratic expressions involving the field operators, and quartic expressions for particles in an external field can easily be calculated in the same way,

$$\begin{aligned} \langle a_m^* a_j^* a_{j'} a_{m'} \rangle &= (\delta_{mm'}\delta_{jj'} \pm \delta_{mj'}\delta_{jm'})(\exp[\beta(\varepsilon_m - \mu)] \mp 1)^{-1} \\ &\quad \times (\exp[\beta(\varepsilon_j - \mu)] \mp 1)^{-1} \\ &= \langle a_m^* a_{m'} \rangle \langle a_j^* a_{j'} \rangle \pm \langle a_m^* a_{j'} \rangle \langle a_j^* a_{m'} \rangle. \end{aligned} \tag{2.34}$$

2.5.25 Remark

If the mean-square deviations of the occupation numbers are calculated in this way, then

$$\langle (a_m^* a_m)^2 \rangle - \langle a_m^* a_m \rangle^2 = \langle a_m^* a_m \rangle (1 \pm \langle a_m^* a_m \rangle).$$

Independent particles would follow a Poisson distribution law $w(n) = \exp(-\bar{n})\bar{n}^n/n!$ for which the mean-square deviation would equal the expectation value of the occupation number. The deviation is greater with Bose statistics and less with Fermi statistics, which can be interpreted as meaning that bosons have a tendency to bunch up and fermions to keep at a distance.

In elementary quantum mechanics a state was characterized by the expectation values of the Weyl operators (cf. (III: 3.1.2; 1)), and likewise now the complete determination of the state requires the expectation value of, say, $\exp[i \int d^3x (a(\mathbf{x}) f^*(\mathbf{x}) + a^*(\mathbf{x}) f(\mathbf{x}))]$ for all $f \in C_0^\infty(\mathbb{R}^3)$. The best way for this to be calculated in the grand canonical ensemble for particles in an external field makes use of coherent states. In Problem 6 it is shown that

$$\frac{\operatorname{Tr} \exp[-\beta\omega a^* a] \exp[i(a^*\alpha + a\alpha^*)]}{\operatorname{Tr} \exp[-\beta\omega a^* a]} = \exp\left[-|\alpha|^2 \left(\frac{1}{2} + \frac{1}{\exp(\beta\omega) - 1}\right)\right] \quad \text{if } [a, a^*] = 1.$$

Therefore:

2.5.26 The Grand Canonical State for Bosons in an External Field

is

$$\left\langle \exp\left[i \sum_m (a_m^* \alpha_m + a_m \alpha_m^*)\right]\right\rangle = \exp\left[-\sum_m |\alpha_m|^2 \left(\frac{1}{2} + \frac{z}{\exp(\beta\varepsilon_m) - z}\right)\right].$$

2.5.27 Example

Free bosons in a cube of volume $V = L^3$, with periodic boundary conditions. Let

$$a_V(\mathbf{k}) = \int_V \frac{d^3x}{L^{3/2}} \exp(-i\mathbf{k} \cdot \mathbf{x}) a(\mathbf{x}),$$

and

$$a_f = \sum_{\mathbf{k} \in ((2\pi/L)\mathbb{Z})^3} L^{-3/2} \tilde{f}(\mathbf{k}) a_V(\mathbf{k}), \quad \tilde{f}(\mathbf{k}) = \int d^3x \exp(i\mathbf{k} \cdot \mathbf{x}) f(\mathbf{x}),$$

for $f \in L^2(V)$. Then because $\omega = |\mathbf{k}|^2$,

$$\langle \exp[i(a_f^* + a_{f^*})] \rangle = \exp\left[- \sum_{\mathbf{k} \in ((2\pi/L)\mathbb{Z})^3} L^{-3} |\tilde{f}(\mathbf{k})|^2 \left(\frac{1}{2} + \frac{z}{\exp(\beta |\mathbf{k}|^2) - z} \right) \right].$$

A more convenient expression in the calculation of ordered products is $\exp[i \sum_m a_m^* \alpha_m] \times \exp[i \sum_m a_m \alpha_m^*]$. Its expectation values can be read off from the formula $\exp(A + B) = \exp A \exp B \exp(\frac{1}{2}[B, A])$, which holds provided that $[A, [A, B]] = [B, [A, B]] = 0$, which in this case is in accordance with the Weyl relations (III: 3.1.2; 1):

2.5.28 The Generating Function for Ordered Products

$$\left\langle \exp\left[i \sum_m a_m^* \alpha_m \right] \exp\left[i \sum_m a_m \alpha_m^* \right] \right\rangle = \exp\left[- \sum_m |\alpha_m|^2 \frac{z}{\exp(\beta \varepsilon_m) - z} \right]$$
$$\equiv E(\alpha_i, \alpha_k^*),$$

which can be written

$$\langle \exp(i a_f^*) \exp(i a_f) \rangle = \exp(-\langle f | \rho_1 f \rangle)$$

with the use of ρ_1 from (2.5.9).

The expectation values of polynomials in the field operators can be obtained by differentiating the generating function by α or α^*. Note that all the factors within a given exponent of (2.5.47) commute, so nothing prevents the exponential functions from being differentiated:

$$\langle a_{m_1}^* \cdots a_{m_n}^* a_{j_1} \cdots a_{j_n} \rangle = (-i)^{n+n'} \frac{\partial}{\partial \alpha_{m_1}} \cdots \frac{\partial}{\partial \alpha_{m_n}} \frac{\partial}{\partial \alpha_{j_1}^*} \cdots \frac{\partial}{\partial \alpha_{j'_n}^*} E \Bigg|_{\substack{\alpha_i = 0 \\ \alpha_k^* = 0}}$$
$$= \delta_{nn'} \sum_P \prod_{i=1}^n \frac{\delta_{m_i j_{P_i}} z}{\exp(\beta \varepsilon_{m_1}) - z},$$

where P stands for any permutation of $(1, 2, \ldots, n)$.

We have been confronted again with a permanent, and it is easy to understand that the analogous expression for fermions contains $(-1)^P$ and thus involves a determinant. The $-z$ in the denominator is also turned into $+z$, but there are no other changes. Linear extension covers the cases of expectation values of products of arbitrary a_f, which are most conveniently written in terms of the one-particle density matrix ρ_1, as before:

2.5.29 The Grand Canonical Expectation Value of an Ordered Product

$$\langle a_{f_1}^* \cdots a_{f_n}^* a_{g_1} \cdots a_{g_{n'}} \rangle = \delta_{nn'} \overset{\text{Per}}{\text{Det}} (\langle f_i | \rho_1 g_i \rangle).$$

This section will conclude with a further investigation into the thermodynamic limit of the grand canonical state of a system of particles in an external field. Such a state will exist under the circumstances in which $\rho_{1,V}$ converges weakly, as for example with free particles, for which:

2.5.30 The Grand Canonical State of an Infinite System

$$\langle a^*_{f_1} \cdots a^*_{f_n} a_{g_1} \cdots a_{g_{n'}} \rangle = \delta_{nn'} \underset{\text{Det}}{\overset{\text{Per}}{}} (\langle f_i | \rho_1 g_j \rangle),$$

$$\langle f | \rho_1 g \rangle = \int \frac{d^3k}{(2\pi)^3} \frac{\tilde{f}^*(\mathbf{k}) \tilde{g}(\mathbf{k}) z}{\exp(\beta |\mathbf{k}|^2) \mp z},$$

where $\beta > 0$, and for bosons, $0 \le z < 1$, or for fermions, $z > 0$.

It was noticed in (2.5.20; 3) that with bosons at $T < T_c = (8\pi^{3/2}\rho/\zeta(\frac{3}{2}))^{2/3}$, the limits $V \to \infty$ and $z \to 1$ have to be taken jointly in order to have a given density ρ. This does not make the sum in (2.5.27) converge to the integral in (2.5.30); rather, if $z = 1 - 1/\rho_0 V$, then the term with $k = 0$ survives separately:

$$\lim_{V \to \infty} \frac{1}{V} \sum_{\mathbf{k} \in ((2\pi/L)\mathbb{Z})^3} \frac{|\tilde{f}(\mathbf{k})|^2 (1 - (1/\rho_0 V))}{\exp(\beta |\mathbf{k}|^2) - 1 + (1/\rho_0 V)}$$

$$\to \rho_0 |\tilde{f}(\mathbf{0})|^2 + \int \frac{d^3k}{(2\pi)^3} \frac{|\tilde{f}(\mathbf{k})|^2}{\exp(\beta |\mathbf{k}|^2) - 1}.$$

This formula is justified if $f \in L^2(\mathbb{R}^3)$ with compact support, which makes $\tilde{f} \in L^2(\mathbb{R}^3) \cap C_0^\infty(\mathbb{R}^3)$, so the integrand remains integrable even at $\mathbf{k} = \mathbf{0}$. Therefore we have:

2.5.31 The Grand Canonical State in Bose Condensation

$$\lim_{V \to \infty} \langle \exp(i a_f^*) \exp(i a_f) \rangle_{\beta, z = 1 - (1/\rho_0 V)}$$
$$= \exp\left[-\rho_0 |\tilde{f}(\mathbf{0})|^2 - \int \frac{d^3k}{(2\pi)^3} \frac{|\tilde{f}(\mathbf{k})|^2}{\exp(\beta |\mathbf{k}|^2) - 1} \right].$$

2.5.32 Remarks

1. If $T < T_c$ then the grand canonical state of the Bose field algebra differs from the canonical state, which can be calculated as

$$\langle \exp(i a_f^*) \exp(i a_f) \rangle = \exp\left[- \int \frac{d^3k}{(2\pi)^3} \frac{|\tilde{f}(\mathbf{k})|^2}{\exp(\beta |\mathbf{k}|^2) - 1} \right]$$
$$\times \int_0^{2\pi} \frac{d\varphi}{2\pi} \exp[2i \sqrt{\rho_0} \operatorname{Re}(\tilde{f}(\mathbf{0}) \exp(i\varphi))]$$

for $T < T_c$ [13].

2. Other than for bosons at $T < T_c$, the representations in the individual factors are the thermal ones (1.4.7). According to Remark (1.4.17;1) the factors are of type III in the infinite system. They form a reducible representation π, the tensor product $\pi_1 \otimes \pi_2$ of two Fock-like representations of the field algebra (cf. (1.4.6)):

$$\pi(a_f) = \pi_1\left(a\frac{\tilde{f}(\mathbf{p})}{\sqrt{\mp\exp[-\beta(|\mathbf{p}|^2-\mu)]+1}}\right)\otimes \mathbf{1} + (-\mathbf{1})^N \otimes \pi_2\left(a^*\frac{\tilde{f}(\mathbf{p})}{\sqrt{\exp[\beta(|\mathbf{p}|^2-\mu)]\mp 1}}\right),$$

where $a_f N = (N+1)a_f$. It is straightforward to verify that

$$\langle a^*_{f_1}\cdots a^*_{f_n} a_{g_1}\cdots a_{g_{n'}}\rangle = \langle \Omega_1\otimes\Omega_2|\pi(a^*_{f_1})\cdots\pi(a_{g_{n'}})|\Omega_1\otimes\Omega_2\rangle.$$

3. For bosons at $T < T_c$ it is not a factor representation; the analogue of the mean magnetization $\mathbf{s}$ (1.4.6: 2) is

$$a_0 \equiv \lim_{V\to\infty} a_0^V, \text{ where } a_0^V \equiv \frac{1}{V}\int_V d^3x a(\mathbf{x}).$$

All bounded functions of a_0 lie in the center of the von Neumann algebra $\pi(\mathcal{A})''$. Now

$$\langle a_0^{*n} a_0^m\rangle = \left(\frac{\partial}{\partial\tilde{f}(\mathbf{0})}\right)^n\left(\frac{\partial}{\partial\tilde{f}^*(\mathbf{0})}\right)^m E|_{f\equiv 0},$$

so for instance $\langle a_0\rangle = 0$, $\langle a_0^* a_0\rangle = \rho_0$. Thus a_0 is not represented as a multiple of the identity.

4. The canonical state (2.5.32;1) is an integral over states ω_φ for which the exponent in the generating function

$$\omega_\varphi\left(\exp(i\lambda a_0^*)\exp(i\lambda a_0)\right) = \exp(2i\lambda\sqrt{\rho_0}\cos\varphi)$$

is linear in $\lambda \in \mathbb{R}$. These states produce factor representations:

$$\pi_\varphi(a_0) = \sqrt{\rho_0}\exp(-i\varphi)\cdot\mathbf{1}.$$

5. If a term $V^\alpha(a_0^V - \sqrt{\rho_0}\exp(-i\varphi))^*(a_0^V - \sqrt{\rho_0}\exp(-i\varphi))$ with $0 < \alpha < 1$ is added to the local Hamiltonian H_V, then the $\mathbf{k} = \mathbf{0}$ component of βH_V becomes $\beta V^\alpha(a_0^V - \sqrt{\rho_0}\exp(-i\varphi))^*(a_0^V - \sqrt{\rho_0}\exp(-i\varphi))$. As will become more apparent below, the thermodynamic functions are unchanged for all $0 < T \le T_c(\rho)$ in the limit $V \to \infty$ if we set $z(V) \equiv 1$ and $\rho_0 = \rho(1-(T/T_c(\rho))^{3/2})$ (cf. (2.5.20;3)). Because

$$\begin{aligned}\mathrm{Tr}\{&\exp[-\beta V^\alpha(a_0^V - \sqrt{\rho_0}\exp(-i\varphi))^*(a_0^V - \sqrt{\rho_0}\exp(-i\varphi))]\\ &\times\exp(i\tilde{f}(\mathbf{0})a_0^{V*})\cdot\exp(i\tilde{f}^*(\mathbf{0})a_0^V)\}\\ &= \mathrm{Tr}\left[\exp(-\beta V^\alpha a_0^{V*}a_0^V)\exp(i\tilde{f}(\mathbf{0})a_0^{V*})\exp(i\tilde{f}^*(\mathbf{0})a_0^V)\right]\end{aligned}$$

$$\times \exp\left(2i\sqrt{\rho_0}\mathrm{Re}(\tilde{f}(\mathbf{0})\exp(i\varphi))\right)$$

and

$$\frac{\mathrm{Tr}[\exp(-\beta V^\alpha a_0^{V*} a_0^V)\cdot\exp(i\tilde{f}(\mathbf{0})a_0^{V*})\cdot\exp(i\tilde{f}^*(\mathbf{0})a_0^V)]}{\mathrm{Tr}\exp(-\beta V^\alpha a_0^{V*} a_0^V)}$$
$$= \exp\left[-|\tilde{f}(\mathbf{0})|^2/\beta V^\alpha + o\left(\frac{1}{V^\alpha}\right)\right]$$

(see Problem 6), in the limit $V \to \infty$ the perturbed grand canonical state reduces to ω_φ, the integrand of the canonical state in the decomposition (2.5.32; 1), since the contribution to the generating function from the components of H_V with $\mathbf{k} \neq \mathbf{0}$ is not affected by the extra term. Since the exponent in this generating function is linear in $\tilde{f}(\mathbf{0})$ and $\tilde{f}^*(\mathbf{0})$,

$$\pi_{\omega_\varphi}(a_0) = \sqrt{\rho_0}\exp(-i\varphi)\cdot\mathbf{1}.$$

This shows that ω_φ is a factor state, and the density of the particles in the ground state is represented by the (dispersionless) multiplication operator $\rho_0 \cdot \mathbf{1}$. Although the assumption that $\alpha > 0$ is essential (the limit state is not changed by perturbations bounded uniformly in V), the bound $\alpha < 1$ only serves to illustrate that a surface effect is enough to single out any given pure phase from a mixture as the limit $V \to \infty$ is taken.

This example appears at first only academic from the physical point of view. Since constant phases of the wave-functions are not observable properties, at least for free particles, the Bose algebra should be replaced with the gauge-invariant subalgebra $\mathcal{E}$, i.e., the subalgebra invariant under the automorphism induced by $f \to \exp(i\varphi)f$. All the states ω_φ are the same on the subalgebra, and the phase mixture of the ground state is not observable. However, these phases do have experimental consequences in the Josephson effect in superconductors.

2.5.33 Problems

1. Calculate the asymptotic forms of $F_{5/2}(z)$ (for $z \to 1$ use $zF'_\sigma(z) = F_{\sigma-1}(z)$, $F_\sigma(1) = \zeta(\sigma)$).

2. Calculate the heat capacity per particle of an ideal Bose gas at constant density, as well as its derivative by the temperature.

3. Verify (2.5.14; 2).

4. Show the maximum properties of (2.5.12; 2) and (2.5.14; 4).

5. Calculate P_B and P_F for particles in a box. Show that the result agrees with (2.22) in the limit $V \to \infty$.

6. Calculate $\mathrm{Tr}\exp[i(a^*\alpha + a\alpha^*)]\exp[-\beta a^* a]/\mathrm{Tr}\exp[-\beta a^* a]$, assuming that $[a, a^*] = 1$.

2.5.34 Solutions

1. $z \to 0$: $F_{5/2}(z) = \sum_{\nu=1}^{\infty} \frac{z^\nu}{\nu^{5/2}} \sim z + \frac{z^2}{2^{5/2}} + \cdots$

$z \to 1$: $F_{5/2}(z) \sim F_{5/2}(1) + (z-1)F'_{5/2}(1) + \cdots = \zeta(\frac{5}{2}) + (z-1)\zeta(\frac{3}{2}) + \cdots$

$z \to \infty$: Let $\alpha = \ln z > 0$

$$\int_0^\infty dt \sqrt{t} \ln(1 + \exp(-t+\alpha)) = \int_{-\alpha}^\infty dt \sqrt{t+\alpha} \ln(1+\exp(-t))$$

$$= \frac{2}{3} \int_{-\alpha}^\infty dt (t+\alpha)^{3/2} (1 + \exp(t))^{-1}$$

$$= \frac{2}{3} \left[\int_0^\alpha dt (\alpha - t)^{3/2} - \int_0^\alpha \frac{dt (\alpha-t)^{3/2}}{1+\exp(t)} + \int_0^\infty \frac{dt (t+\alpha)^{3/2}}{1+\exp(t)} \right]$$

$$= \frac{2}{3} \left[\int_0^\alpha dt (\alpha - t)^{3/2} + \int_0^\infty \frac{dt ((t+\alpha)^{3/2} - |t-\alpha|^{3/2})}{1+\exp(t)} \right] + O(\exp(-\alpha));$$

because

$$|(\alpha+t)^{3/2} - |\alpha - t|^{3/2} - 3t\alpha^{1/2}| \le 2t^2 \alpha^{-1/2}$$

and

$$\int_0^\infty \frac{dt t^{\sigma-1}}{1+\exp(t)} = (1 - 2^{1-\sigma})\Gamma(\sigma)\zeta(\sigma),$$

with $\zeta(2) = \pi^2/6$, $\Gamma(2) = 1$, it follows that

$$\int_0^\infty dt \sqrt{t} \ln(1+\exp(-t+\alpha)) = \frac{2}{3}\left[\frac{2}{5}\alpha^{5/2} + \alpha^{1/2}\frac{\pi^2}{4}\right] + O(\alpha^{-1/2}).$$

2.

$$\varepsilon = \begin{cases} \frac{3}{2} T^{5/2} \frac{1}{8\pi^{3/2}} F_{5/2}(z), & T > T_c, \text{ i.e.,} 0 < z < 1, \\ \frac{3}{2} T^{5/2} \frac{1}{8\pi^{3/2}} \zeta(\frac{5}{2}), & T \le T_c, \text{ i.e.,} z = 1, \end{cases}$$

which implies

$$\gamma \equiv \lim_{N\to\infty} \frac{C_V}{N} = \begin{cases} \frac{15}{4} \frac{1}{8\pi^{3/2}\rho} T^{3/2} F_{5/2}(z) - \frac{9}{4} \frac{F_{3/2}(z)}{F_{1/2}(z)}, & T > T_c, \text{ i.e.,} 0 < z < 1, \\ \frac{15}{4} \frac{T^{3/2}}{8\pi^{3/2}\rho} \zeta(\frac{5}{2}), & T \le T_c, \text{ i.e.,} z = 1, \end{cases}$$

because of the formula $F_{3/2}(z) = 8\pi^{3/2}\rho T^{-3/2}$ for $T > T_c$. The function γ is continuous at $T = T_c$ and equals $(15/4)\zeta(\frac{5}{2})/\zeta(\frac{3}{2}) = 1.93$, and as $T \to \infty$, $F_\sigma(z) \sim z \sim 8\pi^{3/2}\rho T^{-3/2}$, and

$$\gamma \sim \frac{15}{4} \frac{1}{8\pi^{3/2}\rho} T^{3/2} z - \frac{9}{4} \sim \frac{15}{4} - \frac{9}{4} = \frac{3}{2}.$$

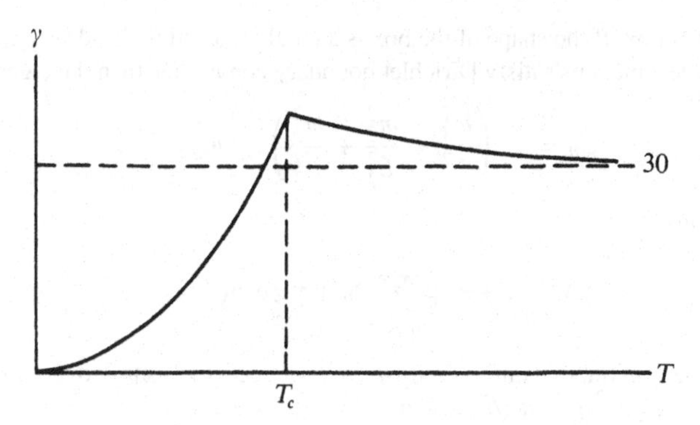

Figure 2.19. Specific heat of an ideal Bose gas.

With the expansion $F_{5/2}(z) = 2.363t^{3/2} + 1.342 - 2.612t - 0.730t^2 \dots$, where $t \equiv -\ln z$, valid for $z \lesssim 1$, and the recursion formula

$$F_{\sigma-1}(\exp(-t)) = -(d/dt)F_\sigma(\exp(-t)),$$

there results

$$\left(\frac{\partial\gamma}{\partial T}\right)_{T=T_c-0} - \left(\frac{\partial\gamma}{\partial T}\right)_{T=T_c+0} = \frac{3.66}{T_c}.$$

3. If the wave-function of $|\mathbf{z}\rangle$ is $\exp(i\mathbf{k}\cdot\mathbf{x})u(\mathbf{x}-\mathbf{q})$ with u real-valued, then $\langle\mathbf{z}||\mathbf{p}|^2|\mathbf{z}\rangle = \int d^3x|i\mathbf{k}u(\mathbf{x}-\mathbf{q}) - \nabla u(\mathbf{x}-\mathbf{q})|^2 = |\mathbf{k}|^2 + \int d^3x|\nabla u|^2$. At the same time, the expectation value of $\int d\Omega_z|\mathbf{z}\rangle\langle\mathbf{z}||\mathbf{k}|^2$ in a normalized ψ equals

$$\begin{aligned}
&\int \frac{d^3q\,d^3k|\mathbf{k}|^2}{(2\pi)^3}\int d^3x\,d^3x'\,\psi^*(\mathbf{x})e^{i\mathbf{k}\cdot\mathbf{x}}u(\mathbf{x}-\mathbf{q})e^{-i\mathbf{k}\cdot\mathbf{x}'}u(\mathbf{x}'-\mathbf{q})\psi(\mathbf{x}')\\
&\quad= \int d^3q\,d^3x\,\nabla(\psi^*(\mathbf{x})u(\mathbf{x}-\mathbf{q}))\cdot\nabla(u(\mathbf{x}-\mathbf{q})\psi(\mathbf{x}))\\
&\quad= \int d^3x|\nabla\psi(\mathbf{x})|^2 + \int d^3q|\nabla u(\mathbf{q})|^2,
\end{aligned}$$

because the mixed terms drop out in the q integration. Therefore,

$$\int d\Omega_z|\mathbf{z}\rangle\langle\mathbf{z}||\mathbf{k}|^2 = |\mathbf{p}|^2 + \int d\Omega_z|\mathbf{z}\rangle\langle\mathbf{z}|\int d^3q|\nabla u|^2.$$

4. Klein's inequality (2.1.4;5) with $K(\rho) = \rho\ln\rho \pm (1\mp\rho)\ln(1\mp\rho)$, $K'(\rho) = -\ln(1/\rho\mp 1)$ and $\bar\rho = [\exp(\beta(h-\mu))\pm 1]^{-1}$ leads to

$$\mathrm{Tr}[K(\rho) - K(\bar\rho) + (\rho-\bar\rho)\beta(h-\mu)] \geq 0,$$

proving (2.5.12;2). In the classical case, i.e., $\rho = \rho(z)$, $h = h(z)$, $\bar\rho = \bar\rho(z)$, all being real,

$$K(\rho(z)) - K(\bar\rho(z)) + (\rho(z) - \bar\rho(z))\beta(h(z)-\mu) \geq 0$$

for all z, and consequently (2.5.14;4).

5. Particles in a box. If the shape of the box is a parallelepiped with sides L_1, L_2, and L_3, and the wave-functions satisfy Dirichlet boundary conditions, then the eigenvalues are

$$\varepsilon_m = \pi^2 \left(\frac{m_1^2}{L_1^2} + \frac{m_2^2}{L_2^2} + \frac{m_3^2}{L_3^2} \right), \quad m_i \in \mathbb{Z}^+.$$

Consequently

$$\beta V P_{\substack{B\\F}}(z) = \mp \sum_{m_i=1}^{\infty} \ln(1 \mp z \exp(-\beta \varepsilon_m)),$$

and in the thermodynamic limit $L_i \to \infty$ the sum over m_i becomes $L_1 \cdot L_2 \cdot L_3 (2\pi)^{-2} \int_0^\infty d\varepsilon \sqrt{\varepsilon} \ldots$, so

$$P_{\substack{B\\F}}(T, z) = \mp T^{5/2} (2\pi)^{-2} \int_0^\infty dt \sqrt{t} \ln(1 \mp z \exp(-t))$$

$$= \pm T^{5/2} \frac{1}{8\pi^{3/2}} F_{5/2}(\pm z).$$

6. Because $\exp A \exp B = \exp(A + B) \exp(\frac{1}{2}[A, B]) = \exp B \exp A \exp[A, B]$ for $[A, B] = c \cdot \mathbf{1}$, the coherent states with $|u\rangle = |0\rangle$, $a|0\rangle = 0$, can be written

$$|z\rangle = \exp\left(\frac{a^* z}{\sqrt{2}}\right) |0\rangle \exp\left(\frac{-|z|^2}{4}\right).$$

As in Remark (III: 3.1.14; 1), with $\exp(-\beta a^* a) f(a^*)|0\rangle = f(a^* \exp(-\beta))|0\rangle$ it follows that

$$\begin{aligned}
&\mathrm{Tr} \exp(\alpha a^*) \exp(-\alpha^* a) \exp(-\beta a^* a) \\
&= \int \frac{dz}{2\pi} \langle 0| \exp\left(\frac{a z^*}{\sqrt{2}}\right) \exp(-\alpha^* a) \exp(-\beta a^* a) \exp(\alpha a^*) \exp\left(\frac{a^* z}{\sqrt{2}}\right) |0\rangle e^{-|z|^2/2} \\
&= \int \frac{dz}{2\pi} \langle 0| \exp\left[a\left(\frac{z^*}{\sqrt{2}} - \alpha^*\right)\right] \exp\left[\exp(-\beta) a^* \left(\frac{z}{\sqrt{2}} + \alpha\right)\right] |0\rangle e^{-|z|^2/2} \\
&= \int \frac{dz}{2\pi} \exp\left[-\frac{|z|^2}{2}(1 - \exp(-\beta)) + \exp(-\beta)\left(\frac{1}{\sqrt{2}}(z^*\alpha - z\alpha^*) - |\alpha|^2\right)\right] \\
&= \exp\left[-|\alpha|^2 \frac{1}{\exp(\beta) - 1}\right] \Big/ (1 - \exp(-\beta)),
\end{aligned}$$

so by changing α to $i\alpha$,

$$\begin{aligned}
\langle \exp[i(a^*\alpha + a\alpha^*)] \rangle &= \langle \exp[\alpha a^* - \alpha^* a] \rangle = \langle \exp(\alpha a^*) \exp(-\alpha^* a) \rangle \exp(-\frac{1}{2}|\alpha|^2) \\
&= \exp\left[-|\alpha|^2 \left(\frac{1}{2} + \frac{1}{\exp(\beta) - 1}\right)\right].
\end{aligned}$$

3
Thermodynamics

3.1 Time-Evolution

Whereas small systems evolve almost periodically in time, large systems appear chaotic and their time-evolution mixes the observables thoroughly.

The framework for this discussion will be an algebra $\mathcal{A}$ of observables with a strongly continuous time-automorphism and a time-invariant state ρ. In the GNS representation the invariant state is made into a vector $|\Omega\rangle$, and the time-automorphism is represented as a unitary group of operators $U = \{\exp(iHt)\}$, $U|\Omega\rangle = |\Omega\rangle$. The time-evolution then extends to the weak closure $\mathcal{A}''$. If the representation is reducible, then it may occur that $U \not\subset \mathcal{A}''$, even if $U_t^{-1}\mathcal{A}U_t \subset \mathcal{A}$. The von Neumann algebra

$$\mathcal{R} \equiv \{\mathcal{A} \cup U\}'', \qquad \mathcal{R}' = \mathcal{A}' \cap U',$$

generated by $\mathcal{A}$ and U is known as the **covariance algebra** and will figure prominently in what follows. If the only invariant elements of $\mathcal{A}'$ are of the form $\alpha \cdot \mathbf{1}$, then it is all of $\mathcal{B}(\mathcal{H})$, as $\mathcal{H} = \alpha \cdot \mathbf{1} \longrightarrow \mathcal{R}'' = \mathcal{R} = \mathcal{B}(\mathcal{H})$.

An initial orientation to the various possibilities can be obtained by looking at some

3.1.1 Examples

1. Classical dynamical systems. The Abelian algebra $\mathcal{A}$ of C^∞ functions $a(p, q)$ on the phase space $T^*(M)$ is a special case of the general schema. If $d\mu$ is

a probability measure on $T^*(M)$, then the elements $a \in \mathcal{A}$ are represented as multiplication operators on the Hilbert-space $L^2(T^*(M), d\mu)$. The advantage of the Hilbert-space approach to classical mechanics is that it ignores exceptional trajectories making up null sets. If a time-invariant measure $d\mu$, such as the Liouville measure $dq_1 \dots dp_{3N}$ is restricted to a time-evolution region Ω of finite volume and normalized, then the time-evolution $a(p, q) \to a(p(t), q(t))$ is represented unitarily on $L^2(\Omega d\mu)$. It can be written formally as $U_t = \exp(-iht)$, where $h = iL_{X_H}$ is the Liouville operator $(I : 2.2.25;\ 1)$, and this unitary group of transformations extends to the von Neumann algebra $\mathcal{A}'' = L^\infty(\Omega, d\mu)$. Of course U_t does not belong to $\mathcal{A}''$, which is maximally Abelian, $\mathcal{A}'' = \mathcal{A}' = \mathcal{Z}$. The algebra $\mathcal{R}$ is all of $\mathcal{B}(\mathcal{H})$ if and only if the system is ergodic, for then the only time-invariant functions are constant almost everywhere, and are thus the constant functions of $L^\infty(\Omega, d\mu)$.

2. A single spin in a magnetic field, cf. (1.1.1):

$$\mathcal{A} = \mathcal{B}(\mathbb{C}^2) = \{\mathbf{1}, \sigma, \sigma^\pm\}'', \qquad \rho(\cdot) = \left\langle \begin{pmatrix} 1 \\ 0 \end{pmatrix} \middle| \cdot \middle| \begin{pmatrix} 1 \\ 0 \end{pmatrix} \right\rangle,$$

$$U_t = \exp(iB(\mathbf{1} - \sigma)t), \qquad \mathcal{A}' = \mathcal{Z} = \mathcal{R}' = \{\alpha \cdot \mathbf{1}\}, \qquad \mathcal{A}'' = \mathcal{A} = \mathcal{R}.$$

Observe that while there is only one invariant vector, there is a second pure invariant state, $\left\langle \begin{pmatrix} 0 \\ 1 \end{pmatrix} \middle| \cdot \middle| \begin{pmatrix} 0 \\ 1 \end{pmatrix} \right\rangle$.

3. A single spin in a magnetic field, in a thermal representation (1.4.6):

$$\mathcal{A} = \{\mathbf{1}, \sigma, \sigma^\pm\}'' \otimes \mathbf{1}, \qquad \rho(\cdot) = \langle \Omega | \cdot | \Omega \rangle,$$

$$\Omega = \sqrt{\frac{1+s}{2}} \begin{pmatrix} 1 \\ 0 \end{pmatrix} \otimes \begin{pmatrix} 1 \\ 0 \end{pmatrix} + \sqrt{\frac{1-s}{2}} \begin{pmatrix} 0 \\ 1 \end{pmatrix} \otimes \begin{pmatrix} 0 \\ 1 \end{pmatrix},$$

$$\mathcal{A}' = \mathbf{1} \otimes \{\mathbf{1}, \tau, \tau^\pm\}'', \qquad U_t = \exp(iB(\tau - \sigma)t),$$

$$\mathcal{A}'' = \mathcal{A}, \quad \mathcal{Z} = \{\alpha \cdot \mathbf{1}\}, \quad \mathcal{R}' = \mathbf{1} \otimes \{\mathbf{1}, \tau\}'', \quad \mathcal{R} = \{\mathbf{1}, \sigma, \sigma^\pm\}'' \otimes \{\mathbf{1}, \tau\}''.$$

This factor representation on $\mathbb{C}^4$ has a two-dimensional invariant subspace and a five-dimensional manifold of invariant states. Two of these are pure states corresponding to nonivariant vectors. Notice that the formal equation $h = B\sigma$ has to be normalized not only with a constant but also by $B\tau \in \mathcal{A}'$, to ensure that $U|\Omega\rangle = |\Omega\rangle$. With a different choice of the basis for $\mathbb{C}^4$, Ω can also be written as $\begin{pmatrix} 1 \\ 0 \end{pmatrix} \otimes \begin{pmatrix} 1 \\ 0 \end{pmatrix}$, which makes the representation π of $\mathcal{A}$ somewhat more complicated (cf. (2.5.32;2)):

$$\pi(\sigma^\pm) = \sqrt{\frac{1+s}{2}}\sigma^\pm \otimes \mathbf{1} - \mathbf{1} \otimes \sqrt{\frac{1-s}{2}}\tau^\mp,$$

$$\pi(\sigma) = \frac{1+s}{2}\sigma \otimes \mathbf{1} - \mathbf{1} \otimes \tau \frac{1-s}{2} + \sqrt{1-s^2}\{\sigma^- \otimes \tau^- + \sigma^+ \otimes \tau^+\}.$$

It is easy to verify the algebraic relationships

$$\pi(\sigma^-)\pi(\sigma^-) \pm \pi(\sigma^-) = \begin{cases} \mathbf{1}. \\ \pi(\sigma), \end{cases} \qquad \pi(\sigma^+)^2 = 0.$$

4. An infinite, interacting spin system. Consider the model of a ferromagnet (2.3.27;2) in the limit $N \to \infty$. It is not hard to discover that the thermal expectation values converge to those with the vector

$$\left(\begin{pmatrix}1\\0\end{pmatrix}\otimes\begin{pmatrix}1\\0\end{pmatrix}\sqrt{\frac{1+s}{2}}+\begin{pmatrix}0\\1\end{pmatrix}\otimes\begin{pmatrix}0\\1\end{pmatrix}\sqrt{\frac{1-s}{2}}\right),$$

as with a type-III representation (1.4.6). The quantities

$$s = \langle\sigma\rangle = -\tanh B_{\text{eff}}\beta, \qquad B_{\text{eff}} = B - 2s,$$

are to be determined self-consistently, for the interaction can be written as

$$\frac{1}{N}\sum_{i,j}\boldsymbol{\sigma}_j\cdot\boldsymbol{\sigma}_j$$
$$= \frac{1}{N}\sum_i(\sigma_i - \langle\sigma_i\rangle)\cdot\sum_j(\sigma_j - \langle\sigma_j\rangle) + 2\langle\sigma\rangle\sum_i\sigma_i + \text{const.}$$

If now $N \to \infty$, the first term on the right describes the fluctuations and becomes negligible compared with $-2\langle\sigma\rangle\sum_i\sigma_i$, and the commutators of H approach those of $B_{\text{eff}}\sum_i\sigma_i$, $B_{\text{eff}} = B - 2\langle\sigma\rangle$ (cf. (1.10)). The time-evolution is accordingly given by

$$U_t = \bigotimes_j \exp(iB_{\text{eff}}(\tau_j - \sigma_j)t).$$

The Hilbert-space $\mathcal{H}$ contains infinitely many invariant vectors, viz., all the ones that differ from Ω in the replacement of finitely many factors with an invariant vector from Example 3. Since B_{eff} depends on β, the time-automorphisms on representations with different β are different. Therefore it is not an automorphism of the algebra $\mathcal{A}$ generated by the σ's on the sum of two representations with different β. Although an isomorphism of $\pi(\mathcal{A})$, as a subalgebra of $\mathcal{B}(\mathcal{H}_\pi)$, is given by

$$\alpha_{-t}(\pi(\mathcal{A})) = U_t^{\beta_1,\beta_2}\pi(\mathcal{A})(U_t^{\beta_1,\beta_2})^{-1}$$

with

$$U_t^{\beta_1,\beta_2} = U_t^{\beta_1}\oplus U_t^{\beta_2}, \qquad \pi = \pi_{\beta_1}\oplus\pi_{\beta_2},$$

it is not an automorphism, since there are times t at which $\alpha_t(\pi(\mathcal{A})) \neq \pi(\mathcal{A})$. The smallest subalgebra of $\mathcal{B}(\mathcal{H}_\pi)$ for which $(\alpha_t)_{t\in\mathbb{R}}$ becomes a group of automorphisms is clearly $\bigcup_t \alpha_t(\pi(\mathcal{A}))$. If $B = 0$ and $T < 2$, then there is such a sum, or even an integral. There are nonzero solutions to the equation $B_{\text{eff}} = 2\tanh\beta B_{\text{eff}}$, but nothing favors any direction. Expectation values are averages over the unit sphere of expectation values with $\mathbf{B}_{\text{eff}} = \mathbf{n}B_{\text{eff}}$, by means of which the representation takes on the form

$$\pi(\mathcal{A}) = \int_{S_2} d\mathbf{n}\pi_{\mathbf{n}}(\mathcal{A}),$$

where $\pi_{\mathbf{n}}$ is specified by (1.4.6) with $\sigma \equiv (\boldsymbol{\sigma} \cdot \mathbf{n})$. The time-evolution on $\pi_{\mathbf{n}}(\mathcal{A})$ is the rotation $\sigma_j^x(t) = (\exp(tR))^{\alpha\beta}\sigma_j^\beta$ having the matrix

$$R = B_{\text{eff}} \begin{pmatrix} 0 & n_3 & -n_2 \\ -n_3 & 0 & n_1 \\ n_2 & -n_1 & 0 \end{pmatrix}.$$

However, as the strong limit of $(1/N)\sum_{j=1}^N \boldsymbol{\sigma}_j$ as $N \to \infty$, $\mathbf{n}$ is contained in $\pi(\mathcal{A})''$ and lies in the center of this algebra but is not a multiple of $\mathbf{1}$. It is constant in time, and the $\mathbf{n}$-dependent time-evolution of the σ's can be viewed as an automorphism of $\pi(\mathcal{A})''$.

5. Free fermions.. The algebra $\mathcal{A}$ is generated by the field operators a_f (1.3.1), and as in (1.3.3:5) the free time-evolution

$$f(\mathbf{p}) \to \exp(-i|\mathbf{p}|^2 t) f(\mathbf{p}) \equiv f_t(\mathbf{p}),$$

provides a group of automorphisms on $\mathcal{A} : a_f \to a_{f_t}$. The thermal state (2.5.30) is clearly invariant in time and leads to a unitary time-evolution $U_t = \exp(-iHt)$. In order to tell the type of the representation, we can write it in a form like the one in Example 3. Let $|\Omega_{1,2}\rangle$ be two Fock vacua and $\pi_{1,2}(a_f)$ be the representations formed with $|\Omega_{1,2}\rangle$. Then with the tensor product

$$|\Omega\rangle = |\Omega_1\rangle \otimes |\Omega_2\rangle$$

we get

$$\pi(a(f)) = \pi_1\left(a\left(\frac{\tilde{f}(\mathbf{p})}{\sqrt{1+\exp(-\beta(|\mathbf{p}|^2-\mu))}}\right)\right) \otimes \mathbf{1}$$
$$+(-\mathbf{1})^N \otimes \pi_2\left(a^*\left(\frac{\tilde{f}^*(\mathbf{p})}{\sqrt{1+\exp(\beta(|\mathbf{p}|^2-\mu))}}\right)\right),$$

where $aN = (N+1)a$ (cf. (1.3.10)). It can be verified that

$$\langle a_{f_1}^* \dots a_{f_n}^* a_{g_1} \dots a_{g_n}\rangle = \langle \Omega|\pi(a_{f_1}^* \dots \pi(a_{f_n}^*)\pi(a_{g_1}) \dots \pi(a_{g_n})|\Omega\rangle,$$

so this representation is equivalent to the thermal representation with infinitely many spins. Consequently, if $T > 0$, then it is a factor of type III. The local field operators in momentum space can be used to write H_π as

$$H_\pi = \int \frac{d^3p}{(2\pi)^3} |\mathbf{p}|^2 \{\pi_1(a^*(\mathbf{p})a(\mathbf{p})) \otimes \mathbf{1} - \mathbf{1} \otimes \pi_2(a^*(\mathbf{p})a(\mathbf{p}))\}.$$

The operator differs from the usual a^*a not only in that the infinite zero-point energy of field theory has been subtracted off, but also in the removal of an operator of $\mathcal{A}'$.

3.1.2 The Time-Evolution of Open Systems

It seems illusory to consider every single local property of a large system as belonging to the algebra of observables. It is certainly true that practically anything can be measured, but not all at once, and putting the system into a state that is dispersionless with respect to a maximally Abelian subalgebra is actually impossible. In reality only fairly small subsystems get measured, so it is of practical interest to divide the total system into the subsystem that is observed, called an "open" system, and all the rest, acting as a reservoir. Accordingly, let $\mathcal{H} = \mathcal{H}_S \otimes \mathcal{H}_R$ and let Tr^{S+R}, Tr^S, and Tr^R be the traces on $\mathcal{H}$, $\mathcal{H}_S$, and $\mathcal{H}_R$. The time-evolution U_t will mix $\mathcal{H}_S$ and $\mathcal{H}_R$, so it does not create an automorphism of $\mathcal{B}(\mathcal{H}_S)$. However, if the initial state postulated can be factorized and written in terms of a density matrix $\rho \otimes \omega$, then a time-evolution $\tau_t : \mathcal{B}(\mathcal{H}_S) \to \mathcal{B}(\mathcal{H}_S)$ can be defined for the open system in the Heisenberg picture, or the dual time-evolution for the density matrices $\tau_t^* : \mathcal{C}_1(\mathcal{H}_S) \to \mathcal{C}_1(\mathcal{H}_S)$ can be defined in the Schrödinger picture. If $a \in \mathcal{B}(\mathcal{H}_s) \otimes \mathbf{1}$, then the time-dependence of the expectation values can be written as

$$\langle a(t) \rangle \equiv \mathrm{Tr}^{S+R}(\rho \otimes \omega) U_{-t} (a \emptyset \mathbf{1}) U_t = \mathrm{Tr}^S \rho \tau_t(a) = \mathrm{Tr}^S \tau_t^*(\rho) a,$$

where by definition

$$\begin{aligned} \tau_t(a) &\equiv \mathrm{Tr}^R (\mathbf{1} \otimes \omega) U_{-t} (a \otimes \mathbf{1}) U_t, \\ \tau_t^*(\rho) &= \mathrm{Tr}^R\, U_t (\rho \otimes \omega) U_{-t}. \end{aligned} \tag{3.1}$$

Note that the states transform with $U_t^* = U_{-t}$ rather than U_t.

3.1.3 Properties of the Time-Evolution of the Subsystem

The operators τ_t and τ_t^* are

(i) one-parameter, strongly continuous families of completely positive linear mappings;

(ii) *not* groups: $\tau_{t_1} \circ \tau_{t_2} \neq \tau_{t_1+t_2}$;

(iii) *not* isomorphisms of the algebra: $\tau_t(a \cdot b) \neq \tau_t(a) \cdot \tau_t(b)$.

Equality holds in (ii) and (iii) only if U_t factorizes.

3.1.4 Gloss

A linear mapping $\Phi : \mathcal{B}(\mathcal{H}) \to \mathcal{B}(\mathcal{H})$ is said to be **n-positive** iff $\Phi \otimes \mathbf{1}$ acting on $\mathcal{B}(\mathcal{H}) \otimes \mathcal{B}(\mathbb{C}^n) : a \otimes M \to \Phi(a) \otimes M$ is positive for all $M \in \mathcal{B}(\mathbb{C}^n)$, i.e., it maps the cone of positive elements of $\mathcal{B}(\mathcal{H}) \otimes \mathcal{B}(\mathbb{C}^n)$ into itself. The mapping Φ is **completely positive** iff it is positive for all n =1,2,.... It can be shown [14] that all completely positive mappings are obtained by taking tensor products of positive operators, composing with unitary operators, and then taking partial traces, just as in the construction of τ_t and τ_t^*. The completely positive mappings form a semigroup with respect to composition.

3.1.5 Examples

1. The classical harmonic oscillator.
   The observables are chosen as the position coordinates q, so

$$\mathrm{Tr}^{S+R} \to \int dpdq, \qquad \mathrm{Tr}^{R} \to \int dp, \qquad \mathrm{Tr}^{S} \to \int dq.$$

Let $\rho(q) = \pi^{-1/2}\exp(-(q-q_0)^2)$ be the probability distribution function of the coordinates and $\omega(p) = \pi^{-1/2}\exp(-(p-p_0)^2)$ be that of the momenta. The time-evolution of the total system, $q(t) = q\cos t + p\sin t$, $p(t) = p\cos t - q\sin t$, induces

$$\tau_t(q) = q\cos t + p_0 \sin t,$$
$$\tau_t^*(\rho) = \pi^{-1/2}\exp[-(q - q_0\cos t - p_0\sin t)^2]$$

on the subsystem. However, τ_t is not an isomorphism,

$$\tau_t(q^2) = (q\cos t + p_0\sin t)^2 + \frac{1}{2}\sin^2 t \neq \tau_t(q)^2,$$

since ω is not free of fluctuations. The choice of equal widths for ρ and ω, as with quantum-mechanical coherent states, causes a rigid oscillation of ρ. If, instead, $\omega(p) = \delta(p - p_0)$, then there would be a periodic focusing and defocusing of ρ.

$$\tau_t^*(\rho) = \frac{\exp[-(q - q_0\cos t - p_0\sin t)^2\cos^{-2} t]}{\sqrt{\pi}\cos t}.$$

2. Quantum-mechanical coupled oscillators.
   Let us return to the chain of oscillators (1.1.2) and take ξ_0 and ξ_1 as the open system. Instead of the pure state (1.18), suppose the system is in a thermal state

$$\left\langle \exp\left[i\sum_{n=-\infty}^{\infty}(\xi_{2n}r_n + \xi_{2n+1}s_n)\right]\right\rangle$$
$$= \exp\left[-\frac{1}{4}\tanh\frac{\eta}{2}\sum_{n=-\infty}^{\infty}(r_n^2+s_n^2) + i\sum_{n=\infty}^{\infty}(r_n s_n' - r_n' s_n)\right].$$

As in (2.5.33;6),

$$\frac{\mathrm{Tr}\exp\left[-\eta\left((p-\bar{p})^2+(q-\bar{q})^2\right)\right]\exp\left[i(pr+qs)\right]}{\mathrm{Tr}\exp\left[-\eta\left((p-\bar{p})^2+(q-\bar{q})^2\right)\right]}$$
$$= \exp\left[-\frac{r^2+s^2}{4}\tanh\frac{\eta}{2} + i(\bar{p}r+\bar{q}s)\right], \tag{3.2}$$

so this state is a Gibbs state with harmonic forces centered at s', $-r'$. Under the time-evolution (1.16), the expectation values of the Weyl operators of the open system are

$$\langle \exp i(r\xi_0(t) + s\xi_1(t))\rangle = \exp\left\{\sum_n \{-\frac{1}{4}\tanh\frac{\eta}{2}\Big[(rJ_{2n} + sJ_{2n+1})^2 + (rJ_{2n+1} + sJ_{2n})^2\Big] + is'_n(rJ_{2n} + sJ_{2n-1}) - -ir'_n(rJ_{2n+1} + sJ_{2n})\}\right\}.$$

At time t the subsystem is in a state of the form (3.2) with

$$s'_0(t) = \sum_n \left(s'_n(0)J_{2n}(t) - r'_n(0)J_{2n+1}(t)\right),$$

$$r'_0(t) = \sum_n \left(r'_n(0)J_{2n}(t) - s'_n(0)J_{2n-1}(t)\right).$$

The terms $\sim rs$ cancel because of (1.1.24). The average values $s'_0(t), r'_0(t)$ move classically as in Example 1. They converge to zero, but not monotonically.

3. Coupled spins.
Consider spin 1 of the chain (1.20) as the open system and the infinitely many others as the thermal reservoir. The coupling constants $\varepsilon(n)$ are chosen as in (1.1.9). The initial state

$$\rho_1 = \frac{1}{2}\left(1 + \sigma_1^+ \exp(-i\alpha) + \sigma_1^- \exp(i\alpha)\right),$$

$$\omega = \prod_{k\neq 1}\frac{1}{2}\left(1 + \sigma_k^+ \exp(-i\alpha) + \sigma_k^- \exp(i\alpha)\right),$$

((1.17) with $s = 0$) evolves as

$$\tau_t^*(\rho) = \frac{1}{2}\left(1 + \frac{\sin^2 t}{t^2}\left[\sigma^+ \exp(-i(\alpha + 2Bt)) + \sigma^- \exp(i(\alpha + 2Bt))\right]\right)$$

if $\eta \to \infty$. The state ρ oscillates as it approaches the equilibrium state $\frac{1}{2}\cdot\mathbf{1}$ as $T \to \infty$. Also the entropy does not converge monotonically.

3.1.6 Remarks

1. The failure of the time-evolution τ or τ^* to be a group is due to the effect of the system on the reservoir and the reaction of the reservoir on the system. The reaction influences the system at later times, so $(\partial/\partial t)\tau_t^*(\rho)$ depends on $\tau_t^*(\rho)$ not only for $s = t$ but for all $s \le t$, i.e., on its whole history. The time-evolution of the density matrix of the reservoir can be written down formally and substituted into the equation for $(\partial/\partial t)\tau_t^*(\rho)$. The resulting **master equation** is an integrodifferential equation for ρ including the memory effects just mentioned.
2. The requirement of complete positivity of the time-evolution is not a mere technicality but a genuine restriction, and it even has some experimentally verifiable

consequences. For instance, its implications for the motion of a spin in a thermal reservoir have been confirmed experimentally [15].

The retrospective effects of (3.1.6; 1) disappear in certain limiting cases, so the time-evolution τ becomes a semigroup. The limits involve the time-scale or the coupling constants. The most understandable case is that of a simplified version of electrodynamic radiative reaction of volume II, $2.4.

3.1.7 **Example** (Model of Brownian Motion)

We modify Example (1.1.2) to take a single harmonic oscillator in three dimensions as the system and represent the rest of the system, functioning as a reservoir, as a continuous scalar field $\Phi(\mathbf{x})$. Suppose initially that the oscillator is coupled to an averaged field $\int d^3x \Phi(\mathbf{x}) c(\mathbf{x})$, $c \in C_0^\infty(\mathbb{R}^3)$, and later take the limit $c(\mathbf{x}) \to \gamma\delta(\mathbf{x})$, $\gamma \in \mathbb{R}$. We shall study the quantum-theoretical time-evolution from the outset; since the equations of motion are linear it agrees with the classical time-evolution. If Q, P and $\Phi(\mathbf{x})$, $\Pi(\mathbf{x})$ are the canonically conjugate coordinate and field variables, then the Hamiltonian is

$$
\begin{aligned}
H_S &= \frac{1}{2}\left(P^2 + \omega_0^2 Q^2\right), \\
H_R &= \frac{1}{2}\int d^3x \left\{\Pi(\mathbf{x})^2 + |\nabla\Phi(\mathbf{x})|^2\right\}, \\
H' &= -\int d^3x c(\mathbf{x})\Phi(\mathbf{x}) Q.
\end{aligned}
$$

The resulting equations of motion,

$$
\begin{aligned}
\left(\frac{\partial^2}{\partial t^2} - \Delta\right)\Phi(\mathbf{x}, t) &= c(\mathbf{x}) Q(t), \\
\left(\frac{\partial^2}{\partial t^2} + \omega_0^2\right) Q(t) &= \int d^3x \Phi(\mathbf{x}, t) c(\mathbf{x}),
\end{aligned}
$$

can be integrated immediately with Green's formula (II:1.2.36). This is the trivial case of a scalar field on $\mathbb{R}^4$, so with the Green function

$$
D(\mathbf{x}, t) = \frac{\delta(r - t)}{4\pi r}
$$

(II:2.2.7), the solution of the initial-value problem is

$$
\begin{aligned}
\Phi(\mathbf{x}, t) = &\int d^3x' (\Phi(\mathbf{x}', 0)\dot{D}(\mathbf{x} - \mathbf{x}', t) + \dot{\Phi}(\mathbf{x}', 0) D(\mathbf{x} - \mathbf{x}', t)) \\
&+ \int d^3x' \int_0^t dt' D(\mathbf{x} - \mathbf{x}', t - t') c(\mathbf{x}') Q(t')
\end{aligned}
$$

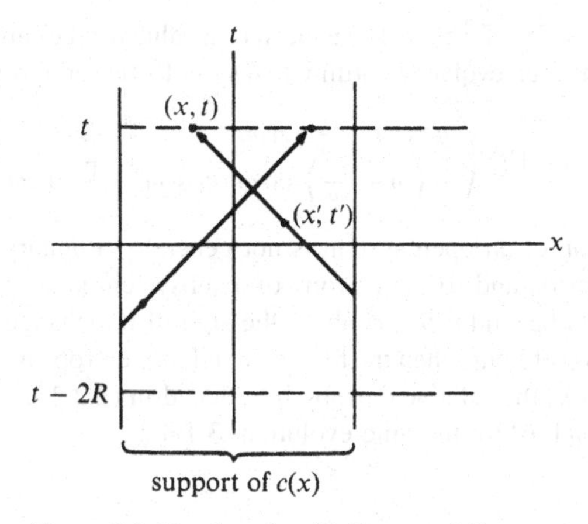

Figure 3.1. The domain of influence of F_{reaction}.

for all $t > 0$, where $\dot{\Phi} = \partial\Phi/\partial t$, etc. Hence the force exerted by the field on the oscillators is

$$\int d^3x \Phi(\mathbf{x}, t) c(\mathbf{x}) = F_{\text{field}}(t) + F_{\text{reaction}}(t),$$

$$F_{\text{field}}(t) = \int d^3x d^3x' c(\mathbf{x})(\Phi(\mathbf{x}', 0)\dot{D}(\mathbf{x}-\mathbf{x}', t) + \dot{\Phi}(\mathbf{x}', 0)D(\mathbf{x}-\mathbf{x}', t))$$

$$F_{\text{reaction}}(t) = \int \frac{d^3x d^3x'}{4\pi|\mathbf{x}-\mathbf{x}'|} c(\mathbf{x})c(\mathbf{x}')Q(t - |\mathbf{x}-\mathbf{x}'|)\Theta(t - |\mathbf{x}-\mathbf{x}'|).$$

In the reaction force $F_{\text{reaction}}(t)$, $Q(t')$ contributes only for $t - 2R \le t' \le t$ if $c(\mathbf{x}) = 0$ for all $\mathbf{x}$ such that $|\mathbf{x}| > R$ (see Figure 3.1).

Now if $c(\mathbf{x}) \to 2\sqrt{\pi}\gamma\delta(\mathbf{x})$, so $R \to 0$, then the retrospective effects disappear, and when the expansion

$$Q(t - |\mathbf{x}-\mathbf{x}'|) = Q(t) - |\mathbf{x}-\mathbf{x}'|\dot{Q}(t) + \frac{1}{2}|\mathbf{x}-\mathbf{x}'|^2 \ldots Q(t) - \ldots$$

is substituted into F_{reaction},

$$F_{\text{reaction}}(t) \to \delta\omega^2 Q(t) - \gamma^2 \dot{Q}(t).$$

The quantity $\delta\omega^2$ is the formally infinite integral $\gamma^2 \int (d^3x d^3x'/|\mathbf{x}-\mathbf{x}'|)\delta(\mathbf{x})\delta(\mathbf{x}')$, so the limit $c(\mathbf{x}) \to \gamma\delta(\mathbf{x})$ must be taken jointly with a change in ω_0^2. If $\bar{\omega}^2 \equiv \omega_0^2 - \delta\omega^2$, then the equation of motion becomes

$$\left(\frac{\partial^2}{\partial t^2} + \bar{\omega}^2 + 2\Gamma\frac{\partial}{\partial t}\right) Q(t) = F_{\text{field}}(t), \qquad \Gamma = \frac{\gamma^2}{2}, t \ge 0.$$

For a thermal state with $\langle\Phi(\mathbf{x}, 0)\rangle = \langle\dot{\Phi}(\mathbf{x}, 0)\rangle = 0$, $\langle F_{\text{field}}(t)\rangle = 0$, and the time-evolution of the expectation value of Q for $t \ge 0$ is

$$\langle Q(t)\rangle = \exp(-\Gamma t)\left(\langle Q(0)\rangle(\cos\omega t + \frac{\Gamma}{\omega}\sin\omega t) + \langle\dot{Q}(0)\rangle\frac{\sin\omega t}{\omega}\right),$$

provided that $\omega^2 \equiv \bar{\omega}^2 - \Gamma^2 > 0$. The expectation values of the canonical variables $\langle Q(t)\rangle$ and $\langle \dot{Q}(t)\rangle$ then evolve according to a symplectic semigroup,

$$\begin{pmatrix} \langle Q(t)\rangle \\ \langle \dot{Q}(t)\rangle \end{pmatrix} = \exp(-\Gamma t) \begin{pmatrix} \cos\omega t + \frac{\Gamma}{\omega}\sin\omega t & \frac{\sin\omega t}{\omega} \\ -\left(\omega + \frac{\Gamma^2}{\omega}\right)\sin\omega t & \cos\omega t - \frac{\Gamma}{\omega}\sin\omega t \end{pmatrix} \begin{pmatrix} \langle Q(t)\rangle \\ \langle \dot{Q}(t)\rangle \end{pmatrix}.$$

The time-evolution of an open system is not generally a unitary transformation of the density matrix, and so the entropy of a subsystem is not necessarily constant. Nothing can be said *a priori* about the sign of the change in entropy: the system might start off hotter than the reservoir and lose entropy as the temperature equalizes. However, the relative entropy introduced in (2.2.20) turns out to be a Liapunov function [16] for the time-evolution (3.1.3).

3.1.8 The Decrease of the Relative Entropy

For the time-evolution τ^* of (3.1.3),

$$S(\tau_t^*(\sigma)|\tau_t^*(\rho)) \leq S(\sigma|\rho).$$

Proof: With Definition (2.2.22) and the unitary invariance,

$$\begin{aligned} &S(\operatorname{Tr}^R U_{-t}\sigma \otimes \omega U_t | \operatorname{Tr}^R U_{-t}\rho \otimes \omega U_t) \\ &\overset{\text{(iv)}}{\leq} S(U_{-t}\sigma \otimes \omega U_t | U_{-t}\rho \otimes \omega U_t) \\ &= S(\sigma \otimes \omega | \rho \otimes \omega) \overset{\text{(iii)}}{=} S(\sigma|\rho). \end{aligned}$$

□

3.1.9 Remarks

1. The relative entropy is always positive, and in the special case of (2.2.21:1), it is β times the difference between the free energy of the state ρ and the free energy at equilibrium. Its decrease reflects the tendency of the system to equilibrium.
2. Monotony in time cannot be claimed if $\tau_{t_1+t_2} \neq \tau_{t_2} \circ \tau_{t_1}$. In Example (3.1.7) friction returned the oscillator monotonically to its rest-point, owing to the semigroup property, which was in turn a consequence of the absence of retrospective effects. The general fact is

3.1.10 Monotony of the Relative Entropy with a Dynamic Semigroup

If $\tau_{t_1+t_2} = \tau_{t_2} \circ \tau_{t_1}$ for all t_1 and $t_2 \geq 0$, then τ_t is said to be a dynamical semigroup. The function $S(\tau_t^*(\sigma)|\tau_t^*(\rho))$ is then a monotonically decreasing function of t.

Proof: This is a direct consequence of (3.1.8). □

3.1.11 Remarks

1. Because $S(\sigma|\rho) \geq 0$, the limit of $S(\tau_t^*(\sigma)|\tau_t^*(\rho))$ as $t \to \infty$ exists.
2. It cannot yet be claimed that the free energy approaches its equilibrium value as $t \to \infty$; $S(\sigma|\rho)$ might stop at some positive value and never fall to zero.
3. The apparent asymmetry in the direction of time comes from the requirement of (3.1.3) that the initial state factorizes. Starting at $t < 0$, the later state at $t = 0$ is factorized, so the relative entropy increases.
4. If the dynamical semigroup is governed by a master equation of the type of (2.1.8:3), then $S(\rho)$ increases monotonically.

That finishes the orientation toward various phenomena connected with the time-evolution. Let us now return to more global questions of time-dependence. The problem, put concisely, is that a finite system the Hamiltonian of which has pure point spectrum $\{\varepsilon_i\}$ has observables whose expectation values $\langle a(t) = \sum_{j,k} a_{jk} \exp(i(\varepsilon_j - \varepsilon_k)t)$ are almost-periodic functions, as superpositions of periodic functions. Only the average over time makes sense: the time-limit exists only for infinite systems the Hamiltonians of which have absolutely continuous spectra. Although in actuality only finite systems come under observation, the recurrence times are so long that they are indistinguishable from infinite systems within the times of relevance to human beings. In any event, the first issue to settle is how to define the time-average of a function $f(t) \in C(\mathbb{R})$, the set of bounded, continuous functions on $\mathbb{R}$. The obvious guesses would be

$$\lim_{T\to-\infty} \frac{1}{2T} \int_{-T}^{T} dt f(t) \qquad \text{or} \qquad \lim_{\varepsilon\to 0} \frac{\varepsilon}{2} \int_{-\infty}^{\infty} dt \exp(-\varepsilon|t|) f(t),$$

but these quantities do not converge for such functions as $\sin(\ln(|t|+1)) \in C(\mathbb{R})$. Any suitable average would have to be linear, positive, and invariant under displacements in time. Every invariant state on the C^* algebra $C(\mathbb{R})$ has the right qualifications, and the existence of many invariant states on $C(\mathbb{R})$ means that there are many possible time-averages. There is thus no question whether a time-average exists, but it is not unique.

3.1.12 The Time-Average of an Observable

Let η be an average over $C(\mathbb{R})$ and $t \to a_t$ be a weakly continuous mapping $\mathbb{R} \to \mathcal{B}(\mathcal{H})$ such that $||a_t|| \leq ||a_0||$ for all t. Then the average $\eta(a)$ is defined by

$$\langle x|\eta(a)|y\rangle = \eta(\langle x|a_t|y\rangle) \qquad \text{for} \quad \text{all} x, y \in \mathcal{H}.$$

3.1.13 Remarks

1. Since $|\eta(\langle x|a_t|y\rangle)| \leq ||x||\cdot||y||\cdot||a_0||$, this sesquilinear form defines a bounded operator $\eta(a)$.

2. In the Schrödinger picture, the average $\eta(\sigma)$ of a state σ on the algebra generated by the operators a_t is defined by $\eta(\sigma)(a) = \eta(\sigma(a_t))$.

3.1.14 Examples

1. If $a_t = \exp(-iHt) \equiv U(t)$, then $\eta(U) = E_0 \equiv$ the projection onto the eigenvectors of H with eigenvector 0.

   Proof:

   (i) $\langle x|E_0\eta(U)y\rangle = \eta\langle E_0 x|U_t y\rangle = \langle x|E_0 y\rangle \Rightarrow E_0\eta(U) = E_0$.

   (ii) $\langle x|U(t_0)\eta(U)y\rangle = \eta\langle x|U(t+t_0)y\rangle = \langle x|\eta(U)y\rangle \Rightarrow$
   $U(t_0)\eta(U) = \eta(U) \Rightarrow E_0\eta(U) = \eta(U) = E_0$ by part (i). □

2. $a_t = U(t)aU^{-1}(t)$, where $U(t)$ has pure point spectrum. If the projections onto the eigenspaces are E_i, then $\eta(a) = \sum_i E_i a E_i$.

   Proof: Take matrix elements with the eigenvectors of $H$and note that $\eta(\exp(i\alpha t)) = 0$ for all η and all $\alpha \in \mathbb{R}$ different from 0. □
3. $\eta(a_t E_0) = E_0 a E_0$, since $\eta(a_t E_0) = \eta(U(t)aE_0) = E_0 a E_0$, as in Example 1.

3.1.15 Remarks

1. In these examples the concrete averages $(1/2T)\int_{-T}^{T} dt \exp(iHt)$ and $(\varepsilon/2) \times \int_{-\infty}^{\infty} dt \exp(-\varepsilon|t|)\exp(iHt)$ converge strongly (Problem 1). Hence E_0 belongs to U'' as well as U'.
2. In the Schrödinger picture the time-average of a vector $|x\rangle$ is defined by $|\eta(x)\rangle \equiv \eta(U(t)|x\rangle) = E_0|x\rangle$. It can be characterized as the vector with the least norm in the convex hull of its trajectory $\{U(t)|x\rangle, t \in \mathbb{R}\}$ (Problem 2). It is not, however, true in general for the state $\sigma(a) = \langle x|a|x\rangle$ formed with $|x\rangle$ that $\eta(\sigma)(a) = \langle\eta(x)|a|\eta(x)\rangle$.
3. There is no definition of $\eta(a)$ independent of the representation; since $\lim_{T\to\infty}(1/T)\int_0^T dt a_t$ belongs only to the weak closure of the algebra, η may send operators out of their C^* algebra. Our representations will usually be such that the time-automorphism α_t can be implemented unitarily, and the image of E_0 will contain a cyclic vector for $\mathcal{A}$. If the averages $\eta(a)$ belong to $\mathcal{A}'$, then they are determined uniquely by

$$\eta(a)E_0 = \lim_{T\to\infty}\frac{1}{T}\int_0^T dt a_t E_0 = \lim_{T\to\infty}\frac{1}{T}\int_0^T U_t a E_0 = E_0 a E_0,$$

   since $E_0\mathcal{H}$ separates $\mathcal{A}'$ (Problem 5). However, as will be seen in (3.1.22;4), $\eta(a)$ in general depends on the representation.
4. The time-average may be nonunique if $f(t)$ converges, as $t \to +\infty$ and $t \to -\infty$, but to different values. This situation is familiar to us from scattering

theory. Whenever the time-average of a function f is unique, it agrees with the "concrete average".

$$\lim_{T\to\infty}\frac{1}{2T}\int_{-T}^{T} dt f(t), \qquad \lim_{\varepsilon\to 0}\frac{\varepsilon}{2}\int_{-\infty}^{\infty} dt \exp(-\varepsilon|t|) f(t),$$

or even

$$\lim_{T\to\infty}\frac{1}{T}\int_{0}^{T} dt f(t).$$

These averages exist in classical ergodic theory, in which the Liouville measure on phase space provides the invariant cyclic vector. Some ergodic systems will be defined later, and for them E_0 is one-dimensional, projecting onto the cyclic vector. This projection is then constant on the energy shell, so the time-average $E_0 a E_0$ equals the average over the energy shell.

5. The point spectrum of H can be turned into a continuum by an arbitrarily small perturbation, so averaging over time focuses unduly on the exact form of H, since η is quite different depending on whether the spectrum is pointlike or continuous: If in the spectral representation of H the operator a on the subspace belonging to σ_{ac} has a continuous integral kernel, then η projects this part of a to 0, and by Remark 2 only its point-spectrum part remains (cf. (1:3.3.4;6)).
6. Pure states of classical systems are points in phase space, and averages over pure states are averages over classical trajectories.
7. If the spectrum of H is pure point and nondegenerate, then every normal, invariant state can be written as the time-average of a pure state. Normal, invariant states are of the form

$$\sigma(a) = \sum_i c_i\langle x_i|a|x_i\rangle, \qquad 0 \le c_i \le 1, \sum_i c_i = 1,\ H|x_i\rangle = \varepsilon_i|x_i\rangle;$$

so

$$\sigma(a) = \langle x|\eta(a)|x\rangle, \qquad x = \sum_i \sqrt{c_i}|x_i\rangle.$$

Although the canonical state $\rho = \exp(-\beta(H - F))$ is an average over the trajectory of a pure state, it is generally not true that every averaged pure state is the canonical state.

Our reasoning until this point has applied indifferently to all sorts of quantum systems, but not all quantum systems exhibit thermodynamic behavior. An isolated atom is rather like a frictionless perpetual-motion machine; only large systems are dissipative. The concept introduced in (1.3.7) of asymptotic commutativity turns out to be a useful characteristic of dissipative systems. If the local observables are asymptotically Abelian with respect to the time-automorphism α_t, that means that local perturbations dissipate through the systems as time passes. Of course, this is possible only if H has continuous spectrum, and hence only if the system is infinite.

We shall remain with Definition (1.3.7), although many of its consequences can be derived with weaker assumptions. Definition (1.3.7) applies to the observables of free fermions, but it has only been possible to prove that weakened versions of it apply to more realistic, interacting systems. It is trivial that classical systems are asymptotically Abelian, and (1.3.7) means roughly that asymptotically Abelian systems behave classically on a macroscopic time scale.

3.1.16 Properties of Asymptotically Abelian Systems

Let $\mathcal{A}$ be an asymptotically Abelian C^* algebra with respect to a group of automorphisms $a \to a_t$, and let ω be an invariant state having a representation on a Hilbert space $\mathcal{H}$ with a cyclic vector $|\Omega\rangle$. Then, abbreviating $\mathcal{A}' = \pi_w(\mathcal{A}')$, etc.

1. *the invariant elements of $\mathcal{A}$ belong to $\mathcal{A}'$;*
2. *the invariant elements of $\mathcal{A}'$ lie in the center (i.e., $\mathcal{R}' = \mathcal{A} \cap U' = \eta(\mathcal{A}')$ is a subalgebra of the center $\mathcal{A}' \cap \mathcal{A}''$), and so $\mathcal{R}' = \eta(\mathcal{A}'')$;*
3. *$E_0 \mathcal{A}'' E_0$ is maximally Abelian in $E_0\mathcal{H}$, where E_0 is the projection onto the invariant vectors of $\mathcal{H}$; and*
4. *if σ produces a factor (i.e., the GNS representations $\pi_\sigma(\mathcal{A})$ and $\pi_\sigma(\mathcal{A}')$ constructed with the cyclic vector Ω_σ generate all of $\mathcal{B}(\mathcal{H})$), then*

$$\lim_{t \to \pm\infty} (\sigma(a_t b) - \sigma(a_t)\sigma(b)) \to 0,$$

even if $\sigma(a_t) \neq \sigma(a)$.

3.1.17 Remarks

1. Neither E_0 nor $E_o \mathcal{A}'' E_0$ necessarily belongs to $\mathcal{A}''$. Moreover, $E_0 \mathcal{A}'' E_0$ may fail to be an algebra, and the somewhat loose phrasing of Property 3 is intended to mean that the algebra generated by $E_0 \mathcal{A}'' E_0$ is the same as its commutant.
2. The point of (3.1.16) is that invariant elements such as time-averages and time-limits form an Abelian algebra, and thus equal its center. Factor states are pure when restricted to the center, and are therefore characters (see Definition(III:2.2.25)), which explains why they factorize in time-limits and time-averages.

Proof:

1. $[a, b] = \lim_{t \to \infty} [a_t, b] = 0$ for all invariant $a \in \mathcal{A}$ and all $b \in \mathcal{A}$.
2. By Property 3, $E_0 \mathcal{R} E_0 = E_0 \mathcal{A}'' E_0$ is maximally Abelian and so equal to $(E_0 \mathcal{R} E_0)' E_0$. Since $E_0 \in \mathcal{R}$, $(E_0 \mathcal{R} E_0)' E_0 = E_0 \mathcal{R}' E_0$ [17], and therefore $E_0 \mathcal{R}' E_0 = E_0 (\mathcal{R}' \cap \mathcal{R}) E_0$. Since $|\Omega\rangle$ separates $\mathcal{A}'$, the equation $E_0 a' E_0 = a' E_0$ determines every $a' \in \mathcal{R}'$ uniquely, so $a' \in \mathcal{R}$. However, $\mathcal{R} \cap \mathcal{R}'$ is $\mathcal{A}'' \cap \mathcal{A} \cap U'$, because $U \cap \mathcal{A}' = \{\mathbf{1}\}$.

3. The set $E_0 \mathcal{A} E_0$ must be Abelian, as otherwise some commutator would fail to vanish as $t \to \pm\infty$:

$$\eta_t[a, b_t] = 0 \longrightarrow \eta_t E_0(aU_t b - bU_{-t}a)E_0 = 0$$
$$\longrightarrow [E_0 a E_0, E_0 b E_0] = 0 \quad \text{for all} a, b \in \mathcal{A}.$$

Hence $E_0 \mathcal{A}'' E_0 = (E_0 \mathcal{A} E_0)''$ is also Abelian, and in fact maximally Abelian, as otherwise $E_0 a E_0$ would be $\sim \mathbf{1}$ on a subspace of dimension greater than one for all $a \in \mathcal{A}$, and $|\Omega\rangle = E_0|\Omega\rangle$ would not be cyclic.

4. For every $b \in \pi_\sigma(\mathcal{A})$ there exist two operators b_1 and b_2 such that $b_2|\Omega_\sigma\rangle = b_1^*|\Omega_\sigma\rangle = 0$ and $b = \mathbf{1}\langle\Omega_\sigma|b|\Omega_\sigma\rangle + b_1 + b_2$. This is obvious for finite matrices:

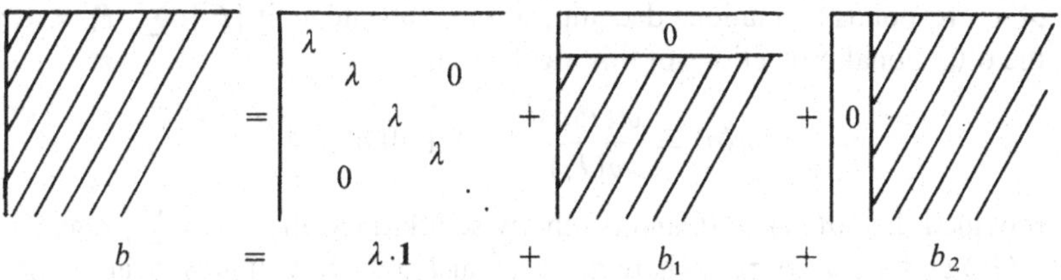

$$b = \lambda \cdot \mathbf{1} + b_1 + b_2$$

and it carries over to $\mathcal{B}(\mathcal{H})$. Then $\sigma(a_t b) - \sigma(a_t)\sigma(b) = \sigma([a_t, b_1])$. If σ produces a factor, then b_1 can be approximated with a finite sum

$$\sum_{i=1}^{n} d_i d_i', \qquad d_i \in \pi_\sigma(\mathcal{A}), \qquad d_i' \in \pi_\sigma(\mathcal{A})',$$

and $\sum_i \sigma([a_t, d_i]d_i')$ tends to 0 as $t \to \pm\infty$ by Definition (1.3.10). Although the subalgebra of $\mathcal{B}(\mathcal{H})$ generated by $\pi_\sigma(\mathcal{A}) \cup \pi_\sigma(\mathcal{A})'$ is only strongly dense, operators with these properties can be approximated even in the norm sense ([18], V.1.4), which justifies these conclusions. □

The set of invariant states is convex and compact, so any invariant state is a convex combination of the extremal points of the set or a limit of such combinations. As the purest among the time-invariant states, the extremal elements deserve a special term:

3.1.18 Definition

An invariant state is **ergodic**, or **extremal invariant**, if it can not be written as a convex combination of other invariant states.

3.1.19 Remarks

1. In classical dynamics an invariant submanifold $\mathcal{N}$ of phase space corresponds to an invariant state (= measure) $\mu_\mathcal{N} = \prod_i dq^i \wedge dp^i_{|\mathcal{N}}$, which is ergodic if

$\mathcal{N}$ cannot be decomposed into invariant pieces with strictly positive measures $\mu_{\mathcal{N}}$.

2. A classical system is said to be ergodic if the surface of the energy shell $\rho(p, q) = \delta(E - H(p, q)) \exp(-S(E))$ corresponds to an ergodic state.
3. Every time-invariant state is a sum or integral of ergodic states, so it is tempting to interpret the ergodic states as the pure phases of the system. Mixtures would then be incoherent superpositions in the sense of quantum theory rather than coexisting, spatially separated phases. With any reasonable definition of pure phases, the decomposition into ergodic states should be unique, and the set of time-invariant states must be a simplex. This is indeed the case for asymptotically Abelian systems, which follows from the observation that $\mathcal{R}' = \mathcal{A}' \cap \{U_t\}'$ is Abelian: As was seen in (1.4.8) and (III:2.3.24;2), every Abelian subalgebra of $\mathcal{A}'$ corresponds to a unique decomposition of a state ω; if $\{P_i\}$, $\sum_i P_i = \mathbf{1}$, are the orthogonal projections of this algebra, and
$$\omega_i(a) = \frac{\omega(P_i a)}{\omega(P_i)} \qquad \text{for all } a \in \mathcal{A},$$
provided that $\omega(P_i) > 0$, and is otherwise arbitrary, then $\omega = \sum_i \lambda_i \omega_i$, $\lambda_i = \omega(P_i)$ and $\pi_\omega = \oplus_i \pi_{\omega_i}$, where π_{ω_i} acts on $P_i \mathcal{H}_\omega$. Now if ω is invariant and is to have a decomposition into other invariant states, then the projections P_i must belong to $\mathcal{A}' \cap \{U_t\}'$, and in fact the external states correspond to the minimal projections. Since $\mathcal{A}' \cap \{U_t\}' \subset \mathbb{Z}$, the decomposition into ergodic states is never as fine as the factor decomposition. Hence if a factor representation is given by the invariant state ω, it is necessarily ergodic.

Ergodicity in fact singles out the desired properties. This is shown by the

3.1.20 Characterization of the Ergodic States

Let $\mathcal{A}$ be an algebra that is asymptotically Abelian in time, ρ an invariant state on $\mathcal{A}$, and $|\Omega\rangle$ the vector of the state ρ in the GNS *representation. Then the following conditions are equivalent:*

1. *ρ is ergodic;*
2. $\mathcal{R}' = \{\alpha \cdot \mathbf{1}\}$;
3. *given any decomposition $\rho = \int \sigma d\mu(\sigma)$ and a μ-measurable mean η, $\eta(\sigma) = \rho$ almost everywhere for μ;*
4. *$\eta(a) = \mathbf{1} \cdot \rho(a)$ for all $a \in \mathcal{A}$ and all invariant means η;*
5. $(\mathcal{A} \cup \mathcal{A}') \cap U' = \{\alpha \cdot \mathbf{1}\}$;
6. $E_0 = |\Omega\rangle\langle\Omega|$;
7. *ρ is a unique, invariant, normal state on $\pi_\rho(\mathcal{A})''$;*
8. *$\eta(\rho(ab_t)) = \rho(a)\rho(b)$ for all a and $b \in \mathcal{A}$ and all invariant means η.*

3.1.21 Remarks

1. If the quantum system is finite, H has pure point spectrum, with eigenvectors $\{|x_i\rangle\}$. As we have learned, the invariant states are of the form $a \rightarrow \sum_i c_i \langle x_i | a x_i \rangle$, so the extremal invariant states are of the form $a \rightarrow \langle x_i | a x_i \rangle$ and therefore pure. If the system is either infinite or classical, the ergodic does not imply pure. For example, the state of free fermions (2.5.30) produces a factor and is therefore ergodic, but $\mathcal{A}'$ is isomorphic to $\mathcal{A}$ and thus different from $\{\alpha \cdot \mathbf{1}\}$. It will be discovered later that this is normal situation for equilibriun states.
2. According to (III:2.3.10;5), Condition 2 means that ρ is a pure state on $\mathcal{R}$, and can also be written as $\mathcal{R} \cap \mathcal{R}' = \{\alpha \cdot \mathbf{1}\}$; in particular, every factor state over $\mathcal{R}$ is ergodic.
3. Condition 3 can be sharpened for classical systems with Birkhoff's ergodic theorem, according to which almost every trajectory fills the energy shell densely. In this case, with the decomposition into pure states, the Cesàro mean exists; $\eta(\sigma)$ is μ-measurable, and the order of η and $\int d\mu$ can be reversed.
4. By Condition 4 the time-average of operators in this situation is unique and a multiple of the identity (we assume $\rho \circ \eta = \eta \circ \rho$). More particularly, the classical time-average of any set of positive ρ-measure is spread out over the whole support of ρ. Hence the time-average of states with a density function equals the equilibrium state. Since averaged observables are multiples of the identity, they exhibit no deviation.
5. The implication of Condition 5 for classical dynamics is that if the system is ergodic, then every measurable, time-independent function is constant on the energy shell. Note that $(\mathcal{A} \cup \mathcal{A}')''$ might contain additional time-invariant operators; for instance, for a factor this set is $\mathcal{B}(\mathcal{H})$ and therefore also contains U.
6. Condition 6 implies that 1 is a simple eigenvalue of U.
7. By Condition 7, all the other eigenvectors of U lead to the same state as ρ. Classically, the eigenfunctions $\varphi(p, q)$ must always have $|\varphi|^2$ constant independently of p and q. Thus ergodicity does not make it impossible for the spectrum to be purely pointlike, but only prevents 0 from being a degenerate eigenvalue of H. The extra word "normal" of Condition 7 is important. In Example (3.1.1;5) of free fermions, equilibrium states at different temperatures from that of the specified representation are invariant in time, but not normal. This means classically that different energy shells have disjoint support.
8. Condition 8 means that the autocorrelation function $\rho(ab_t) - \rho(a)\rho(b)$ has time-average 0. Also, according to Condition 4 the expectation values of operators in states of the form $a|\Omega\rangle$ have the same time-averages as those with the state ρ. Since the states $a|\Omega\rangle$ are dense, the time-average of every normal state is ρ. This is a sort of converse to Condition 3, in so far as $\eta(\sigma) = \rho$ for all σ's that are pure and normal (as states on $\pi_\rho(\mathcal{A})''$). It may happen that the set of such σ's is empty (cf. (1.4.17;3)), and some non-normal, pure states converging to something other than the equilibrium state will make their appearance later.

Proof: 1 ⇒ 2: Let $t \in \mathcal{R}'$, $0 < t < 1$; then the vector $|\Omega_\rho\rangle$ associated with ρ in the GNS representation is cyclic for $\mathcal{R}$ and therefore separates $\mathcal{R}'$. With $|\Omega_\rho\rangle$,

$$0 < ||t^{1/2}\Omega_\rho\rangle|^2 = \langle\Omega_\rho|t\Omega_\rho\rangle \equiv \lambda < 1,$$

so if

$$\rho_1(a) = \frac{1}{\lambda}\langle\Omega_\rho|at\Omega_\rho\rangle,$$

and

$$\rho_2(a) = \frac{1}{1-\lambda}\langle\Omega_\rho|a(1-t)\Omega_\rho\rangle \qquad \text{for all } a \in \mathcal{A},$$

then $\rho = \lambda\rho_1 + (1-\lambda)\rho_2$ has a genuine decomposition into invariant states.

2 ⇒ 1: Let $\rho = \lambda\rho_1 + (1-\lambda)\rho_2$, where $0 < \lambda < 1$. Then according to (III:2.3.24;2) there exists a $t \in \pi_\rho(\mathcal{A})'$ such that $0 \le t \le 1$ and $\rho_1(a) = \langle\Omega_\rho|t\Omega_\rho\rangle^{-1}\langle\Omega_\rho|at\Omega_\rho\rangle$ for all $a \in \mathcal{A}$. If ρ_1 is invariant, then t is in $\mathcal{R}'$, and it follows from Condition 2 that $\rho = \rho_1 = \rho_2$.

2 ⇒ 4: $\mathcal{R}' \supset \{\eta(a) : a \in \mathcal{A}\}$. (Cf. (3.1.16;2).)

1 ⇒ 3: The state $\rho = \int \sigma d\mu(\sigma)$ is invariant in time, so $\rho(a) = \int d\mu(\sigma)\eta(\sigma(a))$. Therefore $\rho = \int d\mu(\sigma)\eta(\sigma)$, and, since ρ is an extremal invariant, it equals the invariant state $\eta(\sigma)$ almost everywhere in μ.

3 ⇒ 1: Suppose that ρ is not ergodic. Then there exist invariant states $\rho_1 \neq \rho_2$ such that $\rho = \lambda\rho_1 + (1-\lambda)\rho_2$. This is a special case of a decomposition with $\rho_i = \eta(\rho_i) \neq \rho$, so Condition 3 would be violated.

2 ⇔ 5: The invariant elements of $\mathcal{A}$ and $\mathcal{A}'$ compose $\mathcal{R}'$.

6 ⇒ 1: Suppose that $\rho = \lambda\rho_1 + (1-\lambda)\rho_2$; then by (III:2.3.24;2), ρ_1 is of the form $\rho_1(a) = \langle\Omega_\rho|t\Omega_\rho\rangle^{-1}\langle t^{1/2}\Omega_\rho|at^{1/2}\Omega_\rho\rangle$ for $a \in \mathcal{A}$, and t is in $\pi_\rho(\mathcal{A})' \cap U'_\rho$ if ρ_1 is invariant. Condition 6 implies that $|t^{1/2}\Omega_\rho\rangle \sim |\Omega_\rho\rangle$, because $|t^{1/2}\Omega_\rho\rangle \in E_0\mathcal{H}$, so $\rho = \rho_1 = \rho_2$.

6 ⇒ 8: $\eta(\rho(ab_t)) = \eta(\langle\Omega|aU_t b|\Omega\rangle) = \langle\Omega|aE_0 b|\Omega\rangle = \rho(a)\rho(b)$.

7 ⇒ 6: If there existed a second invariant vector $|\Omega'\rangle$, then all vectors $\sqrt{\alpha}|\Omega\rangle + \sqrt{1-\alpha}|\Omega'\rangle$ for $0 \le \alpha \le 1$ would give rise to the same state, but by Property (3.1.16;3), since the algebra is maximally Abelian on the subspace, this would mean that $|\Omega\rangle = |\Omega'\rangle$.

4 ⇒ 7 and 8: ω invariant ⇒ $\omega = \eta(\omega) \Rightarrow \eta(\omega)(a) = \rho(a)$.

8 ⇒ 4: From $\eta([b_t, c]) = 0$ it follows that $\rho(ac)\rho(b) = \eta(\rho(acb_t)) = \eta(\rho(ab_t c))$, so the matrix elements of $\rho(b)\cdot\mathbf{1}$ and $\eta(b)$ are equal on a dense set. □

3.1.22 Examples

1. The only possible ergodic states on classical systems are those concentrated on $\delta(E - H(p,q))$; otherwise $\mathcal{A}$ would contain the additional invariant $F(H)$, contradicting Condition 4. Let us examine a chain of N coupled oscillators (1.1.2). The Hamiltonian can be written in terms of action and angle variables

K_i (see (I:3.3.3) and (I:3.3.14)) and $\varphi_i \in T^1$ as

$$H = \sum_{i=1}^{N} \omega_i K_i,$$

and the time-evolution is $\varphi_i \to \varphi_i + \omega_i t$. If $N > 1$, the state $\sim \delta(E - H)$ is not ergodic, although the state $\sim \prod_i \delta(K - i - c_i)$ concentrated on T^N is, provided that the angular velocities ω_i are rationally independent (cf.(I:3.3.3)). To understand why, observe that the operator h on $L^2(T^N)$ introduced in (3.1.1;1) arises when K_i is interpreted as the displacement operator, the eigenvalues of which are $2\pi n$, $n \in \mathbb{Z}$. The spectrum of H is therefore purely pointlike, with eigenvalues $2\pi \sum_i \omega_i n_i$. If the ω_i are rationally independent, then the eigenvalue 0 (all $n_i = 0$) is nondegenerate and otherwise it is degenerate. According to (3.1.20;6) this is a criterion for ergodicity. This example is also useful for illustrating the other criteria. For instance, Condition 4 states that every invariant L^∞ function is constant almost everywhere on T^N. Roughly speaking, a function assuming one value on half the trajectories and a different value on the other half is not measurable.

2. Of the quantum-mechanical examples of (3.1.1), only the free fermions (3.1.1;5) fall within the category covered by (3.1.20), as the others are not asymptotically Abelian. Since (3.1.1;5) has a factor state, it is ergodic according to Condition 5. If we go through the other criteria, we notice that Condition 8 holds in the sharpened form $\lim_{t\to\pm\infty} \rho(ab_t) = \rho(a)\rho(b)$ for all a and $b \in \mathcal{A}$. This means that normal states approach ρ not only in the mean, but also actually in the limit $t \to \pm\infty$. The situation is as described intuitively in §1.1, where the states converge to the equilibrium state.

Even though Example 1 is ergodic, it does not exhibit the sort of behavior appropriate for a thermodynamic system. The time-evolution is a rigid displacement in T^N, and this submanifold does not get thoroughly mixed. States like those given by pieces of T^N do not converge as $t \to \infty$; only their means converge. Example 2 conforms better to the notion of a thermodynamic system, which suggest sharpening some of Criteria (3.1.20) as much as possible, by replacing the time-average with the time-limit.

3.1.23 Definition

An invariant state on an asymptotically Abelian system is called **mixing** iff one of the following equivalent conditions is satisfied:

4. $\text{w-}\lim_{t\to\pm\infty} \pi_\rho(a_t) = \mathbf{1} \cdot \rho(a)$ for all $a \in \mathcal{A}$ (the weak limit is that of the GNS representation);

6. $U \overset{t\to\pm\infty}{\rightharpoonup} |\Omega\rangle\langle\Omega|$;

8. $\lim_{t\to\pm\infty} \rho(ab_t) = \rho(a)\rho(b)$.

3.1.24 Remarks

1. By Condition 4, every operator converges to its equilibrium value and its deviation goes to zero. Hence, in the Schrödinger picture every normal state approaches the equilibrium state ρ. In classical dynamics probability distributions of normal states are described by functions – i.e., not by δ-distributions – and so they spread out through all of ρ.
2. Criterion 6 is satisfied if the spectrum of U is absolutely continuous other that the eigenvalue associated with $|\Omega\rangle$. In any case, $|\Omega\rangle$ must be the only eigenvector.
3. Concerning Condition 8, we have learned that for a factor the correlation functions vanish automatically as $t \to \pm\infty$. Therefore, for factors ergodic is equivalent to mixing. In general it is only true that mixing implies ergodic. It is also not true to say that mixing implies a factor, since there are classical mixing systems. However, it will be shown in the next sections that in quantum theory equilibrium states are mixing iff the algebra is a factor. In the case of free particles with the spatial translations, as the group of automorphisms with respect to which their algebra of observables is asymptotically Abelian, this reasoning implies that the spatial correlation function goes to zero for factors.
4. If a state is a limit of pure states, then it is mixing: If σ is pure and $\sigma_t \to \rho$ then $\rho(ab_tc) - \lim_{s\to\infty} \sigma(a_s b_{s+t} c_s) + \lim_{s\to\infty} \sigma(a_s c_s)\sigma(b_{t+s}) - \rho(ac)\rho(b) = 0$. A pure state is a factor state, so (3.1.16;4) applies, showing that $\rho(ab_tc) \to \rho(ac)\rho(b)$. The converse is not true in general, since the pure states into which ρ is decomposed need not converge as $t \to \pm\infty$. For example, the pure states for classical systems are points in phase space, which will keep moving forever.

Proof of the Equivalence in (3.1.25)

$8' \Leftrightarrow \rho(ab_tc) = \rho(a[b_t, c]) + \rho(acb_t) \to \rho(ac)\rho(b) \Leftrightarrow 4'$, and $\rho(a_tb) = \rho(aU_tb)$, hence $6' \Leftrightarrow 8'$. □

Classical systems that mix are of necessity complicated, and it requires a rather demanding example to show that the concept of (3.1.23) is not empty:

3.1.25 Motion of Surface of Constant, Negative Curvature

The ergodic system (3.1.24;1) is not mixing; the spectrum of U_t is purely discrete. This agrees with the perception that displacements in T^2 do not mix its parts together:

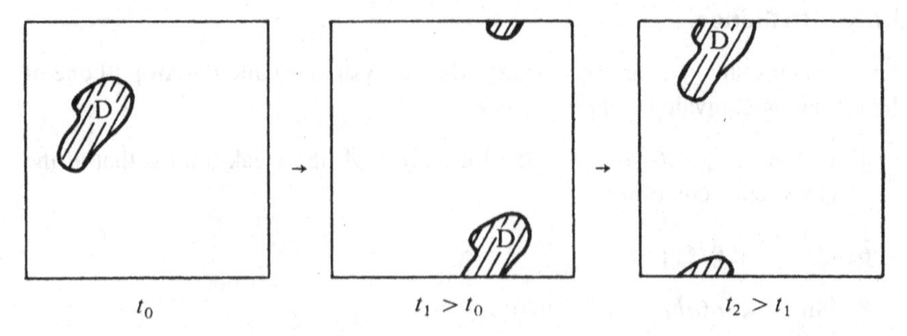

To produce mixing we need a somewhat geometrically irregular configuration; fortunately, as will now be demonstrated, it suffices to have a surface of constant negative curvature. The construction of the example makes use of the following more abstract reformation of (3.1.22;1). Treat $\mathbb{R}^2$ as a two-dimensional group and the trajectory as a one-dimensional subgroup, and consider its image in the quotient space $T^2 = \mathbb{R}^2/\mathbb{Z}^2$. Conservation of angular momentum gets lost, and the trajectory can be dense in T^2. The present example will have an energy shell that is diffeomorphic to the Lorentz group SO(2,1), and the trajectory will be a one-parameter subgroup. In order to destroy the other constants of the motion and have an energy shell of finite volume, map the space to SO(2,1)/$\mathbb{Z}$, where $\mathbb{Z}$ is a discrete subgroup of SO(2,1). The dynamics furnishes a unitary representation $U_t = \exp(mt)$ of a one-parameter subgroup of SO(2,1), but, unlike with $\mathbb{R}^2$, U has only absolutely continuous spectrum other than the point 1, and so the system is mixing by (3.1.23;6).

We realize these ideas in a classical system the Lagrangian of which is quadratic in the velocities. The motion thus proceeds in the absence of forces, but the invariance under SO(2,1) brings about some unusual signs. The extended configuration space is the submanifold of $\mathbb{R}^3$ for which

$$(x|x) \equiv x_1^2 + x_2^2 - x_0^2 = -1. \tag{3.3}$$

If $\dot{x}$ denotes the derivative of x by the proper time t, then the Lagrangian is

$$L = \frac{1}{2}(\dot{x}|\dot{x}).$$

The constraint (3.3) enters into the Euler-Lagrange equations through a Lagrange multiplier,

$$\ddot{x}_i = \lambda x_i, \tag{3.4}$$

and there are the following constants:

$$(x|x) = -1, \qquad (\dot{x}|x) = 0, \qquad (\dot{x}|\dot{x}) = 1 \tag{3.5}$$

(which normalize t). The three-dimensional manifold defined by the constants corresponds to the energy shell (recall that the configuration space is two-dimensional and the phase space is four-dimensional), and on it is the SO(2,1)-invariant Liouville measure

$$d\Omega = d^3x d^3\dot{x} \delta((\dot{x}|x))\delta((x|x) + 1)\delta((\dot{x}|\dot{x}) - 1)\Theta(x_0)\,. \tag{3.6}$$

There are also three constants associated with the angular momentum,

$$l_i = \varepsilon_{ikm} x_k \dot{x}_m, \tag{3.7}$$

which are connected by an algebraic relationship,

$$(l|l) = -(x|x)(\dot{x}|\dot{x}) = 1.$$

One dimension is left for the trajectory. Because $(l_i|x) = 0$, the projection of the trajectory onto configuration space is the intersection of the hyperboloid (3.3)

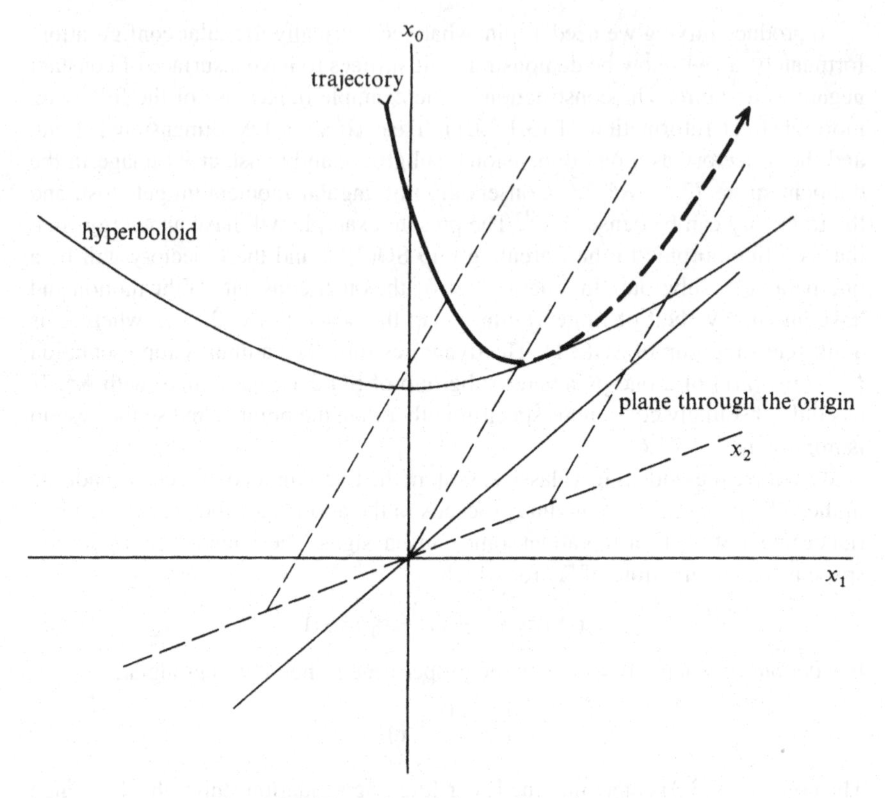

Figure 3.2. The trajectory in configuration space.

with a plane passing through the origin and making an angle less than 45° with the x_0-axis (see Figure 3.2).

The energy is only apparently indefinite; x_0 can be eliminated, and then

$$L = \frac{\dot{x}_1^2 + \dot{x}_2^2 + (\dot{x}_1 x_2 - \dot{x}_2 x_1)^2}{x_1^2 + x_2^2 + 1}$$

describes motion in the $(x_1 - x_2)$-plane without forces, but with a positive effective mass that depends on the position.

The indefinite scalar product $(\cdot|\cdot)$ and consequently also the formalism that has been developed are invariant under SO(2,1). The group SO(2,1) acts transitively on the energy shell (3.5), and every point can be written

$$\{x, \dot{x}\} = \{M(1, 0, 0), M(0, 1, 0)\} \tag{3.8}$$

for some $M \in \mathrm{SO}(2, 1)$. It is easy to see that M is determined uniquely, and this creates the diffeomorphism between the energy shell and SO(2,1) that was mentioned above. Accordingly, every trajectory can be obtained by

$$M(t) = \begin{bmatrix} \cosh t & \sinh t & 0 \\ \sinh t & \cosh t & 0 \\ 0 & 0 & 1 \end{bmatrix}.$$

The most convenient construction of the discrete subgroup makes use of the isomorphism between SO(2,1) and $\mathrm{SL}(2,\mathbb{R})/\{1,-1\}$, since 2×2 matrices are easier to handle than 3×3 matrices. The source of this isomorphism, like that of $\mathrm{SO}(3)=\mathrm{SU}(2,\mathbb{C})/\{1,-1\}$, lies in the observation that

$$\begin{pmatrix} \alpha & \beta \\ \gamma & \delta \end{pmatrix} \in \mathrm{SL}(2,\mathbb{R}), \quad \text{i.e., } (\alpha,\beta,\gamma,\delta) \in \mathbb{R}^4, \alpha\delta - \beta\gamma = 1, \tag{3.9}$$

produces the Lorentz transformation $x \to x'$ by

$$\begin{pmatrix} x_0' + x_2' & x_1' \\ x_1' & x_0' - x_2' \end{pmatrix} = \begin{pmatrix} \alpha & \beta \\ \gamma & \delta \end{pmatrix} \begin{pmatrix} x_0 + x_2 & x_1 \\ x_1 & x_0 - x_2 \end{pmatrix}. \tag{3.10}$$

It is necessary to take the quotient by the center $\{1,-1\}$, since the Lorentz transformations corresponding to the matrices $m \in \mathrm{SL}(2,\mathbb{R})$ and $-m$ are the same. It is not hard to come up with discrete subgroups of SL(2,$\mathbb{R}$), such as

$$\mathcal{Z} = \left\{ \begin{pmatrix} \alpha & \beta \\ \gamma & \delta \end{pmatrix} \in \mathrm{SL}(2,\mathbb{R}) : \alpha,\beta,\gamma,\delta \quad \text{integers} \right\}.$$

Now let us investigate the motion on the quotient space $\Omega_0 = SO(2,1)/\mathcal{Z} \simeq SL(2,\mathbb{R})/(\{1,-1\} \otimes \mathcal{Z})$. Unlike the case of T^2, the quotient space is not a group, since $\mathcal{Z}$ is not a normal divisor, though for our purposes this does not matter. Thus Ω_0 is the energy shell (3.5), on which points are identified if they are transformed into each other by $\mathcal{Z}$. For the trajectory this means that if it goes out one end of the domain of periodicity it reappears at the other. Conservation of angular momentum breaks down, leaving the possibility that the trajectory fills Ω_0 densely.

To get a clearer picture of Ω_0 we have to find out what corresponds to the square $0 \le \varphi_1, \varphi_2 \le 1$ of the earlier example, that is, a region containing no points equivalent under $\mathcal{Z}$, but for each boundary point of which there is a $z \neq 1$ of $\mathcal{Z}$ mapping it to another boundary point. The subgroup $\mathcal{Z}$ is generated by the matrices

$$\begin{pmatrix} 1 & 1 \\ 0 & 1 \end{pmatrix}, \quad \begin{pmatrix} 0 & 1 \\ -1 & 0 \end{pmatrix},$$

the latter of which is the reflection $(x_1, x_2) \to (-x_1, -x_2)$. It is therefore possible to restrict attention to the upper half plane $\{x_2 > 0\}$ in configuration space and choose a region symmetric about the x_2-axis. The boundary curves can be obtained by transforming the x_2-axis with the matrices

$$\begin{pmatrix} 1 & \pm 1/2 \\ 0 & 1 \end{pmatrix}$$

of SL(2,ℝ). They have the parametric representation

$$h_{\pm} = \left\{x_1' : \begin{pmatrix} x_0' + x_2' & x_1' \\ x_1' & x_0' - x_2' \end{pmatrix}\right. \tag{3.11}$$
$$= \begin{pmatrix} 1 & \pm\frac{1}{2} \\ 0 & 1 \end{pmatrix} \begin{pmatrix} \sqrt{1+x_2^2}+x_2 & 0 \\ 0 & \sqrt{1+x_2^2}-x_2 \end{pmatrix} \begin{pmatrix} 1 & 0 \\ \pm\frac{1}{2} & 1 \end{pmatrix}, x_2 > 0\Big\};$$

note that

$$\begin{pmatrix} 1 & 1 \\ 0 & 1 \end{pmatrix} \cdot h_- = h_+,$$

so $x_2' = \pm\frac{1}{4}(1/x_1' - 3x_1')$. The projection of Ω_0 onto configuration space looks as depicted in Figure 3.3, where the lines A indicate the identifications.

The identification of the boundary points by $\binom{1\ 1}{0\ 1}$ means that if the trajectory leaves through one side, it reappears at the corresponding point of the other side (see Figure 3.4).

Now we are in a position to verify that the measure of Ω_0 with $d\Omega$ (3.6) is actually finite. This follows from

$$\int d^3\dot{x}\delta[(\dot{x}|x)]\delta[(\dot{x}|\dot{x}) - 1] \equiv F(x,x) < \infty$$

and

$$F(-1)\int d^3x\delta[(x|x)+1] < \infty,$$

where the integral runs over the region bounded by (3.11).

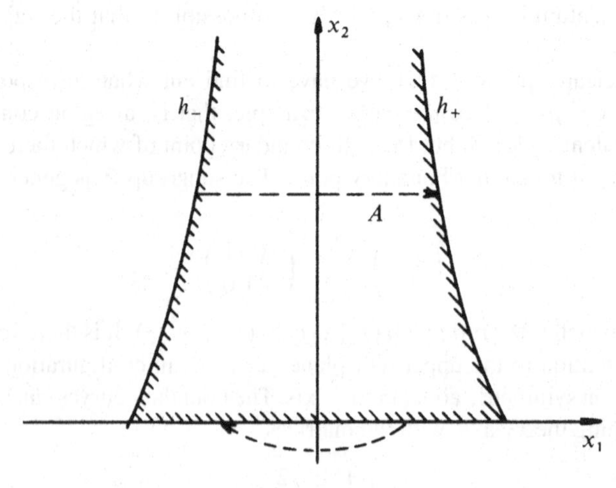

Figure 3.3.

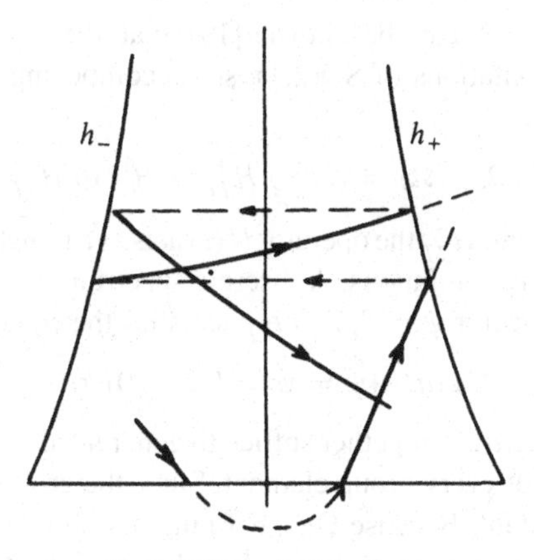

Figure 3.4.

The time-evolution is controlled by the unitary group

$$U_t = \exp(mt), \qquad m = \frac{\partial M}{\partial t}\Big|_{t=0},$$

where the anti-Hermitian operator m is one of the generators of SO(2,1). If the other two generators are combined into $m_\pm \equiv m_1 \pm m_2$, then $m_\pm$ satisfy the commutation relations

$$[m, m_\pm] = \pm m_\pm \qquad \text{and} \qquad [m_+, m_-] = 2m.$$

Note that in contradinstinction to SO(3), this time $(m_\pm)^* = -m_\pm$. This fact will be crucial, since the generators of SO(3) have purely discrete spectra. Instead of SO(2,1), let us now examine the simpler two-parameter subgroups

$$U_\pm(a, t) = \exp(am_\pm)\exp(tm)$$

with the multiplication law

$$U_\pm(a, t)U_\pm(a', t') = U_\pm(a + \exp(\pm t)a', t + t').$$

Because $[m_+, m_-] = 2m$, the operators $U_+(a, 0)$ and $U_-(a, 0)$ generate the whole group, and $U(t) = u_+(0, t) = U_-(0, t)$.

Next consider the representation (3.1.1;1) of classical dynamics on $\mathcal{H} = L^2(\Omega_0, d\Omega)$. Not just U_t, but in fact all of SO(2,1) is represented unitarily on $\mathcal{H}$ by $f(x) \to f(Mx)$, and we shall now reduce this representation according to the irreducible representations of the subgroups $U_\pm$. We start by observing that $U_\pm(a, 0)$ is a normal divisor, and the factor groups $U_\pm(a, t)/U_\pm(a, 0)$ are isomorphic to $\mathbb{R}$. Hence there are irreducible, one-dimensional representations of the type

$$I : U_\pm(a, t) = \exp(i\lambda t), \qquad \lambda \in \mathbb{R}.$$

In addition it is readily seen that $U_\pm$ can also be represented on $L^2(\mathbb{R}, dx)$ by

$$II : [U_+(a, t)\psi](x) = \exp(iae^x)\psi(x + t), \qquad \psi \in L^2(\mathbb{R}, dx),$$

and similarly for U_-. It can be shown [19] that these possibilities exhaust the irreducible representations of SO(2,1), so, decomposing into the irreducible representations of $U_\pm$,

$$L^2(\Omega_0, d\Omega) = \mathcal{H}_I^+ \oplus \mathcal{H}_{II}^+ = \mathcal{H}_I^- \oplus \mathcal{H}_{II}^-.$$

On the subspaces $\mathcal{H}_{II}^+$ and $\mathcal{H}_{II}^-$ the operator $U(t)$ acts as a translation on $L^2(\mathbb{R}, dx)$, and thus its spectrum is continuous. A discrete spectrum could only be found on $\mathcal{H}_I^+ \cap \mathcal{H}_I^-$, but every vector ψ of $\mathcal{H}_I^+ \cap \mathcal{H}_I^-$ satisfies the equation

$$U_+(a,0)\psi = \psi = U_-(a,0)\psi.$$

Since $U_+(a,0)$ and $U_-(a,0)$ together suffice to generate all of SO(2,1), ψ is invariant under the action of every group element. Since the group acts transitively on Ω_0, ψ must be a constant. Because Ω_0 has finite measure, any constant function belongs to $L^2(\Omega_0, d\Omega)$, so the situation is like that of (3.1.24;2). Unless the quotient by $\mathcal{Z}$ is taken, U has no point spectrum, as constant functions would not be integrable. In sum the argument is that the system is mixing because the spectrum of U consists of a single nondegenerate eigenvalue 1 and an absolutely continuous portion. This is in contrast to the motion on the torus, for which the spectrum of U_t was purely discrete, and the system was only ergodic, not mixing.

3.1.26 Example

The quantum-mechanical example of an infinite system of free fermions was seen to be mixing. Despite the absence of interaction, a local perturbation spreads out to infinity through the diffusion of free wave-packets. From among the characterizations of ergodicstates (3.1.20), let us look in particular at the third. It holds in the sharper form (3.1.24;4); the grand canonical state (2.5.30) is the time-limit of a pure state. The proof of this fact uses the transformations

$$a_\uparrow(f) = b_\uparrow(\beta f) + b_\downarrow^*(\sqrt{1-|\beta|^2} f^*)$$

and

$$a_\downarrow(f) = b_\downarrow(\beta f) - b_\uparrow^*(\sqrt{1-|\beta|^2} f^*). \tag{3.12}$$

We have directly taken up the realistic case of spin-$\frac{1}{2}$ fermions, where $\uparrow$ and $\downarrow$ indicate the direction of the spin that the field operator describes. In Fourier-transformed space β is a function $\mathbf{k} \to \beta(\mathbf{k}) : \mathbb{R}^3 \to \{z \in \mathbb{C} : |z|^2 \le 1\}$, and βf is the function $\beta(\mathbf{k})f(\mathbf{k})$. In x-space β is a convolution. It is straightforward to verify that the a's satisfy the usual commutation relations (1.3.2;2),

$$\begin{aligned}
[a_\uparrow(f), a_\uparrow^*(g)]_+ &= [a_\downarrow(f), a_\downarrow^*(g)]_+ = (f|g), \qquad (3.13)\\
[a_\uparrow(f), a_\uparrow(g)]_+ &= [a_\uparrow(f), a_\downarrow(g)]_+ = [a_\uparrow(f), a_\downarrow^*(g)]_+ \\
&= [a_\downarrow(f), a_\downarrow(g)]_+ = 0,
\end{aligned}$$

supposing that the b's satisfy the commutation relations. Clearly the a's and the b's generate the same C^* algebra. The expectation values of the a's in the Fock state $|0\rangle$ (1.3.2) for the b's: $b_\uparrow(f)|0\rangle = b_\downarrow(f)|0\rangle = 0$, are

$$\langle 0|a_\uparrow(f)a_\uparrow^*(g)|0\rangle = \langle 0|a_\downarrow(f)a_\downarrow^*(g)|0\rangle = \int \frac{d^3\mathbf{k}}{(2\pi)^3}|\beta(\mathbf{k})|^2 f^*(\mathbf{k})g(\mathbf{k}),$$

$$-\langle 0|a_\uparrow(f)a_\downarrow(g)|0\rangle = \langle 0|a_\downarrow(f)a_\uparrow(g)|0\rangle$$

$$= \int \frac{d^3\mathbf{k}}{(2\pi)^3} f^*(\mathbf{k})g^*(\mathbf{k})\beta(\mathbf{k})\sqrt{1-|\beta(\mathbf{k})|^2};$$

$$\langle 0|a_\uparrow(f)a_\uparrow(g)|0\rangle = \langle 0|a_\uparrow(f)a_\downarrow^*(g)|0\rangle = 0. \tag{3.14}$$

The state $|0\rangle$ was seen to be pure in (1.3.16;1). Under the time-evolution $f(\mathbf{k}) \to \exp(-it|\mathbf{k}|^2)f(\mathbf{k})$, the quantity $-\langle 0|a_\uparrow a_\downarrow|0\rangle = \langle 0|a_\downarrow a_\uparrow|0\rangle$ goes to 0 as $t \to \pm\infty$ by the Riemann-Lebesque lemma. If

$$\beta(\mathbf{k}) = (1+\exp(-\beta(|\mathbf{k}|^2-\mu)))^{-1/2},$$

then in the limit $t \to \pm\infty$ the generalization of the state (2.5.49) for spin $\frac{1}{2}$ is all that is left over.

3.1.27 Remarks

1. The limit of a pure state is clearly not always an equilibrium state; other functions could be chosen for $\beta(\mathbf{k})$.
2. Since the thermal representation of free fermions (3.1.1;5) is a factor of type III, the pure state $|0\rangle$ associated with the thermal representation cannot be normal (cf. (1.4.16;3)). Likewise, any other states of the latter formed with different $\beta(\mathbf{k})$ are not normal because of (3.1.20;7), even though they are invariant.
3. The state given by $|0\rangle$ is not invariant in time, and in this representation the time-evolution is certainly not a unitary group (cf. (1.3.13;7)). If it were, then the time displacement $\tau_t : a \to a_t$ would be weakly continuous and hence extensible to $\pi(\mathcal{A})''$, which would lead to a contradiction: $\mathcal{A}_e$ is asymptotically Abelian with respect to the spatial translation T_x, so in the representation with the translation-invariant state $|0\rangle$, $\lim_{|x|\to\infty} T_x a = \mathbf{1}\cdot\langle 0|a|0\rangle$ for all $a \in \mathcal{A}_e$. Since T_x commutes with τ_t, it would follow that $\lim_{x\to\infty} T_x\tau_t(a) = \mathbf{1}\cdot\langle 0|a_t|0\rangle = \lim_{x\to\infty}\tau_t T_x(a) = \mathbf{1}\cdot\langle 0|a|0\rangle$, which would then imply that the state $\langle 0|\cdot|0\rangle$ would be invariant in time.

3.1.28 Problems

1. (i) Prove von Neumann's statistical ergodic theorem, $(1/2T)\int_{-T}^{T}\exp(iHt)dt \to E_0$. (Show that on all vectors of the form $x = \exp(iHs)y - y$, $y \in \mathcal{H}$, $s \in \mathbb{R}$, we have $(1/2T)\int_{-T}^{T}\exp(iHt)x dt \to 0$. Let $\mathcal{H}_1$ be the closed linear hull of these vectors, and note that the same fact applies to all $x \in \mathcal{H}_1$. Finally, show that $\mathcal{H}_1^\perp = \{x : \exp(iHs)x = x \text{ for all } s\} = E_0\mathcal{H}$).
   (ii) Show similarly that $(\varepsilon/2)\int_{-\infty}^{\infty}\exp(-\varepsilon|t|)\exp(iHt)dt \to E_0$.
2. Show that in the Schrödinger picture the time-average of a vector x has the following characterization: $\eta(x)$ is the vector of least norm of the norm-closed, convex hull of

$\{U(t)x\}$, denoted $\mathcal{K}$. (Hint: see the example given earlier for $\eta(x) \in \mathcal{K}$. Show (i) that $\mathcal{K}$ contains a unique vector ξ of least norm; (ii) that ξ is invariant under all $U(t)$; and (iii) that $\mathcal{K}$ contains no other fixed point).

3. Show that $\mathcal{Z} = \{\alpha \cdot \mathbf{1}\}$ iff $w(ab) = w(a)w(b)$ for all $w \in \mathcal{A}^*$, $a \in \mathcal{A}$, and $b \in \mathcal{Z}$.
4. Show that for a classical system, if there exists a constant $f(p, q)$ not of the form $\alpha \cdot \mathbf{1}$, then ρ is not ergodic.
5. Show that a set $E \subset \mathcal{H}$ is a totalizer for $\mathcal{A}$ iff E separates $\mathcal{A}'$. (Cf. (III:2.3.4); a totalizer is a set E such that $\mathcal{A}E$ is dense in $\mathcal{H}$, and separating means that $a'E = 0 \Rightarrow a' = 0$.)
6. Boson states of the form (2.5.49) with $\langle f|\rho g\rangle = \int d^3\kappa\rho(\mathbf{k})\tilde{f}^*(\mathbf{k})\tilde{g}(\mathbf{k})$, $0 \le \rho(\mathbf{k})$, are factor states and consequently mixing. Express such a state as a time-limit of a pure state (cf. (3.1.26)).

3.1.29 Solutions

1. (i) If $x = \exp(iHs)y - y$, then

$$\left\| \frac{1}{2T} \int_{-T}^{T} \exp(iHt)x dt \right\| = \left\| \frac{1}{2T} \left\{ \int_{T}^{T+s} \exp(iHt)y dt - \int_{-T}^{-T+s} \exp(iHt)y dt \right\} \right\|$$
$$\le \frac{|s| \|y\|}{T} \to 0.$$

Because $\|(1/2T)\int_{-T}^{T} \exp(iHt)dt\| \le 1$, this holds for all $x \in \mathcal{H}_1$.

$$x \in \mathcal{H}_1^\perp \Leftrightarrow (x|\exp(iHs)y - y) = (\exp(-iHs)x - x|y) = 0 \quad \text{for all} \quad y \in \mathcal{H}$$
$$\Leftrightarrow \exp(iHs)x = x \quad \text{for all} \quad s \Leftrightarrow E_0 x = x$$

by the spectral theorem.

(ii) It suffices to show that $\varepsilon \int_0^\infty \exp(-\varepsilon t)\exp(iHt)dt \to E_0$, which will follow if $\varepsilon \int_0^\infty \exp(-\varepsilon t)\exp(iHt)x dt \to 0$ for vectors $x = \exp(iHs)y - y$. This integral equals

$$\varepsilon \exp(\varepsilon s) \int_s^\infty \exp(-\varepsilon t)\exp(iHt)y dt - \varepsilon \int_0^\infty \exp(-\varepsilon t)\exp(iHt)y dt$$

$$(\exp(\varepsilon s) - 1)\varepsilon \int_s^\infty \exp(-\varepsilon t)\exp(iHt)y dt - \varepsilon \int_0^\infty \exp(-\varepsilon t)\exp(iHt)y dt \to 0,$$

since $\|\varepsilon \int_0^\infty \exp(-\varepsilon t)\exp(iHt)y dt\| \le \|y\|$.

2. (i) Let $\lambda = \inf\{||x|| : x \in \mathcal{K}\}$. There exists a sequence $\{x_n\}$ in $\mathcal{K}$ such that $||x_n|| \to \lambda$. By the parallelogram law,

$$\left\| \frac{x_n - x_m}{2} \right\|^2 + \left\| \frac{x_n + x_m}{2} \right\|^2 = \frac{1}{2}(\|x_n\|^2 + \|x_m\|^2),$$

x_n is a Cauchy sequence, so it has a limit ξ. If $\|x\| = \|\xi\|$, then

$$\left\| \frac{x - \xi}{2} \right\|^2 = \frac{1}{2}(\|x\|^2 + \|\xi\|^2) - \left\| \frac{x + \xi}{2} \right\|^2 \le 0,$$

which implies that $x = \xi$.

(ii) $\|U(t)\xi\| = \|\xi\| \Leftrightarrow U(t)\xi = \xi$.
(iii) Suppose that η is a second fixed point. For all $\varepsilon > 0$, there exist $\lambda_1, \ldots, \lambda_n$ and $\lambda'_1, \ldots, \lambda'_m$ such that $\sum_i \lambda_i = \sum_i \lambda'_i = 1$, with $\lambda_i, \lambda'_i \geq 0$, and there exist $t_1, \ldots, t_n$ and $t'_1, \ldots, t'_m$ such that if $V \equiv \lambda_1 U(t_1) + \ldots + \lambda_n U(t_n)$, and $W \equiv \lambda'_1 U(t'_1) + \ldots + \lambda'_m U(t'_m)$, then $\|Vx - \xi\| < \varepsilon$, and $\|Wx - \eta\| < \varepsilon$. However, then

$$\begin{aligned}\|\xi - \eta\| &\leq \|\xi - VWx\| + \|VWx - \eta\| = \|W\xi - VWx\| + \|VWx - V\eta\| \\ &\leq \|W\|\|Vx - \xi\| + \|V\|\|Wx - \eta\| < 2\varepsilon,\end{aligned}$$

so $\xi = \eta$.
Remark: The strong and weak closures of a convex set are identical.

3. $\Rightarrow$: This part is trivial.
$\Leftarrow$: Let P_1 and P_2 be projections in $\mathcal{L}$, such that $P_1 \perp P_2$, and let $w_i(\cdot) = w(P_i \cdot)$, $a_i = P_i a$, $b_i = P_i b$ for $i = 1, 2$. If $w = \alpha w_1 + (1 - \alpha) w_2$, then

$$\begin{aligned}w(ab) &= \alpha w(a_1) w(b_1) + (1 - \alpha) w(a_2) w(b_2) \\ &\neq (\alpha w(\alpha_1) + (1 - \alpha) w(a_2))(\alpha w(b_1) + (1 - \alpha) w(b_2)).\end{aligned}$$

4. Let $\bar{f}(p, q) = \inf(1, |f(p, q)|)$ (if necessary multiply f by a suitable constant to ensure that $\bar{f}$ is not identically 1). Then $d\rho$ is the sum of two invariant states,

$$d\rho = \frac{1}{2}(1 + \bar{f}) d\rho + \frac{1}{2}(1 - \bar{f}) d\rho.$$

5. $\Rightarrow$: Let $a' \in \mathcal{A}'$. $a'E = 0 \Leftarrow a'\mathcal{A}E = 0 \Leftarrow a' = 0$ on a dense set, which implies that $a' = 0$.
$\Leftarrow$: Let $E_\perp$ be the orthogonal complement of $\mathcal{A}E$. Then $\mathcal{A}E_\perp = E_\perp$, so the projection $P_\perp$ onto $E_\perp$ belongs to $\mathcal{A}'$, but $P_\perp E = 0$, so E does not separate $\mathcal{A}'$.

6. In a Fock representation of the free fields b, $b(k)|0\rangle = 0$, write

$$a(\mathbf{k}) = \sqrt{\rho(\mathbf{k})} b^*(\mathbf{k}) + \sqrt{1 + \rho(\mathbf{k})} b(\mathbf{k}),$$

and

$$a^*(\mathbf{k}) = \sqrt{\rho(\mathbf{k})} b^*(\mathbf{k}) + \sqrt{1 + \rho(\mathbf{k})} b(\mathbf{k}).$$

These operators a likewise satisfy the commutation relations

$$a^*(\mathbf{k}) a^*(\mathbf{k}') - a^*(\mathbf{k}') a(\mathbf{k}) = \delta(\mathbf{k} - \mathbf{k}'),$$

and

$$\begin{aligned}\langle 0|a(\mathbf{k}) a^*(\mathbf{k}')|0\rangle &= \delta(\mathbf{k} - \mathbf{k}') \rho(\mathbf{k}), \\ \langle 0|a(\mathbf{k}) a(\mathbf{k}')|0\rangle &= \delta(\mathbf{k} - \mathbf{k}') \sqrt{\rho(\mathbf{k})} \sqrt{1 + \rho(\mathbf{k})}.\end{aligned}$$

Hence

$$\begin{aligned}\langle a_{f_t} a^*_{g_t}|0\rangle &= \int dk \rho(\mathbf{k}) \bar{f}^*(\mathbf{k}) \bar{g}(\mathbf{k}), \\ \langle 0|a_{f_t} a_{g_t}|0\rangle &= \int dk \exp(2i|\mathbf{k}|^2 t) \sqrt{\rho(\mathbf{k})} \sqrt{1 + \rho(\mathbf{k})} \bar{f}^*(\mathbf{k}) \bar{g}^*(\mathbf{k});\end{aligned}$$

this last integral goes to zero as $t \to \pm\infty$ by the Riemann–Lebesque lemma, and therefore its time-average is zero. The analogous fact holds for the higher correlation functions, so the time-average of the pure Fock state $|0\rangle$ is of the form (2.5.30).

3.2 The Equilibrium State

In the course of time the Maxwell–Boltzmann distribution has proved more and more fundamental, and has become deeply rooted in the mathematical description of infinite quantum systems.

With a certain normalization of H the canonical state has the form $w(a) = \operatorname{Tr}\exp(-\beta H)a$, as we have seen. The appearance of the Hamiltonian H in both the time-evolution and the state creates all sorts of important connections between them. To avoid technical complications at first we shall concentrate only on the finite-dimensional case. The commutativity of the trace gives rise to a symmetry between the representation of the algebra and its commutant.

3.2.1 The GNS Representation of $\mathcal{B}(\mathbb{C}^n)$ with a Faithful State

Let $\mathcal{A} = \mathcal{B}(\mathbb{C}^n)$ be given the inner product $\langle a|b\rangle = \frac{1}{n}\operatorname{Tr} a^*b$ so that it becomes a Hilbert space isomorphic to $\mathbb{C}^{n^2}$, and define

$$\pi : \mathcal{A} \to \mathcal{B}(\mathbb{C}^{n^2}) : \pi(a)|b\rangle = |ab\rangle,$$
$$\pi' : \mathcal{A} \to \mathcal{B}(\mathbb{C}^{n^2}) : \pi(a)|b\rangle = |ba^*\rangle,$$
$$J : \mathbb{C}^{n^2} \to \mathbb{C}^{n^2} : J|b\rangle = |b^*\rangle.$$

Then

(i) *π is a factor representation (*-isomorphism):*

(ii) *π' is a *-antiisomorphism, i.e.*

$$\pi'(ab) = \pi'(a)\pi'(b), \qquad \pi'(\lambda a) = \bar{\lambda}\pi'(a), \qquad \pi'(a^*) = (\pi'(a))^*,$$
$$\pi'(a+b) = \pi'(a) + \pi'(b) \quad with \quad \pi'(\mathcal{A}) = \pi(\mathcal{A})';$$

(iii) *the conjugate-linear operator J preserves norms and $J^2 = \mathbf{1}$;*

(iv) $J\pi(a)J = \pi'(a)$, $J\pi'(a)J = \pi(a)$;

(v) *let w be a faithful state, that is, if $a > 0$, then $w(a) > 0$, so by (2.1.5(ii)), $w(a) = \operatorname{Tr}\rho a = \langle\sqrt{\rho}|a|\sqrt{\rho}\rangle$, $\rho > 0$, $\operatorname{Tr}\rho = 1$. The vector $|\sqrt{\rho}\rangle$ is cyclic and separating for π and π', i.e. $\pi(a)|\sqrt{\rho}\rangle = 0 \Rightarrow a = 0$. Hence the* GNS *representation using w is unitarily equivalent to π.*

Proof: The isomorphism and antiisomorphism properties are obvious. Note $\langle ba^*|ba^*\rangle \leq \langle b|b\rangle\|a\|^2$ so $\pi'(a)$ defines an operator as $|b\rangle = 0$, $\pi'(a)|b\rangle \neq 0$ is impossible.

(ii) $\pi'(a)\pi(b)|c\rangle = \pi'(a)|bc\rangle = |bca^*\rangle = \pi(b)\pi'(a)|c\rangle$, and therefore $\pi'(\mathcal{A}) \subset \pi(\mathcal{A})'$. On the other hand, if $B \in \pi(\mathcal{A})'$, then $B|1\rangle$ is $|b^*\rangle$ for some $b \in \mathcal{A}$. Hence

$$B|a\rangle = B\pi(a)|1\rangle = \pi(a)B|1\rangle = \pi(a)|b^*\rangle = \pi(a)\pi'(b)|1\rangle = \pi'(b)|a\rangle$$

for all $a \in \mathcal{A}$, so $B = \pi'(b)$ and $\pi'(\mathcal{A}) = \pi(\mathcal{A})'$.

(i) Let $\pi(a) \in \pi(\mathcal{A})'$. Then by part (ii) it equals $\pi'(b^*)$ for some b. Hence $\pi(a)|c\rangle = |ac\rangle = \pi'(b^*)|c\rangle = |cb\rangle$, so $ac = cb$ for all $c \in \mathcal{A}$, and therefore $a = b = \alpha \cdot \mathbf{1}$. Thus $\pi(\mathcal{A})$ is a factor.

(iii) $\|J|a\rangle\|^2 = \mathrm{Tr}\, aa^* = \mathrm{Tr}\, a^*a = \||a\rangle\|^2$, and $J^2 = \mathbf{1}$ since $b^{**} = b$.

(iv) $J\pi(a)J|b\rangle = J\pi(a)|b^*\rangle = J|ab^* = |ba^*\rangle = \pi'(a)|b \Rightarrow J\pi(a)J = \pi'(a) \Rightarrow \pi(a) = J\pi'(a)J$, because $J^2 = \mathbf{1}$.

(v) Since ρ^{-1} exists, $|a\rangle$ may be written as $|b\sqrt{\rho} >= \pi(b)|\sqrt{\rho} >, b = a\rho^{-1/2}$, which shows that $\sqrt{\rho}$ is cyclic for π. If $\rho_i > 0$ are the eigenvalues of ρ, then in the diagonal representation of ρ,

$$\|\pi(a)|\sqrt{\rho}\|^2 = \mathrm{Tr}\, \rho a^* a = \sum_{i,k} \rho_i |a_{ik}|^2 = 0,$$

which implies that $a_{ik} = 0$, and similarly for π'. By (III:2.3.10;6) π_ρ is equivalent to π. □

3.2.2 Remarks

1. An anti-isomorphism came up once before, in the reversal of the motion (III:3.3.17), and J is like the conjugate-linear operator Θ' (3.3.19;2).
2. The representation π, being a finite-dimensional factor of type I, is of the form $\pi(a) = a \otimes \mathbf{1}_{|\mathbb{C}^n}$, so $\pi'(a)$ is $\mathbf{1}_{\mathbb{C}^n} \otimes a^*$.

Consider next how to represent the time-evolution $a \to a_t = \exp(iht)a \times \exp(-iht)$. At first thought it might be represented by $\exp(i\pi(h)t)$, but this would not leave the cyclic vector $|\sqrt{\rho}\rangle$ invariant. The correct way to proceed is as in Example (3.1.1;3).

3.2.3 The Time-Evolution on $\mathcal{B}(\mathbb{C}^n)$

The unitary representation (1.3.3) of the time-evolution $a \to a_t$ on the invariant state $a \to \mathrm{Tr}\, \rho a$, $\rho = \exp(-\beta h)$, is given by $U_t = \exp(-iHt)$, $H = \pi(h) - \pi'(h)$. It satisfies the following:

(i) $JHJ = -H, JU_tJ = U_t$;

(ii) $U_{-i\beta/2}\pi(a)|\sqrt{\rho}\rangle = J\pi(a^*)|\sqrt{\rho}\rangle$;

(iii) $\langle\sqrt{\rho}|\pi(a)\pi(b)|\sqrt{\rho}\rangle = \langle\sqrt{\rho}|\pi(b)\pi(a_{i\beta})|\sqrt{\rho}\rangle$.

Proof: It is immediately clear that $\exp(iHt)\pi(a)\exp(-iHt) = \pi(a_t)$. Moreover, $\exp(iHt)|\sqrt{\rho}\rangle = |\exp(iht)\exp(-\beta h)\exp(-iht)\rangle = |\sqrt{\rho}\rangle$.

(i) This follows from (3.2.1(iv)).

(ii) $U_{-i\beta/2}\pi(a)|\sqrt{\rho}\rangle = U_{-i\beta/2}|a\exp(-\beta h/2)\rangle = |\exp(-\beta h)/2)a\rangle$ $= J|a^*\exp(-\beta h/2)\rangle = J\pi(a^*)|\sqrt{\rho}\rangle$.

(iii) $\operatorname{Tr}\exp(-\beta h)ab = \operatorname{Tr}\exp(-\beta h)a\exp(\beta h)\exp(-\beta h)b = \operatorname{Tr}\exp(-\beta h)ba_{i\beta}$.

□

3.2.4 Remarks

1. The density matrix ρ was written simply as $\exp(-\beta h)$ under the assumption that h had been redefined by the addition of a multiple of the identity so that $\operatorname{Tr}\exp(-\beta h) = 1$. This affects neither the time-evolution nor H.
2. Note that J does not reverse the direction of time.
3. The operator $\rho = \exp(-h)$ is always positive. Conversely, if $\rho > 0$ (i.e., all eigenvalues $\rho_i > 0$), then $\ln\rho = -h$ is well defined. This shows that groups of automorphisms and faithful states are bijectively related. There is a special term for their relationship.

3.2.5 The Modular Automorphism

For each faithful state w on $\mathcal{B}(\mathbb{C}^n)$ there is a unique one-parameter group of automorphisms $\tau_t : a \to a_t$ such that

(i) w is invariant in the sense that $w(a_t) = w(a)$.

(ii) w satisfies the **Kubo-Martin-Schwinger** (KMS) condition, $w(ab) = w(ba_i)$.

(iii) there exists an anti-isomorphism $\pi_w(\mathcal{A}) \to J\pi_w(\mathcal{A})J$ onto $\pi_w(\mathcal{A})'$ such that

$$U_{-i/2}\pi(a)|\Omega\rangle = J\pi(a^*)|\Omega\rangle,$$

where $|\Omega\rangle$ is the cyclic vector and U_t is the unitary operator representing τ_t in the GNS representation with w.

If the dimension of the Hilbert space is now infinite, but the state is still given by a density matrix $\rho = \exp(-\beta h)$, then there are a few technical difficulties to clear up.

3.2.6 The Temporal Correlation Functions of Finite Quantum Systems

If the time is made complex, then in general

$$a_{x+iy} \equiv \exp((ix - y)h)\exp(-(ix - y)h)$$

is unbounded, and hence does not belong to the algebra. However, we shall continue to use this notation, as this operator will never act on anything outside its domain of definition.

(i) Continuity in the strip $-\beta \leq \Im t \leq 0$. $w(a_t b) = \langle\Omega|a \exp(-iHt)b|\Omega\rangle$, and if t is complex, then by (3.2.3(ii)), $b|\Omega\rangle$ is in the form domain of $\exp(yH)$ for $y \geq -\beta$. In a spectral representation it is apparent that the vector $\exp(yH/2)b|\Omega\rangle$ is norm-continuous in y, so $\rho(a_t b)$ is norm-continuous in t.

(ii) Boundedness in the strip $-\beta \leq \Im t \leq 0$. Let $H = \pi(h) - \pi'(h)$ as in (3.2.3), so $H|\Omega\rangle = 0$. Because

$$\begin{aligned} a_{x+i} &= \exp((ix - y)H)\exp(-(ix - y)H), \\ |w(a_{x+iy}b)|^2 &= |\langle\Omega|a_x \exp(yH)b|\Omega\rangle|^2 \\ &\leq \langle\Omega|a_x \exp(yH)a_x^*|\Omega\rangle\langle\Omega|b^* \exp(yH)b|\Omega\rangle. \end{aligned}$$

The function $\langle\Omega|a \exp(yH)a^*|\Omega\rangle$ is positive and, because

$$\frac{\partial^2}{\partial y^2}\langle\Omega|a \exp(yH)a^*|\Omega\rangle = \|H \exp(yH/2)a^*|\Omega\rangle\|^2 \geq 0,$$

convex, achieving its maximum at $y = 0$ or $y = -\beta$. It is clear that $w(aa^*) \leq \|a\|^2$, but even at the lower edge it is bounded, as shown by

$$\begin{aligned} w(a_{i\beta/2}a^*_{-i\beta/2}) &= \mathrm{Tr}\exp(-\beta h)\exp(\beta h/2)a^*\exp(-\beta h)a\exp(\beta h/2) \\ &= \mathrm{Tr}\exp(-\beta h)aa^* \leq \|a\|^2, \end{aligned}$$

since $\mathrm{Tr}\exp(-\beta h) = 1$. Therefore

$$|w(a_t b) \leq \|a\|\|b\| \qquad \text{for} \quad -\beta \leq \Im t \leq 0.$$

(iii) Analyticity in the strip $-\beta < \Im t < 0$. The function $w(a_t b)$ is not differentiable on the real axis for generic a's, but only for complex times within the strip. The proof is similar to that of (2.4.6) and will not be repeated here. The relationship $w(ab) = w(ba_{i\beta})$, named for Kubo, Martin and Schwinger, which follows from the invariance of the trace, can be continued analytically to the strip: The functions $w(a_t b)$ and $w(ba_t)$ are analytic respectively in $-\beta < \Im t < 0$ and $0 < \Im t < \beta$, where they satisfy the KMS condition $w(a_t b) = w(ba_{t+i\beta})$, which determines the value of $w(a_t b)$ at $y = -\beta$ as $w(ba)$ (see Figure 3.5).

(iv) The physical significance of the KMS condition. For a finite system the canonical state with $\rho = \exp(-\beta H)$ is not an eigenstate of the energy. The modular Hamiltonian (also denoted H) has $|\Omega\rangle$ as an eigenvector, $H|\Omega\rangle = 0$. This operator H is not generally bounded below; however, the KMS condition distinguishes positive energies because of the positive sign of β. The energy spectrum of $\pi(a)|\Omega\rangle$ for $a = a^* \in \mathcal{A}$ consists predominately of positive energies

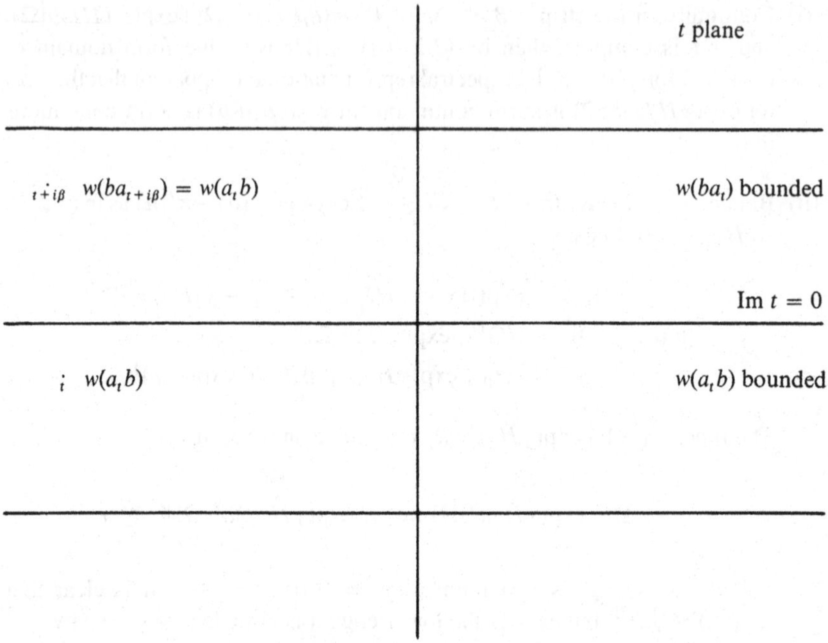

t plane

$t+i\beta$ $w(ba_{t+i\beta}) = w(a_t b)$ — $w(ba_t)$ bounded

Im $t = 0$

t $w(a_t b)$ — $w(a_t b)$ bounded

Figure 3.5. The connection between $w(ba_t)$ and $w(a_t b)$ on their domain of analyticity.

$$
\begin{aligned}
f(E) \equiv \langle \Omega | \pi(a) \delta(H - E) \pi(a) | \Omega \rangle &= \int_{-\infty}^{\infty} \frac{dt}{2\pi} \exp(iEt) \rho(a_t a) \\
&= \int_{-\infty}^{\infty} \frac{dt}{2\pi} \exp(iEt) \rho(a a_{t+i\beta}) \\
&= \exp(\beta E) \langle \Omega | \pi(a) \delta(H + E) \pi(a) | \Omega \rangle,
\end{aligned}
$$

and therefore

$$
\frac{f(E)}{f(-E)} = \exp(\beta E).
$$

It is thus not possible to remove arbitrary amounts of energy from a system in equilibrium, even though $|\Omega\rangle$ is not its ground state.

(v) Analytic operators.. If the dimension of the space is finite, the mapping $t \to a_t$ is analytic, and thus so is $t \to w(a_t b)$. If it is only known that h is semibounded, this is not necessarily the case, and the question arises of which a's are analytic in t. One way to construct such elements of $\mathcal{A}$ is to average over time,

$$a(f) \equiv \int_{-\infty}^{\infty} dt' a(t') f(t').$$

If the Fourier transform $\tilde{f} \in C^2$, and supp $\tilde{f} \subset [-\alpha, \alpha]$, then $f(t)$ is analytic and satisfies the estimate

$$|f(x+iy)| \leq \frac{\exp(\alpha|y|)}{(1+x^2)}\gamma, \quad \text{where} \quad \gamma = (2\pi)^{-1/2}(\|\tilde{f}\|_1 + \|\tilde{f}''\|_1).$$

The time-translate of $a(f)$,

$$\tau_t(a(f)) = \int_{-\infty}^{\infty} dt' a(t') f(t'-t),$$

is then an entire function in t such that $\|\tau_{x+iy}(a(f))\| \leq \pi\gamma\|a\| \exp(\alpha|y|)$. It is easy to see from the continuity of τ_t that the set $\tilde{\mathcal{A}}$ of such regularized a's (for variable f and α) is dense in $\mathcal{A}$ in norm. Within the set $\tilde{\mathcal{A}}$ it is always possible to continue analytically with controlled growth.

If we now think about an infinite system, the density matrix

$$\exp(-\beta H)/\operatorname{Tr}\exp(-\beta H)$$

no longer makes sense. However, the characterization of certain states made in (3.2.5(ii)) may continue to work in the infinite limit.

3.2.7 Definition

Given a C^* algebra $\mathcal{A}$ with a continuous time-automorphism $a \to a_t$, a faithful state w on the algebra is called a **KMS state** with respect to temperature $1/\beta$ whenever the functions $t \to w(a_t b)$ and $t \to w(ba_t)$ can be continued analytically to the strips $-\beta < \Im t < 0$ and, respectively, $0 < \Im t < \beta$, and are continuous on the closures of the strips, where they satisfy the condition

$$w(a_t b) = w(ba_{t+i\beta}).$$

3.2.8 Examples

1. *Free fermions*. The grand canonical state (2.5.30) is KMS with respect to the combination of free time-evolution and gauge transformations,

$$\tau_t : a_f \to a_{f_t}, \qquad \tilde{f}_t(\mathbf{k}) = \exp[it(|\mathbf{k}|^2 - \mu)]\tilde{f}(\mathbf{k}).$$

First, note that clearly

$$\rho(a_f a^*_{g_{i\beta}}) = \int \frac{d^3k}{(2\pi)^3} \tilde{f}^*(\mathbf{k})\tilde{g}(\mathbf{k}) \exp[-\beta(|\mathbf{k}|^2 - \mu)] \times \left(1 - \frac{1}{\exp[\beta(|\mathbf{k}|^2 - \mu)] + 1}\right) = \rho(a^*_g a_f),$$

and likewise

$$\rho(a^*_g a_{f_{i\beta}}) = \int \frac{d^3k}{(2\pi)^3} \tilde{f}^*(\mathbf{k})\tilde{g}(\mathbf{k}) \exp[\beta(|\mathbf{k}|^2 - \mu)] \left(\frac{1}{\exp[\beta(|\mathbf{k}|^2 - \mu) + 1}\right) = \rho(a_f a^*_g).$$

(If f and g are arbitrary functions in L^2, then in general f_t and $\rho(a_f a^*_{g_t})$ have maximal analytic continuations only into the upper half-plane $\{z = t + iy : y > 0\}$, and $\rho(a^*_g a_{f_t})$ only into the region $\{z = t + iy : y < \beta\}$. However, if either $\tilde{f}$ or $\tilde{g}$ has compact support, for example, then the maximal analytic continuation of any of the expressions above is in fact an entire function.) The proof of the KMS property of ρ for arbitrary elements of the algebra will not be given here, because of the amount of combinatorics it requires. The gauge transformation makes an appearance because of the extension of the state to the whole field algebra. If one deals only with the gauge-invariant algebra of observables $\mathcal{A}''_E$ (1.3.14), then the automorphism τ does not depend on μ, so it is identical to the free time-evolution.

2. *Free bosons.* Let ω_φ be the equilibrium state of the field algebra of the free Bose gas at temperature $1/\beta$ and density ρ (see (2.5.32;4)), which appears as the integrand in the decomposition of the canonical limiting state in (2.5.32;1). (The decomposition is nontrivial iff $\rho > \rho_c(\beta)$ – see also (2.5.20;3).) The field algebra of the bosons is generated by the operators

$$W_f \equiv \exp[i(a^*_f + a_f)]; \qquad W_f W_g = \exp[-i\Im(f|g)]W_{f+g},$$

and the free time-evolution of the observables will be extended to the field algebra by $W_f \to W_{f_t}$,

$$\tilde{f}_t(\mathbf{k}) = \exp[it(|\mathbf{k}|^2 - \mu)]\tilde{f}(\mathbf{k}).$$

(The quantity $\mu = \mu(\rho)$ is a unique but not invertible function.) Then $A(f, g, t) \equiv \omega_\varphi(W_f W_{g_t})$ is the continuous boundary value of an analytic function of $z = t + iy$ on the strip $0 < y < \beta$, $t \in \mathbb{R}$, *viz.*,

$$\hat{A}(f, g, z) \equiv \exp\left\{-\int \frac{d^3k}{(2\pi)^3}\left[(|\tilde{f}(\mathbf{k})|^2 + |\tilde{g}(\mathbf{k})|^2 \times \left(\frac{1}{2} + \frac{1}{\exp[\beta(|\mathbf{k}|^2 - \mu)] - 1}\right) + \tilde{f}^*(\mathbf{k})\tilde{g}(\mathbf{k}) \exp[iz(|\mathbf{k}|^2 - \mu)] \left(1 + \frac{1}{\exp[\beta(|\mathbf{k}|^2 - \mu)] - 1}\right)\right.\right.$$

$$+ \tilde{g}^*(\mathbf{k})\tilde{f}(\mathbf{k})\exp[-iz(|\mathbf{k}|^2-\mu)]\left(\frac{1}{\exp[\beta(|\mathbf{k}|^2-\mu)]-1}\right)\Bigg]\Bigg\}$$

$$\times \exp\{2i\sqrt{\rho-\rho_c(\beta)}\Theta(\rho-\rho_c(\beta))\Re[(\tilde{f}(\mathbf{0})+\tilde{\mathbf{g}}(\mathbf{0}))\exp(\mathbf{i}\varphi)]\},$$

and the KMS condition is satisfied: $w(ab_{-t}) = w(a_t b) = w(ba_{t+i\beta})$,

$$\lim_{y\to+\beta} \hat{A}(f,g,t+iy) = \omega_\varphi(W_{g_t}W_f) = \omega_\varphi(W_g W_{f_{-t}}) = A(g,f,-t)$$

$$= \lim_{y\to+0} \hat{A}(g,f,-t+iy).$$

It follows from $\rho < \rho_c(\beta)$ that $\mu(\rho) < 0$, so in this situation f and g can be arbitrary elements of L^2. However, $\mu(\rho) = 0$ for all $\rho \geq \rho_c(\beta)$, so ω_φ must be restricted, for example, to the algebra generated by the W_f with $f \in L^1 \cap L^2$. For general f and g it is not possible to extend $\hat{\alpha}(f,g,z)$ analytically beyond the strip described above. However, if the support of either f or g is compact, then $\hat{A}(f,g,z)$ is an entire function of z.

3.2.9 Properties of a KMS state w

1. *A* KMS *state w is invariant in time.*
2. *When extended to $\pi_w(\mathcal{A})''$, w remains* KMS.
3. *If w is faithful (as a positive functional), then π_w is faithful, and vice versa.*
4. $\mathbb{Z} = \pi_w(\mathcal{A})] \cap \pi_w(\mathcal{A})''$ *consists of time-invariant elements.*
5. *The* KMS *states for any fixed β form a weak-* compact, convex set.*
6. *If w is an extremal* KMS *state, then π_w is a factor.*
7. *For any w faithful, there exists a unique time-evolution under which w is a* KMS *state.*

3.2.10 Remarks

1. According to (1.3.3), if w is invariant in time, then on π_w we can write $a_{-t} = U_t a U_t^{-1}$, and the time-evolution, when extended to $\pi_w(\mathcal{A})''$, transforms this algebra into itself: $a_n \to a \Rightarrow a_n(-t) = U_t a_n U_t^{-1} \to U_t a U_t^{-1} \in \pi_w(\mathcal{A})''$.
2. Of course, the extension of w to $\pi_w(\mathcal{A})''$ with cyclic vector $|\Omega\rangle$ is $w(a'') = \langle\Omega|a''|\Omega\rangle$ for all $a'' \in \pi_w(\mathcal{A})''$. Property 2 means that this state is KMS with respect to the time-evolution defined earlier on $\pi_w(\mathcal{A})''$.
3. According to (III:2.3.10;3),

$$\begin{aligned}\text{Ker}\, w &= \{a \in \mathcal{A} : w(a) = 0\}\\ &\supset N \equiv \{a \in \mathcal{A} : w(a^*a) = 0\}\\ &\supset \text{Ker}\,\pi_w = \{a \in \mathcal{A} : w(b^*a^*ab) = 0 \quad \text{for} \quad \text{all} \quad b \in \mathcal{A}\},\end{aligned}$$

and the statement that w is faithful means that $N = \{0\}$. Property 3 thus means that if $\operatorname{Ker} \pi_w = \{0\}$, then $N = \{0\}$, so $|\Omega\rangle$ is a separating vector for $\pi_w(\mathcal{A})$: $\pi_w(a)|\Omega\rangle \neq 0$ for all $\pi_w(a) \neq 0$. (Speaking field-theoretically, no operator annihilates the vacuum.) If the algebra is simple, and hence has only faithful representations, then all KMS states are also faithful.

4. If the system is asymptotically Abelian, then $\mathcal{R}' = \mathbb{Z}$. The center $\mathbb{Z}$ contains the macroscopic observables, which are therefore constant in time in this case.
5. By Property 5, convex combinations and weak limits of KMS states (at a given β) are KMS states.
6. In a finite system, with $\mathcal{A} = \mathcal{B}(\mathcal{H})$, $U_t = \exp(iHt)$, there is only one normal KMS state. At $t = 0$ the condition is that

$$\operatorname{Tr} \rho ab = \operatorname{Tr} \rho b \exp(-\beta H) a \exp(\beta H) = \operatorname{Tr} \exp(-\beta H) a \exp(\beta H) \rho b$$

for all b, which means that $\rho a = \exp(-\beta H) a \exp(\beta H) \rho$ for all a, so $\exp(\beta H)\rho \in \mathcal{A}'$, and thus $\rho = \exp(-\beta h)$. Since the convex set of KMS states is compact, any KMS state may be decomposed into extremal KMS states. According to Property 6, if the system is asymptotically Abelian, a decomposition into extremal KMS states is the same as a decomposition into elements of the center (defined as a decomposition into factors (1.4.9)), which is the same as a decomposition into extremal invariant states. In the characterization of ergodic states (3.1.20;2) we learned that a factor state is not decomposable into invariant states, and thus *a fortiori* not decomposable into KMS states. Conversely, it is now being claimed that it is always possible to decompose a KMS state w further into other, extremal KMS states, if π_w is not a factor. This means that the extremal KMS states are ergodic and, as factors, even mixing. Since the decomposition by the center is unique, so is the decomposition into extremal KMS states. Hence the set of extremal KMS states is a simplex.
7. If the time-evolution is given, then there can be one or more KMS states (see Problem 2). In contrast, by Property 7, if w is given, then there is a unique time-evolution for which it is KMS.

Proof of (3.2.9)

1. Let $b = 1$; the function $\rho(a_t) = \rho(a_{t+i\beta})$ can be continued analytically to all of $\mathbb{C}$ and is periodic in $\Im t$. Since it is bounded in a strip, it is bounded throughout $\mathbb{C}$ and therefore constant. It follows that ρ is time-invariant.

2. This proposition follows from a more general one to be stated later (3.2.13).

3. If $a \in \mathcal{N}$, then $w(a^*a) = 0$, which implies that for all b, $w(ba) = 0$ (by Cauchy-Schwarz), which means that for all b and c, $0 = w(c_{-i\beta}ba) = w(bac)$, and therefore $a \in \operatorname{Ker} \pi_w$.

4. Suppose $c \in \mathbb{Z} : w(a_t c) = w(a_{t+i\beta} c)$. As in Proposition 1, it can be concluded that $w(a_t c)$ is constant in t. If a is replaced with ab, it follows that $w(a_t c b_t) = \langle \Omega | a U_t c U_{-t} b | \Omega \rangle$ is constant for all a and b, so c is constant.

5. Convexity is trivial. If w_n converges in the weak*- sense to w, then for all $a \in \tilde{\mathcal{A}}$, $b \in \mathcal{A}$ and $t \in \mathbb{C}$, the quantities $w_n(a_t b)$ converge to $w(a_t b)$ and are dominated by $\pi\gamma \|a\| \|b\| \exp \alpha |\Im t|$. Consequently, the limit is holomorphic throughout $\mathbb{C}$ and satisfies $w(a_{t-i\beta} b) = w(b a_t)$. As in Problem 1, this relationship remains valid for norm-limits of a's in the strip $0 \le \Im t \le \beta$, and can thus be extended to all of $\mathcal{A}$ (and, by Property 2, to all of $\mathcal{A}''$).

6. Unless π_w is a factor, $\mathbb{Z}$ contains a nontrivial projection P. Therefore w can be decomposed into a combination of $w_1(a) = w(Pa)/w(P)$ and $w_2(a) = w((1-P)a)/w(1-P)$, and both w_i are KMS states: $w(Pa_t b) = w(a_t Pb) = w(Pba_{t+i\beta})$.

7. Suppose that τ_t and $\tilde{\tau}_t$ are distinct automorphisms under which w is a KMS state. Then if a is entire with respect to τ, and b is entire (see(3.2.6(v))) with respect to $\tilde{\tau}$, it follows that

$$\begin{aligned} F(t) &\equiv w(\tilde{\tau}_{-t}(\tau_t(a)) \cdot b) = w(\tau_t(a) \cdot \tilde{\tau}_t(b)) = w(\tilde{\tau}_t(b) \cdot \tau_{t+i\beta}(a)) \\ &= w(\tau_{t+i\beta}(a) \cdot \tilde{\tau}_{t+i\beta}(b)) = F(t+i\beta). \end{aligned}$$

This fact implies that F is constant, so τ and $\tilde{\tau}$ have the same action on $\tilde{\mathcal{A}}$ and hence on $\mathcal{A}$. □

The foregoing conclusions suggest an interpretation of the decomposition into extremal KMS states as a decomposition of an equilibrium state into its pure phases. Yet it will be apparent from examples that these pure phases are not necessarily identical to physical phases. Property 6 together with Remarks (3.1.24) ensures that these states have mixing properties, meaning that local perturbations eventually die out, and equilibrium gets reestablished. The canonical states were characterized earlier as the states of greatest entropy at a given energy, and the evolution towards them can be thought of as a tendency toward greater entropy. On the other hand, if the system is infinite, it is not the total entropy that is finite, but rather the average entropy, which is unaffected by local perturbations. If a state is normal when restricted to a local algebra (1.3.2;6), then it is possible to define the local entropy, which will then tend to its equilibrium value. It is not, however, claimed that it increases monotonically to that value.

The diagram in Figure 3.6 collects together the various properties of asymptotically Abelian systems in invariant states and shows their connection with the time-evolution. It will be shown later (3.3.13) that the spectrum of H is ordinarily the whole real line $(-\infty, \infty)$. The spectral properties stated then include the supposition that the systems that we shall be concerned with have neither dense point spectrum nor singular continuous spectrum.

3.2.11 Examples

1. *Free fermions.* Consider a system of n kinds of free fermions, described by the field operators $a_{\alpha,f}$, $\alpha = 1, \ldots, n$. The algebra $\mathcal{A}_E$ of observables will be taken to consist only of polynomials containing an equal number of a_α and a_α^*

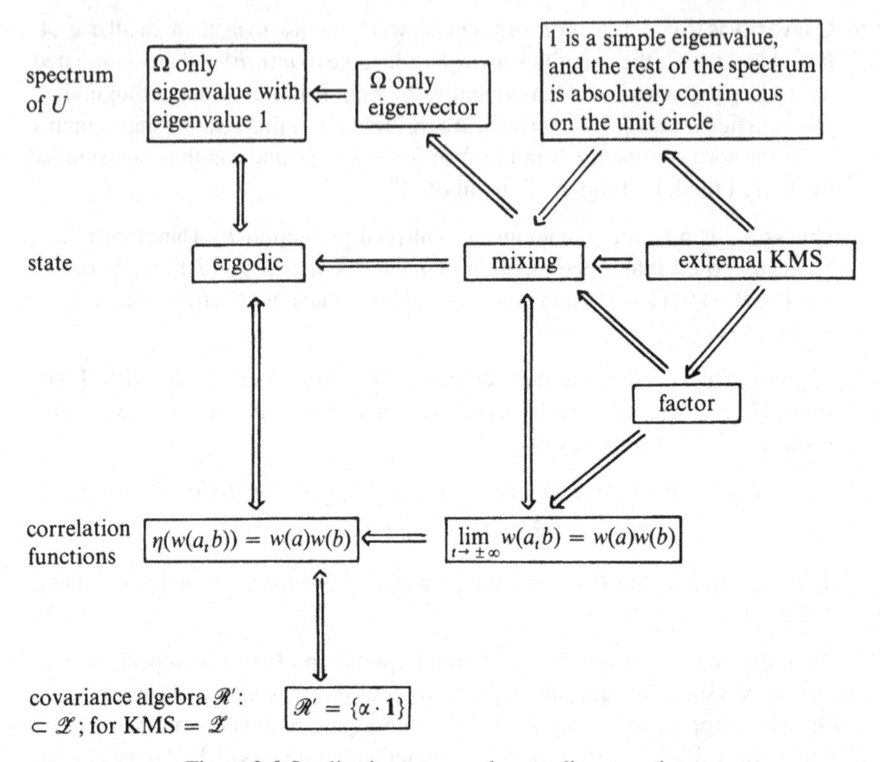

Figure 3.6. Implications among the ergodic properties.

for any α, in accordance with Definition (1.3.5). In other words it contains the densities and currents of the particles. The state is taken as the product of the grand canonical states (2.5.49), i.e.,

$$\langle a^*_{1,f^1_1} \cdots a_{1,f^1_{m_1}} a_{1,g^1_1} \cdots a_{1,g_{m^1_1}} a^*_{2,f^2_1} \cdots a^*_{2,f^2_{m_2}} a_{2,g^2_1} \cdots$$
$$a_{2,g^1_{m_2}} a^*_{3,f^3_1} \cdots a_{n}, g^n_{m_n} \rangle = \prod_\alpha \operatorname{Det}\langle g^\alpha_i | \rho^\alpha_1 f^\alpha_j \rangle,$$

$$\langle f | \rho^\alpha_1 g \rangle = \int \frac{d^3k}{(2\pi)^3} \frac{\tilde{f}^*(\mathbf{k})\tilde{g}(\mathbf{k})}{\exp[\beta(|\mathbf{k}|^2/2m_\alpha - \mu_\alpha)] + 1}.$$

It is KMS with respect to the automorphism $a_{\alpha,f_\alpha} \to a_{\alpha,f_\alpha(t)}$,

$$\tilde{f}_\alpha(t) = \exp\left(\frac{it|\mathbf{k}|^2}{2m_\alpha}\right)\tilde{f}_\alpha.$$

Observe that for this automorphism of the algebra of observables there is an n-parameter family of KMS states. They can be parametrized by the chemical potentials μ_α, and, as factor states, they are extremal. A general KMS state at a given β is an integral over them with some probability measure on the μ_α, which corresponds to the mixture of phases posited in the usual procedure known as

Gibbs's phase rule. As remarked in (2.3.41), with a variable β and n types of matter having only one phase, there is an $n+1$-dimensional manifold of states.

2. *Bose condensation.* If $\rho > \rho_c(\beta)$, then the canonical state (2.5.51;1) may be written as an integral $\int_0^{2\pi}(d\varphi/2\pi)w_\varphi$ over the factor states

$$w_\varphi(\exp(ia_f^*)\exp(ia_f)) = \exp\left[-\int \frac{d^3k}{(2\pi)^3}\frac{|\tilde{f}(\mathbf{k})|^2}{\exp(\beta|\mathbf{k}|^2)-1} + 2i\sqrt{\rho_0}\Re(\tilde{f}(\mathbf{0})\exp(i\varphi f))\right].$$

These states are KMS with respect to the transformation

$$\tilde{f}(\mathbf{k}) \to \exp(i|\mathbf{k}|^2 t)\tilde{f}(\mathbf{k}),$$

and are consequently extremal KMS states. They describe the coexistence of two phases, the normal phase with particle density $\int d^3k[\exp(\beta(\mathbf{k})^2)-1]^{-1}(2\pi)^{-3}$ and a condensed phase of density ρ_0. The latter phase still depends parametrically on the argument φ of a_0, and so for fixed β there are two parameters, ρ_0 and φ, to specify the extremal KMS states. These extremal KMS states are not the same as the phases of Gibbs's phase rule. Although different phases of a substance are coexisting if $\mu = 0$ and $0 < T < T_c$, the condensed phase makes its appearance not a single, pure phase, but rather as combination of infinitely many pure phases, differing in their values of the "hidden parameter" φ, which has no effect on the thermodynamic functions (2.5.20;3). In this way the decomposition into extremal KMS states is finer than the phase decomposition of (2.3.30) into extremal points of the concave function $\sigma(\varepsilon, \rho)$. If the field algebra is confined to its even part $\mathcal{A}_E$ (in the Fock representation, $\mathcal{A}_E = \mathcal{A}_B \cap \{N\}'$), then all the w_φ become the same state. This is apparent when it is observed that gauge transformations $\tau_\varphi : W_f \to W_{\exp(i\varphi)f}$ transform the w_φ into one another: $(w_\varphi \circ \tau_{\varphi'})(W_f) = w_{\varphi+\varphi'}(W_f)$. The restriction to $\mathcal{A}_E$ makes $\tau_{\varphi'}$ the identity, so $w_\varphi = w_{\varphi+\varphi'}$. Recall that for asymptotically Abelian systems the decomposition into extremal KMS states is unique according to (3.2.10;6); the extremal states form a simplex. In contrast, we were not able to adduce any theoretical reasons for why the flat pieces of $\sigma(\varepsilon, \rho)$ had the structure of a simplex.

3. *A model of a ferromagnet.* The time-evolution of Example (2.3.27;2) was investigated in (3.1.1;4). We found that if $B = 0$ and $T < 2$, it was no longer an automorphism of the spin algebra $\mathcal{A}\{\sigma_i\}$, but rather of the strong closure $\pi(\mathcal{A})''$. The state

$$\langle\sigma_1^{\alpha_1}\dots\sigma_m^{\alpha_m}\rangle = \int_{S_2} d\mathbf{n}s^m n_{\alpha_1}\dots n_{\alpha_m}, \qquad s = \tanh(2\beta s),$$

is KMS with respect to this time-evolution. In each of the factors $\pi_\mathbf{n}$ it is a rotation about the axis $\mathbf{n}$ at angular velocity $4s$. For example, if $\mathbf{n}$ point in the z-direction, then $\sigma^+(t) = \exp(-4ist)\sigma^+$ and

$$\langle\sigma^+\sigma^-\rangle = \frac{\langle 1+\sigma\rangle}{2} = \frac{1+s}{2} = \langle\sigma^-\sigma^+_{i\beta}\rangle = \exp(4\beta s)\langle\sigma^-\sigma^+\rangle$$
$$= \exp(4\beta s)\frac{1-s}{2},$$

because $s(1+\exp(4\beta s)) = \exp(4\beta s) - 1$. The individual factors $\pi_{\mathbf{n}}$ thus give rise to extremal KMS states, corresponding to spontaneous magnetization in the direction **n**. Again, from the physical point of view this model would be described as having one magnetized phase, whereas the decomposition into extremal KMS states would distinguish among different directions of **n**, and treat magnetization in each direction as a distinct phase. Notice that the phase transition at $T = 2$ is connected with a change of the type of factor; if $T < 2$ the integral runs over factors of type III, while if $T > 2$, the factors are of type II_1.

3.2.12 Remarks

1. There are many different possible reasons for the existence of several KMS states. One is that the center of the algebra of observables $\mathcal{A}$ might be nontrivial. Unitary elements of the center generate transformations, which, like gauge transformations, leave each element of the algebra invariant. Therefore it is possible to combine the action of these transformations with that of time-evolution τ and study the KMS states with respect to the resulting automorphisms. When restricted to $\mathcal{A}$, these automorphisms are identical to the time-evolution, so all such states are also τ-KMS for $\mathcal{A}$ (cf. Problem 2).
2. Many "degeneracies" of KMS states go away upon enlargement of the algebra of observables. If in Example 1 the particle number is also allowed to vary, for instance by a chemical reaction $(1) \rightleftharpoons (2) + (2)$, then noneven elements like $a_1^* a_2 a_2$ are introduced into the algebra of observables. They are not separately invariant under gauge transformations of the different types of particles, but are invariant only under certain combinations, e.g., if the generator of the transformation has the form $2N_1 + N_2$ in the Fock representation. Consequently, the KMS condition with the free time-evolution makes the chemical potentials satisfy a linear equation such as $\mu_1 - 2\mu_2 = 0$. Similarly, if two condensed Bose systems as in Example (3.2.11;2) are coupled, the relative phase φ becomes observable (the Josephson effect).
3. It is possible that a symmetry is broken, which means that the extremal KMS states w are not invariant under some group σ of automorphisms that commute with τ. This is illustrated in Example (3.2.11;2) with the gauge transformations and in (3.2.11;3) with the rotations. If the symmetry is broken, then $w \circ \sigma_s$ is once again τ-KMS; thus with continuous groups there are even infinitely many KMS states.
4. The theoretical justification of Gibbs's phase rule for continuous systems is still an open problem (cf. [20]).

5. So far we have been considering β as fixed. KMS states with different β's are disjoint, i.e., if $w = (w_{\beta_1} + w_{\beta_2}).2$, then $\pi_w = \pi_{\beta_1} \oplus \pi_{\beta_2}$. In this case the temperature β^{-1} becomes an observable belonging to the center of $\pi_w(\mathcal{A})''$.

As discussed in § 1.1, the ergodic property of a system has been an important ingredient of the justification of statistical mechanics throughout its history. Even though today ergodicity is no longer viewed as the central requirement, it can still be a noteworthy property of realistic systems, so it can still be valuable to have a formulation of ergodicity for infinite quantum systems. In a classical system, if there existed additional constants of the motion beyond H, it would be impossible for the trajectory of almost every point to wind densely throughout the energy shell. However, constants such as momentum or angular momentum are infinite for infinite systems, so ergodicity can not be defined as the absence of additional constants of the motion. But recall that classically constants of the motion also generate diffeomorphisms that commute with the flow of time (see I, §3.3). This property carries over to infinite systems, and even the notions of indecomposable time-invariant surfaces and of dense trajectories have analogies.

In order to characterize ergodic systems, it is only necessary to generalize (3.2.5) to infinite systems. A coarser notion will be discussed in § 3.4.

3.2.13 Modular Automorphisms of a von Neumann Algebra

Let $\mathcal{M}$ be a von Neumann algebra of operators on a Hilbert space H. For every vector $|\Omega\rangle$ that is both cyclic and separating (i.e., $\mathcal{M}|\Omega\rangle = \mathcal{H}$, and if $a|\Omega\rangle = 0$ for any $a \in \mathcal{M}$, then $a = 0$), there exists a unique one-parameter group of automorphisms $a \to \tau_t(a)$ and a conjugate-linear operator J such that

(i) $w(a) \equiv \langle\Omega|a|\Omega\rangle$ *is* τ-KMS (*with* $\beta = 1$);

(ii) $J^2 = \mathbf{1}$, $J\mathcal{M}J = \mathcal{M}'$; *and*

(iii) $U_{-i/2}a|\Omega\rangle Ja^*|\Omega\rangle$, *where* $\tau_t(a) = U_{-t}aU_t$.

3.2.14 Remarks

1. The idea of the proof follows that of (3.2.3), but with additional technical complications, for which reason the reader is referred to [21].
2. Properties (3.2.6) of the correlation functions hold also in the general case. Specifically, (iii) means that $\mathcal{A}|\Omega\rangle \subset D(\exp(-H/2))$, where $U_t = \exp(-iHt)$, from which it follows that $\mathcal{A}|\Omega\rangle \subset D(\exp(-yH))$ for $0 \le y \le \frac{1}{2}$, and $w(a^* \exp(-H)a) = w(aJ^2a^*) \le \|a\|^2$. The proofs of the other properties can be repeated verbatim.
3. It is clear that a further generalization to arbitrary C^* algebras will not work. The state in Example (3.2.11;3) is obviously faithful on the σ's, so it is a candidate for w. However, we have found that the related automorphism under which w is a KMS state maps the C^* algebra generated by the σ's out of itself, leaving only the von Neumann algebra $\pi_w(\mathcal{A})''$ invariant.

4. Suppose that w is a KMS state on the algebra $\mathcal{A}$ with respect to the time-evolution τ_t. By Property (3.2.9;3) the vector $|\Omega\rangle$ given in the GNS representation π_w is cyclic and separates $\pi_w(\mathcal{A})''$, even if w fails to be faithful, and the representation of τ_t is identical to the modular automorphism.
5. Whereas (3.2.13) asserts that for any faithful state there is exactly one automorphism for which it is KMS in the opposite direction there cannot be an analogous statement. A given automorphism may have many KMS states or none whatsoever.

3.2.15 Ergodic Quantum Systems

Let τ be the time-evolution under which the C^* algebra $\mathcal{A}$ of observables is asymptotically Abelian, and let $\mathcal{T}$ be the set of faithful states w with the property that the normal extension of w to $\pi_w(\mathcal{A})''$ is also faithful. Then the following two properties are equivalent:

(i) *A state $w \in \mathcal{T}$ is ergodic if and only if it is an extremal* KMS *state; and*

(ii) *There is no $w \in \mathcal{T}$ such that its modular automorphism σ differs from τ, but $[\sigma, \tau] = 0$.*

If a system has these properties, we shall call it **ergodic**.

Proof that (i) $\Leftrightarrow$ (ii): Not (ii) $\Rightarrow$ not (i). Let w be the σ-KMS state. Since σ and τ commute, $\rho \equiv \eta_t(w \circ \tau_t)$ is also σ-KMS, so our strategy will be to use it to construct a τ-ergodic state. Think of ρ as decomposed in two separate ways, on the one hand into τ-ergodic states and on the other into extremal σ-KMS states. By Remark (3.2.10;6) the latter decomposition is the same as the decomposition into factors, whereas according to Remark (3.1.19;3) the τ-ergodic decomposition is coarser than the factor decomposition. This means that the τ-ergodic components of ρ are combinations of extremal σ-KMS states, but not vice versa. Hence any such component is τ-ergodic but not τ-KMS, since it is not possible for it to be KMS with respect to σ and $\tau \neq \sigma$ at the same time.

Not (i)$\Rightarrow$ not (ii). Suppose that $w(a) = \langle\Omega|a|\Omega\rangle$ is τ-ergodic, and let σ denote the modular automorphism of $\pi_w(\mathcal{A})''$. Since w is invariant under τ and σ, both groups have unitary representations on π_w. Let $\exp(iHt)$ and $\exp(iGs)$ denote their representations. Since w is also σ-KMS, given any a and $b \in \mathcal{A}$,

$$\langle\Omega|\tau_t(a)\sigma_i(b)|\Omega\rangle = \langle\Omega|b\tau_t(a)|\Omega\rangle = \langle\Omega|\tau_{-t}(b)a|\Omega\rangle = \langle\Omega|\sigma_{-i}(a)\tau_{-t}(b)|\Omega\rangle,$$

so

$$\langle\Omega|a \exp(-iHt)\exp(-G)b|\Omega\rangle = \langle\Omega|a \exp(-G)\exp(-iHt)b|\Omega\rangle.$$

Since the vectors of the form $a|\Omega\rangle$ are dense, it follows that $[\exp(-G), \exp(-iHt)] = 0$, so $[\tau, \sigma] = 0$. However, if w is not KMS with respect to τ, then the groups of automorphisms must be different, since w is KMS with respect to σ. □

3.2.16 Remarks

1. Unfortunately, no examples of ergodic quantum systems are known. Although the grand canonical state (2.5.49) of free particles is mixing, there are ergodic states that fail to be KMS: The momentum distribution $[\exp(\beta(|\mathbf{k}|^2-\mu))\pm 1]^{-1}$ would just have to be replaced with some other positive, integrable function. The state would then be time invariant and, as a factor state, ergodic, but not KMS. The hope is that when interactions are switched on, states of this kind will turn into equilibrium states (see §3.3).
2. Property (3.2.15(ii)) forbids the existence of additional constants of the motion. In finite quantum systems, in addition to the Hamiltonian H there are also the constants of the form $f(H)$. If H is nondegenerate, then this accounts for all the constants, because $\{H\}'$ is generated by $f(H)$ and the unitary transformations of the degeneracy space. If the system is infinite, then H exists only in representations π_w of invariant states w, and does not belong to $\pi_w(\mathcal{A})$. It can be shown [22] that only linear functions $f(H)$ produce automorphisms of $\pi_w(\mathcal{A})$. However, the function $H \to cH$ does nothing more than change the scale of time, and we consider scaled time-evolutions as identical.
3. If particle numbers are conserved, then gauge transformations $a_f \to \exp(i\alpha)a_f$, $a \in \mathbb{R}$, certainly commute with time-evolution, and the system is not ergodic as defined by (3.2.15). Yet the corresponding KMS states w are of the form (2.5.30) with infinite temperature but $\beta\mu = 1$,

$$w(a_f a_g^*) = \int \frac{d^3k}{(2\pi)^3} \frac{\tilde{f}^*(\mathbf{k})\tilde{g}(\mathbf{k})}{e+1}.$$

The particle density in this state is infinite, $w(a(\mathbf{x})a^*(\mathbf{x})) = \delta(\mathbf{0})/(1+e)$, however, so it is not even locally normal. This shows that in a nonergodic infinite system it may happen that the states that are ergodic but not KMS never actually occur, so the system behaves ergodically anyway. On the other hand, there is no similar objection to this state on a lattice system, for which $\mathbf{k}$ varies only over a compact region.
4. If an infinite system is homogeneous and isotropic, then translations and rotations commute with τ. The KMS states of these automorphisms have the same defect as that of Remark 3, that the local particle density is infinite.
5. Since under the measurability assumptions of (3.1.22;3) ergodic states are time-averages of pure states, the same will be true of the extremal KMS states of ergodic systems. This is the fulfillment of the hope of classical ergodic theory that the equilibrium state can be obtained as the closure of a single trajectory.

If we wish to conceive of ergodicity roughly as the absence of constants of motion other than $f(H)$, then it is useful to make a table of the implications of this for equilibrium states of systems of various types. As can be seen below, the KMS states of infinite quantum systems inherit the good properties of the canonical and microcanonical states of finite systems.

system / state	Finite, classical There are no additional constants of the motion	Finite, quantum-mechanical H is nondegenerate	Infinite, quantum-mechanical There exists no KMS σ such that $\sigma \neq \tau$, $[\sigma, \tau] = 0$
Microcanonical	Ergodic Time-average of pure states Not faithful	Ergodic Time-average of pure states Not faithful	
Canonical	Not ergodic Faithful	Not ergodic Time-average of pure states Faithful	
Extremal KMS			Ergodic Time-average of pure states Faithful

3.2.17 Problems

1. Consider a sequence of states w_N on a C^* algebra $\mathcal{A}$ converging to w (in the weak*-sense). Show that if the modular automorphism $\tau_{N,t}(a)$ is a norm-convergent sequence in $\mathcal{A}$ for all $a \in \mathcal{A}$ and $t \in \mathbb{R}$, then the $\tau_{N,t}$ converge to the modular automorphism belonging to w.

2. Find an example of an algebra $\mathcal{A} \subset \mathcal{R}(\mathbb{C}^4)$ such that some nontrivial automorphism has many KMS states.

3. Construct the KMS states for translation and rotation of a system of free fermions..

4. In both classical and quantum mechanics, study the automorphisms of the anisotropic oscillator $H = \frac{1}{2}(p_1^2 + p_2^2 + \omega_1^2 q_1^2 + \omega_2^2 q_2^2)$, with ω_1/ω_2 irrational, that commute with the time-evolution. Is the system ergodic?

5. Show that for two faithful states $\omega_{1,2}$ with density matrices $e^{-H_{1,2}}$ the relative entropy $S(\omega_1|\omega_2)$ is given by $-\omega_1(\ln \Delta)$ where the "relative modular operator" Δ (we write Δ without refering to ω_1 and ω_2) is defined by $\omega_2(ab) = \omega_1(b\Delta a)$ $\forall a, b \in \mathcal{A}$.

6. Use 5,

$$\ln \lambda = \int_0^\infty \frac{dt}{t}\left(\frac{\lambda}{\lambda+t} - \frac{1}{1+t}\right)$$

and

$$\frac{\lambda}{\lambda+t} = \inf_{\alpha \in C}\left(\frac{\lambda}{t}|\alpha|^2 + |1-\alpha|^2\right), \quad \lambda, t \in \mathbb{R}^+$$

to prove Kosakis formula (2.2.20). Argue that after having verified it for states represented by density matrices one can use this expression as general definition for $S(\omega_1|\omega_2)$.

3.2.18 Solutions

1. Consider the limits of the correlation functions $w_N(\tau_{N,t}(a(N,f))b)$, where

$$a(N,f) \equiv \int dt \tau_{N,t}(a) f(t),$$

and f is as in (3.2.6(v)), and let $\tau_t(a) = \lim \tau_{N,t}(a)$. The norm-limit of $\tau_{N,t}(a(N,f))$ is $\tau_t(a(f))$ by the dominated convergence theorem, even for complex t, since $\int |f(t+iy)|dt \le \pi\gamma \exp(\alpha|y|)$. The first term of $[w(\tau_t(a(f))b) - w_N(\tau_t(a(f))b)] + w_N(\tau_t(a(f)) - \tau_{N,t}(a(N,f))b)$ goes to zero because of the weak*- convergence $w_N \to w$, and the second term goes to a_f as a consequence of the norm-convergence of $a(N,f)$ to a_f. Therefore, for all $a \in \mathcal{A}$ and $t \in \mathbb{C}$,

$$w_N(\tau_{N,t}(a(N,f))b) \to w(\tau_t(a(f))b).$$

These holomorphic functions converge pointwise and are uniformly bounded on every compact set in $\mathbb{C}$, because they are $\le \|a\|\|b\|\pi\gamma\exp(\alpha|y|)$; the limit is therefore holomorphic and identical to $w(b\tau_{t+i}(a(f)))$.

This means that the KMS condition holds for all $a \in \tilde{\mathcal{A}}$, and of course boundedness in the strip (3.2.6(ii)) is preserved in limits. Passing by norm-limits $a_n \to a$ to general $a \in \mathcal{A}$, if $-1 \le \Im t \le 0$, then $w(\tau_t(a_n)b)$ converges uniformly to $w(\tau_t(a)b)$, which is consequently continuous on the strip and holomorphic in its interior.

It is trivial to see that the identity $w(\tau_t(a)b) = w(b\tau_{t+i}(a))$ continues to hold for limits, as do the group property $\tau_{t+s} = \tau_t \circ \tau_s$ and the invariance of $w : w \circ \tau_t = w$. The GNS construction can now be carried out, so that τ_t is represented unitarily on π_w as U_t. If $\pi(a_n)$ converges weakly to $b \in \pi(\mathcal{A})''$, then $U_{-t}\pi(a_n)U_t$ converges weakly to $U_{-t}bU_t \equiv \tau_t(b)$. Therefore τ_t maps $\pi(\mathcal{A})''$ into itself, and is identical to the modular automorphism according to (3.2.9;7) and (3.2.14;4).

2. Let $\mathcal{A}$ be spanned by $(\mathbf{1}, \tau) \otimes (\mathbf{1}, \sigma_3)$, and let the time-evolution be $\tau^{\pm}(t) = \exp(\pm i\omega t)\tau^{\pm}(0)$, with τ_3 and σ_3 constant. For a given β the density matrices of the form

$$\rho = \frac{\exp(-\beta\tau_3 - \alpha\sigma_3)}{\operatorname{Tr}\exp(-\beta\tau_3 - \alpha\sigma_3)}$$

yield KMS states for all real α.

3. They have the same structure as in (2.5.49), with

$$\langle f|\rho|g\rangle = \int \frac{d^3k}{(2\pi)^3} \frac{\tilde{f}^*(\mathbf{k})\tilde{g}(\mathbf{k})}{1+\exp(k_1)}$$

for translations in the 1-direction, and

$$\int_0^\infty r^2 dr \sum_{l,m} \frac{\hat{f}^*_{lm}(r)\hat{g}_{lm}(r)}{1+\exp m}$$

for rotations about the 3-axis, where $\hat{f}_{lm}$ denote the expansion coefficients of f in spherical harmonics.

4. Classically, $H_i = \frac{1}{2}(p_i^2 + \omega_i^2 q_i^2)$ are two independent constants of the motion, and generate flows that commute with time-evolution. The system is not ergodic in the sense of Table I. Quantum mechanically, H has the eigenvalues $(n_1 + \frac{1}{2})\omega_1 + (n_2 + \frac{1}{2})\omega_2$ and is thus nondegenerate. All constants are of the form $f(H)$, and the system is ergodic in the sense of Table I.

5. With the notation (3.2.1) and $\Pi'(H) = H'$ we get

$$\omega_2(ab) = \mathrm{Tr}\, be^{-H_2}a = \langle b|e^{-H_2}|a\rangle$$
$$= \langle e^{-H_1/2}b|e^{-H_2+H_1'}|ae^{-H_1/2}rangle = \omega_1(be^{-H_2+H_1'}a)$$
$$\Rightarrow \Delta = e^{-H_2+H_1'}.$$

Now

$$H_1'|e^{-H_1/2}\rangle = |e^{-H_1/2}H_1\rangle = |H_1 e^{-H_1/2}\rangle = H_1|e^{-H_1/2}\rangle,,$$

thus

$$-\omega_1(\ln \Delta) = \langle e^{-H_1/2}|H_2 - H_1'|e^{-H_1/2}\rangle = \mathrm{Tr}\,\rho_1(\ln \rho_1 - \ln \rho_2) = S(\omega_1|\omega_2).$$

6.

$$-\omega_1(\ln \Delta) = \int_0^\infty \frac{dt}{t}\left(\frac{1}{1+t} - \langle\Omega_1|\frac{\Delta}{\Delta+t}|\Omega_1\rangle\right)$$

Now

$$\langle\Omega_1|\frac{\Delta}{\Delta+t}|\Omega_1\rangle = \inf_{x\in\mathcal{A}}\left(\langle\Omega_1|x^*\frac{\Delta}{t}x|\Omega_1\rangle + \langle\Omega_1|(1-x)^*(1-x)|\Omega_1\rangle\right),$$

thus

$$S(\omega_1|\omega_2) = -\omega_1(\ln \Delta) = \sup_{x+y=1}\int_0^1 \frac{dt}{t}\left(\frac{1}{1+t} - \frac{1}{t}\omega_2(xx^*) - \omega_1(yy^*)\right).$$

3.3 Stability and Passivity

The distinguishing feature of the equilibrium state is that it does not change abruptly when subjected to a local perturbation. The second law of thermodynamics can be proved in a version stating that a system prevents energy from being extracted by a cyclic perturbation only if the system is in equilibrium.

The next part of the general theory that will be investigated will be the influence of local perturbations on equilibrium. In the mathematical treatment local perturbations play the role of the speck of dust invoked in the traditional theory of statistical mechanics to convert stationary states, not yet in equilibrium, into equilibrium states. As a matter of fact, what makes the KMS states special in the mathematical theory is that they have certain stability properties – they change continuously when the Hamiltonian is perturbed slightly. This is certainly not true of all stationary states, and can even be used to characterize the extremal KMS states of an infinite system; they are precisely the set of states that turn continuously into the unperturbed states as a certain family of perturbations tends to zero. Mixed KMS states represent quantum-mechanical mixtures of phases, and lead to a nontrivial center of the algebra. If an observable from the center is added to H, the time-automorphism is unchanged, but the KMS states do change. Mixtures of KMS states exhibit a kind of instability in that they do not remain unchanged under the influence of a family of perturbations moving spatially off to infinity, and hence entering the center of the algebra.

A second important characteristic of KMS states is their passivity, which is the requirement that the energy of the system at time t can only have increased if the Hamiltonian depends on time and has returned to its initial form at time t. This condition also fixes the sign of β and means that no energy can be removed from a KMS state having $\beta > 0$, just as a periodic process can extract no energy from the ground state. This property does not constitute a kind of stability, and sheds no light on why Nature chiefly produces KMS states. However, it does show the most important empirically familiar feature of equilibrium.

As usual, the study of a finite system will provide us with a first exposure to the effects of perturbations. Its time-evolution will be caused by a self-adjoint operator, which also determines the equilibrium state w by $a_t = \exp(iHt)a\exp(-iHt)$, $w(a) = \operatorname{Tr}\exp(-\beta H)a/\operatorname{Tr}\exp(-\beta H)$. If H is subjected to a bounded, self-adjoint perturbation h, the effects can be written down as norm-convergent series. A simple generalization of (III:3.4.10;3) shows that

$$\begin{aligned}&\exp(i(H+h)t)a\exp(-i(H+h)t)\\&= a_t + \sum_{n\geq 1} i^n \int_{0\leq t_1\leq t_2\ldots\leq t_n\leq t} dt_1 dt_2\ldots dt_n [h_{t_1},[h_{t_2},\ldots,[h_{t_n},a_t]\ldots]],\end{aligned} \tag{3.15}$$

$$\begin{aligned}\exp(-H-h) &= R_h\exp(-H), \qquad \exp(-H+h)/2) = S_h\exp(-H/2),\\ R_h &\equiv \mathbf{1} + \sum_{n\geq 1}(-1)^n \int_{0\leq s_1\leq\ldots\leq s_n\leq 1} ds_1\ldots ds_n h_{is_1}\ldots h_{is_n},\\ S_h &\equiv \sum_{n\geq 0}(-1)^n \int_{0\leq s_1\leq\ldots\leq s_n\leq 1/2} ds_1\ldots ds_n h_{is_1}\ldots h_{is_n}.\end{aligned} \tag{3.16}$$

3.3.1 Remarks

1. Initially, h_{is} is well defined only if h is analytic in time (3.2.6(v)), but since such operators are dense in $\mathcal{A}$ in norm, the formulas it appears in extend to $\mathcal{A}$ by continuity.
2. Inequalities (2.1.5;3) and (2.1.8;7) yield the estimates

$$\begin{aligned}\exp(-\|h\|) &\leq \exp\left(\frac{-\operatorname{Tr}\exp(-H)h}{\operatorname{Tr}\exp(-H)}\right) \leq \frac{\operatorname{Tr}\exp(-H-h)}{\operatorname{Tr}\exp(-H)}\\ &= \frac{\operatorname{Tr} R_h\exp(-H)}{\operatorname{Tr}\exp(-H)} \leq \min\{\|R_h\|, \|\exp(-h)\|\}.\end{aligned}$$

Equation (3.3.1) can now be extended to cover infinite systems, for which H has continuous spectrum, as follows.

3.3.2 Perturbation of the Time-Evolution and KMS State

Let $a \to a_t$ be an automorphism of a C^* algebra $\mathcal{A}$, and let $\tilde{\mathcal{A}}$ be the subalgebra that is analytic in time and w be a KMS state. Assume $\beta = 1$. If $h \in \tilde{\mathcal{A}}$ is self-adjoint, then a perturbed automorphism $a \to \tau_t^h(a)$ and perturbed state are defined by

$$\tau_t^h(a) = a_t + \sum_{n\geq 1} i^n \int_{0\leq t_1\leq \ldots \leq t_n \leq t} dt_1 dt_2 \ldots dt_n [h_{t_2}, \ldots, [h_{t_n}, a_t]\ldots]],$$

$$w_h(a) = \frac{w(aR_h)}{w(R_h)} = \frac{w(R_h^* a)}{w(R_h)} = \frac{w(S_h^* a S_h)}{w(R_h)},$$

where R_h and S_h are defined as in (3.3.2).

3.3.3 Remarks

1. The operator h exists as a local perturbation on a purely algebraic level, whereas H exists only in certain representations. For that reason it is not possible to define $\tau_t^h(a)$ simply as $\exp(i(H+h)t)a\exp(-i(H+h)t)$. As in (3.3.2), for finite times the sums converge in norm.
2. If the system is asymptotically Abelian sufficiently strongly, then the limits as $t \to \pm\infty$ of $\tau_t^h \circ \tau_{-t}^0$ exist. However, such a limit may fail to be an automorphism; like the Möller transformations it might not be surjective. If it is surjective, its inverse transforms w into the perturbed state
$$w_h = \lim_{t\to\pm\infty} w \circ \tau_{-t}^0 \circ \tau_t^h.$$
3. See Problem 1 for the equivalence of the definitions of w_h.
4. $(\partial/\partial t)\tau_t^h(a) = \tau_t^h((\partial/\partial s)a_s|_{s=0}) + i\tau_t^h([h, a])$.
5. The function $\mathcal{A} \to \mathcal{A} : h \to \tau_t^h(a)$ is continuous for all $t \in \mathbb{R}$ and $a \in \mathcal{A}$, if $\mathcal{A}$ has either the strong or the norm topology.
6. The state w_h is KMS with respect to τ_t^h for $\beta = 1$: As shown by (3.3.1), for the domains D of the unbounded operators we have $D(\exp(-H-h)) = D(\exp(-H))$ in the representation using π_w, and because $\exp(H) = \exp(H+h)R_h$, the domains of definition of $\exp(H+h)$ and $\exp(H)$ are also identical. Hence for all a and $b \in \mathcal{A}$,
$$w_h(\tau_{-i}^h(a)b) = \frac{w(R_h^* \exp(H+h)a\exp(-H-h)b)}{w(R_h)}$$
is well defined. From (3.3.1) and the KMS condition for w,
$$w_h(\tau_{-i}^h(a)b) = \frac{w(\tau_{-i}(aR_h)b)}{w(R_h)} = w_h(ba).$$

7. There is an analogue of the variational principle for the free energy, which generalizes (2.1.8;3) for infinite systems. It is a consequence of the convexity of the function $h \to \ln w(R_h)$, which can be proved as follows: From Duhamel's formula (cf. the proof of (III:3.3.15)),

$$\frac{d}{d\lambda} \exp(-(H + \lambda a))$$

$$= - \int_0^1 ds \exp(-s(H + \lambda a)) \exp(-(1-s)(H + \lambda a)),$$

it can be calculated that

$$\frac{d}{d\lambda} w(R_{h+\lambda a})|_{\lambda=0} = \int_0^1 w(\tau_{is}^h(a) R_h) ds = w(a R_h).$$

The second part of the equality makes use of the invariance of w_h under τ^h, which follows from the KMS condition shown above. Likewise,

$$\frac{d^2}{d\lambda^2} w(R_{h+\lambda a})|_{\lambda=0} = \int_0^1 ds w(a \tau_{is}^h(a) R_h),$$

and

$$\frac{d^2}{d\lambda^2} \log w(R_{h+\lambda a})|_{\lambda=0} = \frac{w(R_{h+\lambda a})''}{w(R_h)} - \left(\frac{w(R_{h+\lambda a})'}{w(R_h)} \right)^2$$

$$\int_0^1 ds w_h((a - w_h(a)) \tau_{is}^h (a - w_h(a))).$$

In (3.2.6(ii)) it was seen that the integrands are positive. As in (3.3.3;2) this fact can be used to show that $w(R_h) \geq \exp(-w(h)) \geq \exp(-\|h\|)$.

If there is a bounded sequence of perturbations $h^{(n)}$ all the commutators of which with $\mathcal{A}$ tend to zero as $n \to \infty$, then the automorphism $\tau_t^{h(n)}$ converges to the unperturbed automorphism because

$$\|\tau_t^h(a) - a\| \geq \exp(2\|h\|t) \int_0^t \|[h, a_{t-s}]\| ds.$$

This state of affairs can arise, for instance, if the algebra is asymptotically Abelian with respect to spatial translations. If Λ_n denotes the region Λ translated by na, $a \in \mathbb{R}^3$, and $h^{(n)} \in \mathcal{A}_{\Lambda_n}$ is the corresponding translate of the operator h, then $\|[h^{(n)}, a]\| \to 0$, and consequently $\tau_t^h(a) \to a_t$. The question of whether the associated KMS states $w_{h(y)}$ likewise converge to the unperturbed w depends on whether the KMS states are extremal. This is illustrated even in the finite-dimensional case by

3.3.4 Example

With the notation of (1.1.1), let $\mathcal{A}$ be generated by $\{\mathbf{1}, \sigma_1, \sigma_1^{\pm}, \sigma_2\}$, and suppose that these observables evolve in time into $\{\mathbf{1}, \sigma_1, \exp(\mp 2it)\sigma_1^{\pm}, \sigma_2\}$. This time-evolution has a unitary representation as $U_t = \exp(it(\sigma_1 + c\sigma_2))$ for all $c \in \mathbb{R}$, so there is a one-parameter family of KMS states with density matrix $\rho = \exp(-\beta(\sigma_1 + \mu\sigma_2))$, which is not extremal, because

$$\exp(-\beta\mu\sigma_2) = \exp(-\beta\mu)\mathbf{1} \otimes \begin{pmatrix} 1 & 0 \\ 0 & 0 \end{pmatrix} + \exp(\beta\mu)\mathbf{1} \otimes \begin{pmatrix} 0 & 0 \\ 0 & 1 \end{pmatrix},$$

and $\exp(-\beta\sigma_1) \otimes \begin{pmatrix} 1 & 0 \\ 0 & 0 \end{pmatrix}$ provides a KMS state.

Although adding $h^{(n)} = (1/n)\sigma_1 + c'\sigma_2$ to the Hamiltonian leads to the same time-evolution as $n \to \infty$, the KMS state is different. Only the extremal KMS states provide two-dimensional representations, for which this can not happen.

Infinite systems generically have the property known as

3.3.5 Spatially Asymptotic Dynamical Stability

Let $\mathcal{A}$ be a quasilocal algebra and w be a locally normal KMS *state on $\mathcal{A}$. The state w is an extremal* KMS *state iff for each sequence $h^{(n)}$ of perturbations such that $\|h^{(n)}\|$ and $\|h_i^{(n)}\|$ are bounded in n and $\tau^{h^{(n)}}(a) \to a_t$ for $n \to \infty$ and all $a \in \mathcal{A}$, the sequence $w^{(n)} \equiv w_{h^{(n)}} \to w$ converges in the weak*- sense to w.*

3.3.6 Remarks

1. The assumption that $\mathcal{A}$ is quasilocal (1.3.2;8) serves to guarantee the existence of suitable sequences $h^{(n)}$.
2. If $\mathcal{A}$ is also asymptotically Abelian in time, then the following propositions are equivalent for KMS states (recall Figure 27):

   (a) w is an extremal KMS state;

   (b) π_w is a factor;

   (c) $\lim_{t\to\infty} w(ab_t) = w(a)w(b)$;

   (d) $w_{h^{(n)}} \to w$ for all $h^{(n)}$ as described in (3.3.5).

Proof:

1. If w is extremal, then $w^{(n)} \to w$: By assumption $\|h_i^{(n)}\|$ are bounded uniformly in n, so the same is true of the norms of $R_{h^{(n)}}$. Since, moreover, $w(R_{h^{(n)}}) \geq \exp(-\|h^{(n)}\|)$,
$$\rho_n = \frac{R_{h^{(n)}}}{w(R_{h^{(n)}})}$$
is a bounded sequence of operators. Bounded sequences of operators are weakly relatively compact ([33], VI; 9.6), and the set of states is weak*- compact (III:2.1.21;2), so there is a subsequence $H^{(k)}$, k in some $L \subset \mathbf{N}$, such that $\bar{w} = \lim_L w^{(k)}$ and $\rho = \lim \rho_k$ exist, and $\bar{w}(a) = w(a\rho)$.
The automorphisms converge by assumption, and by Problem (3.2.17;1) $\bar{w}$ is τ-KMS. But this means that ρ belongs not only to $\pi_w(\mathcal{A})''$ (by construction), but also to $\pi_w(\mathcal{A})'$ and thus belongs to the center:
$$w(a\rho b) = w(b_{-i}a\rho) = \bar{w}(b_{-i}a) = \bar{w}(ab) = w(ab\rho)$$
and
$$w(a\rho bc) = w(ab\rho c).$$
However, π_w is a factor, so $\rho = 1$, and since $\bar{w} = w$ is the only point of accumulation it is the limit of $w^{(n)}$.

2. Suppose now that w is not extremal. There is a nontrivial invariant element $z = z^*$ in the center of $\pi_w(\mathcal{A})''$. By Kaplansky's theorem [4] the unit ball of $\mathcal{A}$ is strongly dense in the unit ball of $\mathcal{A}''$, so z belongs to the closure of a bounded set of self-adjoint operators h of $\mathcal{A}$. Because of the locality assumption the closure of $\mathcal{A}_\Lambda \|\Omega\rangle$ is a separable subspace of
$$\mathcal{H} = \overline{\mathcal{A}|\Omega\rangle} = \overline{\bigcup_n \mathcal{A}_{\Lambda(n)}|\Omega\rangle} \qquad (\Lambda(n) \to \mathbb{R}^3),$$
so $\mathcal{H}$ is also separable. As a consequence the strong topology on bounded sets of operators is metrizable, so z is actually the limit of some sequence $h^{(n)}$ in $\bigcup_n \mathcal{A}$. According to (3.3.4) τ_t^h converges to $\tau_t^z = \tau_t^0$. As in (3.2.6(v)) ρ_n can be constructed with the $h^{(n)}(f)$, as they converge to $z_t = z(f) = z$, just like $h_t^{(n)}(f)$ and $h_{is}^{(n)}(f)$. By the dominated convergence theorem it follows that
$$\lim_{n\to\infty} R_{h^{(n)}(f)} = R_z = \exp(-z),$$
and therefore
$$\lim w_{h^{(n)}(f)}(a) = \frac{w(\exp(-z)a)}{w(\exp(-z))}$$
is a KMS state different from w. □

The next topic is that of stability properties that can distinguish the extremal KMS states from other stationary states giving rise to factors. As shown by (3.3.4), if there is an extremal KMS state, then for all $h \in \mathcal{A}$ there exists a state that is stationary under the time-evolution including h as a perturbation, and which transforms continuously into the unperturbed state as $h \to 0$. It is not obvious that such a "linear-response theory" is possible. In fact, we learned (I, §3.3) that even in classical physics there are constants of motion that are not continuous in a parameter of the Hamiltonian. A density in phase space that is a function of such a constant will be unstable when perturbed, no matter by how little. This phenomenon is illustrated in quantum mechanics by the trivial

3.3.7 Example

$\mathcal{H} = \mathbb{C}^2$, $H = 0 \in \mathcal{B}(\mathbb{C}^2)$. Every density matrix ρ corresponds to a stationary state, but with the perturbation $h = \mathbf{n} \cdot \boldsymbol{\sigma}$ the only stationary density matrices are $\rho = \mathbf{1}/2 + \lambda \mathbf{n} \cdot \boldsymbol{\sigma}, \lambda < |\mathbf{n}|/2$. This shows that only the density matrix $\rho = \mathbf{1}/2$ goes continuously into a density matrix that is stationary under all possible perturbed time-evolutions.

The example illustrates that only density matrices of the form $f(H)$, which are proportional to the identity in each degeneracy space of H, adapt themselves well to arbitrary perturbations. Despite the possibility of diagonalizing any stationary density matrix simultaneously with H, there is no telling from stationariness alone how it might vary within a degeneracy space. A requirement that two independent systems be stable would impose an additional restriction on the function f such that $w = f(H)$. The existence of two subsystems shows up mathematically as a tensor product, so if $H = H_1 \otimes \mathbf{1} + \mathbf{1} \otimes H_2$, then we would require that $f(H_1 \otimes \mathbf{1} + \mathbf{1} \otimes H_2) = f(H_1) \otimes f(H_2)$. Since H_1 and H_2 commute, both H_i may be regarded as ordinary numbers in their common spectral representation. Since the only reasonable functions satisfying $f(x+y) = f(x)f(y)$ are of the form $f(x) = \exp(-\beta x)$, we are led to the canonical density matrix, if the H_i may have arbitrary real spectral values. Since our infinite systems are asymptotically Abelian with respect to translations, and thus come to resemble tensor products of independent systems, it is a reasonable expectation that the condition of stability for such systems characterizes the KMS states. It will now be seen that this is the case, given some assumptions.

3.3.8 Local Dynamical Stability

Suppose that the algebra $\mathcal{A}$ is asymptotically Abelian with respect to τ^0, and let w be a stationary factor state, and hence mixing. The question is whether for any perturbed automorphism τ^h it is possible for there to be a unique state w_h that is invariant under τ^h and turns into w as $h \to 0$. The states

$$w_{\pm} = \lim_{t \to \pm\infty} w \circ \tau_t^h$$

are reasonable candidates for w_h. If the limits exist, they would be invariant under τ^h, and the uniqueness of w_h means that the limits are equal. If τ^h is expanded as in (3.3.4) and we use the invariance of w under τ^0, we obtain the

3.3.9 Stability Condition to First Order in *h*

If an invariant factor state w on an algebra $\mathcal{A}$ asymptotically Abelian in time is stable against arbitrary perturbations in the sense stated above, then for all h and $a \in \mathcal{A}$,

$$\int_{-\infty}^{\infty} dt w([h, a_t]) = 0.$$

3.3.10 Remarks

1. The assumption that $h \in \mathcal{A}$ means that we consider only local perturbations. The requirement that $\mathcal{A}$ be asymptotically Abelian makes the commutator $[h, a_t]$ vanish as $t \to \pm\infty$. Condition (3.3.9) requires, roughly speaking, that $w(i[h, a_t])$ is equally often positive and negative. Norm asymptotic abelianess is not really required but we understand by stability that all the integrals are well defined.
2. The physical significance of (3.3.9) is that to first order in h the scattering transformation is the identity in the representation π_w. This can be interpreted as meaning that w is a locally perturbed equilibrium state with respect to the time-automorphism τ^h and should become the equilibrium state as $t \to \pm\infty$, so there is no net change between $t = -\infty$ and $t = +\infty$. In the kinetic theory of gases this is reflected in the argument that collisions do not alter the equilibrium distribution.
3. One can show that (3.3.8) is equivalent to adiabatic stability which means that if h depends on t and is slowly switched on and off then the state regains its original form [56].

Let us introduce the abbreviations

$$F_{ab}(t) = w(ba_t) - w(a)w(b)$$

and

$$G_{ab}(t) = w(a_t b) - w(a)w(b) \tag{3.17}$$

in order to exploit (3.3.11) more fully.

3.3.11 Consequences for the Correlation Functions

Condition (3.3.9) makes

$$\int_{-\infty}^{\infty} dt(F_{ab}(t) - G_{ab}(t)) = 0.$$

Under the assumptions of (3.3.10) we know that F and G tend to zero as $t \to \pm\infty$. In order to ensure that this integral and others to follow make sense, it will be assumed that the correlation functions F and G are integrable in time from $-\infty$ to $+\infty$, at least for a dense set $\mathcal{G} \subset \mathcal{A}$. Since they are bounded, they belong to all $L^p(\mathbb{R})$ for $1 \leq p \leq \infty$. The assumption holds, for example, for free fermions. It will also be assumed that the higher correlation functions decrease rapidly enough for elements of $\mathcal{G}$ that integrals and limits may be interchanged.

If the state is a factor state, then as $u \to \pm\infty$, $w(ab_u c_t d_{t+t_1+u} - c_t d_{t+t_1+u} ab_u)$ tends to $w(ac_t)w(bd_{t+t_1}) - w(c_t a)w(d_{t+t_1} b)$. Therefore

$$\int_{-\infty}^{\infty} dt(F_{ca}(t)F_{db}(t+t_1) - G_{ca}(t)G_{db}(t+t_1)) = 0$$

for all a, b, c and $d \in \mathcal{G}$, $t_1 \in \mathbb{R}$. Similarly, from considering what happens to $w([ab_u c_v, d_t e_{u+t+t_1} f_{v+t+t_2}])$ as $u \to \infty$ and $v \to \infty$,

$$\int_{-\infty}^{\infty} dt(F_{da}(t)F_{cf}(t+t_1)F_{be}(t+t_2) - G_{da}(t)G_{cf}(t+t_1)G_{be}(t+t_2)) = 0$$

for all a, b, c, d, e, and $f \in \mathcal{G}$, $t_{1,2} \in \mathbb{R}$. Because F and G belong to L^1, their Fourier transforms $\tilde{F}$ and $\tilde{G}$ exist and are continuous.

In order to arrive at the KMS condition in Fourier-transformed space, $\tilde{F}_{ab}(E) = \exp(\beta E)\tilde{G}_{ab}(E)$, information about the supports of $\tilde{F}$ and $\tilde{G}$ is needed. It is at least clear that they are contained in the spectrum of H: Let $a_t = U_t^{-1} a U_t$, $U_t = \exp(-iHt)$, writing H as in (1.3.3) in the representation determined by w. Then

$$w(ba_t) = \langle b^*\Omega | U_t^{-1} | a\Omega \rangle,$$

so if $E \neq 0$, then

$$\tilde{F}_{ab}(E) = \tilde{F}_{b^*a^*}(E)^* = \tilde{G}_{ba}(-E) = \langle b^*\Omega | \delta(E-H) a\Omega \rangle. \tag{3.18}$$

This expression is to be interpreted in the spectral representation of H, in which the functions depend continuously on E when a and $b \in \mathcal{G}$.

In order to draw more far-reaching conclusions from these relationships, more information is needed about the energy spectrum. It would simply be additive if the Hamiltonian were the tensor product of Hamiltonians of independent systems: If H_1 and H_2 have eigenvalues $e_n^{(1)}$ and $e_n^{(2)}$, then $H^{(1)} \otimes \mathbf{1} + \mathbf{1} \otimes H^{(2)}$ has eigenvalues $e_n^{(1)} + e_m^{(2)}$. This fact generalizes to an infinite system provided that the system is asymptotically Abelian with respect to an automorphism, such as the translations, that commutes with the time-evolution. Similarly the above integral relations hold if u and v correspond to a translation which leaves w invariant.

3.3.12 The Additivity of the Spectrum of H

Let H generate a time-evolution τ on a factor state w, and suppose that the system is asymptotically Abelian with respect to an automorphism σ such that $[\sigma, \tau] = 0$

and $w \circ \sigma = w$. If H has the spectral values E_1 and E_2, then $E_1 + E_2$ also belongs to the spectrum of H.

Proof: Given any neighborhoods U_i of E_i, $i = 1, 2$, by assumption there exist f_i such that

$$a_{f_i}|\Omega\rangle \equiv \int_{-\infty}^{\infty} dt a_t f_i(t)|\Omega\rangle \neq 0,$$

where the Fourier transforms $\tilde{f}_i$ have their supports in U_i. Since by Property (3.1.16;4) $\|\sigma_s(a_{f_i})a_{f_2}|\Omega\}^2$ approaches

$$\|a_{f_2}|\Omega\rangle\|^2 \|a_{f_1}|\Omega\rangle\|^2 \neq 0$$

as $s \to \infty$, there must be a sufficiently large s that this vector is nonzero. Since the vector is supported in $E_1 + E_2 + U_1 + U_2$ in the spectral representation of H for all s, there are spectral values in every neighborhood of $E_1 + E_2$. Since the spectrum is closed, $E_1 + E_2$ itself belongs to the spectrum. □

3.3.13 Remark

If the system is asymptotically Abelian with respect to τ, then of course it is possible to take $\tau = \sigma$. Since w provides a factor, according to Table I in this case $|\Omega\rangle$ is the only eigenvector, and H has no eigenvalues other than 0. Since the spectrum is additive, it is either $0 \cup [\pm c, \pm\infty)$ for some $c \geq 0$, or else $(-\infty, \infty)$. In the first case there is a ground (or ceiling) state; we shall be concerned only with the second possibility.

3.3.14 Derivation of the KMS Condition

Taking the Fourier transform of the relations for $F(t)$, $G(t)$ we get

$$\int dE e^{-iEt_1} \left(\tilde{F}_{ca}(E)\tilde{F}_{db}(-E) - \tilde{G}_{ca}(E)\tilde{G}_{db}(-E)\right) = 0, \forall t_1 \in \mathbb{R}$$

and since the Fourier transform is injective

$$\tilde{F}_{ca}(E)/\tilde{G}_{ca}(E) = \tilde{G}_{db}(-E)/\tilde{F}_{db}(-E) \quad \forall a, b, c, d \in \mathcal{A}.$$

Thus there must be universal function $\Phi(E) = \Phi^{-1}(-E)$ such that $\tilde{F}_{ca}(E) = \Phi(E)\tilde{G}_{ca}(E)$. The relation with threefold products tells us in Fourier space

$$\tilde{G}_{da}(E_1)\tilde{G}_{cf}(E_2)\tilde{G}_{be}(-E_1 - E_2)\left(\Phi(E_1)\Phi(E_2)\Phi(-E_1 - E_2) - 1\right) = 0$$

or

$$\Phi(E)\Phi(E') = \Phi(E + E'),$$

and since Φ is continuous on $\mathbb{R}$ or $[\pm c, \pm\infty)$ it therefore has the functional form

$$\Phi(E) = \exp(\beta E) \quad \text{for} \quad \text{some} \quad \beta \in \mathbb{R}.$$

This shows the KMS condition for the dense set $\mathcal{G}$. However, since it can be written with the aid of (3.18) in the form

$$\langle b^*\Omega|f(-H)a\Omega\rangle = \langle a^*\Omega|f(H)\exp(-\beta H)b\Omega\rangle$$

for any bounded, continuous $f(H)$, it clearly suffices to derive it on a dense set.

In sum, the foregoing argument has shown the

3.3.15 Equivalence of Dynamical Stability and the KMS Condition

Suppose that the algebra $\mathcal{A}$ is asymptotically Abelian with respect to an automorphism commuting with the time-evolution and that w is a stationary state creating a factor representation. If for all $h \in \mathcal{A}$ there exists a normal state w_1 for $\pi_w(\mathcal{A})''$ to first order in h, such that w and w_1 are both stationary to first order under the perturbed time-evolution, and if w has an absolutely integrable correlation function, then either w is a KMS state, or else the spectrum of H is $\{0\} \cup [\pm c, \pm\infty)$, in which case w is the ground state.

3.3.16 Remarks

1. It does not follow from this argument that $\beta > 0$. This fact did not even emerge from our argument with the tensor product of finite systems.
2. It is hard to tell how much the result suffers from the sharpening of the hypothesis of asymptotic commutativity. All the hypotheses are satisfied by a system of free fermions, but with a Coulomb interaction it is not even known if they hold in weakened forms. To a certain extent our assumptions about decrease at infinity and the interchangeability of limits belong to the realm of unproven hopes.
3. This shows that stability to first order in h implies KMS. Conversely, we have seen that KMS implies stability to every order in h, which means that the higher orders contribute no new information in this respect.
4. If w is not a factor there is no hope to deduce KMS from stability. Take a mixture of two extremal KMS-states to different temperatures, then they satisfy stability but are not KMS for a definite temperature.

Whereas all the perturbations considered until now have been independent of time, we shall now turn our attention to perturbations $h(t)$ depending explicitly on time; they would be due to interference from outside the system. The time-evolution will not have the group property, but it will still be a one-parameter family of automorphisms. Let us, as usual, start by studying finite systems, for which the automorphisms are implemented by the unitary transformations

$$U_t = T\exp\Big[-i\int_0^t dt'(H + h(t'))\Big] \tag{3.19}$$

(cf. (III:3.3.6)).

The most important quality of a passive state for our purposes will be that a system in a passive state will have gained energy when the perturbation has been switched off.

3.3.17 The Passivity of a State

Let us suppose that a finite system evolves under the influence of $H + h(t)$, where by definition $h(0) = h(\tau) = 0$. The Hamiltonian generates a unitary time-evolution (3.19), so the change in energy from $t = 0$ to $t = \tau$ in the state w is given by $\operatorname{Tr} \rho(U_\tau H U_\tau^{-1} - H)$. A state is said to be **passive** if the change in energy is positive for all self-adjoint $h \in \mathcal{B}(\mathcal{H})$, in which case $\operatorname{Tr} \rho U H U^{-1} \geq \operatorname{Tr} \rho H$ for all $U = U^{*-1} \in \mathcal{B}(\mathcal{H})$.

3.3.18 Examples

1. The canonical density matrix. Let $\rho = \exp(-\beta H)/\operatorname{Tr}\exp(-\beta H)$ and $\sigma = U^{-1}\rho U$. From (2.2.20(i)) we know that

$$0 \leq \operatorname{Tr} \sigma(\ln \sigma - \ln \rho) = \operatorname{Tr}(\rho - \sigma) \ln \rho = -\beta \operatorname{Tr}(\rho - U^{-1}\rho U)H$$

   so the system is passive.
2. Negative temperatures. Let ρ be as above, but $\beta < 0$. In order for $\operatorname{Tr}\exp(-\beta H)$ to be finite, H must be bounded from above; this would be realistic for a spin system. The inequality is then reversed, $\operatorname{Tr}(\rho - U^{-1}\rho U)H > 0$, so the system is not passive.

3.3.19 Remarks

1. If it is desired to keep the energy $E = F + TS$ from increasing, the best tactic is to keep S constant (when $T > 0$). Our unitary time-evolution manages this automatically, and so the change in the energy E equals the change in the free energy F. Since the free energy is minimized with the canonical density matrix ρ, in the state ρ the only possibility is for E to increase, so ρ is passive.
2. Obviously, passivity requires the states of lower energy to be more densely occupied, so that the system is ready to gain energy. This is not the case when $\beta < 0$, in which circumstances the system would prefer to give energy away.

3.3.20 The General Form of Passive Density Matrices for Finite Systems

A density matrix ρ on a finite system corresponds to a passive state if and only if

(i) $[\rho, H] = 0$, *and*

(ii) *if ρ_i and e_i designate respectively the ordered eigenvalues of ρ and H, then*

$$(e_i - e_k)(\rho_i - \rho_k) \leq 0.$$

3.3.21 Remarks

1. The condition on the eigenvalues means that if the kth eigenvalue of H is greater than the ith, then the kth eigenvalue of ρ must be less than or equal to the ith. However, it is not necessary for ρ to be simply a function of H, since in a degeneracy space for which $e_i = e_k$ it may happen that $\rho_i \neq \rho_k$.
2. The physical implication of the monotony is that lower-lying states are more densely occupied. On the other hand it implies nothing for the values of ρ where H does not vary:

$$H = \begin{pmatrix} 0 & & \\ & 0 & \\ & & 1 \end{pmatrix}, \quad \rho = \begin{pmatrix} \frac{1}{4} & & \\ & \frac{1}{2} & \\ & & \frac{1}{4} \end{pmatrix},$$

is passive.

Proof: (i) and (ii) $\Rightarrow$ passive $\Leftrightarrow$ $\mathrm{Tr}\,\rho H \leq \mathrm{Tr}\,\rho U H U^{-1}$.

Let U be given in a matrix representation in the common eigenvectors of H and ρ as U_{ik}. The matrix $|U_{ik}|^2$ is doubly stochastic and therefore a convex combination of permutation matrices or a limit of such matrices (cf.(2.1.8;3)). For any such matrix,

$$\mathrm{Tr}\,\rho U H U^{-1} = \sum_{i,k} e_i \rho_k \|U_{ik}\|^2 = \sum_P c_P \sum_i e_i \rho_{P_i},$$

where $\sum_P c_P = 1$, $c_P \geq 0$, and $\{P_i\}$ is a permutation of the $i \in \mathbb{Z}^+$. If $e_i < e_k$ implies that $\rho_i \geq \rho_k$, then for any permutation, $\sum_i e_i \rho_{P_i} \geq \sum_i e_i \rho_i = \mathrm{Tr}\,\rho H$.

Passive $\Rightarrow$(i) and (ii). Suppose that $\mathrm{Tr}\,\rho U H U^{-1}$ has its minimum at $U = \mathbf{1}$, and write $U = \mathbf{1} + M_1 + M_2 + \ldots$, where $\|M_k\| < \varepsilon^k$ for sufficiently small ε. Then $\mathrm{Tr}\,\rho U H U^{-1} = \mathrm{Tr}\,\rho H + \mathrm{Tr}([H, \rho] M_1) + O(\varepsilon^2)$. The operator M_1 only needs to satisfy the condition that $M_1^* = -M_1$, and since $[\rho, H]$ is anti-Hermitian, it must equal zero, as otherwise the energy could be lowered. In order to prove (ii), choose U to have the form

$$u = \begin{pmatrix} 0 & 1 \\ -1 & 0 \end{pmatrix}$$

on the subspace spanned by v_i and v_k, the eigenvectors with eigenvalues e_i, ρ_i and e_k, ρ_k. Then

$$\mathrm{Tr}\,\rho U H U^{-1} - \mathrm{Tr}\,\rho H = -(e_i - e_k)(\rho_i - \rho_k),$$

which is positive only if $(e_i - e_k)(\rho_i - \rho_k) \leq 0$. □

In order to progress beyond the monotonic property to the statement that the function is exponential we must investigate infinite systems. We may either construct the infinite system by taking tensor products of copies of finite systems or go directly to the analysis of some asymptotically Abelian system. As before, the limiting case $\beta = \infty$, i.e., the ground state, would require a special treatment, which we shall not go into. Assuming therefore that β is finite, we can state the main proposition on the

3.3.22 Passivity of Infinite Systems

Within the set of faithful factor states w on a C^*-algebra with a time-automorphism τ and another automorphism commuting with τ and under which w is invariant and asymptotically Abelian, the passive states are precisely the KMS states, for any $\beta \geq 0$.

3.3.23 Remarks

1. Translations of a homogeneous infinite system commute with the time-evolution. Since the local field algebra is asymptotically Abelian with respect to translations, as before this theorem can be used even if it is not known whether the time-evolution is asymptotically Abelian.
2. The sign of β is fixed by passivity, though of course its value is not.
3. To ensure that H is well-defined, assume that the time-evolution can be represented unitarily; then passivity is equivalent to the property that $w(U^{-1}HU - H) \geq 0$ for all unitary $U \in \mathcal{A}$.
4. Since the condition for passivity is linear in w, the passive states form a convex set. Passivity does not single out the extremal KMS states. We shall consider only factor states, which can not be decomposed further into invariant states, as shown in § 3.1.

Proof: Passive $\Rightarrow$ KMS. If the condition of passivity for an infinite systems is written as $w(UHU^{-1}) \geq w(H)$, and we choose $U = \exp(i\varepsilon a)$ for a self-adjoint, then the first two terms of the expansion in powers of ε lead to

(i) $w([a, H]) = 0$ for all $a \in \mathcal{A}$, and

(ii) $w([a, [H, a]]) \geq 0$ for all $a \in \mathcal{A}$.

Equation (i) means that $(\partial/\partial t)w(a_t) = 0$, so w is stationary. In order to deduce the KMS condition from (ii) we employ the modular automorphism of w – call its generator $\bar{H}$ can be used to write (ii) as

$$\begin{aligned} 0 &\leq \langle \Omega | 2aHa - Ha^2 - a^2 H | \Omega \rangle \\ &= \langle \Omega | 2aHa - a\exp(-\bar{H})Ha - aH\exp(-\bar{H})a | \Omega \rangle \\ &= 2\langle \Omega | aH(\mathbf{1} - \exp(-\bar{H}))a | \Omega \rangle . \end{aligned}$$

In the last step we used the fact that $[H, \bar{H}] = 0$, in accordance with our assumption. Since the inequality holds for a dense set of $a = a^* \in \mathcal{A}$, it follows that $H(\mathbf{1} - \exp(-\bar{H})) \geq 0$. This means that in the common spectral representation of H and $\bar{H}$ the spectrum is restricted to the hatched region of the $(H, \bar{H})$-plane shown in Figure 3.7. Now the existence of the commuting, asymptotically Abelian automorphisms comes into play. According to (3.3.12), this implies that the

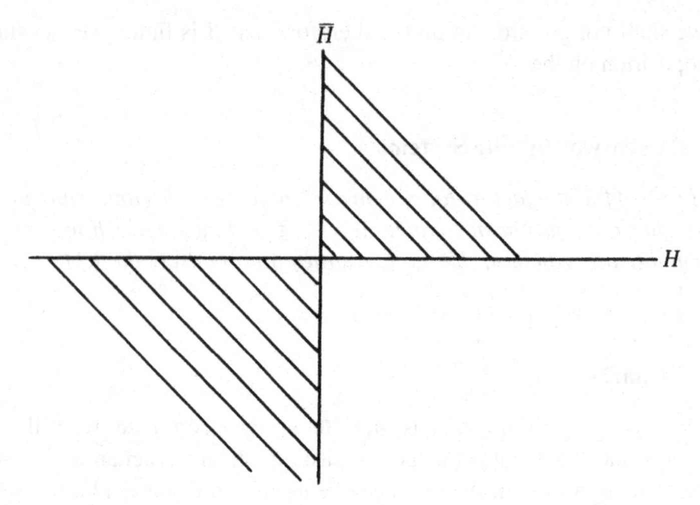

Figure 3.7. Possible location of the spectra of H and $\bar{H}$.

spectrum is additive, i.e., if $(h_1, \bar{h}_1)$ and $(h_2, \bar{h}_2)$ are in the spectrum, then so is $(h_1+h_2, \bar{h}_1+\bar{h}_2)$. As a consequence the spectrum can at most be on a line through $(0, 0)$, so $\bar{H} = \beta H$ for some $\beta > 0$.
KMS $\Rightarrow$ passive. Since $x \geq 1 - \exp(-x)$,

$$w(UHU^{-1}) \geq w(UU^{-1}) - w(U \exp(-H)U^{-1})$$
$$= w(UU^{-1}) - w(U^{-1}U) = 0. \qquad \square$$

3.3.24 Remarks

1. The last inequality proved above is only the first of a whole family of inequalities that the expectation values in KMS states satisfy, and which completely characterize the KMS states [24]. They generalize trace inequalities, which are not directly applicable to infinite systems, since $\exp(-\beta H)$ is not trace-class.
2. Example (3.3.18;1) showed that for finite systems, passivity follows from thermodynamic stability, or, in other words, from the minimum property of the free energy. This fact generalizes to infinite systems, for many of which the implication goes both ways, KMS $\Leftrightarrow$ thermodynamic stability, for instance for lattice systems with finite-range interactions. For these systems KMS is equivalent to global thermodynamic stability, provided that only translation-invariant states are considered, and that the free energy is interpreted as the free-energy density. However, for systems with long-range forces there exist KMS states that do not minimize the free energy; they are instead metastable, minimizing the free energy only on some reduced set of comparison states. Since the free energy is a convex functional on the states, it can not have a relative minimum on the set of all states that fails to be absolute.
3. The state $w_{\beta_1} \otimes w_{\beta_2}$ of two independent systems at different temperatures $T_1 > T_2$ is KMS with respect to the automorphism generated by $\beta_1 H_1 + \beta_2 H_2$. A

perturbation $h(t)$ can cause the temperatures to equalize, and it may happen that the first system will have given up a positive amount of energy $\Delta E_1 \equiv E_1(0) - E_1(\tau) > 0$ by the end of the period. However, because the state is passive, $\beta_1 \Delta E_1 + \beta_2 \Delta E_2 \leq 0$, and the change in the total energy $\Delta E = \Delta E_1 + \Delta E_2$ is bounded by $\Delta E/\Delta E_1 \leq (T_1 - T_2)/T_1$. Since the total entropy remains constant under the unitary time-evolution, ΔE is the amount of energy provided by the total system, and this inequality is Carnot's classical bound on the thermal efficiency.

Another way to characterize the KMS states of an infinite system is known as reservoir stability, and it further justifies the physical interpretation of β as the reciprocal of the temperature. In outline it means that the KMS states are precisely the states that are suitable for thermal reservoirs, allowing the temperature $1/\beta$ to be defined. A more careful formulation states that if the reservoir is coupled to a finite system in the canonical state w, then in the weak-coupling limit w is invariant under the resulting semigroups (cf. (3.1.10)) for a reasonable class of couplings iff the reservoir is in a KMS state [24].

3.3.25 Problems

1. Show that $w(R_h^* a) = w(aR_h) = w(S_h^* a S_h)$.
2. Estimate the length of time for which the "linear-response theory" remains valid; i.e., estimate

$$\|\tau_t^h(a) - a_t - i\int_0^t dt_1 [h_{t_1}, a_t]\| .$$

3. Use the methods of § 2.1 to conclude from $e_i > e_j \Rightarrow \rho_i \leq \rho_j$ that

$$\sum_i e_i \rho_i \leq \sum_i e_i \rho_{P_i}$$

for every permutation P.

3.3.26 Solutions

1. Since H exists in the GNS representation with w, Equations (3.3.1) are applicable. The invariance of Ω holds also for complex z,

$$\exp(zH)|\Omega\rangle = |\Omega\rangle, \qquad R_h|\Omega\rangle = \exp(-H-h)|\Omega\rangle.$$

Now use the KMS condition for w in the form $w(ab) = \langle \Omega | b \exp(-H) a | \Omega \rangle$:

$$w(aR_h) = \langle \Omega | R_h \exp(-H) a | \Omega \rangle = \langle \Omega | \exp(-H-h) a | \Omega \rangle = w(R_h^* a).$$

It is also true in this representation that $S_h \exp(-H) S_h^* = \exp(-H-h)$, so

$$w(R_h^* a) = \langle \Omega | S_h \exp(-H) S_h^* a | \Omega \rangle = w(S_h^* a S_h).$$

2. Apply Taylor's formula $\|f(\alpha) - f(0) - \alpha f'(0)\| \le \| \int_0^1 d\xi (1-\xi) f''(\alpha\xi)\alpha^2 \|$ to $f : [0,1] \to \mathcal{B}(\mathcal{H}), \alpha \to \tau_t^{\alpha h}(a)$. According to (3.3.4),

$$\left\| \frac{\partial^2}{\partial \alpha^2} \tau_t^{\alpha h}(a) \right|_{\alpha=0} \left\| = \left\| 2 \int_0^t dt_2 \int_0^{t_2} dt_1 [h_{t_1}, [h_{t_2}, a_t]] \right\| \le 4t^2 \|h\|^2 \|a\|. \right.$$

This is also true when $\alpha \in [0,1]$; the only change when $\alpha > 0$ is that the time-evolution $a, h \to a_t, h_t$ becomes $a, h \to \tau_t^{\alpha h}(a), \tau_t^{\alpha h}(h)$, which does not affect the norms. Consequently the answer is that $\| \dots \| \le (2t\|h\|)^2 \|a\|/2$. Recall that if $\|h\|$ is on the order of a Rydberg, then $t\|h\| \ll 1$ when $t \ll 10^{-15}$ sec. Therefore this *a priori* estimate guarantees only that the linear approximation remains valid for times on the atomic scale, and not for times measured in seconds. To go further would require knowing that the commutators go to zero for longer than macroscopic times.

3. Order e_i and ρ_i; then

$$e_1\rho_1 + e_2\rho_2 + e_3\rho_3 + \cdots = (e_1 - e_2)\rho_1 + (e_2 - e_3)(\rho_1 + \rho_2) + (e_3 - e_4)(\rho_1 + \rho_2 + \rho_3) + \cdots.$$

All the summands are positive, and permuting the ρ_i can at most make the summands larger.

3.4 Quantum Ergodic Theory

A Reminder

Boltzmann's ideas about the time evolution of large systems has developed in two mathematical disciplines, topological dynamics and measure theoretic ergodic theory. In the former one considers a dynamical system consisting of a C^*-algebra $\mathcal{A}$ (the "observables") and an automorphism τ of $\mathcal{A}$ (the "time evolution" for unit time) whereas in the latter one studies in addition τ-invariant states over $\mathcal{A}$. In both cases one is interested in the long time behaviour ($\lim_{n\to\infty} \tau^n$). This framework can be taken over directly to quantum mechanics where $\mathcal{A}$ is not commutative. (We add as standing assumption $\tau \neq id$ and $\mathbf{1} \in \mathcal{A}$.) However, before generalizing the various classical theorems we have to discuss to which physical situation this framework applies. The system we are interested in is bulk matter which consists of electrons and nuclei. Electrons are fermions and since the difference between various isotopes are of no relevance for us we might as well consider odd mass number isotopes so that all particles are fermions which makes the mathematics much easier. Since we are interested in large systems which are spatially unlimited but have a finite particle density the appropriate description is a quantized fermi field. Thus a minimal algebra $\mathcal{A}$ will be the CAR algebra generated by the annihilation operators $\alpha_f = \int d^3x \bar{f}(x)\alpha(x)$, $f \in L^2(\mathbb{R}^3)$, i.e. the norm closure of the polynomials

$$a = x + \sum_{n,m} \alpha^*_{f_1} \dots \alpha^*_{f_n} \alpha_{g_1} \dots \alpha_{g_m}. \tag{3.20}$$

From the CAR relations

$$[\alpha_f, \alpha_g^*]_+ = \langle f|g\rangle \tag{3.21}$$

one infers $\|\alpha_f\| = \|f\|$. The observable algebra will be the subalgebra where $n = m$. $\mathcal{A}$ is a pleasant C^*-algebra, it is simple (i.e. it has no proper twosided ideal) and it is UHF (i.e. the norm closure of an increasing sequence of matrix algebras). States are positive linear functionals ω over $\mathcal{A}$ and by the GNS construction they give a representation Π_ω of $\mathcal{A}$ in a Hilbert space $\mathcal{H}_\omega$. Popular states are the equilibrium states Φ_ω for a quasifree time evolution $\widetilde{f}(k) \to e^{i\varepsilon(k)t}\widetilde{f}(k)$. They are defined by

$$\Phi_\beta(\alpha_f^* \alpha_g) = \int \frac{d^3k}{(2\pi)^3} \frac{\widetilde{f}(k)\widetilde{g}(k)}{1 + e^{\beta(\varepsilon(k)-\mu)}}$$

and vanishing reduced $n \geq 3$ point functions. The have the following features:

1. Π_{Φ_β} is a factor (i.e. if $\Pi_\omega(\mathcal{A})'$ is the commutant of $\Pi_\omega(\mathcal{A})$, $\Pi_\omega(\mathcal{A})''$ the commutant of $\Pi_\omega(\mathcal{A})'$, then the center $\Pi_\omega(\mathcal{A})' \cap \Pi_\omega(\mathcal{A})''$ is trivial for a factor).
2. For $\beta \neq \infty$ Φ_β is faithful (i.e. $\Phi_\beta(a^*a) = 0 \leftrightarrow a = 0, a \in \mathcal{A}$).
3. Φ_0 is tracial (i.e. $\Phi_0(ab) = \Phi_0(ba)$.
4. Φ_∞ (the Fock state) is pure (i.e. Π_{Φ_∞} is irreducible).

To describe the dynamics of particles interacting via a potential v one usually writes down a Hamiltonian

$$\begin{aligned} H &= \tfrac{1}{2m} \int dx \nabla\alpha^*(x)\nabla\alpha(x) \\ &+ \int dx dx' \alpha^*(x)\alpha^*(x')v(x-x')\alpha(x')\alpha(x) =: T + V \end{aligned} \tag{3.22}$$

and defines a time evolution by

$$\tau^t(\alpha(x)) = e^{iHt}\alpha(x)e^{-iHt}. \tag{3.23}$$

Indeed in the Fock representation Π_{Φ_∞} where one has sectors with finite particle number $N = \int dx\alpha^*(x)\alpha(x)$ there is no doubt that for reasonable v's the restriction H_N will define a time evolution in these sectors. However, in general this will not be a one parameter automorphism group of $\mathcal{A}$ but only of $\Pi_{\Phi_\infty}(\mathcal{A})''$. Having an automorphism of $\mathcal{A}$ means a dynamics independent of the state and this can in general not be expected. Generically potentials are not stable in the sense that there exists $c \in \mathbb{R}^+$ such that (see § 4.3)

$$H_N > -cN \qquad \forall N. \tag{3.24}$$

This condition ensures that the energy per particle stays finite for $N \to \infty$. If it does not hold the motion gets faster and faster as N increases [36] and a limiting dynamics exists only after rescaling the time. Thus one can hope for an automorphism of $\mathcal{A}$ only for stable potentials. Unfortunately, the situation is worse because of the following argument. $K(\alpha_f) = \alpha_{\bar{f}}^*$ is an (anti)automorphism of $\mathcal{A}$, $K^2 = \mathbf{1}$. $K\tau^t K$ would again be an automorphism of $\mathcal{A}$ generated by KHK. Since $\alpha(x)$ is not bounded we should express H by the a_f's otherwise the time derivative $i[H, a]$ will lead out of $\mathcal{A}$. But regularizing H somehow in the form:

$$T = \sum_f a^*_{\nabla f} a_{\nabla f}, \qquad V = \sum_{f,g} a^*_f a_f v(f, g) a^*_g a_g \tag{3.25}$$

we see $KTK = -T + c_1$, $KVK = V + c_2 N + c_3$, $c_i \in \mathbb{R}$. Removing the regularization the constants c_i become infinite but they have no influence on the time development of gauge invariant quantities. For the existence of the N-independent dynamics we need either $H_N > -cN$ or $H_N < cN$. Since $0 < T \not< cN$ we need therefore

$$T + V > -cN, \qquad T - V > -cN$$

that is

$$-cN - T < V < cN + T. \tag{3.26}$$

In general, there is no such operator inequality since $V \sim N^2$. For fermions there is a way out by observing that the particle density in configuration space $\alpha^*(x)\alpha(x)$ is unbounded in spite of the exclusion principle because by going to high momenta we can pack arbitrarily many fermions in any volume. However, the fermion density in phase space $\rho(z)$ is bounded by 1 (in units $\hbar = 1$) if

$$z = (x, p) \in \mathbb{R}^6, \qquad \rho(z) = a^*_z a_z, \qquad a_z = \int \frac{d^3x}{\pi^{3/4}} e^{ipy} e^{-(x-y)^2/2} \alpha(y).$$

We have $0 < \rho(z) < 1$ and if we use a velocity dependent potential which cuts off high momenta, then (3.26) should be satisfied. Indeed we have

3.4.1 Proposition

The potential $V = \int d^6z d^6z' a^*_z a^*_{z'} v(z-z') a_{z'} a_z$ satisfies $-\|v\|_1 N \leq V \leq \|v\|_1 N$, $\|v\|_1 = \int d^6z |v(z)|$.

Proof: We have the operator inequality

$$a^*_z a^*_{z'} a_{z'} a_z \leq a^*_z a_z \|a^*_{z'} a_{z'}\| = a^*_z a_z$$

and thus

$$V \leq \int dz dz' |v(z - z')| a^*_z a_z = \|v\|_1 N$$

and similarly the left inequality.

Once the obstacle of instability is removed we get indeed an automorphism of the CAR algebra. This is stated by the next theorem, the proof of which is based on expansion in the coupling constant and clearly too involved to be given here [37].

3.4.2 Theorem

$H = T + V$, v from (3.4.1) with $\|v\|_1 < \infty$ defines by $\tau^t a_f = e^{iHt} a_f e^{-iHt}$ a one parameter group of automorphisms of the CAR algebra which is norm continuous and Galilei invariant.

3.4.3 Remarks

1. Norm continuous means $\lim_{t\to 0} \|\tau^t a - a\| = 0\, \forall a \in \mathcal{A}$. There is no uniformity, $\lim_{t\to 0} \sup_{a\in\mathcal{A}, \|a\|=1} \|\tau^t a - a\| = 0$ would imply bounded H.
2. Galilei invariant means that together with the shift σ: $\sigma^x a_f = a_{f_x}$, $f_x(y) = f(x+y)$, the boost γ: $\gamma^p a_f = a_{e^{ipx} f}$, and the gauge transformation (first kind) ν: $\nu^\alpha a_f = e^{i\alpha} a_f$ the time evolution τ satisfies
$$\sigma^x \circ \nu^\alpha = \nu^\alpha \circ \sigma^x, \gamma^p \circ \nu^\alpha = \nu^\alpha \circ \gamma^p, \gamma^p \circ \sigma^x = \sigma^x \circ \gamma^p \circ \nu^{-px}$$
$$\tau^t \circ \nu^\alpha = \nu^\alpha \circ \tau^t, \tau^t \circ \sigma^x = \sigma^x \circ \tau^t, \tau^t \circ \gamma^p = \tau^t \circ \sigma^{-pt} \circ \nu^{-p^2 t/2}.$$
This expresses that τ, σ, γ, ν give a realization of the central extension of the Galilei group by norm continuous automorphisms of the CAR algebra.
3. A local potential means v is p-independent in which case $\|v\|_1 = \infty$ and (3.4.2) does not apply. However, there is no limit on the momentum cut-off and one can hope that for stable systems where high momenta do not occur in a reasonable subset of states, then (3.4.2) gives a physically acceptable description of the time evolution.

When we develop the general theory in the next subsection we shall always have this model in mind since it exists mathematically and is relevant for physics.

Topological Dynamics

The knowledge of an observer about the system is contained in the state. Thus the properties of $(\mathcal{A}, \tau)$ are objective in the sense that they do not depend on any observer. On the other hand, $(\mathcal{A}, \tau, \omega)$ reflects how the situation appears to a particular observer. Similar elements from $\Pi_\omega(\mathcal{A})'' \setminus \Pi_\omega(\mathcal{A})$ are extrapolations from some observer and are different for different ω [38]. Thus we shall first study properties of $(\mathcal{A}, \tau)$ alone and in the next section the additional information contained in invariant states. It turns out that there is a close interplay between these points of view. From some states we can draw conclusions about $(\mathcal{A}, \tau)$ and sufficient algebraic structure of $(\mathcal{A}, \tau)$ implies some feature of all invariant states.

3.4.4 Proposition

Between the properties of a dynamical system $(\mathcal{A}, \tau)$

(i) all τ-invariant elements of $\mathcal{A}$ are $\sim \mathbf{1}$,

(ii) all τ-quasiperiodic elements of $\mathcal{A}$ are $\sim \mathbf{1}$,

(iii) $(\mathcal{A}, \tau)$ is a K-system,

(iv) $\mathcal{A}$ has only trivial τ-invariant closed subalgebras,

there are the implications

$$\begin{array}{ccc} \text{(iv)} & \Longrightarrow & \text{(i)} \\ & & \Uparrow \\ \text{(iii)} & \Longrightarrow & \text{(ii)} \end{array}$$

3.4.5 Explanations

ad (i) This is usually called ergodicity and is too weak a property since it does not imply mixing. Nevertheless it excludes systems where τ is an inner automorphism (i.e. $\tau(a) = U^{-1}aU, U \in \mathcal{A}$) and in particular finite quantum systems (where $U = e^{iH}$, $\forall H \in \mathcal{A}$). It also excludes finite τ-invariant subalgebras $\mathcal{A}_0$ since τ restricted to $\mathcal{A}_0$ would be an automorphism of $\mathcal{A}_0$ and they always have invariant elements.

ad (ii) Quasiperiodicity for an element $a \in \mathcal{A}$ means $\forall \varepsilon > 0$, $N \in \mathbb{Z}^+ \; \exists |n| > N$ with $\|\tau^n a - a\| < \varepsilon$. This property is of interest in connection with Poincaré's recurrence theorem which says that classically almost all orbits keep coming arbitrarily close to their origin. Nevertheless classical systems may have property (ii) since the observables are smooth functions on phase space and they may never regain their original form.

ad (iii) K-system [39] means that $\mathcal{A}$ has a C^*-subalgebra $\mathcal{A}_0$ with the properties
(a) $\tau^n \mathcal{A}_0 \supset \mathcal{A}_0 \; \forall n \in \mathbb{Z}^+$
(b) $\bigvee_{n=0}^{\infty} \tau^n \mathcal{A}_0 = \mathcal{A}$ ($\bigvee$ means norm completion of algebra generated by the $\tau^n \mathcal{A}_0$)
(c) $\bigwedge_{n=0}^{\infty} \tau^{-n} \mathcal{A}_0 = c \cdot \mathbf{1}$ ($\bigwedge$ is the intersection).

Rephrased it means that τ^{-1} gives an isomorphism between $\mathcal{A}_1 := \tau \mathcal{A}_0$ and its proper subalgebra $\mathcal{A}_0$ but this isomorphism has no proper invariant subalgebra (this would remain in the "tail" $\bigwedge_{n=0}^{\infty} \tau^{-n} \mathcal{A}_0$). (b) might be considered as definition of $\mathcal{A}$ such that the isomorphism $\tau : \mathcal{A}_n \to \mathcal{A}_{n+1}$ extends to an automorphism $\mathcal{A} \to \mathcal{A}$.

ad (iv) This is equivalent to requiring that $\forall z \cdot \mathbf{1} \neq a \in \mathcal{A}$ the C^*-algebra generated by $\{\tau^n a, n \in \mathbb{Z}\}$ is all of $\mathcal{A}$. For our purposes it is too strong a condition since the particle number conserving time evolutions (3.4.10) of the CAR algebra leave the subalgebras

$$\mathcal{A}_{n_0} = \{z + \sum_{n > n_0} a^*(f_1)a^*(f_2)\ldots a^*(f_n)a(g_1)\ldots a(g_n)\}$$

$\forall n_0 \in \mathbb{Z}^+$ invariant. (iv) is not related to (iii): the tensor product of two K-systems $(\mathcal{A}, \tau_1)$ and $(\mathcal{A}, \tau_2)$ can be endowed naturally with a K-structure but $(\mathcal{A}_1 \otimes \mathcal{A}_2, \tau_1 \otimes \tau_2)$ has $(\mathcal{A}_1 \otimes \mathbf{1})$ and $(\mathbf{1} \otimes \mathcal{A}_2)$ as invariant subalgebras, thus (iii) $\not\Rightarrow$ (iv). Furthermore (iv) is too weak to imply (ii) since there are trivial examples of periodic systems with only trivial subalgebras.

Proof of 3.4.4:
(ii) $\Rightarrow$ (i). Invariant elements are quasiperiodic.
(iv) $\Rightarrow$ (i). The invariant elements form a closed invariant subalgebra.
(iii) $\Rightarrow$ (ii). Let b be a quasiperiodic element. Because of property (b) $\forall \varepsilon > 0$ we can find an $n(\varepsilon) \in \mathbb{Z}^+$ with $\inf_{a \in \mathcal{A}_{n(\varepsilon)}} \|b - a\| < \varepsilon$, $\mathcal{A}_n = \tau^n \mathcal{A}_0$. As the norm is invariant under τ we have $\forall N \in \mathbb{Z}^+$ $\inf_{a \in \mathcal{A}_{n(\varepsilon)-N}} \|\tau^{-N} b - a\| < \varepsilon$ and because b is quasiperiodic $\inf_{a \in \mathcal{A}_{n(\varepsilon)-m}} \|b - a\| \leq 2\varepsilon$ for some $m > N$ and thus $\forall N \in \mathbb{Z}^+$ $\inf_{a \in \mathcal{A}_{n(\varepsilon)-N}} \|b - a\| \leq 2\varepsilon$. This means $\forall \varepsilon > 0$, $n \in \mathbb{Z}$, $\inf_{a \in \mathcal{A}_n} \|b - a\| < \varepsilon$ and since the $\mathcal{A}_n$ are norm-closed $b \in \mathcal{A}_n$ $\forall n \in \mathbb{Z}$. Thus the property (c) requires $b = c \cdot \mathbf{1}$, $c \in C$.

In classical topological dynamics there is the important notion of topologically mixing [40] which we generalize to the noncommutative case as follows [41,42,43,44]:

3.4.6 Definition

A dynamical system $(\mathcal{A}, \tau)$ is called

(i) weakly mixing if $\forall a, b \in \mathcal{A}$ $\exists N, \varepsilon > 0$, such that $\|a\tau^n b\| > \varepsilon \|a\| \|b\|$ $\forall n > N$,

(ii) strongly mixing if $\lim_{n \to \infty} \|a\tau^n b\| = \|a\| \|b\|$.

3.4.7 Remarks

1. The intuitive meaning of mixing is visualized best in the classical case where a and b are continuous functions on phase space. Even if they have disjoint supports such that $ab = 0$ if one is time translated their supports will eventually overlap. (i) says that b is dispersed so finely that eventually its support keeps overlapping with the one of a forever. (ii) says that even the part close to the maximum of b will eventually meet the maximum of a and keep overlapping it.
2. Classically (i) and (ii) are equivalent since (i) must hold for all functions supported near the maxima of a and b. Obviously (ii) $\Rightarrow$ (i) but whether generally (i) $\Rightarrow$ (ii) is an open question.
3. Since $\|a\tau^n b\| = \|b^* \tau^{-n} a^*\|$ there is no distinction between $n \to \infty$ and $n \to -\infty$.

3.4.8 Proposition

3.4.6.(i) $\Rightarrow$ 3.4.4.(ii).

Proof: Suppose b is a quasiperiodic element $\neq c \cdot \mathbf{1}$. Then either $B = b + b^*$ or $B = i(b - b^*)$ has at least two different spectral values and is hermitian and also quasiperiodic. Suppose $f_\pm$ are two smooth functions with disjoint support having their maxima at the two spectral values, respectively. Then $f_\pm(B)$ are $\neq 0$ and $f_+(B)f_-(B) = 0$. Suppose there were some ε with $\|f_+(B)\tau^n f_-(B)\| > \varepsilon\|f_+\|\|f_-\|\ \forall n > N$. Since $f_- f_-^*$ is quasiperiodic there is some $m > N$ with $\|\tau^m f_- f_-^* - f_- f_-^*\| \leq \varepsilon^2\|f_-\|^2/4$ and we have

$$\|f_+\tau^m f_-\| = \|f_+(\tau^m(f_-f_-^*) - f_-f_-^*)f_+^*\|^{1/2} \leq \frac{\varepsilon}{2}\|f_+\|\|f_-\|$$

contradicting our assumption.

Representations by Invariant States:

Each invariant state ω gives a GNS representation Π_ω of $(\mathcal{A}, \tau)$ in a Hilbert space $\mathcal{H}_\omega = \overline{\Pi_\omega(\mathcal{A})|\Omega\rangle}$ where τ is unitarily implemented by U, $U\Pi_\omega(a)|\Omega\rangle = \Pi_\omega(\tau a)|\Omega\rangle$ [21]. The ergodic properties of $(\mathcal{A}, \tau)$ have their counterpart in properties for ω. The following proposition reminds the reader that in a pure quantum situation automatically one gets mixing.

3.4.9 Proposition

Let ω be a τ-invariant state of $(\mathcal{A}, \tau)$ (i.e. $\omega \circ \tau = \omega$) such that ω is faithful on $\Pi_\omega(\mathcal{A})''$ (i.e. $\langle\Omega|a^*a|\Omega\rangle = 0 \Leftrightarrow a = 0\ \forall a \in \Pi_\omega(\mathcal{A})''$) and $\mathcal{Z}_\omega = \Pi''_\omega(\mathcal{A}) \cap \Pi_\omega(\mathcal{A})' = c \cdot \mathbf{1}$.

Then the following properties are equivalent:

(i) *ω is mixing,*

(ii) $w - \lim_{n\to\infty} U^n = |\Omega\rangle\langle\Omega|$.

(iii) $w - \lim_{n\to\infty} \Pi_\omega(\tau^n a) = \omega(a) \cdot \mathbf{1} \forall a \in \Pi_\omega(\mathcal{A})$,

(iv) *$(\mathcal{A}, \tau, \omega)$ is weakly asymptotic abelian.*

3.4.10 Remarks

1. Faithful states are dense and so this requirement does not leave us with exceptions only. There are examples of states faithful on $\Pi_\omega(\mathcal{A})$ and not on $\Pi_\omega(\mathcal{A})''$ but they do not seem to be important in physics. The advantage of faithfulness is that they supply the modular automorphism σ_t^ω which satisfies the KMS condition $\langle\Omega|ab|\Omega\rangle = \langle\Omega|b\sigma_i^\omega(a)|\Omega\rangle$. If ω is τ-invariant σ^ω and τ commute.
2. The center $\mathcal{Z}_\omega$ is the classical part of $\Pi_\omega(\mathcal{A})''$ and may be nontrivial even if $\mathcal{A}$ is simple and therefore its center is trivial. Our requirement means that Π_ω is a factor representation, there the classical observables have fixed values.

3.4.11 Explanations

ad (i) Mixing means $\lim_{n\to\infty}\omega(a\tau^n(b)c) = \omega(ac)\omega(b)$. Since ω is τ-invariant the limits $n \to \infty$ and $n \to -\infty$ are equivalent:

$$\omega(a\tau^n(b)c) = \omega(b\tau^{-n}(c\sigma_i^\omega(a))) = \omega(d^{-1}b\tau^{-n}(c\sigma_i^\omega(a))\sigma_i^\omega d).$$

ad (ii) w-lim means weak limit in the Hilbert space $\overline{\Pi_\omega(\mathcal{A})|\Omega\rangle}$. Strong limit is impossible since strong limits of unitaries must be isometries.

ad (iii) Again strong limits are impossible:

$$\begin{aligned} 0 &= \text{s-}\lim \Pi_\omega(\tau^n(a^* - \omega(a^*))) \cdot \text{s-}\lim \Pi_\omega(\tau^n(a - \omega(a))) \\ &= \text{s-}\lim \Pi_\omega(\tau^n(a^* - \omega(a^*))(a - \omega(a)) = \omega((a^* - \omega(a))(a - \omega(a)). \end{aligned}$$

Faithfulness requires $a = \omega(a)$ and thus strong convergence holds only for multiples of $\mathbf{1}$ (which are invariant under τ).

ad (iv) Weak asymptotic abelianness means

$$\lim_{n\to\infty} \omega(a[b, \tau^n c]d) = 0 \quad \forall a, b, c, d \in \mathcal{A}.$$

It is not possible for inner automorphisms $\tau(a) = U^{-1}aU$ since $\tau(U) = U$ and thus the condition would make $[b, U] = 0\ \forall b \Leftrightarrow \tau = id$.

Proof of 3.4.9 See 3.1. □

3.4.12 Proposition

If **3.4.9** holds $(\mathcal{A}, \tau)$ is weakly mixing.

Proof: $\exists l$ such that $\forall n > N$

$$\begin{aligned} \|a\tau^n(b)\| &= \|a\tau^n(bb^*)a^*\|^{1/2} \geq (\omega(a\tau^n(bb^*)a^*))^{1/2} \\ &\to \omega(aa^*)^{1/2}\omega(bb^*)^{1/2} > l > 0 \end{aligned}$$

since ω is faithful.

In some cases pure states are more convenient and even if they are not faithful $\Pi_\omega(\mathcal{A})$ will be if $\mathcal{A}$ is simple. In this case a strengthening of **3.4.9.(i)** gives an even stronger result.

3.4.13 Proposition

If $\mathcal{A}$ is simple and Π_ω is hyperclustering, $(\mathcal{A}, \tau)$ is strongly mixing.

3.4.14 Explanations

1. Simplicity implies that the center of $\Pi_\omega(\mathcal{A})$ is trivial because a nontrivial twosided ideal can be constructed with nontrivial elements of the center. This does not yet mean that $\Pi_\omega(\mathcal{A})''$ has trivial center.

2. Hyperclustering means
$$\omega(a\tau^n(b)d\tau^n(c)) \to \omega(ad)\omega(bc).$$
It implies $\omega(c[a, \tau^n b][a^*, \tau^n b^*]d) \to 0$ and thus strengthens (3.1,iii) to
$$\text{s-}\lim_{n\to\infty} [\Pi_\omega(a), \Pi_\omega(\tau^n b)] = 0$$
(strong asymptotic abelianness). Together with mixing this implies conversely hyperclustering.

Proof of 3.4.13: For simplicity we write $\tau^n a = a_n$, etc. Consider the operator inequality
$$cd_n ab_n b_n^* a^* d_n^* c^* \le \|ab_n\|^2 cd_n d_n^* c^*$$
which gives us
$$\frac{\omega(cd_n ab_n b_n^* a^* d_n^* c^*)}{\omega(cd_n d_n^* c^*)} \le \|ab_n\|^2.$$
Let n tend to infinity and remember that commutators go strongly to zero. Thus the left hand side tends to
$$\frac{\omega(caa^*c^*)}{\omega(cc^*)}\frac{\omega(dbb^*d^*)}{\omega(dd^*)}.$$
Now take the sup over d and c and take into account that faithfulness of Π_ω means
$$\sup_{c\in\mathcal{A}} \frac{\omega(caa^*c^*)}{\omega(cc^*)} = \|a\|^2$$
thus $\lim_{n\to\infty} \|ab_n\| \ge \|a\|\|b\|$, but generally $\|ab_n\| \le \|a\|\|b\|$ and thus
$$\lim_{n\to\infty} \|a\tau^n b\| = \|a\|\|b\|.$$

Corollary

Let $\mathcal{A}$ be an UHF algebra and $(\mathcal{A}, \tau)$ norm asymptotic abelian, then $(\mathcal{A}, \tau)$ is strongly mixing.

Proof: The unique tracial state ϕ_0 is invariant under all automorphisms α ($\phi_0 \circ \alpha$ would be another tracial state and thus equals ϕ_0). It is known to be faithful over $\Pi_{\phi_0}(\mathcal{A})''$ which is a type II_1 factor and has trivial center. Thus by **3.4.4** ϕ_0 is mixing and even hyperclustering since it is norm asymptotic abelian. Thus **3.4.13** applies.

3.4.15 Corollary

Let τ of the CAR algebra $\mathcal{A}$ be Galilei invariant. Then $(\mathcal{A}, \tau)$ is weakly mixing.

Proof: The tracial state ϕ_0 qualifies for **3.4.4** and from the representation theory of the Galilei group one knows that apart from the invariant vector the U_t which generates the time translation has absolutely continuous spectrum [44]. One knows that $|\Omega\rangle$ is the only translation invariant vector in $\mathcal{H}_{\phi_0}$ and therefore the only Galilei invariant vector. Thus $U_n \rightharpoonup |\Omega\rangle\langle\Omega|$ and the conditions of **3.4.4** are satisfied. □

3.4.16 Remarks

1. Even without appeal to the representation theory of the Galilei group one can show that ϕ_0 is mixing using only its multiplication law [45]. One uses the fact that in faster and faster moving Galilei frames the time translation looks more and more like a space translation and the latter is mixing.
2. For the Galilei invariant interactions mentioned in (3.4.2) one can show strong asymptotic abelianness in the Fock representation [44,46]. Thus (3.4.14.2) tells us that $(\mathcal{A}, \tau)$ is even strongly mixing.

So far we have used properties of particular invariant states to deduce state independent features of $(\mathcal{A}, \tau)$. They do not guarantee ergodic properties of other τ-invariant states. However, with stronger algebraic structures of $(\mathcal{A}, \tau)$ one can make statements for all invariant states [47,48].

3.4.17 Proposition

For a K-system all faithful invariant states are mixing.

Proof: Denote by $P_n \in \Pi_\omega(\mathcal{A}_n)'$ the orthogonal projector projecting onto $\overline{\mathcal{A}_n|\Omega\rangle}$. It may happen, though $\mathcal{A}_n$ is a closed proper subalgebra of $\mathcal{A}$, that $P_n = 1$. In this case the proof requires a more refined argument [49]. We shall here consider only the case where $P_{n+1} \supset P_n$ is a proper inclusion. Since weak convergence of projections to projections implies strong convergence the K-properties **3.4.5.(iii)** are equivalent to

(a) $U_{-n'} P_n U_{n'} = P_{n+n'} \supset P_n \; \forall n' \in N, n \in \mathbb{Z}$,

(b) $\text{s-lim}_{n\to\infty} P_n = \mathbf{1}$,

(c) $\text{s-lim}_{n\to\infty} P_n = |\Omega\rangle\langle\Omega|$.

Thus, if $b \in \mathcal{A}_0$ and using $U_n|\Omega\rangle = |\Omega\rangle$, we have

$$\begin{aligned}
|\omega(a\tau^n b) - \omega(a)\omega(b)| &= |\langle\Omega|aU_{-n}(P_0 - |\Omega\rangle\langle\Omega|)b|\Omega\rangle| \\
&= |\langle\Omega|a(P_n - |\Omega\rangle\langle\Omega|)U_{-n}b|\Omega\rangle| \\
&\le \|\langle\Omega|a(P_n - |\Omega\rangle\langle\Omega|)\| \cdot \|b|\Omega\rangle\|
\end{aligned}$$

and consequently (c) tells us $\forall \varepsilon \exists N$ such that

$$|\omega(a\tau^n b) - \omega(a)\omega(b)| < \varepsilon\omega(b^* b) \quad \forall n < -N,\ b \in \mathcal{A}_0.$$

Clearly the same holds for b from any $\mathcal{A}_n$ and since the $\mathcal{A}_n$ are norm dense in $\mathcal{A}$ we have $\forall a, b \in \mathcal{A}, \varepsilon \exists N$ such that

$$|\omega(a\tau^n b) - \omega(a)\omega(b)| < \varepsilon \qquad \forall n < -N.$$

Faithfulness comes in when we want to treat $n \to \infty$. Since $\omega(a\tau^n b) = \omega(\tau^{-n}(a)b) = \omega(\sigma^\omega_{-i}(b)\tau^{-n}a)$ and the image of σ^ω_{-i} is norm dense in $\mathcal{A}$ we reach the same conclusion for $n > N$.

3.4.18 Corollary

If $\mathcal{A}$ is UHF and $(\mathcal{A}, \tau)$ a K-system, then for the tracial state all conditions **3.4.4** *are satisfied and $(\mathcal{A}, \tau)$ is weakly mixing.*

3.4.19 Corollary

If $\mathcal{A}$ is UHF, then **3.4.4** *can be sharpened to: $(\mathcal{A}, \tau)$ is a K-system $\Rightarrow$ $(\mathcal{A}, \tau)$ is weakly mixing $\Rightarrow$ all quasiperiodic elements of $\mathcal{A}$ are $\sim$* **1**, *$\Rightarrow$ all invariant elements of $\mathcal{A}$ are $\sim$* **1**.

3.4.20 Remark

K-systems are not only mixing but even K-mixing which is the maximally possible uniformity in mixing. Wheras complete uniformity in clustering in the sense that $\forall a \in \mathcal{A}, \|a\| = 1, \varepsilon > 0 \exists N$ such that $|\omega(a\tau^n b) - \omega(a)\omega(b)| < \varepsilon\|b\|\ \forall b \in \mathcal{A}$, $n > N$, is for faithful ω impossible (take $b = \tau^{-n}a^*$) we found in the proof **3.4.17** $\forall a \in \mathcal{A}, m \in \mathbb{Z}, \varepsilon > 0 \exists N$ such that

$$|\omega(a\tau^n b) - \omega(a)\omega(b)| < \varepsilon\|b\| \quad \forall b \in \mathcal{A}_m,\ n < -N.$$

Since the $\mathcal{A}_m$ are dense that is the best one can hope for. It is related to the strong convergence of the P_n and is equivalent to saying that all states when restricted to one of the "strictly local" algebras $\mathcal{A}_m$ converge for $n \to \infty$ strongly to the equilibrium ω. Such states φ can be written $\varphi(b) = \omega(ab)$, $a \in \mathcal{A}$, $\omega(a) = \mathbf{1}$, and we have

3.4.21 Proposition

In the Schrödinger representation $\forall \varepsilon > 0$, $m \in \mathbf{Z}^+$ $\exists N \in \mathbb{Z}^+$ such that

$$\|\varphi_{n|\mathcal{A}_m} - \omega_{|\mathcal{A}_m}\| = \sup_{b \in \mathcal{A}_m, \|b\|=1} |\omega(a\tau^n b) - \omega(b)| < \varepsilon \quad \forall n < -N.$$

The Increase of Entropy with Time

One of the key formulas of quantum statistical mechanics is von Neumann's expression for the entropy of a state over $\mathcal{B}(\mathcal{H})$ given by a density matrix ρ

$$S = -\operatorname{Tr} \rho \ln \rho$$

This relation has been generalized for states over arbitrary von Neumann algebras $\mathcal{A}$ [50] but one always meets with the same difficulties when one discusses the question of entropy increase.

1. All these expressions have to be invariant under automorphisms and therefore cannot change with time. Furthermore, for infinite system S is infinite.
2. The entropy of the state restricted to a subalgebra $\mathcal{A}_0$, $\tau\mathcal{A}_0 \neq \mathcal{A}_0$ may change with time but it may increase or decrease. Furthermore, if the state is time invariant – $\omega \circ \tau = \omega$ – even for a subalgebra the entropy does not change:

$$S(\omega_{|\tau\mathcal{A}_0}) = S(\omega \circ \tau_{|\mathcal{A}_0}) = S(\omega_{|\mathcal{A}_0}).$$

The following two ways around these difficulties have been proposed:

1. Dynamical Entropy. For some infinite algebras $\mathcal{A}_0$ one can give a well defined meaning for the entropy increase per unit time which at face value

$$S(\omega_{|\tau\mathcal{A}_0}) - S(\omega_{|\mathcal{A}_0}) = \infty - \infty = ?$$

If $\mathcal{A}_0 = \bigvee_{n=-\infty}^{0} \tau^n \mathcal{B}_0$, $\mathcal{B}_0 =$ finite, then classically

$$h_\omega(\mathcal{B}_0, \tau) \equiv S(\omega_{|\tau\mathcal{A}_0}) - S(\omega_{|\mathcal{A}_0}) = \lim_{m\to\infty} (S(\omega_{|\bigvee_{n=-m}^{1} \tau^n \mathcal{B}_0}) - S(\omega_{|\bigvee_{n=-m}^{0} \tau^n \mathcal{B}_0})$$

is well defined, ( ) is ≥ 0 and decreasing in m. The dynamical (or Kolmogorov-Sinai [51]) entropy is the maximal increase (per unit time) of the entropy of these algebras

$$h_\omega(\tau) = \sup_{\mathcal{B}_0=\text{finite}} h_\omega(\mathcal{B}_0, \tau). \tag{3.27}$$

In quantum theory the union $\bigvee_{n=-m}^{0} \tau^n \mathcal{B}_0$ may be too big and S is not monotonic so the definition (3.27) is no good. Here one needs a more refined theory which has been elaborated in the past years [52]. One now disposes of a dynamical entropy of an automorphism of an arbitrary C^*-algebra [53] for which almost all the desired properties have been demonstrated. The theory is too extensive to be given here and several other proposals have emerged [54,55].

2. Relative Entropy. We all have learned in thermodynamics that for part of a system which exchanges energy with the rest, it is not the entropy which tends to a maximum but the free energy which tends to a minimum. Again the free energy of an infinite system may be infinite but the difference between energies of a locally disturbed equilibrium ν and the equilibrium state ω may be finite.

The relative entropy $S(\nu|\omega)$ is precisely this difference. For finite quantum systems where states can be described by density matrices it is

$$S(\nu|\omega) = \text{Tr } \nu(\ln \nu - \ln \omega) \qquad (3.28)$$

and if ω is a canonical state (for $\beta = 1$) $\omega = e^{-H+F_\omega}$, F_ω = equilibrium free energy, we found in (2.2.21;4)

$$S(\nu|\mu) = \text{ Tr } \nu(\ln \nu + H - F_\omega) = \langle H\rangle_\nu - S(\nu) - F_\omega \equiv F_\nu - F_\omega.$$

Kosakis formula gives a general definition of $S(\nu|\omega)$ for arbitrary states of C^*-algebras. For infinite systems $S(\nu|\mu)$ is finite if $\nu < c\omega$ and ν is a local perturbation of ω. Again the definition is invariant under automorphisms so that $S(\nu|\omega)$ does not change with time. For subsystems it may change but not necessarily monotonically. We shall see that this monotonicity is a special feature of K-systems.

3.4.22 Theorem

Let $(\mathcal{A}, \mathcal{A}_0, \tau)$ be a K-system, ω an invariant state and φ a normal state for $\Pi_\omega(\mathcal{A})''$ with $\lambda_1\omega \le \varphi \le \lambda_2\omega$, $\lambda_i \in \mathbb{R}^+$. Then for any $m \in \mathbb{Z}$, $\mathcal{A}_m = \tau^m \mathcal{A}_0$,

$$\Delta F(n) = S(\varphi \circ \tau^n_{|\mathcal{A}_m} | \omega_{|\mathcal{A}_m})$$

converges monotonically for $-\infty < n < \infty$ from 0 to $S(\varphi|\omega)$.

3.4.23 Remarks

1. Even in finite dimensions $S(\varphi|\omega)$ is not continuous and becomes infinite where ω is zero but φ greater zero. Thus we need a condition on the states.
2. We have seen that for mixing systems any normal state converges weakly towards equilibrium. However, $\varphi \rightharpoonup \omega$ does not imply $S(\varphi|\omega) \to 0$ since $S(\varphi|\omega) > \frac{1}{2}\|\varphi - \omega\|^2$ [54]. Thus it is only the strong convergence (3.4.27) of the restriction of φ which yields the convergence of the free energy of the subsystems $\mathcal{A}_n$.
3. For K-systems (3.4.4) is strengthened to the statement that it actually approaches its equilibrium value.
4. The thoughtful reader will be perplexed by the fact that ΔF converges away from its equilibrium value 0 to its maximal value $S(\varphi|\omega)$. However, it is not yet said whether τ or τ^{-1} is the real time devlopment. The physical K-systems are actually time reversal invariant in the sense that there is an antiautomorphism K, $K^2 = 1$, with $K\tau K = \tau^{-1}$. Then for $K(\mathcal{A}_0)$ the free energy decreases if it increases for $\mathcal{A}_0$. For time reversal invariant systems there are necessarily as many subsystems with increasing free energy as there are with decreasing free energy.

Proof of 3.4.22:

(i) Monotonicity

$$\varphi \circ \tau^n{}_{|\mathcal{A}_m} = \varphi_{|\mathcal{A}_{m+n}}$$

and for $n < n'$ $\mathcal{A}_{m+n} \subset \mathcal{A}_{m+n'}$. The relative entropy has the monotonicity ((2.2.22) and generally from Kosakis formula)

$$S(\varphi_{|\mathcal{A}}|\omega_{|\mathcal{A}}) \leq S(\varphi_{|\mathcal{B}}|\omega_{|\mathcal{B}}) \quad \text{if } \mathcal{A} \subset \mathcal{B}.$$

(ii) The limit $n \to -\infty$
We have seen in (3.4.21) that

$$\lim_{n \to -\infty} \|\varphi_{|\mathcal{A}_n} - \omega_{|\mathcal{A}_n}\| \to 0$$

and for states satisfying the hypothesis [54] of (3.4.22) this implies

$$S(\varphi_{|\mathcal{A}_n}|\omega_{|\mathcal{A}_n}) \to 0 \quad \text{for } n \to -\infty.$$

(iii) S can be written as sup over an expression linear in φ and ω and therefore weak*-continuous in φ and ω [52]. Thus it is weak*-lower semicontinuous and this implies

$$\lim_{n \to \infty} S(\varphi_{|\mathcal{A}_n}|\omega_{|\mathcal{A}_n}) \geq S(\varphi|\omega).$$

However, monotonicity insures the opposite inequality.

4 Physical Systems

4.1 Thomas–Fermi Theory

Among the best examples of large quantum systems are atoms and molecules with highly charged nuclei. Classical features arise in the limit $Z \to \infty$, $N \to \infty$, except that the Fermi statistics continue to have an important effect.

Matter around us and within us consists of electrons and atomic nuclei, which are governed by the laws of quantum mechanics. Relativistic effects arise only in the fine details (cf. III, § 1), so the forces of primary relevance are electrostatic and, for cosmic bodies, gravistatic (nonrelativistic). Moreover, the precise nature of the atomic nuclei is of little consequence on the macroscopic scale, so they can be considered as point charges. In order to understand the gross features of matter we shall study a Hamiltonian

$$H_{\text{mat}} = \sum_{i=1}^{M} \frac{|\mathbf{p}_i|^2}{2m_i} + \sum_{i>j} \frac{(e_i e_j - k m_i m_j)}{|\mathbf{x}_i - \mathbf{x}_j|} \tag{4.1}$$

for ordinary matter. The first important issue to confront is that of why macroscopic bodies behave classically; in what sense is the thermodynamic limit $N \to \infty$ equivalent to the classical limit $h \to 0$? There are a variety of ways to pass to the limit $N \to \infty$. In this section we begin by letting the nuclear charge Z and the nuclear masses both tend to infinity, while continuing to neglect gravity. This will permit a rather explicit mathematical treatment, as the action is determined by an average field, and the single-particle model becomes exact. The same will be true in § 4.2 when we deal with cosmic bodies, for which gravitation predominates. However, macroscopic bodies on the scale of humans are far from these limits: nuclear charges are for the most part small, and yet gravitation is of little importance. In this intermediate range of normal matter it would be too much to hope for an explicit solution. Section 4.3 will discuss this case, but the results will be confined to general existence theorems and rather crude bounds on the values of observables of physical interest.

Let us consider now what happens to electrons in the field of fixed point charges. In order not to be distracted from the most important facts by physical constants, we shall use units in which $\hbar = 2m = e = k = 1$, so that (4.1.1) becomes

4.1.1 The Hamiltonian for Normal Matter

$$H_N = \sum_{i=1}^{N} |\mathbf{p}_i|^2 - \sum_{i=1}^{N}\sum_{k=1}^{N} \frac{Z_k}{|\mathbf{x}_i - \mathbf{X}_k|} + \sum_{i>j} \frac{1}{|\mathbf{x}_i - \mathbf{x}_j|} + \sum_{k>l} \frac{Z_k Z_l}{|\mathbf{X}_k - \mathbf{X}_l|} + \sum_{i=1}^{N} W(\mathbf{x}_i).$$

4.1.2 Remarks

1. The notation follows that of (III:4.6.6), that is, $\mathbf{x}_i$ and $\mathbf{p}_i$ are the position and momentum of the ith electron, $\mathbf{X}_k$ and Z_k are the position and charge of the kth fixed nucleus, N is the number of electrons, and M is the number of nuclei.
2. The Hamiltonian H operates on an n-fold antisymmetrized tensor product of $L^2(\mathbb{R}^3) \otimes \mathbb{C}^2$ = configuration space ø spin of a given electron. The nuclear coordinates $\mathbf{X}_i$ commute with everything, and are to be regarded as ordinary 3-vectors of numbers.
3. It is usually most convenient to study the many-particle system in the framework of the field algebra (1.3.2). If $a_\alpha(\mathbf{x})$, $\alpha = 1, 2$, denote the annihilation operators of electrons with spin up ($\alpha = 1$) and spin down ($\alpha = 2$), then (4.1.1) reads

$$H = \sum_{\alpha} \int d^3x \big[\nabla a_\alpha^*(\mathbf{x}) \cdot \nabla a_\alpha(\mathbf{x}) + (\sum_{k=1}^{M} \frac{-Z_k}{|\mathbf{x} - \mathbf{X}_k|} + W(\mathbf{x}))a_\alpha^*(\mathbf{x})a_\alpha(\mathbf{x})\big] + \sum_{\alpha,\beta} \frac{1}{2} \int d^3x\, d^3x' \frac{a_\alpha^*(\mathbf{x})a_\beta^*(\mathbf{x}')a_\beta(\mathbf{x}')a_\alpha(\mathbf{x})}{|\mathbf{x} - \mathbf{x}'|} + \sum_{k>l} \frac{Z_k Z_l}{|\mathbf{X}_k - \mathbf{X}_l|}.$$

4. If the temperature is finite, then the attraction of the nuclei is not strong enough to prevent the electrons from escaping to infinity, and the system must be imagined confined to a box. The box can be represented by a potential W, adding a term $\sum_\alpha \int d^3x\, W(\mathbf{x})a_\alpha^*(\mathbf{x})a_\alpha(\mathbf{x})$ to H. The wall potential W will be chosen to be the v_L of (2.5.17).

Most interesting systems are approximately neutral, so N is assumed to be about $\sum_{k=1}^{M} Z_k$. The thermodynamic limit $N \to \infty$ can consist either in $M \to \infty$ or $Z_k \to \infty$. For the moment consider the latter case; the limit $M \to \infty$ will be studied in § 4.3. The first step is to bound the grand canonical partition function in terms of the grand canonical partition function of a theory with free electrons in an external field. This means that the bounds of (III, § 4.5) for the energies have to be generalized for arbitrarily complex systems at nonzero temperatures. After that we shall show that the upper and lower bounds coalesce (when properly scaled) as $Z_k \to \infty$, so the partition function can be calculated exactly in the thermodynamic limit. Finally, the limit of the grand canonical state will be analyzed.

4.1.3 Upper Bounds for the Partition Function

These correspond to lower bounds for the Hamiltonian like those derived in (III:4.5.10). The inequality (III:4.5.4), though, is not well suited to our current purposes, and must be replaced with a variant, which will appear as a by-product of Thomas–Fermi theory in (4.1.40;2). In it the Coulomb repulsion of the electrons is replaced by their energy in an external field:

$$\sum_{i>j=1}^{N} |\mathbf{x}_i - \mathbf{x}_j| \geq \sum_{i=1}^{N} \int \frac{d^3x n(\mathbf{x})}{|\mathbf{x}_i - \mathbf{x}|} - \frac{1}{2} \int \frac{d^3x\, d^3x'}{|\mathbf{x} - \mathbf{x}'|} n(\mathbf{x}) n(\mathbf{x}') - 3.68 \gamma N$$

$$- \frac{3}{5\gamma} \int d^3x n^{5/3}(\mathbf{x}), \quad \forall \mathbf{x}_i \in \mathbb{R}^3, \gamma > 0, n \in L^1(\mathbb{R}^3) \cap L^{5/3}(\mathbb{R}^3). (4.2)$$

This yields a bound on the expression in (4.1.2;3), which is quartic in the a's, in terms of a quadratic expression,

$$\frac{1}{2} \sum_{\alpha,\beta} \int \frac{d^3x d^3x'}{|\mathbf{x} - \mathbf{x}'|} a_\alpha^*(\mathbf{x}) a_\beta^*(\mathbf{x}') a_\beta(\mathbf{x}') a_\alpha(\mathbf{x}) \geq \sum_\alpha \int d^3x a_\alpha^*(\mathbf{x}) a_\alpha(\mathbf{x})$$

$$\times \left[\int \frac{d^3x' n(\mathbf{x}')}{|\mathbf{x} - \mathbf{x}'|} - 3.68\gamma \right] - \frac{1}{2} \int \frac{d^3x d^3x'}{|\mathbf{x} - \mathbf{x}'|} n(\mathbf{x}') n(\mathbf{x}) - \frac{3}{5\gamma} \int d^3x n^{5/3}(\mathbf{x}).$$

Consequently, H is bounded by a

4.1.4 Hamiltonian with an Effective Field

$$H - \mu N \geq H_n - C_n + \sum_{k>l} \frac{Z_k Z_l}{|\mathbf{X}_k - \mathbf{X}_l|},$$

where

$$H_n \equiv \sum_\alpha \int d^3x \{ \nabla a_\alpha^*(\mathbf{x}) \cdot \nabla a_\alpha(\mathbf{x}) + a_\alpha^*(\mathbf{x}) a_\alpha(\mathbf{x})$$

$$\times \left[- \sum_k \frac{Z_k}{|\mathbf{x} - \mathbf{X}_k|} + \int \frac{d^3x' n(\mathbf{x}')}{|\mathbf{x} - \mathbf{x}'|} + W(\mathbf{x}) - \mu - 3.68\gamma \right] \}$$

$$-\frac{1}{2}\int \frac{d^3xd^3x'}{|\mathbf{x}-\mathbf{x}'|}n(\mathbf{x})n(\mathbf{x}'),$$

and

$$C_n = \frac{3}{5\gamma}\int d^3xn^{5/3}(\mathbf{x}).$$

4.1.5 Remarks

1. Although Inequality (4.1.4) holds for any $n(\mathbf{x})$, the optimal choice identifies it with the electron density.. Thus the effective potential in the square brackets [...] consists of the attraction to the nuclei, the repulsion from other electrons, and the chemical potential. However, this interpretation counts the electron repulsion twice, as in $\sum_{i\neq k}|\mathbf{x}_i-\mathbf{x}_k|^{-1}$. The last term in H_n corrects this overcounting.
2. The correlations among the electrons due to their Fermi statistics have the effect of reducing their repulsion. Also, H_n contains the self-energy of the individual electrons. The constant C_n and $-3.68\gamma N$ serve to control any possible effect from these corrections.
3. In (III:4.5.14) we had a correction term $\sim\int n^2$ which now is improved to $\int n^{5/3}$. The best possible result $\sim\int\rho^{4/3}$ has been derived by Lieb and Oxford [55].

The monotonic property (2.1.4;4) translates (4.1.4) into an inequality for the partition function. Then with the aid of the maximum principle of (2.5.12;2) the inequality can be expressed as the supremum of an expression linear in n.

4.1.6 The Partition Function with an Effective Field

$$\Xi(H-\mu N)\equiv T\ln\operatorname{Tr}\exp[-\beta(H-\mu N)]\leq\Xi\Big(H_n-C_n+\sum_{k>l}\frac{Z_kZ_l}{|\mathbf{X}_k-\mathbf{X}_l|}\Big)$$

$$\leq\Xi(H_n)+C_n-\sum_{k>l}\frac{Z_kZ_l}{|\mathbf{X}_k-\mathbf{X}_l|},$$

$$\Xi(H_n)=\operatorname{tr}2T\ln(\mathbf{1}+\exp(-\beta h_n))+\frac{1}{2}\int d^3xd^3x'\frac{n(\mathbf{x})n(\mathbf{x}')}{|\mathbf{x}-\mathbf{x}'|}$$

$$=\sup_{\rho_1}2\operatorname{tr}\{T(-\rho_1\ln\rho_1-(\mathbf{1}-\rho_1)\ln(\mathbf{1}-\rho_1))-\rho_1h_n\}$$

$$+\frac{1}{2}\int d^3xd^3x'\frac{n(\mathbf{x})n(\mathbf{x}')}{|\mathbf{x}-\mathbf{x}'|},$$

$$h_n=|\mathbf{p}|^2-\sum_k\frac{Z_k.}{|\mathbf{x}-\mathbf{X}_k|}+\int\frac{d^3xn(\mathbf{x}')}{|\mathbf{x}-\mathbf{x}'|}+W(\mathbf{x})-\mu-3.68\gamma.$$

4.1.7 Remarks

1. The Hamiltonian h_n of one particle in the effective field acts on the space $\mathcal{H}_1 = L^2(\mathbb{R}^3)$. Spin is accounted for by the factor 2, and tr denotes the trace on $\mathcal{H}_1$.
2. As in Remark (2.5.12;2), $\sup_{\rho_1}$ denotes the supremum over one-particle density matrices ρ_1 such that

$$0 \leq \rho_1 \leq 1, \quad 2 \operatorname{tr} \rho_1 = N \equiv \Big\langle \sum_\alpha \int d^3 x a_\alpha^*(\mathbf{x}) a_\alpha(\mathbf{x}) \Big\rangle.$$

3. There exist $c_i > 0$ such that $h_n \geq c_1 |\mathbf{p}|^2 + W(\mathbf{x}) - c_2$. This ensures that $\operatorname{tr} \ln(\mathbf{1} + \exp(-\beta h_n)) < \infty$.

The next task is to optimize the upper bound. The infimum over n of $\Xi(H_n)$ is in fact achieved. This is a consequence of the

4.1.8 Properties of the Functional $\Xi(H_n)$

The mapping $n \to \Xi(H_n)$ from $\mathcal{N}$ to $\mathbb{R}^+$, where $\mathcal{N}$ is the real Hilbert space of measurable functions $\mathbb{R}^3 \to \mathbb{R}$ finite in the norm

$$\|n\|_c^2 = \langle n|n\rangle_c = \int \frac{d^3 x d^3 x' n(\mathbf{x}) n(\mathbf{x}')}{|\mathbf{x} - \mathbf{x}'|}$$

is

(i) *weakly lower semicontinuous;*

(ii) *strictly convex; and*

(iii) *greater than $\frac{1}{2}\|n\|_c^2$.*

Proof:

(i) In the second version $\Xi(H_n)$ depends on n through $\operatorname{tr} \rho_1 h_n$ and $\|n\|_c$. The norm is $\sup_{n' \in \mathcal{N}, \|n'\|_c \leq 1} \langle n'|n\rangle_c$, and $\operatorname{tr}(\rho_1 \int n(\mathbf{x}') d^3 x' / |\mathbf{x} - \mathbf{x}'|)$ is weakly continuous for

$$\rho_1 \in C_M \equiv \Big\{\rho_1 : \int \frac{d^3 x_1 d^3 x_2}{|\mathbf{x}_1 - \mathbf{x}_2|} \langle \mathbf{x}_1|\rho_1|\mathbf{x}_1\rangle \langle \mathbf{x}_2|\rho_1|\mathbf{x}_2\rangle < M \in \mathbb{R}^+\Big\}.$$

The supremum is attained when $\rho_1 = (\exp(\beta h_n) + \mathbf{1})^{-1}$, which belongs to some C_M. Hence $\sup_{\rho_1}$ may be written as $\sup_{M \in \mathbb{R}^+} \sup_{\rho_1 \in C_M}$. In this way $\Xi(H_n)$ is expressed as the supremum over continuous functions, which is always lower semicontinuous.

(ii) This follows in the first version of $\Xi(H_n)$, when it is observed that $h \to \operatorname{tr} \ln(\mathbf{1} + \exp(-\beta h))$ is convex, $n \to h_n$ is linear, and $n \to \|n\|_c^2$ is strictly convex.

(iii) This follows in the first form of $\Xi(H_n)$, since $\operatorname{tr} \ln(\mathbf{1} + \exp(-\beta h)) \geq 0$. □

4.1.9 Corollaries

1. Because of Property (iii), the infimum over n lies in a weakly compact region where $\|n\|_c < C$. Property (i) means that it is attained at some n_0, which is unique because of (ii).

2. Because of the convexity, we know that the function $\mathbb{R} \to \mathbb{R}^+ : t \to \Xi(H_{n_0+tn_1})$ has a right derivative everywhere, and the minimum is attained at n_0 if and only if

$$\lim_{t \downarrow 0} t^{-1}(\Xi(H_{n_0+tn_1}) - \Xi(H_{n_0})) \geq 0 \quad \text{for all} \quad n_1 \in \mathcal{N}.$$

Although convexity does not imply the existence of a derivative, analyticity can be proved by a variant of Theorem (2.4.6). Granting that, the formal rules for differentiating tr $\ln(1 + A)$ are justified:

$$\frac{d}{dt} \operatorname{tr} \ln(\mathbf{1} + \exp(-\beta h_{n_0+tn_1}))|_{t=0} = -\operatorname{tr} \int \frac{d^3x' n_1(\mathbf{x}')}{|\mathbf{x} - \mathbf{x}'|} \frac{\beta}{\exp(\beta h_{n_0}) + \mathbf{1}}.$$

Therefore the minimum at n_0 is characterized by

$$\int \frac{d^3x' d^3x}{|\mathbf{x} - \mathbf{x}'|} n_0(\mathbf{x}) n_1(\mathbf{x}') = 2 \operatorname{tr} \int \frac{d^3x' n_1(\mathbf{x}')}{|\mathbf{x} - \mathbf{x}'|} \frac{1}{\exp(\beta h_{n_0}) + \mathbf{1}} \quad \text{for all} n_1 \in \mathcal{N}.$$

If n_1 is made to tend to $\Delta\delta(\mathbf{x} - \mathbf{x}_0)$, then there results an equation for $n_0(\mathbf{x}_0)$. Since the integral kernel $K(\mathbf{x}, \mathbf{x}')$ of $(\exp(\beta h_n) + \mathbf{1})^{-1}$ is analytic for $\mathbf{x}, \mathbf{x}' \neq \mathbf{X}_k$, even though $\Delta\delta$ does not belong to $\mathcal{N}$, we have the

4.1.10 Existence of the Self-Consistent Field

The equation

$$n_0(\mathbf{x}) = 2\langle \mathbf{x} | (\exp(\beta h_{n_0}) + 1)^{-1} | \mathbf{x} \rangle$$

has a unique solution, which minimizes $\Xi(H_n)$.

4.1.11 Remarks

1. Since $2\langle \mathbf{x} | \exp(\beta h_{n_0}) + \mathbf{1})^{-1} | \mathbf{x} \rangle$ equals $\sum_\alpha \langle a_\alpha^*(\mathbf{x}) a_\alpha(\mathbf{x}) \rangle_{n_0}$, it is the mean electron density in the state determined by the one-particle Hamiltonian h_{n_0}.
2. The ease with which the existence of the solution of the generalized Hartree equation (4.1.10) was proved depended on the wall potential W. In an infinite space without W there fails to be a solution when $N > \sum_k Z_k$, even at absolute zero temperature – the electrons escape to infinity, and the infimum is never attained. This is a reflection of the general mathematical fact that a strictly convex function need not achieve its infimum on a noncompact region; for example $1/x$ never reaches the value 0 on $[1, \infty)$.
3. A convex function on a finite-dimensional space is continuous on the interior of its domain of definition. This is not always the case when the dimension

of the space is infinite, and $\|n\|_c^2$ is in fact not weakly continuous: The norms $\|\ \|_c$ of the charge distributions $n_R(\mathbf{x}) = R^{-5/2}\Theta(R - |\mathbf{x}|)$ are all equal, but $\int d^3x n_R(\mathbf{x}) \to 0$ as $R \to 0$. Consequently $\langle n_R|n\rangle_c \to 0$ for all n, if

$$V_n(x) \equiv \int \frac{d^3x' n(\mathbf{x}')}{|\mathbf{x} - \mathbf{x}'|} \in L^\infty(\mathbb{R}^3).$$

Since the n's such that $V_n \in L^\infty$ are dense in $\mathcal{N}$, $n_R \rightharpoonup 0$, even though $\|n_R\|_c \not\to 0$. There even exist convex functions that fail to be lower semicontinuous, for example the functional of (III:2.1.15;2). Of course the function $n \to \|n\|_c^2$ is continuous in norm, but this finer norm topology can not be used, because we need the compactness of bounded sets.

4. At the minimum (4.1.12), it is indeed true that $n(\mathbf{x}) > 0$ and $\int d^3x n(\mathbf{x}) = N$.

4.1.12 Lower Bounds for $\Xi(H)$

In (III, § 4.5) upper bounds on the energy were provided by the min-max principle, the generalization of which for nonzero temperatures is the Peierls–Bogoliubov inequality (2.1.8;3) with $\Xi = -F$. Because

$$\begin{aligned}
&\Big\langle H - \bar{\mu}N - \sum_{k>l} \frac{Z_k Z_l}{|\mathbf{X}_k - \mathbf{X}_l|} - H_{n_0}\Big\rangle_{n_0} \\
&= \sum_{\alpha,\beta} \frac{1}{2} \int \frac{d^3x d^3x'}{|\mathbf{x} - \mathbf{x}'|} \big(\langle a_\alpha^*(\mathbf{x}) a_\beta^*(\mathbf{x}') a_\beta(\mathbf{x}') a_\alpha(\mathbf{x})\rangle_{n_0} - n_0(\mathbf{x}) n_0(\mathbf{x}')\big) \\
&= -\sum_{\alpha,\beta} \frac{1}{2} \int \frac{d^3x d^3x'}{|\mathbf{x} - \mathbf{x}'|} |\langle a_\alpha^*(\mathbf{x}) a_\beta(\mathbf{x}')\rangle_{n_0}|^2 \equiv -A(n_0) < 0,
\end{aligned}$$

where $\bar{\mu} = \mu - 3.68\gamma$, it implies that

$$\Xi(H - \bar{\mu}N) \geq \Xi(H_{n_0}) + A(n_0) - \sum_{k>l} \frac{Z_k Z_l}{|\mathbf{X}_k - \mathbf{X}_l|}.$$

When this is combined with (4.1.6), it yields

4.1.13 Two-Sided Bounds for Ξ

$$\begin{aligned}
0 \leq A(n_0) &\leq \Xi(H - \bar{\mu}N) + \sum_{k>l} \frac{Z_k Z_l}{|\mathbf{X}_k - \mathbf{X}_l|} - \Xi(H_{n_0}) \\
&\leq \frac{3}{5\gamma} \int d^3x n_0^{5/3}(\mathbf{x}) + 3.68\gamma N.
\end{aligned}$$

4.1.14 Remarks

1. This means that the true partition function exceeds the partition function with an effective field by more than A but less than C.
2. In particular (4.1.13) states for the exchange energy that $0 \leq A \leq C$. If Z is large, then n_0 approaches the electron density in Thomas–Fermi Theory, and we shall discover that $\int n^{5/3} \sim Z^{7/3}$. If γ is chosen $\sim (Z^{7/3}/N)^{1/2}$, then C and the additional term $3.68\gamma N$ in μ become $\sim N^{1/2}Z^{7/6}$. Since H goes as $Z^{7/3}$, if $N \sim Z$, then the relative error is $O(Z^{-2/3})$.

4.1.15 The Classical Limit

The next topic of study is the way in which $\Xi(H_n)$ approaches the classical phase-space integral (2.5.13) as $Z \to \infty$. According to the general considerations of (1.2.2) the interesting limit would be expected to be that in which the system shrinks as $Z^{-1/3}$. Consider, therefore, a sequence of Hamiltonians H_Z in which not only do the nuclear charges increase as $Z_k = Zz_k$, $\sum_k z_k = 1$, z_k fixed, but also the nuclear coordinates are scaled by changing $\mathbf{X}_k$ into $Z^{-1/3}\mathbf{X}_k$ and the wall potential varies at the same time:

$$
\begin{aligned}
H_Z \equiv \sum_\alpha \int d^3x \big[&\nabla a_\alpha^*(\mathbf{x}) \cdot \nabla a_\alpha(\mathbf{x}) \\
&+ a_\alpha^*(\mathbf{x}) a_\alpha(\mathbf{x}) \big(-Z \sum_{k=1}^{M} \frac{z_k}{|\mathbf{x} - \mathbf{X}_k Z^{-1/3}|} + Z^{4/3} W(Z^{1/3}\mathbf{x}) \big) \big] \\
&+ \frac{1}{2} \sum_{\alpha,\beta} \int a_\alpha^*(\mathbf{x}) a_\beta^*(\mathbf{x}) a_\beta(\mathbf{x}') a_\alpha(\mathbf{x}') \frac{d^3x d^3x'}{|\mathbf{x} - \mathbf{x}'|} + \sum_{k>l} \frac{z_k z_l}{|\mathbf{X}_k - \mathbf{X}_l|} Z^{7/3}, \\
H_{Z,n} \equiv \sum_\alpha \int d^3x \Big\{ &\nabla a_\alpha^*(\mathbf{x}) \cdot \nabla a_\alpha(\mathbf{x}) \\
&+ a_\alpha^*(\mathbf{x}) a_\alpha(\mathbf{x}) \Big[-Z \sum_{k=1}^{M} \frac{z_k}{|\mathbf{x} - \mathbf{X}_k Z^{-1/3}|} + \int \frac{d^3x' n^Z(\mathbf{x}')}{|\mathbf{x} - \mathbf{x}'|} \\
&+ Z^{4/3}(W(\mathbf{x}Z^{1/3}) - \mu) - 3.68\gamma \Big] \Big\} \\
&- \frac{1}{2} \int \frac{d^3x d^3x'}{|\mathbf{x} - \mathbf{x}'|} n^Z(\mathbf{x}) n^Z(\mathbf{x}'); \\
n^Z = Z^2 n(Z^{1/3}\mathbf{x}).&
\end{aligned}
$$

In order always to work in a fixed volume and see what happens in the limit $Z \to \infty$, use a canonical transformation to convert the electron coordinates $\mathbf{x}$ into $Z^{-1/3}\mathbf{x}$ and $\mathbf{p}$ into $Z^{1/3}\mathbf{p}$ at the same time – this entails $a(\mathbf{x}) \to Z^{-1/2}a(Z^{-1/3}\mathbf{x})$ as well. Since the number of electrons also grows as Z, the mean momentum of the electrons grows as $Z^{2/3}$, and every kind of energy per particle, such as T or μ, will depend in the same way on Z. Thus if we calculate $\mathrm{Tr}\exp[-\beta_Z(H_Z-\mu_Z N)]$ with $\beta_Z = Z^{-4/3}\beta$, and $\mu_Z = Z^{4/3}\mu$, and scale n appropriately, we are led to $\mathrm{tr}\ln(\mathbf{1}+\exp(-\beta h_n))$ with

$$h_n = Z^{-2/3}|\mathbf{p}|^2 - \sum_j \frac{z_j}{|\mathbf{x}-\mathbf{X}_j|} + \int \frac{d^3x' n(\mathbf{x}')}{|\mathbf{x}-\mathbf{x}'|} + W(\mathbf{x}) - \mu_\gamma$$

and

$$\mu_\gamma = \mu + Z^{-4/3}\cdot 3.68\gamma.$$

Observe that $Z^{-1/3}$ occurs in the position of $\hbar$, making the limit $Z\to\infty$ equivalent to the classical limit $\hbar\to 0$. Now use the bound (2.5.13) with

$$u^2(\mathbf{x}) = \frac{\kappa^3}{8\pi}\exp(-\kappa r).$$

The Fourier transform of this density is

$$\tilde{u}^2(\mathbf{k}) = \frac{\kappa^4}{(|\mathbf{k}|^2+\kappa^2)^2}.$$

Consequently $\int |\nabla u|^3 d^3x = \kappa^2$, and if $v = 1/r$, then

$$v_u(q) \equiv \int d^3x \frac{1}{|\mathbf{x}|}|u(\mathbf{x}-\mathbf{q})|^2 = \frac{1}{q} - \frac{\exp(-\kappa q)}{q} - \frac{\kappa}{2}\exp(-\kappa q) \equiv \frac{1}{q} - v_s(q).$$

4.1.16 The Classical Upper Bound

Since $1/r$ can not be represented as a smeared potential, v^u makes no sense. Thus it is first necessary to remove v_s, the short-range, singular part of $1/r$, and handle it separately. It can be neglected as $\kappa\to\infty$, and if the smeared remainder is unsmeared, we recover $1/r$:

$$\frac{1}{r} = v_u + v_s, \qquad (v_u)^u = \frac{1}{r}.$$

Let h_c be like the h_n of (4.1.17), but with u_u in place of $1/r$. Then

$$h_n = h_c + V_s,$$

$$h_c = \int \frac{d^3q d^3p}{(2\pi)^3}|\mathbf{q},\mathbf{p}\rangle\langle\mathbf{q},\mathbf{p}|\Big(Z^{-2/3}(|\mathbf{p}|^2-\kappa^2) - \sum_j \frac{z_j}{|\mathbf{q}-\mathbf{X}_j|}$$

$$+ \int \frac{d^3xn(\mathbf{x})}{|\mathbf{q}-\mathbf{x}|} + W^u(\mathbf{q}) - \mu_\gamma\Big),$$

$$V_s = -\sum_j z_j v_s(\mathbf{x}-\mathbf{X}_j) + \int d^3yn(\mathbf{y})v_s(\mathbf{x}-\mathbf{y}).$$

In the x-representation, $|\mathbf{q}, \mathbf{p}\rangle$ is

$$\left(\frac{\kappa^3}{8\pi}\right)^{1/2} \exp(i\mathbf{p}\cdot\mathbf{x}) \exp(-\kappa|\mathbf{x}-\mathbf{q}|),$$

and we let $W^u(\mathbf{x})$ be the unsmeared wall potential W of (2.5.17). Convexity can be appealed to bound the influence of V_s since for convex f, $\alpha \geq 1$ we have $f(1) \leq f(0) + (f(\alpha) - f(0))/\alpha$:

$$\begin{aligned}\operatorname{tr}\ln(\mathbf{1}+\exp(-\beta h_n)) \leq \operatorname{tr}\ln(\mathbf{1}+\exp(-\beta h_c)) \\ +\alpha^{-1}\operatorname{tr}[\ln(\mathbf{1}+\exp(-\beta(h_c+\alpha V_s))) \\ -\ln(\mathbf{1}+\exp(-\beta h_c))] \quad \text{for all} \quad \alpha \geq 1.\end{aligned}$$

The number α will be picked so large that the addition to the first term on the right side goes away in the limit $Z \to \infty$. By (2.5.13), the second term is bounded by

$$\begin{aligned}\operatorname{tr}\ln(\mathbf{1}+\exp(-\beta h_c)) &\leq \int \frac{d^3q d^3p}{(22\pi)^3} \ln\Big(\mathbf{1}+\exp\Big[-\beta\Big(Z^{-2/3}(|\mathbf{p}|^2-\kappa^2 2) \\ &\quad -\sum_j \frac{z_j}{|\mathbf{q}-\mathbf{X}_j|} + \int \frac{d^3xn(\mathbf{x})}{|\mathbf{q}-\mathbf{x}|} + W^u(\mathbf{q}) - \mu_\gamma\Big)\Big]\Big) \\ &= Z \int \frac{d^3qd^3p}{(2\pi)^3} \ln\Big(\mathbf{1}+\exp\Big[-\beta\Big(|\mathbf{p}|^2 - \sum_j \frac{z_j}{|\mathbf{q}-\mathbf{X}_j|} \\ &\quad + \int \frac{d^3xn(\mathbf{x})}{|\mathbf{q}-\mathbf{x}|} + W^u(\mathbf{q}) - \mu_\gamma - Z^{-2/3}\kappa^2\Big)\Big]\Big).\end{aligned}$$

The additional part containing V_s can be taken care of because even for a singular potential $V(\mathbf{x}) \in L^{5/2}(\mathbb{R}^3)$ there is a bound of this form weakened by a factor $c \simeq 7$:

$$\begin{aligned}&\operatorname{tr}\ln(\mathbf{1}+\exp[-\beta(|\mathbf{p}|^2+V(\mathbf{x}))]) \\ &\quad \leq c\int \frac{d^3pd^3q}{(2\pi)^3} \ln(1+\exp[-\beta(|\mathbf{p}|^2+V(\mathbf{q}))]).\end{aligned} \tag{4.3}$$

The derivation of this formula is left for Problems 1 and 2. In this case it leaves us with

$$\begin{aligned}&\operatorname{tr}\ln(\mathbf{1}+\exp[-\beta(h_c+\alpha V_s)]) \\ &\quad \leq cZ \int \frac{d^3qd^3p}{(2\pi)^3} \ln\Big(1+\exp\Big[-\beta(|\mathbf{p}|^2 + W^u(\mathbf{q}) \\ &\quad -\sum_j z_j\Big(\frac{1}{|\mathbf{q}-\mathbf{X}_j|} + (\alpha-1)v_s(\mathbf{q}-\mathbf{X}_j)\Big) \\ &\quad + \int d^3yn(\mathbf{y})\Big(\frac{1}{|\mathbf{q}-\mathbf{y}|} + (\alpha-1)v_s(\mathbf{q}-\mathbf{y})\Big) - \mu_\gamma - Z^{-2/3}\kappa^2\Big)\Big]\Big).\end{aligned}$$

It remains to be shown that α and κ can be sent to infinity with Z in such a way that the additions to the classical one-particle potential in the effective field become negligible. To this end assume that W^u tends to infinity outside some compact set K containing the $\mathbf{X}_i$ so rapidly that the contribution to the integral over the complement CK is insignificant, that is, $\int_K d^3q \ln(\ldots) > \int_{CK} d^3q \ln(\ldots)$ for all $\alpha > 0$. Then it suffices to estimate the integral over K, which can be done in terms of the L^p norms of the potential on K, i.e., $\|V\|_p = (\int_K d^3q|V(\mathbf{q})|^p)^{1/p}$. If $|x|_- \equiv |x|\Theta(-x)$, then

$$\ln(1+\exp(-x)) = |x|_- + \ln(1+\exp(-|x|)) \leq |\mathbf{x}|_- + \exp(-|x|),$$

and if

$$\mathbf{q} \in K_- \equiv \{\mathbf{q} \in K : V(\mathbf{q}) < 0\},$$

then with $\varepsilon = |V(\mathbf{q})|\eta$,

$$\begin{aligned} I &\equiv \int_0^\infty d\varepsilon\sqrt{\varepsilon}\ln(1+\exp[-\beta(\varepsilon+V(\mathbf{q}))]) \\ &< \int_0^\infty d\eta\sqrt{\eta}(\beta|\eta-1|_-|V(\mathbf{q})|^{5/2} + |V(\mathbf{q})|^{3/2}\exp[-\beta|V(\mathbf{q})||\nu-1|]), \end{aligned}$$

and if $\mathbf{q} \in K_+ \equiv \{\mathbf{q} \in K : V(\mathbf{q}) > 0\}$, then

$$I < \int_0^\infty d\eta\sqrt{\eta}V(\mathbf{q})^{3/2}\exp[-\beta V(\mathbf{q})(\eta+1)].$$

Because $|\eta-1| \leq \eta+1$ for all $\eta \geq 0$ and

$$\int_0^\infty d\varepsilon\sqrt{\varepsilon}\ln(1+\exp(-\beta\varepsilon)) < \beta^{-3/2}\sqrt{\pi}/2,$$

if $K' = K_+ \cup K_-$, then

$$\begin{aligned} &\int_K d^3q \int_0^\infty d\varepsilon\sqrt{\varepsilon}\ln(1+\exp[-\beta(\varepsilon+V(\mathbf{q}))]) \\ &\quad< \int_{K'} d^3q \int_0^\infty d\eta\sqrt{\eta}(\beta|V(\mathbf{q})|_-^{5/2}|\eta-1|_- \\ &\qquad+|V(\mathbf{q})|^{3/2}\exp[-\beta|V(\mathbf{q})||\eta-1|]) + \beta^{-3/2}\frac{\sqrt{\pi}}{2}. \end{aligned}$$

The required bound now follows from

$$\int_0^\infty d\varepsilon\sqrt{\varepsilon}\exp(-\gamma|\varepsilon-1|) \leq \int_0^1 d\varepsilon\sqrt{\varepsilon} + \int_0^\infty d\varepsilon(\sqrt{\varepsilon}+1)\exp(-\gamma\varepsilon)$$

$$= \frac{2}{3} + \gamma^{-1} + \frac{\sqrt{\pi}}{2}\gamma^{-3/2},$$

for

$$\int_0^\infty d\varepsilon \sqrt{\varepsilon} \int_K d^3q \ln(1 + \exp[-\beta(\varepsilon + V(\mathbf{q}))])$$

$$\leq \int_K d^3q [\frac{4\beta}{15}|V|_-^{5/2} + \frac{2}{3}|V|^{3/2} + \beta^{-1}|V|^{1/2} + \sqrt{\pi}\beta^{-3/2}].$$

In the case at hand, since $\| V_s \|_p \sim \kappa^{1-3/p}$ and

$$\int d^3x |V + (\alpha - 1)V_s|^p \leq (\|V\|_p + (\alpha - 1)\|V_s\|_p)^p, \qquad p = \frac{5}{2}, \frac{3}{2},$$

or, respectively,

$$\int d^3x |V + (\alpha - 1)V_s|^{1/2} \leq \|V\|_{1/2}^{1/2} + \sqrt{\alpha - 1}\|V_s\|_{1/2}^{1/2},$$

it follows that $\mathbf{1} + \exp[-\beta(h_c + \alpha V_s)]$ remains bounded in the limit as α and $\kappa \to \infty$ when $\alpha \sim \kappa^{1/5}$. If κ goes as $Z^{1/3-\varepsilon}$, $0 < \varepsilon < \frac{1}{3}$, then the correction $Z^{-2/3}\kappa^2$ to the kinetic energy tends to zero, and all corrections to the classical one-particle phase-space integral with the effective field are smaller than this quantity by a factor $Z^{-1/15+\varepsilon/5}$. The quantity μ_γ is no trouble at all, since it approaches μ, provided that $\gamma Z^{-4/3} \to 0$. Likewise, $W^u(\mathbf{q})$ and $W_u(\mathbf{q})$ approach $W(\mathbf{q})$ in the limit $\kappa \to \infty$.

4.1.17 The Classical Lower Bound

For the classical bound (2.5.13), the $1/r$ occurring in the classical phase-space integral has to be replaced with $v_u = 1/r - v_s$. As before, convexity is useful for estimating the influence of the v_s, except that this time the convexity of f for $\alpha > 0$,

$$f(-1) \geq f(0) + \frac{f(0) - f(\alpha)}{\alpha},$$

is used for the other side of the equation. The result is

$$\begin{aligned} &\operatorname{tr} \ln(\mathbf{1} + \exp(-\beta h_n)) \\ &\leq Z \int \frac{d^3q dP}{(2\pi)^3} \ln\big(1 + \exp[-\beta\{|\mathbf{p}|^2 + Z^{-2/3}\kappa^2 - \mu_\gamma + W_u(\mathbf{q}) \\ &- \sum_j z_j (\frac{1}{|\mathbf{q} - \mathbf{X}_j|} - v_s(\mathbf{q} - \mathbf{X}_j)) \\ &+ \int d^3x n(\mathbf{x}) (\frac{1}{|\mathbf{q} - \mathbf{x}|} - v_s(\mathbf{q} - \mathbf{x}))\}]\big) \end{aligned}$$

$$\geq Z \int \frac{d^3q d^3p}{(2\pi)^3} [\ln(1 + \exp[-\beta\{|\mathbf{p}|^2 + Z^{-2/3}\kappa^2 - \sum_j \frac{z_j}{|\mathbf{q} - \mathbf{X}_j|}$$

$$+ \int \frac{d^3x n(\mathbf{x})}{|\mathbf{q} - \mathbf{x}|} + W(\mathbf{q}) - \mu_\gamma\}])(1 + \frac{1}{\alpha})$$

$$- \frac{1}{\alpha} \ln(1 + \exp[-\beta\{|\mathbf{p}|^2 + Z^{-2/3}\kappa^2$$

$$- \sum_j z_j(|\mathbf{q} - \mathbf{X}_j|^{-1} + \alpha v_s(\mathbf{q} - \mathbf{X}_j))$$

$$+ \int d^3x n(\mathbf{x})(|\mathbf{q} - \mathbf{x}|^{-1} + \alpha v_s(\mathbf{q} - \mathbf{x})) + W_u(\mathbf{q}) - \mu_\gamma\}])].$$

The integrals that show up are the same as for the upper bounds, so with $\alpha = \kappa^{1/5}$, $\kappa = Z^{1/3-\varepsilon}$, $0 < \varepsilon < \frac{1}{3}$, the corrections to the classical expression vanish as $Z \to \infty$. The $n(\mathbf{x})$ considered earlier was Z independent, while that defined by (4.1.10) depends on Z. However, it is shown in Problem 4 that the minimum values also converge, so our bounds prove the

4.1.18 Classical Limit of the Partition Function

$$\lim_{Z\to\infty} Z^{-1} \ln \operatorname{Tr} \exp[-\beta(Z^{-4/3} H_Z - \mu N)]$$

$$= \lim_{Z\to\infty} Z^{-1} \ln \operatorname{Tr} \exp(-\beta Z^{-4/3} H_{Z,n}) - \beta \sum_{k>l} \frac{z_k z_l}{|\mathbf{X}_k - \mathbf{X}_l|}$$

$$= 2 \int \frac{d^3p d^3q}{(2\pi)^3} \ln(1 + \exp[-\beta(|\mathbf{p}|^2 - \sum_j \frac{z_j}{|\mathbf{q} - \mathbf{X}_j|}$$

$$+ \int d^3y \frac{n(\mathbf{y})}{|\mathbf{q} - \mathbf{y}|} + W(\mathbf{q}) - \mu)])$$

$$- \beta \sum_{k>l} \frac{z_k z_l}{|\mathbf{X}_k - \mathbf{X}_l|} + \frac{\beta}{2} \int \frac{d^3x d^3y}{|\mathbf{x} - \mathbf{y}|} n(\mathbf{x}) n(\mathbf{y}).$$

According to Remark (2.5.4;4), the optimal density for this formula satisfies

$$n(\mathbf{x}) = 2 \int \frac{d^3p}{(2\pi)^3} \left[\exp(|\mathbf{p}|^2 - \sum_j \frac{z_j}{|\mathbf{x} - \mathbf{X}_j|} \right.$$

$$\left. + \int d^3y \frac{n(\mathbf{y})}{|\mathbf{q} - \mathbf{y}|} + W(\mathbf{x}) - \mu) + 1 \right]^{-1}. \tag{4.4}$$

4.1.19 Remarks

1. The classical functional also has Properties (4.1.8), which ensure the existence and uniqueness of a solution of (4.4).

2. As yet unproved conjectures [11] imply that Equation (4.3) holds even with $c = 1$. If that turns out to be true, then many of the proofs given here can be simplified.

4.1.20 The Density in Phase Space

Now that Ξ has been shown to converge, we can study the limiting behavior of the expectation values of a suitable subalgebra of observables. The densities on classical phase space would seem to be an appropriate subalgebra, since in the classical limit $Z \to \infty$ it ought to make sense to speak of position and momentum simultaneously. As mentioned above (cf. (1.2.2)) position goes as $Z^{-1/3}$ while momentum goes as $Z^{2/3}$, so the product of their relative meansquare deviations would be expected to go as $Z^{-2/3}$, and as $Z \to \infty$ the physics should become classical. This rather airy argument can be made mathematically substantial, and we shall discover that in convenient units, fermions distribute themselves in phase space according to

$$\rho(\mathbf{q}, \mathbf{p}) = \left[\exp \beta \left\{ |\mathbf{p}|^2 - \sum_j \frac{z_j}{|\mathbf{q} - \mathbf{X}_j|} + \int \frac{d^3 x n(\mathbf{x})}{|\mathbf{x} - \mathbf{q}|} + W(\mathbf{q}) - \mu \right\} + 1 \right]^{-1}.$$

Particularly interesting is the observation that fermions behave more classically than bosons. The latter have $[\exp\{\} - 1]$ in the denominator, so $\rho(\mathbf{q}, \mathbf{p})$ becomes negative when $\mathbf{q} = \mathbf{X}_j$, and thus can not turn out to be a probability density on phase space. To make the connection with (2.2.8;5) we define creation and annihilation operators at the point $(\mathbf{q}, \mathbf{p})$ in phase space, and choose u as a sufficiently smooth, decreasing function such that $\|u\|_2 = 1$, like the function of (4.1.15):

4.1.21 The Field Algebra on Phase Space

The operators

$$a_{\mathbf{q},\mathbf{p};\alpha} = Z^{3\varepsilon/2} \int d^3 x a_\alpha(\mathbf{x}) \exp(i Z^{2/3} \mathbf{p} \cdot \mathbf{x}) u(Z^\varepsilon(\mathbf{x} - Z^{-1/3}\mathbf{q})),$$
$$\tfrac{1}{3} < \varepsilon < \tfrac{2}{3}, \qquad u^* = u,$$

satisfy the commutation relations

$$[a_{\mathbf{q},\mathbf{p};\alpha}, a^*_{\mathbf{q}',\mathbf{p}';\beta}]_+ = \delta_{\alpha\beta} \int d^3 x \exp(i Z^{2/3-\varepsilon} \mathbf{x} \cdot (\mathbf{p} - \mathbf{p}')) u(Z^\varepsilon(\mathbf{x} - Z^{\varepsilon-1/3}\mathbf{q}) \times u(\mathbf{x} - Z^{\varepsilon-1/3}\mathbf{q}').$$

If $\mathbf{q} = \mathbf{q}'$ and $\mathbf{p} = \mathbf{p}'$, then the right side is $\delta_{\alpha\beta}$, and otherwise it goes to zero as $Z \to \infty$. Hence $\rho_{\mathbf{q},\mathbf{p}} = \sum_\alpha a^*_{\mathbf{q},\mathbf{p};\alpha} a_{\mathbf{q},\mathbf{p};\alpha}$ are bounded above and below by $0 \le \rho_{\mathbf{p},\mathbf{q}} \le 2$, and generate an algebra that is Abelian in the limit $Z \to \infty$. Defining $d\Omega \equiv d^3 q d^3 p/(2\pi)^3$, we calculate

$$\int d\Omega \rho_{\mathbf{q},\mathbf{p}} F(\mathbf{q}) = Z^{-1} \sum_\alpha \int d^3 x a^*_\alpha(\mathbf{x}) a_\alpha(\mathbf{x}) Z^{3\varepsilon} |u(Z^\varepsilon(\mathbf{x} - \mathbf{x}'))|^2$$

$$\times F(Z^{1/3}\mathbf{x}')d^3x',$$

$$\int d\Omega\rho_{\mathbf{q},\mathbf{p}}|\mathbf{p}|^2 = Z^{-7/3}\sum_\alpha \int d^3x(\nabla a^*_\alpha(\mathbf{x})\cdot\nabla a_\alpha(\mathbf{x}) + a^*_\alpha(\mathbf{x})a_\alpha(\mathbf{x})Z^{2\varepsilon}\int d^3y|\nabla u(\mathbf{y})|^2),$$

$$\int d\Omega d\Omega' \frac{\rho_{\mathbf{q},\mathbf{p}}\rho_{\mathbf{q}',\mathbf{p}'}}{|\mathbf{q}-\mathbf{q}'|} = Z^{-7/3}\left[\sum_\alpha \int d^3x a^*_\alpha(\mathbf{x})a_\alpha(\mathbf{x})v_{uu}(\mathbf{0}) + \sum_{\alpha,\beta}\int d^3x d^3x' a^*_\alpha(\mathbf{x})a^*_\beta(\mathbf{x}')a_\beta(\mathbf{x}')a_\alpha(\mathbf{x})v_{uu}(\mathbf{x}-\mathbf{x}')\right], \tag{4.5}$$

where $F \in L^\infty(\mathbb{R}^3)$ and

$$0 \le v_{uu}(\mathbf{x}-\mathbf{x}') \equiv \int d^3q d^3q' \frac{|u(Z^\varepsilon(\mathbf{x}-\mathbf{q}))|^2|u(Z^\varepsilon(\mathbf{x}'-\mathbf{q}'))|^2}{|\mathbf{q}-\mathbf{q}'|}Z^{6\varepsilon} < \frac{1}{|\mathbf{x}-\mathbf{x}'|}.$$

4.1.22 Remarks

1. As $Z\to\infty$, $Z^{3\varepsilon}|u(Z^\varepsilon(\mathbf{x}-\mathbf{x}'))|^2$ approaches $\delta(\mathbf{x}-\mathbf{x}')$. It is not hard to convince oneself that when the classical Hamiltonian with ρ or $\rho\cdot\rho$ is integrated, the result is H to order $Z^{7/3}$.
2. For neutral states, i.e., $\sum_\alpha\langle N_\alpha\rangle = Z$, it follows that $\langle\int d\Omega\rho_{\mathbf{q},\mathbf{p}}\rangle = 1$.

The convexity of the partition function (2.4.6) can be used to calculate an expectation value by allowing it to be written as the derivative of the partition function by a perturbation parameter. We shall show that the perturbed Ξ still converges as $Z\to\infty$, which will simultaneously prove that the foregoing results are stable against small variations in H. The limit will turn out to be likewise convex and differentiable in the perturbation parameters, so by Problem (2.4.18;3) the limit of the derivative is the derivative of the limit. Since our real aim is to prove that the expectation value of $\rho_{\mathbf{q},\mathbf{p}}$ approaches the Thomas–Fermi density and that the deviations of $\rho_{\mathbf{q},\mathbf{p}}$ vanish, we will perturb H both linearly and quadratically in ρ. To an accuracy of $Z^{-2/3}$ we can by-pass the intermediate steps (4.1.13), so we shall not require the more refined inequality (4.2). Thus we get by with a somewhat simpler effective Hamiltonian.

4.1.23 The Perturbed Hamiltonian

$$H_\lambda \equiv H_Z + \lambda_1 Z^{7/3}\int d\Omega\rho_{\mathbf{q},\mathbf{p}}f(\mathbf{q},\mathbf{p}) + \lambda_2\frac{Z^{7/3}}{2}\left(\int d\Omega\rho_{\mathbf{q},\mathbf{p}}f(\mathbf{q},\mathbf{p})\right)^2,$$

$$H_{\lambda,n} \equiv \int d^3x\sum_\alpha\left\{\nabla a^*_\alpha(\mathbf{x})\cdot\nabla a_\alpha(\mathbf{x}) + a^*_\alpha(\mathbf{x})a_\alpha(\mathbf{x})\right.$$

$$\times \left[-Z \sum_{k=1}^{M} \frac{z_k}{|\mathbf{x} - Z^{-1/3}\mathbf{X}_k|} + Z^{4/3}(W(Z^{1/3}\mathbf{x}) - \mu) \right] \Bigg\}$$

$$+Z^{7/3} \int \frac{d\Omega d\Omega'}{|\mathbf{q} - \mathbf{q}'|} (\rho_{\mathbf{q},\mathbf{p}} - \frac{1}{2} n(\mathbf{q}, \mathbf{p})) n(\mathbf{q}', \mathbf{p}')$$

$$+Z^{7/3}(\lambda_1 + \lambda_2 g) \int d\Omega \rho_{\mathbf{q},\mathbf{p}} f(\mathbf{q}, \mathbf{p}) - Z^{7/3} \lambda_2 g^2/2,$$

where $\lambda_i \in \mathbb{R}$ and $f \in C_0^\infty$. We shall choose $n(\mathbf{q}, \mathbf{p})$ as $< \rho_{\mathbf{q},\mathbf{p}} >$ and let $g \equiv \int d\Omega n(\mathbf{q}, \mathbf{p}) f(\mathbf{q}, \mathbf{p})$. With the idea of (4.1.20), because $0 \leq v_{uu}(\mathbf{x}) \leq 1/|\mathbf{x}|$,

$$H_\lambda - Z^{4/3} \mu N - Z^{7/3} \sum_{k>l} \frac{z_k z_l}{|\mathbf{X}_k - \mathbf{X}_l|} - H_{\lambda,n}$$

$$= \frac{1}{2} \int \frac{d^2x d^2x'}{|\mathbf{x} - \mathbf{x}'|} a_\alpha^*(\mathbf{x}) a_\beta^*(\mathbf{x}') a_\beta(\mathbf{x}') a_\alpha(\mathbf{x})$$

$$- Z^{7/3} \int \frac{d\Omega d\Omega'}{|\mathbf{q} - \mathbf{q}'|} (\rho_{\mathbf{q},\mathbf{p}} - \frac{1}{2} n(\mathbf{q}, \mathbf{p})) n(\mathbf{q}', \mathbf{p}')$$

$$+ Z^{7/3} \frac{\lambda_2}{2} \Big(\int d\Omega \rho_{\mathbf{q},\mathbf{p}} f(\mathbf{q}, \mathbf{p}) - g \Big)^2$$

$$\geq \frac{Z^{7/3}}{2} \int d\Omega d\Omega' \big(\rho_{\mathbf{q},\mathbf{p}} - n(\mathbf{q}, \mathbf{p}) \big) \big(\rho_{\mathbf{q}',\mathbf{p}'} - n(\mathbf{q}', \mathbf{p}') \big)$$

$$\times \left[\frac{1}{|\mathbf{q} - \mathbf{q}'|} + \lambda_2 f(\mathbf{q}, \mathbf{p}) f(\mathbf{q}', \mathbf{p}') \right] - \frac{N}{2} v_{uu}(\mathbf{0}).$$

4.1.24 Remarks

1. Since the Fourier transform in the q variables, $\tilde{f}(\mathbf{k}, \mathbf{p})$, decreases in $\mathbf{k}$ faster than any power, $|\mathbf{k}|^2 + \lambda_2 \tilde{f}(\mathbf{k}, \mathbf{p}) \tilde{f}(\mathbf{k}, \mathbf{p}')$ is positive for sufficiently small $|\lambda_2|$. The expression in square brackets $[\cdots]$ is then of positive type, and the inequality extend to the statement that

$$H_\lambda - Z^{4/3} \mu N - Z^{7/3} \sum_{k>m} \frac{z_k z_m}{|\mathbf{X}_k - \mathbf{X}_m|} - H_{\lambda,n} \geq -\frac{N}{2} v_{uu}(\mathbf{0}).$$

It is easy to calculate that $v_{uu}(\mathbf{0}) \sim Z^\varepsilon$, so the right side is dominated by $Z^{7/3}$, and in the limit as $Z \to \infty$,

$$Z^{-1} \Xi(Z^{-4/3} H_\lambda - \mu N) \leq Z^{-1} \Xi(Z^{-4/3} H_{\lambda,n}) - \sum_{k>m} \frac{z_k z_m}{|\mathbf{X}_k - \mathbf{X}_m|}.$$

2. According to (4.1.20),

$$Z^{7/3} \int d\Omega d\Omega' \frac{\rho_{\mathbf{q},\mathbf{p}} n(\mathbf{q}', \mathbf{p}')}{|\mathbf{q} - \mathbf{q}'|} = \sum_\alpha \int d^3x d^3x' a_\alpha^*(\mathbf{x}) a_\alpha(\mathbf{x}) v_u(\mathbf{x} - \mathbf{x}') n(\mathbf{x}'),$$

where

$$n(\mathbf{x}) = \int \frac{d^3p}{(2\pi)^3} n(\mathbf{x}, \mathbf{p}).$$

Therefore the Coulomb repulsion of the electrons in the Hamiltonian $H_{\lambda,n}$ of (4.1.28) is reduced by $v_s = 1/r - v_u$. As in (4.1.14) the Hamiltonian $H_{\lambda,n}$ with v_u in place of $1/r$ furnishes a lower bound for Ξ. On the other hand, it was shown in (4.1.16) and (4.1.17) that the effect of v_s on $\Xi(H_{\lambda,n})$ was negligible as $Z \to \infty$. Moreover, $\int d\Omega \rho_{\mathbf{q},\mathbf{p}} f(\mathbf{q}, \mathbf{p})$ is the second quantization of the one-particle operator $\int d\Omega |\mathbf{q}, \mathbf{p}\rangle\langle \mathbf{q}, \mathbf{p}| f(\mathbf{q}, \mathbf{p})$, $|\mathbf{q}, \mathbf{p}\rangle = \exp(i\mathbf{p} \cdot \mathbf{x}) u(\mathbf{x} - \mathbf{q})$, the expectation value of which in the state $|\mathbf{q}', \mathbf{p}'\rangle$ reduces to $f(\mathbf{q}', \mathbf{p}')$ in the limit $Z \to \infty$. The generalization of (4.1.18) is consequently

$$\begin{aligned}
&\lim_{Z\to\infty} Z^{-1} \ln \operatorname{Tr} \exp[-\beta(Z^{-4/3} H_\lambda - \mu N)] \\
&= 2\int \frac{d^3q d^3p}{(2\pi)^3} \ln\left\{1 + \exp\left[-\beta\left(|\mathbf{p}|^2 - \sum_k \frac{z_k}{|\mathbf{q} - X_k|}\right.\right.\right. \\
&\quad \left.\left.\left. + \int \frac{d^3q' n_\lambda(\mathbf{q}')}{|\mathbf{q} - \mathbf{q}'|} + W(\mathbf{q}) + f(\mathbf{q}, \mathbf{p})(\lambda_1 + \lambda_2 g_\lambda) - \frac{\lambda_2}{2} g_\lambda^2 - \mu\right)\right]\right\} \\
&\quad + \beta\left(\frac{1}{2}\int \frac{d^3q d^3q'}{|\mathbf{q} - \mathbf{q}'|} n_\lambda(\mathbf{q}) n_\lambda(\mathbf{q}') - \sum_{k>l} \frac{z_k z_l}{|\mathbf{X}_k - \mathbf{X}_l|}\right),
\end{aligned} \tag{4.6}$$

where

$$n_\lambda(\mathbf{q}) = \int \frac{d^3p}{(2\pi)^3} n_\lambda(\mathbf{q}, \mathbf{p}),$$

$$\begin{aligned}
n_\lambda(\mathbf{q}, \mathbf{p}) = 2\left\{\exp\beta\left[|\mathbf{p}|^2 - \sum_k \frac{z_k}{|\mathbf{q} - \mathbf{X}_k|} + \int \frac{d^3q' n_\lambda(\mathbf{q}')}{|\mathbf{q} - \mathbf{q}'|}\right.\right. \\
\left.\left. + W(\mathbf{q}) + f(\mathbf{q}, \mathbf{p})(\lambda_1 + \lambda_2 g_\lambda) - \frac{\lambda_2}{2} g_\lambda^2 - \mu\right] + 1\right\}^{-1},
\end{aligned}$$

$$g_\lambda = \int d\Omega n_\lambda(\mathbf{q}, \mathbf{p}) f(\mathbf{q}, \mathbf{p}),$$

and $|\lambda_2|$ is sufficiently small.

Differentiation by λ_1 and λ_2 at $\lambda_1 = \lambda_2 = 0$ and an optimization of $f \in C_0^2(\mathbb{R}^6)$ reveal the

4.1.25 Convergence of the Expectation Values

$$\lim_{Z\to\infty} \langle \rho_{\mathbf{q},\mathbf{p}} \rangle_Z \equiv \lim_{Z\to\infty} \frac{\operatorname{Tr}(\rho_{\mathbf{q},\mathbf{p}} \exp[-\beta(Z^{-4/3} H_Z - \mu N)])}{\operatorname{Tr} \exp[Z^{-4/3} H_Z - \mu N]}$$

$$= 2\{\exp[\beta(|\mathbf{p}|^2 - \sum_k \frac{z_k}{|\mathbf{q}-\mathbf{X}_k|}$$

$$+ \int \frac{d^3q' n_0(\mathbf{q}')}{|\mathbf{q}-\mathbf{q}'|} + W(\mathbf{q}) - \mu)] + 1\}^{-1} = n_0(\mathbf{q}, \mathbf{p}),$$

$$\lim_{Z\to\infty} \langle \rho_{\mathbf{q},\mathbf{p}}, \rho_{\mathbf{q}',\mathbf{p}'} \rangle_Z = n_0(\mathbf{q}, \mathbf{p}) n_0(\mathbf{q}', \mathbf{p}').$$

4.1.26 Remarks

1. Since f is not arbitrary, but assumed in $C_0^2(\mathbb{R}^6)$, the limit converges only in the sense of distributions. The C^* algebra $\mathcal{A}_Z$ generated by the "smeared" densities on phase space, $\rho \equiv \int d\Omega g(\mathbf{q}, \mathbf{p})\rho_{\mathbf{q},\mathbf{p}}$, together with the identity becomes Abelian in the "weak" limit $Z \to \infty$. Hence, according to the Gel'fand isomorphism (III:2.2.28), if $Z = \infty$, then $\mathcal{A}_Z$ can be represented as the set of continuous functions on a compact Hausdorff space. The space of characters of an Abelian C^* algebra $\mathcal{A}$, i.e., *-homomorphisms from $\mathcal{A}$ to $\mathbb{C}$, is the same as the set $\mathcal{E}$ of pure states and is a compact Hausdorff space in the (relative) weak*- topology. With the identification $[a](\omega) = \omega(a) \in \mathbb{C}$ for all $a \in \mathcal{A}$ and $\omega \in \mathcal{E}$, $\mathcal{A}$ is equivalent to the C^* algebra of the continuous functions with the supremum norm on the set $\mathcal{E}$, given the weak*- topology. In our case, $\mathcal{E} = \{n \in L^\infty(\mathbb{R}^6) | n \geq 0 \text{a.e.}, \|n\|_\infty \leq 2\}$, with the weak*- topology with respect to the linear functionals belonging to the predual $L^1(\mathbb{R}^6)$. (Since $C_0^2(\mathbb{R}^6)$ is dense in $L^1(\mathbb{R}^6)$ in norm, the corresponding weak*- topologies agree on $L^\infty(\mathbb{R}^6)$.) Since $\mathcal{E}$ is the intersection of the cone of the functions that are nonnegative a.e., which is a weak*- closed set, with a multiple of the unit cube of L^∞, it is weak*- compact. The Gel'fand isomorphism correlates ρ_g with the mapping $[\rho_g](n) = \int ng d\Omega$, and since $\|[\rho_g] - [\rho_{g'}]\|_\infty \leq 2\|g - g'\|_1$, the completion contains for instance all ρ_g such that $g \in L^1(\mathbb{R}^6)$. The set of all states on the algebra is the weak-* closure of the convex combinations of characters and can be represented as a set of probability measures; pure states correspond to point measures. If the state is mixed, $\alpha\langle\rangle_{n_1} + (1-\alpha)\langle\rangle_{n_2}$, then the two-point function can not be factorized:

$$\alpha\langle\rho_{z_1}\rho_{z_2}\rangle_{n_1} + (1-\alpha)\langle\rho_{z_1}\rho_{z_2}\rangle_{n_2} = \alpha n_1(z_1)n_1(z_2) + (1\alpha)n_2(z_1)n_2(z_2)$$
$$= (\alpha\langle\rho_{z_1}\rangle_{n_1} + (1-\alpha)\langle\rho_{z_1}\rangle_{n_2})(\alpha\langle\rho_{z_2}\rangle_{n_1} + (1-\alpha)\langle\rho_{z_2}\rangle_{n_2}) \quad \text{for all } z_1, z_2$$
$$\Rightarrow n_1(z) = n_2(z) \qquad \text{for all } z = (\mathbf{q}, \mathbf{p}).$$

Hence it follows from (4.1.25) that the limiting state is a character, and consequently pure.

2. Although the system acts classically on a distance scale $\sim Z^{-1/3}$, it would be expected to behave like a free Fermi gas on the scale $Z^{-2/3} \sim$ the average distance between particles $\sim$ reciprocal of momentum. If the microscopic field operators

$$a_{\mathbf{q}}(\boldsymbol{\xi}) = Z^{-1}a(Z^{-1/3}\mathbf{q} + Z^{-2/3}\boldsymbol{\xi}), \qquad [a_q(\boldsymbol{\xi}), a_q^*(\boldsymbol{\xi}')]_+ = \delta(\boldsymbol{\xi} - \boldsymbol{\xi}')$$

are introduced, it can be seen from (4.1.25) that its expectation value for free fermions is

$$\int \frac{d^3p}{(2\pi)^3}\exp(i\mathbf{p}\cdot\boldsymbol{\xi})\rho_{\mathbf{q},\mathbf{p}}$$
$$= \int \frac{d^3p}{(2\pi)^3}\exp[i\mathbf{p}\cdot(Z^{2/3}(\mathbf{x}-\mathbf{x}')+\boldsymbol{\xi})]a^*(\mathbf{x})a(\mathbf{x}')\frac{Z^{2/3}}{\pi^{3/2}}$$
$$\times\exp[-\frac{Z}{2}(|\mathbf{x}-Z^{-1/3}\mathbf{q}|^2+|\mathbf{x}'-Z^{-1/3}\mathbf{q}|^2)]d^3x\,d^3x'$$
$$= \int d^3z\,a^*_{\mathbf{q}+\mathbf{x}}\Big(-\frac{\boldsymbol{\xi}}{2}\Big)a_{\mathbf{q}+\mathbf{x}}\Big(\frac{\boldsymbol{\xi}}{2}\Big)\exp\left[-Z^{1/3}|\mathbf{x}|^2-\frac{Z^{-1/3}|\boldsymbol{\xi}|^2}{4}\right]\frac{Z^{1/2}}{\pi^{3/2}},$$

where the chemical potential is determined by the potential $V(\mathbf{q})$ at the point $\mathbf{q}$, and we set $\varepsilon=\frac{1}{2}$, $u=\pi^{-3/4}\exp(-|\mathbf{x}|^2/2)$,

$$\int \frac{d^3p}{(2\pi)^3}\exp(i\mathbf{p}\cdot\boldsymbol{\xi})\rho_{\mathbf{q},\mathbf{p}}$$
$$= \int \frac{d^3p}{(2\pi)^3}\exp[i\mathbf{p}\cdot(Z^{2/3}(\mathbf{x}-\mathbf{x}')+\boldsymbol{\xi})]a^*(\mathbf{x})a(\mathbf{x}')\frac{Z^{2/3}}{\pi^{3/2}}$$
$$\times\exp[-\frac{Z}{2}(|\mathbf{x}-Z^{-1/3}\mathbf{q}|^2+|\mathbf{x}'-Z^{-1/3}\mathbf{q}|^2)]d^3x\,d^3x'$$
$$= \int d^3x\,a^*_{\mathbf{q}+\mathbf{x}}\Big(-\frac{\boldsymbol{\xi}}{2}\Big)a_{\mathbf{q}+\mathbf{x}}\Big(\frac{\boldsymbol{\xi}}{2}\Big)\exp\left[-Z^{1/3}|\mathbf{x}|^2-\frac{Z^{-1/3}|\boldsymbol{\xi}|^2}{4}\right]\frac{Z^{1/2}}{\pi^{3/2}},$$

Therefore

$$\int\frac{d^3p\exp(i\mathbf{p}\cdot\boldsymbol{\xi})(2\pi)^{-3}}{\exp[\beta(|\mathbf{p}|^2-V(\mathbf{q}))]+1}$$
$$=\lim_{Z\to\infty}\int d^3x\frac{Z^{1/2}}{\pi^{2/3}}\exp[-Z^{1/3}|\mathbf{x}|^2-\frac{Z^{-1/3}|\boldsymbol{\xi}|^2}{2}]$$
$$\times\Big\langle a^*_{\mathbf{q}+\mathbf{x}}\Big(-\frac{\boldsymbol{\xi}}{2}\Big)a_{\mathbf{q}+\mathbf{x}}\Big(\frac{\boldsymbol{\xi}}{2}\Big)\Big\rangle_{n_0}\equiv\Big\langle a^*_{\mathbf{q}}\Big(-\frac{\boldsymbol{\xi}}{2}\Big)a_{\mathbf{q}}\Big(\frac{\boldsymbol{\xi}}{2}\Big)\Big\rangle_{\infty},$$

$$V(\mathbf{q}) = -\sum_k\frac{z_k}{|\mathbf{q}-\mathbf{X}_k|}+\int\frac{d^3x\,n_0(\mathbf{x})}{|\mathbf{q}-\mathbf{x}|}+W(\mathbf{q})-\mu.$$

3. Results have also been obtained concerning the time-evolution in the limit $Z\to\infty$ [26], but they have only been proved for regularized potentials v_u and not for $1/r$, so they will not be presented here. At any rate the time-evolution of $\omega(a_t)$, where the nonstationary state ω has the same scaling properties as the grand canonical state ρ, is the free time-evolution, as is that of $\rho(a_t b)$, when only the microscopic observables (4.1.26;2) are considered. The equation for the expectation values of the macroscopic observables $\rho_{\mathbf{q},\mathbf{p}}$ is known as the **Vlasov equation**; it describes

$$\frac{dn}{dt}=\mathbf{p}\cdot\frac{\partial n}{\partial\mathbf{q}}-\frac{\partial V}{\partial\mathbf{q}}\cdot\frac{\partial n}{\partial\mathbf{p}},$$

where the potential itself depends on the particle density,

$$V(\mathbf{q}) = -\sum_k \frac{Z_k}{|\mathbf{q} - \mathbf{X}_k|} + \int \frac{d^3q'd^3p'}{(2\pi)^3} \frac{n(\mathbf{q}', \mathbf{p}')}{|\mathbf{q} - \mathbf{q}'|}.$$

Thomas–Fermi theory thus reduces the quantum-mechanical many-body problem to the solution of the integral equation (4.4). Although (4.1.22) is much simpler than the original Schrödinger equation, it can still be solved with reasonable numerical effort and skill only in the radially symmetric case. Despite that, some valuable relationships and properties can be obtained just from the maximum property.

4.1.27 The Relationships among the Contributions to Ξ

Write

$$-\lim Z^{-1}\Xi(Z^{-4/3}H_Z - \mu N) - \sum_{k>l} \frac{z_k z_l}{|\mathbf{X}_k - \mathbf{X}_l|}$$

$$= \inf_{0 \le n \le 2} \int \frac{d^3q d^3p}{(2\pi)^3} \left\{ 2T \left[\frac{n(\mathbf{q}, \mathbf{p})}{2} \ln \frac{n(\mathbf{q}, \mathbf{p})}{2} + \left(1 - \frac{n(\mathbf{q}, \mathbf{p})}{2}\right) \ln\left(1 - \frac{n(\mathbf{q}, \mathbf{p})}{2}\right) \right] \right.$$

$$\left. + n(\mathbf{q}, \mathbf{p})\left(-\mu + |\mathbf{p}|^2 - \sum_{j=1}^{M} \frac{z_j}{|\mathbf{q} - \mathbf{X}_j|} + \frac{1}{2} \int \frac{d^3q'd^3p'}{2(\pi)^3} \frac{n(\mathbf{q}', \mathbf{p}')}{|\mathbf{q} - \mathbf{q}'|} + W(\mathbf{q})\right) \right\}$$

$$= -TS - \mu\lambda + K - A + R + W,$$

where

$$\lambda = \int \frac{d^3qd^3p}{(2\pi)^3} n(\mathbf{q}, \mathbf{p}) = \lim_{Z \to \infty} \frac{N}{Z},$$

K is the kinetic energy of the electrons, A is the potential attracting the electrons to the nuclei, and R is the interelectronic repulsion. Then for the values of μ at which the infimum is attained as a minimum (at a given phase-space density n_0),

(i) $-3(TS + \mu\lambda) + 5K - 3A + 6R + 3W = 0$; *and*

(ii) *if an atom is isolated and in the ground state, i.e., $M = 1$, $X_1 = 0$, $W = 0$, $T = 0$, then*

$$-3\mu\lambda + 3K - 2A + 5R = 0.$$

Proof:

(i) Take the infimum over n' of the form $n_0(q, \gamma_1^{-1} p)$. A change of the variables of integration $\mathbf{p} \to \gamma_1 \mathbf{p}$ converts (4.1.33) into

$$-\gamma_1^3(TS + \mu\lambda + A - W) + \gamma_1^5 K + \gamma_1^6 R.$$

This has its minimum at $\gamma_1 = 1$ when condition (i) holds.

(ii) Now dilate $\mathbf{q}$ so that $n(\mathbf{q}, \mathbf{p}) = n_0(\gamma_2^{-1}\mathbf{q}, \mathbf{p})$, and proceed as before; then

$$\frac{d}{d\gamma_2}\left[\gamma_2^3(K - \mu\lambda) - \gamma_2^3 A + \gamma_2^5 R\right]\Big|_{\gamma_2=1} = 0$$

yields Relationship (ii). □

4.1.28 Corollary

In case (ii) with $\mu = 0$, the three contributions to the energy stand in the ratio

$$K : A : R = 3 : 7 : 1.$$

4.1.29 Remarks

1. The dilatation required for (ii) affects the nuclear coordinates (other than $\mathbf{X}_1 = \mathbf{0}$) and the wall. The reason for setting $T = 0$ was to avoid problems connected with the latter.
2. Since A, K, and R are positive, the second derivatives at $\gamma = 1$ are automatically positive.
3. If $T = \mu = 0$, then $-\Xi$ becomes the minimum of the energy without fixed particle number. We shall learn that the minimum is achieved by a neutral system in Thomas–Fermi theory, and that in case (ii)

$$\int \frac{d^3q d^3p}{(2\pi)^3} n_0(\mathbf{q}, \mathbf{p}) = z_1.$$

The comparison densities $n(\gamma^{-1}\mathbf{q}, \mathbf{p})$ and $n(\mathbf{q}, \gamma^{-1}\mathbf{p})$ correspond to different numbers of particles, and the mystical numbers in (4.1.28) reflect the stability of neutral atoms against spontaneous ionization.

In the limit $T \to 0$, the quantity $(\exp[\beta(\varepsilon - a)] + 1)^{-1}$ approaches $\Theta(a - \varepsilon)$. In that case W may be chosen identically zero, and the integration over $\mathbf{p}$ becomes elementary. The computation yields

4.1.30 The Electron Density in Configuration Space

$$\rho(\mathbf{x}) \equiv \int \frac{d^3p}{(2\pi)^3} n_0(\mathbf{x}, \mathbf{p}) = \frac{1}{3\pi^2}|\Phi(\mathbf{x}) + \mu|_+^{3/2}, \qquad |z|_+ \equiv |z|\Theta(z),$$

$$\Phi(\mathbf{x}) \equiv \sum_j \frac{z_j}{|\mathbf{x} - \mathbf{X}_j|} - \int \frac{d^3x' \rho(\mathbf{x}')}{|\mathbf{x} - \mathbf{x}'|}.$$

The kinetic-energy density is

$$\int \frac{d^3p}{(2\pi)^3} |\mathbf{p}|^2 n_0(\mathbf{x}, \mathbf{p}) = \frac{3}{5}(3\pi^2)^{2/3} \rho^{5/3}(\mathbf{x}).$$

(Since the particles have spin 1/2, the factor $(6\pi^2)^{2/3}$ of (2.28) has become $(3\pi^2)^{2/3}$.) This reveals

4.1.31 The Range of Values of μ and $\Phi(x)$

(i) *μ takes on the values $-\infty < \mu \leq 0$; and*

(ii) *Φ takes on the values $0 \leq \Phi < \infty$.*

Proof: We shall only demonstrate the impossibility of $\mu > 0$ and $\Phi < 0$; Problem 3 will assure us that a minimizing ρ exists for all $\mu \leq 0$, and it can be seen directly that $\Phi(\mathbf{x})$ ranges over $[0, \infty)$ as $\mathbf{x}$ ranges over $\mathbb{R}^3$.

(i) Since $\rho(\mathbf{x})$ must be integrable, $\Phi(\mathbf{x}) \to 0$ as $|\mathbf{x}| \to \infty$. If $\mu > 0$, then $\rho(\mathbf{x})$ would have to approach $\mu^{3/2}/3\pi^2$ as $|\mathbf{x}| \to \infty$, which would contradict integrability.

(ii) The set $A \equiv \{\mathbf{x} \in \mathbb{R}^3 : \Phi(\mathbf{x}) < 0\}$ is open and does not contain $\mathbf{x}_i$. Because $\mu \leq 0$, the density ρ vanishes identically on A, so $\Delta\Phi(\mathbf{x}) = 0$ holds throughout A. Since Φ equals zero on the boundary of A and at infinity and is harmonic, it would have to equal zero on A, because its maximum would be attained either on ∂A or at infinity. However, this contradicts the definition of A, so A must be empty.

□

The quantity $\lambda \equiv \int d^3x\rho(\mathbf{x}) = \lim_{Z\to\infty} N/Z$, where N is the number of electrons and Z is the sum of the nuclear charges, is more intuitively understandable than μ. By expressing the energy as a function of λ, we can find the limits of the observables studied in (III: §4.5).

4.1.32 Properties of the Thomas–Fermi Functional at $T = 0$

Let

$$K(\rho) = \frac{3}{5}(3\pi^2)^{2/3} \int d^3x\rho^{5/3}(\mathbf{x}),$$

$$A(\rho) = \sum_j z_j \int \frac{d^3x\rho(\mathbf{x})}{|\mathbf{x} - \mathbf{X}_j|},$$

$$R(\rho) = \frac{1}{2} \int \frac{d^3x d^3x'}{|\mathbf{x} - \mathbf{x}'|}\rho(\mathbf{x})\rho(\mathbf{x}'),$$

$$E(\rho) \equiv K(\rho) - A(\rho) + R(\rho),$$

and

$$S_\lambda\{\rho : \rho(\mathbf{x}) \geq 0, \int d^3x\rho(\mathbf{x}) = \lambda\}, \quad \sum_j z_j = 1.$$

Then $E[\lambda] = \inf_{\rho\in S_\lambda} E(\rho)$ satisfies

(i)

$$E[\lambda] = -\inf_{\mu}\left(\Xi_\infty(\mu, 0) - \mu\lambda + \sum_{k>l} \frac{z_k z_l}{|\mathbf{X}_k - \mathbf{X}_l|}\right),$$

$$\Xi_\infty(\mu, T) = \lim_{Z\to\infty} Z^{-1} T \ln \operatorname{Tr} \exp[-\beta(Z^{-4/3} H - \mu N)];$$

(ii) $\partial E/\partial\lambda = \mu$ *if* $\lambda \leq 1$, *and =0 if* $\lambda > 1$;

(iii) $E[\lambda]$ *is a nonpositive, convex, decreasing function of* λ; *and*

(iv) *in the atomic case* $z_1 = 1$, *all other* $z_i = 0$, $-\lambda^{-1/6}(-E[\lambda])^{1/2}$ *is a concave, increasing function of* λ

Proof:

(i) Observe first that $E[\lambda]$ is convex, since the convexity of $E(\rho)$ as a function of ρ means that $E[\alpha\lambda_1 + (1-\alpha)\lambda_2] \leq E(\alpha\rho_1 + (1-\alpha)\rho_2) \leq \alpha E[\lambda_1] + (1-\alpha)E[\lambda_2]$, in which $E[\lambda_i] = E(\rho_i)$, because $\alpha\rho_1 + (1-\alpha)\rho_2 \in S_{\alpha\lambda_1+(1-\alpha)\lambda_2}$. As remarked in (2.4.14;2(i)), the Legendre transformation

$$-\Xi_\infty(\mu, 0) = \inf_{\lambda} \inf_{\rho\in S_\lambda} \left(E(\rho) - \mu\lambda + \sum_{k>m} \frac{z_k z_m}{|\mathbf{X}_k - \mathbf{X}_m|}\right)$$

can be inverted for the concave function $-E[\lambda]$, yielding (i).

(ii) The formula $dE/d\lambda = \mu$ will follow from Property (i) once $E[\lambda]$ has been shown to be differentiable. Let ρ_λ denote the minimizing ρ (4.1.30). A calculation shows that

$$\frac{\partial}{\partial t} E(\rho_\lambda(1+t))|_{t=0} = \mu\lambda,$$

so $E[(1+t)\lambda] - E[\lambda] \leq t\mu\lambda + o(t)$ and $E[(1-t)\lambda] - E[\lambda] \leq -t\mu\lambda + o(t)$. In the limit $t \to 0$, this becomes $dE/d\lambda = \mu$. It remains to show that $\lambda < 1 \Leftrightarrow \mu < 0$ and $\lambda = 1 \Rightarrow \mu = 0$, which $\Rightarrow \lambda \geq 1$. Note that Φ goes asymptotically as $(1-\lambda)/r$. If μ were 0, then

$$\rho^{r\to\infty} \overset{r\to\infty}{\sim} \left(\frac{1-\lambda}{r}\right)^{3/2},$$

which would not be integrable; thus μ must be negative when $\lambda < 1$. When $\lambda > 1$, there is no minimum, since if there were, then Φ would be negative as $r \to \infty$, which is impossible because of (4.1.31). However, the infimum has to be $E(1)$, since for $\lambda > 1$ and for any $\varepsilon > 0$ a ρ can be constructed such that $E(\rho) < E(1) + \varepsilon$; start with a ρ_1 with $\lambda = 1$ and compact support, and such that $E(\rho_1) \leq E(1) + \varepsilon/2$, and then let

$$\rho_k = \rho_1 + \frac{1}{k}\chi_k, \qquad k \in \mathbf{N},$$

where the characteristic functions χ_k satisfy $\chi_k\rho_1 \equiv 0$ and $\|\chi_k\|_1 = k(\lambda - 1)$ to ensure that $\rho_k \in S_\lambda$. Then $\|\rho_1 - \rho_k\|_p \to 0$ for all $p > 1$, and it is easy to verify that $E(\rho_k) \to E(\rho_1)$. This accords with the intuitive feeling that a thin electron cloud at a great distance affects the energy only slightly. It means that $E[\lambda]$ decreases while $0 < \lambda < 1$, and becomes constant thereafter.

(iii) This follows from the proof of (i) and (ii), since $\mu \leq 0$.

(iv) Make both of the scaling transformations of (4.1.27) simultaneously and define

$$\inf_{\rho \in S_1} \big(K(\rho) - ZA(\rho) + \alpha R(\rho)\big) = Z^2 \inf_{\rho \in S_1} \big(K(\rho) - A(\rho) + \frac{\alpha}{Z} R(\rho)\big)$$
$$\equiv Z^2 f\big(\frac{\alpha}{Z}\big).$$

This is the infimum of a set of linear functions and consequently concave in (Z, α). The condition that

$$\frac{\partial^2}{\partial Z^2} \frac{\partial^2}{\partial \alpha^2} \geq \big(\frac{\partial^2}{\partial Z \alpha}\big)^2$$

implies that $2f'' \leq f'^2/f$, so $-\sqrt{-f}$ is concave. Because $f' = R(\rho) > 0$, the function f is increasing. With still another scaling transformation, with $\rho(\mathbf{x}) = \lambda \bar{\rho}(\lambda^{2/3}\mathbf{x})$,

$$f(\lambda) = \inf_{\rho \in S_1} \big(K(\rho) - A(\rho) + \lambda R(\rho)\big) = \lambda^{-1/3} \inf_{\bar{\rho} \in S^\lambda} \big(K(\bar{\rho}) - A(\bar{\rho}) + R(\bar{\rho})\big)$$
$$= \lambda^{-1/3} e[\lambda].$$

The at first sight contradictory properties (iii) and (iv) determine the form of $E[\lambda]$ rather narrowly for an atom, making it almost linear:

4.1.33 Properties of $f(\lambda) = \lambda^{-1/3} E[\lambda]$ for an Atom

(i) $0 \leq f' \leq -\frac{1}{\lambda} f$;

(ii) $\frac{2f}{9\lambda^2} - \frac{2f'}{3\lambda} \leq f'' \leq \frac{f'^2}{2f}$.

Proof:

(i) This follows from $E' < 0$ and $f' = \lambda^{-4/3} R(\rho_\lambda) = R(\rho_1) > 0$, where ρ_λ and ρ_1 are the minimizing densities of S_λ and S_1.

(ii) This follows from $E'' \geq 0$ and the concavity of $-\sqrt{-f}$. □

4.1.34 Consequence

1. With the aid of the virial theorem, $2K = A - R$, which follows from (4.1.33) for any μ, Property (i) may be rewritten as $7R(\rho_\lambda) < A(\rho_\lambda)$, $0 \leq \lambda < 1$. This generalizes Corollary (4.1.28), which held for $\lambda = 1 \Rightarrow \mu = 0$, to the statement that $7R = A$, provided that $0 \leq \lambda < 1$.

2. It is not hard to calculate analytically that $f(0) = -0.572$ and $f'(0) = 0.2424$ (Problem 4); computer analysis of the Thomas–Fermi equation has shown that $f(1)$ is -0.384, and by (4.1.38(ii)) and (4.1.28), $f'(1) = -f(1)/3$. Integrating Property (ii) leads to the bounds

$$\max\{-\lambda^{-1/6}|f(1)|^{1/2}, -\lambda|f(1)|^{1/2} - (1-\lambda)|f(0)|^{1/2}\} \leq -|f(\lambda)|^{1/2}$$
$$\leq \min\{-|f(0)|^{1/2}(1 + \frac{\lambda}{2}\frac{f'(0)}{f(0)}), -|f(1)|^{1/2}\frac{7-\lambda}{6}\}$$

(cf. (III:4.3.20)). The concave hull of the left side can be taken, in which case the greatest difference between the bounds is <2% (see Figure 4.1).

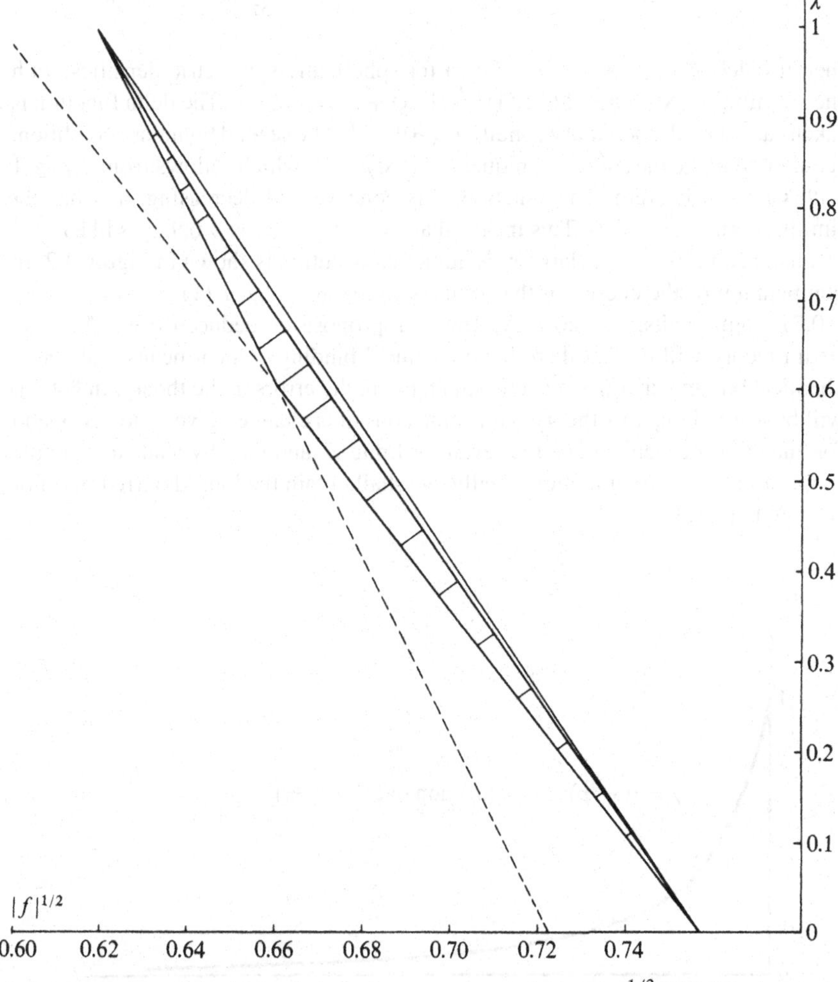

Figure 4.1. The bounds (4.1.34;2) from the concavity of $f(\lambda) = \lambda^{-1/3}E(\lambda)$. The hatched region is allowed.

Since this is already better accuracy than that of the Thomas–Fermi theory itself, there is no point in making fancy numerical calculations of $E[\lambda]$.

If from (4.1.30) we now deduce

4.1.35 The Thomas–Fermi Equation

In the form

$$\Delta\Phi(\mathbf{x}) = -4\pi\delta^3(\mathbf{x}) + 4\pi\rho(\mathbf{x}) = -4\pi\delta^3(\mathbf{x}) + \frac{4}{3\pi}|\mu + \Phi(\mathbf{x})|_+^{3/2},$$

then it reduces to $\sqrt{\xi}\chi'' = \chi^{3/2}\Theta(\chi)$ for spherically symmetric densities, with the substitution $|\mathbf{x}| = r = \xi(3\pi/4)^{2/3}$, $\Phi(\mathbf{x}) + \mu = \chi(\xi)/r$. The delta function is taken care of by the boundary condition $\chi(0) = 1$. The second boundary condition, required to make the solution unique, is $\chi'(\infty) = \mu$, which follows from $\int \rho \leq 1$ with Gauss's theorem. The function χ is concave and decreasing, and has the limiting forms for $\mu = 0$. This means that for neutral atoms ρ behaves like $r^{-3/2}$ at small r, and like r^{-6} at large r. A numerical solution is shown in Figure 4.2. In computation of the energy of the solution yields the value $E(1) = -0.384$, i.e., -0.77 atomic units, or -20.7 eV. The final proposition deduced from Thomas–Fermi theory will be that there is no chemical binding, which means that actual chemical binding energies must be smaller than the errors in the theory. In § 4.3 it will be learned that this theory with some constants changed gives a lower bound for quantum-mechanical energies even for finite Z, and thereby leads to a simple proof of the stability of matter. Finally, we shall obtain the long-deferred proof of Inequality (4.1.3).

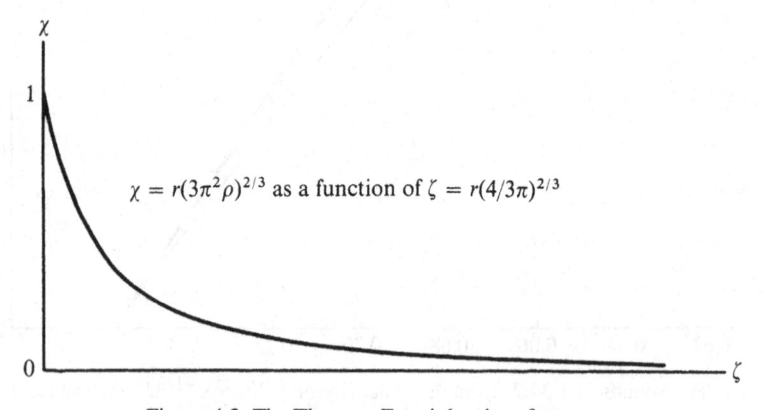

Figure 4.2. The Thomas–Fermi density of an atom.

4.1.36 Monotony of the Thomas–Fermi Potential with Respect to the Nuclear Charges

Let $\Phi_{1,2}$ and $\rho_{1,2}$ be the solutions of the Thomas–Fermi equation with $\mu = 0$ and nuclear charges $z_k^{(1,2)}$. If $z_k^{(1)} \geq z_k^{(2)}$ for all k, then $\Phi_1(\mathbf{x}) \geq \Phi_2(\mathbf{x})$ and $\rho_1(\mathbf{x}) \geq \rho_2(\mathbf{x})$ for all $\mathbf{x}$.

4.1.37 Remarks

1. The normalization $\sum_k z_k^{(i)} = 1$ has of course been dropped.
2. The condition $\mu = 0$ means $\int d^3x\rho_1(\mathbf{x}) = \sum_k z_k^{(1)} \geq \int d^3x\rho_2(\mathbf{x}) = \sum_k z_k^{(2)}$.
3. This can be interpreted as showing how increasing all the nuclear charges causes the configuration with lower energy to have a higher electron density.

Proof: As in the proof of (4.1.31(ii)), let $A \equiv \{\mathbf{x} \in \mathbb{R}^3 : \Phi_1(\mathbf{x}) < \Phi_2(\mathbf{x})\}$. Then A is open and contains none of the $\mathbf{X}_k$, and on it $\psi(\mathbf{x}) \equiv \Phi_1(\mathbf{x}) - \Phi_2(\mathbf{x})$ is negative, continuous, and satisfies

$$\Delta\psi|_A = (\Phi_1^{3/2} - \Phi_2^{3/2})|_A < 0.$$

Hence ψ approaches its infimum on A either on the boundary or at infinity where it vanishes. Since it then vanishes throughout A, the set A must be empty.

The next fact to show is that molecular energies are always greater then those of the isolated atoms. This will require the

4.1.38 Feynman–Hellmann Formula of Thomas–Fermi Theory

Let $E(Z) = \inf_\rho(K(\rho) - ZA(\rho) + R(\rho))$. Then $\partial E/\partial Z = -A(\rho_Z)$, where ρ_Z is the density that minimizes $E(Z)$.

Proof: The function $E(Z)$ is concave, and its right and left derivatives are $\lim_{\varepsilon\downarrow 0}(-A(\rho_{Z\pm\varepsilon}))$, another consequence of the interplay between the concavity of $E(Z)$ and the convexity of the functional in the variable ρ as in (4.1.38(ii)). Since, as shown in Problem 3, for any Z there exists a unique minimizing ρ_Z on a certain compact set, the densities ρ_Z depend continuously on Z. In fact the individual contributions to $E(Z)$ are continuous in Z as well as $E(Z)$ itself. Therefore both the right and the left derivative coincide with $-A(\rho_Z)$. □

Let us now start treating E as a function of each of the nuclear charges, so

$$E(z_1, \ldots, z_M) = \inf_{\rho\in S}\left\{\frac{3}{5}(3\pi^2)^{2/3}\int \rho^{5/3} - \sum_{k=1}^{M} z_k \int \frac{\rho(\mathbf{x})}{|\mathbf{x} - \mathbf{X}_k|} + \frac{1}{2}\int \frac{\rho(\mathbf{x})\rho(\mathbf{y})}{|\mathbf{x} - \mathbf{y}|} + \sum_{k>i} \frac{z_k z_l}{|\mathbf{X}_k - \mathbf{X}_l|}\right\},$$

and define

$$\begin{aligned} E(Z) &= E(Zz_1, Zz_2, \ldots, Zz_M), \\ E_1(Z) &= E(Zz_1, \ldots, Zz_j, 0, 0, \ldots), \\ E_2(Z) &= E(0, \ldots, 0, Zz_{j+1}, \ldots, Zz_M). \end{aligned}$$

Let $\rho \geq \rho_{1,2}$ and $\Phi \geq \Phi_{1,2}$ be the solutions of the appropriately subscripted Thomas–Fermi equations. Then

$$\begin{aligned} \frac{\partial E_1}{\partial Z} &= \sum_{k=1}^{j} z_k \{ Z \sum_{i \neq k} \frac{z_i}{|\mathbf{X}_i - \mathbf{X}_k|} - \int \frac{d^3x \rho(\mathbf{x})}{|\mathbf{x} - \mathbf{X}_k|} \} \\ &= \sum_{k=1}^{j} z_k \lim_{\mathbf{x} \to \mathbf{X}_k} \left(\Phi_1(\mathbf{x}) - \frac{Z z_k}{|\mathbf{x} - \mathbf{X}_k|} \right), \end{aligned}$$

and likewise for E_2. The difference between the energy of the total system and the sum of the energies of the subsystems is easily found to satisfy

$$\frac{\partial E}{\partial Z} - \frac{\partial E_1}{\partial Z} - \frac{\partial E_2}{\partial Z} = \sum_{k=1}^{j} z_k (\Phi(\mathbf{x}_k) - \Phi_1(\mathbf{x}_k)) + \sum_{k=j+1}^{M} z_k (\Phi(\mathbf{x}_k) - \Phi_2(\mathbf{x}_k)) \geq 0.$$

Since E and $E_{1,2}$ become zero when $Z = 0$, this calculation proves the

4.1.39 Instability of Molecules in Thomas–Fermi Theory

$$E(z_1, \ldots, z_M) \geq E(z_1, \ldots, z_j) + E(z_{j+1}, \ldots, z_M).$$

4.1.40 Remarks

1. In the absence of nuclear repulsion the inequality is reversed; in that case

$$\frac{\partial E}{\partial Z} - \frac{\partial E_1}{\partial Z} - \frac{\partial E_2}{\partial Z} = \sum_{k=1}^{j} z_k \int \frac{\rho_1(\mathbf{x}) - \rho(\mathbf{x})}{|\mathbf{x} - \mathbf{X}_k|} + \sum_{k=j+1}^{M} \int \frac{\rho_2(\mathbf{x}) - \rho(\mathbf{x})}{|\mathbf{x} - \mathbf{X}_k|} \leq 0.$$

Although Thomas–Fermi theory predicts some attraction between the nuclei, it is weaker than their Coulomb repulsion. It can even be shown that if the nuclear coordinates are scaled by $X_k \to RX_k$, then E is a convex, decreasing function of R. Thus Thomas–Fermi theory predicts positive pressure and compressibility. However, the molecular energy is not a sum of pair potentials, but contains many-body potentials with alternating signs [34].

2. An alternative version of this theorem reads

$$\sum_{k>i} \frac{z_k z_i}{|\mathbf{X}_k - \mathbf{X}_i|} \geq \sum_{k=1}^{M} z_k \int \frac{\rho(\mathbf{x})}{|\mathbf{x} - \mathbf{X}_k|} - \frac{1}{2} \int \frac{\rho(\mathbf{x})\rho(\mathbf{y})}{|\mathbf{x} - \mathbf{y}|} - \frac{3}{5}(3\pi^2)^{2/3} \int \rho^{5/3} + \sum_k E(z_k)$$

for all $\mathbf{X}_k \in \mathbb{R}^3$ and $\rho \in S$. If $K(\rho)$ is replaced with $(1/\gamma)K(\rho)$, then, because of the way dilatations affect single atoms, $E(z_k)$ becomes $\gamma E(z_k)$. The computed value $E(1) = -0.384$ then leads to Equation (4.2), provided that $\mathbf{X}_k$ are interpreted as the coordinates of the electrons.

3. The proof of (4.1.39) works the same way for a Yukawa potential $\exp(-\mu r)/r$ in place of $1/r$. Because $\Delta \exp(-\mu r)/r + 4\pi\delta^3(\mathbf{x}) = \mu^2 \exp(-\mu r)/r > 0$, the argument with subharmonicity likewise works: $\Delta\psi|_A = \Phi_1^{3/2} - \Phi_2^{3/2} + \mu^2(\Phi_1 - \Phi_2) < 0$, which implies that A must be empty.

4.1.41 Problems

1. Let $H = |\mathbf{p}|^2 + V(\mathbf{x})$ act on $L^2(\mathbb{R}^3)$, and assume that $|V|_- \in L^{5/2}(\mathbb{R}^3)$ and let e_i be the negative eigenvalues of H. Use the bound of Ghirardi and Rimini (III:3.5.35;2) to show that

$$\sum_i |e_i| \leq \frac{4}{15\pi} \int d^3x |V(\mathbf{x})|_-^{5/2}$$

and derive Inequality (4.3) from this fact.

2. Use Problem 1 to prove the inequality

$$T \equiv \langle\psi| - \sum_{i=1}^{N} \Delta_i |\psi\rangle \geq \frac{3}{5}\left(\frac{3\pi}{4}\right)^{2/3} \int d^3x \rho^{5/3}(\mathbf{x})$$

for spin-$\frac{1}{2}$ fermions, where

$$\rho(\mathbf{x}_1) = N \sum_{\alpha_i} \int d^3x_2 \ldots d^3x_N |\psi(\mathbf{x}_1, \mathbf{x}_2, \ldots, \mathbf{x}_N; \alpha_2, \ldots, \alpha_N)|^2,$$

α being the spin index. (Hint: use $\rho^{2/3}$ as the potential in Problem1.)

3. Show that the sts $S \equiv \{\rho \in L^1 \cap L^{5/3} : \rho \geq 0, \|\rho\|_1 \leq N, \|\rho\|_{5/3} \leq K\}$ are compact in the weak $L^{5/3}$ topology, and that the functional $S \to \mathbb{R}$:

$$\varepsilon(\rho) = \frac{3}{5}(3\pi^2)^{2/3} \int d^3x \rho^{5/3}(\mathbf{x}) - \int d^3x \rho(\mathbf{x}) \Big(\sum_k \frac{z_k}{|\mathbf{x} - \mathbf{X}_k|} + \mu\Big) + \frac{1}{2} \int \frac{d^3x d^3x'}{|\mathbf{x} - \mathbf{x}'|} \rho(\mathbf{x})\rho(\mathbf{x}') + \sum_{k>j} \frac{z_k z_j}{|\mathbf{X}_k - \mathbf{X}_j|}$$

has Properties (4.1.8) if $\mu \leq 0$: It is

(i) weakly $L^{5/3}$ lower semicontinuous;

(ii) strictly convex; and

(iii) $\geq \frac{3}{5}(3\pi^2)^{2/3}\|\rho\|_{5/3}^{5/3} - 3(\frac{2}{3})^{5/6}(8\pi)^{1/3}\|\rho\|_{5/3}^{5/6} + |\mu|\|\rho\|_1$.

Conclude that the infimum is attained, and in fact precisely with the ρ of (4.1.36).

4. Solve the Thomas–Fermi equation without Coulomb repulsion, compare with (III:4.5.10), and conclude that the next correction is $O(N^{6/3})$. Use the solution to calculate $f(0)$ and $f'(0)$ of (4.1.34;2).

5. Minimize the functional

$$E(\rho) = \int d^3x \Big(\frac{\rho^2(\mathbf{x})2\pi}{\mu^2} - \sum_i \frac{z_i}{|\mathbf{x} - \mathbf{X}_i|} r(\mathbf{x}) + \frac{1}{2} \int \frac{d^3y \rho(\mathbf{x})\rho(y)}{|\mathbf{x} - \mathbf{y}|} \Big) + \sum_{i>j} \frac{z_i z_j}{|\mathbf{X}_i - \mathbf{X}_j|},$$

and use the result for a new derivation of (III:4.5.14):

$$\sum_{i>j} |\mathbf{X}_i - \mathbf{X}_j|^{-1} \geq \sum_{j=1}^{N} \int \frac{\rho(\mathbf{x})}{|\mathbf{x} - \mathbf{X}_i|} - \frac{1}{2} \int \frac{\rho(\mathbf{x})\rho(\mathbf{y})}{|\mathbf{x} - \mathbf{y}|} - \frac{2\pi}{\mu^2} \int \rho^2 - \frac{\mu N}{2}$$

for all $\mathbf{X}_i \in \mathbb{R}^3$, $\rho \in L^1 \cup L^2$.

4.1.42 Solutions

1. Let $N_E(V)$ denote the number of eigenvalues less than or equal to E. According to (III:3.5.35;2), for all $\alpha > 0$,

$$\begin{aligned} N_{-\alpha}(V) &\leq N_{-\alpha/2}\Big(\Big|V + \frac{\alpha}{2}\Big|_-\Big) \leq \operatorname{tr}\Big[\Big(|\mathbf{p}|^2 + \frac{\alpha}{2}\Big)^{-1/2}\Big|V + \frac{\alpha}{2}\Big|_-\Big(|\mathbf{p}|^2 + \frac{\alpha}{2}\Big)^{-1/2}\Big]^2 \\ &= \frac{1}{(4\pi)^2}\int d^3x d^3y \Big|V(\mathbf{x}) + \frac{\alpha}{2}\Big|_- \frac{\exp(-\sqrt{2\alpha}|\mathbf{x} - \mathbf{y}|)}{|\mathbf{x} - \mathbf{y}|^2}\Big|V(\mathbf{y}) + \frac{\alpha}{2}\Big|_- \\ &\leq \frac{1}{4\pi\sqrt{2\alpha}}\int d^3x \Big|V(\mathbf{x}) + \frac{\alpha}{2}\Big|_-^2. \end{aligned}$$

The last step used Young's inequality, $\|f \cdot (v * g)\|_1 \leq \|v\|_1\|f\|_2\|g\|_2$. Now simply think about what $N_E(V)$ means (see Figure 4.3).

$$\begin{aligned} \sum_j |e_j(V)| &= \int_0^\infty d\alpha N_{-\alpha}(V) \leq \frac{\sqrt{2}}{8\pi}\int d^3x \int_0^{2|V(\mathbf{x})|_-} \frac{d\alpha}{\sqrt{\alpha}}\Big(V(\mathbf{x}) + \frac{\alpha}{2}\Big)^2 \\ &= \frac{4}{15\pi}\int d^3x |V(\mathbf{x})_-^{5/2}. \end{aligned}$$

If $|V|_- \in L^{5/2}$, then the negative part of the spectrum of H is discrete, and we may also write

$$\operatorname{Tr} ||\mathbf{p}|^2 + V(\mathbf{x})|_- \leq 4\pi \int \frac{d^3x d^3p}{(2\pi)^3} ||\mathbf{p}|^2 + V(\mathbf{x})|_-.$$

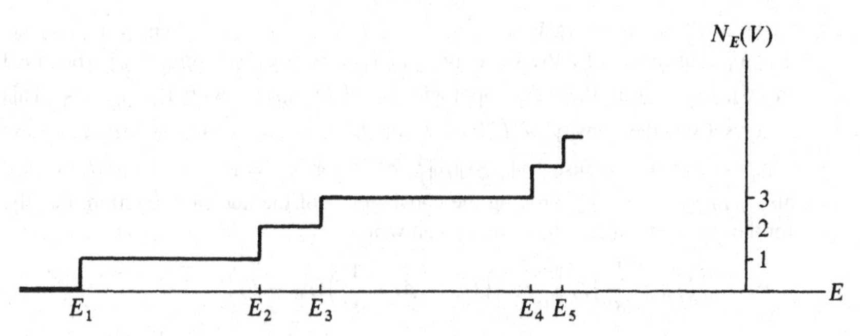

Figure 4.3. The dependence of the function $N_E(V)$ on E.

The partition function can be bounded with the observation that

$$\ln(1+\exp(-\beta H)) = \int_{-\infty}^{\infty} dE\,\Theta(E-H)\beta(1+\exp(\beta E))^{-1} \Rightarrow$$

$$\operatorname{tr}\ln(1+\exp[-\beta(|\mathbf{p}|^2+V(\mathbf{x}))]) \le 4\pi \int \frac{d^3x d^3p}{(2\pi)^3} \ln(1+\exp[-\beta(|\mathbf{p}|^2+V(\mathbf{x}))]).$$

2. Let

$$\rho_\pm(\mathbf{x}_1) = N \sum_{\alpha_2,\ldots,\alpha_N} \int d^3x_2 \cdots d^3x_N |\psi(\mathbf{x}_1,\mathbf{x}_2,\ldots,\mathbf{x}_N;\pm,\alpha_2,\ldots,\alpha_N)|^2$$

be the densities of the electrons with spin $\pm\frac{1}{2}$ and $K = T(\int \rho_+^{5/3} + \int \rho_-^{5/3})^{-1}$. Because of Problem 1 and the min-max principle the lowest energy E_0 of the Hamiltonian

$$H = \sum_i \left(|\mathbf{p}_i|^2 - \frac{5K}{3}[\pi_{i,+}\rho_+^{2/3}(\mathbf{x}_i) + \pi_{i,-}\rho_-^{2/3}(\mathbf{x}_i)] \right),$$

where $\pi_{i,\pm}$ are the spin projections, satisfies the inequalities

$$-\frac{4}{15\pi}\left(\frac{5K}{3}\right)^{5/2}\left(\int \rho_+^{5/3} + \int \rho_-^{5/3}\right) \le E_0 \le \langle\psi|H|\psi\rangle$$
$$= T - \frac{5K}{3}\left(\int \rho_+^{5/3} + \int \rho_-^{5/3}\right).$$

This implies that $K \ge \frac{3}{5}(3\pi/2)^{2/3}$, and then the convexity of the function $x \to x^{5/3}$ yields the inequality for $\rho = \rho_+ + \rho_-$.

3. Since

$$\|\rho\|_{5/3} = \sup_{\|\rho'\|_{5/2}=1} |\langle\rho'|\rho\rangle| \quad \text{and} \quad \|\rho\|_1 = \sup_{\substack{\rho' \in L^{5/2}\cup L^\infty \\ \|\rho'\|_\infty = 1}} |\langle\rho'|\rho\rangle|$$

are suprema over weakly continuous functions, they are weakly lower semicontinuous, so S is weakly compact.

(i) This proposition is equivalent to the statement that $\rho_n \rightharpoonup \rho \Rightarrow \underline{\lim}\,\varepsilon(\rho_n) \ge \varepsilon(\rho)$. First note that $\|\rho\|_{5/3}$ is weakly lower semicontinuous, i.e., $\lim \|\rho_n\|_{5/3} \ge \|\rho\|_{5/3}$. Moreover, $\lim_{n\to\infty} \int \rho_n(1/|\mathbf{x}|) = \int \rho(1/|\mathbf{x}|)$. If the potential $1/|\mathbf{x}|$ is broken up as

$1/|\mathbf{x}| = V_1 + V_2$, where $V_1 \in L^{5/2}$, $V_2 \in L^p$, $3 < p \leq \infty$, then by assumption $\int \rho_n V_1$ converges to $\int \rho V_1$. Since $\sup_n \|\rho_n\|_1 < \infty$ (by assumption $\{\rho_n\}$ is bounded in L^1), $\rho_n \to \rho$ in the weak topologies of all L^q spaces with $1 < q \leq \frac{5}{3}$. This follows from the density of $L^{5/2} \cap L^s$ in L^s for $s \geq \frac{5}{2}$, $1/s + 1/q = 1$, and $\sup_n \|\rho_n\| < \infty$, because $\|\rho\|_q \leq \|\rho\|_p^\alpha \|\rho_r^{1-\alpha}$ for $1/q = \alpha/p + (1-\alpha)/r$. Hence also $\int \rho_n V_2 \to \int \rho V_2$, proving the convergence of the nuclear attraction. Finally, for the repulsion of the electrons we can write

$$\left\| \left(\rho_n * \frac{1}{|\mathbf{x}|}\right)\rho_n \right\|_1 = \left\| \left((\rho_n - \rho) * \frac{1}{|\mathbf{x}|}\right)(\rho_n - \rho) \right\|_1 + 2\left\| \left(\rho * \frac{1}{|\mathbf{x}|}\right)\rho_n \right\|_1 - \left\| \left(\rho * \frac{1}{|\mathbf{x}|}\right)\rho \right\|_1.$$

By Young's inequality, if V is broken up as above and $\rho \in L^1$, then $\rho * V_1 \in L^{5/2}$, $\rho * V_2 \in L^p$, $3 < p \leq \infty$, so the mixed term on the right converges to $2\|(\rho * 1/|\mathbf{x}|)\rho\|_1$, while the first term is positive. Therefore

$$\underline{\lim} \left\| \left(\rho_n * \frac{1}{|\mathbf{x}|}\right)\rho_n \right\| \geq \left\| \left(\rho * \frac{1}{|\mathbf{x}|}\right)\rho \right\|.$$

(ii) $\rho^{5/3}$ is strictly convex, $\int \rho(1/|\mathbf{x}|)$ is linear, and

$$\int \frac{d^3x \rho(\mathbf{x})\rho(\mathbf{y})}{|\mathbf{x} - \mathbf{y}|} = c \int \frac{d^3k}{|\mathbf{k}|^2} |\tilde{\rho}(\mathbf{k})|^2,$$

$c > 0$, is strictly convex.

(iii) The proof of semiboundedness on S will require the following refinements of our earlier estimates. Let $R > 0$,

$$I_+ \equiv \int_{|\mathbf{x}| \geq R} d^3x \frac{\rho(\mathbf{x})}{|\mathbf{x}|}, \quad I_- \equiv \int_{|\mathbf{x}| \leq R} d^3x \frac{\rho(\mathbf{x})}{|\mathbf{x}|} \quad \text{and} \quad f(\mathbf{x}) = \frac{\delta(|\mathbf{x}| - R)}{4\pi}.$$

It follows from

$$\int \frac{d^3x d^3y}{|\mathbf{x} - \mathbf{y}|} \Big(\rho(\mathbf{x})\Theta(|\mathbf{x}| - R) - f(\mathbf{x})\Big)\Big(\rho(\mathbf{y})\Theta(|\mathbf{y}| - R) - f(\mathbf{y})\Big) \geq 0,$$

$$I_+ = \int_{\|\mathbf{y}\| \geq R} \frac{d^3x d^3y}{|\mathbf{x} - \mathbf{y}|} f(\mathbf{x})\rho(\mathbf{y}),$$

and

$$\int \frac{d^3x d^3y}{|\mathbf{x} - \mathbf{y}|} f(\mathbf{x}) f(\mathbf{y}) = \frac{1}{R}$$

that

$$\frac{1}{2} \int_{\substack{|\mathbf{x}| \geq R \\ |\mathbf{y}| \geq R}} \frac{d^3x d^3y}{|\mathbf{x} - \mathbf{y}|} \rho(\mathbf{x})\rho(\mathbf{y}) - I_+ \geq -\frac{1}{2R},$$

and by Hölder's inequality

$$|I_-| \leq \left\| \frac{\Theta(R - |\mathbf{x}|)}{|\mathbf{x}|} \right\|_{5/2} \|\rho\|_{5/3} = (64\pi^2 R)^{1/5} \|\rho\|_{5/3}.$$

If R is chosen as $R = \frac{5}{2}((8\pi)^{2/5}\|\rho\|_{5/3})^{-1}$, then with $\sum_k z_k \leq 1$,

$$\varepsilon(\rho) \geq \frac{3}{5}(3\pi^2)^{2/3}\|\rho\|_{5/3}^{5/3} - 3\left(\frac{2}{3}\right)^{5/6}(8\pi)^{1/3}\|\rho\|_{5/3}^{5/6} + \sum_{k>j}\frac{z_k z_j}{|\mathbf{X}_k - \mathbf{X}_j|},$$

and the function $ax^2 - bx + c$ is bounded below on $\mathbb{R}$ for non-negative a, b, and c. If $\mu < 0$, then because of (iii) the infimum is attained for a ρ in the interior of one of the compact sets S, and ρ must satisfy the Thomas–Fermi equation (4.1.36) by the same argument as in (4.1.12). If $\mu = 0$, then there is also the possibility that the infimum is attained on the boundary $\|\rho\|_1 = N$ of every set S. In that event it would still satisfy the Thomas–Fermi equation with some μ as the Lagrange multiplier for the constraint $\|\rho\|_1 = N$. However, if $N > \sum_i z_i \leq 1$, then there is no such solution, as otherwise $\Phi(\mathbf{x})$ would be negative for large $|\mathbf{x}|$, contradicting (4.1.37(ii)). Therefore, if $\mu = 0$, then the infimum still lies in the interior of some set S.

4. Use units such that $e = \hbar = 2m = 1$, and suppose there is spin. Then

$$E = \int d^3x\left[\frac{3}{5}(3\pi^2)^{2/3}\rho^{5/3} - \frac{Z}{r}\rho\right].$$

From the Thomas–Fermi equations,

$$(3\pi^2\rho)^{2/3} - \frac{Z}{r} + \mu = 0 \Rightarrow \rho = \frac{Z^{3/2}}{3\pi^2}\left(\frac{1}{r} - \frac{1}{R}\right)^{3/2}, \qquad \mu = Z/R.]$$

$$N = \frac{Z^{3/2}}{3\pi^2}4\pi\int_0^R r^2 dr\left(\frac{1}{r} - \frac{1}{R}\right)^{3/2} = \frac{(ZR)^{3/2}}{12} \Rightarrow R$$

$$= N^{-1/3}\frac{N}{Z}4\left(\frac{3}{2}\right)^{2/3},$$

$$-V = Z^{5/2}\frac{4\pi}{3\pi^2}\int_0^R r dr\left(\frac{1}{r} - \frac{1}{R}\right)^{3/2} = \frac{6NZ}{R},$$

$$T = \frac{3}{5}Z^{5/3}\frac{4\pi}{3\pi^2}r^2 dr\left(\frac{1}{r} - \frac{1}{R}\right)^{5/2} = -\frac{V}{2},$$

and

$$E = -T = \frac{3NZ}{R} = -\frac{1}{2}\left(\frac{3}{2}\right)^{1/3}Z^2 N^{1/3}$$

(in units with $2m = 1$; twice this if $m = 1$), so

$$f(0) = -\frac{1}{2}\left(\frac{3}{2}\right)^{1/3} = -0.572,$$

$$f'(0) = \frac{1}{2}\int\frac{d^3x d^3y}{|\mathbf{x} - \mathbf{y}|}\rho(\mathbf{x})\rho(\mathbf{y}) = \left(\frac{4}{3\pi}\right)^2 \int_{0^1 \frac{dr}{\sqrt{r}}(1-r)^3/2} \int_0^r dr'\sqrt{r'}(1 - r')^{3/2}\frac{(ZR)^3}{R}$$

$$= 0.24244, \qquad \text{if} \quad Z = N = 1.$$

If we read the exact ground-state energy off from (III:4.5.9), then to $O(N^{-1/3})$,

$$E_{\text{exact}}/E_{\text{Thomas–Fermi}} = 1 - N^{-1/3}\frac{1}{2}\left(\frac{3}{2}\right)^{1/3}.$$

Thus the Thomas–Fermi energy is below the actual ground-state energy.

5. The density that minimizes E is

$$\rho_0(\mathbf{x}) = \frac{\mu^2}{4\pi} \sum_j z_j \frac{\exp(-\mu|\mathbf{x} - \mathbf{X}_j|)}{|\mathbf{x} - \mathbf{X}_j|},$$

with which

$$E(\rho_0) = -\frac{N\mu}{2} + \sum_{i>j} z_i z_j \Big(\frac{1}{|\mathbf{X}_i - \mathbf{X}_j|} - \mu \exp[-\mu|\mathbf{X}_i - \mathbf{X}_j|] \Big)$$
$$> -\frac{N\mu}{2} \qquad \text{for all } \mathbf{X}_i \in \mathbb{R}^3.$$

If $z_i = 1$, this reduces to (III:4.5.14). In this variant of Thomas–Fermi theory the electron cloud creates an attractive potential $-\mu \exp(\mu r)$ between the nuclei, which is also weaker than their $1/r$ Coulomb repulsion.

4.2 Cosmic Bodies

The Thomas–Fermi theory of stars is thermodynamically more interesting than that of atoms, since it predicts an unusual phase transition

In the year 1926 great discoveries about the laws of matter appeared in rapid succession. Shortly after E. Schrödinger published the equation named after him E. Fermi discovered the distribution law (2.5.16;1) governing particles that satisfy the exclusion principle. This inspired L. Thomas's ingenious idea that the electron cloud of a large atom should satisfy equation (4.1.36) at $T = 0$. Then, still in the year 1926, R. Fowler realized that the stability of cosmic matter is ensured by the zero-point energy of the electrons, and that a cosmic body is closely analogous to a "gigantic molecule in the ground state". Yet it has taken considerably longer to found this vision in mathematics and derive everything from the Schrödinger equation. Today, however, the derivation goes through without gaps, and the Thomas–Fermi theory of atoms and stars is the only many-body problem with realsitic forces to have succumbed, in the appropriate thermodynamic limit, to mankind's attempt at calculation. Yet the zero-point energy guarantees stability only in so far the speeds of the electrons remain slow in comparison with light. If they enter the regime of relativistic kinematics, for which the kinetic energy $\sim c|\mathbf{p}|$, then the zero-point energy goes as $N(N/V)^{1/3}$, whereas the gravitational energy goes as $-\kappa N^2/V^{1/3}$. If $N > (\kappa m_p^2)^{-3/2} \sim 10^{57}$, then the latter predominates, and as V becomes smaller and smaller, the total energy goes to $-\infty$. We shall avoid this instability by remaining within the framework of nonrelativistic kinematics, considering only stars of masses somewhat smaller than that of the sun. Then, according to the estimates (1.2.11;3), if $N > 10^{54}$, the minimum energy occurs when $V \sim N^{-1}$. The situation is again like that of Thomas–Fermi theory, which leads to the hope that the many-body problem can be solved in the limit

$N \to \infty$ with the Hamiltonian

$$H = \sum_{i=1}^{N} \frac{|\mathbf{p}_i|^2}{2m_i} + \sum_{i>j} \frac{e_i e_j - \kappa m_i m_j}{|\mathbf{x}_i - \mathbf{x}_j|}. \tag{4.7}$$

In this limit the system becomes a highly compressed plasma, so the average gravitational field would be expected to be so dominant that the Thomas–Fermi equation is valid. Of course, the total charge of the system must be zero, or, more exactly, the possible excess charge ΔQ is bounded by $(\Delta Q)^2 \le \kappa m_p^2 N_p^2$, so for gravity to predominate, $\Delta Q < 10^{-19} N_p$. Indeed, these conjectures can be derived mathematically for all three ensembles:

4.2.1 The Asymptotic Forms of the State Functions

Let $H_{N,V}$ be the Hamiltonian (4.7) for N_1 positive and N_2 negative fermions of masses $M_{1,2}$, charges e_1 and $e_2 = -e_1$, and spin $\frac{1}{2}$ in a volume V. Let N denote the pair (N_1, N_2). Then the limits

$$E(N, S, V) = \lim_{\lambda\to\infty} \lambda^{-7/3} \inf_{\mathcal{H}_{\lambda S}} \mathrm{Tr}_{\mathcal{H}_{\lambda S}} H_{\lambda N, \lambda^{-1} V},$$

$$F(N, \beta, V) = -\lim_{\lambda\to\infty} \beta^{-1}\lambda^{-1} \ln \mathrm{Tr} \exp(-\beta\lambda^{-4/3} H_{\lambda N, \lambda^{-1} V}). \tag{4.8}$$

exist.

The grand-canonical function Ξ is not defined as in (4.1.6), as now the finiteness of the sum $\sum_N$ requires a factor $N^{-2/3}$ in the interaction if $V \sim N$ [27] (see (4.2.10;4)). With the solution of the Thomas–Fermi equation

$$\rho_\alpha(\mathbf{x}) = 2 \int \frac{d^3p}{(2\pi)^3} [1 + \exp(\beta(\frac{|\mathbf{p}|^2}{2M_\alpha} + W_\alpha(\mathbf{x}) - \mu_\alpha))]^{-1}, \tag{4.9}$$

$$W_\alpha(\mathbf{x}) = \sum_\beta \int_V d^3x' \frac{e_\alpha e_\beta - \kappa M_\alpha M_\beta}{|\mathbf{x} - \mathbf{x}'|} \rho_\beta(\mathbf{x}'), \quad \alpha, \beta = 1, 2, \tag{4.10}$$

$$\int_V d^3x \rho_\alpha(\mathbf{x}) = N_\alpha, \tag{4.11}$$

these quantities are found to be

$$E(N, S, V) = \sum_\alpha \int_V d^3x \left\{ \frac{1}{2} \rho_\alpha(\mathbf{x}) W_\alpha(\mathbf{x}) + 2 \int \frac{d^3p}{(2\pi)^3} \frac{|\mathbf{p}|^2/2M_\alpha}{1 + \exp[\beta(|\mathbf{p}|^2/2M_\alpha + W_\alpha(\mathbf{x}) - \mu_\alpha)]} \right\}, \tag{4.12}$$

$$F(N, \beta, V) = \sum_\alpha \{ -\int_V d^3x \frac{1}{2} \rho_\alpha(\mathbf{x}) W_\alpha(\mathbf{x}) + N_\alpha \mu_\alpha - 2T \int_V d^3x \int \frac{d^3p}{(2\pi)^3} \times \ln(1 + \exp[-\beta(\frac{|\mathbf{p}|^2}{2M_\alpha} + W_\alpha(\mathbf{x}) - \mu_\alpha)])\}, \tag{4.13}$$

and

$$\Xi(\mu_1, \mu_2, \beta, V) = \sum_\alpha \{ \int_V d^3x \frac{1}{2} \rho_\alpha(\mathbf{x}) W_\alpha(\mathbf{x}) + 2T \int_V d^3x \int \frac{d^3p}{(2\pi)^3}$$

$$\times \ln\left(1 + \exp\left[-\beta\left(\frac{|\mathbf{p}|^2}{2M_\alpha} + W_\alpha(\mathbf{x}) - \mu_\alpha\right)\right]\right)\}. \tag{4.14}$$

4.2.2 Gloss

1. For $\lambda S \in \ln \mathbb{Z}^+$, $\mathcal{H}_{\lambda S}$ is an $\exp(\lambda S)$-dimensional subspace of $\mathcal{H}$.
2. The thermodynamic limit has been taken in the sense discussed in (1.2.9), i.e., $E \sim N^{7/3}$, $V \sim N^{-1}$, $S \sim N$, and $T \sim E/S \sim N^{4/3}$. The energies E and F are accordingly neither per particle nor per volume; these specific energies and energy densities do not have thermodynamic limits.
3. The quantity $S = \beta(E - F)$ is extensive for $\beta \sim N^{-4/3}$ and $E - F \sim N^{7/3}$.
4. If one insists on the usual relationships $E \sim N$, $V \sim N$, $S \sim N$, with T constant, then according to (1.2.9) the interaction has to be taken as

$$N^{-2/3} \sum_{i>j} \frac{e_i e_j - \kappa m_i m_j}{|\mathbf{x}_i - \mathbf{x}_j|}.$$

   This means that the system is imagined as getting larger and larger with an ever weaker interaction; all such problems are mathematically equivalent because of the scaling law of (1.2.1). Physically relevant systems are large but finite and have weak, but still nonzero, gravitational interaction. The question of how reasonable the thermodynamic limit is depends only on whether the physical object is sufficiently like the limiting system. If so, the convergence of the thermodynamic quantities (4.2.1) guarantees that the relevant observables of the finite system will have values fairly near those of the infinite idealization.

5. Since ρ_α is a strictly monotonic function of μ_α, the normalization (4.2.6) is an implicit equation for μ_α.

6. We shall soon discover that for certain values of β, N, and V there is more than one solution of the Thomas–Fermi equations. The question of which solutions are the correct limits (4.2.1) is decided by the minimum principles for the thermodynamic potentials (2.3.3;4), (2.2.21;1), and (2.5.3), which survive the limit $\lambda \to \infty$ in the following manner (cf.(4.1.18)): The functionals for energy, entropy, and the phase-space densities n_α are

$$E(n) = -\frac{1}{2} \sum_{\alpha,\beta} \int d^3x d^3x' \frac{d^3p d^3p'}{(2\pi)^6} n_\alpha(\mathbf{x}, \mathbf{p}) n_\beta(\mathbf{x}', \mathbf{p}') \frac{e_\alpha e_\beta - \kappa M_\alpha M_\beta}{|\mathbf{x} - \mathbf{x}'|}$$

$$+ \sum_\alpha \int d^3x \frac{d^3p}{(2\pi)^3} \frac{|\mathbf{p}|^2}{2M_\alpha} n_\alpha(\mathbf{x}, \mathbf{p}),$$

$$S(n) = -2\sum_\alpha \int d^3x \frac{d^3p}{(2\pi)^3}\left[\frac{n_\alpha}{2}\ln\frac{n_\alpha}{2} + \left(1-\frac{n_\alpha}{2}\right)\ln\left(1-\frac{n_\alpha}{2}\right)\right],$$

$$N_\alpha(n) = \int d^3x \frac{d^3p}{(2\pi)^3} n_\alpha(\mathbf{x}, \mathbf{p}).$$

The correct Thomas–Fermi densities are those that minimize the energy for given N_α and S. The variational derivative with T and μ_α as Lagrange multipliers leads to the Thomas–Fermi equations (4.9) — (4.11) again, with

$$\rho_\alpha(\mathbf{x}) = \int \frac{d^3p}{(2\pi)^3} n_\alpha(\mathbf{x}, \mathbf{p}),$$

for the solution of

$$\frac{\delta}{\delta n_\alpha(\mathbf{x}, \mathbf{p})}(E - TS + \mu_1 N_1 + \mu_2 N_2) = 0.$$

However, this equation is also satisfied by merely local extrema and by saddle points. At the minimizing density, $E(n) = E(N_1, N_2, S, V)$.

7. The ensembles are equivalent only in the region where the convex hull of the function $E(S)$ is the same as $E(S)$.

8. We have written E and F as functions of three variables, but it is clear from the definition that they depend on only two ratios. This is reflected in the Thomas–Fermi equation by its scaling behavior when $\mathbf{x} \to \lambda\mathbf{x}$.

Proof: If the only force is gravitation, as in a neutron star ($e = 0$), the methods of § 4.1 are applicable; the lower bound for Ξ is trivial, and Inequality (2.1.5;3) can be used for the upper bound. However, since it requires knowledge of the expectattion value of H, it is necessary to estimate the norm of the quantum fluctuations. If e and κ differ from zero the estimate is much more difficult than that of § 4.1, and can not be given in detail here. The strategy is as follows.

1. Regularization of the potential
   Since one expects the motion of the particles to be determined by an average field, the singular part of the $1/r$ potential should first be cut off, so that the influence of a near-by particle will not be stronger than that of the average field. There are also good physical grounds to insist that the important part of the potential is its long range rather than the singularity, as in reality the singularity is smoothened out with some form factor. By "long range" is meant a length comparable to the diameter of the star, which shrinks to zero as $\lambda \to \infty$. Hence the cut-off length has to be reduced while λ increases, or alternatively one can work in the scaled system (4.2.2;4). It is thus useful to show that changing the potential by, say, $1/r \to (1 - \exp(-\lambda^{1/3} sr))/r$ makes little difference for large s in comparison with the main contribution to the energy, which is $\sim N^{7/3}$. This fact can be shown by an argument similar to the estimate (1.33) and making use of the bound (III:4.5.9) on the number of bound states of a short-range potential.

2. Replacing the potential with a step function
   Since Thomas–Fermi theory is oriented toward free particles in a box, it is useful to divide the volume V into cells inside of which the potential is made constant. The proof that changing the potential to a step function has only a slight effect is trivial, since the continuous function $(1 - \exp(-sr))/r$ can be approximated uniformly on any compact set by a step function.

3. Insertion of walls
   In each of the cells of constant potential the Schrödinger equation reduces to the force-free equation, if they are separated by impenetrable walls. It is thus useful to show that inserting walls will not alter the result much. It is clear that the effect will be to raise all the energy levels. The min-max priniciple can be called upon to show that they do not rise by too much. The presence of the walls means that the wave-function vanishes at their positions, which costs kinetic energy. It is possible to patch together wave-functions for the system without walls so that they vanish at the positions of the walls, and the expectation value of the kinetic energy in such a state is not increased by too much. It is important that the number of walls in this procedure remain constant in the limit $N \to \infty$ so that their effect can be neglected in comparison with $N^{7/3}$.

4. Filling the boxes
   The foregoing manipulations leave the particles in separated boxes moving in constant potentials, which, however, depend on the distribution of the particles. One now finds that the thermodynamic functions of (4.8) are dominated by the contribution from a certain distribution of the particles among the boxes, which is determined by a self-consistent equation, namely the Thomas–Fermi equation for the step potential with walls.

5. Continuity of the Thomas–Fermi equation
   Since we wish to end up with the Thomas–Fermi equation for a $1/r$ potential, we still need to show that the approxiamtions made above for the $1/r$ potential do not change the energy of the solution much. Otherwise, if the solution depended discontinuously on the potential, it would be worthless; the Thomas–Fermi equations can not be solved analytically, and a numerical solution on a computer approximates the potential by a step function. It is thus indispensible, but fortunately also possible, to show that the Thomas–Fermi energy has the required continuity with respect to the potential. □

The structure of the Thomas–Fermi equation is different for stars than for atoms. The energy loses the properties of convexity and weak lower semicontinuity. Consequently the solution is not guaranteed to be unique and there is a possibility of a phase transition, which will be discussed at the conclusion of this section. Meanwhile, we prepare by proving a general

4.2.3 Virial Theorem

The pressure

$$P \equiv -\frac{\partial}{\partial V}F(N,\beta,V),$$

kinetic energy

$$E_k = \sum_\alpha 2\int_V d^3x \int \frac{d^3p}{(2\pi)^3}\frac{|\mathbf{p}|^2/2M_\alpha}{1+\exp[\beta(|\mathbf{p}|^2/2M_\alpha + W_\alpha(\mathbf{x}) - \mu_\alpha)]},$$

and potential energy

$$E_p = \sum_\alpha \int d^3x \frac{1}{2}\rho_\alpha(\mathbf{x})W_\alpha(\mathbf{x})$$

are connected by

$$3PV = 2E_k + E_p.$$

Proof: We start by convincing ourselves of the usual thermodynamic relationships

$$\frac{\partial F}{\partial N_\alpha} = \mu_\alpha \quad \text{and} \quad \beta\frac{\partial F}{\partial \beta} = E - F, \tag{4.15}$$

which follow directly from differentiating (4.13). For this purpose note that ρ depends on β and N, and thus implicity so does W, but that this dependence does not show up when the Thomas–Fermi equations are satisfied. Next rewrite (4.2.8) by integrating by parts in the variable $\mathbf{p}$. The

$$\frac{|\mathbf{p}|^2}{2M_\alpha} = \varepsilon, \qquad \int_0^\infty d\varepsilon\sqrt{\varepsilon}\ln(1+\exp[-\beta(\varepsilon+c)]) = \frac{2}{3}\beta\int_0^\infty \frac{d\varepsilon\varepsilon^{3/2}}{1+\exp[\beta(\varepsilon+c)]},$$

and we conclude that

$$F = \sum_\alpha N_\alpha\mu_\alpha - \frac{2}{3}E_k - E_p. \tag{4.16}$$

Finally, the dilatation relationship mentioned earlier implies that $F(N,\beta,V) = \lambda^{-7/3}F(\lambda N, \lambda^{-4/3}\beta, \lambda^{-1}V)$ for all $\lambda \in \mathbb{R}^+$. With reference to (4.15), the derivative by λ produces

$$0 = -\frac{7}{3}F + \sum_\alpha N_\alpha\mu_\alpha - \frac{4}{3}(E-F) + PV,$$

which concludes the proof of the theorem when combined with (4.16). □

The local densities in phase space,

$$n_\alpha(\mathbf{x},\mathbf{p}) = 2\left[\exp(\beta(\frac{|\mathbf{p}|^2}{2m} + W_\alpha(\mathbf{x}) - \mu_\alpha)) + 1\right]^{-1},$$

have the same significance as in § 4.1. They are stationary solutions of the Vlasov equation (4.1.26;3),

$$\sum_j \frac{\mathbf{p}_j}{M_\alpha} \cdot \frac{\partial}{\partial \mathbf{x}_j} n_\alpha(\mathbf{x}, \mathbf{p}) - \frac{\partial}{\partial \mathbf{p}_j} n_\alpha(\mathbf{x}, \mathbf{p}) \cdot \frac{\partial}{\partial \mathbf{x}_j} W_\alpha(\mathbf{x}) = 0. \tag{4.17}$$

In this equation quantum mechanics enters only through the initial condition $|n_\alpha(\mathbf{x}, \mathbf{p})| \leq 1$. In fact, as a classical equation it is the basis of stellar dynamics [35]. When reduced to configuration space, the local densities describe the hydrostatic equilibrium between the pressure of the matter and of gravitation, in the spherically symmetric case. Since the fermions behave like free particles on the microscopic level, one would expect from (2.28) that

$$\begin{aligned} P(\mathbf{x}) &= \frac{2}{3} E_k(\mathbf{x}) \\ &\equiv \frac{2}{3} \sum_\alpha 2 \int \frac{d^3 p}{(2\pi)^3} \frac{|\mathbf{p}|^2/2M_\alpha}{1 + \exp[\beta(|\mathbf{p}|^2/2M_\alpha + W_\alpha(\mathbf{x}) - \mu_\alpha)]} \end{aligned} \tag{4.18}$$

functions as the pressure, and in fact if (4.2.14) is multiplied by $\mathbf{p}_i$, integrated by $d^3 p$ by parts, and one replaces $\mathbf{p}_i \cdot \mathbf{p}_j \to (|\mathbf{p}|^2/3)\delta_{ij}$, then

$$\nabla P(\mathbf{x}) = -\sum_\alpha \rho_\alpha(\mathbf{x}) \nabla W_\alpha(\mathbf{x}), \tag{4.19}$$

which is the equilibrium condition mentioned above. If the geometry is spherically symmetric, i.e., V is a sphere of radius R and the local observables depend only on $|\mathbf{x}| = r$, then (4.19) can be written as the nonrelativistic **Tolman–Oppenheimer equation**

$$\begin{aligned} \frac{d}{dr} \frac{2}{3} E_k(r) &= -\sum_\alpha \frac{\rho_\alpha(r) M_\alpha(r)}{r^2} \\ M_\alpha(r) &= -\sum_\beta (e_\alpha e_\beta - \kappa M_\alpha M_\beta) \int_0^r dr' 4\pi r'^2 \rho_\beta(r') \end{aligned} \tag{4.20}$$

(cf.(II:4.20)). The electric and gravitational forces have been expressed in terms of the charges and masses within the sphere.

4.2.4 The Connection Between the Local and Global Pressures

By integrating (4.2.17) by $(4\pi/3) \int_0^R dr r^3$ one gets

$$V \frac{2}{3} E_k(R) - \frac{2}{3} E_k = \frac{4\pi}{3} \int_0^R dr r^3 \frac{d}{dr} \frac{2}{3} E_k(r)$$

$$= \sum_{\alpha,\beta} \frac{e_\alpha e_\beta - \kappa M_\alpha M_\beta}{3} \int_0^R dr r 4\pi \rho_\alpha(r) \int_0^r dr' r'^2 4\pi \rho_\beta(r')$$

$$= \frac{E_p}{3},$$

so with the virial theorem (4.2.11) the thermodynamic pressure becomes simply the local pressure at the boundary

$$P = P(R).$$

We see that Thomas–Fermi theory, which begins with the Scrödinger equation, leads eventually to the concepts of classical physics. A more accurate evaluation of the state functions (4.8) requires numerical solutions of Equations (4.9) through (4.11). In order to lend more physical plausibility to those numbers, let us extend the intuitive arguments of § 1.2 to finite temperatures. Since the theory is only valid if gravity is the dominant force, let us simplify by considering only one type of neutral fermion such as neutrons (without nuclear forces). If there were protons and electrons, then the former would provide most of the gravitational force and the latter most of the pressure. This would increase all lengths compared with a system of neutrons by a factor of the ratio of the mass of the neutron to that of the electron, about 2000. Thus, if 10^{57} neutrons are found to have a radius of about 30 km, a similar system made of hydrogen would have a radius of about 6000 km, i.e., that of a white dwarf. We begin with the observation that at a nonzero temperature there is a thermal contribution

$$N\frac{3}{2}T = N\frac{3}{2}\left(\frac{N}{V}\right)^{2/3} \exp\left(\frac{2S}{3N} - 1\right)$$

in addition to the zero-point energy $N(N/V)^{2/3}$. At high temperatures this is exactly the classical expression. In order to interpolate to intermediate temperatures, we shall simply combine the zero-point energy with the classical expression. This turn out to approximate the energy of free fermions (2.28) to within about 20%. It remains to add in the gravitational energy. If the mass of the particles is $\frac{1}{2}$, then up to geometric factors we get

$$\frac{E}{N} = \left(\frac{N}{V}\right)^{2/3}\left(1 + \frac{3}{2e}\exp\left(\frac{2S}{3N}\right)\right) - \frac{\kappa N}{V^{1/3}}$$

in natural units. In checking the properties (2.3.10) of the microcanonical energy density, it becomes readily apparent that, in agreement with (4.2.10;4),

$$\rho^{-1}\varepsilon = \rho^{2/3}\left(1 + \frac{3}{2e}\exp(2\sigma/3\rho)\right) - \kappa N^{2/3}\rho^{1/3}$$

is independent of N only if $\kappa \sim N^{-2/3}$. Although ε increases as a function of σ, conditions (2.3.10(ii)) and (2.3.8(iii)) are not always satisfied; our ansatz does not do justice to the subadditivity (2.3.5). The reason becomes apparent when it is observed that the pressure

$$P = -\frac{\partial E}{\partial V}\Big|_{S,N} = \frac{2}{3}\Big(\frac{N}{V}\Big)^{5/3}\Big(1 + \frac{3}{2e}\exp\Big(\frac{2S}{3N}\Big)\Big) - \frac{\kappa N^2}{3V^{4/3}} = \frac{E - E_p/2}{3V/2},$$

$$E_p = -\frac{\kappa N^2}{V^{1/3}},$$

consists of three parts, from the zero-point, thermal, and gravitational energies. The first two are positive and the last one is negative, and may dominate in the intermediate regime of average densities. However, a negative pressure is physically impossible; the system does not adhere to the walls and pull them inward. What happens is that the system shrinks itself down to such a small radius, $V_0 = (\kappa N^2/(-2E))^3$, that it reaches $P = 0$. A better ansatz consists in replacing V with V_0 in E when $P < 0$,

$$\frac{E}{N} = \Big(\frac{N}{V}\Big)^{2/3}\left(1 + \frac{3}{2e}\exp\Big(\frac{2S}{3N}\Big) - \frac{\kappa N}{V^{1/3}}\right)\Theta_+ - \frac{\kappa^2 N^{4/3}/2}{2 + 3\exp((2S/3N) - 1)}\Theta_-,$$

$$\Theta_\pm = \Theta\Big(\pm(2 + 3\exp\Big(\frac{2S}{3N} - 1\Big) - \kappa(NV^{1/3})\Big).$$

The function $\Theta_\pm$ is also equal to $\Theta(\pm(E + \kappa N^2/2V^{1/3}))$, implying that if the total energy is sufficiently negative, then the system condenses into a volume V_0. As in Example (2.3.27;1) this brings about a phase transition with negative specific heat: The calculation

$$\begin{aligned}\frac{3}{2}T &= \frac{3}{2}\frac{\partial E}{\partial S}\Big|_{V,N} = \frac{3}{2e}\Big(\frac{N}{V}\Big)^{2/3}\exp\Big(\frac{2S}{3N}\Big)\Theta_+ \\ &\quad + \frac{3}{2e}\frac{\kappa^2 N^{4/3}\exp(2S/3N)}{(2 + 3\exp(2S/3N - 1))^2}\Theta_- \\ &= \Big[\frac{E}{N} - \Big(\frac{N}{V}\Big)^{2/3} + \frac{\kappa N}{V^{1/3}}\Big]\Theta_+ + \Big[-\frac{E}{N} - \Big(\frac{2E}{\kappa N^{5/3}}\Big)^2\Big]\Theta_-\end{aligned}$$

reveals that the classical linear dependence of T on E becomes parabolic in the condensation region (see Figure 4.4). The temperature begins to rise again as E

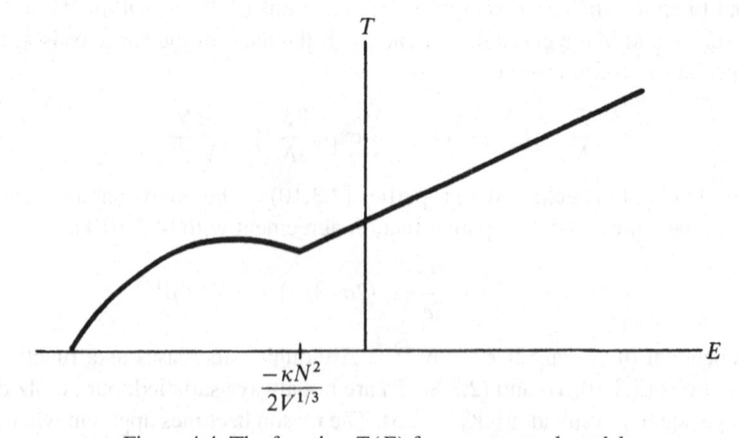

Figure 4.4. The function $T(E)$ for a conceptual model.

decreases, and afterwards, when the zero-point energy gets larger than the gravitational energy, it falls to zero. It is in fact observed by astrophysicists that large gaseous masses contract under the influence of gravity, thereby heating up and radiating the gravitational energy that has been set free. This activity, which indicates a range of values for which $S(E)$ is convex and hence microcanonically a negative specific heat, is a direct consequence of the virial theorem and the theorem of equipartition: energy = −kinetic energy = $-(3N/2)$ temperature. Yet this is true only in the intermediate region, since it ignores the external virial (the pressure) and the equipartition theorem is not valid for degenerate gases. This also becomes visible in the computer solution of the Thomas–Fermi equation, as shown in Figures 4.5 and 4.6. At the smaller radius R=30 km the zero-point energy predominates and the star acts normally, whereas an intermediate region of negative specific heat shows up at R=100 km. This phenomenon can not arise in the canonical ensemble, so our next topic will be what the situation is like in that ensemble. In the transition region the Thomas–Fermi equation has many solutions for a given β, and the analysis leading to (4.2.1) shows that the right solution to choose is the one with the smallest value of F. The existence of many different values of F in

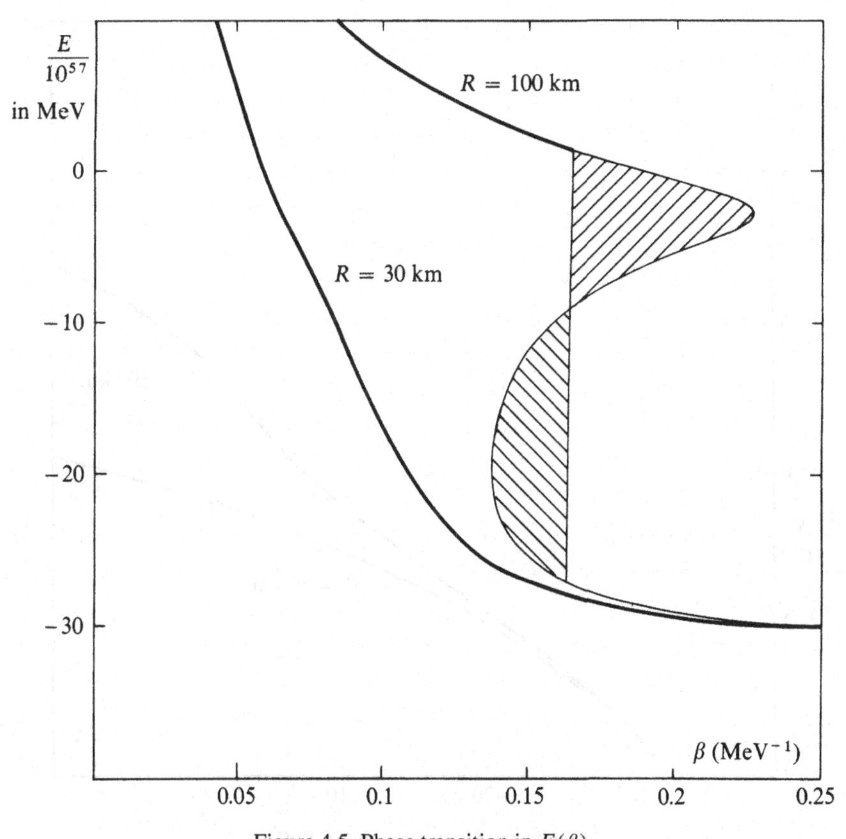

Figure 4.5. Phase transition in $E(\beta)$.

this situation (for a fixed β) follows from the change in the sign of $P = \partial F/\partial V$ (see Figure 4.7). The computed dependence of $-F$ on β is shown in Figure 4.8. If R=100 km, then F has a sharp bend at some transition temperature; in Figure 4.5 it shows up as the lines that divide the surface $E(\beta)$ into two equal parts (Problem 1). At this transition temperature the system in the canonical ensemble rises from one branch of the curve to the other. The energy has a nonzero jump ($\sim$30 MeV per particle) at the transition; in the canonical ensemble the region of negative specific heat is bridged over by a phase transition. Computers have also been used to solve for the local observable $\rho(r)$, which is shown in Figure 4.9 at various temperatures and with R=100 km. At the transition temperature (1/0.165) MeV an almost homogeneous density becomes strongly concentrated at the center. The picture that emerges is of a star with a rather definite surface and a central density about 10^6 times the density of the atmosphere. At still lower temperatures the atmosphere also condenses, but it only increases the density of the star a tiny bit. The radius of a neutron star is only about 10 km at low temperature, which is why at first hardly any difference shows up in S in Figure 4.6 between the systems at R =30 km and R =100 km. Only after the transition energy does the star spread out so as to make

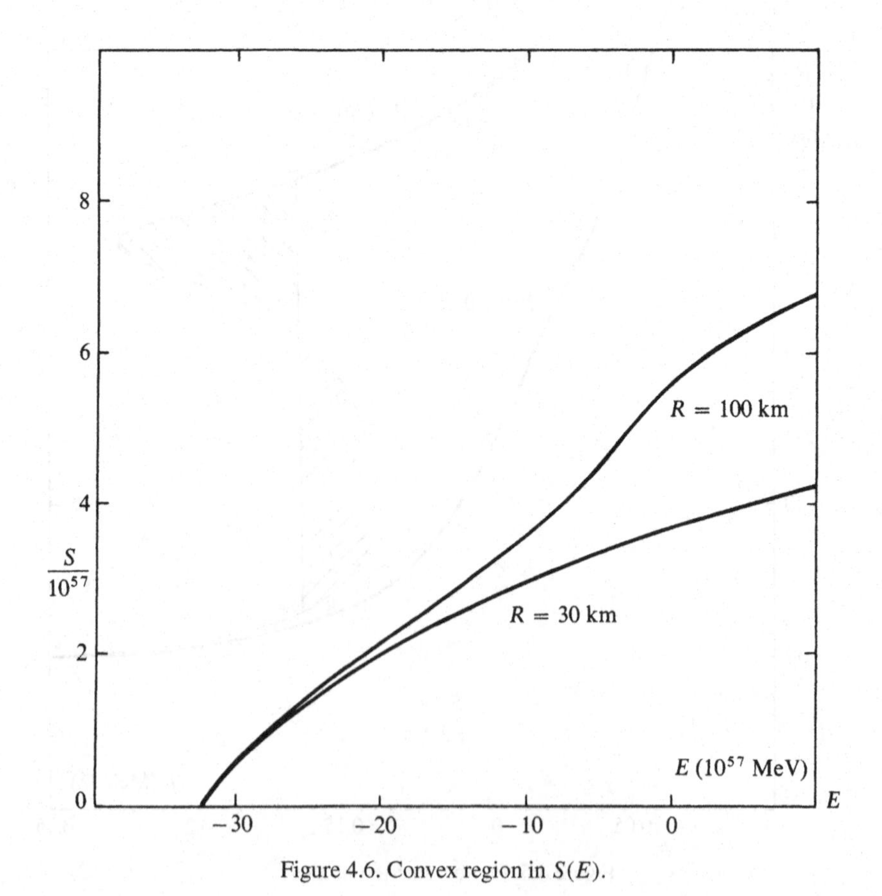

Figure 4.6. Convex region in $S(E)$.

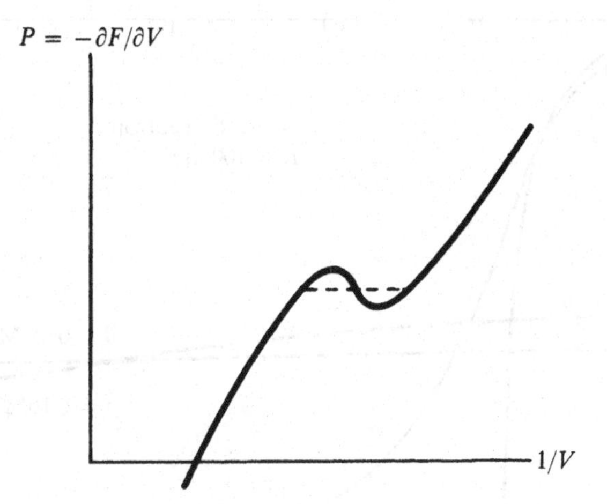

Figure 4.7. Phase transition with negative pressure.

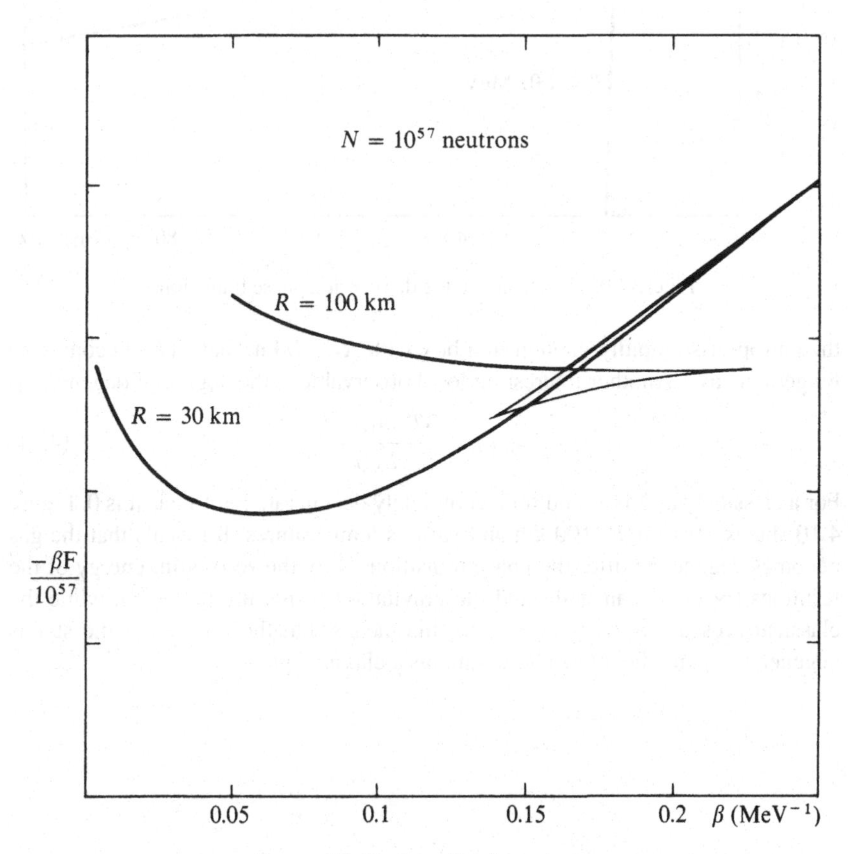

Figure 4.8. The negative free energy.

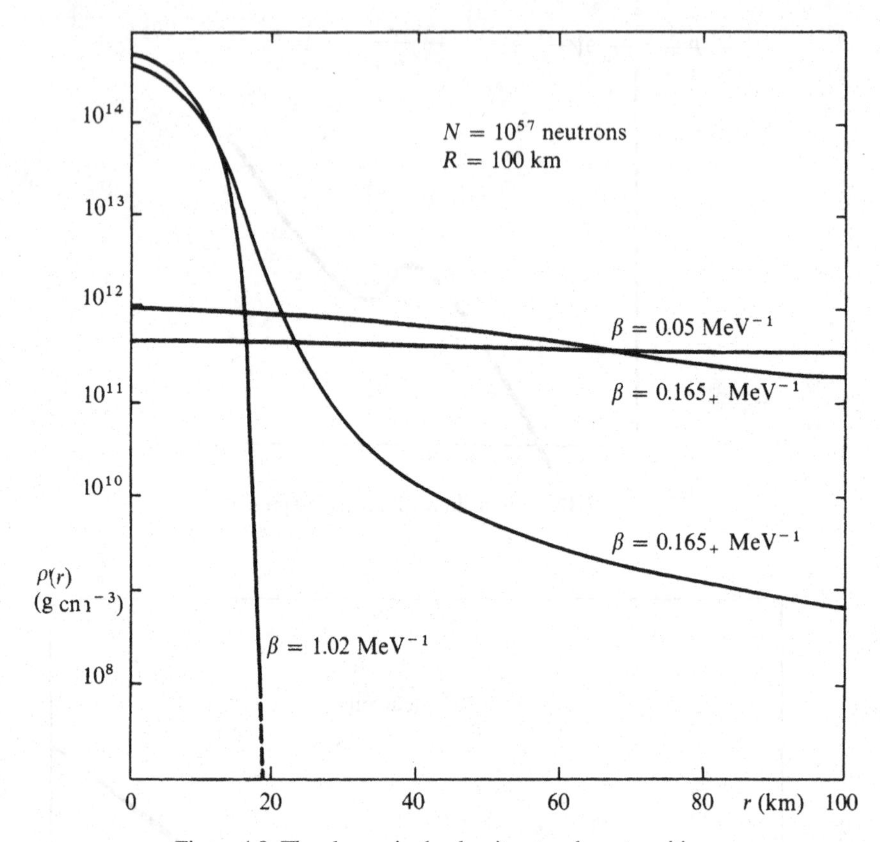

Figure 4.9. The change in the density at a phase transition.

the entropy rise rapidly enough in a box with R=100 km that $S(E)$ becomes no longer concave. Another interesting local observable is the degree of degeneracy

$$\xi(r) = \frac{3T}{2} \frac{\rho(r)}{E_k(r)}. \tag{4.21}$$

For a classical gas ξ is 1, and for a completely degenerate Fermi gas it is 0. Figure 4.10 shows $\xi(r)$ for R=100 km and various temperatures. It reveals that the gas becomes degenerate after the phase transition. Only the zero-point energy of the fermions ($\sim \rho^{5/3}$) can withstand the gravitational pressure ($\sim \rho^{4/3}$), while the classical pressure is weaker ($\sim \rho$). This means that the interior of the star is degenerate, while the atmosphere remains a classical gas.

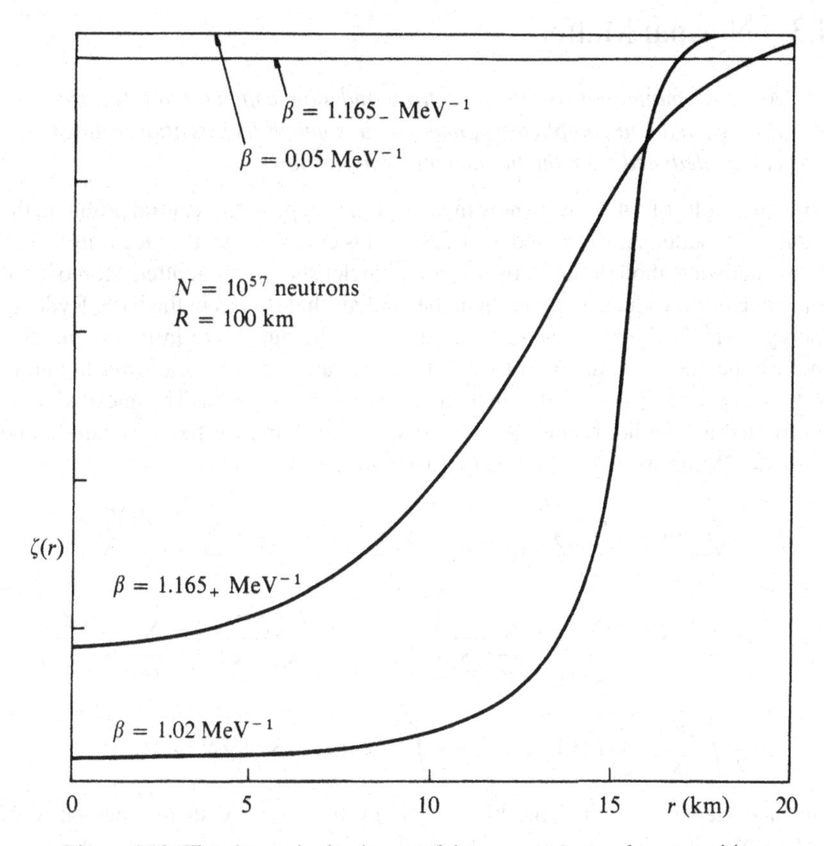

Figure 4.10. The change in the degree of degeneracy ξ at a phase transition.

4.2.5 Problem

Show that the reciprocal β_c of the transition temperature for the canonical ensemble is determined by

$$0 = \int_{E_2}^{E_1} dE(\beta(E) - \beta_c), \quad \beta(E_1) = \beta(E_2) = \beta_c.$$

4.2.6 Solution

Since $\beta = dS/dE$, the condition implies

$$S(E_1) - S(E_2) - \beta_c(E_1 - E_2) = \beta_c(F(E_2) - F(E_1)) = 0.$$

At β_c the two branches of the curves $F(E)$ cross, and the canonical ensemble always selects the lower branch.

4.3 Normal Matter

Although matter consisting of electrons and atomic nuclei exhibits extremely varied and complicated phenomena, some of its essential features can be deduced from the fundamental physics laws.

With the results of § 4.1 we are now in a position to cope with a central problem, the stability of matter. As discussed in (1.2.8;2), it is essential that the electrons follow Fermi statistics, though the statistics of the nuclei should not matter. Moreover, it is the mass of the electron rather than the nucleus that occurs in the basic Rydberg energy $e^4m^2/2$. We shall therefore assume that the nuclei are infinitely massive and use the Hamiltonian H_N of (4.1.2); at any rate it provides a lower bound to (4.1) with $\kappa = 1$. The wall W can then also be dispensed with. The question to be confronted is whether bound $H_N > -AN$ can be found for fixed Z_k but M and $N \to \infty$. With this in mind, write (4.1.6) with $\mu = W = 0$ as

$$H_N = \sum_{i=1}^{N} |\mathbf{p}|^2 - \sum_{i=1}^{N}\sum_{k=1}^{M} \frac{Z_k}{|\mathbf{x}_i - \mathbf{X}_k|} + \sum_{i>j} \frac{1}{|\mathbf{x}_i - \mathbf{x}_j|} + \sum_{k>l} \frac{Z_k Z_l}{|\mathbf{X}_k - \mathbf{X}_l|}$$

$$\geq \sum_{i=1}^{N} |\mathbf{p}_i|^2 + \sum_{i=1}^{N}\left[-\sum_{k} \frac{Z_k}{|\mathbf{x}_i - \mathbf{X}_k|} + \int \frac{d^3x' n(\mathbf{x}')}{|\mathbf{x}_i - \mathbf{x}'|}\right] + \sum_{k>l} \frac{Z_k Z_l}{|\mathbf{x} - \mathbf{x}'|}$$

$$-\frac{1}{2}\int \frac{d^3x d^3x'}{|\mathbf{x} - \mathbf{x}'|} n(\mathbf{x}) n(\mathbf{x}') - \frac{3}{5\gamma}\int d^3x n(\mathbf{x}) - 3.68\gamma N \equiv H_n. \qquad (4.22)$$

The first step is to bound the kinetic energy by $\int \rho^{5/3}$ with the inequality of (4.1.41;2) and set $n = \rho$. This is a bound for every expectation value with spin-$\frac{1}{2}$ fermions, so, again with the aid of (4.1.40;2), we obtain

$$\langle\psi|H_N|\psi\rangle \geq \frac{3}{5}\left(\left(\frac{3\pi}{4}\right)^{2/3} - \frac{1}{\gamma}\right)\int d^3x n^{5/3}(\mathbf{x}) - \int d^3x \sum_{k=1}^{M} \frac{Z_k n(\mathbf{x})}{|\mathbf{x} - \mathbf{X}_k|}$$

$$+\frac{1}{2}\int \frac{d^3x d^3x'}{|\mathbf{x} - \mathbf{x}'|} n(\mathbf{x}) n(\mathbf{x}') + \frac{1}{2}\sum_{k} \frac{Z_k Z_m}{|\mathbf{X}_k - \mathbf{X}_m|} - 3.68\gamma N$$

$$\geq -3.68\left(\gamma N + \frac{1}{(3\pi/4)^{2/3} - 1/\gamma}\sum_{k=1}^{M} Z_k^{7/3}\right) \quad \text{for} \|\psi\|^2 = 1. \qquad (4.23)$$

If this is optimized in γ, it shows the

4.3.1 Stability of Matter

$$H_N \geq -2.08N\left[1 + \left(\sum_{k=1}^{M} \frac{Z_k^{7/3}}{N}\right)^{1/2}\right]^2.$$

4.3.2 Remarks

1. If there were q kinds of electrons instead of the two spin orientations, then the right side would be multiplied by $(q/2)^{2/3}$. Thus there is a bound $\sim N^{5/3}$ independently of the statistics of the electrons.
2. The solution of the Thomas–Fermi equation describes a neutral system, and accordingly the bound is $MZ^{7/3}$ if all Z_k equal $Z = N/M$. The bound is certainly not optimal if $N \ll MZ$, for one would expect $\sim NZ^2$. However, (4.3.3) suffices for our purposes, as we are concerned only with the neutral case.
3. Inequality (4.1.41;2) is presumably not optimal; on the right the constant should be increased by a factor $(4\pi)^{2/3}$ to $3(3\pi^2)^{2/3}/5$. If this conjecture were proved, then (4.3.1) would be improved by the same factor, reading

$$H \geq -0.385 \sum_{k=1}^{M} Z_k^{7/3}(1 + O(Z_k^{-7/6})).$$

However, by more refined methods (4.1.41;2) could be sharpened by 2π. So by now the best proved factor in (4.3.1) is about -0.6 which is not so bad since it holds for any arrangement of atoms of arbitrary complexity. If $Z_k \to \infty$ this approaches the sum of the Thomas–Fermi energies of the atoms. Such an optimal inequality can in fact be proved, although only in the form [28]

$$H \geq -0.385 \sum_{k=1}^{M} Z_k^{7/3}(1 + O(Z_k^{-2/33})).$$

4. Inequality (4.3.1) holds *a fortiori* for a system in a finite volume.
5. Since the kinetic energy of the nuclei was not used, they may follow either Bose or Fermi statistics.
6. The important property of the Coulomb potential for stability is that $1/r$ is a function of positive type, i.e., $\tilde{v} \geq 0$. The Yukawa potential $v(r) = \exp(-\mu r)/r$ similarly satisfies $\tilde{v} > 0$, and stability can be proved analogously. In contrast the potential $v(r) = (a + br)\exp(-\mu r)$ with $b > a\mu > 0$, $\mu > 0$, which is even finite and of short range, does not lead to stability for the Hamiltonian $\sum_{i=1}^{N} |\mathbf{p}_i|^2 + \sum_{i>j} e_i e_j v(\mathbf{x}_i - \mathbf{x}_j)$, even for fermions: There is an $r_0 > 0$ such that $v(r_0) > v(0)$ (this would be impossible if $\tilde{v} > 0$), so let us confine $N/2$ positive and negative particles to separated balls of radius $r_0\varepsilon$, $\varepsilon \ll 1$, arrayed at a distance r_0 from one another. Then the interaction between the balls, $-e^2 v(r_0)N^2/4$, wins out over the repulsive energy of the like-charged particles within the balls, $\sim e^2 v(0)N(N-2)/4$, and also wins out over the kinetic energy $\sim N^{5/3}(r_0\varepsilon)^2$ as $N \to \infty$. Thus the total energy goes to $-\infty$ as $-N^2$ when $N \to \infty$. This shows that the problem of the stability of matter has nothing to do with the long range of the Coulomb potential. The proof with the Yukawa potential is not any simpler; in a way it is more difficult, since stability with a Yukawa potential immediately implies stability with a Coulomb potential – as remarked in (1.2.8;5) the difference produces stability – but not vice versa. However, as we have just seen, the $1/r$ singularity is not the only danger for

stability; even regular potentials v with energies $\sum_{i\leq j} e_i e_j v(\mathbf{x}_i - \mathbf{x}_j)$ that take on both signs can lead to instability. This shows the superficiality of the common opinion that stability is not a real physical problem, since actual potentials do not become singular.

4.3.3 The Extensivity of the Volume

If $H > -cN$ and the expectation value of H in a state is nonpositive, $\langle H\rangle \leq 0$, then no volume $\Omega \leq \varepsilon N$ contains more that $N(20c/3)^{3/5}(4\varepsilon/3\pi)^{2/5}$ particles.

Proof: Let $H = T + V$. Since the energy is proportional to the mass in a Coulomb system, $\frac{1}{2}\langle T\rangle \leq -\langle \frac{1}{2}T + V\rangle \leq 2cN$. Then it follows from

$$\langle T\rangle \geq \frac{3}{5}\Big(\frac{3\pi}{4}\Big)^{2/3} \int \rho^{5/3}$$

that

$$\int_\Omega \rho(\mathbf{x})d^3x \leq \Big(\int_\Omega \rho^{5/3}d^3x\Big)^{3/5}\Big(\int_\Omega d^3x\Big)^{2/5} \leq \Big(\frac{5}{3}\Big(\frac{4}{3\pi}\Big)^{2/3} 4cN\Big)^{3/5}(\varepsilon N)^{2/5}$$
$$\leq N\Big(\frac{20}{3}c\Big)^{3/5}\Big(\frac{4\varepsilon}{3\pi}\Big)^{2/5}. \qquad \square$$

4.3.4 Remarks

1. If Ω is a ball, then it is possible to derive bounds of the form $\langle r^\nu\rangle \geq cN^{\nu/3}$, in analogy with (III:4.5.18).
2. The material up to this point does not allow upper bounds of the form $r \sim N^{1/3}$ to be proved. Neutrality does not enter in an important way, and with an excess of electrons the Coulomb potential would cause the system to swell out to infinity. In other words, it has been proved that matter is stable in the sense that it does not implode, but it might stil explode.

4.3.5 The Existence of the Thermodynamic Functions

We are now faced with the question of how to define the energy density when $N \to \infty$ [30]. It clearly follows from (4.3.1) that $(1/V)E(V_\sigma, V, \rho V) > -\rho \cdot constant$ for all V, and since it is easy to show that E/V remains bounded above, $\underline{\lim}_{V\to\infty}(1/V)E(V\sigma, V, \rho V)$ could be regarded as $\varepsilon(\rho, \sigma)$ (by definition, $\underline{\lim}_{n\to\infty} a_n = \sup_{n'} \inf_{n>n'} a_n$). This cheap way out is physically unsatisfying, however; one would hope that the limit exists and that the energy density becomes independent of V as the system is made infinitely large. This means that the sequence should be proved monotonic, as was done in (2.3.4). Unfortunately, the inductive procedure followed there, of imagining each cube to consist of smaller cubes, does not work in this case, since it is difficult to estimate the Coulomb interaction between cubes. Balls can be used instead of cubes, however, as their interactions are as if the charges were concentrated at their centers, according to a theorem dating from Newton. In particular, if they have a spherically symmetric charge distribution and are overall neutral, then they do not interact with charges placed outside them. Of course, spheres do not fill space as densely as cubes, but by the use of spheres of different radii the unfilled volume can be made arbitrarily small. The convergence proof consequently proceeds by three steps.

(a) We must first show that the interaction between the spheres is not positive, in order to prove monotony.

(b) It must be shown that the radii of the balls can be chosen so that the fraction of volume outside them goes to zero in the limit.

(c) The distribution of particles in this procedure must lead to a homogeneous density in the limit.

4.3.6 The Interaction Between Balls

We consider

$$H = \sum_{i=1}^{N}\Big(|\mathbf{p}_i|^2 - \sum_{k=1}^{M}\frac{Z_k}{|\mathbf{x}_i - \mathbf{X}_k|}\Big) + \sum_{i>j}|\mathbf{x}_i - \mathbf{x}_j|^{-1} + \sum_{k>l}\frac{Z_k Z_l}{|\mathbf{X}_k - \mathbf{X}_l|} + \sum_{k=1}^{M}\frac{|\mathbf{p}_k|^2}{2M_k} \tag{4.24}$$

in a ball B, such that $\psi|_{\partial B} = 0$, and examine the neutral case with only one kind of nucleus: $N = MZ$, $N_t = N(1 + 1/Z) =$ the total number of particles. The eigenvalues $e_i(V, N_t)$, i=1,2,..., of H depend on the volume V of B and on N_t, and the microcanonical energy is given by

$$E(S, V, N_t) = \exp(-S) \sum_{i=1}^{\exp(S)} e_i(V, N_t),$$

where E and E_m have been identified in accordance with (2.3.11;2). Now put k disjoint balls B_α of volumes V_α into B,

$$B \supset \bigcup_{\alpha+1}^{k} B_\alpha,$$

and form a system of trial functions ψ_i by taking tensor products of the eigenfunctions of H_α, defined as H for N_α particles in B_α:

$$\psi_i = \psi_{i_1} \otimes \psi_{i_2} \otimes \cdots \otimes \psi_{i_k}.$$

The trial functions then have to be antisymmetrized in the electron variables and either symmetrized or antisymmetrized in the nuclear coordinates, depending on the nuclear statistics. Yet since ψ_{i_α} and ψ_{i_β} have disjoint support, there are no cross terms in their interaction, and the expectation values are the same as those with the unsymmetrized ψ_i. (The subscript i is to be treated as a multi-inedx $i_1, \ldots, i_k$.) We always choose the first $\exp(S_\alpha)$ eigenfunctions of the operators H_α (and denote the eigenvalues $e_{\alpha,i}$), so

$$\sum_{i=1}^{\exp(S)} = \sum_{i_1=1}^{\exp(S_1)} \cdots \sum_{i_k=1}^{\exp(S_k)},$$

where $S = \sum_{\alpha=1}^k S_\alpha$, $N = \sum_{\alpha=1}^k N_\alpha$, $Z = \sum_{\alpha=1}^k Z_\alpha$ and $N_\alpha/Z_\alpha + 1$ is an integer. Then each B_α can be filled with whole atoms, becoming neutral. As in (2.3.5), with the min-max principle (III:3.5.19),

$$\begin{aligned} E(S,V,N) &\leq \exp(-S)\sum_{i=1}^{\exp(S)} \langle \psi_i | H \psi_i \rangle = \sum_{\alpha=1}^{k} \exp(-S_\alpha) \sum_{i_\alpha=1}^{\exp(S_\alpha)} e_{\alpha,i_\alpha}(V_\alpha, N_\alpha) + U \\ &= \sum_{\alpha=1}^{k} E_\alpha(S_\alpha, V_\alpha, N_\alpha) + U, \end{aligned} \tag{4.25}$$

but this time there is an energy of the interaction between the balls,

$$U = \sum_{\alpha>\beta} \exp(-S_\alpha - S_\beta) \sum_{i_\alpha=1}^{\exp(S_\alpha)} \sum_{i_\beta=1}^{\exp(S_\beta)} U_{i_\alpha i_\beta},$$

$$U_{i_\alpha i_\beta} \equiv \sum_{j=1}^{N_\alpha} \sum_{m=1}^{N_\beta} e_j e_m \int \frac{d^{3N_\alpha}x\, d^{3N_\beta}y}{|\mathbf{x}_j - \mathbf{y}_m|} |\psi_{i_\alpha}(\mathbf{x}_1, \ldots, \mathbf{x}_{N_\alpha})|^2 |\psi_{i_\beta}(\mathbf{y}_1, \ldots, \mathbf{y}_{N_\beta})|^2.$$

Because of the spherical symmetry of B_α and H_α, the functions ψ_{i_α} can be ordered according to the eigenvalues l_α of the total angular momentum L_α about the center of B_α. The eigenvalues $e_{\alpha,i}$ do not depend on the z-component of the angular momentum (which has quantum numbers m_α, $-l_\alpha \leq m_\alpha \leq l_\alpha$), and

$$\rho_\alpha(\mathbf{x}) = \sum_{i_\alpha} \int d^3x_2 \cdots d^3x_{N_\alpha} |\psi_i(\mathbf{x}, \mathbf{x}_2, \ldots, \mathbf{x}_{N_\alpha})|^2$$

will be spherically symmetric if the sum runs over a full L-shells, then U would equal zero by Newton's theorem. It will now be shown that the partially filled shells can be chosen to make U negative. Let μ_α and ν_α be the indices nearest to $\exp(S_\alpha)$ corresponding to filled shells, such that $\mu_\alpha \leq \exp(S_\alpha) \leq \nu_\alpha$. Thus

$$\sum_{i_\alpha=1}^{\exp(S_\alpha)} \sum_{i_\beta=1}^{\exp(S_\beta)} U_{i_\alpha i_\beta} = \sum_{i_\alpha=\mu_\alpha}^{\exp(S_\alpha)} \sum_{i_\beta=\mu_\beta}^{\exp(S_\beta)} U_{i_\beta i_\alpha},$$

and the interaction energy can be written as

$$U = c \sum_{i_1=\mu_1}^{\exp(S_1)} \sum_{i_2=\mu_2}^{\exp(S_2)} \cdots \sum_{i_k=\mu_k}^{\exp(S_k)} U_{i_1,\dots,i_k}, \qquad c > 0,$$

$$U_{i_1,\dots,i_k} = \left\langle \psi_i \right| \sum_{j>m}{}' \frac{e_j e_m}{|\mathbf{x}_j - \mathbf{x}_m|} \left| \psi_i \right\rangle.$$

$\sum'$ indicates that x_j and x_m have to belong to different balls. We know that

$$\sum_{i_1=\mu_1}^{\nu_1} \sum_{i_2=\mu_2}^{\nu_2} \cdots \sum_{i_k=\mu_k}^{\nu_k} U_{i_1,\dots,i_k} = 0,$$

and since the eigenvalues e_{i_α} are degenerate if $\mu_\alpha \le i_\alpha \le \nu_\alpha$, it is possible to select $\exp(S_1) - \mu_1$ indices i_1 such that

$$\sum_{i_2=\mu_2}^{\nu_2} \cdots \sum_{i_k=\mu_k}^{\nu_k} \sum_{i_1=\mu_1}^{\exp(S_1)} U_{i_1,\dots,i_k} \le 0$$

without changing the first sum in (4.25). We now proceed inductively and choose $\exp(S_2) - \mu_2$ indices i_2 such that

$$\sum_{i_2=\mu_2}^{\exp(S_2)} \cdots \le 0$$

and so forth, until finally $U \le 0$. This proves the

4.3.7 Monotony of the Energy

If $B \supset \bigcup_{\alpha=1}^{k} B_\alpha$, $N_t = \sum_{\alpha=1}^{k} N_\alpha$, and $N_\alpha/Z + 1$ is integral, $S = \sum_{\alpha=1}^{k} S_\alpha$, and E is as defined in (4.3.8), then

$$E(S, V, N_t) \le \sum_{\alpha=1}^{k} E_\alpha(S_\alpha, V_\alpha, N_\alpha).$$

4.3.8 Remarks

1. The B_α are required only to be disjoint; how well they fill B does not affect the validity of the equation.
2. All but one of the B_α have to be spherical and electrically neutral, but one of them need not be.
3. The theorem holds regardless of the statistics of the particles, which can affect it only by ensuring the existence of a bound on E/N.

The question of how completely B can be filled by the B_α is a purely geometrical one. It is answered [30] by the

4.3.9 Swiss Cheese Theorem

Let $R_j = (1+p)^j R_0$, $p \in \mathbb{Z}^+$, $1+p \geq 27$, be the radii of the balls of a given size indexed by j and let B_m be a ball of size m. Then for all $m > 0$, B_m contains the union from $j = 1$ to $m-1$ of ν_j disjoint balls of size j, where

$$\nu_j = \frac{(1+p)^{3(m-j)}}{p} \left(\frac{p}{1+p}\right)^{m-j} \in \mathbb{Z}^+.$$

4.3.10 Remarks

1. This theorem makes more precise the fact, clear at the intuitive level, that a large ball can be filled extremely well by smaller ones if their radii are chosen suitably. The total volume of the small balls is

$$\sum_{j=0}^{m-1} R_j^3 \nu_j = ((1+p)^m R_0)^3 \left(1 - \left(\frac{p}{1+p}\right)^m\right),$$

so that the unfilled fraction is only $(p/(1+p))^m$, which tends to zero as $m \to \infty$.
2. Of course, the filling of a ball uses more small balls than large ones, but the fraction of volume filled by the balls of size j is $(1/p)(p/(1+p))^{m-j}$, as the larger balls are much more voluminous.

Proof: See Problem 1. □

4.3.11 The Homogeneity of the Density

The next step in § 2.3 was to consider a sequence of larger and larger cubes, all of which had the same entropy and particle density. Nothing like that is possible in this situation, since to compensate for the gaps some of the balls will have greater densities than the average density overall. Since the unfilled volume gets smaller and smaller, however, it suffices to impose relatively large densities on the balls of size 0 and assign equal densities to all the others. Let us thus choose $N_\alpha / V_\alpha = \rho(p+1) \equiv \rho_0$ for $\alpha = 1, 2, \ldots, \nu_0$, so for the balls of size 0, $N_\alpha / V_\alpha = \rho$ for all $\alpha > \nu_0$. If ρ_j is the density in a ball of size j, and we let $\rho_1, \ldots, \rho_m = \rho$, then the ρ_j satisfy a recursion formula

$$\rho_m = \sum_{j=0}^{m-1} \rho_j \nu_j \left(\frac{R_j}{R_m}\right)^3 = \frac{\rho_0}{p}\left(\frac{p}{p+1}\right)^m + \frac{\rho}{p} \sum_{j=1}^{m-1} \left(\frac{p}{p+1}\right)^{m-j} = \rho, \text{ forall } m \geq 1.$$

In the same way the entropy is distributed so that the entropy density σ_j in the balls of size j satisfies

$$\sigma_0 = \sigma(p+1), \quad \sigma_1 = \sigma_2 = \cdots = \sigma_m = \sigma = \frac{1}{p} \sum_{j=0}^{m-1} \sigma_j \left(\frac{p}{p+1}\right)^{m-j}.$$

If $V_0 = 4\pi R_0^3/3$ and E_j is the energy and ε_j the energy density of the balls of size j, then Proposition (4.3.7) specializes for this particular filling to

$$E_k(S, N) \leq \sum_{j=0}^{k-1} E_j(S_j, N_j)\nu_j,$$

$$\begin{aligned}\varepsilon_k(\sigma_k, \rho_k) &= [(1+p)^{3k} V_0]^{-1} E_k(S_k, N_k) \\ &\leq \frac{1}{V_0 p} \sum_{j=0}^{k-1} \left(\frac{p}{1+p}\right)^{k-j} (1+p)^{-3j} E_j(S_j, N_j) \\ &= \frac{1}{p} \sum_{j=0}^{k-1} \left(\frac{p}{1+p}\right)^{k-j} \varepsilon_j(\sigma_j, \rho_j).\end{aligned}$$

This is a modification of (2.6) and similarly allows the convergence of $\varepsilon_k \equiv \varepsilon_k(\sigma_k, \rho_k)$ to be demonstrated: There exist numbers $c_k \leq 0$ such that

$$\varepsilon_k = c_k + \frac{1}{p} \sum_{j=0}^{k-1} \left(\frac{p}{1+p}\right)^{k-j} \varepsilon_j.$$

The recursion formula has the solution

$$\varepsilon_k = c_k + \frac{1}{1+p}\left(\varepsilon_0 + \sum_{j=0}^{k-1} c_j\right). \tag{4.26}$$

Since the sequence $\{\varepsilon_k\}$ is bounded from below, $\sum_j c_j$ must converge, so $\lim_{k\to\infty} c_k = 0$. Since $\varepsilon_k - c_k$ decreases monotonically as a function of k by (4.26), ε_k must tend to a limit. If $k > 0$, then all the densities had the same values (σ, ρ), and we arrive at the

4.3.12 Existence of the Thermodynamic Limit

For the H of (4.3.6), *the limit* $\varepsilon(\sigma, \rho) \equiv \lim_{V\to\infty}(1/V)E(\sigma V, \rho V)$ *exists.*

4.3.13 Remarks

1. The theorem has been proved for spherical volumes, but it generalizes to other shapes with a reasonable relationship between volume and surface area.
2. Although the theorem and proof are given here for strictly neutral systems, it is clear that a small excess charge ΔQ can be allowed as long as its electrostatic energy $\sim (\Delta Q)^2/V^{1/3}$ can be neglected in comparison with E.
3. Although we have assumed there was only one kind of nucleus, the case of any number of kinds of nuclei can be covered simply by generalizing notation.
4. Since $\varepsilon_k - c_k$ is a monotonic sequence, Dini's theorem guarantees that ε_k converges uniformly on compact sets in (σ, ρ); to use this argument it is necessary to extend the definition of the function ε_V, which was initially defined for finite

V on a discrete set, to make it continuous. The continuity of ε will follow from the convexity to be proved below.

5. The Hamiltonian (4.3.6) includes the kintetic energy of the nuclei. Strangely, the existence of the thermodynamic limit (4.3.12) has not been proved in the apparently simpler case where $M_k = \infty$.

The existence of the limit means that all systems characetrized by N have the same dependence on the averaged quantity ε provided that they are large enough. But does the theory predict a reasonable dependence? The temperature, pressure, specific heat, and compressibility should at least be positive in accordance with our experience. The positivity of the temperature and pressure are ensured by our definition of entropy and by the boundary conditions. With the aid of (2.3.24), the positivity of the other observables is a consequence of the convexity of the function $(\sigma, \rho) \to \varepsilon(\sigma, \rho)$, which, however, does not follow directly from the definitions – recall that the preceding chapter illustrated this with a counter example. Yet it is possible to formulate a theorem on the

4.3.14 Thermodynamic Stability of Coulomb Systems

The mapping $\mathbb{R} \times \mathbb{R}^+ \to \mathbb{R} : (\sigma, \rho) \to \varepsilon(\sigma, \rho)$ *is*

(i) *convex;*

(ii) *nondecreasing in* σ*;*

(iii) *bounded below by* $c\rho$ *for* $c \in \mathbb{R}^+$*;*

(iv) *such that* $\rho^{-1}\varepsilon(\sigma\rho, \rho)$ *is an increasing function of* ρ*.*

Proof:

(i) Let ρ be an odd integer, so that $\nu_j = (1+p)^{2(k-j)}p^{k-j-1}$ is even for $0 \le j \le k-1$, and fill half of the balls of a given size with densities ρ, σ (or, respectively, $\rho_0 = \rho(1+p)$, $\sigma_0 = \sigma(1+p)$) and the other half with ρ', σ' (or, respectively, $\rho'_0 = \rho'(1+p)$, $\rho'_0 = \sigma'(1+p\ )$. Then, since the energy is monotonic as in (4.3.11),

$$\varepsilon_k(\bar{\sigma}_k, \bar{\rho}_k) \le \frac{1}{2p}\sum_{j=0}^{k-1}\left(\frac{p}{1+p}\right)^{k-j}[\varepsilon_j(\sigma_j, \rho_j) + \varepsilon_j(\sigma'_j, \rho'_j)],$$

$$\bar{\sigma}_k = \frac{1}{2p}\sum_{j=0}^{k-1}\left(\frac{p}{1+p}\right)^{k-j}(\sigma_j + \sigma'_j),$$

and

$$\bar{\rho}_k = \frac{1}{2p}\sum_{j=0}^{k-1}\left(\frac{p}{1+p}\right)^{k-j}(\rho_j + \rho'_j),$$

which implies that

$$\varepsilon\Big(\frac{1}{2}(\sigma+\sigma'),\frac{1}{2}(\rho+\rho')\Big)\leq\frac{1}{2}\big(\varepsilon(\sigma,\rho)+\varepsilon(\sigma',\rho')\big)$$

as $k\to\infty$. Now note that ε is monotonic in σ and $\rho^{-1}\varepsilon(\sigma\rho,\rho)$ is monotonic in ρ, so according to (2.3.9;1) ε is convex not just with coefficient $\frac{1}{2}$ but with all $\alpha\in[0,1]$. Hence it is continuous on the interior of $\mathbb{R}^+\times\mathbb{R}^+$.

(ii) See Remark (2.3.3;3).

(iii) This follows from the estimate (4.3.1) showing the stability of matter.

(iv) From the monotonicity (2.3.4) of the energy, $\partial E/\partial V|_{S,N=\text{const}}\leq 0$.

Since ε has the right sort of convexity, one of the assumptions needed to prove the existence of the thermodynamic limit of the canonical ensemble is satisfied. More information about the function $\varepsilon(\sigma,\rho)$ is needed to verify the other hypotheses made in Theorem (2.4.13). In particular it needs to be shown that ε increases rapidly enough with σ that the σ_0 introduced in (2.3.9;4) is finite, and $\lim_{\sigma\to\infty}\varepsilon/\sigma=\infty$. This is shown by the

4.3.15 Lower Bound for the Energy Density

If $H=H_\alpha\equiv K+\alpha\sum_{i>j}e_ie_j|\mathbf{x}-\mathbf{x}_j|^{-1}$ and ε_α are the corresponding energy densities, then

$$\varepsilon_\alpha(\sigma,\rho)\geq\lambda\varepsilon_0(\sigma,\rho)-\frac{c\rho\alpha^2}{1-\lambda}\quad for\quad all\quad 0\leq\lambda<1,$$

where

$$c=2.08(1+Z^{2/3})^2.$$

Proof: According to (2.3.3;4), ε_α is concave in α, and $\varepsilon_\alpha\geq\lambda\varepsilon_0+(1-\lambda)\varepsilon_{\alpha/(1-\lambda)}$. However, by (4.3.1), $-c\rho\alpha^2$ is a lower bound for all ρ and σ. □

4.3.16 Corollaries

1. Since it was shown in (2.5.17) that in the case of one kind of particle, $\varepsilon_0(\sigma,\rho)=c'\rho^{5/3}\exp(2\sigma/3\rho)$, $c'>0$, is the limit as $\sigma\to\infty$, it follows that $\lim_{\sigma\to\infty\rho\text{fixed}}\varepsilon(\sigma,\rho)/\sigma=\infty$.

2. Even for a finite volume $-c\rho\alpha^2$ is a lower bound, which makes it easy to verify that there exists a function $s(\varepsilon,\rho)$ dominating σ for all volumes, and satisfying $\lim_{\varepsilon\to\infty}s/\varepsilon=0$.

3. In (4.3.29;2) we shall find an upper bound on the ground-state energy density, of the form $c_1\rho^{5/3}-\alpha c_2\rho^{4/3}$. When combined with (4.3.15) it yields an upper bound for the σ_0 of (2.3.9;4) at which $\varepsilon(\sigma)$ starts to move up.

4.3.17 Thermodynamic Wish List

1. $\sigma_0 = 0$.
2. $\partial\varepsilon/\partial\sigma|_{\sigma=\sigma_0} = 0$.
3. $\lim_{\sigma\to\infty}(\partial\varepsilon/\partial\sigma) = \infty$.
4. The function ε is continuously differentiable.
5. The function ε is strictly convex for large σ and is linear on certain intervals in σ when σ is small.

Open Questions for the Wish List

1. Statement 1 is a strong formulation of the third law of thermodynamics, and is unproved for Coulomb systems. Although there is an upper bound on σ_0 in (4.3.16;3), it is not sharp enough to show that $\sigma_0 = 0$.
2. The second statement implies that the system does not fall into its ground state if the temperature is higher than absolute zero, and our bounds are likewise too crude to prove it.
3. The third statement means that there is no maximum temperature, and is proved by (4.3.16;1).
4. Kinks in the graph of ε would correspond to "anti-phase-transitions" at which either the temperature or the pressure shows a discontinuity while the energy remains continuous. The specific heat and the compressibility would be zero at such a point. Such things do not appear to happen in reality, though the arguments we have made do not exclude them.
5. It is known empirically that there are no phase transitions at high temperatures, only at low temperatures. However, this fact has not been proved in the theory.

The equivalence with the canonical ensemble requires only the positivity of the specific heat, which is guaranteed by (4.3.14). The assumptions of Theorems (2.4.14) are fulfilled because of (4.3.13;4), (4.3.14(i)), and (4.3.16;2), so it leads to the

4.3.18 Thermodynamic Limit of the Canonical Ensemble

The limit

$$\lim_{V\to\infty}\Big(-\frac{T}{V}\ln \operatorname{Tr}\exp(-\beta H)\Big) = \inf_{\varepsilon}(\varepsilon - T\sigma(\varepsilon,\rho)) = \varphi(T,\rho)$$

exists.

4.3.19 Remarks

1. The properties of the free-energy density listed in (2.4.15) are also proved.

2. It is possible to prove the existence of the limit as $V \to \infty$ directly, but that is not enough to show the equivalence with the microcanonical ε. In particular it does not show that ε is convex in σ.

Finally, consider the grand canonical ensemble, supposing there are N_e electrons and N_k nuclei with chemical potentials μ_e and μ_k. The function to investigate is

$$P(T, \mu_e, \mu_k) \equiv \lim_{V\to\infty} \frac{T}{V} \ln \operatorname{Tr} \exp[-\beta(H - N_e\mu_e - N_k\mu_k)]. \qquad (4.27)$$

One difficulty with (4.3.25) is that the trace contains the sum over all possible numbers of particles, and not only the neutral configuration for which $N_e = ZN_k$. Fortunately, it turns out that the non-neutral contributions have such large Coulomb energies that they play no role. Stated without proof [30], here is the resulting proposition on the

4.3.20 Thermodynamic Limit of the Grand Canonical Ensemble

The limit (4.3.25) *exists, and*

$$P(T, \mu_e, \mu_k) = \sup_{\rho_e = Z\rho_k} (\mu_e\rho_e + \mu_k\rho_k - \varphi(T, \rho)),$$

$$\rho = \frac{N_e + N_k}{V} = \left(1 + \frac{1}{Z}\right)\rho_e.$$

4.3.21 Remarks

1. Although the supremum is *a priori* over all density configurations, it is attained in the neutral sector.
2. Roughly speaking, to generalize this to cover arbitrarily many components it is only necessary to treat μ and ρ as "isovectors".

4.3.22 Bounds for $\varepsilon(\sigma, \rho)$

The question that now arises is to what extent the qualitative propositions that have been derived about $\varepsilon(\sigma, \rho)$ can be sharpened and made quantitative. For instance, it would be desirable to find an upper bound to complement the lower bound (4.3.15); upper bounds are always easy to discover, since with the min-max principle it is only necessary to devise some good trial functions. In the limit $\rho \to 0$ an obvious upper bound for the ground-state energy is the sum of the energies of the individual atoms. If the density is finite, then one would think of using the ground state of the kinetic energy K in the variational principle, and the result is the first-order perturbation-theoretic approximation to $H_\alpha = K + \alpha V$.

4.3.23 Remarks

1. It is impossible for the expansion in powers of α to converge in the thermodynamic limit; if $\alpha < 0$, then the electrons would attract one another, as would the nuclei, whereas the nuclei would repel the electrons. The ground-state energy of fermions with an attractive $1/r$ potential goes as $-N^{7/3}$, and that of bosons goes as $-N^3$ (see (1.2.10) and (1.2.11;3)). If a trial function is constructed with all the electrons on one side of the container and all the nuclei on the other, then the expectation value of the energy is greater than $-N^{7/3} + N^2/R \to -N^{7/3}$, so E/N does not remain bounded from below. On the other hand, the convergence of a series in the limit $N \to \infty$ would imply that $\lim_{N\to\infty} E/N$ would be finite on the whole disc of convergence, which would include some negative values of α. In fact the explicit calculation reveals that even the second-order contribution becomes infinite as $N \to \infty$. Even so, the first-order result is useful as an upper bound.
2. According to (III:3.5.19) the min-max principle applies to finite σ other than the ground state, but it is more difficult to calculate the microcanonical expectation values than the grand canonical ones. Hence, for nonzero temperatures it is better to use (2.1.8;3) to bound the grand canonical partition function with $-P_\alpha \leq -P_0 + \mathrm{Tr}\, V\rho_{GC}$.

4.3.24 The Ground State

The simplest case is $T = 0$, so let us see how far we can get with the easiest methods. Take the expectation value of (4.22) in the ground state of the electrons; if they are confined in a box Λ with periodic boundary conditions, the ground state is a plane wave, producing a constant electron density ρ_e. If the nuclear charges are all Z and the nuclear masses are all μ, that leaves

$$\begin{aligned}\langle H\rangle = &\sum_{k=1}^{M} \frac{|\mathbf{p}_k|^2}{2\mu} + Z^2 \sum_{k>j} |\mathbf{X}_k - \mathbf{X}_j|^{-1} - \sum_k \int_\Lambda \frac{d^3x \rho_e Z}{|\mathbf{x} - \mathbf{X}_k|} \\ &+ \frac{1}{2}\int_\Lambda \frac{d^3x d^3y \rho_e^2}{|\mathbf{x} - \mathbf{y}|} \\ &+ \langle \sum_i |\mathbf{p}_i|^2 \rangle + \langle \sum_{i>k} |\mathbf{x}_i - \mathbf{x}_k|^{-1} - \frac{1}{2}\int_\Lambda \frac{d^3x d^3y \rho_e^2}{|\mathbf{x} - \mathbf{y}|}\rangle \\ &+ \sum_k \langle \int \frac{d^3x \rho_e}{|\mathbf{x} - \mathbf{X}_k|} - \sum_j \frac{1}{|\mathbf{x}_j - \mathbf{X}_k|}\rangle.\end{aligned} \tag{4.28}$$

The first line of this equation is the Hamiltonian H_J of jellium (1.2.6) in the nuclear variables. If we therefore add the ground-state energy of jellium to the other expectation values, we get an upper bound on the ground-state energy of H, corresponding to a trial function consisting of the tensor product of the ground state of H_J with the electron wave-function. The zero-point energy of the electrons is the next term in (4.3.31), followed by what is referred to as the exchange energy, and the final expectation value is zero. By (2.5.32), if the spin is $\frac{1}{2}$, the zero-point energy goes as

$$\langle \sum_{i=1}^{N} |\mathbf{p}_i|^2 \rangle = N\frac{3}{5}(3\pi^2\rho_e)^{2/3} = N\frac{2.2}{r_s^2}, \quad r_s = \left(\frac{3}{4\pi\rho_e}\right)^{1/3}, \tag{4.29}$$

as $N \to \infty$, and with only a little difficulty the exchange energy can be calculated as

$$\langle \sum_{i>j} |\mathbf{x}_i - \mathbf{x}_j|^{-1} - \frac{1}{2}\int \frac{d^3x d^3x' \rho_e^2}{|\mathbf{x} - \mathbf{x}'|} \rangle = -0.458\frac{N}{r_s} \tag{4.30}$$

(Problem 3). It expresses the effect of the correlations among the electrons owing to their having to avoid each to satisfy the exclusion principle. The result is to lower the Coulomb energy in comparison with that of a homogeneous charge distribution.

4.3.25 The Ground State of Jellium

As for H_J, an upper bound can be obtained by using plane waves as trial functions, for which $\langle H_J \rangle$ once again consists of zero-point energy and the minimum of the potential (1.2.10), and when combined they bound E_J according to

$$\frac{2.2}{2\mu r_s^2} - \frac{0.9}{r_s} \le \frac{E_J}{N} \le \frac{2.2}{2\mu r_s^2} - \frac{0.458}{r_s} \tag{4.31}$$

if $Z = 1$ and the spin is $\frac{1}{2}$. If the density is large ($r_s \to 0$), then the bounds are close together, but they spread out if the density is small. At small densities it is better to array the nuclei on a lattice; give them wave-functions $\sim \sin(\pi r a)/r$, where r is the distance from the lattice site if it is less than a and otherwise let the wave-function be 0, and take a small enough that the wave-functions will not overlap, and will thus be orthogonal. The most convenient configuration is a body-centered cubic lattice, which consists of two simple cubic lattices, one of which has been displaced along a diagonal so that its corners are at the centers of the other. If the density is 2, i.e., the lattice constant of the simple cubic lattice is 1, then a must be less than $\sqrt{3}/4$ in order that the balls of radius a do not intersect; in terms of r_s, the distance between nuclei,

$$a \le \left(\frac{8\pi}{3}\right)^{1/3}\frac{\sqrt{3}}{4}r_s. \tag{4.32}$$

If the nuclei were concentrated at the points of the lattice, then the Coulomb energy per particle would be – $0.896/r_s$ according to (1.2.11;2). Provided that they do not overlap, the repulsion between the nuclei will be the same even if they are somewhat spread out. On the other hand, their interaction (per particle) with the background would be affected by

$$\frac{\rho}{2}\int_0^a dr r^2 \sin^2\frac{r\pi}{a} \Big/ \int_0^a dr \sin^2\frac{r\pi}{a} = \frac{\rho}{2}a^2\left(\frac{1}{3}-\frac{1}{2\pi^2}\right). \tag{4.33}$$

If this is added to the kinetic energy $(\pi/a)^2$ (for mass $\frac{1}{2}$), then the minimum

$$\frac{E}{N} = \left(\frac{\pi}{2}-\frac{3}{4\pi^2}\right)^{1/2} r_s^{-3/2} - \frac{0.896}{r_s} = 1.15 r_s^{-3/2} - \frac{0.896}{r_s}$$

is attained when

$$a = \left[\frac{\rho}{2\pi^2}\left(\frac{1}{3}-\frac{1}{2\pi^2}\right)\right]^{-1/4} = r_s^{3/4}\left[\frac{3}{8\pi^3}\left(\frac{1}{3}-\frac{1}{2\pi^2}\right)\right]^{-1/4}.$$

Condition (4.3.36) means that

$$r_s \geq \frac{8^3\pi^4}{3(2\pi^2-3)}\left(\frac{3}{8\pi}\right)^{1/3} \simeq 489. \tag{4.34}$$

If r_s is smaller, then a must be taken as $(8\pi/3)^{1/3}(\sqrt{3}/4)r_s$, which costs some kinetic energy, $12.75/r_s^2$, and raises the Coulomb interaction above that due to the background by $0.026/r_s$. The figures become more favorable, however, when it is recalled that wave-functions of nuclei with opposite spins do not need to be spatially orthogonal to avoid incurring exchange energy. Suppose that the nuclei have spin $\frac{1}{2}$, as with protons, and put nuclei with spin up on one of the simple cubic lattices and nuclei with spin down on the other. Then the spheres are required only not to overlap with other spheres on the same simple cubic lattice. This weakens the bound (4.3.36) to

$$a \leq \left(\frac{8\pi}{3}\right)^{1/3}\frac{r_s}{2},$$

which weakens the lower bound on r_s (4.3.38) by a factor $\frac{9}{16}$, so

$$r_s \geq 275, \tag{4.35}$$

and also diminishes the zero-point energy by $\frac{3}{4}$ to $9.54/r_s^2$ and increases the interaction with the background by the same factor. The Coulomb repulsion between neighboring nuclei decreases, but only by an insignificant amount $10^{-3}/r_s$. The net effect is to produce

4.3.26 Bounds on the Ground-State Energy of Spin-$\frac{1}{2}$ Jellium

(i) $$\leq \frac{2.2}{r_s^2} - \frac{0.458}{r_s}$$

(ii) $$\frac{2.2}{r_s^2} - \frac{0.9}{r_s} \leq \frac{E}{N} \leq \frac{9.58}{r_s^2} - \frac{0.85}{r_s}$$

(iii) $$\leq \frac{1.15}{r_s^{3/2}} - \frac{0.89}{r_s} \quad \text{if} \quad r_s > 275,$$

where $e = 2\mu = 1$. (See Figure 4.11).

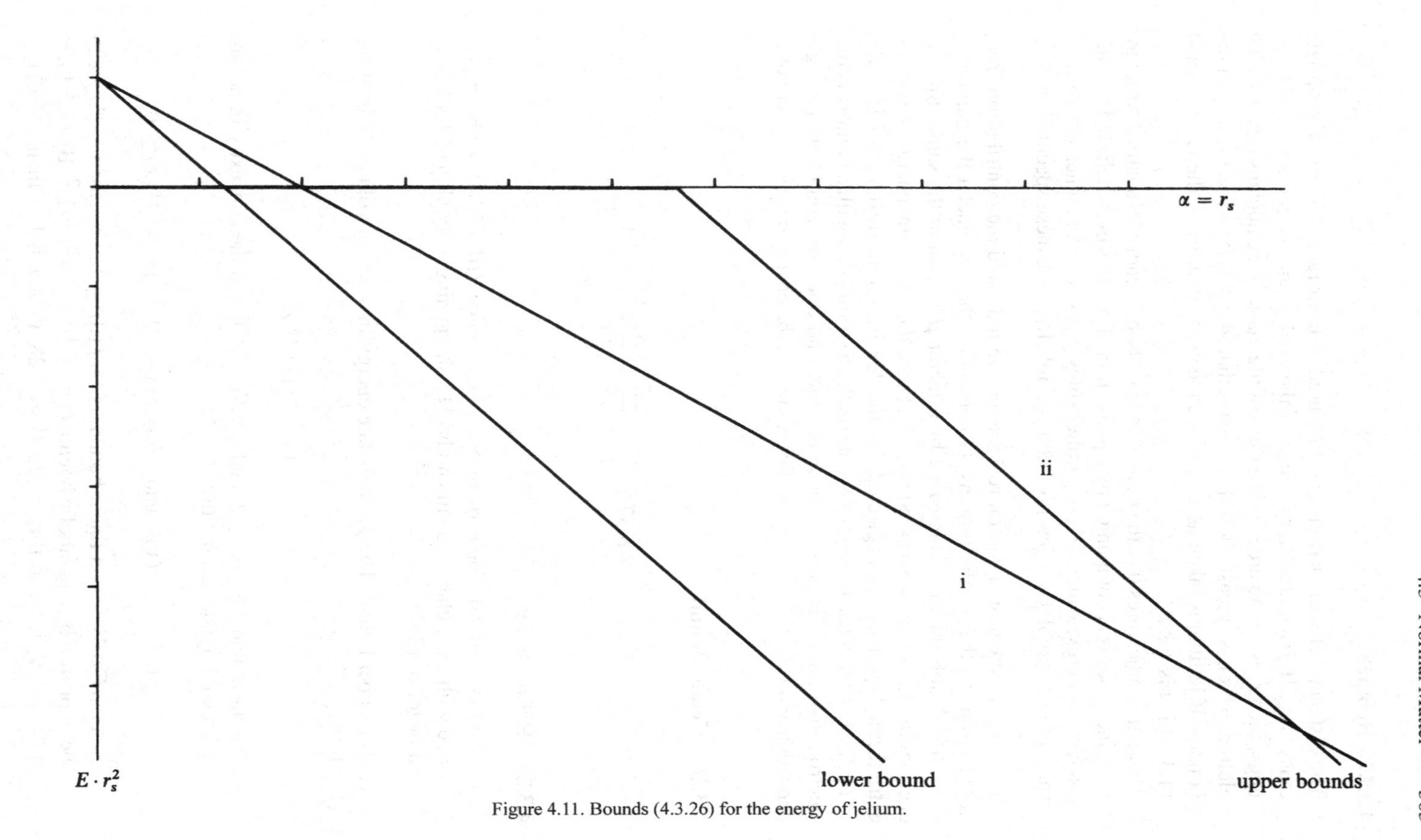

Figure 4.11. Bounds (4.3.26) for the energy of jelium.

4.3.27 Remarks

1. The distance between particles as measured in Bohr radii with the appropriate mass is r_s. If H_J is the Hamiltonian of the nuclei, and the pressure is not too huge, then r_s is on the order of the ratio of the mass of the nucleus to that of the elctron, which is at least 2000. This means that (4.3.40(i)) will be the best of the bounds. If jellium is taken as a model of electrons in a metal, then $r_s \sim 1$, and (4.3.40(ii)) is best.
2. There are conjectures that the transition from homogeneity to a lattice structure as r_s increases is accompanied by a phase transition. It is even believed that the exchange energy, which favors parallel spins, causes ferromagnetism. Despite the simple form of H_J it has not been possible to prove these speculations.

If we focus attention again on real matter, we must add the contribution from the electrons to that of the protons. Observe first that for nuclei the parameter $r_s \sim \rho^{-1/3}$/Bohr radius is increased by a factor μZ^2, but at the same time the energies in (4.3.40) are multiplied by μZ^2. Since the zero-point energy obtains an extra factor $1/\mu$, it can be neglected. For the densities of interest, $r_s > 275/\mu Z^2$, so (4.3.40(iii)) applies to nuclei. Of course, the trial function with a homogeneous electron distribution is poor when Z is large, and does not contribute the right dependence on Z. If $Z = 1$, our earlier results on the energy per electron are only

4.3.28 Crude Bounds

$$-8.32 \leq \frac{E}{N} \leq \frac{2.2}{r_s^2} - \frac{1.34}{r_s}.$$

4.3.29 Refinements

1. The lower bound. The Birman–Schwinger bound (III:3.5.34) can be improved with more elaborated methods [31], sharpening Inequality (4.1.47;2) by a factor of $(2\pi)^{2/3}$.

2. The upper bound. The ground-state energy in a box of volume V is of the form

$$E = V^{-2/3} f(V^{1/3}\alpha).$$

The facts that $\partial E/\partial V \leq 0$ and $\partial^2 \varepsilon/\partial\rho^2 \geq 0$ and the convexity in $n\alpha$ are expressed by the inequalities

$$f(x) \geq \frac{x}{2} f'(x) \quad \text{and} \quad 6xf'(x) - 10f(x) \leq x^2 f''(x) \leq 0.$$

Since $\partial E/\partial V \leq 0$, a linear bound $f(x)/f(0) \leq 1 - \gamma x$ for $x > 2/\gamma$ can be improved by a parabolic bound $f(x)/f(0) \leq -x^2(\gamma/2)^2$. By (4.3.43;1) $\gamma^{-1} = 2.2/1.34$, so if $r_s > 2\gamma^{-1} = 3.28$, then f is less than $-f(0)x^2 \cdot$

$1.34/4(2.2)^2$. It follows that

$$\frac{E}{N} \leq \begin{cases} 2.2/r_s^2 - 1.34/r_s, & \text{if } r_s < 3.28 \\ -0.204, & \text{if } r_s \geq 3.28. \end{cases}$$

These bounds are far from satisfactory. Not only do they fail to allow finer details to be discerned, but indeed they do not even prove that hydrogen holds together at $T = 0$ rather than breaking up into separated atoms. In these units the energy of a separated hydrogen atom is $-\frac{1}{4}$, i.e., less than the upper bound, which only shows how large a territory still remains open to exploration with exact methods in physics.

4.3.30 Problems

1. Prove the Swiss cheese theorem (4.3.13): For any region $\Lambda \subset \mathbb{R}^3$ and any real number h let $\Lambda_h = \{\mathbf{x} \in \Lambda : d)\mathbf{x}, \Lambda^c) < h\}$, if $h > 0$ and $\Lambda_h = \{\mathbf{x} \in \Lambda^c : d(\mathbf{x}, \Lambda) \leq -h\}$, if $h \leq 0$, and denote the volume of Λ_h by $V(h, \Lambda)$.
   Then prove the following two lemmas: (i) Suppose Λ is covered by closed cubes of side l, the interiors of which do not intersect, and let v be the number of cubes entirely contained in Λ. Then the volume of Λ not covered by these v cubes is at most $V(l\sqrt{3}, \Lambda)$. (ii) Let $B \subset \mathbb{R}^3$ be an open ball of radius R and y a number satisfying the inequality $R \geq 2\sqrt{3}y \geq 0$. Then $V(2\sqrt{3}y, B) \leq V(-2\sqrt{3}y, B) \leq 56\pi R^2 y\sqrt{3}$. Finish the proof of the theorem by covering B_0 with a cubic lattice of spacing $2R_1$, and in each cube of the lattice place a ball of radius R_1, then cover the balls with a cubic lattice of spacing R_2, etc. Use the lemmas to estimate v_j and the fraction of volume taken up by the balls of size j.

2. Use Inequalities (III:4.5.14) and (4.1.5) to find a lower bound for the potential energy of jellium,

$$U = \sum_{j>k} |\mathbf{x}_j - \mathbf{x}_k|^{-1} - \sum_i d^3x\rho(\mathbf{x})|\mathbf{x} - \mathbf{x}_i|^{-1} + \frac{1}{2}\int \frac{d^3x d^3y}{|\mathbf{x} - \mathbf{y}|}\rho(\mathbf{x})\rho(\mathbf{y})$$

   and compare with (1.2.10). (Let ρ be constant in any ball.)

3. Calculate

$$\lim_{V\to\infty} V^{-1}\Big\langle\psi\Big|\Big(\sum_{i>k} |\mathbf{x}_i - \mathbf{x}_k|^{-1} - \frac{1}{2}\int_\Lambda \frac{d^3x d^3y}{|\mathbf{x} - \mathbf{y}|}\rho_e^2\Big)\Big|\psi\Big\rangle$$

   if ψ is the ground state of a system of free electrons in a box of volume V. (The momentum states in both spin orientations are occupied up to a maximum momentum p such that $p^3/3\pi = N/V = 3/4\pi r_s^3$.)

4. Verify that the concavity of E as a function of $(1/m, \alpha)$ is no more severe a restriction than the concavity of f in (4.3.43;2).

4.3.31 Solutions

1. (i) If Λ is covered by cubes of length l, but all cubes intersecting Λ^c are removed, then the uncovered portion of Λ is contained in $\Lambda_{l\sqrt{3}}$. (Hence the number ν_{2l} of cubes of length $2l$ that can be packed entirely into Λ is at least $(2l)^{-3}[V(\Lambda) - V(2\sqrt{3}l, \Lambda)]$.)
 (ii) If $0 \le h \le R$, then

$$V(h, B) = \frac{4\pi}{3}[R^3 - (R-h)^3] \le V(-h, B) = \frac{4\pi}{3}[(R+h)^3 - R^3].$$

The lemma is then a consequence of the convexity of the function $f(\varepsilon) \equiv (1+\varepsilon)^3 - 1$, which implies that $f(\varepsilon) \le f(0) + \varepsilon[f(1) - f(0)] = \varepsilon[2^3 - 1] = 7\varepsilon$.
Proof of the packing estimates. For simplicity assume that $R_0 = 1$, and let $\nu_j = p^{j-1}(1+p)^{2j}$. If a unit ball is covered by cubes of length $2R_1 = 2(1+p)^{-1}$, then it contains ν_1 cubes, as we shall show. If we then cover the unit ball with a lattice of length $2R_2$, then there are ν_2 cubes contained in the unit ball and not intersecting the first ν_1 balls of size 1. The general fact will follow by induction. Therefore it needs to be shown that when the ball has been filled with smaller balls up to size j, it is still possible to pack ν_{j+1} balls of radius R_{j+1} into the remaining space $B - \bigcup_{k=1}^{j}$ (balls of size k)$\equiv \Omega_j$:

$$V(\Omega_j) = \frac{4\pi}{3}(1 - \sum \nu_k R_k^3) = \frac{4\pi}{3}\left(\frac{p}{p+1}\right)^j.$$

$V(2\sqrt{3}R_{j+1}, \Omega_j) \le M_j$, defined as the sum of $V(-2\sqrt{3}R_{j+1}, B)$ for all balls of size $\le j$ and $V(2\sqrt{3}R_{j+1}, B)$, where B is the unit ball. Because of (ii) and the inequality $2\sqrt{3}R_{j+1} < R_j$,

$$\begin{aligned} V(2\sqrt{3}R_{j+1}, \Omega_j) \le M_j &\le \frac{56\pi}{\sqrt{3}} R_{j+1}[1 + \sum \nu_k R_k^2] \\ &= (p' + p - 2)(p-1)^{-1}(1+p)^{-(j+1)}\frac{56\pi}{\sqrt{3}} \equiv \tilde{M}_j. \end{aligned}$$

Therefore it suffices to show that

$$(2R_{j+1})^3 \nu_{j+1} \le [V_j - \tilde{M}_j] \le [V(\Omega_j) - V(2\sqrt{3}R_{j+1}, \Omega_j)],$$

i.e.,

$$1 \le \frac{\pi}{6}\left[p + 1 - 14\sqrt{3}\frac{1 + p^{-j}(p-2)}{p-1}\right].$$

Since $p^{-j}(p-2) \le (p-2)$, this reduces to

$$1 \le \frac{\pi}{6}[p + 1 - 14\sqrt{3}],$$

which is true when $p + 1 \ge 27$. The fraction of the volume taken up by the balls of radius R_j is

$$\frac{p^{j-1}}{(1+p)^j},$$

which shows that the packing fills the original ball exponentially fast.

2. From (III;4.5.14),

$$U \geq -\frac{3}{4}\Big[8\pi N^2 \int d^3x\rho^2(\mathbf{x})\Big]^{1/3} = -1.35N/r_s,$$

and from (4.1.5),

$$U \geq -2\Big[3.68N\frac{3}{5}\int \rho^{5/3}\Big]^{1/2} = -1.84N/r_s.$$

3. As N and $V \to \infty$, make the replacements

$$\frac{1}{V}\sum_{|k|\leq p} \to \int_{|k|\leq p} \frac{d^3k}{(2\pi)^3}$$

$$v(\mathbf{k}) \equiv \frac{1}{V}\int \frac{d^3x d^3x'}{|\mathbf{x}-\mathbf{x}'|}\exp[i\mathbf{k}\cdot(\mathbf{x}-\mathbf{x}')] \to 4\pi/|\mathbf{k}|^2,$$

to find that

$$\sum_{\substack{|\mathbf{k}|\leq p\\ |\mathbf{k}'|\leq p}} = \int \frac{d^3k}{(2\pi)^3}\frac{4\pi}{|\mathbf{k}|^2}\int \frac{d^3q}{(2\pi)^3}\Theta(p-|\mathbf{q}|)\Theta(p-|\mathbf{k}-\mathbf{q}|)$$

$$= \frac{2}{\pi}\int_0^{2p} dk\frac{p^3}{6\pi^2}\Big[1-\frac{3k}{4p}+\frac{k^3}{16p^3}\Big] = \frac{p^3}{3\pi^2}\frac{p}{\pi}\frac{3}{4}$$

$$= \frac{N}{V}\Big(\frac{9\pi}{4}\Big)^{1/3}\frac{1}{r_s}\frac{3}{4\pi} = \frac{N}{V}\frac{0.458}{r_s}.$$

In order to justify this formal calculation, make a convolution so that

$$v(\mathbf{k}) = \frac{4\pi}{|\mathbf{k}|^2} * F(\mathbf{k}),$$

where

$$F(\mathbf{k}) = \frac{1}{V}\int_{\mathbf{x}\in V,\mathbf{x}'\in V} d^3xd^3x'\exp[i\mathbf{k}\cdot(\mathbf{x}-\mathbf{x}')]$$

$$= L^{-3}\Big(\frac{\sin k_1L/2}{L/2}\Big)^2\Big(\frac{\sin k_2L/2}{L/2}\Big)^2\Big(\frac{\sin k_3L/2}{L/2}\Big)^2$$

is the Fourier transform of the characteristic function of the box, and use Lebesgue's dominated convergence theorem to show that the integrals have the limits given above.

4. With $1/m = \nu : E = \nu f(\alpha/\nu)$,

$$E_{,\alpha\alpha} = \frac{1}{\nu}f'', \qquad E_{,\nu\nu} = \frac{\alpha^2}{\nu^3}f'',$$

$$E_{,\nu\alpha} = -\frac{\alpha}{\nu^2}f'', \qquad E_{,\alpha\alpha}E_{,\nu\nu} - (E_{,\nu\alpha})^2 = 0.$$

Bibliography to Part I

[1] J. Dieudonné. Foundations of Modern Analysis, in four volumes. New York, Academic Press, 1969–1974.

[2] P.R. Halmos. Measure Theory. New York, Springer, 1974.

[3] M. Reed and B. Simon. Methods of Modern Mathematical Physics, in four volumes. New York, Academic Press, 1972–1979.

[4] D.B. Pearson. An Example in Potential Scattering Illustrating the Breakdown of Asymptotic Completeness. *Commun. Math. Phys.* **40**, 125–146, 1975.

[5] F. Weinhold. Criteria of Accuracy of Approximate Wavefunctions. *J. Math. Phys.* **11**, 2127–2138, 1970.

[6] E.H. Lieb and W. Thirring. Inequalities for the Moments of the Eigenvalues of the Schrödinger Hamiltonian and their Relation to Sobolev Inequalities. In: Studies in Mathematical Physics, Essays in Honor of Valentine Bargmann, A.S. Wightman, E.H. Lieb, and B. Simon, eds. Princeton, Princeton Univ. Press, 1976.

[7] Ph. Martin. On the Time-Delay of Simple Scattering Systems. *Commun. Math. Phys.* **47**, 221–227, 1976.

[8] H. Narnhofer. Continuity of the S-Matrix. *Il Nuovo Cim.*, ser. 2, **30B**, 254–266, 1975.

[9] L. Hostler. Coulomb Green's Functions and the Furry Approximation. *J. Math. Phys.* **5**, 591–611, 1964. J. Schwinger. Coulomb Green's Function. *J. Math. Phys.* **5**, 1606–1608, 1964.

[10] M. Abramowitz and I.A. Stegun, eds. Handbook of Mathematical Functions. Applied Mathematics Series, vol. 55, Washington, National Bureau of Standards, 1964.

[11] R. Lavine. Scattering Theory for Long Range Potentials. *J. Func. Anal.* **5**, 368–382, 1970; Absolute Continuity of Positive Spectrum for Schrödinger Operators with Long-Range Potentials. *Ibid.* **12**, 30–54, 1973. C.R. Putnam. Commutation Properties of Hilbert Space Operators and Related Topics. Ergebnisse der Mathematik und ihrer Grenzgebiete, vol. 36. Berlin–Heidelberg–New York, Springer, 1967.

[12] J.E. Avron and I.W. Herbst. Spectral and Scattering Theory of Schrödinger Operators Related to the Stark Effect. *Commun. Math. Phys.* **52**, 239–254, 1977.

[13] I.W. Herbst and B. Simon. The Stark Effect Revisited. *Phys. Rev.* **41**, 67–69, 1978. L.Benassi, V. Grecchi, E. Harrell and B. Simon. The Bender–Wu Formula and the Stark Effect in Hydrogen. *Phys. Rev. Lett.* **42**, 704–707, 1979; erratum, *Ibid.*, p.1430.

[14] T. Kinoshita. Ground State of the Helium Atom. *Phys. Rev.* **105**, 1490–1502, 1957; Ground State of the Helium Atom, II. *Ibid.* **115**, 366–374, 1959.

[15] R.N. Hill. Proof that the H^- Ion has Only One Bound State. Details and Extension to Finite Nuclear Mass. *J. Math. Phys.* **18**, 2316–2330, 1977.

[16] B. Simon. Resonances and Complex Scaling: A Rigorous Overview. *J. Quantum Chem.* **14**, 529–542, 1978.

[17] H. Behncke and F. Sommer. Theorie der analytischen Funktionen einer komplexen Veränderlichen. Die Grundlehren der mathematischen Wissenschaften in Einzeldarstellungen, vol. 77. Berlin–Heidelberg–New York, Springer, 1972, p. 178. W. Rudin. Real and Complex Analysis. New York, McGraw-Hill, 1966, p. 316.

[18] E.H. Lieb and B. Simon. The Hartree–Fock Theory for Coulomb Systems. *Commun. Math. Phys.* **53**, 185–194, 1977.

[19] J.M. Combes. The Born–Oppenheimer Approximation. In: The Schrödinger Equation. *Acta Phys. Austr. Suppl.*, **17**, W. Thirring and P. Urban, eds. Vienna and New York, Springer, 1977, p. 139.

[20] C.A. Coulson. The Van der Waals Force between a Proton and a Hydrogen Atom. *Proc. Roy. Soc. Edinburgh* **A 61**, 20–25, 1941.

[21] W. Kolos. Accurate Theoretical Determination of Molecular Energy Levels. In: The Schrödinger Equation, *op. cit.*, p. 161.

[22] P. Perry, I. Sigal, and B. Simon. Absence of Singular Continuous Spectrum in N-Body Quantum Systems. *Bull. Amer. Math. Soc.* **3**, 1019–1023, 1980.

[23] E. Harrell. Double Wells. *Commun. Math. Phys.* **75**, 239–261, 1980. R.J. Damburg and R.Kh. Propin. On Asymptotic Expansions of Electronic Terms of the Molecular Ion H_2^+. *J. Phys. B*, ser. 2, **1**, 681–691, 1968.

[24] T. Hoffmann-Ostenhof. A Lower Bound to the Decay of Ground States of Two-Electron Atoms. *Phys. Lett.* **77A**, 140–142, 1980.

[25] G.V. Rosenbljum. The Distribution of the Discrete Spectrum for Singular Differential Operators. *Sov. Math. Dokl.* **13**, 245–249, 1972. M. Cwickel. Weak Type Estimates and the Number of Bound States of Schrödinger Operators. *Ann. Math.* **106**, 93–100, 1977. E.H. Lieb. The Number of Bound States of One-Body Schrödinger Operators and the Weyl Problem (unpublished).

Further Reading

Chapter 2, General

N.I. Akhiezer and I.M. Glazman. Theory of Linear Operators in Hilbert Space, in two volumes. New York, F. Ungar, 1961–1963.

N. Dunford and J.T. Schwartz. Linear Operators, in three volumes. New York, Interscience, 1958–1971.

E. Hille and R.S. Phillips. Functional Analysis and Semigroups. Amer. Math. Soc. Colloquium Publications, vol. 31. Providence, American Mathematical Society, 1957.

K. Jörgens and J. Weidmann. Spectral Properties of Hamiltonian Operators, Lecture Notes in Mathematics, vol. 313. Heidelberg–New York, Springer, 1973.

T. Kato. Perturbation Theory for Linear Operators. Die Grundlehren der mathematischen Wissenschaften in Einzeldarstellungen, vol. 132. Berlin–Heidelberg–New York, Springer, 1966.

G. Köthe. Topological Vector Spaces, vol. 1. Die Grundlehren der mathematischen Wissenschaften in Einzeldarstellungen, vol. 159. Berlin–Heidelberg–New York, Springer, 1969.

J. von Neumann. Mathematical Foundations of Quantum Mechanics. Princeton, Princeton University Press, 1955.

F. Riesz and B.Sz.-Nagy. Functional Analysis. New York, F. Ungar, 1965.

A.P. Robertson and W.J. Robertson. Topological Vector Spaces. Cambridge, At the University Press, 1964.

J. Weidmann. Lineare Operatoren in Hilberträumen. Stuttgart, B.G. Teubner, 1976.

K. Yôsida. Functional Analysis. Die Grundlehren der mathematischen Wissenschaften in Einzeldarstellungen, vol. 123. Berlin–Heidelberg–New York, Springer, 1978.

Chapter 2, Specific

(2.2.34)

J.M. Jauch. Foundations of Quantum Mechanics. Reading, Mass., Addison-Wesley, 1968.

G. Ludwig. Die Grundlagen der Quantenmechanik. Die Grundlehren der mathematischen Wissenschaften in Einzeldarstellungen, vol. 70. Berlin–Heidelberg–New York, Springer, 1954.

G. Mackey. The Mathematical Foundations of Quantum Mechanics. New York, Benjamin, 1963.

P. Mittelstaedt. Philosophical Problems of Modern Physics. Boston Studies in the Philosophy of Science, vol. 18. Boston, D. Reidel, 1976.

C. Piron. Foundations of Quantum Physics. New York, Benjamin, 1976.

Section 2.3

O. Bratteli and D.W. Robinson. Operator Algebras and Quantum Statistical Mechanics, vol. I: C^* and W^* Algebras. Symmetry Groups. Decomposition of States. Texts and Monographs in Physics. Berlin–Heidelberg–New York, Springer, 1979.

J. Dixmier. C^* Algebras. Amstredam, North-Holland, and New York, Elsevier North Holland, 1977.

M.A. Naimark. Normed Algebras. Groningen, Wolters-Noordhoff, 1972.

S. Sakai. C^* and W^* Algebras. Ergebnisse der Mathematik und ihrer Grenzgebiete, vol. 60. Berlin–Heidelberg–New York, Springer, 1971.

H. Trotter. On the Product of Semigroups of Operators. *Proc. Amer. Math. Soc.* **10**, 545–551, 1959.

F. Dyson. The Radiation Theories of Tomonaga, Schwinger, and Feynman. *Phys. Rev.* **75**, 486–502, 1949.

(2.5.14)
H. Narnhofer. Quantum Theory for $1/r^2$-Potentials. *Acta Phys. Aust.* **40**, 306–322, 1974.

(2.5.20)
G. Flamand. Applications of Mathematics to Problems in Theoretical Physics. In: Cargèse Lectures in Theoretical Physics, F. Lurçat, ed. New York, Gordon and Breach, 1967, p. 247.

Chapter 3, General

C. Cohen-Tannoudji, B. Diu, and F. Laloë. Quantum Mechanics. New York, Wiley, 1979.
A. Galindo and P. Pascual. Mecánica Cuántica. Madrid, Alhambra, 1978.
G. Grawert. Quantenmechanik. Wiesbaden, Akademie Verlagsgesellschaft, 1977.
R. Jost. Quantenmechanik, in two volumes. Zurich, Verlag des Vereins der Mathematiker und Physiker an der ETH, 1969.
A. Messiah. Quantum Mechanics, in two volumes. Amsterdam, North-Holland, 1961–1962.
F.L. Pilar. Elementary Quantum Chemistry. New York, McGraw-Hill, 1968.
L.I. Schiff. Quantum Mechanics. New York, McGraw-Hill, 1968.
B. Simon. Quantum Mechanics for Hamiltonians Defined as Quadratic Forms. Princeton, Princeton University Press, 1974.
H. Weyl. Gruppentheorie und Quantenmechanik. Leipzig, S. Hirzel, 1931.

(3.1.13)
A.M. Perelomov. Coherent States for Arbitrary Lie Groups. *Commun. Math. Phys.* **26**, 222–236, 1972.

(3.1.16)
R. Jost. The General Theory of Quantized Fields. Providence, American Mathematical Society, 1965.
R.F. Streater and A.S. Wightman. PCT, Spin, Statistics, and All That. New York, Benjamin, 1964.

Section 3.2

H. Weyl, *op. cit.*
A.R. Edmonds. Angular Momentum in Quantum Mechanics. Princeton, Princeton University Press, 1974.

(3.3.1)
P. Ehrenfest. Bemerkung über die angenäherte Gültigkeit der klassischen Mechanik innerhalb der Quantenmechanik. *Z. Phys.* **45**, 455–457, 1927.

(3.3.4)
T. Kato. On th Eigenfunctions of Many-Particle Systems in Quantum Mechanics. *Commun. on Pure and Appl. Math.* **10**, 151–177, 1957.

(3.3.13)
K. Hepp. The Classical Limit for Quantum Mechanical Correlation Functions. *Commun. Math. Phys.* **35**, 265–277, 1974.

(3.3.18)
See references for (3.1.16).

(3.4.6)
W.O. Amrein, Ph.A. Martin, and B. Misra. On the Asymptotic Condition of Scattering Theory. *Helv. Phys. Acta* **43**, 313–344, 1970.

(3.4.11)
T. Kato. Wave Operators and Similarity for some Non-Selfadjoint Operators. *Math. Ann.* **162**, 258–279, 1966.
S. Agmon. Spectral Properties of Schrödingher Operators and Scattering Theory. *Ann. Scuola Norm. Sup. Pisa, Cl. di Sci.*, ser. IV, **2**, 151–218,1975.

(3.4.19)
P. Deift and B. Simon. A Time-Dependent Approach to the Completeness of Multiparticle Quantum Systems. *Commun. on Pure and Appl. Math.* **30**, 573–583, 1977.

(3.5.21)
A. Weinstein and W. Stenger. Methods of Intermediate Problems for Eigenvalues: Theory and Ramifications. New York, Academic Press, 1972.

(3.5.28)
M.F. Barnsley. Lower Bounds for Quantum Mechanical Energy Levels. *J. Phys.* **A11**, 55–68, 1978.
B. Baumgartner. A Class of Lower Bounds for Hamiltonian Operators. *J. Phys.* **A12**, 459–467, 1979.
R.J. Duffin. Lower Bounds for Eigenvalues. *Phys. Rev.* **71**, 827–828, 1947.

(3.5.30)
H. Grosse, private communication.

(3.5.31)
See references for (3.5.21).
P. Hertel, H. Grosse, and W. Thirring. Lower Bounds to the Energy Levels of Atomic and Molecular Systems. *Acta Phys. Austr.* **49**, 89–112, 1978.

(3.5.36)
B. Simon. An Introduction to the Self-Adjointness and Spectral Analysis of Schrödinger Operators. In: The Schrödinger Equation, *op. cit.*, p. 19.

(3.5.38;1)
V. Glaser, H. Grosse, A. Martin, and W. Thirring. A Family of Optimal Conditions for the Absence of Bound States in a Potential. In: Studies in Mathematical Physics, *op. cit.*, p.169.
See also reference [6].

Section 3.6

W.O. Amrein, J.M. Jauch, and K.B. Sinha. Scattering Theory in Quantum Mechanics: Physical Principles and Mathematical Methods. Lecture Notes and Supplements in Physics, vol. 16. New York, Benjamin, 1977.

M.L. Goldberger and K.M. Watson. Collision Theory. New York, Wiley, 1964.

R.G. Newton. Scattering Theory of Waves and Particles. New York, McGraw-Hill, 1966.

W. Sandhas. The N-Body Problem. *Acta Phys. Austr. Suppl.*, Vol. 13, Vienna and New York, Springer, 1974.

J.R. Taylor. Scattering Theory. New York, Wiley, 1972

Chapter 4, General

K. Osterwalder, ed. Mathematical Problems in Theoretical Physics. Proc. Int. Conf. on Math. Phys. (Lausanne, Switzerland, Aug. 1979). Berlin–Heidelberg–New York, Springer, 1980.

Chapter 4, Specific

Section 4.1

M.J. Englefield. Group Theory and the Coulomb Problem. New York, Interscience, 1972.

(4.1.18)

H. Grosse, H.-R. Grümm, H. Narnhofer, and W. Thirring. Algebraic Theory of Coulomb Scattering. *Acta Phys. Austr.* **40**, 97–103, 1974.

(4.2.2)

See reference [3], section X.5.

E. Nelson. Time-Ordered Operator Products of Sharp-Time Quadratic Forms. *J. Func. Anal.* **11**, 211–219, 1972.

W. Faris and R. Lavine. Commutators and Self-Adjointness of Hamiltonian Operators. *Commun Math. Phys.* **35**, 39–48, 1974.

(4.2.11)

R. Lavine. Spectral Densities and Sojourn Times. In: Atomic Scattering Theory, J. Nuttall, ed. London, Ontario, University of Western Ontario Press, 1978.

(4.2.13)

R. Lavine and M. O'Carroll. Ground State Properties and Lower Bounds for Energy Levels of a Particle in a Uniform Magnetic Field and External Potential. *J. Math. Phys.* **18**, 1908–1912, 1977.

(4.2.15)

See reference [13].

(4.2.19)

H. Narnhofer and W. Thirring. Convexity Properties for Coulomb Systems. *Acta Phys. Austr.* **41**, 281–297, 1975.

(4.3.27)

See the first reference for (3.5.38;1).

(4.3.36)

T. Kinoshita. Ground State of the Helium Atom. *Phys. Rev.* **105**, 1490–1502, 1957.

C.L. Pekeris. Ground State of Two-Electron Atoms. *Phys. Rev.* **112**, 1649–1658, 1958.

K. Frankowski and C.L. Pekeris. Logarithmic Terms in the Wave Function of the Ground State of Two-Electron Atoms.*Phys. Rev.* **146**, 46–49, 1966.

(4.3.38)

R. Ahlrichs, M. Hoffmann-Ostenhof, T. Hoffmann-Ostenhof, and J.D. Morgan. Bounds on the Decay of Electron Densities with Screening. *Phys. Rev.* **A23**, 2106, 1981.

(4.3.39–40)

S. Agmon. Lectures on exponential decay of solutions of second-order elliptic equations: bounds on eigenfunctions of N-body Schrödinger operators. Mathematical Notes, vol. 29. Tokyo, University of Tokyo Press, 1982.

(4.3.45;2)

See reference for (3.3.4).

S. Fournais, M. Hoffmann-Ostenhof, T. Hoffmann-Ostenhof, and T.O. Sørensen. The electron density is smooth away from the nuclei. *Commun. Math. Phys.* **228**, 401–415, 2002.

(4.3.50)

See reference for (4.3.43).

W. Faris. Inequalities and Uncertainty Principles. *J. Math. Phys.* **19**, 461–466, 1978.

Section 4.4

D.B. Pearson. Singular Continuous Measures in Scattering Theory. *Commun. Math. Phys.* **60**, 13–36, 1978.

L. Faddeev. Mathematical Aspects of the Three-Body Problem in the Quantum Scattering Theory. Translation of Trudy Steklov Math. Inst., vol. 69, 1963. Jerusalem, Israel Program for Scientific Translation. 1965.

J. Ginibre and M. Moulin. Hilbert Space Approach to the Quantum Mechanical Three-Body Problem. *Ann. Inst. H. Poincaré* **21A**, 97–145, 1974.

I.M. Sigal. Mathematical Foundations of Quantum Scattering Theory for Multiparticle Systems. *Memoirs of the Amer. Math. Soc.* **16**, no. 209. Providence, American Mathematical Society, 1978.

R. Blankenbecler and R. Sugar. Variational Upper and Lower Bounds for Multichannel Scattering. *Phys. Rev.* **136B**, 472–491, 1964.

L. Spruch and L. Rosenberg. Upper Bounds on Scattering Lengths for Static Potentials. *Phys.Rev.* **116**, 1034–1040, 1959.

L. Rosenberg and L. Spruch. Subsidiary Minimum Principles for Scattering Parameters. *Phys.Rev.* **A10**, 2002–2015, 1974.

(4.4.5)
W. Hunziker. On the Spectra of Schrödinger Multiparticle Hamiltonians. *Helv. Phys. Acta* **39**, 451–462, 1966.
C. Van Winter. Theory of Finite Systems of Particles I. The Green Function. *Mat-Fys. Scr. Danske Vid. Selsk.* **2**, No. 8, 1, 1964.
G.M. Zhislin. Issledovaniye Spektra Operatora Shredingera dlya Sistemy Mnogikh Chastits. *Trudy Mosk. Mat. Obs.* **9**, 81–120, 1960.
S. Weinberg. Systematic Solution of Multiparticle Scattering Problems. *Phys. Rev.* **133B**, 232–256, 1964.
V. Enss. A Note on Hunziker's Theorem. *Commun. Math. Phys.* **52**, 233–238, 1977.

(4.4.10)
E. Balslev and J.M. Combes. Spectral Properties of Many-Body Schrödinger Operators with Dilatation-Analytic Interactions. *Commun. Math. Phys.* **22**, 280–294, 1971.

(4.4.13)
See reference [3], section XII.6.
B. Simon. Resonances in n-Body Quantum Systems with Dilation Analytic Potentials and the Foundations of Time-Dependent Perturbation Theory. *Ann. Math.* **97**, 247–274, 1973.

(4.4.14)
B. Simon. N-Body Scattering in the Two-Cluster Region. *Commun. Math. Phys.* **58**, 205–210, 1978.
V. Enss. Two-Cluster Scattering of N Charged Particles. *Commun. Math. Phys.* **65**, 151–165, 1979.
J.M. Combes and H. Narnhofer, private communication for the proof used here.

(4.4.18)
See references for section 3.6.
H. Grosse, H. Narnhofer, and W. Thirring. Accurate Determination of the Scattering Legth of Electrons on $\mu^- p$ Atoms. *J. Phys.* **B12**, L189–L192, 1979.

(4.5.20)
P. Hertel, E.H. Lieb, and W. Thirring. Lower Bound to the Energy of Complex Atoms. *J. Chem. Phys.* **62**, 3355–3356, 1975.

(4.5.27;1)
E.H. Lieb and B. Simon. The Thomas–Fermi Theory of Atoms, Molecules and Solids. *Adv. Math.* **23**, 22–116, 1977.

(4.5.27;2)
See the second reference for (3.5.31).

(4.5.28)
See reference [6].

(4.6.4)
See reference for (4.2.19).

(4.6.12)
S.T. Epstein. Ground-State Energy of a Molecule in the Adiabatic Approximation. *J. Chem. Phys.* **44**, 838–839, 1966; erratum, *ibid.*, p. 4062.
See reference [19].
J.M. Combes and R. Seiler. Regularity and Asymptotic Properties of the Discrete Spectrum of Electronic Hamiltonians. *Int. J. Quantum Chem.* **14**, 213–229, 1978.

(4.6.18)
See reference for (4.2.19).

(4.6.24)
E.H. Lieb and B. Simon. Monotonicity of the Electronic Contribution to the Born–Oppenheimer Energy. *J. Phys.* **B11**, L537–L542, 1978.

(4.6.27)
See the second reference for (3.5.31).

(4.6.28;2)
A subsequent justification of the formal argument of C.A. Coulson; see R. Ahlrichs. Convergence Properties of the Intermolecular Force Series ($1/R$-Expansion). *Theor. Chim. Acta* **41**, 7–15, 1976.

(4.6.28;5)
J.D. Power. Fixed Nuclei Two-Centre Problem in Quantum Mechanics. *Phil. Trans. Roy. Soc. London* **A274**, 663–697, 1973.

(4.6.28;7)
See reference [19].

(4.6.31;1)
R.F. Alvarez-Estrada and A. Galindo. Bound States in Some Coulomb Systems. *Il Nuovo Cim.* **44B**, 47–66, 1978.

Literature Added to Part I of This Edition

1. H. Grosse, L. Pittner. On the Number of Unnatural Parity States of the H^--Ion. *J. Math. Phys.* **24**, 1142–1147, 1983.
2. H.L. Cycon, R.G. Froese, W. Kirsch, and B. Simon. Schrödinger Operators and Applications to Quantum Mechanics and Global Geometry. Berlin–Heidelberg, Springer, 1987.
3. I.M. Sigal, A. Soffer. The N-Particle Scattering Problem: Asymptotic Completeness for Short Range Systems. *Ann. Math.* **126**, 35–108, 1987.

4. F. Schwabl. Quantenmechanik. New York, Springer, 1988.
5. D.R. Yafaev. Quasi-classical asymptotics of the scattering amplitude and of the scattering cross-section. In: IXth International Congress on Mathematical Physics (17–27 July 1988, Swansea, UK). Bristol, IOP Publishing Ltd, 1989.
6. G.-M. Graf. Asymptotic completeness for N-body short-range quantum systems: A new proof. *Commun. Math. Phys.* **132**, 73–101, 1990.
7. V. Bach. Error bound for the Hartree–Fock energy of atoms and molecules. *Commun. Math. Phys.* **147**, 527–548, 1992.
8. A. Martin, J.M. Richard, T.T. Wu. Stability of three-unit-charge Systems. *Phys. Rev.* **A46**, 3697–3703, 1992.
9. V. Bach. Accuracy of mean field approximations for atoms and molecules. *Commun. Math. Phys.* **155**, 295–310, 1993.
10. V. Bach, R. Lewis, E.H. Lieb, and H. Siedentop. On the number of bound states of a bosonic N-particle Coulomb system. *Math. Z.* **214**(3), 441–460, 1993.
11. B. Baumgartner. Postulate for Time Evolution in Quantum Mechanics. *Found. Phys.* **24**, 855–872, 1993.
12. D. Yafaev. New channels in three-body long-range scattering. In: Equations aux derivees partielles, Seminairre 1993–1994, Ecole Polytechnique.
13. G.-M. Graf and J.-P. Solovej. A correlation estimate with application to quantum systems with Coulomb interactions. *Rev. Math. Phys.* **6**, 977–997, 1994.
14. W.O. Amrein, A.B. de Monvel, and V. Georgescu. C_0-Groups, Commutator Methods and Spectral Theory of N-Body Hamiltonians. Basel–Boston–Berlin, Birkhäuser Verlag, 1996.
15. J. Dereziński and C. Gérard. Scattering Theory of Classical and Quantum N-Particle Systems. Berlin, Springer, 1997.
16. E.H. Lieb and M. Loss. Analysis. Graduate Studuies in Mathematics, vol. 14. Providence, American Mathematical Society, 1997 (2001, Second Edition).
17. E.H. Lieb. Inequalities. Selecta. M. Loss and M.B. Ruskai, eds. Berlin–Heidelberg, Springer, 2002.

Bibliography to Part II

[1] F.J. Dyson and A. Lenard. Stability of Matter, I. *J. Math. Phys.* **8**, 423-433, 1967; Stability of Matter, II. *Ibid.* **9**, 698-711, 1968.
[2] E.H. Lieb. The $N^{5/3}$ Law for Bosons. *Phys.Lett.* **70A**, 71-73, 1979.
[3] R.A. Goldwell-Horstall and A.A. Maradulin. Zero-Point Energy of an Electron Lattice. *J. Math. Phys.* **1**, 395-404, 1960.
[4] J. Dixmier. Les Algèbres d'Opérateurs dans l'Espace Hilbertien. Paris, Gauthier-Villars, 1969.
[5] A. Wehrl. General Properties of Entropy. *Rev. Mod. Phys.* **50**, 221-260, 1978. E.H. Lieb. Convex Trace Functions and the Wigner–Yanase–Dyson Conjecture. *Adv. Math.* **11**, 267-288, 1973. B. Simon. Trace Ideals and their Applications. London and New York, Cambridge Univ. Press, 1979. A. Uhlmann. Relative Entropy and the Wigner–Yanase–Dyson–Lieb Concavity in an Interpolation Theory. *Commun. Math. Phys.* **40**, 147-151, 1975; Sätze über Dichtematrizen. *Wiss. Z. Karl-Marx-Univ. Leipzig* **20**, 633, 1971; The Order Structure of States. In: Proc. Int. Symp. on Selected Topics in Statistical Mechanics. JINR-Publ. D17-11490. Dubna USSR, 1978.
[6] M.B. Ruskai. Inequalities for Traces on von Neumann Algebras. *Commun. Math. Phys.* **26**, 280-289, 1972. M. Breitenecker, H.-R. Grümm. Note on Trace Inequalities. *Commun. Math. Phys.* **26**, 276-279, 1972. K. Symanzik. Proof and Refinements of an Inequality of Feynman. *J. Math. Phys.* **6**, 1155-1156, 1965.
[7] J. Aczel, B. Forte, and C.T. Ng. Why the Shannon and Hartry Entropies are ''Natural". *Adv. Appl. Prob.* **6**, 131-146, 1974.
[8] E.H. Lieb and B. Ruskai. Proof of the Strong Subadditivity of Quantum-Mechanical Entropy. *J. Math. Phys.* **14**, 1938-1941, 1973. H. Araki and E.H. Lieb. Entropy Inequalities. *Commun. Math. Phys.* **18**, 160-170, 1970.
[9] P.C. Martin and J. Schwinger. Theory of Many-Particle Systems, I. *Phys. Rev.* **115**, 1342-1373, 1959.
[10] R. Peierls. Surprises in Theoretical Physics. Princeton, Princeton Univ. Press, 1976.

[11] E.H. Lieb and W. Thirring. Inequalities for the Moments of the Eigenvalues of the Schrödinger Hamiltonain and their Relation to Sobolev Inequalities. In: Studies in Mathematical Physics, Essays in Honor of Valentine Bargmann, A.S. Wightman, E.H. Lieb, and G.N. Simon, eds. Princeton, Princeton Univ. Press. 1976.
[12] E.T. Whittaker and G.N. Watson. A Course of Modern Analysis. Cambridge, at the University Press, 1969.
[13] J.T. Cannon. Infinite Volume Limits of the Canonical Free Bose Gas States on the Weyl Algebra. *Commun. Math. Phys.* **29**, 89-104, 1973.
[14] G. Lindblad. On the Generators of Quantum Dynamical Semigroups. *Commun. Math. Phys.* **48**, 119-130, 1976.
[15] A. Kossakowski and E.C.G. Sudarshan. Completely Positive Dynamical Semigroups of N-Level Syatems. *J. Math. Phys.* **17**, 821-825, 1976.
[16] T.L. Saaty and J. Bram. Nonlinear Mathematics. New York, McGraw-Hill, 1964.
[17] D. Ruelle. Statistical Mechanics, Rigorous Results. New York, Benjamin, 1969.
[18] A. Guichardet. Systèmes Dynamiques non Commutatifs. *Astérisque* **13-14**, 1974.
[19] I.M. Gel'fand, R.A. Minlos, and Z.Ya. Shapiro. Representations of the Rotation and Lorenz Group and their Applications. Oxford, Pergamon Press, 1963.
[20] R.B. Israel, ed. Convexity in the Theory of Lattice Gases. Princeton, Princeton Univ. Press, 1979.
[21] O. Bratteli and D.W. Robinson. Operator Algebras and Quantum Statistical Mechanics, in two volumes. New York, Springer, 1979, 1980.
[22] H. Narnhofer. Kommutative Automorphismen und Gleichgewichszustände. *Acta Phys. Austriaca* **47**, 1-29, 1977.
[23] H. Narnhofer. Scattering Theory for Quasi-Free Time Automorphisms of C^*-Algebras and von Neumann Algebras. *Rep. Math. Phys.* **16**, 1-8, 1979.
[24] H. Araki and G.L. Sewell. KMS Conditions and Local Thermodynamic Stability of Quantum Lattice Systems. *Commun. Math. Phys.* **52**, 103-109, 1977.
[25] G.L. Sewell. Relaxation, Application, and the KMS Conditions. *Ann. Phys.*(N.Y.) **85**, 336-377, 1974.
[26] H. Narnhofer and G.L. Sewell. Vlasov Hydrodynamics of a Quantum Mechanical Model. *Commun. Math. Phys.* **79**, 9-24, 1981.
[27] J. Messer. The Pressure of Fermions with Gravitational Interaction. *Z. Phys.* **B33**, 313-316, 1979.
[28] W. Thirring. A Lower Bound with the Best Possible Constant for Coulomb Hamiltonians. *Commun. Math. Phys.* **79**, 1-7, 1981.
[29] E.H. Lieb. The Stability of Matter. *Rev. Mod. Phys.* **48**, 553-569, 1976.
[30] J. Lebowitz and E.H. Lieb. The Constitution of Matter: Existence of Thermodynamics for Systems Composed of Electrons and Nuclei. *Adv. Math.* **9**, 316-398, 1972.
[31] E.H. Lieb. The Number of Bound States of One-Body Schrödinger Operators and the Weyl Problem. *Proc. Symposia in Pure Math.* **36**, 241-252, 1980.
[32] E.H. Lieb. Proof of an Entropy Conjecture of Wehrl. *Commun. Math. Phys.* **62**, 35-41, 1978.
[33] N. Dunford and J.T. Schwartz. Linear Operator, part I. New York, Wiley-Interscience, 1967.
[34] E.H. Lieb. Thomas–Fermi and Related Theories of Atoms and Molecules. *Rev. Mod. Phys.* **53**, 603-641, 1981.
[35] S. Chandrasekhar. An Introduction to the Study of Stellar Structure. New York, Dover, 1967.

[36] H. Posch, H. Narnhofer, W. Thirring. Dynamics of Unstable Systems. *Phys. Rev.* **A42**, 1880-1890, 1990.
[37] H. Narnhofer, W. Thirring. Quantum Field Theories with Galilei-Invariant Interactions. *Phys. Rev. Lett.* **64**, 1863-1866, 1990.
[38] R. Haag. Local Quantum Field Theory. Berlin, Springer, 1992.
[39] G. Emch. Generalized K-Flows. *Commun. Math. Phys.* **49**, 191-215, 1976.
[40] P. Walters. An Introduction to Ergodic Theory. New York, Springer, 1982.
[41] R. Longo, C. Peligrad. Noncommutative Topological Dynamics and Compact Actions on C^*-Algebras. *J. Funct. Anal.* **58**, 157-174, 1984.
[42] A. Kishimoto, D. Robinson. Dissipations, Derivations, Dynamical Systems, and Asymptotic Abelianess. *J. Op. Theor.* **13**, 237-253, 1985.
[43] O. Bratteli, G. Elliott, D. Robinson. Strong Topological Transitivity and C^*-Dynamical Systems. *J. Math. Soc. Jap.* **37**, 115-133, 1985.
[44] H. Narnhofer, W. Thirring. Mixing Properties of Quantum Systems. *J. Stat. Phys.* **57**, 811-825, 1989.
[45] H. Narnhofer, W. Thirring. Galilei-Invariant Quantum Field Theories with Pair Interactions. *Int. Journ. Mod. Phys.* **A6**, 2937-2990, 1991.
[46] C.D. Jäkel. Asymptotic Triviality of the Moller Operators in Galilei Invariant Quantum Field Theories. *Lett. Math. Phys.* **21**, 343-350, 1991.
[47] W. Schröder. In: Quantum Probability and Applications, Lecture Notes in Mathematics, vol. 1055, L. Accardi, A. Gorini, eds. Berlin, Springer, 1984.
[48] H. Narnhofer, W. Thirring. Algebraic K-systems. *Lett. Math. Phys.* **20**, 231-250, 1990.
[49] H. Narnhofer, W. Thirring. Clustering for Algebraic K-Systems. *Lett. Math. Phys.* **30**, 307-316, 1994.
[50] H. Narnhofer, W. Thirring. From Relative Entropy to Entropy. *Fizika* **17**, 258-265, 1985.
[51] A. Kolmogorov. A New Metric Invariant of Transitive Systems and Automorphisms of Lebesgue Spaces. *Dokl. Akad. Nauk* **119**, 861-864, 1958.
[52] A. Connes, H. Narnhofer, W. Thirring. Dynamical Entropy of C^*-Algebras and von Neumann Algebras. *Commun. Math. Phys.* **112**, 691-719, 1987.
[53] J. Sauvageot, J. Thouvenot. Une nouvelle définition de l'entropie dinamique des systèmes non commutatifs. *Commun. Math. Phys.* **145**, 411-423, 1992.
[54] M. Ohya, D. Petz, Quantum Entropy and its Use. New York, Springer, 1993.
[55] E.H. Lieb, S. Oxford. Improved Lower Bound on the Indirect Coulomb Energy. *Int. J. Quant. Chem.* **19**, 427-439, 1981.
[56] H. Narnhofer, W. Thirring. Adiabatic Theorem in Quantum Statistical Mechanics. *Phys. Rev. A* **26**, 3645-3651, 1982.

Further Reading

Section 1.1

H. Wergeland. Irreversibility in Many-Body Systems. In: Irreversibility in the Many-Body Problem, J. Biel and J. Rae, eds. New York and London, Plenum, 1972.

(1.1.1)
G. Emch. Non-Markovian Model for the Approach to Equilibrium. *J. Math. Phys.* **7**, 1198-12206, 1966.

(1.1.13)
E. Schrödinger. Zur Dynamik elastisch gekoppelter Punktsysteme. *Ann. Phys.* (Leipzig) **44**, 916-934, 1914.
I. Prigogine and G. Klein. Sur la Mécanique Statistique des Phénomènes Irréversibles, III. *Physica* **19**, 1053-1071, 1953.

(1.2.10)
E.H. Lieb and H. Narnhofer. The Thermodynamic Limit for Jellium. *J. Stat. Phys.* **12**, 291-310, 1975.

Section 1.3

F.A. Berezin. Method of Second Quantization. New York, Academic Press, 1966.
M. Reed and B. Simon. Methods of Modern Mathematical Physics, vol. II: Fourier Analysis, Self-Adjointness. New York, Academic Press, 1975.

(1.4.2)
J. von Neumann. On Infinite Direct Products. *Compositio Math.* **6**, 1-77, 1939.

(1.4.9)
O. Bratteli and D.W. Robinson. Operator Algebras and Quantum Statistical Mechanics, vol. I. New York, Springer, 1987.

Section 2.1

A. Wehrl. How Chaotic is a State of a Quantum System? *Rep. Math. Phys.* **6**, 15-28, 1974.
A. Uhlmann. Endlichdimensionale Dichtermatrizen, I. *Wiss. Z. Karl-Marx-Univ. Leipzig* **21**, 421, 1972; Endlichdimensionale Dichtermatrizen, II. *Ibid.* **22**, 139, 1973.

(2.1.7)
E.H. Lieb. Some Convexity and Subadditivity Properties of Entropy. *Bull. Amer. Math. Soc.* **81**, 1-13, 1975.
B. Simon. Trace Ideals and their Applications. London and New York, Cambridge Univ. Press, 1979.

(2.2.9)
B. Baumgartner. Classical Bounds on Quantum Partition Functions. *Commun. Math. Phys.* **75**, 25-41, 1980.

(2.2.11)
F.A. Berezin. Wick and Anti-Wick Operator Symbols. *Math. USSR Sbornik* **15**, 577-606, 1971. (Translation of Vikovskie i antivikovskie simboly operatorov. *Mat. Sbornik* **86(128)**, 578-610, 1971.)

(2.2.22)

G. Lindblad. Entropy, Information, and Quantum Measurements. *Commun. Math. Phys.* **33**, 305-322, 1973.

U. Umegaki. Conditional Expectation in an Operator Algebra. *Kodai Math. Seminar. Rep.* **14**, 59-85, 1962.

H. Araki. RIMS preprint 190, Kyoto, 1975.

Section 2.3

R. Griffiths. Microcanonical Ensemble in Quantum Statistical Mechanics. *J. Math. Phys.* **6**, 1447-1461, 1965.

(2.3.39)

A.S. Wightman. Convexity and the Notion of Equilibrium State in Thermodynamics and Statistical Mechanics. In: Convexity in the Theory of Lattice Gases, R. Israel, ed. Princeton, Princeton Univ. Press, 1979.

(2.4.7)

H.D. Maison. Analyticity of the Partition Function for Finite Quantum Systems. *Commun. Math. Phys.* **22**, 166-172, 1971.

(2.4.9)

E.H. Lieb. The Classical Limit of Quantum Spin Systems. *Commun. Math. Phys.* **31**, 326-341, 1973.

(2.4.15;2)

A.S. Wightman, *op. cit.* in (2.3.39)

(2.5.15)

W. Thirring. Bounds on the Entropy in Terms of One Particle Distributions. *Lett. Math. Phys.* **4**, 67-70, 1980.

(2.5.26)

K. Huang. Statistical Mechanics. New York, Wiley, 1963.

F. Reif. Fundamentals of Statistical and Thermal Physics. New York, McGraw-Hill, 1965.

Section 3.1

D. Ruelle. Statistical Mechanics, Rigorous Results. New York, Benjamin, 1969.

G. Emch. Algebraic Methods in Statistical Mechanics and Quantum Field Theory. New York, Wiley, 1971.

(3.1.2)

E.B. Davies. Quantum Theory of Open Systems. New York, Academic Press,1976.

(3.1.4)

V. Gorini, A. Frigerio, M. Verri, A. Kossakowski, and E.C.G. Sudarshan. Properties of Quantum Markovian Master Equations. *Rep. Math. Phys.* **13**, 149-173, 1978.

P. Martin. Modèls en Méchanique Statistique des Processus Irréversibles. New York and Berlin, Springer, 1979.

(3.1.12)

G. Lindblad. Completely Positive Maps and Entropy Inequalities. *Commun. Math. Phys.* **40**, 147-151, 1975.

A. Uhlmann. Relative Entropy and the Wigner–Yanase–Dyson–Lieb Concavity in an Interpolation Theory. *Commun. Math. Phys.* **54**, 21-322, 1970.

(3.1.14)

F. Greenleaf. Invariant Means on Topological Groups. New York, Van Nostrand, 1966.

A. Guichardet. Systémes Dynamiques non Commutatifs. *Astérisque* **13-14**, 1974.

(3.1.25)

V.I. Arnol'd and A. Avez. Ergodic Problems of Classical Mechanics. New York, Benjamin, 1968.

(3.2.1)

N.M. Hugenholtz. In: Mathematics of Contemporary Physics, R. Streater, ed. New York and London, Academic Press, 1972.

(3.2.6)

R. Haag, N.M. Hugenholtz, and M. Winnink. On the Equilibrium States in Quantum Statistical Mechanics. *Commun. Math. Phys.* **5**, 215-236, 1967.

(3.2.13)

M. Takesaki. Tomita's Theory of Modular Hilbert Algebras and Its Applications, Lecture Notes in Mathematics, vol. 128. New York and Berlin, Springer, 1970.

(3.3.7)

H. Narnhofer and D.W. Robinson. Dynamical Stability and Pure Thermodynamic Phases. *Commun. Math. Phys.* **41**, 89-97, 1975.

(3.3.10)

R. Haag, D. Kastler, and E.B. Trych-Pohlmeyer. Stability and Equilibrium States. *Commun. Math. Phys.* **38**, 173-193, 1974.

(3.3.22)

W. Pusz and S.L. Woronovicz. Passive States and KMS States for General Quantum Systems. *Commun. Math. Phys.* **58**, 273-290, 1978.

A. Lenard. Thermodynamical Proof of the Gibbs Formula for Elementary Quantum Systems. *J. Stat. Phys.* **19**, 575-586, 1978.

Section 4.1

E.H. Lieb and B. Simon. The Thomas–Fermi Theory of Atoms, Molecules, and Solids. *Adv. Math.* **23**, 22-116, 1977.

H. Narnhofer and W. Thirring. Asymptotic Exactness of Finite Temperature Thomas–Fermi Theory. *Ann. Phys.* (N.Y.) **134**, 128-140, 1981.

B. Baumgartner. The Thomas–Fermi Theory as Result of a Strong-Coupling Limit, *Commun. Math. Phys.* **47**, 215-219, 1976.

Section 4.2

P. Hertel, H. Narnhofer, and W. Thirring. Thermodynamic Functions for Fermions with Gravostatic and Electrostatic Interactions. *Commun. Math. Phys.* **28**, 159-176, 1972.

P. Hertel and W. Thirring. In: Quanten und Felder. H. Dürr, ed. Brunswick, Vieweg, 1971.

J. Messer. Temperature Dependent Thomas–Fermi Theory, Lectures Notes in Physics, vol. 147. New York and Berlin, Springer, 1979.

B. Baumgartner. Thermodynamic Limit of Correlation Functions in a System of Gravitating Fermions. *Commun. Math. Phys.* **48**, 207-213, 1976.

Section 4.3

E.H. Lieb. The Stability of Matter. *Rev. Mod. Phys.* **48**, 553-569, 1976.

W. Thirring. Stability of Matter. In: Current Problems in Elementary Particle and Mathematical Physics, P. Urban, ed. *Acta Phys. Austriaca Suppl.* **XV**, 337-354, 1976.

(4.3.22)

R. Griffiths. Microcanonical Ensemble in Quantum Statistical Mechanics. *J. Math. Phys.* **6**, 1447-1461, 1965.

(4.3.40)

H. Narnhofer and W. Thirring. Convexity Properties for Coulomb Systems. *Acta Phys. Austriaca.* **41**, 281-297, 1975.

Literature Added in the Second Edition

Chapter 3

1. J. Bisognano, E.H. Wichmann. On the Duality Condition for a Hermitian Scalar Field. *Journ. Math. Phys.* **16**, 985-1007, 1975.
2. R. Figari, R. Hoegh-Krohn, C.R. Nappi. Interacting Relativistic Boson Fields in the de Sitter Universe with Two Space-Time Dimensions. *Commun. Math. Phys.* **44**, 265-278, 1975.
3. J. Bisognano, E.H. Wichmann. On the Duality Condition for Quantum Fields. *Journ. Math. Phys.* **17**, 303-312, 1976.
4. G.G. Emch. Generalized K-Flows. *Commun. Math. Phys.* **49**, 191-215, 1976.

5. L. Accardi, A. Frigerio, J.T. Lewis. Quantum Stochastic Processes. *Publ. RIMS Kyoto* **18**, 97-133, 1982.
6. H. Kosaki. Extension of Jones Theory on Index to Arbitrary Factors. *Journ. Funct. Analysis* **66**, 123-140, 1986.
7. M. Pimsner, S. Popa. Entropy and Index for Subfactors. *Ann. Sci. Ec. Norm. Sup.* **19**, 57-106, 1986.
8. A. Connes, H. Narnhofer, W. Thirring. Dynamical Entropy of C*-Algebras and von Neumann Algebras. *Commun. Math. Phys.* **112**, 691-719, 1987.
9. R. Longo. Simple Injective Subfactors. *Advances in Mathematics* **63**, 152-171, 1987.
10. L. Price. Shifts of II_1 Factors. *Canad. J. Math.* **39**, 492-511, 1987.
11. B. Kümmerer. Survey on a Theory of non commutative stationary Markov processes, Quantum probability and applications III, L. Accardi, W. v. Waldenfeld eds., Berlin, Springer, 1988, p.228-244.
12. M. Pimsner, S. Popa. Iterating the Basic Construction. *Trans. Amer. Math. Soc.* **310**, 127-133, 1988.
13. R.T. Powers. An Index Theory for Semigroups of *-endomorphisms of $B(H)$ and Type II_1 Factors. *Canad. J. Math.* **40**, 86-114, 1988.
14. H. Narnhofer, W. Thirring. Quantum K-Systems. *Commun. Math. Phys.* **125**, 565-577, 1989.
15. H. Narnhofer, A. Pflug, W. Thirring. Mixing and Entropy Increase in Quantum Systems. *Scuola Normale Superiore Pisa* 597-626, 1989.
16. R. Longo. Index of Subfactors and Statistics of Quantum Fields. *Commun. Math. Phys.* **130**, 285-309, 1990.
17. H. Narnhofer, W. Thirring, Algebraic K-Systems. *Lett. Math. Phys.* **20**, 231-250, 1990.
18. F. Benatti. Deterministic Chaos in Infinite Quantum Systems, SISSA Notes in Physics, Berlin-Heidelberg-New York, Springer, 1993.
19. H.J. Borchers. On Modular Inclusion and Spectrum Condition. *Lett. Math. Phys.* **27**, 311-324, 1993.
20. H.W. Wiesbrock. Halfsides Modular Inclusions of von Neumann Algebras. *Commun. Math. Phys.* **157**, 83-92, 1993; *ibid.* **184**, 683-685, 1997.
21. R. Alicki, H. Fannes. Defining Quantum Dynamical Entropy. *Lett. Math. Phys.* **32**, 75-82, 1994.
22. G.G. Emch, H. Narnhofer, G.L. Sewell, W. Thirring. Anosov Actions on Non-Commutative Algebras. *J. Math. Phys.* **35**/11, 5582-5599, 1994.
23. H. Narnhofer, W. Thirring. Clustering for Algebraic K-Systems. *Lett. Math. Phys.* **30**, 307-316, 1994.
24. H. Narnhofer, E. Størmer, W. Thirring. C*-dynamical systems for which the tensor product formula for entropy fails. *Ergod. Th. and Dyn. Systems* **15**, 961-968, 1995.
25. D.V. Voiculescu. Dynamical Approximation Entropies and Topological Entropy in Operator Algebras 2. *Commun. Math. Phys.* **170**, 249-282, 1995.
26. H. Narnhofer, I. Peter, W. Thirring. How Hot is the de Sitter Space. *Int. J. Mod. Phys. B* **10**, 1507-1520, 1996.
27. H. v. Beijeren, J.R. Dorfmann, H.A. Posch, Ch. Dellago. Kolmogorov–Sinai entropy for dilute gases in equilbrium. *Phys. Rev. E* **56**/5, 5272-5277, 1997.
28. H.J. Borchers. On the Revolutionization of Quantum Field Theory by Tomita's Modular Theory. Preprint ESI-773-1999 (160 pages, 148 references).
29. D. Shlyakhtenko. Free Quasifree States. *Pac. Journ. of Math.* **177**, 329-368, 1997.
30. H.W. Wiesbrock. Symmetries and Modular Intersections of von Neumann Algebras. *Lett. Math. Phys.* **39**, 203-212, 1997.

31. G.G. Emch, I. Peter. Quantum Anosov Flows: A New Family of Examples. *Journ. Math. Phys.* **39**/9, 4513-4539, 1998.
32. V.Ya. Golodets, S.V. Neshveyev. Non Bernoullian Quantum K-Systems. *Commun. Math. Phys.* **195**, 213-232, 1998.
33. V.Ya. Golodets, E. Størmer. Entropy of C*-Dynamical Systems Defined by Bitstreams. *Ergod. Th. and Dynam. Systems* **18**, 1-16, 1998.
34. H. Narnhofer, W. Thirring. Equivalence of Modular K and Anosov Dynamical Systems. Preprint UWThPh-1998-xx, to be publ. in *Rev. Math. Phys.*.
35. H.W. Wiesbrock. Modular Intersections of von Neumann Algebras in Quantum Field Theory. *Commun. Math. Phys.* **193**, 269-285, 1998.
36. H.J. Borchers. On Thermal States of (1+1)-Dimensional Quantum Systems. Preprint ESI-788-1999.
37. H.J. Borchers, J. Yngvason. Modular Groups of Quantum Fields in Thermal States. *J. Math. Phys.* **40**, 601-624, 1999.
38. M. Choda. Entropy on Cuntz's Canonical Endomorphism. *Pacif. Journ. of Math.* **190**, 235-245, 1999.
39. H. Narnhofer. Time Reversibility for Modular K-Systems. Preprint (1999), ESI 716, to be publ. in *Rep. Math. Phys.*.
40. H. Narnhofer, W. Thirring. Realizations of Two-Sided Quantum K-Systems. Preprint (1999), ESI 717, to be publ. in *Rep. Math. Phys.*.

Chapter 4

1. H. Brézis. Some Variational Problems of the Thomas–Fermi Type. In: Variational Inequalities and Complementary Problems, Cottle, Giannessi, and J.-L. Lions, eds. new York, Wiley, 1980, 53-73.
2. H. Brézis, R. Benguria, and E.H. Lieb. The Thomas–Fermi–von Weizsäcker Theory of Atoms and Molecules. *Commun. Math. Phys.* **79**, 167-180, 1981.
3. E.H. Lieb. Thomas–Fermi and Related Theories of Atoms and Molecules. *Rev. Mod. Phys.* **53**, 603-641, 1981.
4. E.H. Lieb. Bound on the Maximum Negative Ionization Energy of Atoms and Molecules. *Phys. Rev. A* **29**, 3018-3028, 1984.
5. R. Benguria, E.H. Lieb. The Most Negative Ion in the Thomas–Fermi–von Weizsäcker Theory of Atoms and Molecules. *J. Phys. B* **18**, 1045-1059, 1985.
6. C. Fefferman. The Thermodynamic Limit for a Cristal. *Commun. Math. Phys.* **98**, 289-311, 1985.
7. C. Fefferman. The Atomic and Molecular Nature of Matter. *Rev. Mat. Iberoam.* **1**, 1985.
8. P.-L. Lions. Solutions of Hartree–Fock Equations for Coulomb Systems. *Commun. Math. Phys.* **109**, 33-97, 1987.
9. J.P. Solovej. Universality in the Thomas–Fermi–von Weizsäcker Model of Atoms and Molecules. *Commun. Math. Phys.* **129**, 561-598, 1990.
10. J. Yngvason. Thomas–Fermi Theory for Matter in a Magnetic Field as a Limit of Quantum Mechanics. *Lett. Math. Phys.* **22**, 107-117, 1991.
11. E.H. Lieb, J.P. Solovej, and J. Yngvason. Heavy Atoms in the strong Magnetic Field of a neutron star. *Phys. Rev. Lett.* **69**, 749-753, 1992.
12. I. Catto, P.-L. Lions. Binding of Atoms and Stability of Molecules in Hartree and Thomas–Fermi Type Theories, parts 1, 2, 3, 4. *Commun. Part. Diff. Eq.* **17, 18**, 1992, 1993.

13. C. Le Bris. Some Results on the Thomas–Fermi–Dirac–von Weizsäcker Model. *Diff. Int. Eq.* **6**, 337-353, 1993.
14. I. Fushiki, E. Gudmundsson, C.J. Pethick, Ö.E. Rögnvaldsson, and J. Yngvason. Thomas–Fermi Calculations of Atoms and Matter in Magnetic Neutron Stars: Effects of Higher Landau Bands. *Astrophys. J.* **416**, 276-290, 1993.
15. E.H. Lieb, J.P. Solovej, and J. Yngvason. Asymptotics of Heavy Atoms in High Magnetic Fields. I : Lowest Landau Band Regions. *Commun. Pure and Appl. Math.* **47**, 513-593, 1994.
16. E. Lieb, J.P. Solovej, and J. Yngvason. Asymptotics of Heavy Atoms in High Magnetic Fields. II: Semiclassical Regions. *Commun. Math. Phys.* **161**, 77-124, 1994.
17. J.P. Solovej. An Improvement on Stability of Mater in Mean Field Theory. *Proceedings of the Conference on PDEs and Mathematical Physics*. University of Alabama, International Press, 1994.
18. C. Fefferman. Stability of Coulomb Systems in a Magnetic Field. *Proc. Natl. Acad. Sci. USA* **92**, 5006-5007, 1995.
19. E.H. Lieb, M. Loss, and J.P. Solovej. Stability of Matter in Magnetic Fields. *Phys. Rev. Lett.* **75**, 985-989, 1995.
20. E. Lieb, J.P. Solovej, and J. Yngvason. The Ground States of Large Quantum Dots in Magnetic Fields. *Phys. Rev. B* **51**, 10646-10665, 1995.
21. C. Fefferman, J. Fröhlich and G.M. Graf. Stability of Nonrelativistic Quantum Mechanical Matter Coupled to the (ultraviolet cutoff) Radiation Field. *Proc. Natl. Acad. Sci. USA* **93**, 15009-15011, 1996.
22. C. Catto, C. Le Bris, and P.-L. Lions. Limite Thermodynamique pour des modèles de type Thomas–Fermi. *C. R. Acad. Sci. Paris* **322**, Série I, 357-364, 1996.
23. F. Nakano. The Thermodynamic Limit of the Magnetic Thomas–Fermi Energy. *J. Math. Sci. Univ. Tokyo* **3**, 713-722, 1996.
24. Y. Netrusov, T. Weidl. On Lieb–Thirring Inequalities for Higher Order Operators with Critical and Subcritical Powers. *Commun. Math. Phys.* **182**, 355-370, 1996.
25. T. Weidl. On the Lieb-Thirring Constants. *Commun. Math. Phys.* **178**, 135-146, 1996.
26. C. Fefferman, J. Fröhlich and G.M. Graf. Stability of Ultraviolet Cutoff Quantum Electrodynamics with Non-relativistic Matter. *Commun. Math. Phys.* **190**, 309-330, 1997.
27. The Stability of Matter: from Atoms to Stars. Selecta of E.H. Lieb. 2nd enl. ed., W. Thirring ed., Berlin-Heidelberg, Springer, 1997.
28. E.H. Lieb, H. Siedentop, and J.P. Solovej. Stability and Instability of Relativistic Electrons in Magnetic Fields. *J. Stat. Phys.* **89**, 37-59, 1997.
29. I. Catto, C. Le Bris, and P.-L. Lions. Mathematical Theory of Thermodynamic Limits: Thomas–Fermi Type Models. Oxford Mathematical Monographs. Oxford, Clarendon Press, 1998.
30. J. Yngvason. Quantum dots. A Survey of Rigorous Results. *Operator Theory: Advances and Applications* **108**, 161-180, 1999.
31. D. Hundertmark, A. Laptev, T. Weidl. New Bounds on the Lieb–Thirring Constants. (to be published in *Acta Mathematica*).
32. A. Laptev, T. Weidl. Sharp Lieb–Thirring Inequalities in High Dimensions. (to be published in *Invent. Mathematicae*).

Index